Reactivity, Mechanism and Structure
in
Polymer Chemistry

Reactivity, Mechanism and Structure
in
Polymer Chemistry

Edited by

A. D. Jenkins

School of Molecular Sciences, University of Sussex

and

A. Ledwith

Donnan Laboratories, University of Liverpool

A Wiley–Interscience Publication

JOHN WILEY & SONS

London · New York · Sydney · Toronto

Library of Congress Catalog Card No 73-2786

ISBN 0 471 44155 4

PRINTED IN NORTHERN IRELAND AT
THE UNIVERSITIES PRESS, BELFAST

Preface

The concept of mechanism, as it is now understood in chemistry, is very modern; it certainly did not concern the earliest experimenters in polymer chemistry, who were almost entirely involved in attempts to elucidate the nature of polymer molecules and to establish the basic empirical facts of the interconversion of familiar well-characterized chemicals and the more indefinite polymers. Isoprene and rubber exemplify the kind of system which attracted attention, and it is salutary to recall that it required something like a century of effort to establish the true high molecular weight nature of such a relatively simple and abundant substance as natural rubber.

Useful thoughts about mechanism could scarcely pre-date the acceptance of the macromolecular hypothesis, so it is natural to look first at the controversy of the 1920's to seek the emergence of mechanistic postulates. Staudinger had at that time developed the chain reaction hypothesis against opposition which clung to a belief in step-wise synthesis. Although both concepts are now known to be valid, this is not the sense in which one currently uses the expression 'mechanism' which perhaps can be more fruitfully defined as a description of the movements undergone by electrons, nuclei and atoms during the period of chemical transformation.

In any search for landmarks in the development of mechanistic understanding certain pioneers or authors of critical experiments or discoveries stand out: chronologically one can place in sequence radical chain reactions (Staudinger, Melville), step-wise reactions (Carothers, Flory), cationic chains (Whitmore, Evans), carbanionic chains (Szwarc) and the co-ordination catalysts (Ziegler, Natta). This list is not intended to be comprehensive; moreover, it deliberately excludes mention of contributors to the more recent literature, but it seems to indicate the course of basic progress, at least in the revelation of important facets of mechanism, especially during the period 1940–1955. Since that time, of course, great progress has been made in raising the level of sophistication and it is our purpose in this book to present, in summary, a statement of the present state of understanding and belief in most of the areas which seem to be of greatest importance.

We have deliberately elected to present not only an overall treatment of reactivity mechanism and structure in the context of polymerization but also to relate the behaviour of the important reactive intermediates in polymerization processes to that displayed by the same entities in the wider compass of general chemical reactions. In this way we hope to avoid any unnecessary dichotomy between polymerization in particular and organic chemistry in general; in fact, we hope to show that it is profitable to keep the relationship between the two constantly in mind. In polymer chemistry, no less than in any other branch of chemistry, discussion of reactivity and mechanism is greatly assisted by detailed consideration of relevant configurational, conformational, and thermodynamic aspects, and this book has been compiled with due regard for all these factors.

A. D. JENKINS
A. LEDWITH

Contributing Authors

BAKER, R.
Department of Chemistry,
University of Southampton, England.

BAMFORD, C. H.
Department of Inorganic, Physical and Industrial
Chemistry, University of Liverpool, England.

BILLINGHAM, N. C.
School of Molecular Sciences,
University of Sussex, England.

CIARDELLI, F.
Institute of Industrial Organic Chemistry of the
University, Pisa, Italy.

COOPER, W.
Dunlop Research Centre,
Birmingham, England.

DART, E. C.
Imperial Chemical Industries Ltd.,
Corporate Laboratory, Runcorn, England.

DE SCHRYVER, F. C.
Department of Chemistry,
University of Louvain, Belgium.

IVIN, K. J.
The Queen's University of Belfast,
Northern Ireland.

JENKINS, A. D.
School of Molecular Sciences,
University of Sussex, England.

LEDWITH, A.
Department of Inorganic, Physical and Industrial
Chemistry, University of Liverpool, England.

LUISI, P. L.
Swiss Federal Institute of Technology, Zürich,
Switzerland.

NORTH, A. M.
Department of Pure and Applied Chemistry,
University of Strathclyde, Scotland.

PARRY, A.
Imperial Chemical Industries Ltd., Corporate
Laboratory, Runcorn, England.

SHERRINGTON, D. C.
Department of Pure and Applied Chemistry,
University of Strathclyde, Scotland.

SMETS, G.
Department of Chemistry, University of Louvain,
Belgium.

TEDDER, J. M.   *Department of Chemistry, University of St. Andrews, Scotland.*

WEALE, K. E.   *Department of Chemical Engineering and Chemical Technology, Imperial College, London, England.*

WITT, D. R.   *Phillips Petroleum Co., Bartlesville, Oklahoma, U.S.A.*

Contents

1. General Aspects of Reactivity and Structure in Polymerization Processes

N. C. BILLINGHAM *and* A. LEDWITH

1.1 The Nature of High Polymers 1
1.2 Step Reaction Polymerization 2
1.3 Chain Reaction Polymerization 6
1.4 Polymerizability of Unsaturated Compounds . . . 7
1.5 Polymerizability of Ring Compounds 15
1.6 Isomerism in Vinyl Polymers 19
    1.6.1 Structural isomerism 19
    1.6.2 Sequence isomerism 20
    1.6.3 Stereoisomerism 21
1.7 Isomerism in Diene Polymers 25
1.8 Isomerism in Polyacetals and Polyethers . . . 27
1.9 Stereo-regular and Stereo-selective Polymerizations . . 27
1.10 Molecular Weights of High Polymers 28
1.11 References 29

2. The Reactivity of Free-Radicals

J. M. TEDDER

2.1 Radical Combination Reactions 32
2.2 Radical Disproportionation Reactions 34
2.3 Radical Transfer Reactions 35
2.4 Radical Addition Reactions 37
2.5 References 50

3. Organometallic Derivatives of Transition Metals as Initiators of Free-Radical Polymerization

C. H. BAMFORD

3.1 Systems Based on Organometallic Derivatives and Organic
     Halides   52
     3.1.1 General kinetic features   55
     3.1.2 Reaction mechanisms   59
     3.1.3 Activities of organometallic derivatives in initiation .   .   73
     3.1.4 Activities of halides in initiation   76
     3.1.5 Some detailed studies   78
     3.1.6 Photoinitiation   85
     3.1.7 Practical applications   91
3.2 Initiation by Chelate Derivatives of Transition Metals   .   .   95
     3.2.1 Chelates alone   95
     3.2.2 Initiation by chelates in the presence of donor additives   .   103
     3.2.3 Photochemical initiation by chelates   .   .   .   112
3.3 References   113

4. The Reactivity of Polymer Radicals in Propagation and Transfer Reactions

A. D. JENKINS

4.1 Definition of Terms   117
     4.1.1 Propagation   117
     4.1.2 Transfer   120
     4.1.3 General   121
4.2 The Inherent Reactivity of Radicals   122
4.3 The Alternating Tendency   123
4.4 Steric Effects   123
4.5 Penultimate Unit Effects   123
4.6 Complexes of Radicals with Salts   124
4.7 Interpretation of Radical Reactivity Data   .   .   .   125
     4.7.1 The Q–e scheme   125
     4.7.2 The patterns of reactivity approach   .   .   .   127
     4.7.3 Other schemes   138
4.8 Comparison of Schemes   139
4.9 Conclusions   140
4.10 References   140

Contents

5. The Influence of Chain Structure on the Free-Radical Termination Reaction

A. M. NORTH

5.1 Introduction 142
5.2 The Diffusive Processes in Termination . . . 143
  5.2.1 The rate-determining step in termination . . 143
  5.2.2 Translational and segmental rearrangement diffusion . 145
  5.2.3 Comparison of rate and diffusion constants . . . 146
  5.2.4 Structure, back-bone rotation and radical termination rate 148
  5.2.5 The effect of copolymer composition 152
  5.2.6 The effect of temperature 154
5.3 Combination and Disproportionation of Radicals . . . 156
5.4 Conclusions 156
5.5 References 156

6. The Influence of Pressure on Polymerization Reactions

K. E. WEALE

6.1 Introduction 158
6.2 Principal Effects of Pressure on Equilibrium and Rate Constants in Liquids. 159
  6.2.1 Pressure and equilibria 159
  6.2.2 Pressure and the rate constants of reactions . . . 160
  6.2.3 Some physical effects of pressure which influence chemical reactivity 162
6.3 General Features of Radical Polymerization at High Pressure. 163
6.4 Reactions of Radicals with Molecules, and Radical-Producing Reactions, under Pressure 166
  6.4.1 Radical addition to vinyl monomers (chain growth) . 166
  6.4.2 Pressure and chain transfer in vinyl polymerization . 167
  6.4.3 Pressure and radical production;the dissociation of initiators 168
6.5 Some Further Effects of Pressure on Radical and Ionic Polymerizations 169
  6.5.1 Ionic polymerization 169
  6.5.2 Polymer-monomer equilibria: pressure and ceiling temperatures 170
  6.5.3 Pressure and the micro-structure of polymers . . . 171
  6.5.4 Phase-separations and phase changes in high pressure polymerization 172
6.6 Conclusion 173
6.7 References 173

7. Emulsion Polymerization
W. COOPER

7.1   Introduction 175
7.2   General Features of Emulsion Polymerization . . . 176
      7.2.1   The emulsifier 176
      7.2.2   Auxiliary stabilizers 180
      7.2.3   Initiators 181
      7.2.4   Chain transfer agents 184
      7.2.5   Other additives 184
      7.2.6   Polymerization 185
7.3   Kinetics of Emulsion Polymerization 186
      7.3.1   Nucleation 189
      7.3.2   Particle number 189
      7.3.3   Free-radicals/polymer particle 195
      7.3.4   Monomer concentration in emulsion polymerization . 198
      7.3.5   Polymer molecular weights 199
7.4   Semi-continuous and Continuous Emulsion Polymerization . 200
7.5   Emulsion Polymer Latices 202
      7.5.1   Surface chemistry of emulsion polymer particles . 203
      7.5.2   Particle size and particle size distribution . . 204
7.6   Inverse Emulsion Polymerization 206
7.7   Polymerization in Emulsion by Non-Radical Processes . 207
7.8   Individual Monomers 208
      7.8.1   Styrene 208
      7.8.2   Vinyl acetate 208
      7.8.3   Vinyl chloride 211
      7.8.4   Conjugated dienes: butadiene, isoprene, 2,3-dimethyl
              butadiene, 1,3-pentadiene, chloroprene, and their copolymers   212
      7.8.5   Vinylidene chloride 216
      7.8.6   Ethylene 217
      7.8.7   Methyl methacrylate 218
      7.8.8   Other monomers: methyl acrylate, 2-methyl vinyl pyridine,
              acrolein, acrylonitrile 218
7.9   Emulsion Polymerization Recipes 219
7.10 References 224
7.11 Addendum 228
      7.11.1 References 228

8. Carbonium Ions
R. BAKER

8.1   Generation and Stability of Carbonium Ions . . . 229
8.2   Neighbouring Group Participation 233

Contents

8.3 Ion-pairs and Solvent Effects   235
8.4 Reactions of Carbonium Ions   239
8.5 Rearrangements in Carbonium Ions   240
8.6 Other Reactions   242
8.7 References   243

9. Reactivity and Mechanism in Cationic Polymerization

A. LEDWITH and D. C. SHERRINGTON

9.1 Introduction   244
9.2 The Nature of the Active Intermediate in Cationic Polymeriza-
     tion   245
     9.2.1 Spectroscopic detection of cationic intermediates   .   246
     9.2.2 Effects of additives and impurities on cationic polymeriza-
            tion   247
     9.2.3 Copolymerization, conductance, and solvent effects .   .   248
     9.2.4 Free ions and ion pairs   250
     9.2.5 Pseudo-cationic systems   257
9.3 Initiation Reactions   262
     9.3.1 Mechanism of the initiation of polymerization of vinyl
            monomers   263
     9.3.2 Mechanism of the initiation of ring-opening polymerization
            (cyclic ethers)   267
     9.3.3 Effect of solvent and temperature on the initiation reaction .   273
9.4 Propagation Reactions   274
     9.4.1 Free ions and ion pairs in propagation reactions   .   .   276
     9.4.2 Pseudo-cationic polymerization   281
     9.4.3 Propagation with concurrent isomerization   .   .   282
     9.4.4 Effect of solvent and temperature on propagation reactions   290
9.5 Chain Breaking Reactions   292
     9.5.1 Transfer reactions   294
     9.5.2 Termination reactions   302
9.6 References   305

10. Carbanions

E. C. DART

10.1 Introduction   310
10.2 pK_a, An Index of Carbanion Stability.   311
      10.2.1 Measurement of pK_a   313
      10.2.2 The relation of carbanion structure to pK_a .   .   314

10.3   Ion Pairing and Reactivity   322
10.4   Preparation of Carbanions   328
    10.4.1   From other carbanions by metalation   .   .   .   328
    10.4.2   Direct metalation   330
    10.4.3   Metal-metal exchange   330
    10.4.4   Halogen-metal interchange   330
    10.4.5   Halogen-metal interconversion   .   .   .   331
    10.4.6   Ether cleavage   331
10.5   Stereochemistry and Configurational Stability   .   .   332
10.6   Reactivity of Delocalized Ions   335
10.7   Reactions of Carbanions   338
10.8   Radical Anions   342
    10.8.1   Generation   342
    10.8.2   Disproportionation   344
    10.8.3   Dimerization   345
    10.8.4   Radical ions as reducing agents   .   .   .   345
10.9   References   346

11. Anionic Polymerization

A. PARRY

11.1   Introduction   350
11.2   Historical Aspects   352
11.3   Initiation Reactions   352
    11.3.1   The use of alkali metal alkyls   .   .   .   352
    11.3.2   Initiation with alkali metals   .   .   .   353
    11.3.3   Initiation by alkali metal complexes   .   .   356
    11.3.4   Initiation by living polymers   .   .   .   357
    11.3.5   Ether cleavage reactions   358
    11.3.6   Electrochemical initiation   359
11.4   Propagation of Anionic Polymerization   .   .   359
    11.4.1   Molecular weight distribution   .   .   .   361
    11.4.2   Kinetics of the propagation reaction   .   .   362
    11.4.3   Ionic species   362
    11.4.4   Polymer structure   366
11.5   Termination Reactions   369
    11.5.1   Reaction with proton donors   .   .   .   369
    11.5.2   Transfer reactions   369
    11.5.3   Isomerization of the polymeric carbanion   .   370
    11.5.4   Reaction with impurities   .   .   .   371
11.6   Design Features for Anionic Polymerization Systems   .   372
11.7   Anionic Copolymerization Reactions   .   .   .   373

Contents

11.7.1 Random copolymerization 373
11.7.2 Block copolymerization 375
11.7.3 Properties of block copolymers 378
11.8 References 380

12. Reactivity and Mechanism in Polymerization by Complex Organometallic Derivatives

A. LEDWITH *and* D. C. SHERRINGTON

12.1 Polymerization of Ethylene and α-olefins by Ziegler-Natta Catalysts. 384
12.1.1 Mechanisms of initiation of propagation . . . 385
12.1.2 Experimental evidence for existing mechanisms . . 398
12.1.3 Oxidation state of the transition metal at the active site . 399
12.1.4 Origin of stereo-regulation 400
12.1.5 Stereo-selective and stereo-elective polymerization . . 406
12.2 Polymerization of Conjugated Di-olefins (Dienes) by Ziegler-Natta and Related Transition Metal Complexes . . . 409
12.2.1 Ziegler-Natta catalysis of the polymerization of conjugated dienes 411
12.2.2 Cobalt and π-allyl-type catalysts for the polymerization of conjugated dienes 416
12.3 References 427

13. Reactivity and Mechanism with Chromium Oxide Polymerization Catalysts

D. R. WITT

13.1 Introduction 431
13.1.1 Description of catalyst and catalyst activation . . 431
13.1.2 Description of polymerization process . . . 432
13.1.3 Reaction of CrO_3 with support 432
13.2 Detailed Study of Co-Treated Catalysts . . . 435
13.3 Reactivity and Mechanism 435
13.4 References 445

14. Interaction of Light with Monomers and Polymers

F. C. DE SCHRYVER *and* G. SMETS

14.1 Introduction 446
14.2 Principles of Photochemical Processes . . . 446

14.3   Photoinitiated Polymerization   449
14.4   Photopolymerization   454
   14.4.1   Photopolymerization in the solid state   .   .   .   454
   14.4.2   Solution photopolymerization   .   .   .   456
14.5   Photochemical Processes in High Polymers   .   .   458
14.6   Conclusions   467
14.7   References   467

15. Configuration and Conformation in High Polymers

P. L. LUISI *and* F. CIARDELLI

15.1   Introduction   471
15.2   Configuration   473
   15.2.1   Evaluation of stereo-regularity type and degree   .   482
   15.2.2   Influence of stereo-regularity on polymer properties   .   482
15.3   Basic Problems in the Investigation of Conformational
         Equilibria   483
15.4   The Conformation of Macromolecules in the Crystalline
         State   485
15.5   Statistical Approach to the Investigation of the Conforma-
         tional Isomerism of Macromolecules in Solution   .   .   489
15.6   Experimental Investigations on the Conformation of Macro-
         molecules in Solution   493
15.7   Optical Activity and Conformation of Macromolecules in
         Solution   498
15.8   Conclusions   504
15.9   References   506

16. Thermodynamics of Addition Polymerization Processes

K. J. IVIN

16.1   Introduction   514
16.2   Types of Equilibria in Polymer-Monomer Systems   .   .   514
16.3   Addition Polymerization Considered as an Aggregation
         Process   514
16.4   Methods for Determining Ceiling Temperatures   .   .   517
16.5   The Physical State of the System; Some Possible Situations
         in Equilibrium Polymerization   520
   16.5.1   Polymer and monomer immiscible   .   .   .   520
   16.5.2   Polymer soluble in monomer   .   .   .   522

Contents

16.5.3   Polymer and monomer in a solvent; effect of concentration 523
16.5.4   1:1 Copolymerization of miscible monomers to give insoluble polymer 527
16.5.5   Effect of pressure 531
16.6   Effect of Monomer Structure on Polymerizability of Unsaturated Compounds 532
16.6.1   Ethylene derivatives 532
16.6.2   Aldehydes and ketones 535
16.6.3   Isocyanates 536
16.6.4   1:1 Copolymerization of alkenes with sulphur dioxide . 536
16.7   Effect of Monomer Structure on Polymerizability of Ring Compounds 538
16.7.1   General effect of ring size and substituents; the cycloalkanes 538
16.7.2   Heterocyclic compounds 539
16.7.3   Ring-chain equilibria involving rings of different sizes . 551
16.8   Copolymerizations Involving Reversible Addition . . 553
16.9   References 559

Author Index   567

Subject Index   591

1

General Aspects of Reactivity and Structure in Polymerization Processes

N. C. Billingham

School of Molecular Sciences, University of Sussex

A. Ledwith

Donnan Laboratories, University of Liverpool

1.1 THE NATURE OF HIGH POLYMERS

Despite the wide variety of polymeric materials which are available, certain fundamental principles apply to all polymers. High polymers are composed of very large molecules formed by linking together large numbers of simple *structural units*. The simplest polymer which can be envisaged consists of the repetition of a single structural unit many times over to form a *linear homopolymer* which can be represented by the formula

$$X—M—M—M \cdots\cdots\cdots M—M—Y$$

in which M is the structural unit and X and Y the *end-groups*. The end-groups of a polymer represent only a very small fraction of its total weight and their presence is frequently ignored; they may nevertheless exert a considerable influence on the properties of the polymer, particularly with respect to its thermal stability. The structural units of the polymer chain are normally closely related to the structure of the monomer from which the polymer is prepared. Linear polymer structures are not restricted to those formed from a single structural unit; the admixture of two or more different units gives rise to linear *copolymers*. The majority of synthetic copolymers contain only two types of structural unit which may occur in a *random* sequence along the chain or in more ordered *block* or *alternating* arrangements.

Clearly the principal requirement for a substance to be capable of forming a structural unit of a polymer chain is that it must be difunctional, i.e. it must possess two sites capable of bonding to other units. By definition an exclusively difunctional monomer can only give rise to a linear polymer. If, however, some of the structural units are polyfunctional, i.e. capable of bonding to more than two other units, then wide variations in molecular geometry are possible giving rise to *branching* and *cross-linking* in homopolymers and copolymers and allowing the synthesis of *graft copolymers*, in which side chains derived from one monomer are attached to a main chain formed from a different monomer.

Difunctionality, which is the essential characteristic of a polymer precursor, may be achieved in many different ways, but polymers are most commonly formed by two main types of growth reaction traditionally referred to as condensation (step reaction) and addition (chain reaction) processes.

1.2 STEP REACTION POLYMERIZATION

Many simple organic reactions are known in which two molecules become joined, typical examples being the condensation of an acid with an alcohol to yield an ester and the similar reaction of an acid with an amine to yield an amide. Reactions of this type may be adapted to the formation of polymers by using molecules which are functionally capable of coupling indefinitely by condensation reactions, that is to say molecules having more than one carboxyl, amine or alcohol function. Difunctionality may be achieved by using a single monomer bearing two different functions as for example in the polymerization of an hydroxy acid to a polyester

$$n \text{ HO—(CH}_2)_x\text{—COOH} \longrightarrow \text{H} \!\!\left[\!\text{O—(CH}_2)_x\text{—}\overset{\displaystyle \overset{\text{O}}{\|}}{\text{C}}\!\right]_{\!n}\!\! \text{O—H} + (n-1)\text{H}_2\text{O}$$

Alternatively the two functional groups may be present on different molecules as for example in the reaction of a diacid with a diamine to yield a linear polyamide

$$n \text{ HOOC—(CH}_2)_4\text{—COOH} + \text{H}_2\text{N—(CH}_2)_6\text{—NH}_2 \longrightarrow$$

$$\text{H—O}\!\!\left[\!\overset{\overset{\text{O}}{\|}}{\text{C}}\text{—(CH}_2)_4\text{—}\overset{\overset{\text{O}}{\|}}{\text{C}}\text{—}\overset{\overset{\text{H}}{|}}{\text{N}}\text{—(CH}_2)_6\text{—}\overset{\overset{\text{H}}{|}}{\text{N}}\!\right]_{\!n}\!\!\text{H} + (n-1)\text{H}_2\text{O}$$

In this latter example the repeat unit of the polymer is produced from both starting materials—despite the fact that two monomers are used the polymer has only a single repeat unit (contained within the brackets in the formula) and is thus formally classified as a homopolymer. The term copolymer is reserved for more complex cases in which the polymer contains more than one repeat unit.

Most of the coupling reactions used for polymer production are condensation reactions, in which the coupling is accompanied by the elimination of a small molecule such as water, hydrogen chloride or methanol. For this reason such reactions are frequently termed *condensation polymerizations;* it must be emphasized that this term refers to a specific polymerization mechanism in which growth occurs by a step-wise coupling of small molecules, irrespective of whether small molecules are eliminated or not. Thus the reaction of a diol with a diisocyanate to produce a polyurethane

$$HO-R-OH + OCN-R'-NCO \longrightarrow \left\{ O-R-O-\overset{\overset{\displaystyle O}{\parallel}}{C}-\overset{\overset{\displaystyle H}{\mid}}{N}-R'-\overset{\overset{\displaystyle H}{\mid}}{N}-\overset{\overset{\displaystyle O}{\parallel}}{C} \right\}_n$$

proceeds without elimination of any small molecule and is therefore not a condensation reaction. Nevertheless both the mechanism of polymerization and the structure of the final product are typical of condensation polymerizations and the reaction is normally considered as such. In many ways the description of these reactions as *step reaction polymerizations* is both preferable and more correct, but the term condensation is now firmly entrenched in the literature of polymer science. In recent years the concept of step reaction polymerization has been utilized to produce a very wide range of polymeric materials having useful thermal and mechanical characteristics. Although there is usually apparent formal elimination of a small molecule during the growth processes, the reaction mechanisms are frequently very much more complex and representative examples of monomers, catalysts, and polymer structures are indicated in Table 1.1.

The production of polymers by intermolecular condensation reactions exhibits a number of features[1] which are very different from those found in addition reactions (see later), since in polymerizations of this type all of the molecules present are functionally capable of reaction at any time. Consider for example the polymerization of a linear ω-hydroxy acid, $HO-R-COOH$. The first stage is the elimination of water between two monomer molecules to produce a dimer:

$$2\ HO-R-COOH \rightleftharpoons HO-R-\overset{\overset{\displaystyle O}{\parallel}}{C}-O-R-COOH + H_2O$$

Polymerization may then continue by step-wise condensation of a monomer molecule with either end of the dimer, successive reactions yielding trimer, tetramer etc. After a short reaction time most of the monomer will have disappeared and the reaction system will contain a mixture of dimer, trimer and other low polymers. Subsequent condensation steps will occur by coupling of these low molecular weight oligomers to give the final polymer. If the concentration of —OH or —COOH at the beginning of the reaction is C_0 and the concentration at any time is C then the degree of polymerization $\bar{P}$, which is equal to the ratio of the total number of monomer molecules at zero time to the total number of molecules at any other time, is C_0/C. Alternatively, if we let ϕ be the fraction of functional groups which have reacted at any time we have:

$$\bar{P} = C_0/C = C_0/C_0(1-\phi) = \frac{1}{1-\phi}$$

Table 1.1 Examples of step reaction polymerization

Monomer	Polymer
$HO-R-COOH \xrightarrow{-H_2O} HO(RCOO)_nH$	Polyester
$H_2N-R-COOH \xrightarrow{-H_2O} H(NHRCO)_nH$	Polyamide
$HO-R-OH + OCNR'NCO \longrightarrow H(OROCONHR'NHCO)_nO-$	Polyurethane
$ClSiR_2Cl \xrightarrow[-HCl]{H_2O} HOSiR_2OH \xrightarrow{-H_2O} HO(SiR_2O)_nH$	Polysiloxane
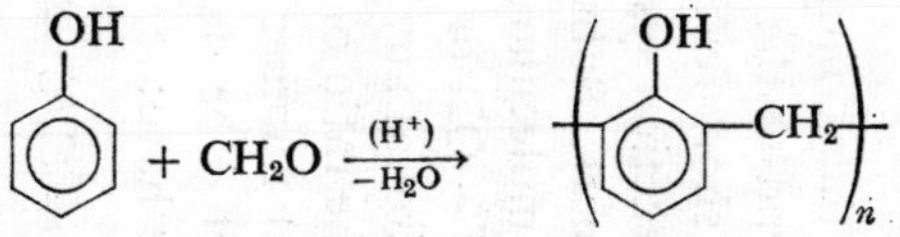	Phenol-formaldehyde
$Cl-R-Cl + Na_2S_x \xrightarrow{-NaCl} (R-S_x)_n$	Polysulphide
$RCHO + HO-R'-OH \xrightarrow{-H_2O} (O-R'-OCHR)_n$	Polyacetal
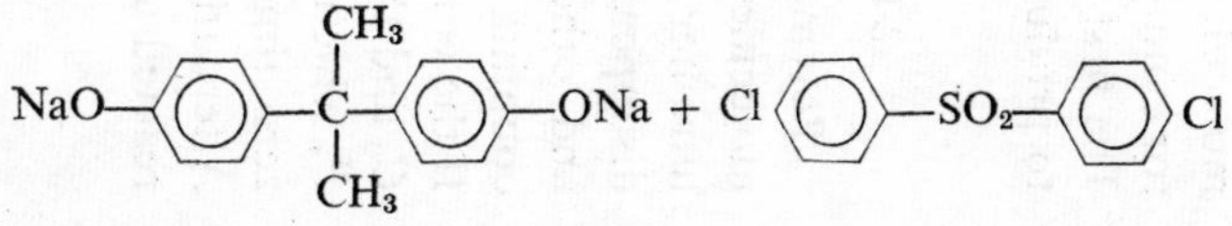	Polysulphone

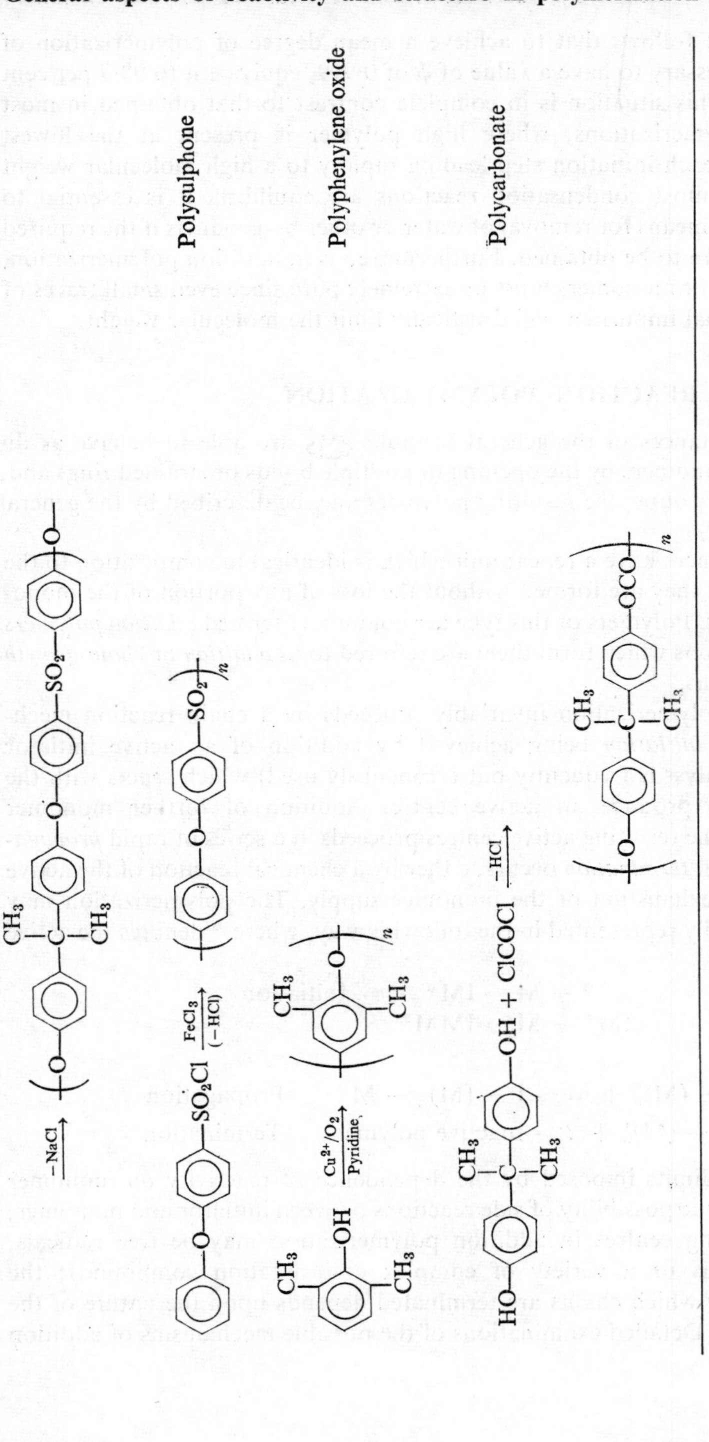

form which it offers that to achieve a given degree of polymerization of 100 it is necessary to attain a value of q of 0.99 or 99 per cent conversion. The situation is fundamentally nearer to that of most addition polymerizations, where high polymer is present at the lowest conversions, so that the molecular mass increased rapidly up to a high molecular weight while, either way, condensation must one way... this is especial, for molecules except for monomers with... conditions must be satisfied... monomers in... cause they even give an even smaller chance of making even small chains of monomers and material... different from the required length.

1.2 CHAIN-GROWTH POLYMERIZATION

Many substances in the form of the nat... are either homo- or heterofunctional molecules as the resultant of polymer bonds produced during, given the end... groups... the materials... producing the polymer... described by the general formula

$$\text{...}$$

A distinct possibility... structures... a condensation... anionic, cationic or free radical addition of a monomer...

Polysulphone

$$\cdots \text{—} \text{SO}_2\text{Cl} \xrightarrow[(-\text{HCl})]{\text{FeCl}_3} \cdots \text{—} \text{SO}_2 \text{—} \cdots$$

Polyphenylene oxide

$$\cdots \text{OH} \xrightarrow[\text{Pyridine}]{\text{Cu}^{2+}/\text{O}_2} \cdots$$

Polycarbonate

$$\cdots \text{OH} + \text{ClCOCl} \xrightarrow{-\text{HCl}} \cdots \text{OCO} \cdots$$

from which it follows that to achieve a mean degree of polymerization of 1000 it is necessary to have a value of ϕ of 0·999, equivalent to 99·9 per cent conversion. This situation is in complete contrast to that obtained in most addition polymerizations, where high polymer is present at the lowest conversions, each initiation step leading rapidly to a high molecular weight chain. Since most condensation reactions are equilibria it is essential to provide some means for removal of water or other by-products if the required conversions are to be obtained. Furthermore, as in addition polymerization, the monomer or monomers must be extremely pure since even small traces of monofunctional impurities will drastically limit the molecular weight.

1.3   CHAIN REACTION POLYMERIZATION

Many substances of the general formula M are able to behave as difunctional monomers by the opening of multiple bonds or strained rings and, ignoring end-groups, the resulting polymers may be described by the general formula $(M)_n$.

These polymers have a repeat unit which is identical in composition to the monomer and they are formed without the loss of any portion of the monomer molecules. Polymers of this type are commonly termed *addition polymers* and the reactions which form them are referred to as *addition* or *chain-growth* polymerizations.

Addition polymerization invariably proceeds by a chain reaction mechanism, chain *initiation* being achieved by addition of an active initiator (the term catalyst is frequently but erroneously used) which reacts with the monomers to produce an active centre. Addition of further monomer molecules to the resulting active centres proceeds in a series of rapid *propagation* steps, until *termination* occurs, either by a chemical reaction of the active centre or by exhaustion of the monomer supply. The polymerization may thus be formally represented in the following way, where * denotes the active centre.

$$I^* + M \rightarrow IM^* \qquad \text{Initiation}$$
$$IM^* + M \rightarrow IMM^*$$

or in general

$$I - (M)_n^* + M \rightarrow I - (M)_n - M^* \qquad \text{Propagation}$$
$$I - (M)_n^* + ? \rightarrow \text{Inactive polymer} \qquad \text{Termination}$$

Within the limits imposed by the dependence of reactivity on monomer structure and the possibility of side reactions between initiator and monomer, the propagating centres in addition polymerization may be free radicals, anions, cations or a variety of complex co-ordination compounds; the mechanism by which chains are terminated depends upon the nature of the active centres. Detailed examinations of the possible mechanisms of addition polymerization are given in later chapters and before discussing the general tendencies of types of monomer to polymerize via the various propagating species, it is worthwhile to examine the essential differences between addition and condensation polymerizations.

We have seen that addition polymerizations may involve a variety of active centres, and although each may lead to rather different polymerization behaviour, all such polymerizations have a number of common features. The production of polymer in each case involves the rapid addition of monomer to a few active centres, the monomer concentration decreasing slowly throughout the reaction. High molecular weight polymer is present even at low conversions, the mixture always consisting of high polymer and unreacted monomer. As with all chain reactions, successful addition polymerization requires highly purified materials to avoid adventitious termination by impurities. Polymers produced by addition polymerization have chains of repeat units identical in composition to the monomers, and which are joined (in the majority of cases) by carbon—carbon bonds. In contrast, condensation polymers have repeat units which differ from the monomer or monomers in the loss of water, hydrogen chloride etc., and which are often joined by polar groupings containing hetero-atoms. The presence of polar groupings in the chain, with the associated possibility of strong interchain forces due to hydrogen bonding and dipolar interactions, accounts for many of the differences in physical properties between typical addition and condensation polymers. By the nature of the polymerization reaction, condensation polymers have regular, simple structures with internal functional groups repeating at regular intervals; many of the problems associated with chain branching and isomerism in addition polymers (see section 1.6) do not arise. Since the distinction between step reaction and chain reaction mechanisms of polymerization is of fundamental importance it is summarized in Table 1.2.

As already noted, whether or not a molecule will undergo addition polymerization depends upon the degree and nature of any unsaturation or ring strain present in the molecule and it is convenient to survey the two main types of monomer activation separately.

1.4 POLYMERIZABILITY OF UNSATURATED COMPOUNDS

The types of unsaturated molecule readily undergoing chain reaction polymerization include olefins, conjugated dienes, acetylenes, and carbonyl compounds with the first two types being of most importance. Representative examples of monomers and corresponding polymers are shown in Table 1.3.

All monosubstituted ethylenes ($RCH{=}CH_2$) are capable of homopolymerization by an appropriate mechanism (see below) to give polymers

Table 1.2

Chain reaction polymerization	Step reaction polymerization
Growth occurs by rapid addition of monomer to a small number of active centres	Growth occurs by coupling of any two species (monomer or polymer)
Monomer concentration decreases slowly during reaction	Monomer concentration decreases rapidly before any high polymer is formed
High polymer is present at low conversion	Polymer molecular weight increases continuously during polymerization; high polymer is present only at very high conversions
Mean degree of polymerization may be very high	Mean degree of polymerization usually fairly low
Rate of polymerization is zero initially, quickly rises to a maximum, as active centres are formed from the initiator, and remains more or less constant during the reaction before falling off when the initiator is consumed	Rate of polymerization is a maximum at the start and decreases continuously during the reaction as the concentration of functional groups decreases

Table 1.3   Mono-olefins polymerizable by chain reaction processes

Monomer	Polymer	Polymerization mechanisms
$CH_2{=}CH_2$	$-CH_2CH_2-$ Polyethylene	Radical, Ziegler-Natta, supported metal oxide
$CH_2{=}CHCH_3$	$-CH_2-CH-$ $\quad\quad\; \mid$ $\quad\quad CH_3$ Polypropylene	Ziegler-Natta, supported metal oxide
$CH_2{=}CHPh$	$-CH_2-CH-$ $\quad\quad\; \mid$ $\quad\quad Ph$ Polystyrene	Radical, anionic, cationic, Ziegler-Natta
$CH_2{=}CHCl$	$-CH_2-CH-$ $\quad\quad\; \mid$ $\quad\quad Cl$ Polyvinylchloride	Radical
$CH_2{=}CHCN$	$-CH_2-CH-$ $\quad\quad\; \mid$ $\quad\quad CN$ Polyacrylonitrile	Radical, anionic
$CH_2{=}CHCOOCH_3$	$-CH_2-CH-$ $\quad\quad\;\;\; \mid$ $\quad\quad COOCH_3$ Polymethylacrylate	Radical, anionic

Monomer	Polymer	Polymerization mechanisms
$CH_2{=}CH{-}OR$	$-CH_2{-}CH-$ $\quad\quad\;\; \mid$ $\quad\quad\; OR$ Polyalkylvinylether	Cationic
$CH_2{=}CHOCOCH_3$	$-CH_2{-}CH-$ $\quad\quad\;\;\; \mid$ $\quad\quad\; OCOCH_3$ Polyvinylacetate	Radical
$\quad\quad CH_3$ $\quad\quad\; \mid$ $CH_2{=}C{-}CH_3$	$\quad\quad CH_3$ $\quad\quad\; \mid$ $-CH_2{-}C-$ $\quad\quad\; \mid$ $\quad\quad CH_3$ Polyisobutene	Cationic
$\quad\quad CH_3$ $\quad\quad\; \mid$ $CH_2{=}C{-}Ph$	$\quad\quad CH_3$ $\quad\quad\; \mid$ $-CH_2{-}C-$ $\quad\quad\; \mid$ $\quad\quad Ph$ Poly-α-methylstyrene	Cationic, anionic
$\quad\; Cl$ $\quad\; \mid$ $CH_2{=}C{-}Cl$	$\quad\quad Cl$ $\quad\quad\; \mid$ $-CH_2{-}C-$ $\quad\quad\; \mid$ $\quad\quad Cl$ Polyvinylidenechloride	Radical
$\quad\quad CH_3$ $\quad\quad\; \mid$ $CH_2{=}C{-}COOCH_3$	$\quad\quad CH_3$ $\quad\quad\; \mid$ $-CH_2{-}C-$ $\quad\quad\; \mid$ $\quad\quad COOCH_3$ Polymethylmethacrylate	Radical, anionic
$CH{=}CH$ $\; \mid\quad\; \mid$ $CO\quad CO$ $\quad \searrow \swarrow$ $\quad\;\; N$ $\quad\;\; \mid$ $\;\; R(Ar)$	$-CH{-}CH-$ $\;\;\; \mid\quad\; \mid$ $\;\; CO\quad CO$ $\quad\; \searrow \swarrow$ $\quad\quad N$ $\quad\quad \mid$ $\quad\; R(Ar)$ Polymaleimides	Radical, anionic
$CH_2{-}CH$ $\; \mid\quad\;\; \parallel$ $CH_2{-}CH$	$-CH{-}CH-$ $\;\;\; \mid\quad\;\; \parallel$ $\;\; CH_2{-}CH_2$ Polycyclobutene	Ziegler-Natta
$F_2C{=}CF_2$	$-CF_2CF_2-$ Polytetrafluoroethylene	Radical

having exclusively carbon–carbon backbones, that is

$$n\ RCH{=}CH_2 \longrightarrow -(CH_2{-}CH)_n{-}$$
$$\text{(R)}$$

Similarly many 1,1-disubstituted ethylenes ($R_1R_2C{=}CH_2$) may be homo-polymerized to high polymer, notable exceptions being 1,1-diarylethylenes which yield dimers exclusively, because of steric effects, for example

$$2\ Ph_2C{=}CH_2 \longrightarrow CH_3{-}\overset{\displaystyle Ph}{\underset{\displaystyle Ph}{C}}{-}CH{=}CPh_2$$

Other structurally isomeric dimers and their derivatives may be formed by attempted polymerization of 1,1-diaryl ethylenes, and these olefins provide useful models for studies of the mechanisms of catalysis and initiation.

In contrast, 1,2-disubstituted ethylenes ($R_1CH{=}CHR_2$) are not normally homopolymerizable to high molecular weight polymer except when the substituents comprise part of a strained ring system or, alternatively, when either R_1 or R_2 is an alkoxy group (RO—). It is well known that alkoxy groups have comparatively poor steric hindering effects because of the bond angle and flexible nature of the C—O—C linkage, confirming the steric nature of the forces inhibiting homopolymerization of 1,2-disubstituted ethylenes. Originally it was thought that this steric interference arose from interactions in the polymeric backbone but inspection of completely eclipsed conformations for homopolymers obtained from the isomers RCH=CHR and $R_2C{=}CH_2$, shows that, if anything, steric interference is less in polymers from the former olefin.

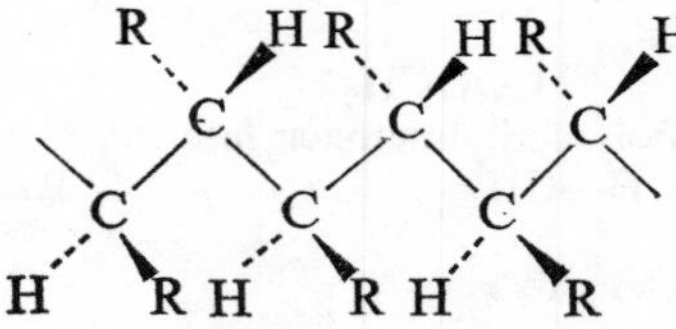

polymer from RCH=CHR

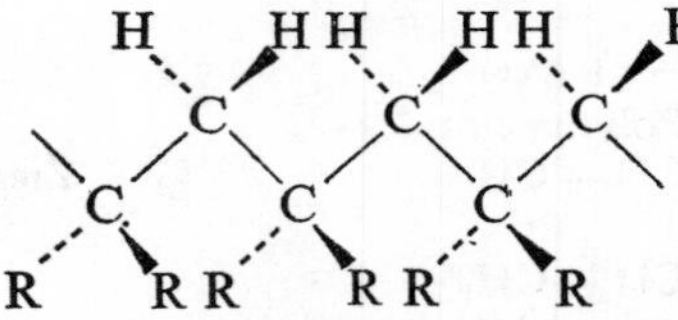

polymer from $R_2C{=}CH_2$

It follows therefore that the steric blocking effects of substituents are reflected in adverse kinetic, rather than thermodynamic factors; this fact being clearly demonstrated by the ready formation of polyethylidene

$$-\!(CH)_{\overline{n}} \qquad -\!(CH)_{\overline{n}}$$
$$\quad\; |\qquad\qquad\; |$$
$$\quad\; CH_3 \qquad\quad\; Ph$$

and polybenzylidene from (highly reactive) diazoethane CH_3CHN_2, and phenyldiazomethane $PhCHN_2$ respectively.[2] It is worth noting that formation of polyalkylidenes from diazoalkanes is an unmistakable chain polymerization although a molecule of nitrogen is eliminated at each propagation step, for example

$$-CH-\overset{+}{N_2} + RCHN_2 \longrightarrow -CH-CH-\overset{+}{N_2} + N_2 \text{ etc.}$$
$$\quad |\qquad\qquad\qquad\qquad\quad |\quad\; |$$
$$\quad R\qquad\qquad\qquad\qquad\quad R\quad R$$

Steric effects completely inhibit homopolymerization of the vast majority of tri- and tetra-substituted ethylenes except in the special case of fluorinated derivatives, notably tetrafluoroethylene which homopolymerizes exceedingly readily to give polytetrafluoroethylene. Presumably both the very small size, and activating influence, of fluorine atoms contribute to this exception.

More detailed consideration of the thermodynamics of chain reaction polymerization is given in Chapter 16 but it should be noted that the generalizations made above relate only to *homopolymerization* and that most alkenes, whatever the substitution pattern, may be *copolymerized* with suitable co-monomers.

A further guide to strain effects in monomers and polymers is provided by appropriate values of heats (enthalpies) of polymerization.[3] The conversion of alkenes $R_1R_2C\!=\!CR_3R_4$ to high polymer is a strongly exothermic process with the specific enthalpy change ΔH_p depending on the actual reaction conditions, e.g. dissolved monomer to dissolved polymer, dissolved monomer to solid polymer, gaseous monomer to solid polymer etc. A selection of typical values of ΔH_p is given in Table 1.4 and the steric effect of substituents is clearly apparent, increased size of substituent or increased numbers of substituents resulting in a decrease in the heat of polymerization. Many important polymers are derived from ethylene and its monosubstituted derivatives, and the high exothermicity of homopolymerization largely precludes the possibility of polymerizing olefins in bulk. The problems of heat dissipation are usually overcome by employing solution, suspension or emulsion procedures, although in recent years bulk polymerization of vinyl chloride has become commercially feasible.

Acetylenes are much more difficult to homopolymerize and copolymerize than corresponding alkenes and in general, only monosubstituted acetylenes $RC\!\equiv\!CH$ may be homopolymerized and then only to comparatively low

Table 1.4 Heats of polymerization[a]

Monomer	$-\Delta H_p$ (kcal mol^{-1})
Ethylene	22·7
Propylene	20·5
Isobutylene	12·3
1,3-Butadiene	17·4
Isoprene	17·8
Styrene	16·7
α-Methyl styrene	8·4
Vinyl chloride	22·9
Vinylidene chloride	18·0
Tetrafluoroethylene	37·2
Methyl acrylate	18·8
Methyl methacrylate	13·5
Vinyl acetate	21·0

a. ΔH_p refers to the conversion of liquid monomer
to amorphous or slightly crystalline polymer.

molecular weight material. Addition polymerization of acetylenes produces a polymer backbone having a completely conjugated structure, that is

$$(n + 2)RC\equiv CH \longrightarrow -\underset{R}{C}=CH\text{-}(\underset{R}{C}=CH)_n\underset{R}{C}=CH-$$

The extensive conjugation confers, after a certain chain length, a high degree of resonance stabilization (and therefore a low reactivity) on the propagating intermediate, whether it be cation, anion, or radical. It is this fact which largely precludes formation of high molecular weight addition polymers although an additional factor is to be found in the facility for chain transfer to monomer, especially for anionic and free radical polymerizations, that is

$$\sim CH=\underset{R}{C}-CH=\underset{R}{C}^* + RC\equiv CH \longrightarrow \sim CH=\underset{R}{C}-CH=\underset{R}{CH} + RC\equiv C^*$$

On the other hand, interest in polymers from acetylenes rests entirely on potential application as organic semiconductors, a property dependent mainly on the degree of conjugation in the backbone. Up to now, polymers from acetylenes which have been properly characterized are derived mainly from phenyl acetylene PhC≡CH and its ring substituted derivatives, and are intensely coloured, or even black, as a result of backbone conjugation.[4]

Conjugated dienes are readily polymerized alone and with other mono-olefins, although here again, 1,4-terminal substitution drastically lowers reactivity. Important examples of conjugated dienes undergoing homo-polymerization include 1,3-butadiene $CH_2\!=\!CHCH\!=\!CH_2$, isoprene

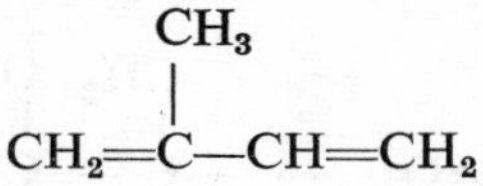

$$
\begin{array}{c}
CH_3 \\
| \\
CH_2\!=\!C\!-\!CH\!=\!CH_2
\end{array}
$$

and chloroprene

$$
\begin{array}{c}
Cl \\
| \\
CH_2\!=\!C\!-\!CH\!=\!CH_2
\end{array}
$$

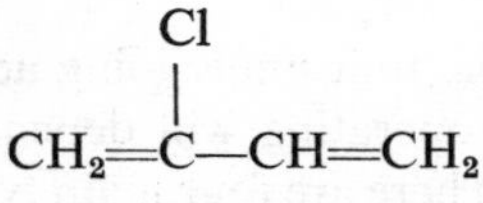

As for all addition reactions of conjugated dienes, there are several possible structural isomers of the diene unit which may be incorporated in a polymer chain, and these are outlined fully in section 1.7.

Carbonyl compounds $R_1R_2C\!=\!O$ are, in theory, capable of addition homopolymerization via the $C\!=\!O$ group but, in reality, readily character-izable high polymers are formed only from aliphatic aldehydes RCHO, notably formaldehyde CH_2O and acetaldehyde CH_3CHO

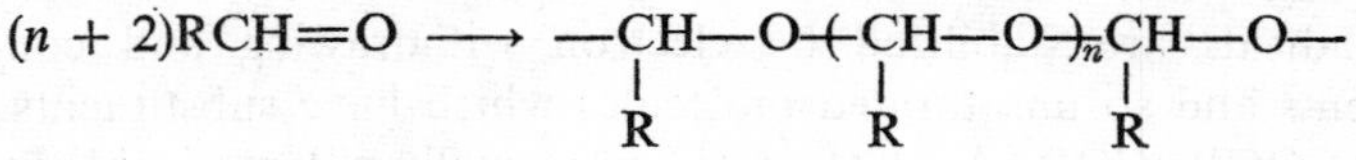

$$
(n+2)RCH\!=\!O \longrightarrow
\begin{array}{c}
-CH\!-\!O\!\!\!-\!\!\!(CH\!-\!O\!\!-\!)_n CH\!-\!O- \\
| \qquad\quad | \qquad\qquad | \\
R \qquad\quad R \qquad\qquad R
\end{array}
$$

The homopolymers are more correctly described as polyacetals rather than polyethers.

There have been reports of the synthesis of a homopolymer of acetone but, perhaps because of its anticipated low ceiling temperature (see Chapter 16), these remain largely unconfirmed although it is interesting that thioacetone $(CH_3)_2C\!=\!S$ readily gives a high molecular weight homopolymer.[5]

Ketenes, notably dimethyl ketene, readily undergo oligomerization and polymerization and, depending upon reaction conditions, the product may be a polyketone formed by addition polymerization of the ketene double bond,

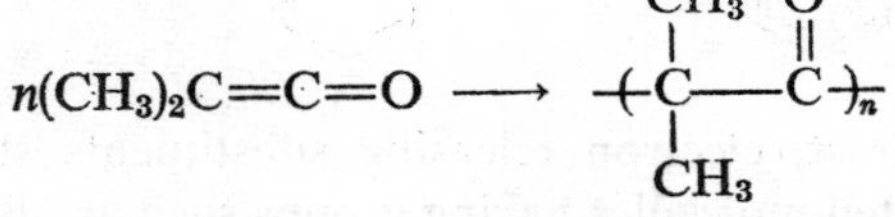

$$
n(CH_3)_2C\!=\!C\!=\!O \longrightarrow
\begin{array}{c}
\quad CH_3 \;\; O \\
\quad | \quad\;\; || \\
-\!\!(C\!-\!C)_n \\
\quad | \\
\quad CH_3
\end{array}
$$

a polyacetal formed by addition polymerization of the carbonyl group,

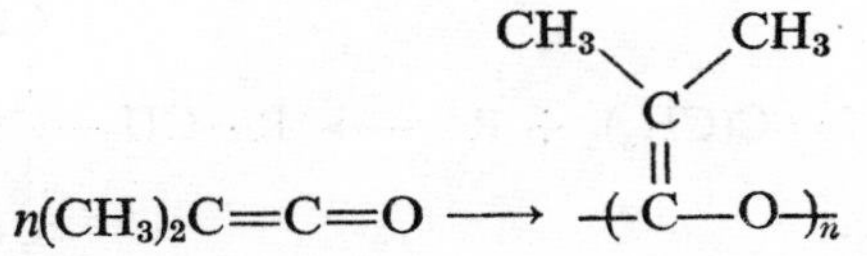

$$
n(CH_3)_2C\!=\!C\!=\!O \longrightarrow
\begin{array}{c}
CH_3\;\;\;\;\;\; CH_3 \\
\diagdown \;\; \diagup \\
C \\
|| \\
-\!\!(C\!-\!O)_n
\end{array}
$$

or, a polyester formed by a remarkable alternation of polymerization through both C=C, and C=O linkages.[6]

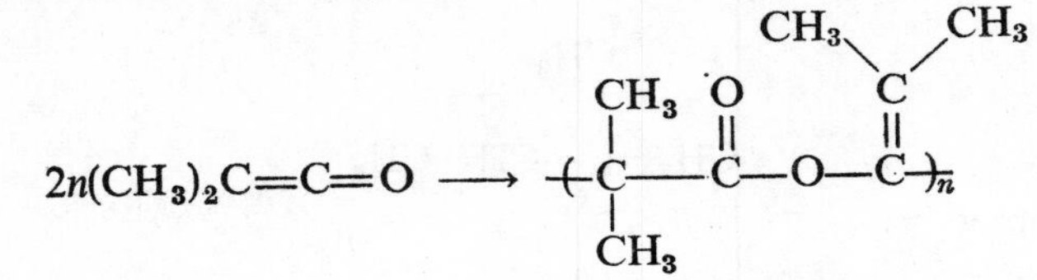

$$2n(CH_3)_2C{=}C{=}O \longrightarrow \left(\begin{matrix} CH_3 \\ | \\ C \\ | \\ CH_3 \end{matrix} \begin{matrix} O \\ \| \\ C \end{matrix} {-}O{-} \begin{matrix} CH_3 \diagdown \diagup CH_3 \\ C \\ \| \end{matrix}\right)_n$$

Whatever the nature of the unit undergoing addition polymerization, the actual reaction mechanism operating will depend upon the substituents at the active propagation site. There are four main types of reactive intermediate encountered in chain reaction polymerizations: radicals, cations, anions and complex organometallic derivatives. Each type of intermediate has characteristic features which are just as much reflected in appropriate chain reaction polymerizations, as they are in the wider compass of general organic reactions, and the major part of this volume is devoted to detailed discussion of reactivity and mechanisms associated with the various propagating species. However, it is perhaps useful at this stage to indicate briefly the factors contributing specific mechanistic activity to polymerizable monomers.

Free-radicals are stabilized by electron withdrawing and conjugating substituents and so unsaturated molecules which have substituents such as Cl, CN, COOR, COR, Ar, CH=CH$_2$, are usually polymerizable by radical processes. The same type of substituent stabilizes a carbanion by resonance delocalization, and these monomers are likewise prone to anionic polymerization, for example

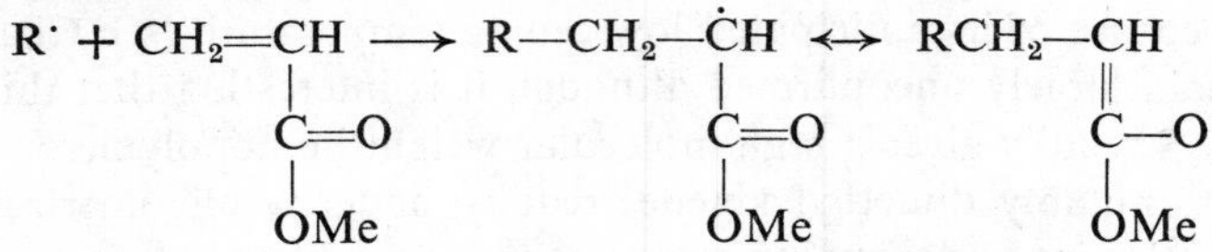

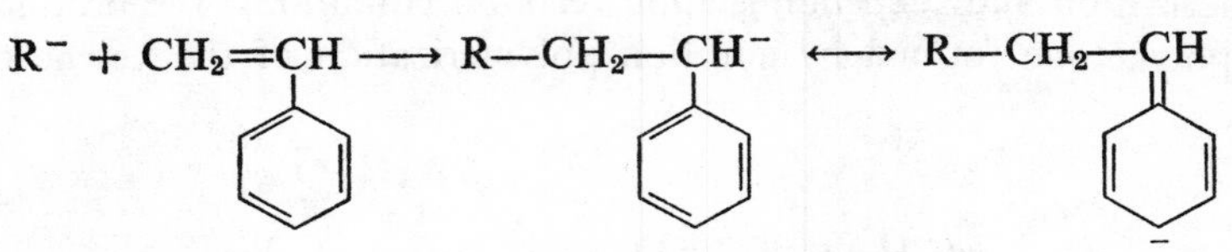

$$R^{\cdot} + CH_2{=}\underset{\underset{\underset{OMe}{|}}{\overset{\|}{C}{=}O}}{CH} \longrightarrow R{-}CH_2{-}\underset{\underset{\underset{OMe}{|}}{\overset{\|}{C}{=}O}}{\overset{\cdot}{C}H} \longleftrightarrow RCH_2{-}\underset{\underset{\underset{OMe}{|}}{\overset{\|}{C}{-}O^{\cdot}}}{CH}$$

$$R^- + CH_2{=}CH \longrightarrow R{-}CH_2{-}CH^- \longleftrightarrow R{-}CH_2{-}CH$$

In complete contrast, electron releasing substituents stabilize carbonium ions and unsaturated molecules having groups such as alkyl, RO, RS, R$_2$N, Ar and CH$_2$=CH—, are most easily polymerized by cationic processes, for example

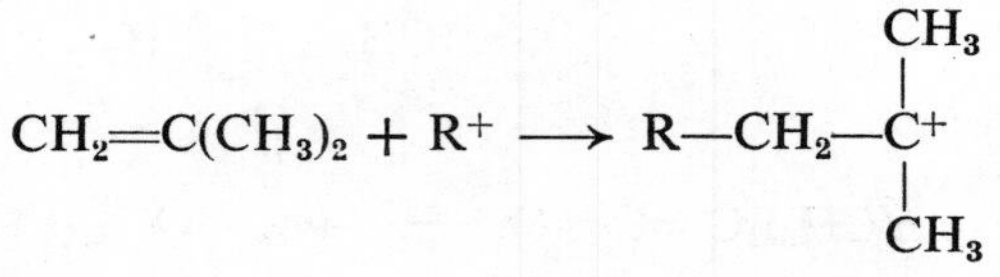

$$CH_2{=}C(CH_3)_2 + R^+ \longrightarrow R{-}CH_2{-}\underset{\underset{CH_3}{|}}{\overset{\overset{CH_3}{|}}{C^+}}$$

$$CH_2{=}CHOR' + R^+ \longrightarrow \underset{OR'}{RCH_2{-}CH^+} \longleftrightarrow \underset{+OR}{RCH_2{-}CH}$$

$$CH_2{=}CHPh + R^+ \longrightarrow RCH_2{-}CH^+ \longleftrightarrow RCH_2{-}CH$$

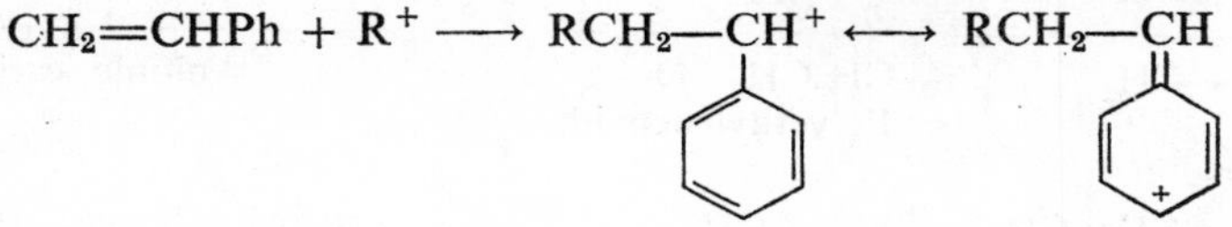

Simple hydrocarbon monomers $RCH{=}CH_2$ are most easily polymerized by complex organometallic initiators where the propagating species may be approximated to a transition-metal-primary-alkyl bond

$$\underset{R'{-}CH{-}CH_2{-}Metal}{\overset{R}{|}}$$

although ethylene is polymerized industrially by a comparatively high pressure, high temperature, free radical reaction; isobutene $CH_2{=}C(CH_3)_2$ in contrast, readily gives high molecular weight polymer by cationic polymerization at temperatures around $-100\,°C$.

Aldehydes and other carbonyl compounds are most easily polymerized by essentially anionic mechanisms on account of the stability of alkoxide ions, that is

$$R^- + R'CH{=}O \longrightarrow R{-}\underset{|}{\overset{R'}{C}}H{-}O^- \text{ etc.}$$

although cationic activity is also shown because of the stabilizing influence of the adjacent oxygen atom, that is

$$R^+ + R'CHO \longrightarrow R{-}O{-}\overset{R'}{\underset{|}{C}}H^+ \longleftrightarrow R{-}\overset{+}{O}{=}\overset{R'}{\underset{|}{C}}H$$

1.5 POLYMERIZABILITY OF RING COMPOUNDS

The tendency of a ring compound to polymerize depends upon the extent of ring strain and the availability of a suitable mechanism for the ring opening process.[7] Ring strain is a thermodynamic property which is caused by either forcing the bonds between ring atoms into angular distortion or by steric interaction of substituents on the ring atoms.

The thermodynamic factors affecting polymerizability of ring compounds were first rationalized[8] by use of semi-empirical methods to calculate free energy changes for the hypothetical polymerization of pure liquid cyclo-alkanes, and, since the bond lengths of C—C and C—O bonds do not differ very much, replacement of a carbon atom in a cycloalkane by a heteroatom such as oxygen, or other hetero units affects the free energy change only

Table 1.5 Ring compounds polymerizable by chain reaction processes

Monomer	Polymer	Polymerization mechanisms
CH_2—CH_2 with O bridge (ethylene oxide)	—CH_2CH_2—O— Polyethyleneoxide	Anionic, cationic
CH_2—CH—CH_3 with O bridge	CH_3 | —CH_2—CH—O— Polypropyleneoxide	Anionic, cationic
CH_2—CH—CH_3 with S bridge	CH_3 | —CH_2—CH—S— Polypropylenesulphide	Anionic, cationic
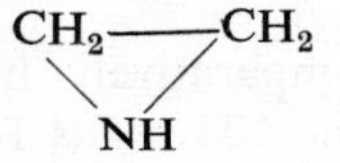 CH_2——CH_2 with NH bridge	—CH_2CH_2—NH— Polyethyleneimine	Cationic
CH_2—O | | CH_2—CH_2	—$CH_2CH_2CH_2$—O— Polytrimethyleneoxide	Cationic
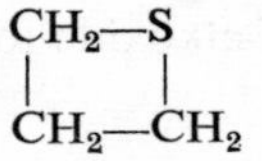 CH_2—S | | CH_2—CH_2	—$CH_2CH_2CH_2$—S— Polytrimethylenesulphide	Anionic
CH_2—CH_2 | | CO—O	—CH_2—CH_2CO—O— Polypropiolactone	Anionic, cationic
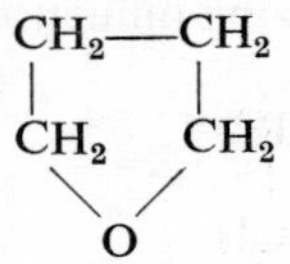 CH_2—CH_2 | | CH_2 CH_2 with O bridge	—$CH_2CH_2CH_2CH_2$—O— Polytetramethyleneoxide	Cationic
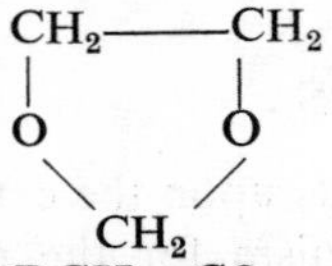 CH_2——CH_2 | | O O with CH_2 bridge	—CH_2CH_2O—CH_2—O— Poly-1,3-dioxolane	Cationic
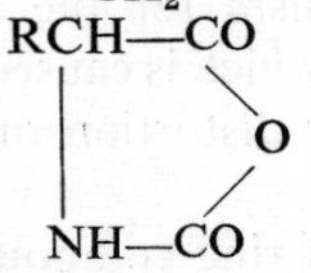 RCH—CO | O | NH—CO	—NH—$CHRCO$— (—CO_2) Poly-N-carboxyacidanhydride	Anionic
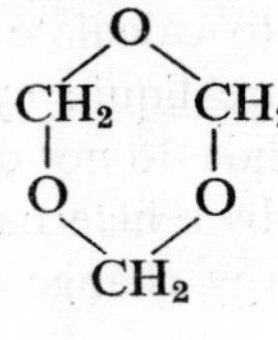 O | CH_2 CH_2 | O O with CH_2 bridge	—OCH_2— Polymethyleneoxide	Cationic

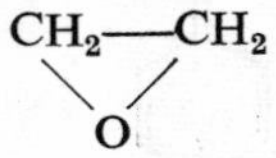

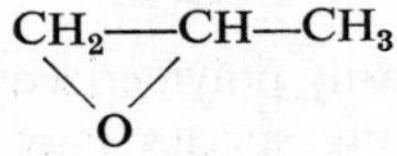

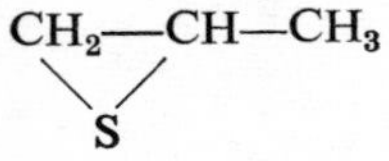

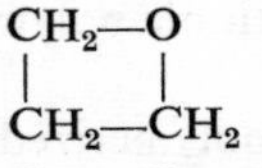

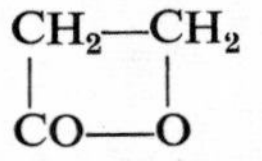

Table 1.5 (*contd.*)

Monomer	Polymer	Polymerization mechanisms
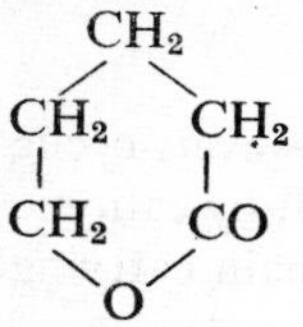 CH_2 / CH_2 CH_2 / CH_2 CO / O (ring structure)	—$CH_2CH_2CH_2CH_2CO$—O— Poly-δ-valerolactone	Anionic, cationic
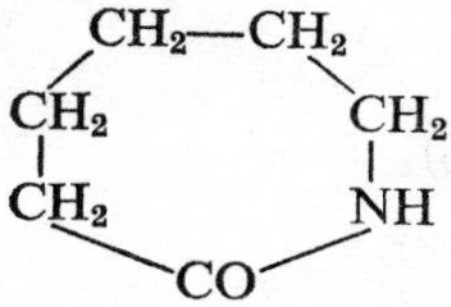 CH_2—CH_2 / CH_2 CH_2 / CH_2 NH / CO (ring structure)	—$CH_2CH_2CH_2CH_2CH_2CONH$— Poly-ε-caprolactam	Anionic

slightly. Data for cycloalkanes can therefore be taken to represent roughly, the free energy changes involved in ring opening polymerization and the following generalizations emerge:

1. In any one group of derivatives, the thermodynamic feasibility of the ring opening process decreases with increasing ring size in the order 3 > 4 > 5 > 6, at which ring size the free energy change has become positive. With further increase in ring size, ring opening again becomes possible, the free energy change favouring the reaction in the order 8 > 7 > 6.

2. For any one size of ring, the thermodynamic feasibility of ring opening decreases in the order unsubstituted derivative > monomethyl derivative > 1,1-dimethyl derivative.

3. The thermodynamic criterion for a spontaneous process, that the free energy changes must be negative, is of importance only in the case of the six-membered rings (unsubstituted, monomethyl-substituted and 1,1-disubstituted), and substituted five-membered rings. In all these cases the attendant free energy change is positive, thereby precluding their participation in the (hypothetical) ring opening polymerization.

Since these calculations were based mainly on cycloalkanes, it is to be expected that exceptions will be noted. However, the general rules apply to a variety of ring opening polymerizations and representative examples are given in Table 1.5. A notable exception is the case of δ-valerolactone (see Table 1.5) which may be homopolymerized by radical, anionic and cationic mechanisms. This reaction is a typical chain-reaction polymerization; no water is eliminated and the polymer structural unit is directly related to the monomer. However, the product is a linear polyester more typical of the products of step reaction polymerization, as indeed are the polyamides produced by ring opening polymerization of cyclic lactams, e.g. ε-caprolactam.

It would be expected that cyclic ethers polymerize readily by simple anionic ring opening chain reactions because of the stability of alkoxide ions, for example

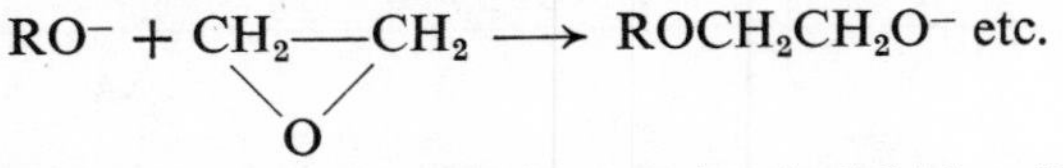

$$RO^- + CH_2\!\!-\!\!CH_2 \longrightarrow ROCH_2CH_2O^- \text{ etc.}$$
$$\underset{O}{\diagdown\diagup}$$

and this is certainly the case for epoxides and episulphides. However, cyclic ethers (and sulphides) with larger ring sizes, polymerize, if at all, by cationic processes in which cyclic oxonium (or sulphonium ions) are chain carrying intermediates, for example

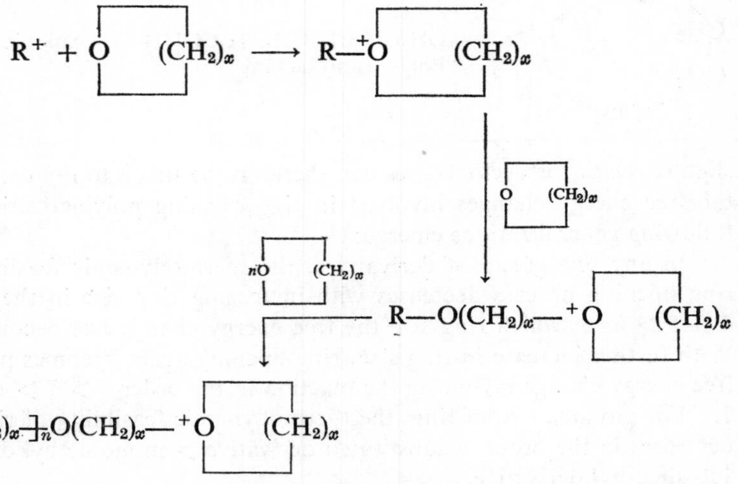

$$R^+ + O\quad(CH_2)_x \longrightarrow R\!\!-\!\!\overset{+}{O}\quad(CH_2)_x$$

$$O\quad(CH_2)_x$$

$$n\,O\quad(CH_2)_x$$

$$R\!\!-\!\!O(CH_2)_x\!\!-\!\!\overset{+}{O}\quad(CH_2)_x$$

$$R\!\!-\!\!\!\left[O(CH_2)_x\right]_{\!n}\!\!O(CH_2)_x\!\!-\!\!\overset{+}{O}\quad(CH_2)_x$$

Epoxides, episulphides and ethylene imines also polymerize via similar cyclic 'onium' ion mechanisms and for the first named class of monomer, polymerization by complex organometallic catalysts is important although whether this type of polymerization should be termed cationic or anionic is not absolutely certain.

Special mention must be made of the type of polymer obtainable by polymerization of cyclic olefins. Regular vinyl addition polymers may be obtained from particularly strained endocyclic double bonds (see Table 1.3) but recent work with complex organometallic systems or activated metal oxide catalysts has shown that commercially important unsaturated hydrocarbon polymers may be obtained by ring opening reactions of cyclopentene but not cyclohexene.[9] Essentially this work follows use of similar catalysts to achieve redistribution reactions of olefins, e.g. disproportionation of 1-butene to ethylene and 3-hexene

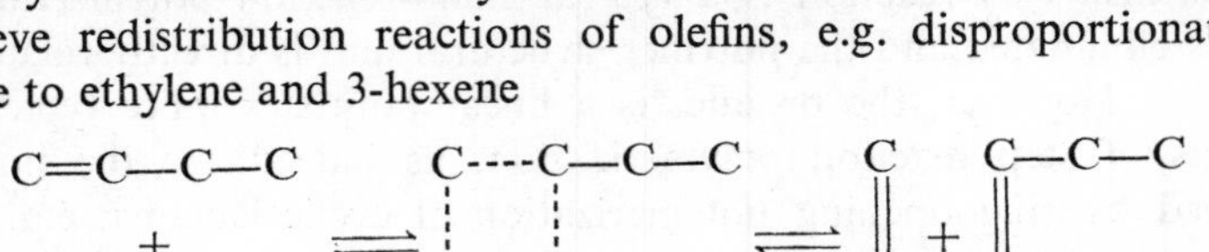

$$
\begin{array}{ccc}
C{=}C{-}C{-}C & C\text{----}C{-}C{-}C & C \quad C{-}C{-}C\\
+ & \rightleftharpoons & \rightleftharpoons \quad \| + \|\\
C{=}C{-}C{-}C & C\text{----}C{-}C{-}C & C \quad C{-}C{-}C
\end{array}
$$

The mechanism of this remarkable transformation remains enigmatic but extension of the reaction to strained cyclic olefins, particularly cyclopentene, gives rise to high molecular weight macrocyclic polyalkenamers

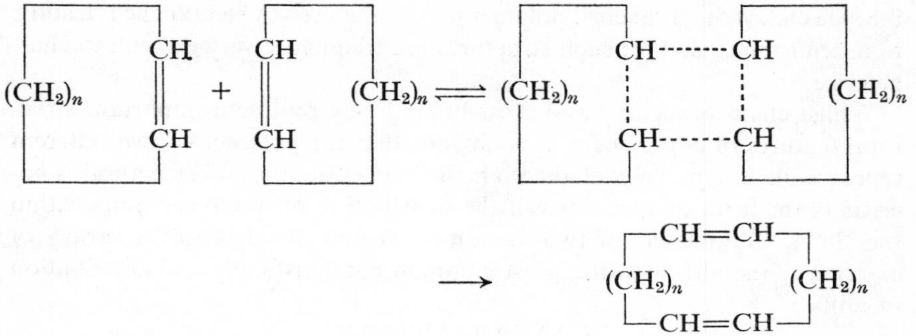

$$(CH_2)_n \quad \begin{matrix} CH \\ \| \\ CH \end{matrix} \quad + \quad \begin{matrix} CH \\ \| \\ CH \end{matrix} \quad (CH_2)_n \rightleftharpoons (CH_2)_n \quad \begin{matrix} CH \text{------} CH \\ \vdots \quad\quad \vdots \\ CH \text{------} CH \end{matrix} \quad (CH_2)_n$$

$$\longrightarrow \quad (CH_2)_n \begin{matrix} \ulcorner CH{=}CH \urcorner \\ \\ \llcorner CH{=}CH \lrcorner \end{matrix} (CH_2)_n$$

etc.

1.6 ISOMERISM IN VINYL POLYMERS

From the discussion of polymerization mechanisms in the preceding paragraph it can be seen that step reaction or condensation polymerization imposes considerable restrictions on the manner in which structural units combine to form the polymer chain. As a result simple condensation polymers are characterized by a high level of structural regularity, and the ease with which factors such as chain branching and copolymer composition may be controlled. Such is not the case in addition polymerization where the nature of the propagation reaction and its interruption by chain transfer allows important variations in the mode of addition of the monomer units to the growing chain; the following discussion is an attempt to summarize some of the more important structural features which may arise in addition polymers formed from vinyl and diene monomers.

1.6.1 Structural isomerism

Structural isomerism in a simple homopolymer may be defined as gross structural variations arising from the insertion of monomer units in arrangements differing from the linear chain expected on the basis of the postulated mechanism. The most common form of structural isomerism in homopolymers is *chain branching*, which may occur in several ways, the most probable of which is a chain transfer reaction, involving preformed polymer molecules, resulting in the formation of an initiating centre on the main chain of the polymer molecule; for a radical polymerization this type of reaction may be formulated as

$$\sim CH{-}CHX\cdot + \sim CH{-}CHX\sim \rightarrow \sim CH{-}CH_2X + \sim CH{-}\dot{C}X\sim$$

The radical so generated is capable of further propagation leading to a branch chain. Chain transfer reactions of this type may be intermolecular leading to long branch chains, or intramolecular (back-biting), leading to short branches. Depending upon the ease with which chain branching occurs the side chains of a branched polymer may in themselves be branched, leading to a dendritic structure; such structures are frequently very difficult to characterize.

Whilst chain branching and cross-linking may represent important structural features of copolymers, it is obvious that the presence of two different repeat units in a polymer chain offers further possibilities for structural isomerism in the form of variations of the distribution of monomer units within the chain. Copolymers of two monomers A and B may be of a variety of extreme types, although the most common contains a random distribution of units

$$-AABABBABBAAB-$$

Suitable choice of monomers and synthetic routes can yield linear copolymers in which the monomer units alternate

$$-ABABABABAB-$$

or are present in the form of regular blocks either in a linear arrangement

$$-AAAAABBBBBB-$$

or in the form of a graft copolymer

1.6.2 Sequence isomerism

In principle the unsymmetric nature of the majority of polymerizable vinyl monomers allows three different modes of linkage between repeat units in the polymer chain, for example

$$-CH_2-CHX-CH_2-CHX- \qquad \text{Head to tail}$$
$$-CH_2-CHX-CHX-CH_2- \qquad \text{Head to head}$$
$$-CHX-CH_2-CH_2-CHX- \qquad \text{Tail to tail}$$

(the substituted group —CHX is referred to as the head of the repeat unit). Generally speaking, head to head and tail to tail polymers will be identical although, in principle, they could be formed by quite different mechanisms.

The proportions of these isomeric sequences occurring in any polymer chain is dependent on the relative rates of the two reactions

$$R^* + CH_2 = CHX \rightarrow R—CH_2—\overset{*}{C}HX \qquad (a)$$

$$R^* + CHX = CH_2 \rightarrow R—CHX—\overset{*}{C}H_2 \qquad (b)$$

where R^* is a propagating chain with any type of active end. The relative rates of reactions (a) and (b) are qualitatively expected to be influenced by electrostatic forces between the growing chain and the approaching monomer molecule (most important where R^* is an ionic species), by steric hindrance to the attack of an active centre on the more heavily substituted carbon atom and by the preferential production of the most stable active centres. Whilst it is difficult to assess the relative importance of these factors, it is evident that, in general, reaction (a) will be favoured by both steric and electronic factors since the steric hindrance to propagation is minimum in this mode, whilst the substituent is in a position to exert its maximum stabilizing effect on the active centre. Thus, qualitatively it is expected that vinyl polymers will contain regular head to tail sequences and such is indeed found to be the case for all vinyl polymers so far studied. Major exceptions to this rule are found only in fluorinated monomers and the subject is discussed in detail in Chapter 2.

1.6.3 Stereoisomerism

It is well known that a carbon atom which bears four different substituents is able to adopt two configurations having a mirror image relationship; these structures are not interconvertible without bond breaking and are termed *stereoisomers* or *enantiomorphs*. If we consider a single repeat unit in the chain of a vinyl polymer R′—CH$_2$—CHX—R it is immediately obvious that the carbon atom of the —CHX— group is an asymmetric centre, since the chain elements R′—CH$_2$— and R— on either side of a given carbon atom will not normally be the same. The —CHX— group is thus capable of adopting two configurations

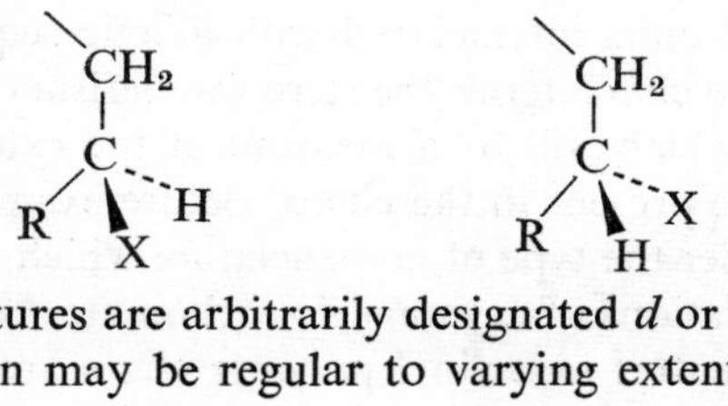

If these two structures are arbitrarily designated *d* or *l* then it can be seen that a polymer chain may be regular to varying extents depending on the

distribution of *d* and *l* configurations in the chain. Polymers displaying a high level of stereo-regularity are said to be *tactic* whilst those with a random distribution of configuration are termed *atactic*. In vinyl polymers we may distinguish two distinct forms of stereo-regularity; if each asymmetric centre has the same configuration, then the polymer is said to be *isotactic* whilst a polymer in which the configurations alternate is termed *syndiotactic*. The stereochemistry of such polymers is most easily visualized by imagining the backbone of the polymer as being stretched out in the form of a zig-zag in the plane of the paper; in an isotactic polymer (Figure 1.1(a)) all of the X

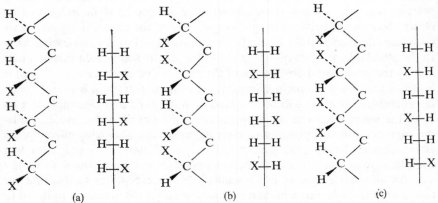

Figure 1.1 Stereo-regular structures from monosubstituted ethylene $CH_2{=}CHX$, (a) isotactic, (b) syndiotactic, (c) atactic (heterotactic)

substituents lie on the same side of this plane and in a syndiotactic polymer (Figure 1.1(b)) they lie alternately on opposite sides. In practice, steric and dipolar forces normally prevent a polymer chain from adopting the planar zig-zag conformation but this has no effect on the tacticity since the configuration of the asymmetric carbon atoms cannot be altered without bond breaking.

Whilst it is possible to prepare polymers with highly tactic structures it must be emphasized that completely tactic and completely atactic structures represent the extremes of stereoisomerism; many polymers exhibit intermediate behaviour, their structures consisting of short sequences of syndiotactic or isotactic units interspersed with atactic sequences. Clearly it is desirable to be able to characterize the stereoisomerism of a polymer in terms of some parameter which will be a measure of the extent to which stereoregular sequences are present in the chain. Before examining this concept it is pertinent to consider the type of nomenclature which should be applied to describe the configurations of asymmetric carbon atoms in a polymer chain.

The asymmetric centres in a vinyl polymer chain are pseudo-asymmetric because the asymmetry is derived from the varying length of the polymer chain on either side of the asymmetric carbon atom rather than from asymmetry of the repeat unit; at some asymmetric carbon atom in a symmetrical chain there must be a plane of symmetry where the two substituent chains are of equal length so that optical activity is impossible. The use of the d and l notation to describe the configuration of asymmetric centres may be criticized on these grounds, and there are even more serious objections to its use. Consider as an example the structure shown in the Fischer projection below

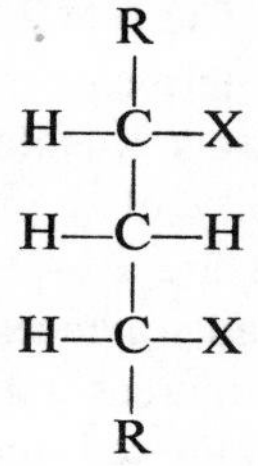

$$
\begin{array}{c}
\text{R} \\
| \\
\text{H--C--X} \\
| \\
\text{H--C--H} \\
| \\
\text{H--C--X} \\
| \\
\text{R}
\end{array}
$$

In the simple case where R is CH_3 this structure represents the *meso-*enantiomorph of a 2,4-disubstituted pentane and its two asymmetric centres would be regarded as being of opposite configuration and designated *dl*. The use of the symbols d and l is purely arbitrary; in principle the R and S chirality nomenclature would be preferable, but its extension to polymers raises some difficulties and, in this Chapter, we have preferred to follow the older convention for clarity. A more detailed and much more exact discussion of these points is given in Chapter 15.

If the substituents R are both polymer chains then the two asymmetric centres would be regarded by the polymer chemist as being in an isotactic arrangement; the two units would then be considered as having the same configuration and designated as *dd* or *ll*. It is this basic difference in approach which creates difficulties in the use of the *d*, *l* convention. In the context of polymer stereochemistry the absolute configuration at any asymmetric carbon atom is clearly unimportant, rather it is the relationship of one asymmetric carbon atom to the adjacent one which determines tacticity; it is thus possible to characterize the stereochemistry of a polymer in terms of the fraction of pairs of monomer units (diads) which may be described as *meso* (*m*) or *racemic* (*r*), a polymer in which all diads are *m* being wholly isotactic whilst a polymer in which all diads are *r* is wholly syndiotactic. Projection formulae for *m* and *r* diads are

$$
m\quad
\begin{array}{c}
| \\
\text{H--C--H} \\
| \\
\text{H--C--X} \\
| \\
\text{H--C--H} \\
| \\
\text{H--C--X} \\
|
\end{array}
\qquad
r\quad
\begin{array}{c}
| \\
\text{H--C--H} \\
| \\
\text{H--C--X} \\
| \\
\text{H--C--H} \\
| \\
\text{X--C--H} \\
|
\end{array}
$$

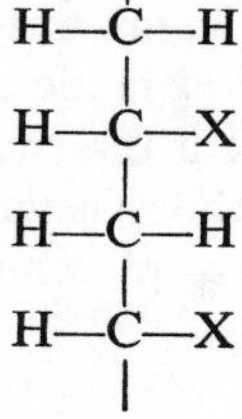

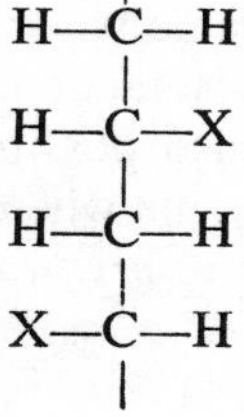

Without doubt, the most powerful method of determining the stereo-regular content of polymers is high resolution nuclear magnetic resonance spectroscopy, a technique which is especially suitable for polymers of relatively low stereo-regularity which are difficult to characterize by any other method.[10] Basically the application of NMR depends upon the fact that chemical shifts and coupling constants for protons in a polymer repeat unit are dependent upon the environment of that unit with respect to its two nearest neighbours in the chain. The NMR method leads most conveniently to characterization of tacticity in terms of triad sequences of monomer units. Three types of triad may be envisaged, isotactic (*mm*), syndiotactic (*rr*), and heterotactic (*mr*) and the tacticity of a polymer may be specified by the fractions of its structure which are iso-, syndio- and heterotactic triads; projection formulae for these triad squences are shown in Figure 1.1. As the resolution of NMR techniques increases the tacticity of longer sequences than diads and triads could be analysed and indeed it is already possible to characterize some polymers in terms of tetrad and pentad sequences.

The simple types of stereo-order in polymer backbones, referred to above, relate to polymers formed from monosubstituted ethylenes or 1,1-disubstituted ethylenes in which the two substituents are different, e.g. methyl-

$$\underset{\displaystyle CH_3}{\overset{}{|}} \qquad\qquad \underset{\displaystyle CH_3}{\overset{}{|}}$$

methacrylate $CH_2{=}C{-}COOCH_3$, but not isobutylene $CH_2{=}C{-}CH_3$.

For polymers from 1,2-disubstituted ethylenes, the same reasoning shows that every carbon atom in the polymer backbone may be regarded as an asymmetric centre resulting in di-tacticity. In order to describe configuration sequences it is conventional to adopt the nomenclature utilized by organic chemists to describe stereospecific addition to *cis*- and *trans*- olefinic linkages

$$\left.\begin{array}{l} trans\text{-addition to }trans\text{-olefin} \\ cis\text{-addition to }cis\text{-olefin} \end{array}\right\} \longrightarrow erythro \text{ structure}$$

$$\left.\begin{array}{l} trans\text{-addition to }cis\text{-olefin} \\ cis\text{-addition to }trans\text{-olefin} \end{array}\right\} \longrightarrow threo \text{ structure}$$

By definition, the adjacent asymmetric centres in an *erythro* structure have the same configuration whereas those in *threo* structure have alternating configurations. Thus a polymer produced from an olefin RCH=CHR′ will be defined as *erythro*-di-isotactic or *threo*-di-isotactic accordingly, as indicated in Figure 1.2. It is also possible to have a di-syndiotactic polymer from a 1,2-disubstituted ethylene where the backbone asymmetric centres are made up of equal amounts of *erythro*- and *threo*- placements, as shown in Figure 1.2.

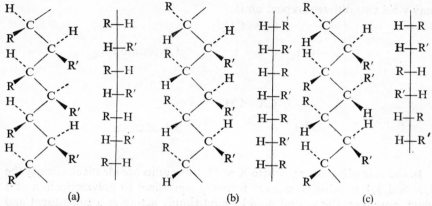

Figure 1.2   Stereo-regular structures from a disubstituted ethylene RCH=CHR′, (a) *threo*-di-isotactic, (b) *erythro*-di-isotactic, (c) disyndiotactic

In reality comparatively few high polymers of 1,2-disubstituted ethylenes are of any significance but the concepts of di-tacticity are nonetheless valuable since deuterium substitution of readily polymerizable mono-olefins (e.g. RCH=CHD) gives rise to *cis-trans*-pairs which are very valuable in mechanistic studies. This follows since, whenever it is possible to characterize the backbone configuration sequences by NMR studies in solution or X-ray crystallographic analysis of solid polymers, it will be immediately apparent whether the mechanism of incorporation of the olefin into the polymer occurs in an exclusively *cis*- or *trans*- fashion.

1.7   ISOMERISM IN DIENE POLYMERS

The preceding discussion of molecular isomerism has been confined to polymers produced from monomers with only one polymerizable double bond. An important group of polymers are those produced from monomers having two conjugated double bonds and this allows several isomeric chain units, depending on which and how many of the double bonds become involved in the polymer chain. If we denote a general monosubstituted diene by the formula

$$\underset{1\qquad2\qquad3\qquad4}{CH_2\!=\!\overset{\overset{\displaystyle X}{\displaystyle |}}{C}\!-\!CH\!=\!CH_2}$$

where X may be Cl or CH_3, it is obvious that simple vinyl type polymerization

may yield two different repeat units

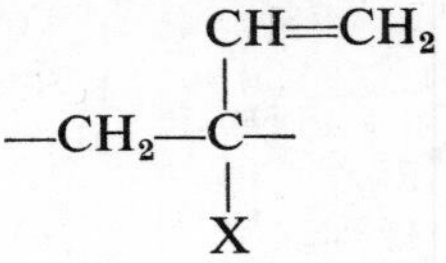

$$
\begin{array}{c}
CH{=}CH_2 \\
| \\
-CH_2-C- \\
| \\
X
\end{array}
\qquad \text{1,2-addition}
$$

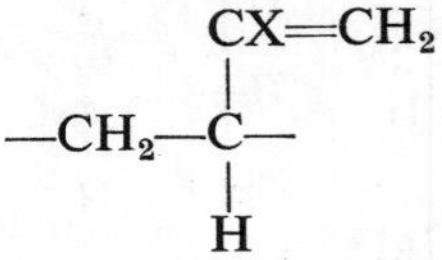

$$
\begin{array}{c}
CX{=}CH_2 \\
| \\
-CH_2-C- \\
| \\
H
\end{array}
\qquad \text{3,4-addition}
$$

In the case of butadiene, where $X = H$, these units are identical. Clearly the 1,2- and 3,4-addition modes are formally equivalent to polymerization of a vinyl monomer, the second double bond simply acting as a substituent and polymers obtained via these modes of addition can exhibit the same isomeric features as simple vinyl polymers.

As well as the 1,2- and 3,4-addition modes, diene monomers can undergo conjugate addition reactions involving both double bonds and leading to 1,4-addition. Under these conditions two forms of isomerism may arise, sequence isomerism and geometric isomerism. Sequence isomerism is similar to that found in vinyl polymers and three structures may be envisaged

$$
\begin{array}{ll}
-CH_2-CX{=}CH-CH_2-CH_2-CX{=}CH-CH_2- & \text{Head to tail} \\
-CH_2-CH{=}CX-CH_2-CH_2-CX{=}CH-CH_2- & \text{Head to head} \\
-CH_2-CX{=}CH-CH_2-CH_2-CH{=}CX-CH_2- & \text{Tail to tail}
\end{array}
$$

In practice, 1,4-sequences in diene polymers are found to be mainly head to tail as in vinyl polymers; however, the proportion of other sequences is sometimes much higher than that in vinyl polymers.

The most important aspect of 1,4-addition polymerization of conjugated dienes is the formation of cis- and trans- isomeric linkages in the polymer backbone, that is

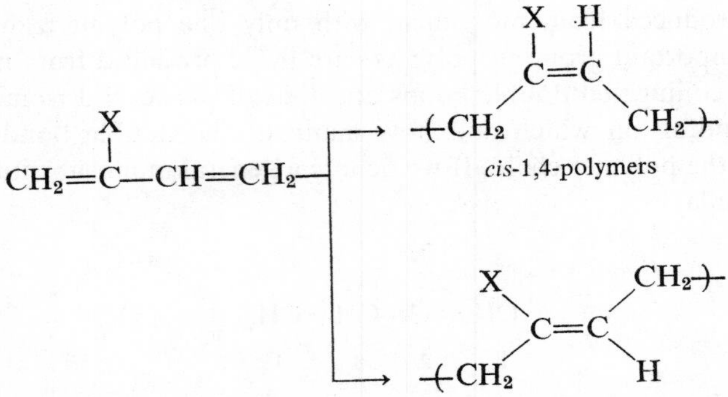

cis-1,4-polymers

trans-1,4-polymer

If the diene is substituted in the 1- or 4-positions then the resulting 1,4-addition polymers may exhibit configurational order of the asymmetric carbon atom in addition to *cis-* or *trans-* ordering of the double bond, e.g.

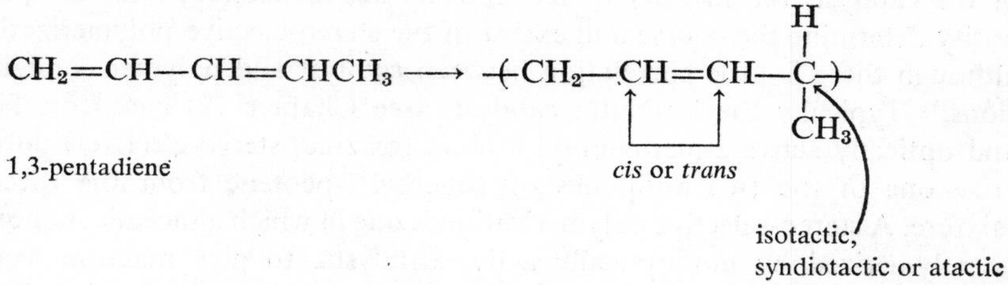

$$CH_2=CH-CH=CHCH_3 \longrightarrow \quad -(CH_2-CH=CH-\overset{\overset{\displaystyle H}{|}}{\underset{\underset{\displaystyle CH_3}{|}}{C}})-$$

1,3-pentadiene

cis or *trans*

isotactic,
syndiotactic or atactic

1.8 ISOMERISM IN POLYACETALS AND POLYETHERS

The concepts of stereoisomerism developed for polymers from mono-olefins may be applied equally well to addition polymers of aldehydes, epoxides and their hetero-analogues. Thus polyacetaldehyde

$$-\left(\overset{\overset{\displaystyle CH_3}{|}}{CH}-O\right)-$$

may exist in atactic, syndiotactic and isotactic forms and, as for polyolefins, each polymer chain is essentially a meso form so that stereo-regular poly-acetals are not normally optically active. On the other hand, whilst the rather similar polymers obtained by addition polymerization of epoxides

$$-(\overset{\overset{\displaystyle R}{|}}{CH}-CH_2-O)-$$

may also be characterized as isotactic, syndiotactic or atactic, in this case the *monomer* (e.g. propylene oxide) possesses an asymmetric centre and every substituted carbon atom of the backbone is truly asymmetric because it is bonded to four different groups. Consequently it is possible to polymerize optically active monomer, with preservation of configuration during chain growth, and to end up with optically active stereo-regular polyether (see Chapter 15).

1.9 STEREO-REGULAR, STEREO-ELECTIVE AND STEREO-SELECTIVE POLYMERIZATIONS

For the purposes of this volume a stereo-regular polymerization is defined as any polymerization process leading to an ordered arrangement of monomer segment *configuration* in the polymer backbone. Thus isotactic and syn-diotactic polymers are stereo-regular as are all *cis*-1,4-polydienes and all *trans*-1,4-polydienes.

A stereo-elective polymerization is one in which, by use of an optically active catalyst, there is exclusive incorporation of one of the two antipodes of a racemic monomer mixture into a polymer chain. The exact composition of the catalyst, the identity of its optically active portion, and its optical purity determine the course and extent of the stereo-elective polymerization, although there is little correlation between catalyst and polymer configurations.[11] Typically Ziegler-Natta catalysts (see Chapter 12) based on $TiCl_3$ and optically active 2-methylbutyl lithium (or zinc) stereo-electively polymerize one of the two antipodes of 3-methyl-1-pentene from the racemic mixture. A stereo-selective polymerization is one in which a racemic monomer is polymerized by non-optically active catalysts, to give macromolecules uniquely composed of one, or the other, monomer antipode. Further details of the configurational aspects of this type of polymerization are given in Chapter 15.

1.10 MOLECULAR WEIGHTS OF HIGH POLYMERS

Irrespective of the mechanism of their formation, high polymers generally differ from small size compounds in that they are polydisperse or heterogeneous in molecular weight. Even when synthesized from rigorously purified monomer, a polymer is not a pure substance in the usually accepted sense. Polymers, in their purest form, are mixtures of molecules of different molecular weight as a result of either the random nature of initiation, termination and transfer processes in chain reaction polymerization or the random interaction of species of all sizes which often occurs in step reaction polymerizations. The polymer sample cannot then be characterized by a single molecular weight, but instead, various molecular weight averages may be defined.

1. The *number average molecular weight* $\bar{M}_n$ is determined by measurement of colligative properties such as freezing point depression, boiling point elevation, osmotic pressure, and vapour pressure lowering. $\bar{M}_n$ is defined as the total weight w of all molecules present in a polymer sample divided by the total number of molecules present

$$\bar{M}_n = \frac{w}{\sum N_x} = \frac{\sum N_x M_x}{\sum N_x}$$

N_x is the number of molecules whose weight is M_x (the summation is from $x = 1$ to $x = \infty$) and the equation can be rewritten as

$$\bar{M}_n = \sum \bar{N}_x M_x$$

where $\bar{N}_x$ is the mole fraction (or number fraction) of molecules of size M_x.

2. The *weight-average molecule weight* $\bar{M}_w$ is obtained from light scattering measurements and defined as

$$\bar{M}_w = \sum w_x M_x = \frac{\sum c_x M_x}{\sum c_x} = \frac{\sum c_x M_x}{c} = \frac{\sum N_x M_x^2}{\sum N_x M_x}$$

where w_x is the weight fraction of molecules whose weight is M_x, c_x is the weight concentration of M_x molecules, and c is the total weight concentration of all the polymer molecules.

3. *The viscosity average molecular weight* $\bar{M}_v$ is obtained from viscosity measurements and defined by

$$\bar{M}_v = [\sum w_x M_x^a]^{1/a} = \left[\frac{\sum N_x M_x^{a+1}}{\sum N_x M_x}\right]^{1/a}$$

where a is a constant.

The viscosity and weight average molecule weights are equal when a is unity. Usually however, $\bar{M}_v$ is less than $\bar{M}_w$ since a lies in the range 0·5–0·9. Inspection of the equations for the three molecular weight averages shows that, for a polydisperse polymer,

$$\bar{M}_w > \bar{M}_v > \bar{M}_n$$

with the differences increasing as the molecular weight distribution broadens. The ratio $\bar{M}_w/\bar{M}_n$ is often taken as a measure of the distribution of molecular weights in a polymer sample and has a value of approximately 2 for polymers having a most probable distribution, e.g. step reaction (condensation) polymers produced at high conversions. In general, chain reaction polymerizations give values for $\bar{M}_w/\bar{M}_n$ markedly greater than 2 according to the nature of the initiation, propagation and termination/transfer processes. In very special circumstances (see Chapter 11) essentially monodisperse polymers ($\bar{M}_w/\bar{M}_n \simeq 1$) can be obtained.

1.11 REFERENCES

1. For an up-to-date survey of all features of step growth polymerization see G. Odian, *Principles of Polymerization*, McGraw-Hill, London, 1970.
2. G. W. Cowell and A. Ledwith, *Quart. Rev. (London)*, **24**, 119 (1970).
3. R. M. Joshi and B. J. Zwolinski, *Vinyl Polymerization*, ed. G. E. Ham, Vol. I, Part I, p. 445, Dekker, London, 1967.
4. B. E. Lee and A. M. North, *Makromol. Chem.* **79**, 135 (1964); C. E. H. Bawn, B. E. Lee and A. M. North, *J. Polymer Sci.*, **B2**, 263, 1964.
5. O. Vogl, *Polymer Chemistry of Synthetic Elastomers*, eds. J. P. Kennedy and E. Tornqvist, Vol. I, p. 419, Interscience, London, 1968.
6. Y. Yamashita and S. Nunomoto, *Makromol. Chem.*, **58**, 244 (1962); G. F. Pregaglia, M. Minaghi and M. Cambini, *ibid*, **67**, 10 (1963).
7. *Ring Opening Polymerization*, eds. K. C. Frisch and S. L. Reegen, Dekker, London, 1969.

8. F. S. Dainton, T. R. E. Devlin, and P. A. Small, *Trans. Faraday Soc.*, **51**, 1710 (1955); P. A. Small, *ibid*, **51**, 1717 (1955).
9. G. Dall'asta and G. Motroni, *Die Angew. Makromol. Chem.*, **16/17**, 51 (1971); N. Calderon, *Accounts Chem. Res.*, **5**, 127 (1972).
10. F. A. Bovey, *Polymer Conformation and Configuration*, Academic Press, London, 1969; *High Resolution NMR of Macromolecules*, Academic Press, London, 1972.
11. P. Pino, *Fortschr. Hochpolymer. Forsch.*, **4**, 399 (1966).

2

The Reactivity of Free-Radicals

J. M. Tedder

Department of Chemistry, University of St. Andrews

Free-radicals commonly undergo five different types of reaction:

(i) $\qquad X\cdot + Y\cdot \rightarrow X\!-\!Y$ Radical *combination*

$\qquad$ (e.g. $CH_3\cdot + CH_3\cdot \rightarrow C_2H_6$)

(ii) $\qquad YZ\cdot \rightarrow Y\cdot + Z$ Radical *fragmentation*

$\qquad$ (e.g. $CH_3CO\cdot \rightarrow CH_3\cdot + CO$)

(iii) $\qquad X\cdot + Y\!-\!Z \rightarrow X\!-\!Y + Z\cdot$ Radical *transfer*

$\qquad$ (e.g. $CF_3\cdot + CH_4 \rightarrow CF_3H + CH_3\cdot$)

(iv) $\qquad X\cdot + Y\!=\!Z \rightarrow X\!-\!Y\!-\!Z\cdot$ Radical *addition*

$\qquad$ (e.g. $CH_3\cdot + CH_2\!=\!CH_2 \rightarrow CH_3CH_2CH_2\cdot$)

(v) $\quad X\cdot + Y\!-\!Z\!-\!W\cdot \rightarrow X\!-\!Y + Z\!=\!W$ Radical *disproportionation*

$\qquad$ (e.g. $CH_3\cdot + CH_3CH_2\cdot \rightarrow CH_4 + CH_2\!=\!CH_2$)

There are, in addition, reactions involving one electron transfer to or from a radical, with the formation of either an anion or a cation. The present article will not be concerned with such processes. Our main attention will be focused on (iv) radical addition, particularly the addition of simple radicals to olefins, however, we will not be able to discuss this topic without paying some attention to (iii) radical transfer reactions and (ii) radical fragmentation, which is often simply the reverse of addition. We must begin however by considering (i) radical combination.

The striking feature of the five types of radical reaction, enumerated above is that only (i) and (v) involve disappearance of radical species. In the other three reactions new radicals replace the initial radicals formed in the reaction processes. In other words reactions (ii), (iii) and (iv) represent chain carrying steps, whereas (i) and (v) are chain termination processes. Some knowledge of the nature and rate of the chain termination steps is essential if the chain propagating steps are to be investigated.

2.1 RADICAL COMBINATION REACTIONS

Atom and radical combination reactions have received much study since they represent one of the simplest possible reactions. At moderate pressures the rate of combination of simple radicals in the gas phase is temperature independent and is in close accord with the predictions of simple collision theory. At low pressures and for atoms, the observed combination rates are very much lower than those predicted. The reason for this is very simple; the energy required to break the bond between two atoms in a diatomic molecule, the bond dissociation energy, is exactly the energy which must be given out when the two atoms come together and combine. If a molecule formed by the combination of two atoms in a gas phase collision, cannot get rid of this energy within one vibration period the atoms will simply dissociate again. This dissociation is a kind of unimolecular decomposition and there is a very close relationship between an understanding of radical combination reactions and unimolecular rate theory. We can regard a molecule formed as the result of the combination of two atoms or radicals as an activated species, that is a molecule containing excess energy which it must seek to lose. In a simple diatomic molecule in the gas phase this will normally occur via collision with the walls of the containing vessel or with other molecules. When radicals combine some of the energy released can be taken up in other modes of vibration and rotation of the new molecule. The rate of radical combination is therefore less pressure dependent than atom combination. In solution the situation is very much more complicated. The solvent molecules are close and readily available to deactivate the excited molecule through collision and the solvent 'cage' will keep the two atoms or radicals in close proximity, so that even if they re-dissociate immediately after their first combination, they will have further opportunities of combining.

As indicated above atom combination rates are very much slower than the collision frequency. This is because the energy released when the bond is formed will cause the atoms to redissociate unless there is a collision with a third body within the period of the first vibration of the new molecule. At normal temperatures and pressures a molecule undergoes a collision about 10^8 to 10^9 times a second, whereas the vibrational frequency of a simple diatomic molecule is approximately 10^{13} s^{-1}. Thus homogeneous combination is very slow and most atom recombination takes place at the walls of the reaction vessel. Chain reactions involving atoms and simple molecules like the chlorination or fluorination of methane have very long chains in the gas phase.

Radical combination at high pressures (above about 10 mm) shows normal second order behaviour and gives rate constants slightly below the range expected from transition state theory. At low pressures the apparent second order rate falls and the reaction approaches third order behaviour. The reason is exactly the same as with atom combination only the life time of the activated molecule is very much longer, so that at high pressures the chance of collisional deactivation increases until virtually every collision between two radicals results in combination.

We are now in a position to consider the experimental data available for radical combination at normal pressures (the so called 'high pressure region'). All the data reported suggest that combination occurs with little or no activation energy.

Table 2.1 Rates of combination of simple alkyl radicals[1]

Radical	log k experimental $cm^3 \, mol^{-1} \, s^{-1}$
$CH_3\cdot + CH_3\cdot \rightarrow C_2H_6$	13·4
$C_2H_5\cdot + C_2H_5\cdot \rightarrow C_4H_{10}$	12·6
$i\text{-}C_3H_7\cdot + i\text{-}C_3H_7\cdot \rightarrow C_6H_{14}$	11·6
$CF_3\cdot + CF_3\cdot \rightarrow C_2F_6$	12·7
$CCl_3\cdot + CCl_3\cdot \rightarrow C_2Cl_6$	12·8
$t\text{-}C_4H_9\cdot + t\text{-}C_4H_9\cdot \rightarrow C_8H_{18}$	~8·6
$c\text{-}C_6H_{11}\cdot + c\text{-}C_6H_{11}\cdot \rightarrow (c\text{-}C_6H_{11})_2$	~8·5

Until recently it was believed that most radicals combined at rates close to that predicted by collision theory and the experimental data seemed to support this belief. New methods of determination now suggest that apart from methyl other simple alkyl radicals combine about one order of magnitude slower. Furthermore some large alkyl radicals, notably t-butyl and cyclohexyl combine three orders of magnitude slower still.

Data have also been obtained for cross-combination, that is the combination of two unlike radicals

$$R\cdot^a + R\cdot^a \xrightarrow{k_1^{aa}} R^a \!-\! R^a \quad \text{(Combination of } R\cdot^a)$$

$$R\cdot^b + R\cdot^b \xrightarrow{k_1^{bb}} R^b \!-\! R^b \quad \text{(Combination of } R\cdot^b)$$

$$R\cdot^a + R\cdot^b \xrightarrow{k_1^{ab}} R^a \!-\! R^b \quad \text{(Cross-combination)}$$

On the basis of simple collision theory $k_1^{ab}/(k_1^{aa} \times k_1^{bb})^{\frac{1}{2}}$ should be equal to 2, assuming there is no disporportionation (see below). As Table 2.2 shows this is almost exactly what has been found for a large number of radicals.

In the solution phase not only is there the question of solvent cage referred to above, but there is the problem associated with the diffusion of radicals through the solvent in order to meet each other and collide. In practice it is found for alkyl and alkoxy radicals that the rate of reaction is often controlled

Table 2.2   Cross-combination ratios

$R \cdot_a$	$R \cdot_b$	$\phi = \dfrac{k_{ab}}{(k_{aa}k_{bb})^{\frac{1}{2}}}$	Temp (K)
$CH_3 \cdot$	$C_2H_5 \cdot$	$2 \cdot 00^2 \ (1 \cdot 92^3)$	298 (288)
$CH_3 \cdot$	$n\text{-}C_3H_7 \cdot$	$2 \cdot 02^2$	298
$C_2H_5 \cdot$	$n\text{-}C_3H_7 \cdot$	$2 \cdot 08^2$	298
$CH_3 \cdot$	$CH_2F \cdot$	$2 \cdot 4^4$	329
$CF_3 \cdot$	$CHF_2 \cdot$	$2 \cdot 4^5$	564
$CF_3 \cdot$	$C_2F_5 \cdot$	$1 \cdot 95^6$	357

entirely by the rate of diffusion. This means that the *apparent* rate of combination in solution may be considerably slower than in the gas phase.

2.2   RADICAL DISPROPORTIONATION REACTIONS

Two radicals may, on collision, react to form not a dimeric molecule but two new molecules, one of which must be unsaturated (e.g. two ethyl radicals, besides yielding butane, also yield ethane and ethylene)

$$C_2H_5 \cdot + C_2H_5 \cdot \quad \begin{array}{l} \overset{k_c}{\nearrow} \quad n\text{-}C_4H_{10} \qquad \textit{Combination} \\[2mm] \underset{k_d}{\searrow} \quad C_2H_6 + CH_2{=}CH_2 \quad \textit{Disproportionation} \end{array}$$

The ratio k_d/k_c hereafter written as Δ, has been determined for a number of radicals. For primary radicals it is usually much less than one but, as the Table 2.3 shows, with highly branched radicals it can be greater than unity.

Table 2.3   Disproportionation-combination ratios of some alkyl radicals at 298 K[6a]

$$R^a + R^b \to R^aH + \text{Olefin} \quad k_d$$
$$R^a + R^b \to R^a - R^b \qquad k_c$$

R^a	R^b	Δ
$CH_3 \cdot$	$C_2H_5 \cdot$	$0 \cdot 036$
$CH_3 \cdot$	$n\text{-}C_3H_7 \cdot$	$0 \cdot 058$
$CH_3 \cdot$	$(CH_3)_2CH \cdot$	$0 \cdot 163$
$C_2H_5 \cdot$	$C_2H_5 \cdot$	$0 \cdot 135$
$C_2H_5 \cdot$	$n\text{-}C_3H_7 \cdot$	$0 \cdot 066$
$C_2H_5 \cdot$	$(CH_3)_2CH \cdot$	$0 \cdot 180$
$n\text{-}C_3H_7 \cdot$	$C_2H_5 \cdot$	$0 \cdot 057$
$(CH_3)_2CH \cdot$	$C_2H_5 \cdot$	$0 \cdot 124$
$n\text{-}C_3H_7 \cdot$	$n\text{-}C_3H_7 \cdot$	$0 \cdot 158$
$(CH_3)_2\dot{C}H$	$(CH_3)_2CH \cdot$	$0 \cdot 687$
$t\text{-}C_4H_9 \cdot$	$t\text{-}C_4H_9 \cdot$	$2 \cdot 30$

The mechanism of disproportionation has been the subject of much discussion. A simple head-to-tail hydrogen abstraction appears to be a reasonable mechanism and is consistent with the observed products from the disproportionation of 1,1-dideuterioethyl radicals

$$2\ CH_3CD_2\cdot \xrightarrow{k_d} CH_3CD_2H + CH_2{=\!\!=}CD_2$$

(ethylene found to be >90 per cent $C_2H_2D_2$)

However this simple interpretation has been questioned and it has alternatively been suggested that the activated complex for disproportionation and combination are the same. It is significant, that in the solution phase the extent of disproportionation of alkyl radicals is less than in the gas phase; this would be consistent with the idea that disproportionation represents a mode of breakdown of the highly energized dimer molecules formed as the result of a combination, since in solution the excess energy would be dissipated by collision before the molecule had time to rearrange and redissociate. However, thermal cracking yields radicals only, that is the reverse of combination, and this suggests that disproportionation and combination involve different activated complexes.

2.3 RADICAL TRANSFER REACTIONS

The most common and important radical transfer is that which involves hydrogen atom abstraction, for example

$$CH_3\cdot + CH_3C_6H_6 \rightarrow CH_4 + C_6H_5CH_2\cdot$$

$$Cl\cdot + CH_4 \rightarrow CH_3\cdot + HCl$$

$$CCl_3\cdot + CH_3CH{=\!\!=}CH_2 \rightarrow CCl_3H + \overset{\frown}{CH_2{-}CH{-}CH_2}$$

A substantial amount of kinetic data is now available for these reactions. The pre-exponential rate factors are close to those predicted by the Transition State Theory, that is around 10^{13} cm^3 mol^{-1} s^{-1} for atoms and 10^{10} cm^3 mol^{-1} s^{-1} for simple alkyl radicals. The observed activation energies show that the relative strengths of the bonds broken and formed play a big part in determining the rate. However the experimental results show that polar effects, familiar to the organic chemist in reactions like aromatic nitration, also have an important influence on radical transfer reactions.

Hydrogen abstraction from an alkane by chlorine atoms is more than 10^5 times faster than abstraction by methyl radicals at normal temperatures. This increase in rate is partly because there is an additional loss of rotation in the transition state for methyl radical (i.e. the difference in the A factors already referred to), and partly because abstraction by a chlorine atom is a slightly exothermic process whereas abstraction by a methyl radical is nearly thermo-neutral. The electrophilic nature of a chlorine radical compared with

Table 2.4 Arrhenius parameters for hydrogen atom abstraction from alkanes by different atoms and radicals

	Activation energies (kcal mol^{-1})			
Abstracting atom or radical	Primary CH$_3$—	Secondary CH$_2$	Tertiary CH—	log A (cm^3 mol^{-1} s^{-1}) Average
Cl·[7]	1·0	0·3	0	14·2
Br·[7]	13·4	10·2	7·5	13·5
CH$_3$O·[8]	7·1	5·2	4·1	11·0
(CH$_3$)$_3$CO·[9]	6·5	5·6	4·9	11·0
CH$_3$·[10]	11·7	10·1	8·0	11·5
CF$_3$·[11]	7·5	5·3	3·0	12·2
CCl$_3$·[12]	14·3	10·7	—	12·6

the nucleophilic character of a methyl radical may also be important. When we look at the relative rates of hydrogen abstraction from propionic acid we find a methyl radical attacks the hydrogen atoms attached to carbon atom 2, nearly eight times faster than the rate at which it attacks hydrogen atoms attached to carbon atom 3. This result is consistent with the expected strengths of the two types of bond. However, the relative rate of hydrogen abstraction by chlorine atoms from the same molecule shows exactly the reverse selectivity, namely hydrogen atoms in the 3-position are abstracted thirty times

Table 2.5 Arrhenius parameters for hydrogen abstraction from hydrogen and halogeno-methanes by different atoms and radicals

$$R—H + X· \xrightarrow{k_2} R· + HX$$

	X							
	Cl[7]		Br[7]		CF$_3$[13]		CH$_3$[7]	
R—H	log A_2	E_2	log A_2	E_2	log A_2	E_2	log A_2	E_2
H$_2$	13·6	5·5	14·1	19·7	12·1	10·71	11·6	10·5
CH$_4$	13·8	3·8	13·8	18·2	12·1	11·2	11·8	14·7 (12·8)
CH$_3$Cl	13·0	3·3	13·1	14·3	12·1	10·6	11·9	9·4*
CH$_3$Cl$_2$	18·1	3·0	12·9	10·9	11·2	7·6	11·5	7·2*
CHCl$_3$	12·8	3·3	12·3	9·3	11·0	6·6	10·8	5·8*
CH$_3$Br	—	—	13·5	15·9	12·0	10·4	12·5	10·1*

* Uncertain values.
Log A in cm^3 mol^{-1} s^{-1}
E in kcal mol^{-1}

more readily than those in the 2-position. This reversal of the expected reactivity is attributed to a polar effect present in the chlorine atom case. Hydrogen chloride is a very polar molecule and it is assumed that, if the carbon atom from which the hydrogen is being transferred is electron deficient, a substantial polar force will have to be overcome in forming the activated complex

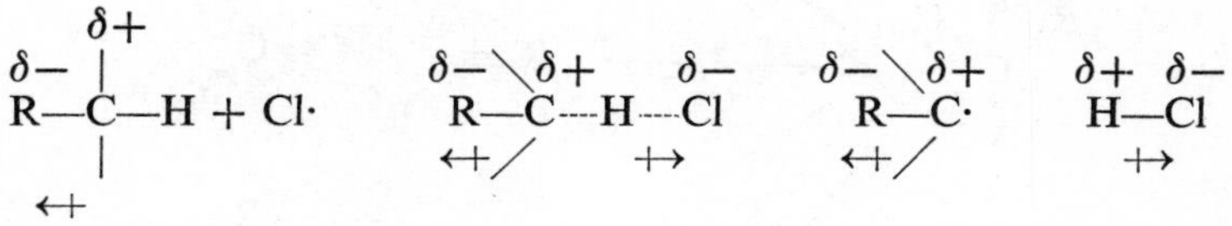

$$
\begin{array}{c c c}
\overset{\delta+}{\underset{\leftarrow+}{\delta-|}} & \delta-\diagdown\delta+ \quad \delta- & \delta-\diagdown\delta+ \quad \delta+ \ \delta- \\
R-\overset{|}{\underset{|}{C}}-H + Cl\cdot & R-C\text{----}H\text{----}Cl & R-C\cdot \quad H-Cl \\
 & \overset{\leftarrow+}{\diagup} \quad \overset{\rightarrow+}{} & \overset{\leftarrow+}{\diagup} \quad \overset{\rightarrow+}{} \\
\text{Reactants} & \text{Activated complex} & \text{Products}
\end{array}
$$

The exact balancing between bond strength and polar effect is hard to predict, although on a qualitative plane most observed results can be explained.

2.4 RADICAL ADDITION REACTIONS

The present chapter is primarily concerned with radical addition reactions in the gas phase. We will start by considering the addition of a radical or atom to an olefinic double bond

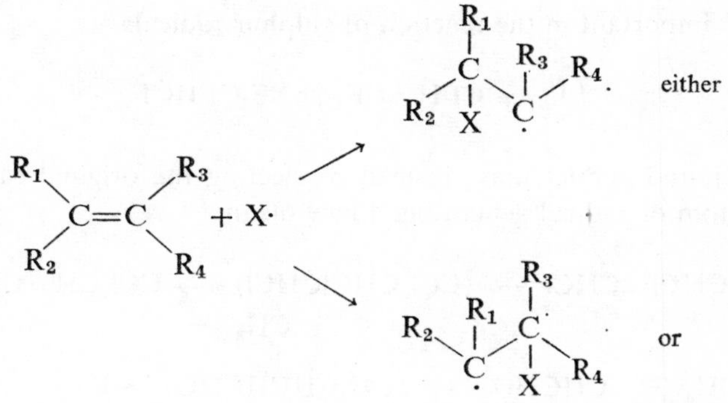

The problems that concern us are, firstly, how the nature of X and of R_1, R_2, R_3, R_4 affect the rate of addition and, secondly, if R_1 and R_2 are different from R_3 and R_4, to which end of the double bond will X become attached? These two questions are clearly going to be inter-related. However, before we can consider these problems, we must first determine the fate of the initial adduct radical. Four processes seem probable

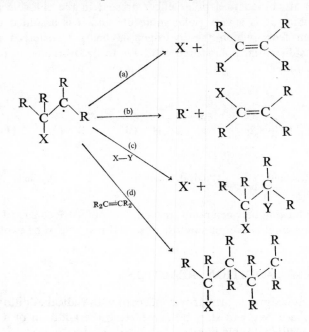

(a)   The first process is simply the reverse of the addition process. This probably occurs to some extent in all radical additions but it is particularly kinetically important in the reaction of sulphur radicals

$$SF_5\cdot + CFH{=}CF_2 \rightleftarrows SF_3CFH\dot{C}F_2$$

(b)   The initial adduct may, instead of ejecting the original radical, eject another atom or radical generating a new olefin

$$CCl_3\cdot + CHCl{=}CHCl \longrightarrow (CCl_3CHCl\dot{C}HCl) \longrightarrow CCl_3CH{=}CHCl + Cl\cdot$$

$$CH_3\cdot + CH_3CH{=}CHCHO \longrightarrow (CH_3\overset{\displaystyle CH_3}{\underset{\displaystyle |}{\dot{C}H}}CHCHO) \longrightarrow$$
$$CH_3CH{=}CHCH_3 + \dot{C}HO$$

(c)   The adduct radical may abstract an atom from another molecule, often so producing a chain process

$$CCl_3\cdot + CH_2{=}CH_2 \rightarrow CCl_3CH_2CH_2\cdot$$
$$CCl_3CH_2CH_2\cdot + CCl_3Br \rightarrow CCl_3CH_2CH_2Br + CCl_3\cdot$$

(d)   Finally, the adduct radical may itself add to an olefin. This can lead to teleomers or to polymers

$$CF_3\cdot + CF_2{=}CF_2 \rightarrow CF_3CF_2CF_2\cdot$$

$$CF_3CF_2CF_2\cdot + CF_2{=}CF_2 \rightarrow CF_3(CF_2CF_2)_2\cdot$$

and will be discussed in detail in later chapters.

There is a considerable difference between the reactions of hydrogen and halogen atoms on the one hand with the reactions of nitrogen, oxygen and sulphur atoms on the other. Hydrogen and halogen atoms add to olefinic double bonds to yield a new adduct radical, as described above; with nitrogen and oxygen atoms, however, the reaction is much more complicated and, in general, we will not be dealing with these latter species.

Probably the most extensively studied gas phase radical addition reaction is that of the addition of methyl radicals to ethylene. The rate of addition is usually measured by the change in the methane/ethane ratio, when a source of methyl radicals (often acetone) is photolysed alone or with ethylene

$$(CH_3)_2CO \xrightarrow{h\nu} CH_3\cdot + (CH_3\dot{C}O \longrightarrow CH_3\cdot + CO) \qquad (1)$$

$$CH_3\cdot + (CH_3)_2CO \xrightarrow{k_2} CH_4 + \text{Acetonyl radical} \qquad (2)$$

$$CH_3\cdot + CH_3\cdot \xrightarrow{k_3} C_2H_6 \qquad (3)$$

When an olefin is added the methyl radicals are consumed in a new reaction

$$CH_3\cdot + \text{Olefin} \xrightarrow{k_4} \text{Alkyl radical} \qquad (4)$$

If k_2 and k_3 are known, k_4 may be computed provided certain assumptions are made about the fates of the acetonyl radical formed in reaction (2) and the alkyl radical formed in reaction (4). Some workers have adjusted their conditions so that the propyl radical formed in reaction (4) mainly abstracts hydrogen, in which case the rate of addition can be correlated with the rate of propane formation (the addition reaction can be shown not to be significantly reversible under the conditions employed). The difficulties in this work are that there are a number of possible subsidiary chain transfer and chain termination steps. Table 2.6 records five different studies and it will be seen that, although the reported rates at 160 °C only differ by a factor of 5, the Arrhenius parameters are very different.

Although the absolute Arrhenius parameters are not very accurate, there is every reason to believe that the trends in a series of olefins reported by any

Table 2.6 The addition of methyl radicals to ethylene

$$CH_3\cdot + CH_2 = CH_2 \xrightarrow{k_2} CH_3CH_2CH_2\cdot$$

$\log k^{164°}$ cm^3 mol^{-1} s^{-1}	$\log A$ cm^3 mol^{-1} s^{-1}	E kcal mol^{-1}	Reference
7·4	10·0	5·5	14
7·5	12·1	8·7	15
7·6	11·5	7·8	16
7·7	11·1	6·8	17
8·0	11·9	7·8	18

one group of workers represents a real change in the rate constants. An example of such a series is that reported by Cvetanovic and Irwin.

Table 2.7 Arrhenius parameters for the addition of methyl radicals to hydrocarbon olefins[18]

$$CH_3\cdot + Olefin \xrightarrow{k} Alkyl\ radical$$

Olefin	$\log k^{164°C}$ cm^3 mol^{-1} s^{-1}	$\log A$ cm^3 mol^{-1} s^{-1}	E kcal mol^{-1}	Relative rate at 164 °C
$CH_2{=}CH_2$	8·0	11·9	7·9	0
$CH_3CH{=}CH_2$	7·8	11·5	7·4	0·8
$CH_3CH{=}CCHCH_3$	7·2	10·9	7·5	0·2
$CH_3CH{=}CHCH_3$	7·5	11·5	8·1	0·3
$(CH_3)_2C{=}CH_2$	8·0	11·5	6·9	1·1

The difficulty about work involving hydrocarbon olefins is that hydrogen abstraction from the allylic position is extremely facile and it is quite impossible to suppress this competing process. The relative rate constants for addition and abstraction from *cis*-but-2-ene reported by another group of workers are[19]

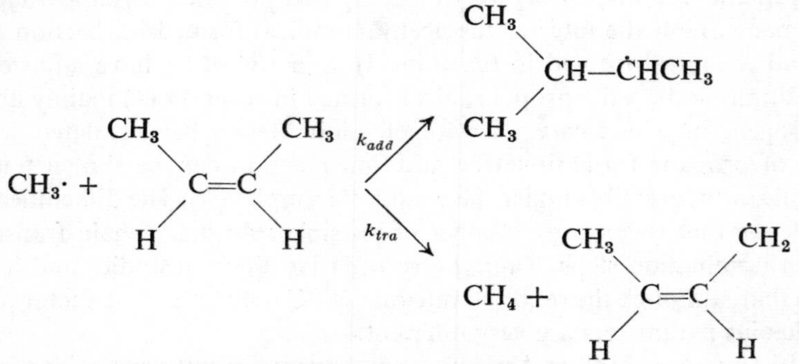

$$k_{add} = 4\cdot5 \times 10^{10} \exp(-7000\ cal/RT)$$
$$k_{tra} = 1\cdot8 \times 10^{10} \exp(-7300\ cal/RT)$$

One of the few studies of the addition of alkyl radicals to olefins with polar substituents so far reported is that of James and his colleauges. The rate constants refer to the total rate of addition to both ends of these unsymmetric

Table 2.8 The addition of ethyl radicals to polar olefins[20,21]

$$C_2H_5\cdot + CH_2{=}CHR \xrightarrow{k}$$

Olefin	log A cm^3 mol^{-1} s^{-1}	E kcal mol^{-1}	Relative rate at 164 °C
$CH_2{=}CHCN$	11·1	3·4	70
$CH_2{=}C(CH_3)CN$	11·7	4·6	70
$CH_3CH^c{=}CHCN$	10·5	5·0	2·8
$CH_3CH^t{=}CHCN$	10·8	5·2	7·1
$CH_2{=}CHOCHOCH_3$	11·2	6·9	1·6
$CH_2{=}CHC_6H_5$	10·9	4·1	20
$CH_2{=}CHC_5H_{11}$	11·1	7·0	1·1
$CH_2{=}CH_2$	11·0[a]	6·9[a]	1

a. Average values from three separate reports.

olefins and no data about the relative orientation were obtained. Probably the addition occurred predominantly at the $CH_2\cdot$ end of all the olefins except the crotonitriles where both adducts may have been formed. Reactivity appears to be enhanced by the substituents in the order NC— > C_6H_5 > $CH_3CO\cdot O$— > Alkyl, and this change in reactivity appears to be mainly due to the changes in the activation energy portion of the rate constants. The effects of the substituents are similar to the effects they would have on a nucleophile in ionic reactions and so it might be argued that these results provided evidence that an alkyl radical showed nucleophilic characteristics.

The relative rates of addition of trifluoromethyl radicals to a number of fluoro-olefins in the gas phase have been studied by Szwarc and his co-workers. Szwarc's technique was to photolyse hexafluoroazomethane in the presence of 2,3-dimethylbutane. From these experiments the ratio $[CF_3H]_f/[N_2]_f$ was obtained, the experiment was then repeated with the addition of an olefin; since trifluoromethyl radicals now are consumed in another reaction (addition to the olefin) the ratio is reduced.

$$CF_3N{=}NCF_3 \xrightarrow{h\nu} 2\,CF_3\cdot + N_2$$

$$CF_3\cdot + C_6H_{14} \xrightarrow{k_2} CF_3H + C_6H_{13}\cdot$$

$$CF_3\cdot + E \xrightarrow{k_3} CF_3E\cdot \qquad (E = \text{Ethylene})$$

No absolute rates can be obtained in this way but the ratio k_3/k_2 can be determined over a range of temperatures. Taking Thynne and Sangster's value for the rate of addition of trifluoromethyl radicals to ethylene in collaboration with Szwarc's results, Table 2.9 can be drawn up.

Table 2.9 The addition of trifluoromethyl radicals to fluoro-olefins (Szwarc et al.)[22,23,24] [Calculated using the value of ethylene of Thynne][25]
$$CF_3 \cdot + E \rightarrow$$

Olefin	$\log A$ $cm^3\,mol^{-1}\,s^{-1}$	E kcal mol^{-1}	Relative rate at 164 °C
$CCl_2{=}CCl_2$	10·9	5·6	0·008
$CHCl^c{=}CHCl$	10·7	4·3	0·02
$CF_2{=}CCl_2$	10·3	3·1	0·04
$CHCl^t{=}CHCl$	10·6	3·5	0·04
$CF_2{=}CF_2$	10·7	2·9	0·11
$CFH{=}CH_2$	11·2	3·8	0·12
$CClH{=}CH_2$	11·4	3·0	0·20
$CH_2{=}CH_2$	11·4	2·4	**1**
$CH_3CH{=}CH_2$	11·5	2·0	2·0
$C_6H_5CH{=}CH_2$	11·6	1·3	5·6
$CH_2{=}CH{-}CH{=}CH_2$	11·5	1·0	10·0

The rates of addition of a number of radicals to ethylene have now been measured and are tabulated below.

Table 2.10 The relative reactivity of different radicals in addition to ethylene
$$X \cdot + CH_2{=}CH_2 \rightarrow XCH_2\dot{C}H_2$$

Radical	$\log A$ $cm^3\,mol^{-1}\,s^{-1}$	E kcal mol^{-1}	Relative rate at 164 °C	Reference
$CH_3 \cdot$	11·1	6·8	**1**	17
$C_2H_5 \cdot$	11·2	7·6	0·5	26
$CH_3\!\diagdown$ $\quad CH \cdot$ $CH_3\!\diagup$	11·2*	6·9	1·0	26
$CH_3(CH_2)_3 \cdot$	10·3*	6·7	0·2	26
$CH_3\!\diagdown$ $\quad CH(CH_2)_2 \cdot$ $CH_3\!\diagup$	10·5*	6·4	0·4	26
$CH_2F \cdot$	10·6	4·3	5·6	27
$CCl_3 \cdot$	11·3	6·3	1·1	28, 29
$CF_3 \cdot$	11·4	2·4	320	30
$SF_5 \cdot$	(8·6?)	1·9	0·9	31
$CF_2Br \cdot$	10·3	2·3	1·2	32

* Assuming $X \cdot + X \cdot \rightarrow X_2\ k + 10^{13}$ (see page 33).

Probably the most surprising feature of this table is that the rate of addition of a wide variety of radicals to ethylene varies so little. Only the rate for trifluoromethyl radicals differs by more than a power of ten from the rate of addition of methyl radicals at 164 °C. On the other hand the Arrhenius parameters do show a fairly considerable variation. The more electronegative the radical, the lower the activation energy. The decrease in activation energy with increasing electronegativity fits in with the apparent effect of the substituents in the olefin on the rate of ethyl radical addition. In other words the polar concepts of the organic chemist have some relevance, albeit considerably modified, in interpreting the results of radical addition.

A very important result is the recent determination of the rate constant for the addition of methyl radicals to tetrafluoroethylene. This result clearly shows the importance of the polar effects in these reactions, for the electrophilic CF_3-radical addition to ethylene is faster than to tetrafluoroethylene whereas for the CH_3-radical the very reverse is observed. This is in line with the suggestion made above that the ethyl-radical (and by presumption also the methyl-radical) is nucleophilic.

Table 2.11 The addition of methyl and trifluoromethyl radicals to ethylene and tetrafluoroethylene

Olefin	CH_3			CF_3		
	$\log k$ cm^3 mol^{-1}s^{-1}	$\log A$ cm^3 mol^{-1}s^{-1}	E kcal mol^{-1}	$\log k$ cm^3 mol^{-1}s^{-1}	$\log A$ cm^3 mol^{-1}s^{-1}	E kcal mol^{-1}
$CH_2{=}CH_2$	8·0	11·9	7·9[18]	10·2	11·4	2·4[25]
$CF_2{=}CF_2$	9·1	11·95	5·7[30]	9·8	10·5	2·9[24]

Attention has also been drawn to the fact that trifluoromethyl radicals are pyramidal, whereas methyl radicals are planar. It is then argued that the pyramidal radical would be more reactive. On the other hand, the pre-exponential terms for $CF_3\cdot$ and $CH_3\cdot$ radicals are nearly the same, which rather suggests that the shape of the radical is less important and this argument is not in accord with the tetrafluoromethylene results.

The problem of orientation of addition goes right back to the beginning of free-radical chemistry. The observation that the orientation of addition of hydrogen bromide to unsymmetric olefins was reversed in the presence of peroxides or of ultra-violet light was explained by Kharasch and Mayo in terms of a free-radical mechanism for 'anti-Markovnikov' addition as opposed to an ionic mechanism for normal Markovnikov addition.

$$RCH{=}CH_2 + H^+ \longrightarrow R\overset{+}{C}HCH_3$$

$$R\overset{+}{C}HCH_3 + Br^- \longrightarrow RCHBrCH_3$$

i.e.

$$RCH{=}CH_2 + HBr \xrightarrow[\text{dark}]{\text{solution}} RCHBrCH_3 \qquad \textit{Markovnikov}$$

$$RCH{=}CH_2 + Br{\cdot} \longrightarrow R\dot{C}HCH_2Br$$

$$R\dot{C}HCH_2Br + HBr \longrightarrow RCH_2CH_2Br + Br{\cdot}$$

i.e.

$$RCH{=}CH_2 + HBr \xrightarrow[\text{uv light}]{\text{gas phase}} RCH_2CH_2Br \qquad \textit{Anti-Markovnikov}$$

The first feature to notice about these two reactions is that the cation and the radical add to the same site. This had led some authors to regard radicals as essentially electrophilic species, but this hardly seems reasonable, they could equally be regarded as nucleophilic.

The explanation of the orientation of ionic addition is usually given in terms of extra canonical forms; thus the intermediate ion in Markovnikov addition of an acid to vinyl chloride is 'resonance stabilized' whereas the reverse adduct is not. It is really assumed that

$$[:\ddot{C}l{-}\overset{+}{C}H{-}CH_3 \leftrightarrow :\overset{+}{C}l{=}CHCH_3]$$

this 'resonance stabilization' of the intermediate means that the electronic energy of the ion is lower than for an 'unstabilized' ion. The electronic theory of organic chemistry has been extraordinarily successful in giving a pictorial explanation of the orientation not only of ionic addition to olefins, but also of electrophilic and nucleophilic aromatic substitution. It is not surprising therefore that attempts should be made to apply the same type of explanation to radical addition. This was first done in an unambiguous manner by Walling and Mayo. Taking the same example, namely vinyl chloride, the intermediate radical in Anti-Markovnikov addition can be written in two canonical forms like the intermediate ion in Markovnikov addition. Since the

$$[:\ddot{C}l{-}\dot{C}H{-}CH_2Br \leftrightarrow :\ddot{C}l{-}\overset{+}{C}H{-}\overset{-}{C}H_2Br]$$

second canonical form required separation of charge one might expect 'resonance stabilization' to be less important than in the ionic addition. Walling and Mayo supported their hypothesis from co-polymerization results, but it was left largely to Haszeldine and his co-workers to provide experimental support with examples of the addition to monomers. Haszeldine studied the addition of trifluoromethyl radicals generated by the photolysis of trifluoromethyl iodide to a large number of unsymmetric fluoro-olefins. This work was entirely qualitative in nature and depended on the analysis of products in experiments in which the reaction had gone a long way towards completion. In the early work, separation of the products depended on distillation and it says a great deal for the quality of the original experimental work that when these results were re-investigated using gas chromatography the earlier work was confirmed (changes were made in the ratio of the two possible adducts, but the major product had invariably been correctly identified). More recent work by this group includes a study of thermally initiated addition of CF_3I as well as the re-investigation of the photochemically initiated reaction. Unfortunately, since the experiments are done under different conditions of temperature, little information is available beyond the identification of the main direction of addition.

Table 2.12 The site of the predominant radical attack by trifluoromethyl radicals in the gas phase (R. N. Haszeldine et al.)

$CH_3CH{=}CH_2$[*][34,35]	$CF_3CF{=}CF_2$[37]
$CHF{=}CH_2$[†][34,35]	$CFCl{=}CF_2$[‡][38]
$CHCl{=}CH_2$[34]	$CHCl{=}CF_2$[35]
$CF_3CH{=}CH_2$[33]	$CHF{=}CF_2$[35]
$CH_3O{\cdot}COCH{=}CH_2$[34]	$CH_3CH{=}CF_2$[39]
$CF_2{=}CH_2$[36]	$CF_3CH{=}CF_2$[∥][,37]

[*]90:10   [†]89:11   [‡]80:20   [§]92:8   [∥]60:40

The one completely anomalous result here is that for chloro-1,1-difluoromethylene and, although the authors make no mention of it, one cannot help wondering whether there was not some loss of chlorine from

$$CF_3\dot{C}HClCF_2$$

radicals. By and large the results appear to fit with the Walling-Mayo type of hypothesis, although certain *ad hoc* assumptions have to be made to accommodate them all.

Before discussing quantitative data now available which throw considerable doubt on the Walling-Mayo hypothesis, it is perhaps worth examining it in further detail. In ionic electrophilic addition reactions, and particularly in aromatic substitution, there is no doubt that the carbonium ion intermediate is of much higher energy than either the reactant arene (or alkene) or the products. It is not unreasonable therefore to assume that the carbonium ion

Ionic addition
$$\begin{cases} X^+ + R_2C{=}CR_2 \rightarrow R_2C\overset{+}{X}CR_2 \\ \qquad\qquad\text{Carbonium ion} \\ R_2CXCR_2 + Y^- \rightarrow R_2CXCYR_2 \\ \qquad\qquad\text{Product} \end{cases}$$

Radical addition
$$\begin{cases} Z{\cdot} + R_2C{=}CR_2 \rightarrow R_2CZ\dot{C}R_2 \\ \qquad\qquad\text{Adduct radical} \\ R_2CZ\dot{C}R_2 + W{-}Z \rightarrow R_2CZCWR_2 + Z{\cdot} \\ \qquad\qquad\text{Final product} \end{cases}$$

is nearer in structure and in energy to the transition state than are either the alkene or the product. It is very important to realize that this is not true in radical addition, in fact the adduct radical will often be of lower energy than the reactant radical and energy-wise further from the transition state.

The formal similarity between ionic addition and radical addition is much less real than the drawing of the equations suggests. Another feature of the Walling-Mayo hypothesis is the emphasis it puts on the siting of the odd electron in the adduct radical, while really it would seem more likely that the site where the new bond is formed is the predominant factor.

No real test of any theory relating to the orientation of radical addition can be obtained unless there are accurate rate data for the addition of a radical to opposite ends of the same olefin. In other words, it is necessary to find a system in which a radical will add irreversibly to both ends of any unsymmetric olefin and where it is possible to determine the relative rates of addition to both ends over a range of temperatures. Kharasch and his co-workers showed that it was possible to add bromotrichloromethane across olefinic double bonds photochemically

$$CCl_3Br \xrightarrow{h\nu} CCl_3\cdot + Br\cdot$$

$$CCl_3\cdot + CH_2{=}CH_2 \longrightarrow CCl_3CH_2CH_2\cdot$$

$$CCl_3CH_2CH_2 + CCl_3Br \longrightarrow CCl_3CH_2CH_2Br + CCl_3\cdot$$

Subsequently Melville and his co-workers measured the rate of addition of trichloromethyl radicals to cyclohexene and vinyl acetate in the liquid phase using the photolysis of bromotrichloromethane as their source of trichloromethyl radicals. The absolute rates were determined using the rotating sector method.

Recently this method has been employed in gas phase studies, and the rate of trichloromethyl radical addition to a number of unsymmetrical fluoro-olefins has been determined.

Table 2.13   The addition of trichloromethyl radicals to olefins

$$CCl_3\cdot + CWX{=}CYZ \xrightarrow{k} CCl_3CWX\dot{C}YZ^{40,28,29}$$

Addition to CH$_2$			Addition to CHF			Addition to CF$_2$		
	log A	E		log A	E		log A	E
CH$_2$=CH$_2$	11·3	6·3	CFH=CH$_2$	11·1	8·4	CF$_2$=CH$_2$	11·4	11·4
CH$_2$=CHF	11·1	6·4				CF$_2$=CHF	12·1	10·2
CH$_2$=CF$_2$	11·4	7·7	CFH=CF$_2$	12·1	9·2	CF$_2$=CF$_2$	12·7	9·2

Log A in cm^3 mol^{-1} s^{-1}
E in kcal mol^{-1}

The first feature to notice about these results is that the trichloromethyl radical adds preferentially to the CH_2-end of vinyl fluoride, not because attack at this end is facilitated (e.g. by resonance stabilization involving the fluorine atom) but because the fluorine atom hinders addition at the CHF-end. It is noticeable that addition to the CH_2-end of vinylidene fluoride is slightly retarded, although there are two fluorine atoms to stabilize the radical. On the other hand addition to the CF_2-end of vinylidene fluoride is greatly retarded although the odd electron is sited on a CH_2-group just as it is in the addition to ethylene. In other words, the results are in complete contradiction to the Walling-Mayo picture in its simplest form.

Very similar results have been obtained for the addition of heptafluoropropyl radicals to fluoro-olefins.

Table 2.14   The addition of heptafluoropropyl radicals to fluoro-olefins[41]
$$C_3F_7\cdot + CWX{=}CYZ \rightarrow C_3F_7CWX\dot{C}YZ$$

	Addition to CH_2			Addition to CFH			Addition to CF_2	
	log A	E		log A	E		log A	E
$CH_2{=}CH_2$	11·5*	3·1*	$CHF{=}CH_2$	11·6	7·3*	$CF_2{=}CH_2$	12·6	11·7
$CH_2{=}CHF$	11·4	3·9				$CF_2{=}CHF$	11·7	7·2
$CH_2{=}CF_2$	12·2	5·8	$CHF{=}CF_2$	11·7	6·2	$CF_2{=}CF_2$	13·1	7·8

* Projected values (other values relative to this).
Log A in cm^3 mol^{-1} s^{-1}
E in kcal mol^{-1}

Again attack occurs preferentially at the CH_2-end of vinyl fluoride because attack is retarded at the CHF-end, not because attack at the CH_2-end is facilitated (i.e. by supposed resonance stabilization).

Results involving other electrophilic radicals trichloromethyl, bromo-difluoromethyl and sulphur pentafluoride radicals all confirm the general picture provided by the trichloromethyl and heptafluoromethyl radical results listed. It is possible to devise a qualitative picture of the results in terms of the strength of the bond formed, i.e. one attempts to predict which will be the stronger of the two possible σ-bonds made when a radical adds to an unsymmetrical olefin. It is fairly easy to see that such an argument will always lead to the same prediction for the orientation of addition to a single olefin as the Walling-Mayo picture. Where it differs is in its predictions in going from one olefin to another and here it is in accord with the results whereas the Walling-Mayo argument is not. For instance Table 2.16 shows the Arrhenius parameters for the addition to the CH_2-end of a wide range of olefins. According to the Walling-Mayo picture, the activation energy should fluctuate very widely since the extent of resonance stabilization can vary widely whereas the results show it to be almost constant, except for two extremely polar molecules $CH_2{=}CF_2$ and $CH_2{=}CHCF_3$.

Table 2.15 Rate data for the addition of trichloromethyl radicals to the CH_2-end of some unsymmetric olefins[28,29,40,42]

Olefin	log A $cm^3 mol^{-1} s^{-1}$	E kcal mol^{-1}
$CH_2=CH_2$	11·3	6·3
$CH_2=CHCH_3$	12·6	6·5
$CH_2=CHF$	11·1	6·4
$CH_2=CFCH_3$	11·8	6·3
$CH_2=CHCl$	12·3	6·5
$CH_2=CF_2$	11·4	7·7
$CH_2=CHCF_3$	12·1	8·0

If we consider the strength of the bond formed to be all important then we would expect the activation energies to be almost all the same. We can accommodate the higher activation energies for the addition to the last two olefins by involving the concept of electronegativity. The trichloromethyl radical is electrophilic and we have already discussed the effect of this in our discussions on the rate of additions to olefins as a whole (p. 41). There is one word of caution needed here, and that is that although the activation energies are almost constant the rates vary. Whether this is due to experimental error in the temperature-dependent term, or whether the pre-exponential terms really do vary cannot be determined at the present time. What is certain, however, is that the orientation of radical addition is principally determined by the temperature-dependent term, i.e. the activation energy. The cause of the small differences in rate for attack at the CH_2-end of the olefins listed is as yet uncertain.

So far all our discussion about orientation of addition has been concerned with electrophilic radicals. We have earlier argued that methyl and ethyl radicals show some nucleophilic character. It would therefore be extremely interesting to compare the orientation of nucleophilic radicals with that of the electrophilic radicals already discussed. The striking feature of Table 2.16 is that the preferred orientation for methyl radicals to trifluoroethylene is the reverse of that for the two electrophilic radicals. The proportion of attack at the carbon atom bonded to fluorine in vinyl fluoride and 1,1-difluoroethylene is greater as well. The ratio of methyl radical attack at $CHF=$ or $CF_2=$ to the attack at $CH_2=$ in the same olefins is an order of magnitude greater than the same ratio for attack by the electrophilic radicals. The same is also true for the ratio of methyl attack at $CF_2=$ to attack at $CHF=$ in trifluoroethylene. These results are consistent with those discussed earlier in Table 2.11 which shows that the rate of addition of methyl radicals to tetrafluoroethylene is ten times faster than addition to ethylene, whereas trifluoromethyl radicals add two and half times faster to ethylene.

Table 2.16 The orientation of radical addition to unsymmetrical fluoro-olefins
Orientation ratios $\alpha:\beta$ at 150 °C

Radical	α β $CH_2{=}CHF$	α β $CH_2{=}CF_2$	α β $CHF{=}CF_2$
$CCl_3\cdot$	1:0·08	1:0·01	1:0·29
$CF_3\cdot$	1:0·09	1:0·03	1:0·42
$C_2F_5\cdot$	1:0·05	1:0·01	1:0·25
$(CF_3)CF\cdot$	1:0·02	1:0·001	1:0·06
$CF_3\cdot$	1:0·09	1:0·03	1:0·42
$CF_2Br\cdot$	1:0·09	1:0·03	1:0·47
$CFBr_2\cdot$	1:0·08	1:0·02	1:0·37
$CH_3\cdot$	1:0·59	1:0·42	1:7·26
C_2H_5	1:0·29	—	1:13·5

It is interesting to note in Table 2.16 that, although heptafluoropropyl radicals are appreciably more reactive than trichloromethyl radicals (the rate of addition of $C_3F_7\cdot$ to ethylene is 10^2 faster than $CCl_3\cdot$ at 150 °C), they are more selective.

Table 2.17 shows the Arrhenius parameters for methyl addition to fluoro-ethylenes, the difference between this table and Tables 2.14 and 2.15 is substantial; there are however important similarities. In all three tables the

Table 2.17 The addition of methyl radicals to fluoroethylenes in the gas phase[44]
$CH_3\cdot + CWX{=}CYZ \rightarrow CH_3CWX'C\dot{Y}Z$

Addition to CH₂			Addition to CHF			Addition to CF₂		
	$\log A$	E		$\log A$	E		$\log A$	E
$CH_2{=}CH_2$	11·2	7·7	$CHF{=}CH_2$	11·5	9·7	$CF_2{=}CH_2$	9·9	8·0
$CH_2{=}CHF$	11·2	8·7				$CF_2{=}CHF$	10·1	6·1
$CH_2{=}CF_2$	11·8	11·1	$CHF{=}CF_2$	10·5	8·4	$CF_2{=}CF_2$	10·5	5·4

Log A in cm³ mol⁻¹ s⁻¹
E in kcal mol⁻¹.

activation energy increases for the addition to a $CH_2{=}$ group as the other terminal atom becomes increasingly substituted with fluorine atoms (i.e. the exact converse of the predictions of the simple Walling–Mayo picture). However there is an exactly opposite trend in the addition to the $CF_2{=}$ group of difluoro, trifluoro and tetrafluoroethylene.

Two transition states can be visualized for the addition of small radicals to olefins. A 'π-transition state' in which the radical is associated with the double bond and a 'σ-transition state' in which the radical is associated with one of the terminal carbon atoms. Probably the reaction passes through a

'π-complex' stage followed by a 'σ-transition state.' The fact that the relative proportions of the two possible adducts from an unsymmetrical olefin vary with temperature shows that the 'σ-complex' represents the top of the potential energy pass. The absence of any appreciable kinetic isotope effect implies that the transition state resembles reactants rather than products, a conclusion supported by the small activation energies. We would expect correlation of the rate data with atom, rather than bond properties, and localization energy $L\mu$ calculated from simple Hückel MO theory gives reasonable correlations with the electrophilic radicals $CCl_3\cdot$ and $C_3F_7\cdot$. The preliminary methyl radical data do not correlate well with simple $L\mu$ but a correlation can be obtained if a charge transfer structure

$$\left(CH_3{-}C{-}C\cdot \leftrightarrow CH_3^+ \quad \ddot{C}{-}\dot{C} \right)$$

is incorporated. At the present therefore there is no single method for correlating all the observed data which is satisfactory. With the development of more sophisticated CNDO methods it is reasonable to hope that we will soon be able to compare theoretical parameters with experimental observations. In these gas phase studies the complications and uncertainties of a solvent can be avoided so that there is a real hope that good correlations will be obtained. At present the gas phase data already show in a qualitative way that the polar concepts of the physical organic chemist are relevant to radical reactions. In subsequent chapters describing radical polymerizations this point will be further emphasized.

I would like to acknowledge the valuable criticisms of Drs J. A. Kerr, J. C. Thynne and E. Whittle, all of whom have kindly read the chapter and made valuable comments.

2.5   REFERENCES

1. R. Hiatt and S. W. Benson, *Int. J. Chem. Kinetics*, **4**, 151, 479, (1972); D. F. McMillen, D. M. Golden and S. W. Benson, *J. Amer. Chem. Soc.*, **94**, 4403, (1972); J. Currie, H. W. Sidebottom and J. M. Tedder, unpublished results.
2. J. O. Terry and J. H. Futrell, *Canad. J. Chem.*, **45**, 2327, (1967).
3. J. Grotewold, E. A. Lissi and M. G. Newmann, *J. Chem. Soc.*, **1968**, A, 375.
4. G. O. Pritchard and R. L. Thommarson, *J. Phys. Chem.*, **71**, 1674, (1969).
5. R. D. Giles, L. M. Quick and E. Whittle, *Trans. Faraday Soc.*, **63**, 662, (1967).
6. G. O. Pritchard and J. T. Bryant, *J. Phys. Chem.*, **72**, 1603, (1968).
6a. J. P. Terry and J. H. Futrell, *Canad. J. Chem.*, **45**, 2327, (1967).
7. G. C. Fettis and J. H. Knox, *Prog. React. Kinetics*, **2**, 1, (1963).
8. T. Berces and A. F. Trotman-Dickenson, *J. Chem. Soc.*, **1961**, 348.
9. J. L. Brokenshire, A. Nechvatal and J. M. Tedder, *Trans. Faraday Soc.*, **66**, 2029, (1970).

10. A. F. Trotman-Dickenson, *Quart. Rev.*, **7**, 198, (1953).
11. P. B. Ayscough and E. W. R. Steacie, *Can. J. Chem.*, **34**, 103, (1956).
12. J. M. Tedder and R. A. Watson, *Trans. Faraday Soc.*, **62**, 1215, (1966). J. A. Kerr and M. Parsonage; H. W. Sidebottom, J. M. Tedder and J. C. Walton, *Int. J. Chem. Kinetics*, **4**, 245, 249, (1972).
13. W. G. Alcock and E. Whittle, *Trans. Faraday Soc.*, **62**, 134, (1966). L. M. Quick and E. Whittle, *Trans. Faraday Soc.*, **67**, 1727, (1971).
14. L. Mandelcorn and E. W. R. Steacie, *Canad. J. Chem.*, **32**, 79, (1954).
15. R. K. Briton, *J. Chem. Phys.*, **29**, 781, (1958).
16. A. M. Hogg and P. Kebarle, *J. Amer. Chem. Soc.*, **86**, 4558, (1964).
17. L. Endrenyi and D. J. LeRoy, *J. Amer. Chem. Soc.*, **71**, 1334, (1967).
18. R. J. Cvetanovic and R. S. Irwin, *J. Chem. Phys.*, **46**, 1694, (1967).
19. N. Yokoyama and R. K. Briton, *Canad. J. Chem.*, **47**, 2987, (1969).
20. D. G. L. James and D. MacCallum, *Canad. J. Chem.*, **43**, 633, (1965).
21. D. G. L. James and T. Ogawa, *Canad. J. Chem.*, **43**, 640, (1965).
22. P. S. Dixon and M. Szwarc, *Trans. Faraday Soc.*, **59**, 112, (1963).
23. J. M. Pearson and M. Szwarc, *Trans. Faraday Soc.*, **60**, 564, (1964).
24. G. E. Owen, J. M. Pearson and M. Szwarc, *Trans. Faraday Soc.*, **61**, 1722, (1965).
25. J. M. Sangster and J. C. J. Thynne, *J. Phys. Chem.*, **73**, 2746, (1969).
26. K. W. Watkins and L. A. O'Deen, *J. Phys. Chem.*, **71**, 1334, (1967).
27. J. M. Sangster and J. C. Thynne, *Trans. Faraday Soc.*, **65**, 2110, (1969).
28. J. M. Tedder and J. C. Walton, *Trans. Faraday Soc.*, **60**, 1769, (1964).
29. J. A. Kerr and M. Parsonage; H. W. Sidebottom, J. M. Tedder and J. C. Walton, *Int. J. Chem. Kinetics*, **4**, 245, 249, (1972).
30. J. M. Sangster and J. C. Thynne, *Int. J. Chem. Kinetics*, **1**, 571, (1969).
31. H. W. Sidebottom, J. M. Tedder and J. C. Walton *Trans. Faraday Soc.*, **65**, 2103, (1969).
32. J. M. Tedder and J. C. Walton, *Trans. Faraday Soc.*, **66**, 1135, (1970), and unpublished results.
33. R. N. Haszeldine, *J. Chem. Soc.*, **1952**, 2504.
34. R. N. Haszeldine and B. R. Steele, *J. Chem. Soc.*, **1953**, 1199.
35. R. N. Haszeldine and B. R. Steele, *J. Chem. Soc.*, **1957**, 2800.
36. R. N. Haszeldine and B. R. Steele, *J. Chem. Soc.*, **1954**, 923.
37. R. N. Haszeldine, *J. Chem. Soc.*, **1953**, 3559.
38. R. N. Haszeldine and B. R. Steele, *J. Chem. Soc.*, **1957**, 2193.
39. R. N. Haszeldine, *J. Chem. Soc.*, **1953**, 3565.
40. J. M. Tedder and J. C. Walton, *Trans. Faraday Soc.*, **62**, 1859, (1966).
41. J. M. Tedder, J. C. Walton and K. D. R. Winton, *J. Chem. Soc. (Faraday I)*, **68**, 160, (1972).
42. D. P. Johari, H. W. Sidebottom, J. M. Tedder and J. C. Walton, *J. Chem. Soc.*, **1971, B**, 95.
43. J. M. Tedder, J. C. Walton and K. D. R. Winton, *Chem. Comm.*, **1971**, 1046; D. S. Ashton, A. F. Mackay, J. M. Tedder, D. C. Tiprey and J. C. Walton, *Chem. Comm.*, **1973**, 496; J. P. Sloan, J. M. Tedder and J. C. Walton, *J. Chem. Soc. (Faraday I)*, **69**, 1143, (1973).
44. J. M. Tedder, J. C. Walton and K. D. R. Winton, *J. Chem. Soc. (Faraday I)*, **68**, 1866, (1972), and unpublished results.

3

Organometallic Derivatives of Transition Metals as Initiators of Free-Radical Polymerization

C. H. Bamford

Department of Inorganic, Physical and Industrial Chemistry,
University of Liverpool

Two types of system which generate free radicals suitable for initiating the polymerization of vinyl monomers are discussed in this chapter. Members of the first group contain an organometallic derivative of a transition metal in a low (often the zeroth) oxidation state (e.g. a metal carbonyl) together with an organic halide, while those of the second consist of a transition metal chelate in which the metal is in a high oxidation state, either alone, or accompanied by a suitable electron-donor additive. All the initiators function thermally under suitable conditions and some are also active photoinitiators. We shall be concerned essentially with homogeneous reactions.

Most of the initiating systems do not participate in chain-transfer or retardation reactions, except under unusual conditions, consequently the molecular weights of the resulting polymers are the same as those obtained with more conventional initiators, for similar rates of initiation, temperature, monomer concentrations and so forth. Molecular weights and distributions may therefore be calculated from standard kinetic equations. However, it will become apparent from our discussion that retardation is a characteristic feature of the behaviour of a few systems; naturally the foregoing remarks do not apply to these.

There is no evidence that the microstructures of the polymeric products are significantly influenced by the initiators under consideration.

3.1 SYSTEMS BASED ON ORGANOMETALLIC DERIVATIVES AND ORGANIC HALIDES

In 1959 and 1961 reports[1,2] by Freydlina and Belyavskii of the use of some metal carbonyls for initiating the telomerization of ethylene in carbon tetrachloride solution prompted Bamford and Finch[3] to explore the possibilities of employing these derivatives as initiators of polymerization.

Although the carbonyls alone were found to be inactive or only feebly active, it soon transpired that in the presence of carbon tetrachloride they are effective initiators. Since 1962 many organometallic derivatives have been studied from this point of view; a list is given in Table 3.1 of those which have been examined in some detail. Others which have received more cursory treatment are referred to by Strohmeier and Grübel[4] and by Bamford and Finch.[4a]

Table 3.1 Thermal initiation by organometallic derivatives. Values of a, where $I = a$[derivative], I (mol $l^{-1}s^{-1}$) being the rate of initiation in bulk monomer at high [halide]. Values refer to initial rates of polymerization of methyl methacrylate in the presence of carbon tetrachloride, unless otherwise indicated (see section 3.13). St = styrene

Group VIA

Derivative	$T\,°C$	$10^6\,a\,s^{-1}$	Reaction type	Reference
$Cr(CO)_6$	80	0·7		4
	60	0·10		4
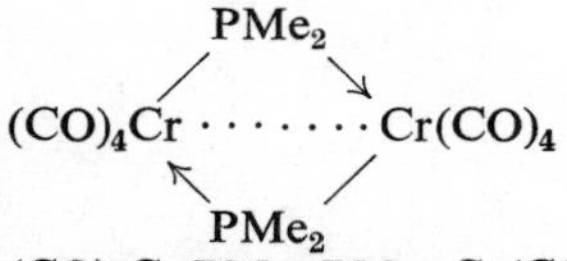 $(CO)_4Cr \cdots Cr(CO)_4$ (with PMe_2 bridges)	60	72		5
$(CO)_5Cr \cdot PMe_2 \cdot PMe_2 \cdot Cr(CO)_5$	60	0·50		5
† $Mo(CO)_6$	80	112	S_N2	6
	80	22·5 (St)		12
	80	290 (CBr_4)	S_N2	7
$Mo(CO)_5Py$	40	1040 (CCl_3COOEt)		8
	40	360		8
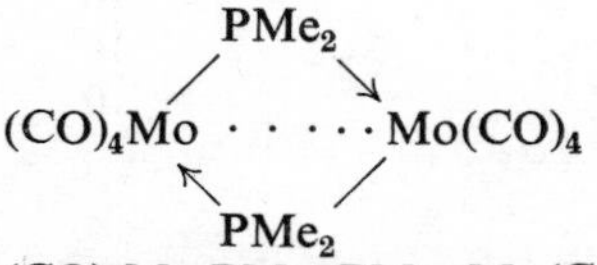 $(CO)_4Mo \cdots Mo(CO)_4$ (with PMe_2 bridges)	60	1200	S_N1	5
$(CO)_5Mo \cdot PMe_2 \cdot PMe_2 \cdot Mo(CO)_5$	60	2·5		5
$(CO)_5Mo \cdot PEt_2 \cdot PEt_2 \cdot Mo(CO)_5$	60	12·0	S_N1 (mainly)	5
$W(CO)_6$	80	9·9		9
$Cr(CNAr)_6$				
$Ar = $ 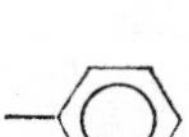 —⬡	80	97	S_N2	10
	25	0·9		10
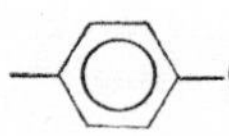 —⬡—OMe	80	865		11

Table 3.1 *Continued*

Group VI A *Continued*

Derivative	$T\,^\circ C$	$10^6\,a\,\mathrm{s}^{-1}$	Reaction type	Reference
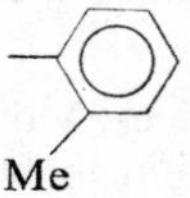 (phenyl with *o*-Me)	80	185	S_N2	11
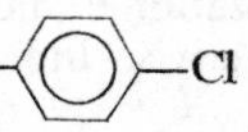 (phenyl with *p*-Cl)	80	119		11
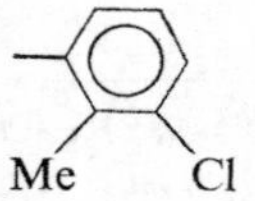 (phenyl with Me, Cl)	80	33·3		11
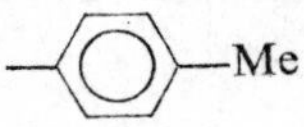 (phenyl with *p*-Me)	80	222	?S_N2	11
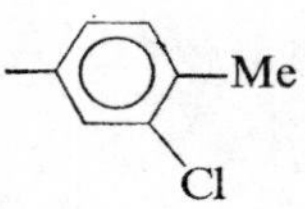 (phenyl with Me, Cl)	80	173		11
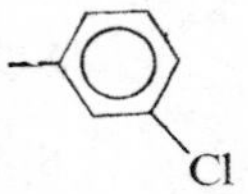 (phenyl with *m*-Cl)	80	121	S_N2	11
$Mo(CNPh)_6$	80	340	S_N2	10
	25	0·8		10
$W(CNPh)_6$	80	140	S_N2	10
	25	1·9		10

Group VII A

$Mn_2(CO)_{10}$	80	3·9	S_N1	12
	80	5·2 (St)		12
$(C_5H_5)Mn(CO)_3$	80	~1·6		13
$Re_2(CO)_{10}$	100	~30		14
	80	~3		14
$Re_2(CO)_8(PPh_3)_2$	80	~4·9		14
$Re(CO)_3(PPh_3)_2$	80	~0·4		14

Group VIII

$Fe(CO)_5$	60	0·70		5
$Fe(CO)_4(PPh_3)$	60	3·0		5
$Fe(CO)_3(PPh_3)_2$	60	6·4		5

Table 3.1 *Continued*

Group VIII *Continued*

Derivative	$T\,°C$	$10^6\,a\ s^{-1}$	Reaction type	Reference
$(CO)_3Fe \cdots\cdots Fe(CO)_3$ (bridged by two PMe_2)	60	18		5
$(CO)_3BrFe \quad\quad FeBr(CO)_3$ (bridged by two PMe_2)	60	350	$?S_N2$	5
$(CO)_4Fe\cdot PMe_2\cdot PMe_2\cdot Fe(CO)_4$	60	22		5
$(CO)_3Fe \cdots\cdots Fe(CO)_2(PPh_3)$ (bridged by two PMe_2)	60	7·9		5
$Co_2(CO)_8$	25	0		15
$(CO)_3Co \quad\quad Co(CO)_3$ (bridged by two PPh_2)	80	2250	$?S_N2$	5
	25	130		5
$Co_2(CO)_6(PhC\equiv CCOOMe)$	25	>1·6		15
$Co_3(CO)_9H(PhC\equiv CH)$	25	>1		15
$Co_4(CO)_{12}$	25	1170	$?S_N2$	15
† $Ni(CO)_4$	25	79·4	S_N2	16
	25	24·5 (St)		16
† $Ni(CO)_2(PPh_3)_2$	25	172	S_N2	17
$(CO)_3Ni\cdot PPh_2\cdot PPh_2\cdot Ni(CO)_3$	60	433		5
† $Ni\{P(OPh)_3\}_4$	25	270	S_N1	18
$Ni\{P(OC_2H_5)_3\}_4$	60	22	S_N1	11
$Ni\{P(OC_2H_4Cl)_3\}_4$	60	19	S_N1	11
$CoH(N_2)(PPh_3)_3$	−10	~0·2*		19
† $Pt(PPh_3)_4$	60	36·5	$?S_N2$	19a
	25	4·5		19a

* From one observation at concentration 10^{-2} mol 1^{-1}.

† See also Figure 3.4.

3.11 GENERAL KINETIC FEATURES

All the systems which have been investigated show characteristic kinetic behaviour, which we now discuss. For brevity we shall subsequently refer to the organometallic and halide components of the initiator as In, H, respectively, and we shall denote the rate of polymerization by ω ($\equiv -d[M]/dt$, where M represents the vinyl monomer).

(i)   At constant [H], ω is proportional to $[In]^{0.5}$ at sufficiently low values of [In] as would be expected for a free radical polymerization. The lowest concentration of In at which departure from the square-root relation becomes apparent depends on the nature of the system. For example, Figure 3.1 shows that with $Mo(CO)_6/CCl_4$ and $W(CO)_6/CCl_4$ deviations occur at relatively

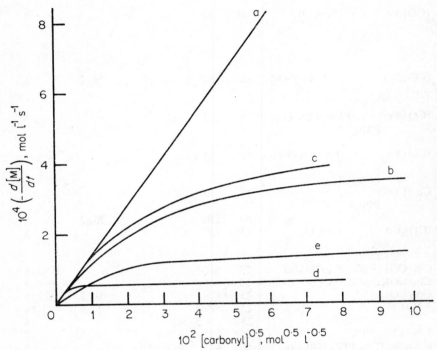

Figure 3.1   Onset of self-inhibition with increasing carbonyl concentration; influence of the halide component. Polymerization of methyl methacrylate at 80 °C. Curves (a), (b), (c), (d) for $Mo(CO)_6$, and halides: (a) CCl_3COOEt, 0·19 mol l^{-1} (b) CCl_3COOEt, 0·014 mol l^{-1} (c) CCl_4, 0·19 mol l^{-1} (d) $CHCl_3$, 0·19 mol l^{-1}. Curve (e) is for $W(CO)_6$; CCl_4, 0·192 mol l^{-1}. (From Bamford, Eastmond and Maltman[20], and Bamford and Finch[9])

low concentrations and at high concentrations become very pronounced, especially with $W(CO)_6$.[9,20] On the other hand, with $Mn_2(CO)_{10}/CCl_4$ the departures are notably smaller,[13] while with $Cr(CNPh)_6/CCl_4$[10] and $Ni\{P(OPh)_3\}_4/CCl_4$[24] the square root relation holds up to the highest concentrations studied (> 10^{-3} mol l^{-1}). In the majority of systems showing these departures from 'ideal' kinetics retardation by the initiator is negligible or small, unless [In] is unusually high; this follows from the observation that the kinetic parameter $k_p k_t^{-0.5}$ (k_p, k_t being the rate coefficients of chain propagation and termination for the vinyl monomer, respectively), evaluated from measurements of ω and the mean kinetic chain length $\bar{\nu}$ by conventional relations, has a 'normal' value for the vinyl monomer at the appropriate temperature. It follows, therefore, that at high [In] inhibition of the initiation process must occur, some species formed under these conditions interfering with the generation of initiating radicals, so that the rate of initiation falls below that required for direct proportionality to [In]. This process, which we shall term 'self-inhibition' will be considered in more detail later (section 3.123). The extent of self-inhibition is influenced by the nature of the halide component and is smaller for highly active halides such as CBr_4, $Cl_3CCOOEt$, than for the less active CCl_4 (Figure 3.1).

(ii) The dependence of ω on [H] at constant [In] has the same form in all cases. The rate of polymerization is essentially zero for [H] = 0, increases with increasing [H], and finally achieves a plateau value. At *high* values of [H] the rate is therefore zero order in [H]. The sharpness of the ω-[H] curve depends on the natures of the organometallic derivative and the halide; it may vary from monomer to monomer and be influenced by the presence of solvent and other additives. Figure 3.2 compares the ω-[CCl_4] dependence for $Mo(CO)_6$ and $Mn_2(CO)_{10}$ in methyl methacrylate at 80 °C and indicates the much sharper dependence with the manganese derivative.[21] Table 3.2 lists approximate halide concentrations required to achieve 90 per cent of the limiting rates of polymerization for a few systems.

In general, the activity of the halide component of the initiator increases with the number of chlorine or bromine atoms joined to a single carbon atom;[21a,b] derivatives of bromine are much more active than the corresponding one of chlorine.[21a,b] This is illustrated by the data in Table 3.2. Introduction of an electron-attracting group often increases the activity;[21a,b] thus CCl_3COOH and CBr_4 have similar activities. N-chlorinated and N-brominated amides and imides are very active components of initiating systems as might be anticipated from the relative weakness of the N-halogen bond and the electron attracting properties of the CO group.[23] (See also section 3.14.)

Saturated fluorine derivatives (which do not contain other halides) are inactive, no doubt on account of the high C—F bond strength, while iodine derivatives are ineffective since they tend to form molecular iodine which is a powerful retarder.[3]

(iii) As a rule the rate of initiation is depressed by addition of the free ligand. Again, the magnitude of the effect depends on the nature of the system. Thus, initiation by $Mn_2(CO)_{10}$ is relatively insensitive to the presence of carbon monoxide, pressures exceeding one atmosphere being necessary to reduce the rate by a factor of two in a typical case.[13] On the other hand, the $Ni(CO)_4/CCl_4$ system is extremely sensitive, 1 mmHg of CO leading to an approximately seven-fold reduction in the rate of initiation.[16] Since carbon monoxide is a product of the initiation reaction, investigation of the kinetics

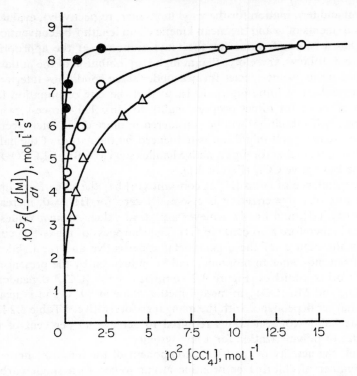

Figure 3.2 Dependence of the initial rate of polymerization of methyl methacrylate at 80 °C on carbon tetrachloride concentration: comparison of different systems. Curves have been normalized (factor f) at high [CCl$_4$]. [Monomer] = 3·3 mol l^{-1}
●   Mn$_2$(CO)$_{10}$, 1·5 × 10^{-3} mol l^{-1}; benzene and ethyl acetate as solvents; f = 2·33 (section 3.122)
○   Mo(CO)$_6$; 3·49 × 10^{-4} mol l^{-1}; ethyl acetate as solvent; f = 1·00
△   Mo(CO)$_6$, 3·49 × 10^{-4} mol l^{-1}; benzene as solvent; f = 1·60. A similar curve (f = 0·38) is obtained for bulk monomer. (From Bamford and Eastmond)[21]

Table 3.2 Approximate halide concentrations (mol l^{-1}) required to obtain 90 per cent of maximum rate of polymerization.[22] Methyl methacrylate as monomer

Derivative	Temperature (°C)	10^2 [CCl$_4$]	10^2 [CBr$_4$]
Mo(CO)$_6$	80	13	0·8
Mn$_2$(CO)$_{10}$	80	0·6	
Ni(CO)$_4$	25	5	0·3
	0	3	
Ni(CO)$_2$(PPh$_3$)$_2$	25	10	
Ni{P(OPh)$_3$}$_4$	25	9	

of such sensitive systems presents obvious difficulties. The effect is not confined to metal carbonyls; as a further example we may cite[18,24] the reduction in rate produced by addition of triphenyl phosphite to polymerizations initiated by $Ni\{P(OPh)_3\}_4$.

Presence of the free ligand often reduces the sharpness of the ω against [H] curve. An example[24] is presented in Figure 3.3; carbon monoxide behaves

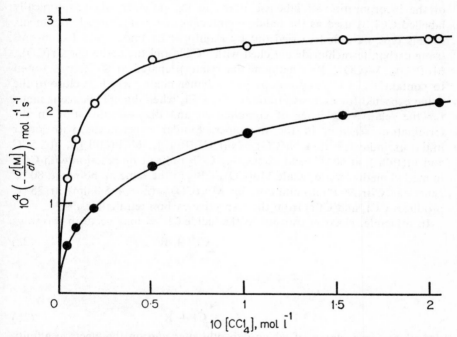

Figure 3.3 Influence of the free ligand on the ω against $[CCl_4]$ relation. Polymerization of methyl methacrylate at 25 °C initiated by $Ni\{P(OPh)_3\}_4/CCl_4$. $[Ni\{P(OPh)_3\}_4] = 7.62 \times 10^{-4}$ mol l^{-1}. ○ No additional $(P(OPh)_3$; ● $P(OPh)_3$ added, concentration 2×10^{-2} mol l^{-1}. (From Bamford and Hargreaves)[24]

similarly with initiators based on the hexacarbonyls of the Group VIA metals.[3,9] More rarely, the shape of the ω against [H] curve remains unchanged in the presence of the additive, the rates being reduced by a constant factor independent of [H]; this is the case[13] with $Mn_2(CO)_{10}$ (section 3.122).

3.12 REACTION MECHANISMS

The process of radical formation in these systems is basically electron transfer from transition metal to halide, the metal assuming a higher oxidation state, and the halide generally splitting into an ion and a radical fragment.

A typical radical-producing reaction is represented in equation (1) **M** being the transition metal.

$$M^0 + \overset{\frown}{C-CCl_3} \longrightarrow M^I Cl^{\ominus} + \dot{C}Cl_3 \qquad (1)$$

It has been amply demonstrated that the initiating radicals generated from CCl_4 are indeed $\dot{C}Cl_3$. The most direct evidence is obtained by determination of the incorporation of labelled atoms in the polymer when isotopically labelled CCl_4 is used as the halide component of the initiator. Experiments of this type were first carried out by Bamford, Eastmond and Robinson[25] using carbon tetrachloride enriched with $^{14}CCl_4$ and the carbonyls $Cr(CO)_6$, $Mo(CO)_6$, $W(CO)_6$. Polymethylmethacrylate prepared at 80 °C was found to contain 1·27 CCl_3 end-groups per polymer chain, which is close to the value expected for exclusive initiation by $\dot{C}Cl_3$ when due allowance is made for the relative incidence of combination and disproportionation in the termination reaction in the monomer. Similar experiments with other initiators indicate[26] that $Ni(CO)_4$, $Ni(CO)_2(PPh_3)_2$, $Ni\{P(OPh)_3\}_4$ at 25 °C and $Pt(PPh_3)_4$ at 60 °C yield exclusively $\dot{C}Cl_3$ radicals on reaction with CCl_4 in methyl methacrylate, while $Mo(CO)_6/CBr_4$ in the same monomer at 80 °C generates $\dot{C}Br_3$.[7b] Photoinitiation by $Mn_2(CO)_{10}$ ($\lambda = 435·8$ nm) at 25 °C produces $\dot{C}Cl_3$ and $\dot{C}Br_3$ from the respective carbon tetrahalides.[7b,26]

In principle, electron transfer to the halide Cl—R may occur in two ways

$$Cl^{\ominus} + \dot{R} \qquad (2a)$$

$$e^{\ominus} + Cl-R$$

$$\dot{Cl} + R^{\ominus} \qquad (2b)$$

the relative importance of which depends *inter alia* on the electron affinity of R. The discussion above shows that equation (2a) is the predominant reaction when $R = CCl_3$. Further, trichloracetic acid derivatives such as $Cl_3CCOOEt$ react with $Mo(CO)_6$, $Mn_2(CO)_{10}$, and $Pt(PPh_3)_4$ according to equation (2a), but in the reactions between these halides and the nickel derivatives $Ni(CO)_4$, $Ni(CO)_2(PPh_3)_2$, $Ni\{P(OPh)_3\}_4$, reactions (2a) and (2b) are of comparable importance.[26] Thus the anion

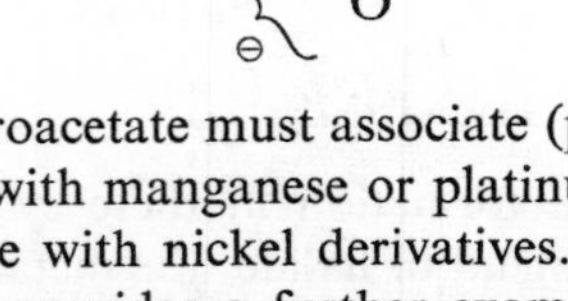

$$(R^{\ominus} = Cl_2C \overset{..}{=} \underset{\underset{\ominus}{\overset{||}{O}}}{C} - OEt)$$

derived from ethyl trichloroacetate must associate (presumably as an enolate ion) with nickel, but not with manganese or platinum, so that reaction (2b) is enhanced in importance with nickel derivatives. The work of Yoshisato and Tsutsumi[27] probably provides a further example of the occurrence of reaction (2b) in the presence of nickel derivatives. These investigators found that β-epoxy ketones are formed in good yield (50–80 per cent) by reaction of nickel carbonyl with α-bromo ketones in N,N-dimethylformamide solution at 30 °C. We may suppose that the first step in a typical case

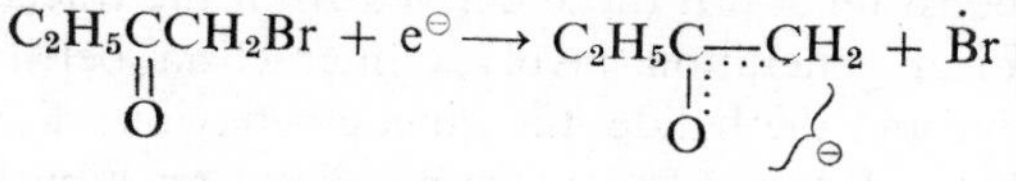

$$C_2H_5\underset{\underset{O}{\|}}{C}CH_2Br + e^\ominus \longrightarrow C_2H_5\overset{\cdot}{C}\text{---}CH_2 + \dot{B}r$$

is followed by addition of the product anion to the carbonyl group of a second ketone molecule

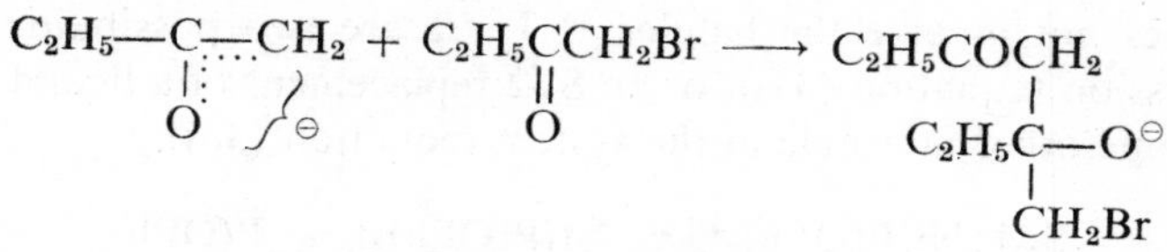

$$C_2H_5\text{---}C\text{---}CH_2 + C_2H_5\underset{\underset{O}{\|}}{C}CH_2Br \longrightarrow C_2H_5COCH_2$$

Cyclization with elimination of $Br^\ominus$ then yields the β-epoxy ketone

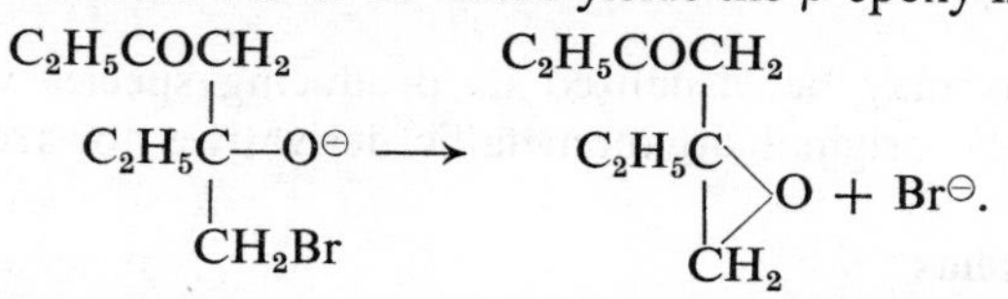

Go Hata[28] has shown that α-bromoacetone in the presence of $Ni\{P(OPh)_3\}_4$ is a good initiator of the polymerization of methyl methacrylate at 25 °C, the limiting (plateau) rate being the same as that obtained with CCl_4. The reaction sequence shown is clearly able to explain the formation of a β-epoxy ketone and the concurrent initiation of free-radical polymerization (by bromine atoms). In the absence of vinyl monomer the bromine atoms enter into oxidation reactions leading ultimately to $NiBr_2$.

Some attention has been paid[23] to the nature of the primary radicals formed from the chloroamide $n\text{-}C_5H_{11}CONCln\text{-}C_3H_7$ by thermal interaction with $Mo(CO)_6$ and $Mn_2(CO)_{10}$, and by photochemical interaction with $Mn_2(CO)_{10}$ ($\lambda = 435\cdot8$ nm). Tracer experiments show that, provided water is rigorously excluded, the radicals generated are mainly $n\text{-}C_5H_{11}CO\dot{N}n\text{-}C_3H_7$; however, a minor contribution from the reaction corresponding to equation (2b) cannot be excluded on the basis of the reported experimental findings. The presence of small quantities of water has a marked effect on initiation by $Mo(CO)_6$/chloroamide at 80 °C; under these conditions there is very little incorporation of radioactivity in the polymer when either of the labelled halides $n\text{-}C_5H_{11}{}^{14}CONCln\text{-}C_3H_7$, $n\text{-}C_5H_{11}CONCl^{14}CH_2C_2H_5$ is used. It has been suggested[23] that hypochlorous acid, formed by reaction between the halide and water, might react with $Mo(CO)_6$ to form $\dot{O}H$ radicals. Systems based on $Mn_2(CO)_{10}$ appear to be much less sensitive to water. There are no reports in the literature of studies of the interaction of N-halo-compounds and nickel derivatives but a significant contribution from route (2b) might be anticipated.

A different technique for investigating the nature of the primary radicals, based on gelation observations, is described briefly in section 3.17. It leads to conclusions consistent with those derived from the tracer method.

Although radical generation involves interaction between the organo-metallic derivative and the halide, the kinetic behaviour discussed in section 3.11 indicates that it does not follow from a direct reaction between the two species. The plateau value of the rate of polymerization reached at high [H] has been taken to imply the existence of a potentially rate-determining step which does not involve the halide.[3,29] There are two possibilities: an S_N1 ligand scission (equation (3a)), or an S_N2 replacement of a ligand by mono-mer or some other molecule in the system (equation (3b)).

$$S_N1 : Ni\{P(OPh)_3\}_4 \rightleftharpoons Ni\{P(OPh)_3\}_3 + P(OPh)_3 \qquad (3a)$$

$$S_N2 : Mo(CO)_6 + M \rightleftharpoons M \cdots Mo(CO)_5 + CO \qquad (3b)$$

These reactions may be visualized as producing species which are more reactive than the original organometallic derivatives towards the halide.

3.121 S_N2 systems

We now consider the reaction scheme originally proposed[6,9] for initiation by $Mo(CO)_6/CCl_4$ (equations (4a-d)), which may be regarded as typical for systems in which the primary process is an S_N2 ligand displacement. Although, as will appear later, the scheme is an idealised one, it is capable of explaining many of the experimental observations.

$$Mo(CO)_6 + M \underset{k_2}{\overset{k_1}{\rightleftharpoons}} \underset{(I)}{M \cdots Mo(CO)_5} + CO \qquad (4a)$$

$$(I) \overset{k_2}{\longrightarrow} \text{inactive products} \qquad (4b)$$

$$(I) + CCl_4 \overset{k_3}{\longrightarrow} (II) + CO \qquad (4c)$$

$$(II) \overset{k_4}{\longrightarrow} \dot{C}Cl_3 + Mo^I \text{ derivative} \qquad (4d)$$

Under stationary conditions, scheme (4) leads to the following expressions for the rates of initiation I and polymerization if all $\dot{C}Cl_3$ radicals formed initiate.

$$I = k_1[Mo(CO)_6][M]\frac{[CCl_4]}{k_2'/k_3 + (k_2/k_3)[CO] + [CCl_4]}, \qquad (5)$$

$$\omega = k_p k_t^{-0.5} k_1^{0.5}[Mo(CO)_6]^{0.5}[M]^{1.5}\left\{\frac{[CCl_4]}{k_2'/k_3 + (k_2/k_3)[CO] + [CCl_4]}\right\}^{0.5}. \qquad (6)$$

Equation (6) is consistent with the available kinetic data on polymerizations initiated by $Mo(CO)_6/CCl_4$ at sufficiently low $[Mo(CO)_6]$. Thus Bamford, Denyer and Eastmond[6] showed that in benzene solution at 80 °C the rate of polymerization is proportional to $[M]^{1.5}$ (M = methyl methacrylate) and that $[M]^3/\omega^2$ is linear in $1/[CCl_4]$ as expected from equation (6). The reduction in the rate of polymerization brought about by addition of carbon monoxide (section 3.11) results from the back reaction in equation (4a); it is therefore a true inhibition process interfering with chain initiation and not with chain growth. In equation (6) it is represented quantitatively by the term in the denominator involving [CO].

The form of the dependence of ω on $[CCl_4]$ in equation (6) seems to be generally applicable; the plateau value of ω is given by

$$\omega_p = k_p k_t^{-0.5} k_1^{0.5} [Mo(CO)_6]^{0.5} [M]^{1.5} \tag{7}$$

and the sharpness of the ω against $[CCl_4]$ curve, which depends inversely on $\{k_2'/k_3 + (k_2/k_3)[CO]\}$, is clearly reduced by addition of carbon monoxide. Inspection of the reaction mechanism shows that this latter behaviour arises from competition between CO and CCl_4 for reaction with the complex (I).

The parameters k_1, k_2'/k_3 and k_2/k_3 may be evaluated from measurements of the dependence of the rate of polymerization on $[CCl_4]$ and [CO] with the aid of equation (6).

Although benzene appears to act as an inert diluent in systems containing $Mo(CO)_6$, it has been shown that some other solvents such as ethyl acetate,[6] dioxan[30] and acetic anhydride[30] play a more active role, and can replace the monomer in the ligand exchange equation (4a). In these circumstances the additional reactions shown in equations (8a-d), in which S represents the solvent, must be included.

$$Mo(CO)_6 + S \underset{k_{2s}}{\overset{k_{1s}}{\rightleftharpoons}} \underset{(I_s)}{S \cdot \cdot Mo(CO)_5} + CO \tag{8a}$$

$$(I_s) \xrightarrow{k_{2s}'} \text{inactive products} \tag{8b}$$

$$(I_s) + CCl_4 \xrightarrow{k_{3s}} (II_s) + CO \tag{8c}$$

$$(II_s) \longrightarrow \dot{C}Cl_3 + Mo^I \text{ derivative} \tag{8d}$$

Bamford, Denyer and Eastmond have shown that the augmented mechanism agrees satisfactorily with experimental data.[6,30] At high $[CCl_4]$ the total rate of initiation is given by

$$I = [Mo(CO)_6]\{k_1[M] + k_{1s}[S]\} \tag{9}$$

so that, for constant [M], I would be expected to be linear in [S], as is observed. Kinetic observations allow the parameters k_{1s}, k_{2s}'/k_{3s}, k_{2s}/k_{3s} to be evaluated; the available data are collected in Table 3.3.

4

Table 3.3 Kinetic parameters at 80 °C:halide-CCl₄

Organometallic derivative	Monomer	Solvent	$10^6 k_1$	$10^3 k_2'/k_3$ mol l⁻¹	k_2/k_3	$10^6 k_{1s}$ mol⁻¹ l s⁻¹	$10^3 k_{2s}'/k_{3s}$	$* 10^6 k_{CO}$
Mo(CO)₆	MMA	none[6,9,20]	13 mol⁻¹ l s⁻¹	30	~960	—	—	11·7 mol⁻¹ l s⁻¹
		benzene[6,20]				0	—	0
		ethyl acetate[6,20]				10·7	2·8	10·3 mol⁻¹ l s⁻¹
		dioxan[20,30]				7·54	6·14	5·57 mol⁻¹ l s⁻¹
		acetic anhydride[30]				18·7	4·34	
		benzonitrile[30]				42		
	St	none[12]	2·8 mol⁻¹ l s⁻¹					
Mn₂(CO)₁₀	MMA	none[12]	3·9 s⁻¹	1·37				4·5 s⁻¹
		benzene[12]	3·9 s⁻¹	1·37		0		4·5 s⁻¹
		ethyl acetate[12]	3·9 s⁻¹	1·37		0		4·5 s⁻¹
		cyclohexane[12]	3·9 s⁻¹			0		
	St	none[12]	5·2 s⁻¹					

* From rates of evolution of carbon monoxide in the absence of halide.

The presence of a solvent which is involved in initiation clearly influences the order of the polymerization reaction in [M]. For example, it follows from the values of k_1, k_{1s} in Table 3.3 that in mixtures containing methyl methacrylate and ethyl acetate, the rate of initiation is not much affected by changes in the monomer/solvent ratio, consequently the order of the polymerization reaction in [M] approximates to unity since the rate of propagation is proportional to [M]. In general,[30] the presence of an active solvent at constant concentration leads to an order in [M] between unity and 1·5.

Table 3.3 shows that k'_{2s}/k_{3s} can be much smaller than k'_2/k_3, so that in the presence of an active solvent the ω against [CCl_4] curve may be sharper than that in bulk monomer. Figure 3.2 illustrates this point for mixtures containing ethyl acetate. This behaviour is further evidence for the formation of a separate species (I_s) from the solvent, especially since ethyl acetate and methyl methacrylate are not very dissimilar chemically and it would be surprising if there were a marked medium effect.

In the absence of marked differences in dielectric effects, the differing reactivities of the solvents which participate in ligand replacements must be attributed to differences in solvent type. These reactions involve electron-donation from the incoming ligand to the empty d-orbitals of the molybdenum. All solvents found to be active possess lone-pairs which could take part in donation, and the measured values of k_{1s} partly reflect the relative donor powers. The similarity in the values for methyl methacrylate and ethyl acetate may indicate that donation occurs predominantly through the ester group in both cases.

We have already remarked that the species (I, I_s) arising from ligand substitution are more reactive than the original carbonyl towards the halide. It is of interest to examine the information provided by the above studies about the stabilities of these species, which we may represent as $Mo(CO)_5L$, L being the incoming ligand. If L is methyl methacrylate, styrene, ethyl acetate, dioxan or acetic anhydride, $Mo(CO)_5L$ is not sufficiently stable to permit isolation; these complexes react directly with CCl_4. If L = pyridine, or a phosphine derivative, the complex is often isolatable, and requires activation (e.g. by reaction with monomer, or dissociation) before reaction with CCl_4 is possible. Such considerations indicate that the reactivity of $Mo(CO)_5L$ towards CCl_4 depends on the nature of L as indicated in the sequence:

$$\text{O-compound} > \text{nitrile} > \text{N-base} > \text{P-compound} > CO$$

This conclusion agrees with the views of Strohmeier, who proposed the reverse order for the stability of the complexes.[31] The rather meagre data available suggest that the dependence of the rate of polymerization on carbon monoxide pressure is less marked in the presence of the more active solvents;[30] this would be anticipated, since a higher value of k_{1s} reflects increased stability of (I_s) and therefore a reduced probability of the back-reaction with $CO(8a)$.

3.122 S_N1 systems

Thermal initiation of polymerization by manganese carbonyl in the presence of carbon tetrachloride provided the first example of an S_N1 primary process, and we now consider this.[12,13]

Polymerizations initiated by $Mn_2(CO)_{10}/CCl_4$ differ kinetically from those considered in section 3.121 in several ways. Typically, the overall reaction is first-order in monomer concentration, as would be expected if the monomer does not enter into initiation in a rate-determining manner. Changes in the nature of the solvent have relatively little effect; thus Figure 3.2 shows that, in the polymerization of methyl methacrylate, the same ω against $[CCl_4]$ relation is obtained with benzene and ethyl acetate as solvent (see also Table 3.3). Further, the sharpness of the ω against $[CCl_4]$ curve is not significantly changed in the presence of carbon monoxide; CO reduces the rate of polymerization to an extent which is not dependent on $[CCl_4]$. Bamford and Denyer[12] proposed the mechanism of initiation given in equations (10a–e)

$$Mn_2(CO)_{10} \underset{k_2}{\overset{k_1}{\rightleftarrows}} \underset{(I)}{Mn_2(CO)_9} + CO \qquad (10a)$$

$$Mn_2(CO)_9 \xrightarrow{k_2'} \text{inactive products} \qquad (10b)$$

$$Mn_2(CO)_9 + CCl_4 \xrightarrow{k_3} (II) + CO \qquad (10c)$$

$$(II) \xrightarrow{k_4} \dot{C}Cl_3 + Mn \qquad (10d)$$

$$(II) + CO \xrightarrow{k_4''} \text{inactive products} \qquad (10e)$$

As written, (I) is an incompletely co-ordinated species; co-ordination could be completed, however, by (non-rate-determining) addition of monomer or solvent molecules.

Assuming as before that $\dot{C}Cl_3$ is 100 per cent efficient in initiating polymerization, we obtain from this scheme

$$I = k_1[Mn_2(CO)_{10}]\left\{\frac{[CCl_4]}{k_2'/k_3 + (k_2/k_3)[CO] + [CCl_4]} \cdot \frac{k_4}{k_4 + k_4''[CO]}\right\}, \quad (11)$$

$$\omega = k_p k_t^{-0.5} k_1^{0.5}[Mn_2(CO)_{10}]^{0.5}[M]$$
$$\times \left\{\frac{[CCl_4]}{k_2'/k_3 + (k_2/k_3)[CO] + [CCl_4]} \cdot \frac{k_4}{k_4 + k_4''[CO]}\right\}^{0.5} \quad (12)$$

The influence of carbon monoxide on the rate of initiation is considered to arise predominantly through reaction (10e), the back-reaction in equation (10a) being relatively unimportant. Thus, under the experimental conditions, the term $(k_2/k_3)[CO]$ in equations (11) and (12) is of little significance; there is no effective competition for reaction with (I) between CO and CCl_4 so that the sharpness of the ω against $[CCl_4]$ curve is little affected by the presence of CO. Deactivation steps involving carbon monoxide similar to equation (10e) have been postulated in several other systems, e.g. $Ni(CO)_4/CCl_4$,[16] $Ni\{P(OPh)_3\}_4/CCl_4$.[24] This type of deactivation is therefore not confined to S_N1 systems, since the experimental evidence favours an S_N2 primary step in $Ni(CO)_4$ (see section 3.13).

On the basis of their earlier work on the thermal decomposition of manganese carbonyl at higher temperatures, Haines and Poë[32] have argued that the most important primary step involves metal–metal bond rupture

$$Mn_2(CO)_{10} \rightarrow 2Mn(CO)_5 \qquad (13a)$$

Subsequently,[33] this was modified to

$$Mn_2(CO)_{10} \rightarrow (CO)_4MnCOMn(CO)_5 \qquad (13b)$$

The sequence (10a-d) was included to allow for the effect of carbon monoxide. This conclusion has been contested.[33a] More recently Wawersik and Basolo[34] have concluded that substitution reactions of $Mn_2(CO)_{10}$ in p-xylene solution take place by the dissociative mechanism in equation (10a), reactions (13a,b) being unsuitable as primary steps.

Of the systems which initiate by a primary S_N1 process those containing $Ni\{P(OPh)_3\}_4$ have been studied in most detail.[18] An account of the reactions of this derivative is presented later (section 3.152).

3.123 Some additional mechanistic features

Entropies of activation for initiation

Entropies of activation for initiation measured at high halide concentration would be expected to be characteristic of the primary process, which is rate-determining under these conditions. Table 3.4 presents the experimental data. As anticipated, $\Delta S^{\ddagger}$ is negative for the S_N2 reactions, the value for $Ni(CO)_2(PPh_3)_2$ being in the 'normal' range; the other values, which are

Table 3.4 Enthalpies and entropies of activation

Derivative	Monomer	Reaction type	$\Delta H^{\ddagger}$ kJ mol^{-1}	$\Delta S^{\ddagger}$ e.u. mol^{-1}
$Ni\{P(OPh)_3\}_4$[18]	MMA, St	S_N1	108	12
$Ni(CO)_4$[16]	MMA	S_N2	52·6	−43
$Ni(CO)_2(PPh_3)_2$[17]	MMA	S_N2	88·5	−11
$Mo(CO)_6$[35]	MMA	S_N2	80·0	−30

markedly more negative, probably indicate increased crowding in the transition states. The molecule of $Ni\{P(OPh)_3\}_4$ shows considerable congestion, which is partly relieved in the transition state (see section 3.152), so that $\Delta S^{\ddagger}$ in this case is more positive than the 'normal' value (zero) for an S_N1 process.

Monomer selectivity

It will be evident that, in the absence of additional complicating features, the rates of initiation by an S_N1 system do not depend on the nature of the monomer, while those obtained with an S_N2 system are generally monomer-selective. Thus Bamford and Denyer[12] have pointed out that the ratio of the (effectively limiting) rates of initiation in methyl methacrylate and styrene (at 80 °C) is close to unity for initiation by $Mn_2(CO)_{10}/CCl_4$ but exceeds a value of four for $Mo(CO)_6/CCl_4$. Tetrakis-triphenyl phosphite nickel (0), which functions by an S_N1 mechanism,[18] initiates the polymerization of styrene and methacrylate at the same rate (section 3.152); nickel carbonyl, operating through an S_N2 process,[16] is more effective with the latter monomer (Table 3.1). The presence of a solvent or other additive capable of independent S_N2 initiation would be expected to reduce the monomer selectivity of a given system.

Rates of carbon monoxide evolution from metal carbonyls

Several measurements of the rates of carbon monoxide evolution from carbonyls have been made under conditions appertaining to those holding in initiating systems.

In the absence of halide, evolution of carbon monoxide at 80 °C is undetectably slow from solutions of $Mo(CO)_6$ in benzene or cyclohexane but is readily measurable from solutions in methyl methacrylate, ethyl acetate or dioxan.[20] Values of k_{CO} calculated from the equation

$$\frac{d[CO]}{dt} = k_{CO}[Mo(CO)_6][S] \tag{14a}$$

are shown in Table 3.3 and indicate that k_{CO} has values similar to k_{1s} (or k_1) in all systems examined. These observations clearly support the reaction mechanisms of equations (4) and (8), which predict an initial rate of CO evolution equal to $k_{1s}[Mo(CO)_6][S]$ (or $k_1[Mo(CO)_6][M]$).

In the presence of CCl_4 (concentration 0·2 mol l^{-1}, approximately) little or no carbon monoxide is evolved if the solvent is of the inert type (e.g. benzene, cyclohexane),[20] a result which provides further evidence against the occurrence of a direct initiation reaction between the carbonyl and halide. When the system contains a monomer or an active solvent, addition of CCl_4 increases the rates of CO evolution above those listed in Table 3.3, e.g. in methyl methacrylate, the rate is approximately doubled.[20] This behaviour is also consistent with schemes (4) and (8). Further discussion of $Mo(CO)_6$ is deferred until the more detailed considerations in section 3.151.

Rates of evolution of carbon monoxide from solutions of manganese carbonyl which do not contain halide are independent of the nature of the solvent,[12] and k_{CO} calculated from the equation

$$\frac{d[CO]}{dt} = k_{CO}[Mn_2(CO)_{10}] \tag{14b}$$

has values similar to k_1 (Table 3.3). Again it is found that addition of CCl_4 almost doubles the rate.[12]

The different medium effects exhibited by S_N1 and S_N2 systems therefore extend to the rates of carbon monoxide evolution. Although the possibility that carbon monoxide arises from a different process not directly connected with radical generation cannot be completely excluded for any given system, this does not seem likely when all the results are considered together.

A few data for other organometallic derivatives are available. No carbon monoxide is evolved from solutions of $Ni(CO)_2(PPh_3)_2$ in methyl methacrylate in the absence of halide, and it has therefore been proposed[17] that the primary step is scission of PPh_3 rather than CO. In the case of the molybdenum derivative

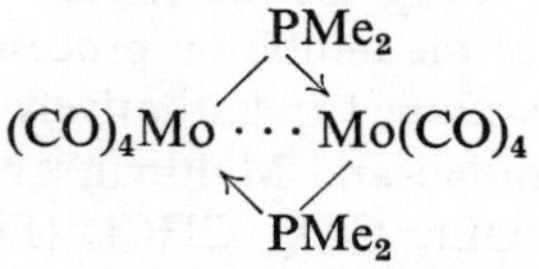

$$(CO)_4Mo \overset{\nearrow \text{PMe}_2 \searrow}{\underset{\nwarrow \text{PMe}_2 \swarrow}{\cdots}} Mo(CO)_4$$

the rate of carbon monoxide evolution measured at 60 °C with 'high' $[CCl_4]$ is almost the same in benzene and methyl methacrylate, probably indicating that the primary reaction proceeds predominantly by S_N1 rupture of a Mo—P bond. The linear phosphine derivatives

$$(CO)_5Mo \cdot PR_2 \cdot PR_2 \cdot Mo(CO)_5 \quad (R = Me, Et)$$

were found to behave similarly.

Rates of evolution of carbon monoxide from solutions of $Mo(CO)_5Py$ in methyl methacrylate are consistent with the view that the primary process is mainly S_N2 displacement of pyridine by monomer with a small contribution from carbon monoxide displacement.[8]

Observations on tetracobalt dodecacarbonyl have been reported;[15] it is difficult to make accurate determinations of initial rates of carbon monoxide evolution in this system since they are greatly reduced by the presence of low pressures of carbon monoxide. Solutions of the carbonyl in carbon tetrachloride at 25 °C do not evolve carbon monoxide at a significant rate, but carbon monoxide is liberated from solutions in methyl methacrylate in the absence of halide, suggesting the occurrence of an S_N2 reaction.

Self-inhibition and retardation

In this discussion we are concerned with the self-inhibition and self-retardation which may occur at relatively high initiator concentrations. Both processes lead to the same kind of kinetic deviation from the ideal linear ω against $[In]^{0.5}$ relation, the order in $[In]$ decreasing below 0·5 as $[In]$ is increased. In principle they are readily distinguishable since the apparent value of the parameter $k_p k_t^{-0.5}$ (evaluated from a knowledge of ω and $\bar{v}$) is not affected by pure inhibition, but is reduced by retardation. However, detection of retardation when accompanied by marked inhibition is not straightforward, since the order in $[In]$ is no longer diagnostic, and estimation of $\bar{v}$ depends on an accurate knowledge of the extent of chain-transfer, which is frequently not available.

It is therefore not surprising that self-retardation is not well-documented in the systems we are considering. Its occurrence at high $[In]$ has been suggested in polymerizations initiated by $Mo(CO)_6$,[9] $Re_2(CO)_{10}$,[14] and $Co_4(CO)_{12}$[15] in the presence of CCl_4, mainly from evidence derived from apparent values of $k_p k_t^{-0.5}$. In general, there is no reason to believe that self-retardation is significant under ordinary conditions.

On the other hand, we have already seen (section 3.11) that self-inhibition may be very marked. Since it becomes more pronounced with increasing $[In]$ the effective order in $[In]$ of the inhibition process must exceed unity.

Self-inhibition has been studied quantitatively only in systems based on $Mo(CO)_6$. Bamford, Eastmond and Maltman[20] compared the behaviour of the three halides CCl_3COOEt, CCl_4, $CHCl_3$ (Figure 3.1). At high halide concentrations (0·19 mol l^{-1}), such that the rates of polymerization correspond effectively to the plateau values (Figure 3.2), CCl_3COOEt, CCl_4 and $CHCl_3$ give the same rates of polymerization for sufficiently low values of $[Mo(CO)_6]$. At higher $[Mo(CO)_6]$, self-inhibition is negligible with CCl_3COOEt (a linear dependence of ω on $[Mo(CO)_6]^{\frac{1}{2}}$ being found) but is appreciable with CCl_4. When $[CCl_3COOEt]$ is reduced to 0·014 mol l^{-1} there is also marked self-inhibition. Chloroform (0·19 mol l^{-1}) gives very powerful self-inhibition for concentrations of $Mo(CO)_6$ as low as 10^{-4} mol l^{-1}. It was proposed[20] that self-inhibition is attributable to interaction between complexes (I) and (II) (scheme (4)):

$$(I) + (II) \xrightarrow{k_5} (III) \qquad (4e)$$

The resulting products, represented by (III), were assumed to be inactive in radical generation. Reaction (4e), occurring at a rate proportional to [I][II], has the required kinetic order. Further, it is clear that if [I] is reduced without compensating increase in [II] the extent of inhibition will be reduced, so that the effects of halide activity and concentration mentioned above are understandable. If, for simplicity, it is assumed that reaction (4b) is negligible, the modified reaction mechanism gives expression (15) for the rate of initiation.

$$I = \frac{k_1}{2\Delta} \left(\Delta[Mo(CO)_6][M] + (k_2/k_3)[CO] + [H] \right)$$

$$\times \left[\left\{ 1 + \frac{4\Delta[Mo(CO)_6][M][H]}{\{\Delta[Mo(CO)_6][M] + (k_2/k_3)[CO] + H\}^2} \right\}^{\frac{1}{2}} - 1 \right] \quad (15)$$

In expression (15)

$$\Delta = k_1 k_5 / k_3 k_4 \quad (16)$$

The parameter Δ determines the extent of self-inhibition in a given system. Chloroform is a relatively inactive halide, and probably both k_3 and k_4 are low, so that a high value of Δ is not surprising. On the other hand, a high value of k_3 would be expected for CCl_3COOEt in view of the electron-attracting powers of —COOEt (section 3.14), and Δ may well be smaller. Bamford, Eastmond and Maltman[20] showed that expression (15) is in fair agreement with the experimental observations at 80 °C if $\Delta(CCl_4) = 15\cdot3$, $\Delta(CCl_3COOEt) = 2\cdot0$ mol^{-1} l. An essentially similar treatment has been applied to initiation by $Mo(CO)_5Py$ at 40 °C, Δ being assigned the values 80, 40 mol^{-1} l for CCl_4, CCl_3COOEt, respectively.[8]

From the preceding discussion self-inhibition at high [H] might be expected to decrease on lowering the temperature, on account of the smaller values of [I] and (probably) [II]. This is found to be so in $Mo(CO)_6$[35] and $Cr(CO)_6$[3] systems. Further, when carbon monoxide is present, [I] is reduced by the enhanced back reaction in equation (4a), and there are indications that for $Cr(CO)_6/CCl_4$ self-inhibition is smaller under these conditions.[3] Similarly, addition of triphenyl phosphite, which closely resembles carbon monoxide in its effects on $Mo(CO)_6$ systems, also causes a decrease in self-inhibition.[36]

The incidence of self-inhibition not only complicates interpretation of the reaction kinetics, but it also restricts the practical value of an initiator. In unfavourable cases, a system which is highly active at low concentrations may not be capable of producing the desired rate of initiation when more concentrated. $Co_4(CO)_{12}/CCl_4$ is an example of such an initiator.[15]

Critique

Most of the information on the systems we are considering comes from kinetic measurements; unfortunately there have been few direct investigations of the chemistry of the component reactions. It is to be hoped that chemical data will be forthcoming in the future, but in the meantime one is justified in enquiring whether any modification of the mechanisms proposed is necessary or desirable.

One of the least satisfactory features is the postulated reaction

$$(I) \rightarrow \text{inactive products} \qquad (17)$$

which appears in all the schemes (equations (4b), (8b) and (10b)) but which is difficult to formulate in chemical terms. Is it possible to dispense with reaction (17)? Some process leading to deactivation of (I) is clearly necessary, otherwise the ω against [H] curves would be infinitely sharp—an extremely low value of [H] sufficing to achieve the maximum rate of initiation. In the absence of reaction (17), would the back-reaction in equations (4a), (8a) or (10a) provide the necessary deactivation? An unequivocal answer to this question can only be given for one system at present. Investigations of the reactions of $Ni\{P(OPh)_3\}_4$ in methyl methacrylate and styrene,[18] to which reference has already been made, allow the contribution of the process

$$Ni\{P(OPh)_3\}_3 + P(OPh)_3 \rightarrow Ni\{P(OPh)_3\}_4$$

(corresponding to the back-reaction in equation (10a)) to be calculated unambiguously (section 3.152). Significantly, it turns out that no step analogous to reaction (17) is necessary. It is virtually impossible to ensure the complete absence of carbon monoxide when examining initiation by metal carbonyls, so that the back-reaction we are considering must always occur to some extent, even in studies purporting to measure initial rates of polymerization. (Although the concentration of carbon monoxide would change during the experiment, the measurements may not have been precise enough to detect the kinetic consequences.) It may be significant that deactivation of (I) by carbon monoxide is particularly small for $Mn_2(CO)_{10}/CCl_4$ (section 3.122), an initiator which has an unusually sharp ω-[H] curve (Figure 3.2).

Bamford, Eastmond and Maltman[20] have enquired into the necessity for including process (17) in the mechanism of $Mo(CO)_6$/halide initiation from a different point of view. These workers noted that equation (15), derived from a reaction scheme which does not include equation (17), leads to an ω against [H] relation resembling that found experimentally, but they concluded that the mechanism without reaction (17) is unsatisfactory in some respects.

Extension of the techniques employed[18] with $Ni\{P(OPh)_3\}_4$ to carbonyl systems is evidently desirable. (Observations of this type recently made on $Mo(CO)_6$/halide systems confirm the reality of the deactivation step (4b), although they do not elucidate its chemistry.[21b]) Even though future developments may lead to partial or complete replacement of reaction (17) by other deactivation steps, it is convenient to retain this step at present in discussion of formal kinetics.

Deactivation reactions between complex (II) and some other species (e.g. carbon monoxide in equation (10e)) are not open to corresponding objections, since they can usually be formulated as ligand exchanges.

3.13 ACTIVITIES OF ORGANOMETALLIC DERIVATIVES IN INITIATION

Table 3.1 presents data on the initiating activities of the organometallic derivatives investigated. The activity is expressed in terms of a defined by the relation

$$I = a[\text{organometallic derivative}];$$

a is therefore the first-order coefficient for initiation at limiting halide concentration. The values in Table 3.1 generally refer to rates at high carbon tetrachloride concentration (0.1–0.2 mol l^{-1}) but for practical purposes they differ little from limiting values. Note that $a = k_1$ only if each molecule of initiator decomposed yields one initiating radical; frequently the radical yield is not known so that values of k_1 cannot be deduced from rates of initiation. The activities in Table 3.1 were mostly deduced from initial rates of polymerization of methyl methacrylate; when another monomer was employed this is indicated. As a rule, concentrations of organometallic derivatives were sufficiently low for the square root relation to hold.

Of the group VIA carbonyls, $Mo(CO)_6$ is by far the most active, in keeping with the generally higher reactivities of molybdenum compounds. The conventional initiator azo-bis-isobutyronitrile, with $10^6 a = 14$ s^{-1} at 60 °C,[36a] has comparable activity at this temperature. A striking feature associated with this group is the high activity of dinuclear carbonyls carrying phosphine bridges. These derivatives appear to undergo primary S_N1 dissociation of a metal-phosphorus bond (section 3.123), the relative weakness of which is presumably responsible for the high activities. Although the phosphine-bridged derivatives have high a values, self-inhibition is pronounced at higher concentrations.[5] The linear phosphine derivatives containing —PR$_2$·PR$_2$— groups are much less active, having a values similar to those of the parent carbonyls. (Probably both metal atoms in these compounds enter into radical formation, so for a true comparison it is necessary to divide the appropriate a values by two.) The metal-CO bonds in these derivatives are stronger than those in the parent carbonyls on account of increased back-donation from the metal, so that the S_N2 mechanism occurs less readily, and the predominant primary process is S_N1 rupture of a metal-phosphorus bond.[5]

Phenylisocyanide resembles carbon monoxide as a ligand, but the extent of back-donation from the filled d-orbitals of the metal is less. The metal-carbon bond is therefore weaker and the ligand is more easily displaced. In accord with this it is found that hexakis(phenylisocyano) derivatives as a class are more active initiators than the corresponding hexacarbonyls, as is apparent from Table 3.1. The molybdenum derivatives are again more active than those of chromium or tungsten (with unsubstituted ligands), although the differences are smaller than with the carbonyls. Rates of initiation are sensitive to substitution in the benzene rings; substituents (such as p-OMe) which increase the electron-density in the isocyanide group also augment the rate of initiation. This effect probably arises from a reduction in the extent of back-donation from the d-orbitals of the metal which weakens the metal–carbon bonds. The low activity of the derivative carrying o-Me and m-Cl substituents is not easy to understand and requires further investigation. Self-inhibition is much less in evidence with the hexakis(arylisocyano) derivatives than with the carbonyls; at 25 °C it is virtually absent at concentrations up to $1\cdot3 \times 10^{-3}$ mol l^{-1} (the highest examined), while at 80 °C it appears in this concentration range only with the molybdenum derivative.

Derivatives of group VIIA metals which have been studied show rather low activities. A low sensitivity towards carbon monoxide appears to be another group characteristic. $Re_2(CO)_8(PPh_3)_2$ is somewhat more active than Re_2-$(CO)_{10}$; since the Re—CO bonds are stronger in the former compound, the rate-determining step is probably the scission of a PPh_3 ligand. The very low activity of the mononuclear derivative, in which the Re—CO bonds are still stronger, may be attributed to the strengthening of all metal–ligand bonds by use of an electron from the Re—Re bond of the parent carbonyl. The metal-CO bonds are also relatively strong in $(C_5H_5)Mn(CO)_3$; since the hydrocarbon ligand is very strongly bound the very low activity of this compound is understandable.

The progressive increase in activity accompanying replacement of CO ligands by triphenyl phosphine is well illustrated by the group VIII derivatives $Fe(CO)_5$, $Fe(CO)_4(PPh_3)$ and $Fe(CO)_3(PPh_3)_2$. For reasons which will be clear from the above discussion it is thought that fission of an Fe—P bond is the primary step in initiation by the last two derivatives. Table 3.1 shows that the dinuclear complexes with phosphine bridges also have increased activities, although the enhancement over the parent $Fe(CO)_5$ is not so great as with the group VIA compounds. The bridged iron complex containing two bromine atoms has the highest recorded activity for any iron derivative. It has been suggested[5] that the decrease in electron-density on the metal atoms brought about by introduction of the bromine atoms may be responsible, since this would favour co-ordination of monomer molecules and encourage an S_N2 reaction. However, it has not been established that the primary process is, in fact, of this type.

Several derivatives of cobalt and nickel are active initiators at room temperatures. The inactivity of $Co_2(CO)_8$ is remarkable; this is the only metal carbonyl examined which does not appear to function as an initiator. It has been demonstrated[15] that at 25 °C $Co_2(CO)_8$ effectively prevents polymer formation from methyl methacrylate containing azo-bis-isobutyronitrile, so that the ineffectiveness of the carbonyl as an initiator is probably connected with its ability to destroy free-radicals. At low concentrations, tetracobalt-dodecacarbonyl is the most active initiator of the type under discussion which has been studied; however, its usefulness is limited by strong self-inhibition at concentrations exceeding 10^{-5} mol l^{-1}.

Initiation by $Ni(CO)_4$ has been examined in some detail, particularly to ascertain whether the rates of initiation are comparable to those of ligand substitution in $Ni(CO)_4$

$$Ni(CO)_4 + L \rightarrow Ni(CO)_3L + CO$$

After considerable controversy[37-39] it is now agreed[40,41] that rates of ligand substitution and of carbon monoxide exchange are similar, and that both processes involve the intermediate $Ni(CO)_3$. The currently accepted mechanism is shown in equations (18a) and (18b).

$$Ni(CO)_4 \rightarrow Ni(CO)_3 + CO \qquad \text{slow} \qquad (18a)$$
$$Ni(CO)_3 + L \rightarrow Ni(CO)_3L \qquad \text{fast} \qquad (18b)$$

Day, Basolo and Pearson[40] have recorded the following activation parameters for substitution of CO by PPh_3 in toluene solution:

$$\Delta H^{\ddagger} = 93 \text{ kJ mol}^{-1} \tag{19a}$$
$$\Delta S^{\ddagger} = 8 \cdot 4 \pm 1 \text{ e.u. mol}^{-1} \tag{19b}$$

The positive entropy of activation is clearly consistent with a rate-determining dissociative process such as equation (18a). The values in equations (19a) and (19b) contrast sharply with those referring to initiation in Table 3.4, and there seems little doubt that the rate-determining reactions for ligand-exchange and initiation are different. In methyl methacrylate at 25 °C initiation is approximately 100 times slower than would be expected if equation (18a) were the rate-determining step. Bamford, Eastmond and Murphy[16] considered several different mechanisms of initiation and concluded that an S_N2 process involving displacement of carbon monoxide by monomer is the most acceptable. Although this mechanism agrees well with the kinetic observations, it encounters some difficulties. For example, formation of $M \cdots Ni(CO)_3$ from monomer and $Ni(CO)_3$ (or from monomer and the cage $[Ni(CO)_3 + CO]$) must be assumed to occur at an insignificant rate under the experimental conditions. Inspection of the kinetic parameters suggests that complex formation between monomer and carbonyl without elimination of carbon monoxide may be a possible alternative primary process which could avoid the difficulty mentioned above, and would also be consistent with the large negative $\Delta S^{\ddagger}$ (Table 3.4).

The relatively low activities of tetrakis(triethyl phosphite)Ni^0 and tetrakis-(β-chloroethyl phosphite)Ni^0 compared to the triphenyl phosphite derivative are the result of greatly reduced crowding in the molecules of these compounds (section 3.152).[11]

Although rates of initiation have been measured for many organometallic derivatives, most determinations have been confined to a single temperature. Figure 3.4 presents in the form of Arrhenius lines most of the data extending over ranges of temperature which are currently available. Comparative data for azo-bis-isobutyronitrile are included for comparison.[36a,42] The Figure shows that initiation by the organometallic initiators studied has a lower

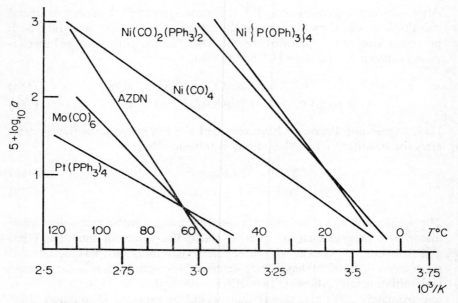

Figure 3.4 Dependence of rate of initiation of methyl methacrylate polymerization on temperature for several initiators; for definition of a see section 3.13. Halide is CCl_4; AZDN = azo-bis-isobutyronitrile. Data from references 16, 17, 18, 19a, 35, 36a, 42

activation energy than initiation by the azo compound. Further, the relatively high activities of nickel derivatives are apparent—the compounds shown in Figure 3.4 initiate at $0\,°C$ almost as rapidly as azo-bis-isobutyronitrile initiates at $60\,°C$.

3.14 ACTIVITIES OF HALIDES IN INITIATION

The radical-generating processes summarized by equations (1) and (2) involve the formation of complex (II) by interaction of (I) with the halide, and the decomposition of (II), equations (4c) and (4d), (8c) and (8d) and (10c) and (10d). In principle, either step may be rate-determining. It seems likely that (II) is a charge-transfer complex, in which case the electron-affinity of the halide would be one important factor determining the rate of a reaction such as equation (4c).

The reactions between sodium atoms and halides, equation (20), which were studied in the classic work of Polanyi, provide an analogy to the processes we are considering.

$$Na + hal\text{-}R \rightarrow Na^+hal^- + R\cdot \tag{20}$$

Wilson and Herschbach[43] have pointed out that there is a strong correlation between the rate constant of equation (20) and the electron absorption co-efficient of the halide hal-R. Such coefficients are known from the investigations of Lovelock and Gregory;[44] strictly their observations provide a measure of the adiabatic electron-affinity of the halide, whereas for our purposes the vertical electron-affinity is appropriate, but the difference between the two quantities is not likely to be significant in our systems. The data indicate that the factors leading to an increase in the rate of capture of thermal electrons by a halide, and in the rate of reaction of the halide with sodium atoms, are as follows: (a) change of halogen atom $F \rightarrow Cl \rightarrow Br \rightarrow I$, (b) presence of an oxygen atom in the halide molecule, (c) multiple substitution, (d) location of the halogen atom at an allylic or benzylic position. Further, the rates are depressed by the presence of neighbouring double bonds in the α position to the halogen atom. Comparison of these factors with those listed in section 3.11(ii) corresponding to high halide activity in initiation reveals a striking similarity and leads to the conclusion that the electron-affinity of a halide plays a dominant role in determining its activity.

It is of interest to compare the activities of halides in initiation and in chain-transfer reactions. Several halides have high reactivities in both processes, e.g. CCl_4, CBr_4. However, there is not always a correlation; for example, ethyl trichloracetate is much more active than CCl_4 in initiation, although its transfer constant towards polystyrene radicals is smaller (the values at 60 °C being $6\cdot5 \times 10^{-3}$ and $9\cdot2 \times 10^{-3}$ for CCl_3COOEt, CCl_4, respectively).[45] Polar factors are known to have an important influence on transition state energies of free-radical reactions in general;[46] the balance of polar and non-polar contributions apparently differs in initiation and chain-transfer processes.

Little systematic work has been carried out on the relations between halide structure and reactivity in initiation. The reaction between complex (I) and halide has so far been studied directly only in the $Ni\{P(OPh)_3\}_4/CCl_4$ system[18] (section 3.152), and has turned out to be far from simple. A more detailed discussion must await the results of further experimental investigations. (Recent determinations of the absolute rate coefficients for reactions between a series of halides and complexes of type (I) in $Mo(CO)_6$ and $Ni\{P(OPh)_3\}_4$

systems are reported in references 21*a* and *b*. The results substantiate and extend the views described above.)

3.15 SOME DETAILED STUDIES

In this section we give brief accounts of recent investigations on thermal initiation by $Mo(CO)_6$ and $Ni\{P(OPh)_3\}_4$. These systems have received more detailed examination than others of the group we are discussing, and the mechanism of initiation by $Ni\{P(OPh)_3\}_4$ is probably the most clearly understood.

3.151 Molybdenum carbonyl

It has been known for some time[6] that reaction between molybdenum carbonyl and a halide in an active solvent produces paramagnetic molybdenum species which give readily observable electron spin resonance (ESR) spectra. Figure 3.5 shows the room-temperature ESR spectrum obtained after reaction with carbon tetrachloride in methyl methacrylate at 80 °C for 1 h.[7a] The central line, situated at $g = 1.947$, arises from the isotopes

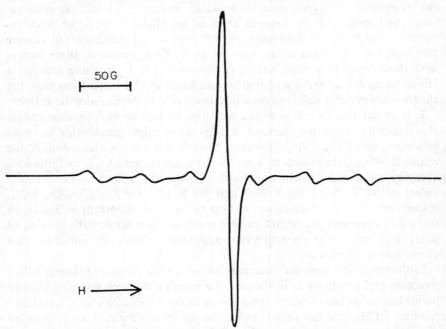

50 G

H ⟶

Figure 3.5 Room-temperature ESR spectrum of reaction products of $Mo(CO)_6$ (3.50×10^{-4} mol l^{-1}) and CCl_4 (0.192 mol l^{-1}) in methyl methacrylate at 80 °C. Reaction time = 1 h. (From Bamford, Eastmond and Fildes)[7a]

^{94}Mo, ^{96}Mo, ^{98}Mo, ^{100}Mo (spin zero), and is flanked by six satellites produced by ^{95}Mo, ^{97}Mo (spin $\frac{5}{2}$). The non-symmetric spectrum shown in Figure 3.6 is obtained from the same specimen at 77 °K and is consistent with a structure in which the molybdenum atom is situated in an axially symmetric ligand field, randomly oriented with respect to the applied magnetic field.

The development of the ESR signal with reaction-time cannot be followed conveniently in methyl methacrylate solution on account of complications

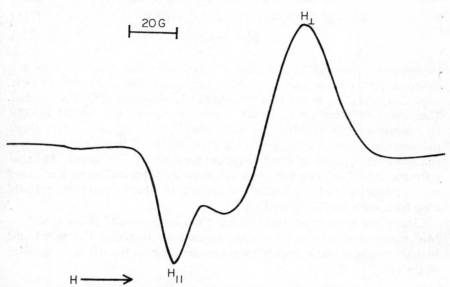

Figure 3.6 77 K ESR spectrum of reaction products of Mo(CO)$_6$ (3·50 × 10^{-4} mol l^{-1}) and CCl$_4$ (0·192 mol l^{-1}) in methyl methacrylate at 80 °C. Reaction time = 1 h. (From Bamford, Eastmond and Fildes)[7a]

arising from the rapidly increasing viscosity, but measurements of this type have been made[7a] in solution in ethyl acetate, which is known to resemble the monomer in its activity in initiation.[6] For reaction-times of 1 h, approximately, the two solvents give virtually identical ESR spectra. When the reaction is carried to completion the intensity of the spectrum reveals that each molybdenum atom in the paramagnetic product carries one free spin, if no diamagnetic molybdenum products are formed. From the nature of the room temperature and 77 K spectra, which are identical with those of MoCl$_5$ in ethyl acetate, and the observations that model derivatives of Mo^{I} and Mo^{III} are oxidized by CCl$_4$ in ethyl acetate solution at 80 °C, it was concluded that the final reaction products contain Mo^{V}.

The interesting observation was made that the curve relating signal strength to reaction time is sigmoid in character; for short reaction times the rate of development of observable free spins is much less than the rate of radical generation calculated from polymerization kinetics.[6] The results indicate that intermediate molybdenum species (e.g. Mo^I) undetected by ESR are first formed, and are subsequently oxidized more slowly to the Mo^V product. Under the conditions of the experiments (high $[CCl_4]$, low $[CO]$) observations are consistent with the following sequence

$$Mo(CO)_6 + S \rightarrow S \cdots Mo(CO)_5 + CO \qquad (21a)$$

$$S \cdots Mo(CO)_5 + CCl_4 \rightarrow Mo^I + \dot{C}Cl_3 \qquad (21b)$$

$$Mo^I \rightarrow Mo^V \qquad (21c)$$

In equations (21a), (21b) and (21c) all secondary oxidation processes are collapsed into a single rate-determining step (21c). The two relatively slow steps (21a) and (21c) account for the sigmoid development of the ESR signal. The first rate-determining process—ligand exchange—is followed rapidly by primary oxidation; this is the only radical-producing step at very short reaction times, and, according to the experiments of Bamford, Eastmond and Maltman,[20] generates one free-radical per $Mo(CO)_6$ consumed. (See also reference 21b.) The secondary oxidation equation (21c) leading to Mo^V, also rate-determining, yields a further two radicals, so that *in toto* three radicals arise from each $Mo(CO)_6$ reacting.

Clearly one or more of the oxidation steps in the overall process $Mo^0 \rightarrow Mo^V$ must occur without free-radical generation; Bamford, Eastmond and Fildes[7a] consider some possible mechanisms, e.g. a two-electron transfer of the type

$$Mo^m + CCl_4 \rightarrow [Mo^{m+2}Cl]^+Cl^-.$$

It should be noted that no positive identification of Mo^I as a long-lived species in the reaction has been achieved, and alternatives are possible, for example a rapid non-radical-forming oxidation of Mo^I to Mo^{III}, and subsequent rate-determining oxidation to Mo^V with formation of two radicals may occur.

At long reaction times (24 h) in ethyl acetate at 80 °C all six carbonyl ligands are released, so that no carbon monoxide is retained in the final products. It has been tentatively suggested[7a] that evolution of 5 molecules of carbon monoxide accompanies primary radical formation. Recent measurements[7a] confirm the earlier findings of Bamford, Eastmond and Maltman[20] that in methyl methacrylate solution at short reaction times only two molecules of carbon monoxide are released for each $Mo(CO)_6$ consumed. Ultimately, however, all six ligands are probably released, so that a higher proportion of carbon monoxide comes off during secondary oxidation under these conditions.

The paramagnetic products decompose on isolation and a black dia-magnetic solid may be isolated from the reaction in ethyl acetate. This material appears to contain Mo^{IV}, and possibly bridging acetate groups derived from reaction of ethyl acetate.

When the polymerization of methyl methacrylate is initiated by $Mo(CO)_6/CBr_4$ at 80 °C the rate constant for carbonyl consumption is reported to be the same as in the corresponding CCl_4 system.[7b] Each $Mo(CO)_6$ consumed gives rise to three radicals, shown by tracer experiments to be $\dot{C}Br_3$. On account of the high reactivity of CBr_4 the secondary oxidation is relatively fast and the two stages of oxidation cannot readily be separated. Thus, the limiting initial rate of polymerization is consistent with effectively simul-taneous generation of all radicals arising from a single $Mo(CO)_6$ molecule. The reaction products formed in ethyl acetate solution at 80 °C contain Mo^V. In certain circumstances the rate of growth of the ESR signal is surprisingly rapid; this has been attributed to a chain oxidation of $Mo(CO)_6$, in which $\dot{C}Br_3$ and a Mo^I derivative are chain carriers. For details the reader is referred to the original paper.[7b]

3.152 Tetrakis(triphenyl phosphite)nickel(0)

In methyl methacrylate[18a] or styrene[18b] solution tetrakis(triphenyl phos-phite)nickel(0) participates in a reversible ligand exchange with the monomer. The reaction has been investigated kinetically,[18a,b] the concentration of the product being monitored spectrophotometrically at $\lambda = 401$ nm; the exchange mechanism is shown for MMA in equations (22a,b), in which P represents triphenyl phosphite.

$$NiP_4 \underset{k_2}{\overset{k_1}{\rightleftharpoons}} NiP_3 + P \qquad (22a)$$

$$NiP_3 + MMA \overset{K}{\rightleftharpoons} NiP_3(MMA) \qquad (\Delta H_K^\circ, \Delta S_K^\circ) \qquad (22b)$$

The process which is initially rate-determining is independent of the nature and concentration of the monomer, and hence is considered to be the S_N1 reaction (22a); the equilibrium in equation (22b) is rapidly established and lies predominantly over to the right. From these kinetic studies k_1, k_2/K and K_0, the equilibrium constant of the overall reaction in equation (23), have been evaluated for each monomer at a series of temperatures.

$$NiP_4 + MMA \overset{K_0}{\rightleftharpoons} NiP_3(MMA) + P \qquad (\Delta H_0^\circ, \Delta S_0^\circ) \qquad (23)$$

The derived activation parameters are collected in Table 3.5 and the enthalpy changes are represented schematically in Figure 3.7. Within the limits of

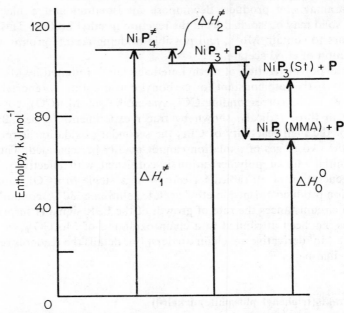

Figure 3.7 Schematic representation of enthalpy changes in reactions involved in ligand exchange between NiP_4 and methyl methacrylate and styrene (equations (22a), (22b) and (23)). (From Bamford and Hughes)[18]

error, $\Delta H_1^{\ddagger}$ and $\Delta S_1^{\ddagger}$ have the same values in methyl methacrylate and styrene, so that there is no pronounced medium effect on the forward reaction in equation (22a). The figure for $\Delta H_1^{\ddagger}$ is close to that reported by Meier, Basolo and Pearson[47] for the dissociation of tetrakis(triethyl phosphite)-nickel (0) in benzene solution (109 kJ mol^{-1}) but the entropy of activation in the present system is higher ($\Delta S_1^{\ddagger} = 1·8 \pm 3·0$ e.u. mol^{-1} for Ni{P(OEt)$_3$}$_4$). These findings reflect the relatively high degree of crowding in the NiP_4 molecule; reduction of this in the transition state of reaction (22a) is responsible for the high entropy of activation.

The data in Table 3.5 do not allow separate evaluation of ΔH_K° or $\Delta H_2^{\ddagger}$, but permit comparisons for the two monomers to be made. It is found that

$$\Delta H_K^{\circ}(MMA) - \Delta H_K^{\circ}(St) \sim -22 \text{ kJ mol}^{-1}; \qquad (24)$$

the Ni-monomer bond is therefore stronger with methyl methacrylate than with styrene, a conclusion which is in line with the higher reactivity of the former monomer in S_N2 reactions with zero-valent metal derivatives (cf. section 3.123). According to Bamford and Hughes[18b] the dissociation energy of the Ni—St bond is less than 20 per cent of that of the Ni—P bond broken in reaction (22a).

Table 3.5 $Ni\{P(OPh)_3\}_4$—reaction and equilibrium parameters

| | K_0 | | | k_1 | | k_2/K | | k_3 | |
	ΔH_0°	ΔS_0°	ΔG_0° (298 K)	$\Delta H_1^\ddagger$	$\Delta S_1^\ddagger$	$\Delta H_2^\ddagger - \Delta H_K^\circ$	$\Delta S_2^\ddagger - \Delta S_K^\circ$	$\Delta H_3^\ddagger$	$\Delta S_3^\ddagger$
MMA	69·4	34·5	26·4	108[a]	11·9	44·4	−18·5	~6	−57
				107[b]	10·6				
St	96·2	55·2	27·3	114[a]	16·2	20·1	−37·3	48·6	−19·5

a. From absorbance measurements.
b. From polymerization kinetics.
ΔH, ΔG values in kJ mol^{-1}, ΔS in e.u. mol^{-1}
Probable errors: $\Delta H \pm 8\cdot5$ kJ mol^{-1}, $\Delta S \pm 5$ e.u. mol^{-1}

Table 3.5 shows that there is a significant difference in ΔS_0° for the two systems. We also see that

$$\Delta S_K^\circ(\text{MMA}) - \Delta S_K^\circ(\text{St}) \sim -20 \text{ e.u. mol}^{-1} \qquad (25)$$

The relatively large decrease in entropy associated with the formation of $\text{NiP}_3(\text{MMA})$ (reaction (22b)) probably indicates a greater loss of rotational degrees of freedom in the case of methyl methacrylate.

Although the equilibrium constant for reaction (23) is small ($K_0 = 2 \cdot 59 \times 10^{-5}$ for methyl methacrylate at 25 °C), under the conditions of concentration holding in polymerizations (low $[\text{NiP}_4]$, bulk monomer) an appreciable fraction of the NiP_4 initially present exists as $\text{NiP}_3(\text{M})$ when equilibrium has been established (in the absence of halide). For example, with $[\text{NiP}_4]_0 = 10^{-3}$ mol l^{-1}, the equilibrium concentration of $\text{NiP}_3(\text{MMA})$ in bulk methyl methacrylate at 25 °C is $3 \cdot 86 \times 10^{-4}$ mol l^{-1}.

The second stage of initiation—the radical-forming reaction between $\text{NiP}_3(\text{M})$ and halide (equation (26))—has been investigated directly[18] by monitoring the changes in absorbance occurring when the halide is added to equilibrium mixtures containing NiP_4 and monomer.

$$\text{NiP}_3(\text{M}) + \text{CCl}_4 \xrightarrow{k_3} n\text{R}\cdot + r\text{P} + \text{Ni}^\text{I}(\text{or Ni}^\text{II}) \qquad (26)$$

As expected, reactions of this type are found to be relatively fast. $\text{NiP}_3(\text{St})$ reacts much more rapidly than $\text{NiP}_3(\text{MMA})$, the second-order rate constants at 25 °C being $0 \cdot 14$, $1 \cdot 19$ mol^{-1} l s^{-1}, respectively. The difference may be attributed to the lower Ni—M bond energy in the former complex. This observation rules out the possibility that the reaction involves the uncomplexed species NiP_3 in both monomers. The higher reactivity of the styrene derivative cannot be explained by a larger equilibrium concentration of NiP_3 compared to that in MMA, since there is only a modest decrease in K (33 per cent), while the rate constant is higher by a factor of 8 at 25 °C. Finally, k_3 appears to be independent of monomer concentration, which would not be so if the uncomplexed species were reacting. Thus there are good reasons for believing that $\text{NiP}_3(\text{M})$ is the active species under the conditions examined.

Values of $\Delta H_3^\ddagger$ and $\Delta S_3^\ddagger$ are given in Table 3.5. The very low $\Delta H_3^\ddagger$ and highly negative $\Delta S_3^\ddagger$ for reaction of the methyl methacrylate complex are remarkable and do not find a parallel in the styrene system. The classical concept of orientation of polar molecules around a polar transition state may be invoked here, but the effects seem rather large to be attributable solely to such a mechanism.

No observations derived from polymerization have been used in the work so far described. Determination of conversion against time curves for polymerizations at high $[\text{CCl}_4]$ have, however, been employed to evaluate the parameters k_1 and n by an elaboration of the technique described by

Bamford, Eastmond and Maltman.[20] In the methyl methacrylate system the values of k_1 so found agree well with those derived from spectrophotometry. It appears that n, equation (26), is slightly in excess of unity; since the final products are derivatives of Ni^{II}, $NiP_3(MMA)$ must be oxidized by (at least) two different routes, one giving two radicals per Ni atom, and one leading directly to Ni^{II} without radical generation. (In a more recent publication[21a] it is suggested that with chloro- and bromo-derivatives the products consist of Ni^{I} and Ni^{II} species, respectively.) Values of $[NiP_4]$, $[NiP_3(MMA)]$, $[P]$ during reaction have been computed by numerical integration of the three simultaneous differential equations derived from the reaction scheme (22a) and (22b) and (26); $[NiP_3(MMA)]$, initially zero, increases initially, then passes through a maximum and declines as NiP_4 is consumed. The height of the maximum, and the time taken to achieve it, depend inversely on the halide concentration. Rates of polymerization have also been computed, with results which are in good agreement with experimental polymerization studies. In particular, 'stationary' rates of polymerization are found to depend on $[CCl_4]$ in a manner similar to that determined experimentally (Figure 3.3).[24] This is an important result since it demonstrates that, in this case at least, the inclusion of an additional reaction (corresponding to equation (17)) leading to the deactivation of $[NiP_3(MMA)]$ is unnecessary; the back reaction in equation (22a) supplies the only deactivation process.

Similar studies with styrene[18b] indicate a difference in the mechanism of radical formation. The conversion against time curves at high $[CCl_4]$ are not consistent with the simple expression of Bamford, Eastmond and Maltman,[20] and a two-stage initiation process, equations (27a) and (27b), in which the second step is relatively slow, appears to be necessary.

$$NiP_3(St) + CCl_4 \rightarrow Ni^{I} + R\cdot \qquad (27a)$$

$$Ni^{I} + CCl_4 \rightarrow Ni^{II} + R\cdot \qquad (27b)$$

It has been suggested on the basis of ESR evidence that the Ni^{I} derivative is dimeric; this might also account for the slowness of equation (27b). With CBr_4 as halide equation (27b) is sufficiently fast for the simpler kinetic treatment to apply.

It will be apparent that the mechanism of initiation by $Ni\{P(OPh_3)\}_4$ is now understood in reasonable detail, although some interesting chemical problems remain for further investigation. Similar studies of systems based on other derivatives, in which the various stages of radical-formation can be investigated separately, would seem to be worthwhile.

3.16 PHOTOINITIATION

A valuable review of the photochemistry of metal carbonyls and other derivatives has been written by Koerner von Gustorf and Grevels.[48] These authors have summarized evidence demonstrating that the primary photo-chemical process with a simple metal carbonyl is dissociation, with formation of a co-ordinatively unsaturated species

$$M(CO)_n + h\nu \rightarrow M(CO)_n^* \rightarrow M(CO)_{n-1} + CO \qquad (28)$$

The product $M(CO)_{n-1}$ may react in several ways, e.g. by combination with carbon monoxide to reform the original carbonyl or by addition of an n- or π-donor

$$M(CO)_{n-1} + S \rightarrow M(CO)_{n-1}S \qquad (29)$$

Other possibilities include reaction with a metal carbonyl in the system or addition of a molecule X—Y with splitting of the X—Y bond.

Reactions (28) and (29) constitute an S_N1 mechanism for ligand replacement. Strohmeier,[49] Strohmeier and Gerlach,[50] Dobson et al.[51] and Dobson[52] demonstrated the occurrence of these processes with group VIA carbonyls and several types of n-donor. More complex carbonyls, especially those containing extended π-systems, may undergo ligand exchange by an S_N2 mechanism, a molecule activated by light absorption reacting directly with the incoming ligand prior to dissociation.[48] We shall not be concerned here with this type of reaction.

If S is a vinyl monomer, reactions (28) and (29) lead to the formation of species which have been designated as (I) in our earlier discussion (cf. equation (4a)); since such species are considered to react readily with suitable halides with generation of free radicals it is not surprising that mixtures of simple carbonyls and halides are photoinitiators of free-radical polymerization. Probably most of the systems listed in Table 3.1 are active in this way. Many have been shown to photoinitiate polymerization, although in the majority of cases the reaction conditions were unfavourable for mechanistic investigations. Koerner von Gustorf and colleagues[53,54] have shown that irradiation of $Fe(CO)_5$ in a vinyl monomer (methyl methacrylate, vinyl chloride, styrene, propylene, vinyl ethyl ether, vinyl acetate) yields a product $MFe(CO)_4$ in which the metal is co-ordinated to the double bond. (Similar adducts may also be prepared by thermal reaction.[53]) In the presence of an organic halide, polymerization is photoinitiated in suitable systems.[53] Initiation also occurs when the halogen component is added in the dark to the complex $MFe(CO)_4$. The reactions shown in equations (30a) and (30b) have been envisaged as a possible mechanism of radical formation.[48] It is reported that the complex $(CO)_4Fe(CHBr_2)Br$ is itself a polymerization catalyst; its mode of action does not appear to have been established but a co-ordination type of propagation has been mentioned.

Bamford, Eastmond and Fildes[55] showed that when CBr_4 is added to a previously-irradiated solution of $Mo(CO)_6$ in ethyl acetate (containing $(EtOAc)Mo(CO)_5$ formed by equations (28) and (29)) a species is produced

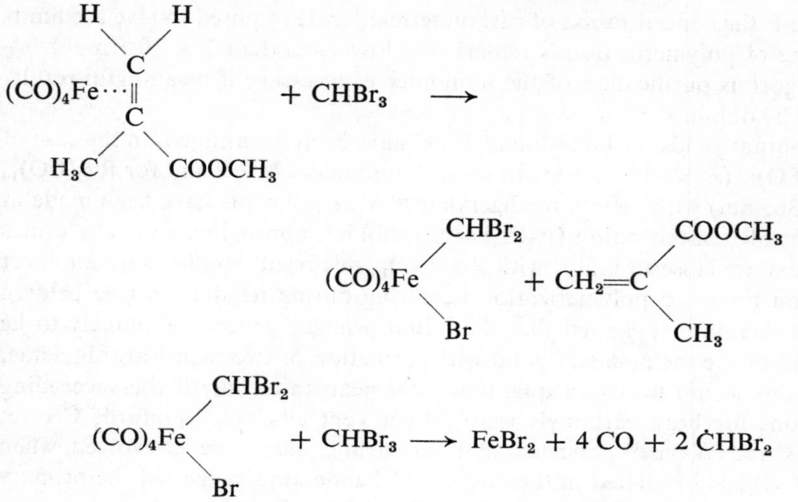

$$
\underset{H_3C}{\overset{H\quad H}{\underset{\diagdown}{\diagup}}}\;C\;\; (CO)_4Fe\cdots\|\;\;\;\underset{COOCH_3}{\overset{C}{\diagup}}\quad + CHBr_3 \longrightarrow
$$

$$
(CO)_4Fe\underset{Br}{\overset{CHBr_2}{\diagup}} \quad + CH_2{=}C\underset{CH_3}{\overset{COOCH_3}{\diagup}} \qquad (30a)
$$

$$
(CO)_4Fe\underset{Br}{\overset{CHBr_2}{\diagup}} \quad + CHBr_3 \longrightarrow FeBr_2 + 4\,CO + 2\,\dot{C}HBr_2 \quad (30b)
$$

slowly in the dark at 25 °C which gives ESR spectra characteristic of the Mo^V derivatives arising during thermal initiation (section 3.151). Further, polymerization occurs at 25 °C when a solution of CBr_4 in methyl methacrylate is added to the ethyl acetate solution of $(EtOAc)Mo(CO)_5$. Photochemical activation of the carbonyl therefore enables a separation to be made of the ligand-exchange and oxidation steps occurring in the thermal reaction (see equations (4) and (8), and section 3.15), and the technique evidently provides an opportunity for detailed mechanistic studies of the type described for $Ni\{P(OPh)_3\}_4$ (section 3.152). (See also reference 21b.)

Strohmeier and Hartmann[56] reported photoinitiation of polymerization of ethyl acrylate by several transition metal carbonyls and related derivatives in the presence of carbon tetrachloride; vinyl chloride has also been polymerized in a similar manner.[4] No detailed mechanisms were discussed, but it seems most likely that initiation proceeds by the route of equations (28), (29) and (4c) and (4d).

Most attention has been paid to initiation photosensitized by manganese and rhenium carbonyls.[57-61] The longwave limits of absorption by these materials are approximately 460 nm and 380 nm, respectively, and photoinitiation occurs up to these wavelengths. Manganese carbonyl is therefore a particularly convenient initiator since it is active in visible light.

The dependence of the rate of polymerization on halide concentration for $Mn_2(CO)_{10}$ is similar to that found for thermal reactions, but the exact relation depends on the light intensity.[58] At higher absorbed intensities the dependence is less sharp, so that a process probably occurs leading to the destruction of active intermediates which is second-order in their concentration. The halide dependence with rhenium carbonyl has not been studied in detail; the concentration of carbon tetrachloride required to give the limiting rate of polymerization is remarkably low—less than 2×10^{-5} mol l^{-1}—and rigorous purification of the monomer is necessary if meaningful results are to be obtained.[60]

Quantum yields for initiation at 25 °C have been determined, in the case of $Mn_2(CO)_{10}$ ($\lambda = 435 \cdot 8$ nm) with several monomers,[58,62,63] and for $Re_2(CO)_{10}$ ($\lambda = 365$ nm) with methyl methacrylate.[60] Measurements have been made at high halide concentration ($[CCl_4] = 0 \cdot 1$ mol l^{-1}, approximately). The values reported are close to unity; with $Re_2(CO)_{10}$ this result applies only for short reaction times, to polymerization occurring during irradiation (see below). It was therefore suggested that the initial primary process is unlikely to be rupture of the metal–metal bond with formation of two identical fragments, since this would lead to a quantum yield near unity only if the succeeding reactions for both carbonyls were 50 per cent efficient. Bamford, Crowe, Hobbs and Wayne[60] reported that no carbon monoxide is evolved when $Mn_2(CO)_{10}$ is irradiated in the absence of halide, and suggested the primary dissociation shown in equation (31) ($M = Mn$ or Re).

$$M_2(CO)_{10} + h\nu \rightarrow M(CO)_4 + M(CO)_6 \qquad (31)$$

Observations on rhenium carbonyl, to be described below, were also adduced in support of equation (31). It was supposed that only the fragment $M(CO)_4$ reacts rapidly with CCl_4 to yield $\dot{C}Cl_3$ radicals. More recently, Kwok[59] has confirmed the unit quantum efficiency for photoinitiation by $Mn_2(CO)_{10}$, and has further shown that two initiating radicals are generated for each $Mn_2(CO)_{10}$ which disappears and that the major product is $Mn(CO)_5Cl$. These findings are consistent with the reaction scheme in equations (32a), (32b) and (32c).

$$Mn_2(CO)_{10} \underset{}{\overset{h\nu}{\rightleftharpoons}} Mn(CO)_4 + Mn(CO)_6 \qquad (32a)$$

$$Mn(CO)_4 + CCl_4 \longrightarrow Mn(CO)_4Cl + \dot{C}Cl_3 \qquad (32b)$$

$$\downarrow +CO$$

$$Mn(CO)_5Cl$$

$$2\,Mn(CO)_6 \longrightarrow Mn_2(CO)_{10} + 2\,CO \qquad (32c)$$

Reaction (32c) is rapid, and regenerates 50 per cent of the $Mn_2(CO)_{10}$ consumed in reaction (32a). In essence, this mechanism closely resembles that advanced by Bamford et al.;[60] the latter authors discussed alternatives to reaction (32c) and preferred one in which CCl_4 is involved. (Further discussion of the mechanism of reactions (32a), (32b) and (32c) is given in reference 63a.)

Polymerizations photoinitiated by rhenium carbonyl show a long-lived after effect, reaction continuing at room temperature for several hours after the light has been cut off.[60] The after effect, which is shown by a variety of monomers, is therefore quite distinct from that associated with the finite rate of normal free-radical decay; nor does it originate in thermal initiation by $Re_2(CO)_{10}$, which is exceedingly slow at 25 °C.[14] The observations were interpreted in terms of a mechanism similar to that of equations (32a), (32b) and (32c), with the difference that mutual interaction of $Re(CO)_6$ fragments (cf. equation (32c)) is relatively slow, so that $Re(CO)_6$ may survive for long periods and enter into rather slow radical-forming reactions with CCl_4, thus giving rise to the after effect. Supporting evidence was provided by the observation that a solution of $Re_2(CO)_{10}$ in methyl methacrylate after a short irradiation ($\lambda = 365$ nm) and subsequent standing in the dark for one hour, produces polymer when added to a mixture of monomer and carbon tetrachloride.[57] (No significant polymerization occurs during the irradiation.) With manganese carbonyl, intermediates formed on irradiation were found to decay much more rapidly, and indeed in this system the after effect is hardly significant. The difference in the behaviour of the two systems is therefore considered to originate from differences in the relative rates of reactions typified by equation (32c).

More recent investigations[64,65] have shown that prolonged after effects can be obtained with manganese carbonyl in the presence of certain additives, notably cyclohexanone and acetyl acetone, although other ketones show some activity. Typical results are presented in Figure 3.8. The after effect occurs only if the additive is present during irradiation; the presence of monomer or halide during illumination is not necessary, and the after effect develops if these components are added to a previously illuminated mixture containing carbonyl and additive. It therefore appears that photochemical reaction between $Mn_2(CO)_{10}$ and the additive (S) produces a species (Z) which generates free-radicals by interaction with the halide. (Z) is probably formed from $Mn(CO)_6$ by a process such as recation (33).[65]

$$Mn(CO)_6 + S \rightarrow S \cdots Mn(CO)_5 + CO \qquad (33)$$
$$(Z)$$

Spectrophotometric observations are consistent with the formation of a labile species with a half-life of several hours at 25 °C during irradiation of $Mn_2(CO)_{10}$ in cyclohexanone; no similar species could be detected in cyclohexane solution.[65]

Kinetic observations are simply explained in terms of a mechanism based on equations (32), (33) and (34); the latter reaction, in which RCl represents the halide component, is the process responsible for producing free-radicals during the after effect.

$$Z + RCl \rightarrow R\cdot + Mn^I + \cdots \qquad (34)$$

The concentration of (Z) formed during irradiation and its rate of decay by reaction (34) may be deduced from the kinetics of the after effect. It has thus been demonstrated that, within experimental error, the maximum value of

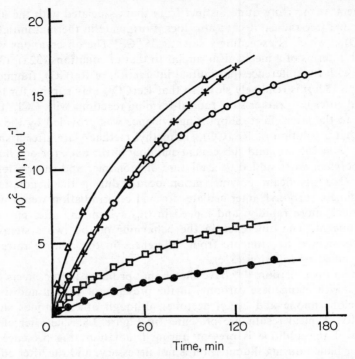

Figure 3.8 After effects in polymerizations at 25 °C photoinitiated by manganese carbonyl. Irradiation by light of $\lambda = 435\cdot8$ nm for 5 min, interrupted at time zero in Figure. ΔM—conversion. $[CCl_3COOEt] = 6 \times 10^{-2}$ mol l^{-1}. Monomers—methyl methacrylate (MMA) and vinyl chloride (VC). Concentrations are given below in mol l^{-1}.

●	$[MMA] = 4\cdot7$,	$[Mn_2(CO)_{10}] = 3\cdot0 \times 10^{-4}$,	solvent-benzene
□	$[MMA] = 4\cdot7$,	$[Mn_2(CO_{610})] = 2\cdot1 \times 10^{-4}$,	solvent-acetone
○	$[MMA] = 4\cdot7$,	$[Mn_2(CO)_{10}] = 3\cdot9 \times 10^{-4}$,	solvent-cyclohexanone
+	$[MMA] = 4\cdot7$,	$[Mn_2(CO)_{10}] = 3\cdot0 \times 10^{-4}$,	solvent-acetylacetone
△	$[VC] = 8\cdot6$,	$[Mn_2(CO)_{10}] = 6\cdot4 \times 10^{-4}$,	solvent-cyclohexanone

(From Bamford and Paprotny)[65]

[Z] is equal to the total concentration of radicals produced by reactions (32a) and (32b) during the period of irradiation, a result which clearly supports equation (32a) as the primary photolytic act. Further, the rate of initiation in the light is not decreased by the presence of cyclohexanone, showing that the latter does not deactivate an intermediate normally participating in photoinitiation. The observations therefore do not favour a symmetrical primary photochemical process leading exclusively to two Mn(CO)$_5$ fragments. If such a reaction is only part of the primary act it is possible, by making certain assumptions, to explain the experimental observations.[65]

Most of the quantitative studies have been made with rather short irradiation periods ($\leqslant$ 5 min); if irradiation is prolonged complications arise from photolysis of (Z). The reader is referred to the original paper for details.[65]

Irradiation of manganese carbonyl in solution in the presence of n- or π-donors yields mono- or di-substitution products,[66,67] the former in rather low yield. These observations in isolation do not provide unambiguous evidence about the nature of the primary process, since they can be reconciled either with rupture of the Mn—Mn bond or with CO scission. We may note that although data on the after effect find a natural explanation in terms of reactions (32)–(34), they are not intrinsically incompatible with primary CO scission. Thus a sequence such as that shown below can be envisaged, in which reaction of $Mn_2(CO)_9$ with halide breaks the Mn—Mn bond and leads to one fragment which, in the absence of a stabilizing additive, regenerates $Mn_2(CO)_{10}$.

$$Mn_2(CO)_9 + CCl_4 \longrightarrow Mn(CO)_5 + Mn(CO)_4Cl + \dot{C}Cl_3$$
$$\downarrow \qquad\qquad \downarrow CO$$
$$\tfrac{1}{2} Mn_2(CO)_{10} \quad Mn(CO)_5Cl$$

The after effect observed in the presence of such an additive could then arise by trapping of $Mn(CO)_5$ by the latter, and subsequent reactions similar to those already described in equation (34). However, such a mechanism may be incompatible with the report that CO is not evolved during irradiation of $Mn_2(CO)_{10}$.[60] Clearly investigation of the nature of the primary act by a different technique, e.g. flash photolysis, is desirable.

3.17 PRACTICAL APPLICATIONS

We conclude our considerations of initiating systems composed of organo-metallic derivatives and organic halides by describing their practical applications.

First, their value as initiators of free-radical polymerization will be evident from the preceding pages and little need be added. Activities are summarized in Table 3.1 and Figure 3.4. The data refer mostly to initiation in methyl methacrylate, and, if there is monomer-selectivity (section 3.123), will not necessarily apply to other monomers. However, they should hold approximately for solutions of other monomers in solvents resembling methyl methacrylate in electron-donating power, e.g. ethyl acetate. The more active systems will initiate satisfactorily at temperatures down to about 0 °C, or perhaps somewhat lower. Nickel derivatives are generally useful for initiation at low temperatures.

The systems under discussion have the advantage of being composed of components which are mostly stable at ordinary temperatures, and therefore convenient to store. $Ni\{P(OPh)_3\}_4$ is a good example; in the presence of

CCl_4 it is active at 0 °C, although the solid organo-nickel derivative does not decompose rapidly below the melting point (~146 °C).

The utility of $Mn_2(CO)_{10}$ as a photoinitiator with a quantum yield of unity in visible light has been mentioned in section 3.16.

Secondly, these initiators provide a convenient route for synthesizing vinyl polymers with a variety of different end-groups. A halide component R-hal generally yields R· radicals as the sole initiating species so that the residue R becomes incorporated in the polymer as a terminal group. (Note that this does not always hold when organo-nickel derivatives are used, section 3.12.) If the termination reaction is radical combination, the polymer molecules carry R groups at both ends, provided transfer is insignificant. The nature of R may be varied over a wide range; for example, polystyrene chains with $-CBr_3$ or $-CHClCOOH$ terminations may be prepared in this way.

An extension to the case in which R is part of a preformed polymer chain leads to a technique for synthesizing block or graft copolymers which is probably the most important practical application of these initiators.[22] Let us suppose that the halide component is a vinyl polymer of a monomer A having one terminal CBr_3 per chain, and that reaction takes place in a monomer B. The primary radicals (IV) will be $A_p\dot{C}Br_2$, and the propagating chains $A_pCBr_2B_q\dot{B}$; if termination occurs exclusively by combination the product will be the three-block polymer $A_pCBr_2B_rCBr_2A_s$ (equation (35)).

$$\sim\!\!A\!\!\sim\!\!CBr_3 \xrightarrow[h\nu]{Mn_2(CO)_{10}} \sim\!\!\sim A\!\!\sim\!\!\dot{C}Br_2$$
$$(IV)$$

$$(35)$$

$$\xrightarrow{B} \sim\!\!\sim\!\!A\!\!\sim\!\!CBr_2B\!\!\sim\!\!\sim\!\!\dot{B} \longrightarrow \sim\!\!\sim A\!\!\sim\!\!CBr_2B\!\!\sim\!\!BCBr_2\!\!\sim\!\!A\!\!\sim$$

If the initial polymer carries two terminal CBr_3 groups per chain, the first product will be (V); further reaction will activate CBr_3 groups in (V) so that the next product will be the seven block copolymer (VI), having four blocks of A and three of B, and so on. Generally, therefore, the product will

$$Br_3C\!\!\sim\!\!A\!\!\sim\!\!CBr_2\!\!\sim\!\!\sim\!\!B\!\!\sim\!\!\sim\!\!CBr_2\!\!\sim\!\!\sim\!\!A\!\!\sim\!\!\sim\!\!CBr_3$$
$$(V)$$

$$Br_3C\!\!\sim\!\!A\!\!\sim\!\!CBr_2\!\!\sim\!\!\sim\!\!B\!\!\sim\!\!\sim\!\!CBr_2\!\!\sim\!\!A\!\!\sim\!\!\sim\!\!CBr_2\!\!\sim\!\!\sim\!\!B$$
$$Br_3C\!\!\sim\!\!A\!\!\sim\!\!CBr_2\!\!\sim\!\!\sim B\!\!\sim\!\!\sim\!\!CBr_2\!\!\sim\!\!A\!\!\sim\!\!\sim\!\!CBr_2\!\!\sim\!\!\sim\!\!B$$
$$(VI)$$

consist of chains of alternating blocks of A and B linked by CBr_2 units and with CBr_3 terminations. Since CBr_2 is much less reactive to organometallic derivatives than CBr_3 (probably by a factor approaching 100) activation of

CBr$_2$ leading to grafting will not be considerable until polymers containing a large number of blocks have been formed. The concentration of reactive halogen groups will usually be low in these preparations, and it is clearly advantageous to employ an organometallic derivative having a sharp ω to [halide] relation (see Table 3.2). Use of manganese carbonyl is convenient for this reason.

Syntheses of block polymers by this route are much more versatile than those based on anionic polymerization since a wider range of monomers may be incorporated into the blocks. Although the mean block-length is controllable through the parameters which normally determine the mean kinetic chain-length in a free-radical polymerization, the molecular weight distributions are, of course, much broader than with anionic polymerization, and the polymers are therefore less well-defined. It is also worthy of note that the syntheses we are discussing have the merit (important in practice) of avoiding homopolymer formation when properly carried out, since all the initial radicals are attached to polymer chains. This is not so with many free-radical syntheses.

Graft polymers are produced when the halide component of the initiator is a pre-formed polymer carrying side-chains with active halide groups. Initiation in a vinyl monomer then leads to synthesis of a network if there is a combination component in the termination reaction.[63,68-70] Equation (36) represents the formation of a cross-link in a typical case. A very wide range

$$\text{(scheme)}$$

$$
\begin{array}{c}
\left.\begin{array}{l}\text{—CBr}_3\\[6pt]\text{Br}_3\text{C—}\end{array}\right\}
\xrightarrow[h\nu]{\text{Mn}_2(\text{CO})_{10}}
\left.\begin{array}{l}\text{—}\dot{\text{C}}\text{Br}_2\\[6pt]\text{Br}_3\text{C—}\end{array}\right\}
\xrightarrow{\text{B}}
\left.\begin{array}{l}\text{—CBr}_2\!\!\sim\!\!\text{B}\!\!\sim\!\!\dot{\text{B}}\\[6pt]\text{Br}_3\text{C—}\end{array}\right\}
\end{array}
$$

$$\tag{36}$$

$$
2\;\left.\begin{array}{l}\text{—CBr}_2\!\!\sim\!\!\text{B}\!\!\sim\!\!\dot{\text{B}}\\[6pt]\text{Br}_3\text{C—}\end{array}\right\}
\longrightarrow
\left.\begin{array}{l}\text{—CBr}_2\!\!\sim\!\!\text{B}\!\!\sim\!\!\text{Br}_2\text{C—}\\[6pt]\text{Br}_3\text{C—}\qquad\qquad\text{—CBr}_3\end{array}\right\}
$$

of polymer networks may be constructed in this manner; in most of the reported work polyvinyltrichloracetate has been the halide component, with styrene,[71] methyl methacrylate[71] and chloroprene[72] as cross-linking units. Polycarbonates and polystyrene with suitable functional groups,[72a] N-halogenated polyamides and polypeptides have also been employed as prepolymers.[23]

The statistics of these networks are readily controllable. The number and distribution of halide groups in the prepolymer chains and the molecular weight of the prepolymer may be varied. Cross-link length may be controlled through the rate of initiation and/or the monomer concentration, and cross-link density in a given system is determined by the extent of initiation, $\int I\,dt$. When termination occurs partly by disproportionation the prepolymer chains in the product carry branches as well as cross-links. Addition of a sufficiently high concentration of a transfer agent suppresses network formation and the products are then simple graft copolymers mixed with the vinyl homopolymer (which originates from the unattached radicals produced by transfer). Lower concentrations of transfer agent permit some cross-linking of the prepolymer chains, but lead to the formation of branches of the vinyl polymer. The ratio cross-links to branches may therefore be varied by altering the incidence of chain transfer. Branching necessarily occurs if any process generates radicals which are not attached to prepolymer chains. Such radicals interfere with cross-link formation by interacting with propagating radicals, equation (37), and also yield homopolymer by obvious reactions. Unattached

$$\text{\Large{}}{-}CBr_2\!\!\diagdown\!\!\!\diagup\!\!\!\diagdown\!\!\!\diagup\!\!\!\diagdown\!\!\dot{B} \;+\; \dot{B}\!\!\diagdown\!\!\!\diagup\!\!\!\diagdown\!\! R \longrightarrow \text{\Large{}}{-}CBr_2\!\!\diagdown\!\!\!\diagup\!\!\!\diagdown\!\!\!\diagup\!\!\!\diagdown\!\! R \qquad (37)$$

radicals arise through chain-transfer, as as mentioned above, and may also be formed in initiation, e.g. if organo-nickel derivatives are employed (section 3.12) or an active non-polymeric halide such as ethyl trichloracetate is present.

For a description of the properties of some networks prepared by this technique the reader may consult references 71–73.

Since cross-link formation depends on the occurrence of combination in the termination reaction, it is clear that observations of the rate of cross-linking permit the incidence of combination in this reaction to be assessed. If all the initial radicals are attached to chains of the polymeric initiator, and chain-transfer is negligible, we may derive the following expression for the gel time t_g from simple gel theory.[68]

$$\frac{1}{t_g} = \frac{I\bar{P}_w}{c}\,\frac{k_{tc}}{k_{tc} + k_{td}} \qquad (38)$$

Here I is the rate of initiation, $\bar{P}_w$, c are the weight-average degree of polymerization and base-molar concentration of the polymeric initiator, and the velocity coefficients k_{tc}, k_{td} refer to combination and disproportionation, respectively. According to equation (38) a plot of $1/t_g$ against I should be a straight line, from the slope of which $k_{tc}/(k_{tc} + k_{td})$ may be evaluated. Bamford et al.[63,68,70] have described observations of this kind and have also specified the conditions under which the simple relation in equation (38) is obeyed; they have also calculated some corrections which are necessary in practice, for example to allow for initiator consumption and chain transfer.[69,70] Combination is effectively the sole termination reaction with styrene, acrylonitrile, methyl-, ethyl- and n-butyl acrylates and chloroprene while methyl, ethyl and n-butyl methacrylates and methacrylonitrile show a mixed termination.[22,63,68,72]

When only a fraction of the initial radicals is attached to chains of the polymeric initiator the gel time is longer than that predicted by equation (38). Determination of the gel time with a monomer for which k_{tc}/k_{td} is known permits estimation of the fraction of attached initial radicals. The required relations have been given by Bamford, Eastmond and Whittle,[70] and the same workers have shown that this technique and the tracer method (section 3.12) give similar results.

3.2 INITIATION BY CHELATE DERIVATIVES OF TRANSITION METALS

Initiation of polymerization by metal chelates has received considerable attention in recent years. Reports in the literature suggest that these derivatives can initiate various types of polymerizations in suitable circumstances; thus Nishikawa and Otsu[74] reported the cationic or co-ordinated cationic polymerization at 80 °C of some vinyl monomers and cyclic ethers by acetylacetonates (particularly $MoO_2(acac)_2$), while Nishikawa, Otsu and Watanuma[75] found that $Mn^{II}(acac)_2$ and $Co^{II}(acac)_2$ bring about the polymerization of chloral at room temperatures, probably by a co-ordinated anionic mechanism.

3.21 CHELATES ALONE

Most interest, however, has centred on the behaviour of chelates as free-radical initiators. This stems from the original work of Arnett and Mendelsohn,[76] who, in the course of oxidation studies, showed that some chelates produce free-radicals on heating. For $Fe^{III}(acac)_3$ they postulated the reactions in equations (39a) and (39b). The peroxide form of the radical in equation (39a) was invoked on the grounds of its assumed non-reactivity towards oxygen. Arnett and Mendelsohn also found that some chelates can initiate the polymerization of styrene at 110 °C and gave the following order of rates of polymerization in the presence of 0·32 mol per cent of various acetylacetonates

$$V^{III} < Cr^{III} = Al^{III} < Fe^{III} < Co^{III} < Ce^{IV} < Mn^{III}$$

Chromium and aluminium chelates did not affect the rate of polymerization, so that the vanadium derivative behaved as an inhibitor.

Kastning et al.[77] also observed the high initiating activities of $Mn^{III}(acac)_3$ and $Co^{III}(acac)_3$ and reported a proportionality between the rate of polymerization of styrene at 110 °C and $[Co^{III}(acac)_3]^{0.5}$. They noted that this result is consistent with a free-radical polymerization and adduced supporting evidence from studies of the copolymerization of methyl methacrylate and styrene. Kastning et al. further remarked that 'it has been established by infrared spectroscopy and by labelling the ligands with ^{14}C that the ligand combines with the polymer,' although they were inclined to doubt whether

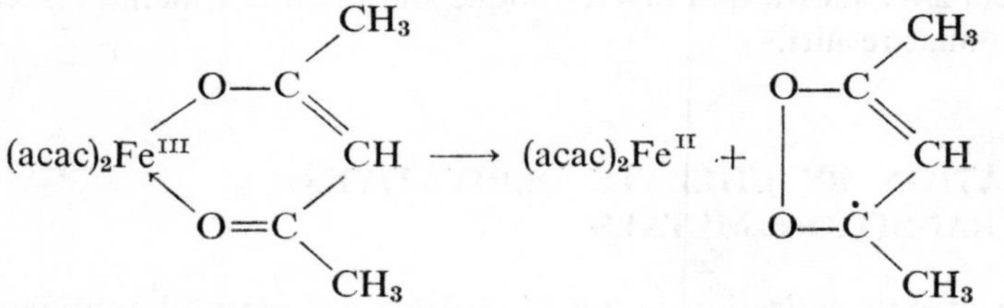

$$(39a)$$

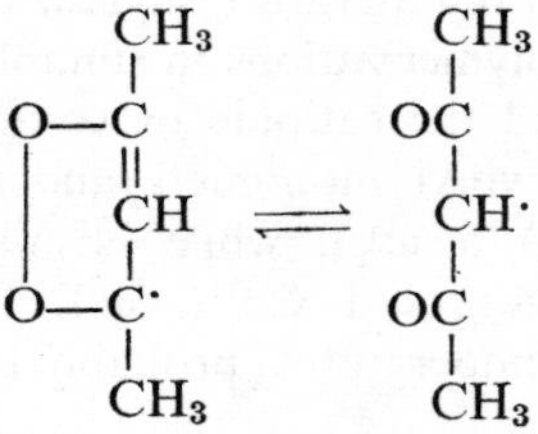

$$(39b)$$

initiation involves a change in oxidation state of the metal atom, as would be implied if reactions (39a) and (39b) were the initiation process.

Bamford and Lind[78] have investigated initiation by $Mn^{III}(acac)_3$ and the related derivatives (VII)–(X).

$Mn^{III}(PhCO:CHCOPh)_3$

(VII)

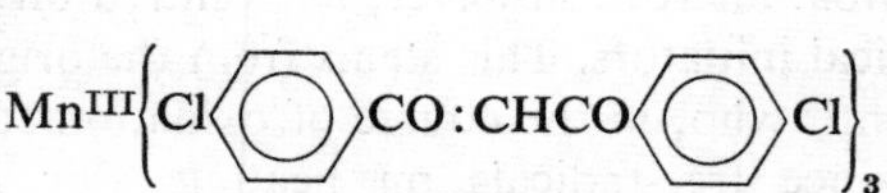

(VIII)

$Mn^{III}\{CH_3CO:CPhCOCH_3\}_3$

(IX)

$Mn^{III}(CF_3CO:CHCOCH_3)_3$
$(Mn^{III}(facac)_3)$

(X)

Subsequently Nishikawa and Otsu[79] repeated the work on initiation of the polymerization of styrene and methyl methacrylate by $Mn(acac)_3$, and showed that $Cu(acac)_2$ is a less effective initiator. These authors also reported that at 80 °C $Mn(acac)_3$ and azo-bis-isobutyronitrile initiate the polymerization of butadiene and isoprene at similar rates, other chelates being much less effective.

Bamford and Lind[78] confirmed that the ω against $[Mn(acac)_3]^{0.5}$ relation observed by Kastning *et al.*[77] holds at 80 °C and demonstrated that the polymers have degrees of polymerization expected for unretarded free-radical reactions in methyl methacrylate and styrene, corresponding to normal values of the parameter $k_p k_t^{-0.5}$. Measurements of specific susceptibilities of the manganese products established that reduction to a Mn^{II} derivative occurs on reaction, so that the essential step in initiation is the scission of a ligand as a free-radical:

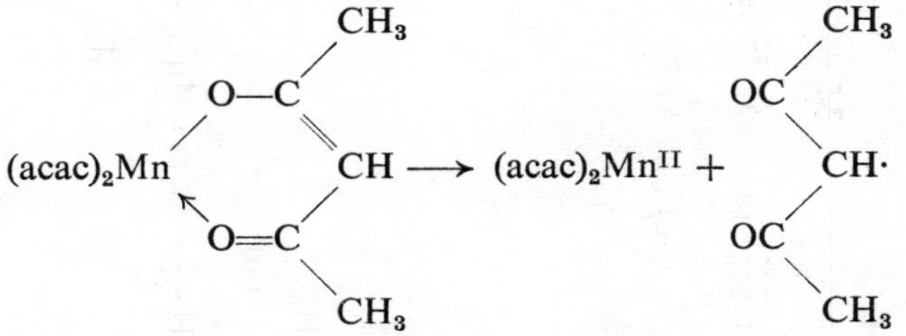

$$(40)$$

Such a process is consistent with the findings of earlier investigators.[76,77] During this work three other features emerged: (i) decomposition of $Mn(acac)_3$ is more rapid in ethyl acetate than in benzene solution, (ii) addition of methyl methacrylate to ethyl acetate increases the rate, and (iii) addition of styrene to benzene increases the rate. Clearly, monomers and ethyl acetate participate in the decomposition of the chelate, and it is not surprising to find that the observed order in monomer for the polymerization exceeds unity. The order depends on the nature of the diluent as set out in Table 3.6. The data are reminiscent of those presented in section 3.121 for

Table 3.6 Initiation by $Mn(acac)_3$ at 80 °C

Monomer	Diluent	Order in [M]
MMA	Benzene	1·5
	Ethyl acetate	1·25
St	Toluene	1·5
	Ethyl acetate	1·2

S_N2 systems and may be interpreted similarly. We therefore conclude that radical generation requires prior co-ordination of a monomer or solvent molecule to the metal, the details of the overall process in equation (40) being shown in equation (41). After the rate-determining steps (a) and (b) co-ordination of the Mn^{II} species is probably completed by rapid addition of

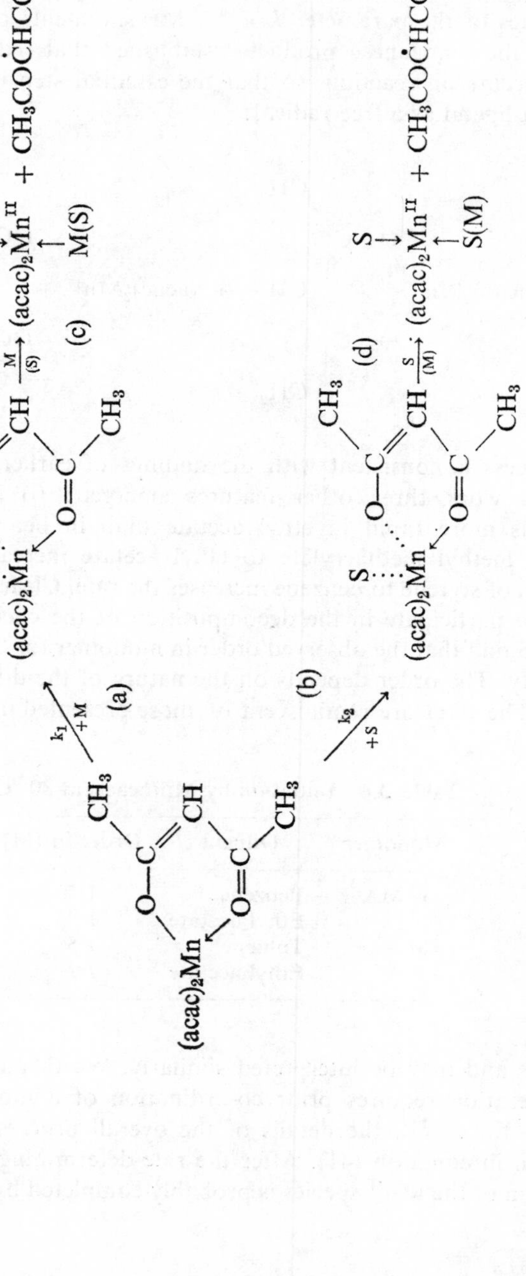

$$(acac)_2Mn \xleftarrow{+M}{k_1} \text{(a)}; \quad (acac)_2Mn \xrightarrow[(S)]{M} (acac)_2Mn^{II} + CH_3COCHCOCH_3 \quad \text{(c)}$$

$$(acac)_2Mn \xrightarrow{+S}{k_2} \text{(b)}; \quad (acac)_2Mn \xrightarrow[(M)]{S} (acac)_2Mn^{II} + CH_3CO\dot{C}HCOCH_3 \quad \text{(d)}$$

(41)

monomer or solvent as indicated in (c) and (d). The total rate of initiation is given by the expression

$$I = [Mn(acac)_3](k_1[M] + k_2[S]) \tag{42}$$

which is of the required kinetic form, and closely analogous to equation (9).

Values of a ($= k_1[M]$) for several chelates are presented in Table 3.7, together with available data on the activation parameters for initiation. For initiation by $Mn(acac)_3$, neither $\Delta H^{\ddagger}$ nor $\Delta S^{\ddagger}$ is significantly affected by changing the monomer from methyl methacrylate to styrene. Corresponding data are not available for chelates (VII)–(IX), but we may note that with (VII)

Table 3.7 Initiation by Mn^{III} chelates

Chelate	$10^6 a$ s^{-1}	$\Delta H^{\ddagger}$ kJ mol^{-1}	$\Delta S^{\ddagger}$ e.u. mol^{-1}
$Mn(acac)_3$	51 (80 °C, bulk MMA)[78]	106[78], 109[80]	−10, −9
	0·047 (25 °C, bulk MMA)[80]		
	50 (80 °C, bulk St)[78]	109[80]	−9
	0·049 (25 °C, bulk St)[80]		
(VII)	33 (80 °C, bulk MMA)[78]		
	43 (80 °C, bulk St)[78]		
(VIII)	~33 (80 °C, bulk MMA)[78]		
(IX)	150 (80 °C, bulk MMA)[78]	51	−39
	140 (80 °C, bulk St)[78]		
$Mn(facac)_3$	380 (80 °C, bulk MMA)[78]	63[78], 82[85]	−31[78], −21[85]
	1·82 (25 °C, bulk MMA)[85]		
	<53, probably effectively zero (80 °C, bulk St)[78]		

and (IX) rates of initiation at 80 °C are similar for the two monomers. Thus there does not appear to be any marked monomer selectivity in these systems. Data for chelates (VII), (VIII) are derived from observed rates of polymerization at very low chelate concentrations; at higher concentrations inhibition or retardation occurs. On the other hand, with chelate (IX) the square root relation holds over the whole concentration range studied (up to 2×10^{-3} mol l^{-1}). The large negative entropy of activation for initiation by (IX) is noteworthy and probably indicates congestion in the transition state of the reaction with monomer, while the low enthalpy of activation may reflect either the stabilization of the radical product by the phenyl substituent, or a relatively unstable chelate ring in (IX).

Manganese 1,1,1-trifluoroacetylacetonate initiates the polymerization of methyl methacrylate more rapidly than $Mn(acac)_3$. Bamford and Lind[78] reported that at 80 °C the rate of polymerization is proportional to $[Mn(facac)_3]^{0.5}$

but later work[80] has shown that at lower temperatures (when rates of initiation are lower, and kinetic chain lengths correspondingly longer) retardation is significant.

Mn(facac)₃ is a powerful retarder of the polymerization of styrene, and in the presence of the chelate the rate of polymerization is lower than the spontaneous thermal rate for pure styrene.[78] When a conventional initiator is also present, the rate of polymerization decreases as [Mn(facac)₃] is increased (see Figure 3.9); benzoyl peroxide was used as initiator in the experiments of Figure 3.9 but may be replaced by azo-bis-isobutyronitrile with similar

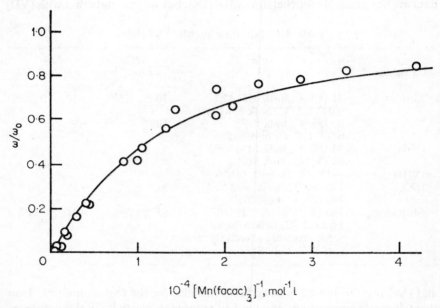

Figure 3.9 Retardation by Mn(facac)₃ of polymerization of styrene initiated by benzoyl peroxide at 80 °C. [Benzoyl peroxide] $= 2 \cdot 0 \times 10^{-3}$ mol l⁻¹; ω_0, ω are the initial rates of polymerization in the absence and presence of chelate, respectively. The curve was calculated from equations (43) and (44) with the following values of the parameters: $a = 5 \cdot 3 \times 10^{-5}$ s⁻¹, $k_t^{0 \cdot 5}/b = 0 \cdot 129$ mol$^{0 \cdot 5}$ l$^{-0 \cdot 5}$ s$^{0 \cdot 5}$, $I_0 = 1 \cdot 9 \times 10^{-7}$ mol l⁻¹ s⁻¹. Experimental points ○. (From Bamford and Lind)[78]

results. Bamford and Lind[78] analysed these data in terms of a mechanism in which both initiation and retardation by chelate occur; this was shown to lead to the relation

$$\frac{\omega}{\omega_0} = \frac{1}{\sigma}\left[\left\{1 + \sigma^2\left(1 + \frac{a[C]}{I_0}\right)\right\}^{0 \cdot 5} - 1\right] \tag{43}$$

In equation (43) ω_0, ω are the rates of polymerization in the absence and presence of chelate C, respectively, I_0 is the (constant) rate of initiation by the conventional initiator, and a, b are the rate constants for initiation (first-order) and retardation (second-order) by the chelate, respectively. Further

$$\sigma = 2(I_0 k_t)^{0.5}/(b[\text{C}]). \tag{44}$$

equation (43) contains only two unknown quantities, namely a, b. At high concentrations the equation takes the form

$$\frac{\omega}{\omega_0} = \frac{a}{b}\left(\frac{k_t}{I_0}\right)^{0.5} + \frac{(I_0 k_t)^{0.5}}{b[\text{C}]}\left(1 - \frac{a^2 k_t}{b^2 I_0}\right) \tag{45}$$

so that under these conditions ω/ω_0 is linear in $[\text{C}]^{-1}$. The intercept on $[\text{C}]^{-1} = 0$ is $(a/b)(k_t/I_0)^{0.5}$, from which a/b may be calculated; measurement of the slope of the line then permits separate evaluation of a and b. The curve in Figure 3.9 has been calculated from equation (43) with $k_t^{0.5}/b = 0.129$ $\text{mol}^{0.5}\,\text{l}^{-0.5}\,\text{s}^{0.5}$, $a = 5.3 \times 10^{-5}\,\text{s}^{-1}$, and $I_0 = 1.9 \times 10^{-7}\,\text{mol l}^{-1}\,\text{s}^{-1}$, the value appropriate to the benzoyl peroxide concentration employed. However, with these values the magnitude of a plays only a minor part in the calculation of ω/ω_0, and a fit with experiment which is almost equally good can be obtained with $a = 0$, and $k_t^{0.5}/b$ close to the value given above. Indeed, it is clear from an inspection of Figure 3.9 that the intercept is indistinguishable from zero, so that a must be very small, the value quoted being an upper limit. Thus although Mn(facac)_3 readily initiates the polymerization of methyl methacrylate at 80 °C, it initiates that of styrene very inefficiently, or not at all—a surprising conclusion in view of the high reactivity of styrene towards free-radicals in general.

The highly monomer-selective character of initiation by Mn(facac)_3, compared to initiation by Mn(acac)_3, is also demonstrated by the data[78] in Table 3.8. Selectivity is most likely to be associated with monomer polarity in

Table 3.8 Comparison of Mn(acac)_3 and Mn(facac)_3[78]

Monomer	Chelate; $10^4 \times$ concentration, mol l^{-1}	T °C	10^6 rate of polymerization, mol l^{-1} s^{-1}
Styrene	Mn(acac)_3; 2·5	80	34
Styrene	Mn(facac)_3; 2·5	80	1·6
Vinyl acetate	Mn(acac)_3; 4·0	60	190
Vinyl acetate	Mn(facac)_3; 4·0	60	38
Methyl methacrylate	Mn(acac)_3; 4·0	80	210
Methyl methacrylate	Mn(facac)_3; 4·0	80	530
Acrylonitrile	Mn(acac)_3; 5·8	60	15
Acrylonitrile	Mn(facac)_3; 5·8	60	1090

these systems; $Mn(facac)_3$ is ineffective with the relatively non-polar mono-mers styrene and vinyl acetate, but is remarkably effective with acrylonitrile. Table 3.8 illustrates the similar behaviour of styrene and vinyl acetate, and of methyl methacrylate and acrylonitrile; the efficiencies of the two chelates as initiators are in a different order for the two monomer pairs. The evidence therefore suggests that the electron-accepting properties of a monomer are an important factor in determining the efficiency of initiation by $Mn(facac)_3$.

The mechanism of initiation in equation (46) is essentially that proposed by Bamford and Lind.[78] As a consequence of electron withdrawal by CF_3, the primary activation of $Mn(facac)_3$ in reaction (46a) differs from that of $Mn(acac)_3$ in reactions (41a) and (41b) in being heterolytic rather than homolytic. This step, like the corresponding one with $Mn(acac)_3$, involves co-ordination of monomer (or an electron-donating solvent) to manganese, but leads to the dipolar species (XI). If the monomer $CH_2 {=} CXY$ has a

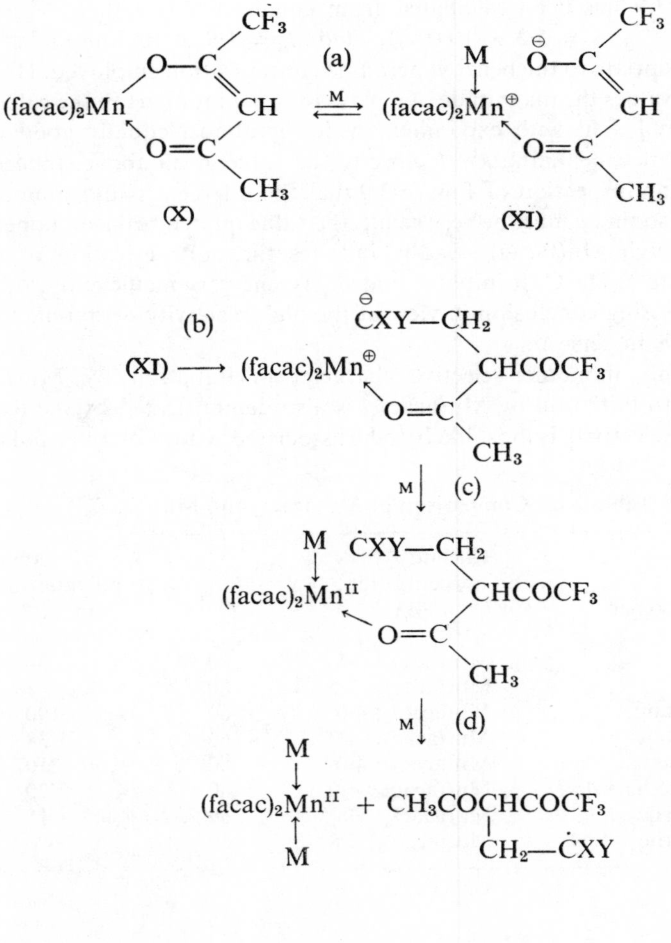

$$(46)$$

sufficiently high electron-accepting capacity (i.e. if it is readily susceptible to anionic polymerization) it may add to the anionic moiety in (XI) to give a monomer anion. This step, represented in reaction (46b), probably involves the co-ordinated monomer molecule, if present; when (XI) is formed from a donor solvent a molecule of monomer must enter into reaction (46b). (In reaction (46b) addition to C3 has been assumed, but this is not essential.) The radical-generating process in reaction (46c) is the oxidation of the monomer anion by the Mn^{III} atom in the same molecule; rapid co-ordination of monomer or solvent in reaction (46d), accompanies the separation of the radical which initiates polymerization. The final manganese product is the Mn^{II} chelate. Monomer selectivity probably arises mainly through the anionic addition in reaction (46b); from what has been written it will be clear that this process is more precisely regarded as monomer insertion. Indeed reactions (46a) and (46b) are essentially the initial stages of an anionic co-ordination polymerization which is prevented from continuing as such by intervention of the electron-transfer in reaction (46c).

In principle, either (46a) or (46b) may be rate-determining. With styrene, (46b) would be expected to be very slow. According to equation (46), in the presence of an inert diluent such as toluene the order in monomer of the polymerization reaction should be 1·5, and this is the observed value for methyl methacrylate.[78] On the other hand, an order between 1 and 1·5 would be expected if the diluent can co-ordinate with the metal, regardless of which of reactions (46a) or (46b) is rate-determining. An order of 1·1 for methyl methacrylate in ethyl acetate has been reported.[78] Since initiation proceeds through relatively polar transition states, the rather large negative entropy of activation for initiation in Table 3.7 is understandable.

3.22 INITIATION BY CHELATES IN THE PRESENCE OF DONOR ADDITIVES

Several groups of workers have reported that the rate of initiation by metal chelates may be increased by the presence of additives. Kastning et al.[77] remarked that a wide range of compounds, including materials as diverse as 1,3-cyclooctadiene and zinc chloride, 'activate' metal chelates, but later investigators[78] have found difficulty in reproducing some of the results of Kastning et al. In 1967 Lind[81] observed that free-radical polymerization of acrylonitrile initiated by $Mn(acac)_3$ occurs readily at 25 °C in dimethyl sulphoxide (30 per cent v/v acrylonitrile) although there is little detectable reaction in the bulk monomer. Korshak, Bevza and Dolgoplosk[82] found that initiation by $Mn(acac)_3$ is accelerated by the presence of acids (CH_3COOH, CCl_3COOH, HCl) and synthesized graft copolymers by initiating the polymerization of styrene with a mixture of $Mn(acac)_3$ and polymethacrylic acid. In Japan, Uehara, Kataoka, Tanaka and Murata[83] reported that pyridine, piperidine and 2,6-dimethylpyridine increase the rates of polymerization of styrene, methyl methacrylate and vinyl acetate initiated by Mn(acac)$_3$ at 70 °C and Kaeriyama[84] observed increased rates of initiation by Mn(acac)$_3$ and Fe(acac)$_3$ at 60 °C in the presence of various carbonyl compounds (e.g. cyclohexanone, caproic aldehyde) and has concluded that the effectiveness of ketones is dependent on their nucleophilicity. Not all these studies readily

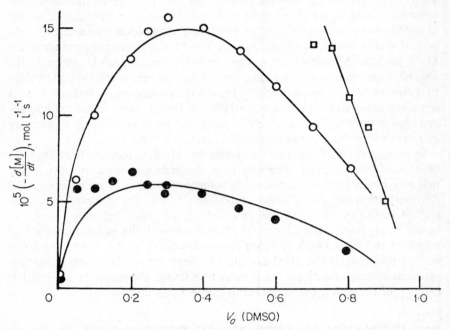

Figure 3.10 Rates of polymerizations at 25 °C initiated by Mn(acac)$_3$ (5×10^{-3} mol l^{-1}) in the presence of dimethylsulphoxide (DMSO), volume fraction v_a. Experimental points: ● = styrene, rates × 5, ○ = methyl methacrylate, □ = acrylonitrile. Curves calculated from rates of initiation given by equations (48), (49a), (49b) and (49c) with parameters in Table 3.11 (see Figure 3.12). (From Bamford and Ferrar)[80]

lead to conclusions of mechanistic significance, since reactions have been carried to high monomer conversions or extensive decomposition of initiators. Bamford and Ferrar[80] have examined the kinetics of initiation by chelates in the presence of electron donors (e.g. dimethylsulphoxide and organic amines) at 25 °C.

Rates of initiation by Mn(acac)$_3$ are greatly augmented, up to a thousand-fold by the presence of donor additives such as dimethylsulphoxide or 1,2-diaminopropane, so that these systems become active initiators at room temperatures. Figures 3.10 and 3.11 show the dependence of the initial rates of polymerization at 25 °C on the volume fractions v_a of dimethylsulphoxide (DMSO) and 1,2-diaminopropane (DAP), respectively. With methyl methacrylate and styrene the rates rise rapidly as the concentration of additive increases, then pass through a maximum and ultimately decline. Polyacrylonitrile is soluble only for v_a(DMSO) > 0·65 and the observations in Figure 3.10 are confined to this range of composition. Evidently DAP is

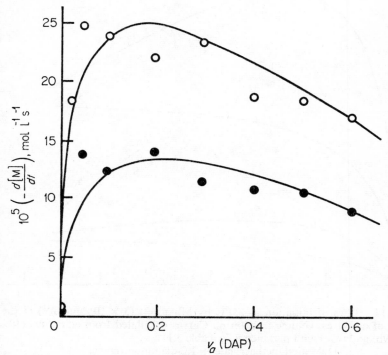

Figure 3.11 Rates of polymerizations at 25 °C initiated by Mn(acac)$_3$ (5 × 10^{-3} mol l^{-1}) in the presence of 1,2-diaminopropane (DAP), volume fraction v_a. Experimental points: ● = styrene, rates × 5, ○ = methyl methacrylate. Curves calculated from rates of initiation given by equations (48), (49a), (49b) and (49c) with parameters in Table 3.11 (see Figure 3.12). (From Bamford and Ferrar)[80]

more effective than DMSO, relatively low concentrations sufficing to achieve the maximum rate. The value of the maximum rate shown for MMA in Figure 3.11 corresponds approximately to 10 per cent conversion per hour for [Mn(acac)$_3$] = 5 × 10^{-3} mol l^{-1}, whereas the rate without additive is negligibly small.

At constant v_a, rates of polymerization are proportional to [Mn(acac)$_3$]$^{0.5}$, indicating the absence of retardation at 25 °C.[80] Kaeriyama[84] has reported non-linear relations between rates and [Mn(acac)$_3$]$^{0.5}$ in the presence of cyclohexanone or caproic aldehyde at 60 °C, but the deviations are not large and seem most pronounced at rather high chelate concentrations. Rates of initiation may be calculated from the results in Figures 3.10 and 3.11 with the aid of the conventional relation[42] between ω and I. (Note that the value of $k_p k_t^{-0.5}$, which is required for these calculations, is a function of composition

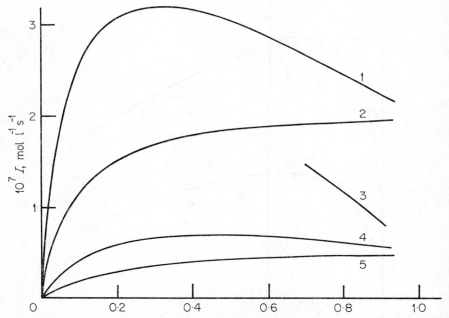

Figure 3.12 Rates of initiation at 25 °C by Mn(acac)$_3$ (5×10^{-3} mol l^{-1}) in the presence of additives, volume fraction v_a. Curves calculated from equations (48), (49a), (49b) and (49c) with parameters in Table 3.11.

         (1) Methyl methacrylate + 1,2-diaminopropane.
         (2) Styrene + 1,2-diaminopropane.
         (3) Acrylonitrile + dimethylsulphoxide.
         (4) Methyl methacrylate + dimethylsulphoxide.
         (5) Styrene + dimethylsulphoxide.
(From Bamford and Ferrar)[80]

of the system. Termination is diffusion-controlled and k_t is inversely proportional to the viscosity;[86] further, in the case of methyl methacrylate, k_p is affected by the presence of certain additives, including DMSO).[80,87,88] Values of I calculated from the curves in Figures 3.10 and 3.11 are presented in Figure 3.12, which reveals the striking enhancement produced by the additives. As an illustration we note that the maximum rate of initiation for MMA in the presence of DAP in Figure 3.12 corresponds to a first-order rate coefficient of initiation of 60×10^{-6} s^{-1}, approximately, while the

Table 3.9 Initiation by $Mn(acac)_3$ ($5 \cdot 0 \times 10^{-3}$ mol l^{-1}) in the presence of different additives at 25 °C.[80] Monomer: methyl methacrylate

Additive	v_a	$10^4 \omega$ mol l^{-1} s^{-1}	$10^7 I$ mol l^{-1} s^{-1}
None	—	0·08	0·002
1,3-diaminopropane	0·1	3·30	3·85
DAP	0·1	2·36	2·52
Hexamethylphosphoramide	0·1	2·23	1·76
1,2-diaminoethane	0·1	2·14	1·63
1,10-phenanthroline	—	1·82	1·25
N,N,N′,N′-tetramethyl- diaminoethane	0·02	1·54	0·93
N,N,N′,N′-tetramethyl- diaminoethane	0·1	1·67	0·99
Pyridine	0·1	0·99	0·346
DMSO	0·1	0·98	0·340
N,N-dimethylformamide	0·1	0·43	0·066
Formamide	0·1	0·42	0·063

coefficient observed with the chelate alone is only $0 \cdot 047 \times 10^{-6}$ s^{-1} (Table 3.7).

The type of behaviour we have described appears to be general for chelates of transition metals and additives capable of co-ordination to the metal atoms. Tables 3.9 and 3.10 summarize results at 25 °C for various additives and chelates.

The nature of the primary radicals formed by $Mn(acac)_3$ in the presence of DMSO and DAP has been investigated by radiochemical labelling of the various components of the system, and measurement of the extents to which activity is incorporated in the polymers.[80] It was concluded that initiating radicals are derived predominantly from the acetylacetonate ligands. In agreement with this conclusion, Uehara et al.[83] could not detect pyridine

Table 3.10 Activation of metal chelates by DAP ($v_a = 0 \cdot 1$) at 25 °C.[80] Monomer: methyl methacrylate. Chelate concentration 5×10^{-3} mol l^{-1}, except for $Mn(facac)_3$; $[Mn(facac)_3] = 2 \cdot 5 \times 10^{-4}$ mol l^{-1}. Subscripts of o and a indicate bulk monomer, and monomer in the presence of additive, respectively. Units for ω and I: mol l^{-1} s^{-1}

Chelate	$10^5 \omega_o$	$10^5 \omega_a$	$10^{10} I_o$	$10^{10} I_a$	I_a/I_o
$Cr(acac)_3$	0·24	2·22	0·22	23·7	103
$Mn(acac)_3$	0·79	23·65	2·36	2520	1072
$Fe(acac)_3$	0	1·03	0	4·8	high
$Co(acac)_3$	0	1·10	0	5·4	high
$Cu(acac)_2$	0·29	3·12	0·40	44·0	110
$Mn(facac)_3$	0·81	2·32	2·50	24·4	10·5

residues in polymers prepared by chelate initiation in the presence of the base.

One of the most interesting features in Figure 3.12 is the monomer selectivity which becomes apparent when DMSO or DAP is present. Thus, with DAP the polymerization of methyl methacrylate is initiated more rapidly than that of styrene, and with DMSO the order is

$$AN > MMA > St$$

This is the same kind of monomer selectivity found with $Mn(facac)_3$ in the absence of additives (cf. Table 3.8). However, the selectivity is less extreme in the present systems, since initiation in styrene is appreciable, and it is possible that, in general, initiation involves both selective and non-selective components. It is also clear from Figure 3.12 that the enhancement of the rate of initiation depends on the nature of the additive; Uehara et al.,[83] Kaeriyama,[84] and Bamford and Ferrar[80] remark that the activity of an additive is primarily determined by its donor power.

The above workers have shown from spectral observations that the chelates and additives interact to form complexes, and we may assume that a process of this type constitutes the first step in initiation. We therefore write the formal mechanism in equations (47a), (47b) and (47c), in which ds is a donor additive.

$$Mn(acac)_3 + ds \overset{K}{\rightleftharpoons} Z \tag{47a}$$

$$Z \xrightarrow{k_1} R\cdot \tag{47b}$$

$$Z + M \xrightarrow{k_2} R\cdot \tag{47c}$$

Complex Z is considered to be formed in the rapidly established equilibrium in equation (47a); the existence of a maximum or plateau in I may be correlated with effectively complete conversion of the chelate into Z at sufficiently high [ds]. Z may decompose spontaneously into radicals as in equation; (47b), or generate radicals by reaction with monomer as in equation (47c); processes (47b) and (47c) are therefore the non-selective and selective components of the initiation reaction, respectively. From equations (47a), (47b) and (47c) it follows that

$$I = K[Mn(acac)_3]_0[ds]\frac{k_1 + k_2[M]}{1 + K[ds]}$$

$$= A[Mn(acac)_3]_0\frac{v_a(1 - Bv_a)}{C + v_a} \tag{48}$$

where

$$A = k_1 + k_2[M]_0 \tag{49a}$$

$$B = k_2[M]_0/(k_1 + k_2[M]_0) \tag{49b}$$

$$C = (K[ds]_0)^{-1} \tag{49c}$$

In these equations $[Mn(acac)_3]_0$ is the initial chelate concentration (i.e. the concentration of manganese in solution) and $[M]_0$, $[ds]_0$ represent concentrations of bulk monomer and additive, respectively.

Equation (48), with values of the parameters shown in Table 3.11, leads to the calculated curves in Figure 3.12; as we have seen, these correspond to the curves in Figures 3.10 and 3.11, and so are generally in reasonable agreement with the experimental findings. The derived kinetic parameters are presented in Table 3.11.

Table 3.11 Experimental values of A, B, C (equations (48) and (49));[80] 25 °C

	DAP			DMSO		
	10^7A, s^{-1}	10^3B	10^3C	10^7A, s^{-1}	10^3B	10^3C
Styrene	430	0	80	120	0	200
Methyl methacrylate	965	555	80	270	560	200
Acrylonitrile	—	—	—	970	878	200

Table 3.12 Kinetic parameters;[80] 25 °C

	DAP			DMSO		
	10^2K mol^{-1}l	10^6k_1 s^{-1}	10^7k_2 mol^{-1}l s^{-1}	10^2K mol^{-1}l	10^6k_1 s^{-1}	10^7k_2 mol^{-1}l s^{-1}
Styrene	106	43	0	36	12	0
Methyl methacrylate	106	43	57	36	12	16
Acrylonitrile	—	—	—	36	12	56

The order of the initiation reaction in [M] predicted by equations (47a), (47b) and (47c) lies between 0 and 1, depending on the relative contributions of equations (47b) and (47c); the overall polymerization would therefore be expected to have an order between 1 and 1·5. We may anticipate that the presence of the additive would lead to orders in [M] which are lower than those observed in an inert solvent with $v_a = 0$ (unless the contribution of equation (47b) were negligible). This appears to be borne out in practice.

An electron-donor would be expected partially to displace an acetyl-acetonate ligand from $Mn(acac)_3$, with production of a complex with a structure containing contributions from forms such as those typified in (XII).

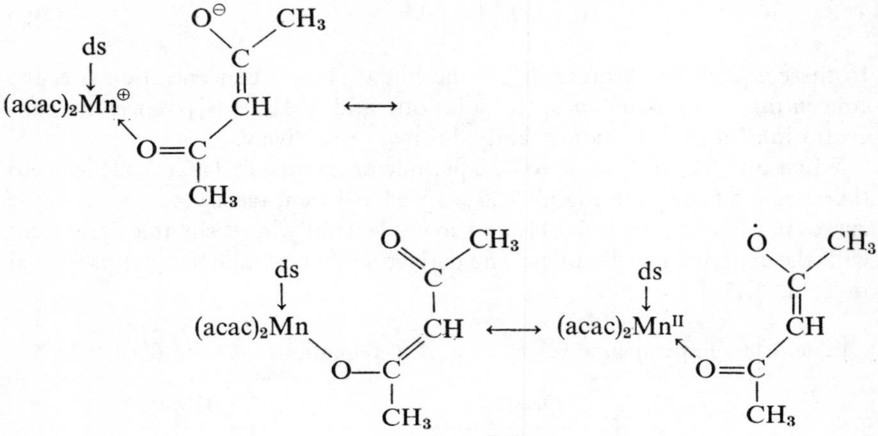

(XII)

If DMSO is the donor we may identify Z with (XII). A chelating additive such as DAP would tend to react further, to give more effective displacement of the acetylacetonate ligand, so that structures such as (XIII) would also contribute. In both cases spontaneous decomposition of Z follows electron

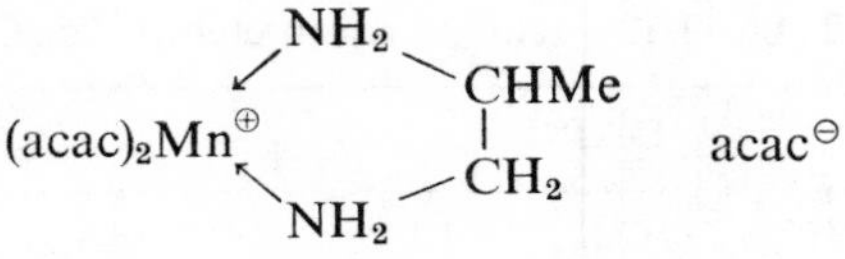

(XIII)

transfer from an acetylacetone residue to the metal atom, thus giving rise to a Mn^{II} derivative and an initiating $CH_3CO\dot{C}HCOCH_3$ radical. This process is the non-selective initiation in equation (47b). An essentially similar mechanism for the enhanced initiation brought about by ketones has been suggested by Kaeriyama.[84] This author believes that with aldehydes the intermediate Z contains two molecules of the additive, the metal atom having eight-fold co-ordination. The monomer-selective reaction in equation (47c) probably follows the path for $Mn(facac)_3$ which has already been described (equation (46)), with the minor difference that co-ordination of monomer in reaction (a) of equation (46) is replaced by more extensive co-ordination of ds. The resulting complex Z (XII) then interacts with a suitable monomer according to equation (50), which is closely analogous to reactions (b), (c) and (d) of equation (46). The final Mn^{II} product arising with a bidentate ligand is the corresponding Mn^{II} heterochelate; such derivatives have been known for many years.[89-91]

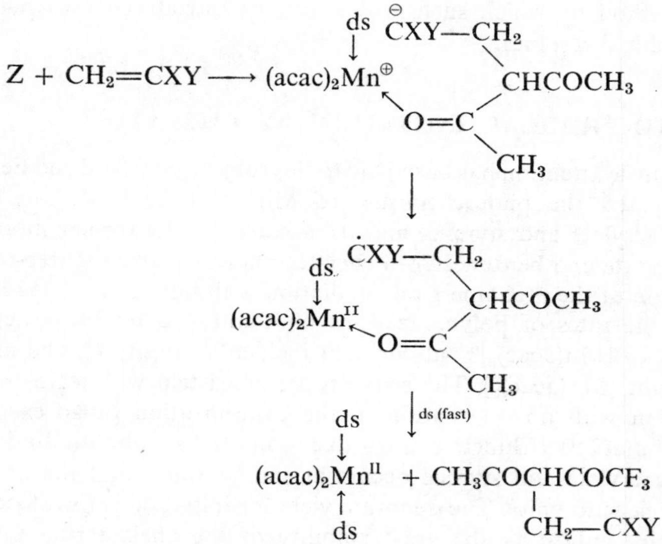

$$Z + CH_2{=}CXY \longrightarrow (acac)_2 \overset{ds}{\underset{}{\overset{\downarrow}{Mn}}}{}^{\oplus} \begin{array}{c} \overset{\ominus}{CXY}{-}CH_2 \\ \diagdown \\ \quad CHCOCH_3 \\ O{=}C \diagup \\ \diagdown \\ \quad CH_3 \end{array}$$

$$\downarrow$$

$$(acac)_2 \overset{ds}{\underset{}{\overset{\downarrow}{Mn}}}{}^{II} \begin{array}{c} \overset{\cdot}{CXY}{-}CH_2 \\ \diagdown \\ \quad CHCOCH_3 \\ O{=}C \diagup \\ \diagdown \\ \quad CH_3 \end{array} \qquad (50)$$

$$\overset{ds}{\underset{}{\overset{\downarrow}{}}} \Big\downarrow \text{ds (fast)}$$

$$\underset{\overset{\uparrow}{ds}}{(acac)_2 Mn^{II}} + \underset{\overset{|}{CH_2{-}\overset{\cdot}{C}XY}}{CH_3COCHCOCF_3}$$

The values of K in Table 3.12 reflect the co-ordinating abilities of the two ligands, DAP being more effective than DMSO. According to Table 3.9 1,3-diaminopropane is more active than DAP, and the activity of N,N,N′,N′-tetramethyldiaminoethane exceeds that of the unsubstituted diamine (on a molar basis). Both observations are probably attributable to differences in co-ordinating power. Comparison of the rate coefficients for the selective component (k_2) for a given additive (Table 3.12) shows that the order of reactivity of the monomers is

$$AN > MMA > St$$

in agreement with views already expressed. It is of interest that the increase in rate of initiation produced by DAP is much less for $Mn(facac)_3$ than for $Mn(acac)_3$ (Table 3.10). However, it must be remembered that the rate of initiation by the fluorochelate is already high in the absence of additive, on account of the potential dipolar structure, and further, the complex Z may be deactivated to some extent by electron-withdrawal by CF_3.

To summarize, we believe that monomer selectivity in initiation by chelates is connected with the existence of intermediates with markedly dipolar structures; these arise either as a result of electron attraction by a substituent in a ligand as with Mn(facac)$_3$, or by electron-donation to the metal by suitable co-ordination.

In conclusion, we may note that since the initiating radicals in all the chelate reactions discussed are derived from the ligands, these processes provide a method by which such groups can be introduced into polymer chains as terminal residues.

3.23 PHOTOCHEMICAL INITIATION BY CHELATES

Relatively little attention has been paid to this subject. Bamford and Ferrar[85] have investigated the photochemistry of Mn(acac)$_3$ and Mn(facac)$_3$ in methyl methacrylate and styrene, and in mixtures of the former monomer with ethyl acetate and benzene. Both chelates photosensitize the free-radical polymerization of the monomers on irradiation with light of $\lambda = 365$ nm.

At 25 °C the rates of polymerizations photoinitiated by Mn(acac)$_3$ are proportional to [Mn(acac)$_3$]$^{0.5}$ at constant incident intensity I_0, and also to $I_0^{0.5}$ at constant [Mn(acac)$_3$]. These results are consistent with a free-radical polymerization with no retardation in the concentration range examined $(< 2 \times 10^{-4}$ mol l^{-1}). Under comparable conditions, photoinitiation in methyl methacrylate and styrene occurs at similar rates, and the order in monomer is close to unity. The quantum yield for initiation is low, 8×10^{-3} approximately, indicating the great stability of the chelate ring towards photolysis. No fluorescence from dilute solutions in benzene or methanol has been observed.

Rates of chelate decay on irradiation have been measured spectrophotometrically. In benzene or ethyl acetate solution rates increase linearly with methyl methacrylate concentration, results for the two solvents being indistinguishable. The rates are identical in bulk methyl methacrylate and styrene and the quantum yield under these conditions is 2×10^{-2} approximately. It is clear therefore that not every molecule of chelate which is decomposed initiates polymerization; apparently excited molecules of Mn(acac)$_3$ may react with monomer to form inactive products.

Photolysis of Mn(acac)$_3$ in benzene and cyclohexane yields biphenyl and bicyclohexyl, respectively, together with the ligand acetylacetone. These results support the view that the primary products are Mn^{II}(acac)$_2$ and acetylacetone radicals; the latter enter into hydrogen abstraction reactions with solvent, or initiate polymerization if a monomer is present.

Quantum yields for photoinitiation and photodecomposition are also low with Mn(facac)$_3$. However, the two quantum yields are equal -1.5×10^{-2} approximately—so that there is no significant interaction between monomer and Mn(facac)$_3^*$ leading to inactive products. It is of interest that no monomer selectivity is encountered, demonstrating that photo- and thermal initiation by Mn(facac)$_3$ proceed through different intermediates. As with Mn(acac)$_3$, the products of photolysis in benzene and cyclohexane show that the primary decomposition is scission of a ligand as a free-radical, with formation of the Mn^{II} chelate. The most unusual feature is the striking dependence of the rate of initiation on the nature of the solvent. Addition of 70 per cent (v/v) of ethyl acetate to methyl methacrylate leaves I practically unchanged, whereas the presence of 20 per cent (v/v) of benzene causes a reduction in I of 30 per cent. These phenomena have been interpreted in terms of exciplex formation between the excited chelate and methyl methacrylate or ethyl acetate; such complexes may subsequently revert to $Mn(facac)_3$ or decompose with radical generation. It is supposed that benzene does not form similar exciplexes, but only leads to deactivation of $Mn(facac)_3^*$.

The reader may consult the original paper[85] for a quantitative discussion of the reaction mechanism.

3.3 REFERENCES

1. R. K. Freydlina and A. B. Belyavskii, *Dokl. Akad. Nauk, SSSR*, **127**, 1027, (1959).
2. R. K. Freydlina and A. B. Belyavskii, *Izvest. Akad. Nauk, SSSR, Otdel. Khim. Nauk*, **1**, 177, (1961).
3. C. H. Bamford and C. A. Finch, *Proc. Roy. Soc.* A, **268**, 553, (1962).
4. W. Strohmeier and H. Grübel, *Z. Naturforsch*, **22B**, 98, 553, (1967).
4a. C. H. Bamford and C. A. Finch, *Z. Naturforsch*, **17B**, 500, (1962).
5. C. H. Bamford and W. R. Maltman, *Trans. Faraday Soc.*, **62**, 2823, (1966).
6. C. H. Bamford, R. Denyer and G. C. Eastmond, *Trans. Faraday Soc.*, **61**, 1459, (1965).
7. C. H. Bamford, G. C. Eastmond and F. J. T. Fildes, (a) *Proc. Roy. Soc.*, A, **326**, 431, (1972); (b) *Proc. Roy. Soc.*, A, **326**, 453, (1972).
8. C. H. Bamford, R. Denyer and G. C. Eastmond, *Z. Naturforsch*, **22b**, 580, (1967).
9. C. H. Bamford and C. A. Finch, *Trans. Faraday Soc.*, **59**, 118, (1962).
10. C. H. Bamford, G. C. Eastmond and K. Hargreaves, *Trans. Faraday Soc.*, **64**, 1611, (1967).
11. K. Hargreaves, Thesis, University of Liverpool, 1966.
12. C. H. Bamford and R. Denyer, *Trans. Faraday Soc.*, **62**, 1567, (1966).
13. C. H. Bamford and C. A. Finch, *Trans. Faraday Soc.*, **59**, 540, (1963).
14. C. H. Bamford, G. C. Eastmond and W. R. Maltman, *Trans. Faraday Soc.*, **61**, 267, (1965).
15. C. H. Bamford, G. C. Eastmond and W. R. Maltman, *Trans. Faraday Soc.*, **60**, 1432, (1964).
16. C. H. Bamford, G. C. Eastmond and P. Murphy, *Trans. Faraday Soc.*, **66**, 2598, (1970).
17. C. H. Bamford and K. Hargreaves, *Trans. Faraday Soc.*, **63**, 392, (1967).
18. C. H. Bamford and E. O. Hughes, (a) *Proc. Roy. Soc.* A, **326**, 469, (1972); (b) *Proc. Roy. Soc.*, A, **326**, 489, (1972).
19. B. A. Casey, A. D. Jenkins, P. S. Lawry and M. G. Rayner, *Polym. Reprints*, **11.2**, 904, (1970).

19a. C. H. Bamford, G. C. Eastmond and K. Hargreaves, *Trans. Faraday Soc.*, **64**, 175, (1968).
20. C. H. Bamford, G. C. Eastmond and W. R. Maltman, *Trans. Faraday Soc.*, **62**, 2531, (1966).
21. C. H. Bamford and G. C. Eastmond, *Pure and Appl. Chem.*, **12**, 183, (1966).
21a. C. H. Bamford and I. Sakomoto, in course of publication, *J. Chem. Soc. Faraday Trans I.*
21b. C. H. Bamford and I. Sakomoto, in course of publication, *J. Chem. Soc. Faraday Trans. I.*
22. C. H. Bamford, *European Polymer J.—Supplement* **1969**, 1.
23. C. H. Bamford, F. J. Duncan, R. J. W. Reynolds and J. D. Seddon, *Polymer Sci.* C. No. **23**, 419, (1968).
24. C. H. Bamford and K. Hargreaves, *Proc. Roy. Soc.*, A, **297**, 425, (1967).
25. C. H. Bamford, G. C. Eastmond and V. J. Robinson, *Trans. Faraday Soc.*, **60**, 751, (1964).
26. C. H. Bamford, G. C. Eastmond and D. Whittle, *Polymer*, **10**, 771, (1969).
27. E. Yoshisato and S. Tsutumi, *J. Amer. Chem. Soc.*, **90**, 4488, (1968).
28. Go Hata, unpublished observations.
29. C. H. Bamford, *J. Polymer Sci.* C, No. **4**, 1571, (1966).
30. C. H. Bamford, R. Denyer and G. C. Eastmond, *Trans. Faraday Soc.*, **62**, 688, (1966).
31. W. Strohmeier, *Angew. Chem. Internat. Edit.* **3**, 730, (1964).
32. L. I. B. Haines and A. J. Poë, *Nature*, **215**, 699, (1967).
33. L. I. B. Haines, D. Hopgood and A. J. Poë, *J. Chem. Soc.*, **2A**, 421, (1968).
33a. C. H. Bamford and P. Denyer, *Nature*, **217**, 59, (1968).
34. H. Wawersik and F. Basolo, *Inorg. Chem. Acta.*, **3**, 113, (1969).
35. C. H. Bamford and P. Emery, 1970, unpublished observations.
36. C. H. Bamford, F. J. T. Fildes and W. R. Maltman, *Trans. Faraday Soc.*, **62**, 2544, (1966).
36a. C. H. Bamford, A. D. Jenkins and R. Johnston, *Trans. Faraday Soc.*, **58**, 1212, (1962).
37. F. Basolo and A. Wojcicki, *J. Amer. Chem. Soc.*, **83**, 520, (1961).
38. R. F. Heck, *J. Amer. Chem. Soc.*, **85**, 657, (1963).
39. L. R. Kangas, R. F. Heck, P. M. Henry, S. Breitschaft, E. M. Thorsteinson and F. Basolo, *J. Amer. Chem. Soc.*, **88**, 2334, (1966).
40. J. P. Day, F. Basolo and R. G. Pearson, *J. Amer. Chem. Soc.*, **90**, 6927, (1968).
41. F. Basolo, *Chem. in Britain*, **5**, 505, (1969).
42. C. H. Bamford, W. G. Barb, A. D. Jenkins and P. F. Onyon, *The Kinetics of Vinyl Polymerization by Radical Mechanisms* Butterworths, London, 1958.
43. K. R. Wilson and D. R. Herschbach, *Nature*, **208**, 182, (1965).
44. J. E. Lovelock and N. L. Gregory in *Gas Chromatography* eds. N. Brenner, J. E. Callen and M. D. Weiss, Academic Press, New York, 1962.
45. R. A. Gregg and F. R. Mayo, *J. Amer. Chem. Soc.*, **75**, 3530, (1953).
46. C. Walling, E. R. Briggs, K. B. Wolfstirn and F. R. Mayo, *J. Amer. Chem. Soc.*, **70**, 1537, (1948); also following papers. For a recent review see A. D. Jenkins this volume, Chapter 4.
47. M. Meier, F. Basolo and R. G. Pearson, *Inorg. Chem.*, **8**, 795, (1969).
48. E. Koerner von Gustorf and F. W. Grevels, *Fortschritte der Chemische Forschung*, **13**, 366, (1969).
49. W. Strohmeier, *Angew. Chem.*, **76**, 873, (1964).
50. W. Strohmeier and K. Gerlach, *Chem. Ber.*, **94**, 398, (1961).
51. G. R. Dobson, M. F. A. El Sayed, I. W. Stolz and R. K. Sheline, *Inorg. Chem.* **1**, 526, (1962).

52. G. R. Dobson, *J. Phys. Chem.*, **69**, 677, (1965).
53. E. Koerner von Gustorf, M. C. Henry and C. Di Pietro, *Z. Naturforsch*, **21B**, 42, (1966).
54. E. Koerner von Gustorf, M-J. Jun and G. O. Schenck, *Z. Naturforsch*, **18B**, 503, (1963).
55. C. H. Bamford, G. C. Eastmond and F. J. T. Fildes, *Chem. Comm.* **1970**, 144.
56. W. Strohmeier and P. Hartmann, *Z. Naturforsch*, 1964, **19B**, 882.
57. C. H. Bamford, J. Hobbs and R. P. Wayne, *Chem. Comm.* **1965**, 469.
58. C. H. Bamford, P. A. Crowe and R. P. Wayne, *Proc. Roy. Soc.* A, **284**, 455, (1965).
59. J. C. Kwok, Thesis, University of Liverpool, 1971.
60. C. H. Bamford, P. A. Crowe, J. Hobbs and R. P. Wayne, *Proc. Roy. Soc.*, A, **292**, 153, (1966).
61. W. Strohmeier and H. Grübel, *Z. Naturforsch*, **22B**, 115, (1967).
62. C. H. Bamford, J. Bingham and H. Block, *Trans. Faraday Soc.*, **66**, 2612, (1970).
63. C. H. Bamford, R. W. Dyson and G. C. Eastmond, *Polymer*, **10**, 885, (1969).
63a. C. H. Bamford, *Pure Appl. Chem.*, **34**, 173, (1973).
64. C. H. Bamford and J. Paprotny, *Chem. Comm.*, **1971**, 140.
65. C. H. Bamford and J. Paprotny, *Polymer*, **13**, 208, (1972).
66. A. G. Osborne and M. H. B. Stiddard, *J. Chem. Soc.*, **1964**, 634.
67. M. L. Ziegler, H. Haas and R. K. Sheline, *Chem. Ber.*, **98**, 2454, (1965).
68. C. H. Bamford, R. W. Dyson and G. C. Eastmond, *J. Polymer Sci.* C, No. **16**, 2425, (1967).
69. C. H. Bamford, R. W. Dyson, C. G. Eastmond and D. Whittle, *Polymer*, **10**, 759, (1969).
70. C. H. Bamford, G. C. Eastmond and D. Whittle, *Polymer*, **10**, 771, (1969).
71. C. H. Bamford, G. C. Eastmond and D. Whittle, *Polymer*, **12**, 247, (1971).
72. J. Ashworth, C. H. Bamford and E. G. Smith, *Polymer*, **13**, 57, (1972).
72a. C. H. Bamford and G. C. Eastmond, *Polym. Preprints Amer. Chem. Soc.*, 14.2, 1973
73. J. Ashworth, C. H. Bamford and E. G. Smith, *Pure Appl. Chem.* **30**, 25, (1972).
74. Y. Nishikawa and T. Otsu, *Makromolek. Chem.*, **128**, 276, (1969).
75. Y. Nishikawa, T. Otsu and S. Watanuma, *Makromolek. Chem.*, **115**, 278, (1968).
76. E. M. Arnett and M. A. Mendelsohn, *J. Amer. Chem. Soc.*, **84**, 3821, 3824, (1962).
77. E. G. Kastning, H. Naarmann, H. Reis and C. Berding, *Angew. Chem. Internat. Ed.*, **4**, 322, (1965).
78. C. H. Bamford and D. J. Lind, *Proc. Roy. Soc.*, A, **302**, 145, (1968).
79. Y. Nishikawa and T. Otsu, *J. Chem. Soc. Japan*, *Ind. Chem. Sect.*, **72**, 1836, (1969).
80. C. H. Bamford and A. N. Ferrar, *Proc. Roy. Soc.* A, **321**, 425, (1971).
81. D. J. Lind, Thesis, University of Liverpool, 1967.
82. Yu. V. Korshak, T. I. Bevza and B. A. Dolgoplosk, *Vysokomolek, Soedin*, **B11**, 794, (1969).
83. K. Uehara, Y. Kataoka, M. Tanaka and N. Murata, *J. Chem. Soc. Japan*, *Ind. Chem. Sect.* **72**, 754, (1969).
84. K. Kaeriyama, *Bull. Res. Inst. Polymers and Textiles*, Yokohama, No. **92**, 1, (1970).
85. C. H. Bamford and A. N. Ferrar, *J. Chem. Soc. Faraday Trans. I*, **68**, 1243, (1972).

86. A. M. North and G. A. Reed, *Trans. Faraday Soc.*, **57**, 853, (1961).
87. C. H. Bamford and S. Brumby, *Makromolek. Chem.*, **105**, 122, (1967).
88. C. H. Bamford and S. Brumby, *Chem. and Ind.*, **1969**, 1020.
89. F. P. Dwyer and A. M. Sargeson, *J. Proc. R. Soc. N.S.W.*, **90**, 29, (1956).
90. A. Syamal, *J. Proc. Inst. Chem. India*, **40**, 131, (1968).
91. A. K. Das and D. V. R. Rao, *Curr. Sci.*, **39**, 60, (1970).

4

The Reactivity of Polymer Radicals in Propagation and Transfer Reactions

A. D. Jenkins

School of Molecular Sciences, University of Sussex

In the previous chapter the manner in which radicals react with double bonds to yield simple addition products has been surveyed, and our present purpose is to complement this discussion with a review of the closely related chain reaction responsible for the production of polymers from vinyl and vinylidene monomers by radical mechanisms. It may be as well at this point to emphasize that there is no fundamental reason for divorcing a discussion of the reactivity of polymer radicals from that of their small counterparts because the sheer size of the former species has no special bearing on their participation in addition processes, except where the physical state of the system is affected. However, it is generally much simpler to obtain data for polymer radicals and in the present context interest must be restricted to polymer-producing reactions.

Of course, the addition process is the essential reaction which produces polymer molecules and it consists simply of the repetitive addition of a radical to double bonds; in any actual polymerization study it is usually necessary to pay adequate regard also to transfer reactions which, as will be seen below, can be treated in reactivity terms in just the same way as propagation steps, and we shall keep both types of reaction in mind in this review. Termination processes are also competitive with propagation but the chemical activation energy for these reactions between two radicals is so low that the rate-determining step is usually a diffusion process; for this reason termination will be omitted from further discussion here, rather it forms the subject of Chapter 5.

4.1 DEFINITION OF TERMS

4.1.1 Propagation

Let us consider the most simple reaction scheme for the radical-propagated polymerization of a single monomer.

$$\text{Initiation} \qquad\qquad \rightarrow R_{\dot{c}} \qquad \text{Rate of initiation} = I$$
$$R_{\dot{c}} + M \rightarrow R_1 \qquad \text{Rate constant } k_1$$
$$\text{Propagation } R_{\dot{r}} + M \rightarrow R_{\dot{r}+1} \qquad\qquad k_p$$
$$\text{Termination } R_{\dot{r}} + R_{\dot{s}} \rightarrow P_{r+s} \qquad\qquad k_t$$
$$\text{or} \qquad \rightarrow P_r + P_s \qquad\qquad k_t'$$

According to this scheme, radicals $R_{\dot{c}}$ are generated by some unspecified means, frequently the decomposition of a radical-generating substance (such as azo-*bis*-isobutyronitrile or benzoyl peroxide), thermally or photolytically, but which may be direct thermal, photochemical or radiation-induced homolysis of the monomer. If the radicals arise directly from the monomer we represent them by the symbol R_1 which denotes that one monomer unit has been incorporated, and which occurs in the second equation in the reaction scheme for the general case where the primary radical $R_{\dot{c}}$ is derived from some added substance.

The propagation reaction consists of the enlargement of the size of the radical by addition of one monomer unit, and the termination reaction involves a bimolecular process in which the radicals may either combine or undergo disproportionation.

Application of the stationary state hypothesis to this reaction scheme gives the two following expressions for the rate of polymerization ω and the number average degree of polymerization $\bar{P}$.

$$\omega = k_p[M]\left[\frac{I}{(k_t + k_t')}\right]^{\frac{1}{2}} \tag{1}$$

$$\bar{P} = k_p[M]\{I(k_t/2 + k_t')\}^{-\frac{1}{2}} \tag{2}$$

From equation (1) it follows that, at a given rate of initiation and a given monomer concentration, the rate of polymerization of a single monomer will be directly proportional to the ratio $k_p/(k_t + k_t')^{\frac{1}{2}}$, a quantity which might be regarded as a measure of reactivity. Although for practical purposes the value of $k_p/(k_t + k_t')^{\frac{1}{2}}$ is useful, too many factors are involved in the various individual rate constants for the compound quantity to have any informative relationship to radical structure and it is this dependence of reactivity on structure which will be regarded in this article as the most important aspect of reactivity.

It thus becomes desirable to seek either a determination of individual velocity constants or some meaningful ratios of velocity constants. Methods are available for the measurement of absolute velocity constants but they are by no means easy to apply; comparatively few determinations of this kind have been made. Fortunately, it is not necessary to have many such data because the vital information regarding reactivity is available in the form of ratios of velocity constants from studies of copolymerization and chain transfer.

Much useful reactivity information is derived from copolymerization studies for which the following basic scheme applies. Consider a reaction mixture containing two monomers A and B so that the growing polymer radicals will have either terminal A or terminal B groups; such radicals are denoted by A· and B· respectively. It can be assumed to a good approximation that (a) the reactivity of the radical is independent of its size, and (b) that reactivity is determined solely by the terminal monomer unit, i.e. radicals ending in ⌁A—A· and ⌁B—A· are assumed to be of equal reactivity in propagation. (See p. 123.)

Four propagation reactions are possible:

$$
\left.\begin{array}{c}
\text{⌁A· + A} \rightarrow \text{⌁A·} \quad k_{p_{AA}} \\
\text{⌁A· + B} \rightarrow \text{⌁B·} \quad k_{p_{AB}} \\
\text{⌁B· + A} \rightarrow \text{⌁A·} \quad k_{p_{BA}} \\
\text{⌁B· + B} \rightarrow \text{⌁B·} \quad k_{p_{BB}}
\end{array}\right\} \tag{3}
$$

For a low conversion (i.e. concentrations of all reactants are assumed to be constant) it follows that the ratio n of the concentrations of A and B units in the polymer is given by equation (4), and by application of the stationary state treatment one obtains the following most important relationship, known as the 'copolymer composition equation' equation (5).

$$
n = \left(\frac{-d[A]}{dt}\right) \Big/ \left(\frac{-d/[B]}{dt}\right) \tag{4}
$$

$$
n = \frac{[A]}{[B]} \left\{\frac{r_A[A] + [B]}{r_B[B] + [A]}\right\} \tag{5}
$$

where

$$
r_A = \frac{k_{p_{AA}}}{k_{p_{AB}}} \quad \text{and} \quad r_B = \frac{k_{p_{BB}}}{k_{p_{BA}}} \tag{6}
$$

In principle, r_A and r_B, the 'monomer reactivity ratios', may be evaluated by analysis of two copolymers prepared from monomer mixtures with different [A]/[B] ratios. In practice, such little information rarely suffices but, even so, a large number of monomer reactivity ratios have been determined. For a list of such data the reader is referred to Young's extensive compilation.[1] Reference to the significance of the symbols employed above shows that r_A expresses directly the relative speeds with which the radical A· reacts with monomers A and B. Evidently, if the relative reactivity of any radical towards the two given olefins A and B were independent of its structure, one would always find that $r_A = r_B^{-1}$ or $r_A r_B = 1$. This criterion defines an 'ideal' copolymerization;[2] ideal copolymerization systems are of academic interest only, since in practice the criterion $r_A r_B = 1$ is rarely satisfied. Broadly, two types of behaviour are encountered:

(i) r_A is greater than unity and r_B is small, tending perhaps to zero. This would be expected to occur if monomer A had a much greater 'inherent reactivity' than monomer B so that both types of propagating radical add preferentially to A.

(ii) r_A and r_B are both less than unity, corresponding to a tendency for each radical to react preferentially with the foreign monomer. This effect has been recognized for many years[3] and is called the 'alternating tendency' since it corresponds to a situation where the monomers tend to enter the growing chain alternately.

4.1.2 Transfer

In the polymerization scheme above radicals are depicted as participating in addition and termination reactions only but, as is well known, they also function by abstraction of atoms (or possibly groups) in metathetical processes such as equation (7). Where the process represented by equation (7) interrupts a chain reaction it is known as a 'chain transfer' step; in a polymerization specifically, it results in the cessation of growth of a particular polymer molecule but may not have any influence on the total rate of reaction, although this depends upon the relative reactivities of R· and S· radicals in addition and termination reactions.

$$\text{R·}_r + \text{S} \rightarrow \text{P}_r + \text{S·} \qquad k_{tr} \tag{7}$$

$$\frac{1}{\bar{P}} = \sum \frac{k_{tr}[\text{S}]}{k_p[\text{M}]} + \frac{\{I(k_t/2 + k_t')\}^{\frac{1}{2}}}{k_p[\text{M}]} \tag{8}$$

Where transfer occurs, the number average degree of polymerization of the polymer produced is given by equation (8) in which the ratio k_{tr}/k_p is the chain transfer constant, usually denoted by C_s. C_s is usually determined by measuring the number average degree of polymerization of polymers prepared at a variety of concentrations of S, although care may be necessary in the adjustment of other variables due to the interdependence of the concentrations of species present. (The summation sign is included in the transfer term to take account of the possibility that transfer may take place with a number of species present, including monomer and polymer.)

C_s is clearly analogous to the monomer reactivity ratio in copolymerization in furnishing a measure of the relative ease of reaction for a radical with a given substrate (relative to its rate of propagation) and in providing a basis for the absolute comparison of reactivities where individual propagation rate constants have been determined.

If the radical S· resulting from the act of chain transfer has a small probability of entering into an addition (i.e. re-initiation) reaction, then there will be a decrease in rate of polymerization, and 'retardation' will be observed. Under extreme conditions where every kinetic chain is interrupted by transfer and the transfer radicals all enter into termination the reaction will be inhibited, that is, it will be unobservably slow. In both cases suitable kinetic observations facilitate the evaluation of C_s. Important atom transfer reactions which are illuminating in the interpretation of radical reactivity include:

Hydrocarbons	$R\cdot + R'H \rightarrow RH + R'\cdot$
Halides	$R\cdot + CBr_4 \rightarrow RBr + \cdot CBr_3$
Mercaptans	$R\cdot + C_4H_9SH \rightarrow RH + C_4H_9S\cdot$
Amines	$R\cdot + (C_2H_5)_3N \rightarrow RH + (C_2H_5)_2N\dot{C}HCH_3$

Polymer radicals also participate in 'electron transfer' reactions with salts of transition metals, for example, ferric chloride, e.g.

$$R\cdot + FeCl_3 \rightarrow RCl + FeCl_2 \quad k_5$$

This reaction, formally involving electron transfer, is actually a termination process since no possible chain carrier is produced.

4.1.3 General

Taking propagation and the two types of transfer reaction together, we have a collection of processes in which a given radical may react with a whole variety of centres, including carbon–carbon double bonds, carbon–hydrogen, carbon–halogen, sulphur–hydrogen and a number of other bonds, as well as with ions or highly polar molecules. The immediate problem is to find some rationalization of radical behaviour over this extensive range in terms of the structures of the radicals and the substrates. Perhaps the most concise statement of the problem is to say that we wish to account for the relative magnitudes of the specific velocity constants for reactions such as the three following in relation to measurable properties of the substituent group X in the radical and the structures of the substrate molecules.

$$
\text{wCH}_2-\dot{\text{C}}\text{H} \left\{
\begin{array}{l}
+CH_2{=}CH \xrightarrow{k_p} \text{wCH}_2-CH-CH_2-\dot{C}H \\
\qquad\quad | \qquad\qquad\qquad | \qquad\qquad | \\
\qquad\quad Y \qquad\qquad\qquad X \qquad\qquad Y \\[2mm]
\qquad\qquad\qquad\qquad\qquad\qquad Cl \\
\qquad\qquad\qquad\qquad\qquad\qquad | \\
+CCl_4 \xrightarrow{k_{tr}} \text{wCH}_2-CH + CCl_3 \\
\qquad\qquad\qquad\qquad\qquad\qquad | \\
\qquad\qquad\qquad\qquad\qquad\qquad X \\
\qquad\qquad\qquad\qquad\qquad\qquad Cl \\
\qquad\qquad\qquad\qquad\qquad\qquad | \\
+FeCl_3 \xrightarrow{k_5} \text{wCH}-CH + FeCl_2 \\
\qquad\qquad\qquad\qquad\qquad\qquad | \\
\qquad\qquad\qquad\qquad\qquad\qquad X
\end{array}
\right.
$$

4.2 THE INHERENT REACTIVITY OF RADICALS

At one time it was believed that one could assess the reactivity of radicals, and the converse order for monomers, by considering the change in stabilization energy for the following process:

$$R\cdot + CH_2{=}\underset{X}{CH} \longrightarrow R{-}CH_2{-}\underset{X}{\dot{C}H}$$

According to this view, the most reactive monomer will yield the least reactive radical, and so on, because of their structural relationship. To some extent this view is supported by data which generally indicate that vinyl acetate is one of the least reactive of monomers while yielding one of the most reactive radicals, precisely the reverse being the case for styrene. Unfortunately, for substances of intermediate reactivity the situation is far less clear cut and it is here that one is forced to conclude that different factors prove dominant in determining reactivity in different reactions. With the more strongly polar transfer agents, such as CBr_4 or $FeCl_3$, the intrusion of polar factors is particularly strong and the concept of a unique order of reactivity cannot possibly be maintained.[4]

A conclusive demonstration of the failure of the unique order of reactivity concept is afforded by an examination of the relative reactivity of polystyrene (S$\cdot$) and polyacrylonitrile (AN$\cdot$) radicals towards a variety of substrates across the polarity spectrum. The data in Table 4.1 indicate that, judged by the

Table 4.1 Relative rate constants for
polystyrene and polyacrylonitrile radicals

Substrate	Ratio of rate constants $\dfrac{S\cdot}{AN\cdot}$
$FeCl_3$	100
$CH_2{=}CH\cdot CN$	2
$CH_2{=}CH\cdot Cl$	0·05
$CH_2{=}CH\cdot\phi$	0·002
$N(C_2H_5)_3$	0·0002

standard of transfer with ferric chloride, the S$\cdot$ radical is one hundred times as reactive as the AN$\cdot$, although transfer with triethylamine leads to the conclusion that the AN$\cdot$ radical is five thousand times as reactive as the S$\cdot$. In fact, one can obtain almost any apparent ratio of reactivities by suitable substrate selection so that bald statements that one radical is more reactive than another are seen to be without meaning.

4.3 THE ALTERNATING TENDENCY

In copolymerization studies the breakdown of the concept of the unique order of inherent reactivity was attributed to the influence of an 'alternating tendency' which became stronger as the polar properties of the two monomers involved became more diverse.[3] This was an early recognition of the importance of polarity in radical processes and one that has been refined in the more precise schemes discussed in more detail below.

4.4 STERIC EFFECTS

Undoubtedly steric effects play an important role in some radical polymerization processes but this is rather in a negative sense in that 1,2-disubstituted monomers are usually almost unable to homopolymerize and may copolymerize without enthusiasm. (This is not true if the substituent is fluorine.) Where some tendency to polymerize is displayed by a 1,2-disubstituted monomer, the *cis-* and *trans*-isomers often differ considerably in ease of reaction, an effect which can hardly be electronic and must therefore be attributed to steric influences.[5]

If discussion is restricted to monomers substituted on only one side of the double bond, the assumption is normally made that steric hindrance is negligible; this assumption is important in all treatments of reactivity which attribute differences in reactivity to activation energy differentials and it is the success of such treatments which justifies the neglect of steric factors.

4.5 PENULTIMATE UNIT EFFECTS

The analysis of copolymerization data sometimes shows a trend in the values of monomer reactivity ratios with the variations in the monomer composition in the working mixtures.[6-9] One interpretation of this effect is that the reactivity of a radical is modified by the nature of the penultimate unit: with monomers A and B there will be a relatively higher concentration of ~AA· radicals compared with ~BA· radicals at high concentrations of A and if the penultimate unit indeed can influence the reactivity of the terminal unpaired electron, the monomer reactivity ratio will vary with [A]/[B]. Some examples of this type of trend are shown in Table 4.2.

There is also some evidence for penultimate unit effects in transfer reactions but, in general, significant dependence of reactivity ratios on co-monomer composition only arises when the polarity differential between the two monomers is large.[10]

Ham[11] has presented a generalized treatment of penultimate and other progressively more remote units where the influence of as many as four

Table 4.2 Effects of penultimate unit on radical reactivity

		Reactivity ratio for radical with terminal structure	
Monomer A	Monomer B	AA·	BA·
Styrene	Maleic anhydride	0·017	0·063
Styrene	Citraconic anhydride	0·07	0·25
Styrene	Fumaronitrile	0·07	1·0
Styrene	Acrylonitrile	0·30	0·45
α-Methylstyrene	Acrylonitrile	0·055	0·093
Chloroprene	Vinylidene cyanide	0	0·057
Methyl methacrylate	Vinylidene cyanide	0·04	0·08
Styrene	Oxygen	4.4×10^{-8}	$3.3 \times 10^{-7} - 1.8 \times 10^{-6}$

units back from the active end can be taken into account, although it is very hard to see how such effects could operate.

4.6 COMPLEXES OF RADICALS WITH SALTS

Kinetic studies have afforded clear evidence that polymerizations carried out in the presence of certain salts, notably lithium chloride and zinc chloride, are affected by complex formation between the propagating radicals and the salt in such a way that the balance of reactivity is altered.

With acrylonitrile in N,N-dimethylformamide solution there is evidence that complex formation takes place with both the lithium and chloride ions, presumably to give structures such as

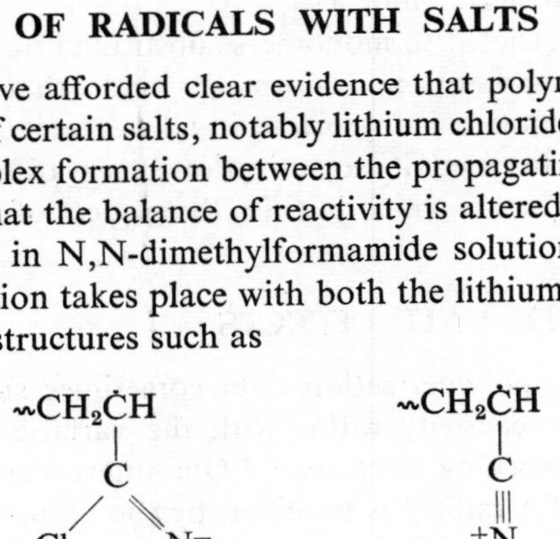

$$\sim CH_2\dot{C}H \qquad \sim CH_2\dot{C}H$$

but the concentration of complexed radicals is very low.[12] This leads to the conclusion that no reduction of activity of complexed radicals would be observable but increases are reflected in changes in both rate of reaction and degree of polymerization.

Anticipating discussion below, it is interesting to note that the velocity constants for transfer with carbon tetrabromide and triethylamine are both increased in the presence of lithium chloride, although the polar effects are expected to operate in opposite senses. Presumably the explanation is that in the former case complexing with chloride ion raises the reactivity whereas in the latter case the lithium ion complex is especially reactive.

In copolymerization the presence of the salt may lead to striking effects, for example, in the copolymerization of acrylonitrile with styrene in the presence of zinc chloride an equimolar copolymer with a regular alternation of the monomer residues is produced spontaneously, irrespective of the ratio of the monomer concentration in the feed.[13] The presence of a radical-generating catalyst accelerates the reaction without affecting the polymer structure.

A deeper examination of this system, including a fractionation of the product into soluble and insoluble components, has suggested a possible re-interpretation of the normal radical copolymerization of these monomers which takes place in the absence of the salt. This new concept is that the reaction is composed of two entirely independent mechanisms, one being a random copolymerization in which the polymer has the same composition as the feed, all monomers being equally reactive, while the other consists of the homopolymerization of a 1:1 charge-transfer complex of the two monomers. If this suggestion proves to have validity, all copolymerizations involving monomers capable of forming a charge-transfer intermediate will require new scrutiny.

4.7 INTERPRETATION OF RADICAL REACTIVITY DATA

4.7.1 The Q-e scheme

In 1945 or thereabouts the generally accepted view of polymer radical reactivity was that the inherent order of radical reactivities (the inverse of the order of stabilization energies) was perturbed to a rather minor extent by the influence of the alternating tendency. Price[14] suggested that the latter could be treated as a polar effect resulting from electrostatic interactions between permanent charges on the reactants, an idea which was shortly developed by Alfrey and Price into a quantitative treatment.[15]

The basic assumption is that two parameters characterize each participant, radical or monomer, according to (i) its 'general reactivity', and (ii) its polarity. For radicals and monomers the former are represented by P and Q, respectively, and the latter by e, the e value for a parent monomer and daughter radical being assumed to be identical. The velocity constant for reaction between a radical X and a monomer Y is then given by the equation

$$k_{XY} = P_X Q_Y e^{-e_X e_Y} \tag{9}$$

and, if we work in terms of the monomer reactivity ratios

$$r_X = \frac{k_{XX}}{k_{XY}}; \quad r_Y = \frac{k_{YY}}{k_{YX}}$$

we have

$$
\left.\begin{aligned}
r_X &= (Q_X/Q_Y)e^{-e_X(e_X-e_Y)} \\
r_Y &= (Q_Y/Q_X)e^{-e_Y(e_Y-e_X)} \\
r_X r_Y &= e^{-(e_X-e_Y)^2}
\end{aligned}\right\} \tag{10}
$$

Copolymerization data (r_X and r_Y) furnish values for ratios of Q factors and differences between e factors so that assignment of individual values of Q and e parameters requires values to be attributed to one of each on an arbitrary basis. The authors of the scheme selected styrene as the reference monomer with $Q = 1.0$ and $e = -0.8$: there have been several attempts to change the latter to $e = 0$ but the original Q-e scheme has become so firmly entrenched that it does not seem possible either to modify it or to replace it.

Looking at the assumptions in the Q-e scheme at this time one sees clearly how simple the approach was and how difficult it would be to justify. Thus the concept of permanent charges on the reactants is hard to sustain especially as transient polarization effects in the transition state are in any case likely to dominate the kinetic behaviour. Further, the permanent charge view leads to a dependence of rate of reaction on the dielectric constant of the medium, a dependence which is simply not observed, the reactivity ratios generally being quite insensitive to the dielectric parameter.

A recital of the theoretical weaknesses of the Q-e scheme, however, must be seen in relation to the impressive success that the treatment enjoys and, taken as an empirical exercise, it must be concluded that it constituted a most important advance in the rationalization of the available polymerization data.

Table 4.3 Q-e parameters

Monomer	e	Q
p-Methoxystyrene	−1·11	1·36
Butadiene	−1·05	2·39
Styrene	(−0·80)[a]	(1·00)[a]
2-Vinylpyridine	−0·50	1·30
Vinyl acetate	−0·22	0·026
Vinyl chloride	0·20	0·044
Methyl methacrylate	0·40	0·74
Methyl acrylate	0·60	0·42
Methacrylic acid	0·65	2·34
Acrylic acid	0·77	1·15
Methacrylonitrile	0·81	1·12
Vinyl trifluoroacetate	1·06	0·50
Acrylonitrile	1·20	0·60
Vinyl fluoride	1·28	0·012
Maleic anhydride	2·25	0·23
Vinylidene cyanide	2·58	20·13

a. Arbitrary reference values.

Standard text-books of polymer science sometimes contain tables of Q-e values, but special reference should be made to the extensive compilation, published in 1961, by Young.[1] A selection of values is reproduced in Table 4.3.

In the foregoing the selection of arbitrary base points was mentioned; undoubtedly it is pushing the scheme far beyond its author's intention to look for a fundamental quantitative significance for the Q and e parameters but there has been a desire to correlate such values with other measurable characteristics of compounds which has prompted several examinations of the 'best' reference values. Kawabata, Tsuruta and Furukawa[16] considered the consequences of altering the e value for styrene to zero and concluded that the revised list was superior. It should be pointed out that this change in the reference e value results in modifications to both e and Q parameters and one consequence of the change was that the revised Q values correlated well with the methyl affinities of the monomers which tends to confirm that they are representative of general reactivities. The corresponding P values for radicals were also more satisfactory since they maintained constant values for a whole series of copolymerization reactions with a large variety of co-monomers.

On the other hand the new e values seemed individually much less plausible, a result which led to the postulate that the combined $e_X e_Y$ product is to be regarded as having physical significance and not its components. The zero point for the Q-e scheme will be touched upon again in a subsequent section (p. 139).

With regard to the assumption that conjugate monomer and radical should have the same e value, it must be said that there is no obvious justification. Following an analysis of data for copolymerizations involving dienes which do not fit the Q-e scheme at all satisfactorily, Wall[17] modified the treatment to a Q-e-e^* scheme so that the monomers and radicals could have different polar parameters but, although this naturally permits a much better fit between theory and observation, it incurs the need for the arbitrary assignment of an e^* reference value and this tends to undermine the utility of the modification.

4.7.2 The patterns of reactivity approach

In an attempt to avoid the criticisms reasonably levelled at the basis of the Q-e scheme while retaining the essential approach, Bamford, Jenkins and Johnston[18-20] made a new interpretation of radical reactivity data in which no need arises for arbitrary reference parameters, rather *experimental* assessments of both general reactivity and polar character are employed in the basic equation.

To begin with the assumption is made that one can find a chemical reaction so little influenced by the polar character of substituents in the reactants that its rate is essentially dependent on general reactivity and constitutes a measure of that quantity. The association chain implied is general reactivity, delocalization, heat of reaction, activation energy, velocity constant and the reaction chosen is that of chain transfer with toluene

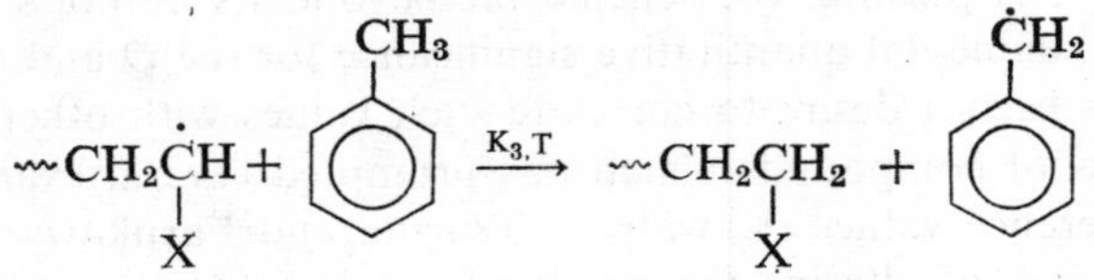

$$\sim CH_2\dot{C}H + \bigcirc^{CH_3} \xrightarrow{K_{3,T}} \sim CH_2CH_2 + \bigcirc^{\dot{C}H_2}$$

The choice of this reaction can be defended and the decadic logarithm of its velocity constant is used as a measure of general reactivity and its dependence on X since the nature of this group is the only variable in the reaction. For six of the most important polymer radicals the values of $k_{3,T}$ are collated in Table 4.4.

Table 4.4 Data for standard radicals at 60 °C[15]

Radical (derived from)	k_p $(l\,m^{-1}\,s^{-1})$	$k_{3,T}$ $(l\,m^{-1}\,s^{-1})$	σ
Styrene	176	$2 \cdot 11 \times 10^{-3}$	$-0 \cdot 01$
Methyl methacrylate	734	$1 \cdot 25 \times 10^{-2}$	$0 \cdot 28$
Methacrylonitrile	201	$2 \cdot 00 \times 10^{-2}$	$0 \cdot 49$
Acrylonitrile	2458	$0 \cdot 785$	$0 \cdot 66$
Methyl acrylate	2090	$0 \cdot 56$	$0 \cdot 45$
Vinyl acetate	3700	$7 \cdot 73$	$0 \cdot 31$

The question of the possible uniqueness of toluene as a reference compound is answered by consideration of similar data for a number of hydrocarbons. If all are equally 'non-polar' then for reagent B (say, benzene) there should be a linear correlation between $\log k_{3,B}$ and $\log k_{3,T}$ for a series of radicals such as the six in Table 4.4. That such a linear relationship does exist is shown by Figure 4.1; not only does the expected proportionality exist but the values of the parameter γ, defined by the equation

$$k_{3,B} = \gamma k_{3,T} \tag{11}$$

fall in the order expected on the basis of general beliefs regarding the relative reactivities of primary, secondary and tertiary aliphatic and aromatic hydrogen atoms, as may be seen from Table 4.5.

Since the Q-e scheme was applied primarily to propagation reactions (there was some development into transfer) while the basis of the Patterns treatment rests upon a transfer process up to the present point, it is of immediate relevance to attempt a correlation of propagation data with transfer to toluene. In order to keep within the hydrocarbon field, styrene

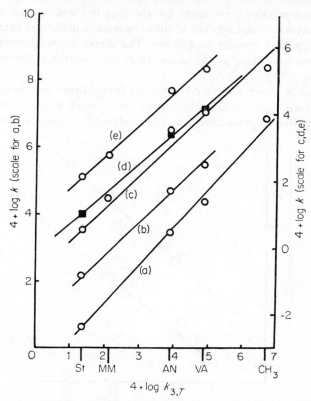

Figure 4.1 Double logarithmic plot of the velocity constants for chain transfer with various hydrocarbons as functions of the velocity constants for chain transfer with toluene. All reactions at 60 °C. (a) cyclohexane; (b) isopropylbenzene; (c) benzene; (d) t-butylbenzene; (e) ethylbenzene. Abbreviations for radicals as in Figure 4.2

Table 4.5 Transfer reactivity parameters

Transfer agent	γ	No. available H atoms			
		Aromatic	Aliphatic		
			1°	2°	3°
Benzene	0·15	6	0	0	0
Toluene	1·00	5	3	0	0
Ethylbenzene	4·20	5	3	2	0
Isopropylbenzene	5·60	5	6	0	1
t-Butylbenzene	0·24	5	9	0	0
Cyclohexane	0·42	0	0	12	0

is the obvious monomer to select for the first test and indeed Figure 4.2 shows that velocity constants for addition to styrene monomer correlate very well with those for transfer to toluene. The direct proportionality persists for these data and also, as the same figure demonstrates, for addition to vinyl acetate.

Moving on to make similar comparisons for addition reactions involving monomers containing polar substituents or transfer reactions with compounds with obvious polar character, one finds a progressive tendency for the

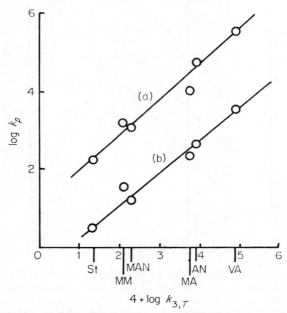

Figure 4.2 Double logarithmic plots of the velocity constants for addition or transfer reactions with the substrate indicated as functions of the velocity constants for chain transfer with toluene. All reactions at 60 °C. (a) Styrene; (b) vinyl acetate. Abbreviations for radicals: S, styrene; MM, methyl methacrylate; MAN, methacrylonitrile; MA, methyl acrylate; AN, acrylonitrile; VA, vinyl acetate

linear correlation to weaken and disappear as the substrate becomes increasingly polar. This trend is clearly shown by Figures 4.3–4.5 with data for methyl methacrylate, n-butyl mercaptan and ferric chloride. Towards the extreme end of this series polar factors have come to dominate and submerge the general reactivity factor but the most significant fact to emerge from this comparison is that the same pattern of points, qualitatively, is observed in all cases.

The problem of quantifying the polar influence in terms of an independent experimentally accessible parameter was solved in the following manner.

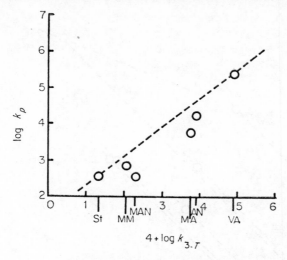

Figure 4.3   Methyl methacrylate

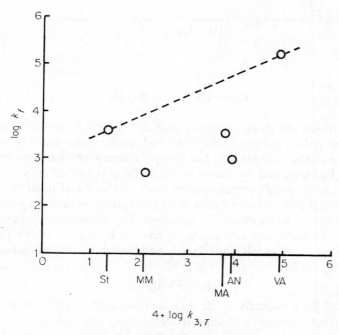

Figure 4.4   n-Butyl mercaptan

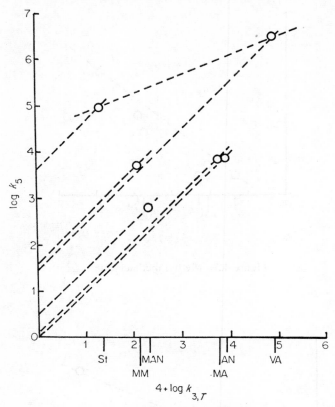

Figure 4.5 Ferric chloride

Take one of the reactivity diagrams, such as Figure 4.5, and draw through each of the points a line of unit slope and measure the positions of their intercepts p on the ordinate axis. For non-polar reactions these lines would be entirely coincident, and the extent to which the p values differ is a measure of the spread of points corresponding to the intrusion of polar influences. The problem is then converted into finding a correlation between p and some other recognized and measurable quantity. The Hammett σ function was chosen for the latter with immediate success; as Figure 4.6 shows for three important reactions, p and σ are directly and linearly related so that one may write the equation

$$p = \alpha\sigma + \beta \tag{12}$$

where α and β are respectively the slope and intercept of a line on this figure. Now, simple trigonometrical considerations will show that

$$p = \log k - \log k_{3,\mathrm{T}} \tag{13}$$

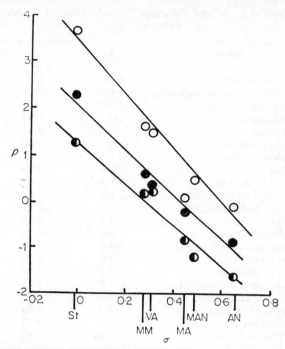

Figure 4.6 Correlation of Hammett σ-functions of radicals and polarity parameters for reactions with ferric chloride $\bigcirc$, n-butyl mercaptan $\bullet$ and carbon tetrabromide $\bullet$

and, by combining this with the previous equation, we have

$$\log k = \log k_{3,T} + \alpha\sigma + \beta \tag{14}$$

This statement is the epitome of the Patterns treatment; it claims that the absolute velocity constant for the reaction of a radical with a substrate can be predicted from a knowledge of

(i) $\log k_{3,T}$ and σ for the radical, both *experimentally* determinable, and

(ii) α and β for the substrate.

α and β present little difficulty since, in principle, it is only necessary to determine the rate constants for reaction of the chosen substrate with two calibrated radicals, i.e. radicals of known $k_{3,T}$ and σ. However, we have six such radicals with known characteristics (Table 4.4) and all can be used for this purpose. This obviates the possible criticism that α and β are not well authenticated. Some α and β values are listed in Table 4.6.

There is one complicating factor which deserves mention at this point, namely that a very small number of substrates (examples are N,N-dimethylformamide and triethylamine) give reactivity patterns different from those

Table 4.6 α- and β-Values for substrates in reactions
with radicals

Substrate	α	β
Triethylamine	+2·4	1·8
Form-N,N-dimethylamide	+1·0	−0·2
p-Methoxystyrene	+1·0	4·86
Butadiene	0	5·03
Styrene	0	4·85
Vinyl acetate	0	3·00
2-Vinylpyridine	−1·0	5·18
Chloroform	−1·4	0·90
Vinyl chloride	−1·5	3·65
Methyl methacrylate	−1·5	4·90
Vinyl trifluoroacetate	−2·0	3·22
Methacrylonitrile	−2·5	5·40
Methyl acrylate	−3·0	5·20
Acrylic acid	−3·0	5·52
Acrylonitrile	−3·0	5·30
Vinyl fluoride	−3·5	3·77
Carbon tetrabromide	−4·3	5·25
Maleic anhydride	−4·5	5·75
n-Butyl mercaptan	−4·8	6·05
Vinylidene cyanide	−5·0	7·22
Ferric chloride	−5·65	7·40

displayed above and exemplified by Figure 4.7. Detailed examination reveals that the departure of each point from the 'non-polar' straight line relation is just opposite to that found in the previous cases, a result which would be expected if the α value appropriate to the substrate is positive rather than negative as heretofore.

The sign of α is regarded as an indication of the direction in which charges tend to be transferred in the transition state, negative values of α corresponding to a tendency to transfer an electron from the radical to the substrate while positive values are associated with partial proton transfer from substrate to radical.

Reactions such as transfer with hydrocarbons or addition to styrene are characterized by an α value of zero for the substrate so that equation reduces to the following form:

$$\log k = \log k_{3,\mathrm{T}} + \beta \tag{15}$$

Comparison with equation (11) requires that

$$\beta = \log \gamma \tag{16}$$

In discussing the Q-e scheme, the postulate of identical values of e for conjugate monomer and radical was regarded as an unjustifiable oversimplification and we should now enquire about the manner in which the

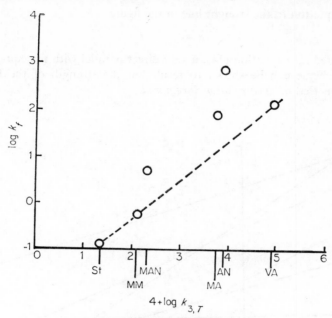

Figure 4.7 Triethylamine

Patterns treatment avoids the same charge. The corresponding assumption in the latter case would be that a direct relation exists between α and σ but, although Figure 4.8 demonstrates that a broad correlation can be found, it is most important to realize that the departure from the straight line is quite outside experimental error and that α and σ are independently known in the majority of cases.

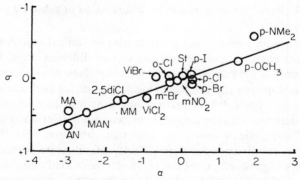

Figure 4.8 Correlation of α-values for monomers and Hammett σ-values for conjugate radicals. The monomers include substituted styrenes and vinyl bromide (ViBr), 2-vinylpyridine (2-VP) and vinylidene chloride (ViCl$_2$)

If the equation to the straight line in the figure

$$\alpha = -5\cdot3\sigma \tag{17}$$

is substituted into equation (14), a very direct parallel with the equations of the *Q-e* scheme can be shown to result but the strength of the Patterns approach is that *no such relation is required*.

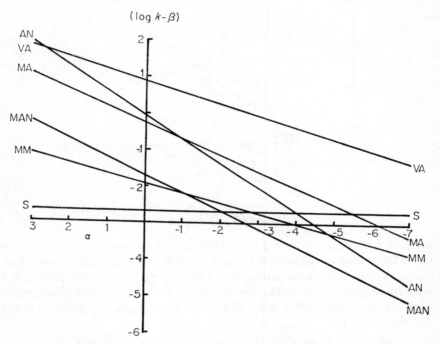

Figure 4.9   Relative radical reactivity as a function of substrate polarity

In order to obtain useful concise summaries of radical reactivity data we should consider equation (14) rearranged in two different ways. In the first $(\log k\text{-}\beta)$ is plotted against α. This process involves constructing straight lines of known slope (σ) and intercept $(\log k_{3,\text{T}})$, one for each radical and the result appears as Figure 4.9. A vertical line drawn at any selected α value gives, by its intercepts with the sloping lines, a direct indication of the order of reactivities of the radicals concerned with that substrate. Seen in this light the variation in reactivity order with polarity of substrate is readily understood. The very high reactivity of the vinyl acetate radical and the rather low reactivity of methacrylonitrile radical keep these two well separated over most of the range but with radicals of intermediate reactivity the polar effect can bring about a striking change in the sequence.

The second rearrangement requires a plot of $(\log k - \log k_{3,T})$ against σ, in other words we plot one straight line for each substrate, the intercept being β and the slope α. A selection of such lines is shown in Figure 4.10.

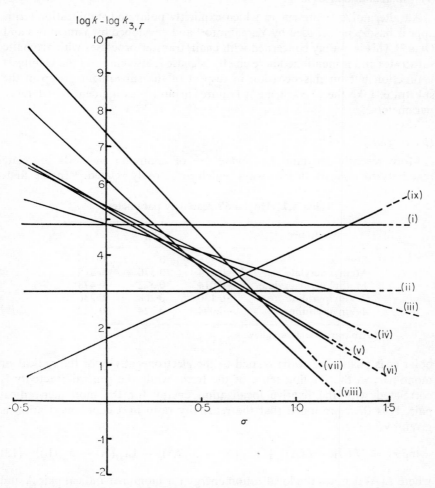

Figure 4.10 Relative substrate reactivity as a function of radical polarity: (i) styrene; (ii) vinyl acetate; (iii) methyl methacrylate; (iv) methacrylonitrile; (v) acrylonitrile; (vi) methyl acrylate; (vii) ferric chloride; (viii) n-butyl mercaptan; (ix) triethylamine

By drawing a vertical line at a chosen σ value, one obtains the relative reactivities towards a given radical of a variety of substrates. Thus, one can rapidly by inspection calculate a reactivity ratio or a transfer constant.

4.7.3 Other schemes

(i) Yamamoto and Otsu

An alternative treatment in which explicitly polar and stabilization terms appear has been proposed by Yamamoto[21] and developed by Yamamoto and Otsu.[22] This is mainly concerned with chain transfer processes with aromatic substrates and is mentioned here chiefly because it also employs the Hammett σ function but on this occasion in respect of the substituent group in the substrate. Like the *Q-e* scheme it requires arbitrary assignment of reference parameters.

(ii) Hoyland

More recently Hoyland has made use of computer methods to relate reactivity to polarity in two ways which are closely related.[23] In the first,

Table 4.7   Hoyland's reactivity parameters

Monomer	L	X_R	X_M
Styrene	0*	0*	0·126
Methyl acrylate	0·574	0·570	0·818
Methyl methacrylate	−0·014	0·522	0·478
2-Vinylpyridine	−0·130	0·268	0·236
4-Vinylpyridine	−0·094	0·324	0·322

* Arbitrary standard values.

polarity is postulated to be related to the electronegativity of the radical or monomer, in the Pauling sense of the term, while the general reactivity is correlated with the relative localization energy for the monomer-radical pair. It is then presumed that the reactivity ratio in a copolymerization is given by

$$\log r_A = [L(B) - L(A)] + [X_R(A) - X_M(A)] - [X_R(A) - X_M(B)] \quad (18)$$

where $L(A)$ denotes the localization energy for monomer-radical pair A and $X_R(A)$, $X_M(A)$ are the electronegativities of radical A and monomer A respectively.

With this equation as a basis, the known reactivity data for five selected monomers (Table 4.7) were used to derive the best values of L, X_R and X_M but, once again, it was necessary to assign two such values arbitrarily, $L = 0$ and $X_R = 0$ being selected for styrene. Having settled the L and X values for the five standard monomers, data for another twelve were processed by computer to obtain the best fit with experiment.

Apart from the undesirability of working with arbitrary reference values, it may be thought that the five standard monomers are not the best possible selection since styrene, 2-vinylpyridine and 4-vinylpyridine are all rather similar and the choice is biased to the non-polar end of the spectrum, nevertheless the use of the computer to fit the data in this way is attractive.

The second treatment is based on a charge transfer rather than an electro-negativity concept; it requires rather more elaborate analysis than the former scheme but the procedure in principle and the results are much the same.

4.8 COMPARISON OF SCHEMES

One of the most searching tests of a copolymerization reactivity scheme was found by consideration of a system of three monomers A, B and C which are reacted together in pairs to give a total of six reactivity ratios: from A + B, r_{AB} and r_{BA}; from B + C, r_{BC} and r_{CB}; from C + A, r_{CA} and r_{AC}. We now consider the function H defined by Mayo[24] such that

$$H = \frac{r_{AB} r_{BC} r_{CA}}{r_{AC} r_{CB} r_{BA}} \tag{19}$$

By substitution of the appropriate expression for the reactivity ratios from either the Q-e scheme (equation (14)) or the Hoyland electronegativity

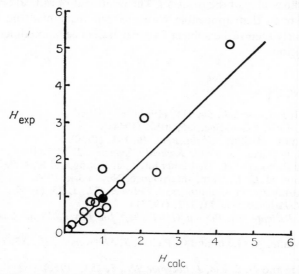

Figure 4.11 Comparison of observed and calculated H-factors, ○ points corre-sponding to H-values calculated from the Patterns treatment: line of unit slope denotes complete agreement between experiment and calculations of this type. Q-e scheme predicts that all points coincide at H (experimental) = H (calcu-lated) = 1, denoted by ●

treatment (equation (18)) it is evident that according to each of these approaches the only possible value for H is unity. If the approximation represented by equation (17) is combined with the Patterns expression (equation (14)), the same result ensues but by use of equation (14) alone, without approximation, one obtains the following result[25]

$$-\log H = \sigma_A(\alpha_B - \alpha_C) + \sigma_B(\alpha_C - \alpha_A) + \sigma_C(\alpha_A - \alpha_B) \qquad (20)$$

which enables H to be calculated *a priori* from the known σ and α values for the three monomers and their derived radicals. This calculated H can then be compared with its experimental counterpart obtained by substitution of the experimental values of the six reactivity ratios in equation (19). The comparison for fourteen such systems, each comprising three monomers, is displayed in Figure 4.11. This demonstrates the manner in which the Patterns treatment gives a remarkably good fit to the data while the alternative approaches fail completely to cope with the situation.

4.9 CONCLUSIONS

It is now possible to offer a highly satisfactory interpretation of the reactivity data for propagation and transfer processes in radical polymerization reactions which is based on the combined influence of stabilization energy and polarity on the course of the reaction. The treatment of radical processes is formally similar to that applicable with charged reactants, and radical processes are clearly seen to be subject to polar influence, sometimes to the extent of domination.

4.10 REFERENCES

1. L. J. Young, *J. Polymer Sci.*, **54**, 411, (1961).
2. F. T. Wall, *J. Amer. Chem. Soc.*, **66**, 2050, (1944).
3. F. R. Mayo and C. Walling, *Chem. Rev.*, **46**, 191, (1950).
4. A. D. Jenkins, in *Advances in Free Radical Chemistry*, **2**, 139, (1967).
5. H. F. Mark, B. Immergut, E. H. Immergut, L. J. Young and K. I. Beynon, in *Copolymerization* ed. G. E. Ham, Interscience, New York, 1964.
6. E. Merz, T. Alfrey and G. Goldfinger, *J. Polymer Sci.*, **1**, 75, (1946).
7. W. G. Barb, *J. Polymer Sci.*, **11**, 117, (1953).
8. F. R. Mayo, *Berichte der Bunsen Gesellscheft für Physikalische Chemie*, **7**, 233, (1966).
9. J. Guillot, J. Vialle, A. Guyot and C. Pichot, *J. Macromol. Sci.*, **A5**, 735, 753, (1971).
10. M. C. De Wilde and G. Smets, *J. Polymer Sci.*, **5**, 253, (1950).
11. G. E. Ham, *J. Polymer Sci.*, **A2**, 4169, (1964) and previous papers.
12. C. H. Bamford, A. D. Jenkins and R. Johnston, *Proc. Roy. Soc.* **A241**, 364, (1957).
13. B. Patnaik, A. Takahashi and N. Gaylord, *J. Macromol. Sci.*, **A4**, 143, (1970).
14. C. C. Price, *J. Polymer Sci.*, **1**, 83, (1946).

15. T. Alfrey and C. C. Price, *J. Polymer Sci.*, **2**, 101, (1947).
16. N. Kawabata, T. Tsuruta and J. Furukawa, *Makromol. Chem.*, **51**, 70, 80, (1962).
17. L. A. Wall, *J. Polymer Sci.*, **2**, 542, (1947).
18. C. H. Bamford, A. D. Jenkins and R. Johnston, *Trans. Faraday Soc.*, **55**, 418, (1959).
19. C. H. Bamford and A. D. Jenkins, *J. Polymer Sci.*, **53**, 149, (1961).
20. C. H. Bamford and A. D. Jenkins, *Trans. Faraday Soc.*, **59**, 530, (1963).
21. T. Yamamoto, *Bull. Chem. Soc. Japan*, **40**, 642, (1967).
22. T. Yamamoto and T. Otsu, *Chem. and Ind.*, **1967**, 787.
23. J. R. Hoyland, *J. Polymer Sci.*, **(A-1)8**, 885, 901, 1863, (1970).
24. F. R. Mayo, *J. Polymer Sci.*, **A-2**, 4207, (1964).
25. A. D. Jenkins, *European Polymer J.*, **1**, 177, (1965).

5

The Influence of Chain Structure on the Free-Radical Termination Reaction

A. M. North

Department of Pure and Applied Chemistry,
University of Strathclyde

5.1 INTRODUCTION

This chapter deals with the way in which the chemical structure of a vinyl polymer chain affects both the mechanism and the rate of the bimolecular termination process in free-radical polymerization.

Free-radicals may be removed from a polymerizing system by a variety of mechanisms, both first and second order in the reacting species. These have been listed in a recent review.[1] In this article attention is restricted to the bimolecular reaction between two polymeric free-radicals, as a result of which inert polymer is formed by either a combination or a disproportionation process.[2-4]

This termination reaction occurs between two reactive species, initially present as a very dilute solution in a condensed phase. Consequently, in discussing the mechanism of the reaction we must start with two molecules initially separated in space and consider all those processes which occur up to the final formation of inert polymer. These processes are both physical and chemical and may be listed as follows. First of all the two macroradicals must diffuse together so that it becomes possible for the two radical chain ends to move into close proximity. This first process, involving movement of the centres of gravity of the two radical chains, is referred to as translational diffusion. Now the locus of radical activity forms a very small volume element on each chain so that although certain segments of each chain are in a 'nearest neighbour' situation, the two free-radical sites may still be separated by several solvent molecules or inert chain segments. In order that the radicals may react they must approach to within the 'nearest neighbour' configuration and this may take place without further movement of the centres of gravity with the whole chains. This process takes place as each chain moves through a variety of conformations by way of rotation (or partial rotation) about the backbone bonds, and is referred to as segmental rearrangement. We have thus considered two physical processes necessary to bring the free-radical sites into a position in which chemical change can occur. Once this has been achieved chemical change may take place by either a combination or a disproportionation process.

The overall reaction thus involves two physical processes which are followed by either or both of two chemical processes. Consequently, it is convenient to divide this discussion in the same way and consider in turn each aspect of the overall reaction. However, before doing this it is useful to consider the relative rates of the physical and chemical processes.

In any bimolecular reaction, when the chemical process is slow compared to the physical diffusive processes, reactant molecules diffuse into and out from 'nearest neighbour' positions many times before chemical reaction takes place. Under these circumstances the reaction is said to be 'chemically-controlled' and the reaction rate is treated in the normal way using either transition state theory or a modified collision theory.[5] On the other hand when the chemical process is very rapid, and the diffusive processes are relatively slow, reaction can take place only as fast as the reactants can diffuse together. Under these circumstances chemical change occurs on almost every occasion that reactants become nearest neighbours and the process is referred to as being 'diffusion-controlled'. Many fast reactions such as those between oppositely charged ions, two free-radicals, or excited electronic states are known to be diffusion-controlled[6] even in mobile liquids.

In the case of the free-radical polymerization termination reaction, whether or not the reaction is diffusion-controlled will have a considerable effect on the kinetic treatment used to discuss the overall polymerization rate. However, it is important to stress that (unlike the analogous situation in photo-chemistry) there will be no effect on the chemical composition of the resulting polymer. Thus, the molecular weight is always lower when disproportionation is operative than when combination occurs, and the polymer composition in a copolymerization is likewise unaffected by which termination process is rate-determining.

5.2 THE DIFFUSIVE PROCESSES IN TERMINATION

5.2.1 The rate-determining step in termination

Since the diffusive steps may affect the rate of formation, but not the chemical composition, of polymer a discussion of these processes has major significance only if the overall reaction is diffusion-controlled. It has been known for many years that the termination reaction is diffusion-controlled in viscous or gel-like systems.[7,8] However, only over the last decade has it become apparent that for most cases the reaction is diffusion-controlled even in the most mobile solvents available.

There are several ways of testing whether the termination reaction is diffusion- or chemically-controlled.[9] The test most commonly applied is to conduct the polymerization in solvents of differing viscosity but of similar solvent power towards polymer. Under conditions of diffusion-controlled termination the termination rate constant should vary inversely as the solvent viscosity (frequently as the first power) and consequently (though less stringently) the overall polymerization rate should increase as the square root of the solution viscosity. In all the systems which have been tested to date (with the possible exception of butyl acrylate[10,11] about which there is some doubt) a dependence of reaction rate on solution viscosity has been observed down to the lowest obtainable solution viscosities. Unfortunately, these studies have not covered a wide range of chemical types, but include acrylamide,[12] methyl methacrylate[10,13,14,15] various alkyl methacrylates,[16] various copolymers of methyl methacrylate[17,18,19] and vinyl acetate.[17]

The general conclusion which can be drawn from these studies is that if a free-radical polymerization is characterized by a termination rate constant with a value 10^6–10^8 l mol^{-1} s^{-1} then it is extremely probable that the reaction is diffusion-controlled. This criterion is sometimes extended to a comment on the Arrhenius activation energy observed for the process. Thus, if the activation energy is comparable with the activation energy for viscous flow of solvent, or is sufficiently small that the reaction rate must be very rapid, then there is a definite probability that the reaction will be diffusion-controlled. This last criterion is not always totally reliable, especially when the internal segmental rearrangement of a polymer chain is a slow or inefficient process. Indeed, in the extreme case of very slow segmental rearrangement of a very stiff chain, the rate of rearrangement (and hence free-radical termination) may be governed entirely by the structural features of the chain and become independent of the external viscosity.[20] Some indications of this are seen when the dependence of the termination rate on solvent viscosity is examined for chains with large side groups. For the three alkyl methacrylates illustrated in Figure 5.1, the dependence on solvent viscosity decreases as the ester side group changes from methyl, to isobutyl, to 3,5,5-trimethyl hexyl. We thus meet the first effect of structure on reaction rate, an influence on the importance of solvent viscosity.

Although the observation of a solvent viscosity effect suggests that some diffusive process is the rate determining step in termination, it does not tell us whether translational diffusion or some segmental rearrangement (which also moves the chain end past several solvent molecules) is the important slow step. One reason for this is that Brownian motion, and so translational diffusion, of a flexible chain can take place by segments of the chain undergoing motion identical in character to that of the chain end in the 'segmental rearrangement process'. Consequently a consideration of the overall reaction as two independent consecutive processes is a very considerable simplification.

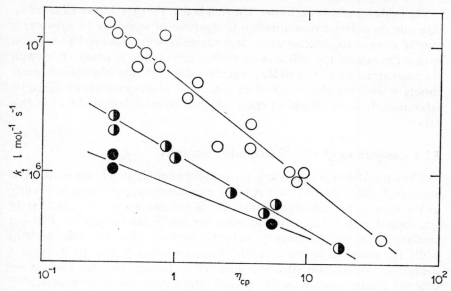

Figure 5.1   Viscosity dependence of termination rate constants at 30 °C. ○ methyl methacrylate; ◐ isobutyl methacrylate; ● 3,5,5-trimethyl hexyl methacrylate. Mixed solvents ethyl acetate, sucrose acetate isobutyrate

5.2.2 Translational and segmental rearrangement diffusion

Although the translational motion of a flexible chain does occur with concomitant segmental rearrangement, it is convenient to separate the two processes when examining the effect of chain structure on chemical reaction rate. Thus the reactive centres on perfectly rigid molecules can be brought together only by translational and gross rotational diffusion. Furthermore the perfectly flexible chain (with infinitely fast segmental reorientation) will search through all possible conformations in a time less than that required for translational separation of molecules, and so again translational diffusion will be rate-determining. Between these two extremes must lie a situation where the relative rates of translation and segmental rearrangement govern whether or not reaction takes place when two macro-radicals diffuse together. Both of these rates, but particularly the rate of segmental rearrangement, will depend on the chemical structure of the chain.

An unambiguous assessment of the importance of segmental rotation is not easy to make, and the evidence presented to date takes the form of inference rather than proof. However, every indication is that in the polymerization systems studied, (and mentioned earlier) segmental rearrangement is the important rate-determining phenomenon.

The evidence takes two forms. The first is that the dependence of termination rate on polymer concentration (conversion of monomer to polymer) is not the same as the concentration-dependence of self-diffusion.[13] The second is that the rate of the diffusion-controlled termination reaction varies with chemical structure over a wider range than does the rate of diffusion, more closely resembling the variation in rotational processes such as dielectric relaxation. It is this chemical effect which is of most interest to us in this article.

5.2.3 Comparison of rate and diffusion constants

When translational diffusion is the rate-determining step in the reaction of small molecules, the observed reaction rate constant varies almost linearly as the sum of the diffusion coefficients of the two species involved.[6,21] In free-radical polymerization the situation is complicated because the diffusion coefficients of the terminating radicals decrease (due to chain growth) while the reactants approach each other, and in any case there exists a distribution of reactant sizes and diffusivities. Despite this, the translation-controlled rate constant in free-radical termination would be expected[23,24] to vary almost linearly as the translational diffusion coefficient of inert polymer with molecular weight average and distribution the same as the active radicals.

The first 'structure' effect we might examine is molecular weight. A very detailed and precise examination of the dependence of the termination rate constant upon final polymer (and hence free-radical) molecular weight has been made for polymethyl methacrylate by Fischer, Mucke and Schulz.[14] These authors covered a twenty-fold range of degree of polymerization (above 2×10^3), yet found that the termination rate constant (at low conversion) was independent of the degree of polymerization. Since the translational diffusion coefficient of methyl methacrylate varies roughly as the inverse square root of chain length for chain sizes of interest here (or even as the inverse first power for very short chains) the inference is that translational diffusion cannot be of major importance in controlling the termination reaction.

A rather elegant examination of the chain length effect in a radical termination reaction has been made by Borgwardt, Schnabel and Henglein.[22] These authors studied the disappearance of polyethylene oxide radicals formed by the pulse radiolysis of dilute aqueous solutions.

$$.OH + \text{\textasciitilde}CH_2CH_2O\text{\textasciitilde} \rightarrow \text{\textasciitilde}CH_2\dot{C}HO\text{\textasciitilde} + H_2O$$

They found that the bimolecular termination rate constant was dependent on the polymer degree of polymerization up to values of 10^3, but did appear to be independent of chain length above this value (Figure 5.2). These observations were made on polymers, some samples of which were almost monodisperse in molecular weight, and of course the radical chain length did not increase during the radical lifetime. The value of 10^3 is considerably larger than the quite arbitrary suggestion[23] of 10^2 made for vinyl polymers.

For chains of degree of polymerization less than 10^3 the reaction rate constant varies almost as the inverse square root of the chain length. This is in accord with a model in which translational diffusion is rate determining, but the collision volume involved is the size of the reactive chain segment,

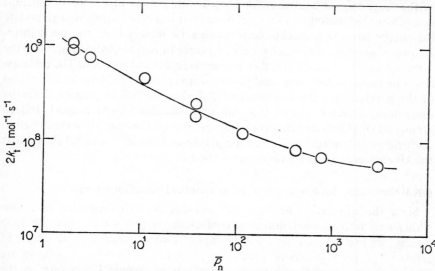

Figure 5.2   Chain length dependence of polyethylene oxide termination rate constants[22]

not the overall polymer chain. In other words the important diffusive motion is migration of the radical segment retarded to the low diffusivity of the whole chain. This, somewhat light-heartedly, has been named[23] the 'ball and chain' model.

Whether the change from translation-controlled to segmental-controlled reaction occurs for all chains at degrees of polymerization of 10^3 cannot yet be stated. However, the polyethylene oxide chain is one of the most flexible (most rapid backbone rotation) known, so perhaps it is not surprising that segmental rearrangement is not so clearly the slow process as in sterically hindered chains like polymethyl methacrylate.

A different structure change, the effect of which has been examined, is the size of the ester group in a series of alkyl methacrylates. The nature of the alkyl group has little effect on the chemical activity of the free-radical or monomer (as evidenced by propagation rate constants[25]) but does affect the termination rate constant.[16] Thus the polybutylmethacrylyl and polynonyl-methacrylyl radicals have termination rate constants respectively one fifth and less than one tenth of the value for polymethylmethacrylyl radicals. The exactly corresponding diffusion coefficients for polybutylmethacrylate and polynonylmethacrylate were not measured, but were estimated from other solution properties as respectively 0·9 and 0·8 of the coefficient for polymethylmethacrylate. Again the inference is that translational diffusion is not the important rate-determining process.

Probably the most convincing circumstantial evidence lies in a comparison of the reaction rate constants for chains with groups of markedly different size substituted directly on the backbone. For example in polymethylacrylate the energy barriers to backbone rotation arise mainly from the steric interactions of neighbouring ester groups, whereas in polymethylmethacrylate the α-methyl group presents a further barrier to segmental rotation. The termination rate constant for the α-methyl substituted chain is 35 times less than that of the acrylate, yet the translational diffusion coefficients cannot differ by more than a factor of two (for similar molecular weight chains). Indeed almost all vinyl polymer chains of molecular weight around 10^5 have diffusion coefficients 2–$3 \times 10^{-7} \text{ cm}^2 \text{ s}^{-1}$, yet a glance at Tables 5.1 and 5.2 shows that the kinetic rate constants vary over a thousandfold range.

5.2.4 Structure, backbone rotation and radical termination rate

Since the correlation between diffusion-controlled termination rate constants and translational diffusion coefficients is not satisfactory for long chains, we are led to search for an explanation of structure-rate effects in terms of polymer backbone rotational phenomena. In other words we are looking for a correlation between the effects of chemical structure on the rotation around substituted carbon–carbon single bonds and upon termination rate constants.

Dielectric relaxation is one technique which has been widely used to obtain information on the rate of segmental rotation in dissolved macromolecules.[27] Obviously when the dipole is attached rigidly to the chain backbone, measurement of dipole motion automatically implies observation of backbone motion. However, even in the case of acrylic polymers reorientation of the carbonyl dipole component perpendicular to the backbone contour requires backbone rotation. In effect it is found[28] that both the parallel and perpendicular dipole components relax together, indicating that

$$
\begin{array}{c}
\text{CH}_3 \\
| \\
-\text{CH}_2-\text{C}- \\
\updownarrow \; | \\
\text{C} \leftrightarrow \\
\diagup \quad \diagdown \\
\text{O} \qquad \text{O} \\
\diagdown \\
\text{CH}_3
\end{array}
$$

movement of the carbonyl group is co-operative with movement of the backbone. An identical phenomenon is observed with polyvinylacetate.

A comparison of termination rate constants and dielectric relaxation frequencies, both measured in dilute solution, is reported in Tables 5.1 and 5.2. It is strikingly apparent that an increase in the size of the groups substituted on the polymer backbone causes a parallel decrease in the rates of both the dielectric and chemical rate processes. The intimate dependence of the termination reaction upon chain segmental rotation thus seems quite un-ambiguous for these polymers.

Table 5.1 Comparison of kinetic rate constants and dipole reorientation frequencies. Polymers with ester side groups

Polymers	Dipole relaxation frequency Hz at 25 °C	Termination rate constant $1 \text{ mol}^{-1} \text{ s}^{-1}$ at 30 °C
$-CH_2-CH-$ $\quad\quad\mid$ $\quad\quad COOCH_3$	2×10^9	$2 \cdot 6 \times 10^8$
$-CH_2-CH-$ $\quad\quad\mid$ $\quad\quad OCOCH_3$	2×10^9	$2 \cdot 0 \times 10^8$
$-CH_2-C(CH_3)-$ $\quad\quad\quad\mid$ $\quad\quad\quad COOCH_3$	$3 \cdot 9 \times 10^7$	$1 \cdot 6 \times 10^7$
$-CH_2-C(CH_3)-\!\!-\!\!-CH-C(CH_3)-$ $\quad\quad\mid\quad\quad\quad\quad CO\quad CO$ $\quad\quad COOCH_3\quad\quad\;\backslash\quad/$ $\quad\quad\quad\quad\quad\quad\quad\quad O$	$1 \cdot 4 \times 10^7$	$3 \cdot 8 \times 10^6$

(2 per cent citraconic anhydride)

Unfortunately, though, the correspondence between the dielectric and chemical processes is not so good when changes are made to groups some atoms removed from the backbone. Thus changing the size of the alcohol residue in a series of polyalkyl methacrylates produces a marked change in chemical reaction rate,[16,29] but little change in the dipole relaxation frequency, Table 5.3. This emphasizes that the two processes are not exactly equivalent. The dielectric phenomenon involves a rather small scale motion which can be accommodated by a series of partial bond rotations close to the dipole without necessarily requiring movement of atoms well removed from the dipole. On the other hand the chemical process requires large scale movement of the free-radical chain end, and this in turn requires all the substituents in a side chain to move in a co-operative fashion.

A further dissimilarity between the chemical and dielectric measurements is that the chemical process involves movement of the chain end, whereas

Table 5.2 Comparison of kinetic rate constants and dipole reorientation frequencies. Polymers with dipoles attached rigidly to the backbone

Polymers	Dipole relaxation frequency Hz at 25 °C	Termination rate constant l mol^{-1} s^{-1} at 30 °C
—CH$_2$—CH— | Cl	2×10^8	5×10^8
—CH$_2$—CH— | Br	3×10^7	$2 \cdot 5 \times 10^8$
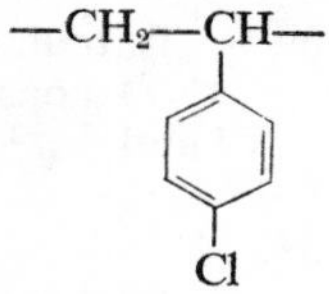 —CH$_2$—CH— (phenyl-Cl)	3×10^7	$7 \cdot 7 \times 10^7$
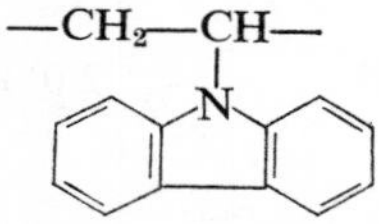 —CH$_2$—CH— (carbazolyl)	9×10^5	$2 \cdot 5 \times 10^5$

the dielectric observation is heavily weighted by the behaviour of groups in the chain 'interior'. An observation of chain end mobility can be made using the technique of fluorescence depolarization. In this experiment a fluorescent group is attached to the chain end (perhaps by way of a radical transfer reaction during polymerization or reaction with 'living' chain ends) and the fluorescence is excited using plane polarized light. A measurement of the depolarization of the emitted fluorescence then can be related to the extent of rotational reorientation of the fluorescent group during its lifetime in the excited state. An independent measurement of excited state lifetime finally

Table 5.3 Comparison of termination rate constants and dipole reorientation frequencies. Alkyl methacrylates

Alkyl group	Dipole relaxation frequency Hz at 25 °C	Termination rate constant l mol^{-1} s^{-1} at 30 °C
Methyl	$3 \cdot 9 \times 10^7$	$1 \cdot 6 \times 10^7$
Butyl	$3 \cdot 5 \times 10^7$	$2 \cdot 5 \times 10^6$
3,5,5-Trimethyl hexyl	$3 \cdot 0 \times 10^7$	$1 \cdot 3 \times 10^6$

yields the rotational relaxation time of the group. There are several experimental problems associated with this technique, not the least being the choice of a fluorescent group, movement of which is a true guide to movement of the natural chain end it replaces. However, measurements have been made on polystyrene,[30] polyacrylamide,[31] polymethylmethacrylate, polybutylmethacrylate and polymethylacrylate,[32] and the conclusions are set forth in Table 5.4. Again the most sterically hindered chains exhibit the longest rota-

Table 5.4 Comparison of chain end rotational relaxation times obtained from fluorescence depolarization and termination rate constants at 30 °C

Polymer	Dye used in transfer reaction	Rotational relaxation time $\times 10^9$ s	Termination rate constant l mol^{-1} s^{-1}
—CH$_2$—CH— COOCH$_3$	Dichlorofluorescein	31	2.6×10^8
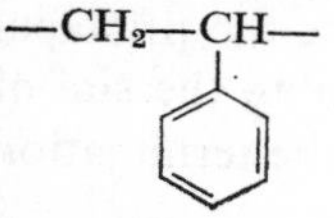 —CH$_2$—CH— (phenyl)	1-Sulphonyldimethyl-5-aminonaphthalene	4.4	2.5×10^7
—CH$_2$—C(CH$_3$)— COOCH$_3$	Dichlorofluorescein	54	1.6×10^7
—CH$_2$—CH— CONH$_2$	Fluorescein	4	1.2×10^7
—CH$_2$—C(CH$_3$)— COOC$_4$H$_9$	Dichlorofluorescein	6.5	2.5×10^6

tional relaxation times and in general the slowest reaction rates. However, the quantitative agreement is not as striking as in the dielectric case. Whether or not this is due to independent rotation of the fluorescent residue is not yet clear.

Obviously much further quantification of the backbone rotational parameters affecting termination is possible. Thus ultrasonic experiments[33] are beginning to yield the energy barriers to backbone rotation, and it should be possible to compare these with activation energies for termination. Unfortunately, there is still considerable uncertainty about the exact values of these latter, which are not easy to measure. Indeed, the activation energies for dielectric relaxation increase as the steric size of the backbone substituent increases, and so a similar trend should be observable for the kinetic rate constant.

A further quantification of the steric effect of chain substituents on motion, and hence reaction rate, might be sought in a correlation with the Taft[34] steric factor, E_s. Strangely enough a correlation with the Taft polar factor, σ^*, has been observed[24] for the termination rate constants of a series of alkyl methacrylates. Why this should be so is puzzling, because the reaction rate is independent of 'chemical' factors. It might be that polar interactions contribute to the barriers to rotation, or that the Taft treatment fails to separate completely steric and inductive effects. Indeed, some correlation between σ^* and E_s does exist in a limited selection of compounds.

5.2.5 The effect of copolymer composition

The final structural feature of a polymer chain which we shall examine is the composition of a copolymer. Obviously the chain structure will affect the diffusion-controlled and chemically-controlled termination processes differently. In the former case the termination rate will be governed by the composition of the whole chain, or at least the composition extending for several monomer units from the chain end, whereas in the latter case the principal feature affecting reactivity will be the monomer unit containing the site of free-radical activity. Again it is important to remember that the termination mechanism does not affect the copolymer composition.

For many years it has been conventional to consider three chemically-controlled termination reactions in a binary copolymerization of monomers A and B.

$$\left.\begin{array}{l} \text{\tiny{v}}\text{A}\cdot + \text{\tiny{v}}\text{A}\cdot \xrightarrow{k_{tAA}} \\[4pt] \text{\tiny{v}}\text{A}\cdot + \text{\tiny{v}}\text{B}\cdot \xrightarrow{k_{tAB}} \\[4pt] \text{\tiny{v}}\text{B}\cdot + \text{\tiny{v}}\text{B}\cdot \xrightarrow{k_{tBB}} \end{array}\right\} \textit{inert polymer}$$

This leads to an expression for the overall rate of copolymerization

$$\frac{-d([A] + [B])}{dt} = \frac{(r_1[A]^2 + 2[A][B] + r_2[B]^2)R_i^{\frac{1}{2}}}{(r_1^2\delta_A^2[A]^2 + 2\phi r_1 r_2\,\delta_A\delta_B[A][B] + r_2^2\,\delta_B^2[B]^2)^{\frac{1}{2}}} \quad (1)$$

where r_1, r_2 are the monomer reactivity ratios and ϕ is a ratio of the cross-termination to homo-termination rate constants, $k_{tAB}/k_{tAA}^{\frac{1}{2}}k_{tBB}^{\frac{1}{2}}$, δ_i is the ratio $(2k_{tii}^{\frac{1}{2}}/k_{pii})$ and R_i is the rate of initiation. This equation has been applied to many copolymerizations and attempts made to rationalize ϕ in terms of polar factors and other structure-reactivity correlations. Unfortunately the systems studied now appear to have termination reactions which are diffusion-controlled, and so equation (1) is not valid. The reason is that, for any given monomer composition, there are not three different termination rates depending on chain end unit, but only a single rate depending on overall chain composition. More precisely there will be a distribution of rates corresponding to a distribution in composition, but this is treated as a single average rate. When the kinetic equation for copolymerization is reinvestigated for such a scheme, there results[17,18]

$$\frac{-d([A] + [B])}{dt} = \frac{(r_1[A]^2 + 2[A][B] + r_2[B]^2)R_i^{\frac{1}{2}}}{(r_1^2\varepsilon_A^2[A]^2 + 2r_1r_2\varepsilon_A\varepsilon_B[A][B] + r_2^2\varepsilon_B^2[B]^2)^{\frac{1}{2}}} \tag{2}$$

Equation (2) bears a superficial resemblance to equation (1) except that there is now no cross-termination factor, ϕ, yielding 'chemical' information, and the homopolymerization constants, δ_i, have been replaced by composition dependent variables, ε_i defined as $(2k_{t(AB)})^{\frac{1}{2}}/k_{pii}$. $k_{t(AB)}$ is the single diffusion-controlled rate constant which is independent of the nature of the end unit, but depends on the nature of the whole chain.

On this basis consider the copolymerization of a sterically hindered monomer with an unhindered monomer. When the copolymer composition is rich in the former, segmental motion will be slow and so the termination rate constant $k_{t(AB)}$ will be low. When the copolymer composition is rich in the 'flexibilizing' monomer, segmental motion will be easier and so the termination rate constant will be higher. The copolymerization of methyl methacrylate and vinyl acetate has been analysed[17] in this fashion and a sensible dependence of $k_{t(AB)}$ on monomer feed composition obtained. (Use of equation (1) yields composition-dependent values of ϕ, a phenomenon not acceptable on the simple theory of end-unit control). Indeed at 60 °C it appears that the flexibilizing effect almost follows an 'ideal' equation combining the termination constants for homopolymerization and the mole fractions of each unit in the polymer.

$$k_{t(AB)} = x_A k_{tAA} + x_B k_{tBB} \tag{3}$$

On this basis, termination may be suppressed not only by the addition of a very hindered comonomer but also by one which exerts dipole-dipole forces that hinder segmental rotation. This has been shown[19] in the effect of citraconic anhydride on methyl methacrylate polymerization (Table 5.1).

Many other interesting correlations can be conjured up,[35] such as a termination rate—composition dependence inversely reflecting a glass transition temperature-composition dependence, or a correlation between termination rates and phase transitions in solution.[36]

The chemically-controlled treatment can be extended[37] to cover the diffusion-controlled case by the incorporation of penultimate and other unit effects. However the mathematical formulation becomes lengthy and does not appear to offer much advantage over the scheme incorporated in equation (2).

5.2.6 The effect of temperature

When the termination rate is controlled by segmental rotation, the temperature dependence of the rate will depend upon the energy barriers to rotation. It has already been shown[27] that the activation energy for

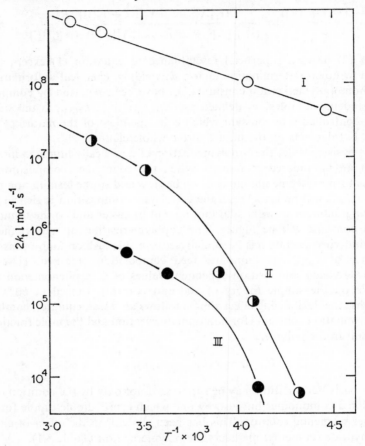

Figure 5.3 Arrhenius diagram of termination rate constants.
○ vinyl bromide   ◓ methyl methacrylate   ● N-vinyl carbazole

dielectric relaxation in solution increases with the size of backbone substituent and a comparable effect should be apparent in termination rates. Although this has not yet been observed (due to the difficulty in obtaining accurate activation energies for termination) an even more remarkable phenomenon occurs in chains with bulky steric groups. This is a catastrophic decrease[47] in the termination rate constants at low temperatures.

In the series vinyl bromide, methyl methacrylate and N-vinyl carbazole, the two latter monomers form hindered chains, and exhibit a marked reduction in termination rate around $-50\ ^{\circ}$C (Figure 5.3). The polyvinyl bromide chain is less hindered, however, and as a result the termination rate is higher and exhibits an Arrhenius temperature-dependence over the whole temperature range studied.

The marked reduction in termination rate has been ascribed to an effective cessation of backbone rotation as the energy barriers to rotation become comparable with kT. The phenomenon then corresponds roughly to a 'glass transition' in a single dissolved chain. Again no quantitative structure 'transition temperature' studies have been made, but the phenomenon might be expected to be most easily observed in the polymerization to those polymers of high bulk glass transition.

5.3 COMBINATION AND DISPROPORTIONATION OF RADICALS

Although it appears that most termination reactions are diffusion-controlled, the molecular weight of the final polymer formed and the nature of the end groups do depend on whether the chemical step is combination or disproportionation of the two free-radicals. Although considerable study of this question was made during the decade 1950–1960, activity now seems to have lessened. However there is still no general theory which successfully and usefully links polymer chain structure with relative extents of combination and disproportionation.

One of the problems in devising such a theory is that it should be based upon a difference of the free energies of the two transition states, one leading to combination and the other leading to disproportionation. However it is by no means certain[38,39] that the two transition configurations are markedly different, especially for primary alkyl radicals.

A 'rule-of-thumb' estimate of the relative importance of disproportionation might be that the steric hindrance of α-disubstituted chain radicals would discourage combination. In support of this hypothesis a table[40] of gas phase disproportionation to combination ratios shows that the former is more favourable for t-butyl + t-butyl or t-butyl + cyclohexadienyl radicals, but not for i-butyl + i-butyl or simpler radicals. Recent solution studies[48] confirm that, in general, disproportionation is favoured for tertiary radicals, but becomes of decreasing importance for secondary and primary radicals. Pursuing these lines of thought it is not surprising that disproportionation is favoured in the case of methyl methacrylate polymerization,[41,42] but not in the case of styrene[41–44] or acrylonitrile[45] polymerization.

Various techniques have been used to determine the relative amounts of disproportionation and combination, and these are reviewed in texts on the kinetics of free-radical polymerization.[46]

5.4 CONCLUSIONS

The correlation of structure and reactivity in the free-radical termination reaction probably has not developed as far as in other branches of organic chemistry. This is due to the fact that physical factors (often difficult to evaluate) control the rate of reaction. However a clear qualitative picture is emerging, and it is to be expected that the next few years will see the appearance of useful quantitative correlations.

With regard to the combination or disproportionation question, the importance of structural effects is little clearer today than twenty years ago. This is disappointing when one considers the effort that has been expended on both macromolecular and small molecule systems.

5.5 REFERENCES

1. A. M. North and D. Postlethwaite, 'The Termination Mechanism in Radical Polymerization, Chapter 4 in T. Tsuruta and K. F. O'Driscoll, *Structure and Mechanism in Vinyl Polymerization*, Marcel Dekker, New York, 1969.
2. G. M. Burnett, *Mechanism of Polymer Reactions*, Interscience, New York London, 1954.
3. C. H. Bamford, W. G. Barb, A. D. Jenkins and P. F. Onyon, *Kinetics of Vinyl Polymerization by Radical Mechanisms*, Butterworths, London, 1958.
4. A. M. North, *The Kinetics of Free Radical Polymerization*, Pergamon, Oxford, London, New York, Toronto, 1966.
5. A. M. North, *The Collision Theory of Chemical Reactions in Liquids*, Methuen, London, 1964.
6. A. M. North, *Quarterly Reviews*, **20**, 421, (1966).
7. R. G. W. Norrish and R. R. Smith, *Nature*, **150**, 566, (1942).
8. E. Tromsdorff, H. Kohle and P. Lagally, *Makromol. Chem.*, **1**, 169, (1947).
9. A. M. North, 'Diffusion-Control of Homogeneous Free Radical Reactions', in *Progress in High Polymers*, Volume 2, ed. J. C. Robb and F. W. Peaker, Heywood, London, 1968.
10. S. W. Benson and A. M. North, *J. Amer. Chem. Soc.*, **81**, 1339, (1959).
11. W. I. Bengough and H. W. Melville, *Proc. Roy. Soc.*, **A249**, 445, (1959).
12. G. K. Oster, G. Oster and G. Prati, *J. Amer. Chem. Soc.*, **79**, 595, (1957).
13. A. M. North and G. A. Read, *Trans. Faraday Soc.*, **57**, 859, (1961).
14. J. P. Fischer, J. Mucke and G. V. Schulz, *Ber. Bunsengesellshaft*, **73**, 154, (1969).
15. K. Yokota and M. Itoh, *J. Polymer Sci.*, **B6**, 825, (1968).
16. A. M. North and G. A. Read, *J. Polymer Sci.*, **A1**, 1311, (1963).
17. J. N. Atherton and A. M. North, *Trans. Faraday Soc.*, **58**, 2049, (1962).
18. A. M. North and D. Postlethwaite, *Polymer*, **5**, 237, (1964).
19. A. M. North and D. Postlethwaite, *Trans. Faraday Soc.*, **62**, 2843, (1966).
20. R. D. Burkhart, *J. Polymer Sci.*, **A3**, 883, (1965).
21. R. M. Noyes *Progress in Reaction Kinetics* ed. G. Porter, Vol. 1, Chapter 5, Pergamon, London, 1961.
22. U. Borgwardt, W. Schnabel and A. Henglein, *Makromol. Chem.*, **127**, 176, (1969).

23. S. W. Benson and A. M. North, *J. Amer. Chem. Soc.*, **84**, 935, (1962).
24. K. Ito, *J. Polymer Sci.*, A-1, **7**, 827, 2247, 2707, 2995, (1969).
25. G. M. Burnett, P. Evans and H. W. Melville, *Trans. Faraday Soc.*, **49**, 1096, 1105, (1953).
26. H. A. Ende, *Polymer Handbook*, ed. J. Brandrup and E. H. Immergut, Section IV-2, Interscience, New York, London, Sydney, 1966.
27. H. Block and A. M. North, *Advances in Molecular Relaxation Processes*, **1**, 309, (1970).
28. A. M. North and P. J. Phillips, *Trans. Faraday Soc.*, **64**, 3235, (1968).
29. K. Yokota, M. Kani, and Y. Ishii, *J. Polymer Sci.*, A-1, **6**, 1325, (1968).
30. M. Frey, P. Wahl and H. Benoit, *J. Chim. Phys.*, **61**, 69, (1964).
31. Y. Nishijima, A. Teramoto, M. Yamamoto and S. Hiratsuka, *J. Polymer Sci.*, A-2, **5**, 23, (1967).
32. A. M. North and I. Soutar, *J. Chem. Soc., Faraday Trans. I*, **68**, 1101, (1972).
33. H. Jl Bauer, H. Hasler and M. Immendorfer, *Disc. Faraday Soc.*, **49**, 238, (1970).
34. R. W. Taft jun., *J. Amer. Chem. Soc.*, **75**, 4534, (1953).
35. A. M. North, *Polymer*, **4**, 134, (1963).
36. K. F. O'Driscoll, W. Wertz and A. Husar, *J. Polymer Sci.*, A-1, **5**, 2159, (1967).
37. G. E. Ham, *Copolymerization*, Interscience, New York, London, 1964.
38. J. N. Bradley, *J. Chem. Phys.*, **35**, 748, (1961).
39. S. W. Benson and W. B. DeMore, *Ann. Rev. Phys. Chem.*, **16**, 397, (1965).
40. J. A. Kerr, *Ann. Rep. Chem. Soc.*, **65A**, 189, (1968).
41. J. C. Bevington, H. W. Melville and R. P. Taylor, *J. Polymer Sci.*, **12**, 449, (1954).
42. C. H. Bamford and A. D. Jenkins, *Nature*, **176**, 78, (1955).
43. G. M. Burnett and A. M. North, *Makromol. Chem.*, **73**, 77, (1964).
44. S. Olive and G. Henrici-Olive, *Makromol. Chem.*, **68**, 120, (1963).
45. C. H. Bamford, A. D. Jenkins and R. Johnston *Trans. Faraday Soc.*, **55**, 179, (1959).
46. See for example A. M. North and D. Postlethwaite, Chapter 4 in *Structure and Mechanism of Vinyl Polymerization* eds. K. F. O'Driscoll and T. Tsuruta, Marcel Dekker, New York, 1969.
47. J. Hughes and A. M. North, *Trans. Faraday Soc.*, **62**, 523, (1966).
48. R. A. Sheldon and J. K. Kochi, *J. Amer. Chem. Soc.*, **92**, 4395, (1970).

6

The Influence of Pressure on Polymerization Reactions

K. E. Weale

Department of Chemical Engineering and Chemical Technology, Imperial College

6.1 INTRODUCTION

The rates and equilibria of many chemical reactions in liquids are altered very considerably by pressures of two or three kilobars, and higher, (1 kb = 100 MN m^{-2}). The effects are often of much greater magnitude than the decrease in volume of the system (typically ten or twenty per cent) which is caused by the pressure, and are not directly related to it. They reflect the more fundamental pressure-dependence of the rate constants and equilibrium constants, which is determined by the detailed molecular structures of the participating species, and by the nature of the reaction mechanism.

In polymerization systems the most striking effects of high pressure are large increases in reaction rates; and the enhancement of 'polymerizability' which occurs when a polymer-monomer equilibrium undergoes marked displacement towards the polymer. Polymers formed under pressure may also have much higher molecular weights, and may show important structural differences from those obtained under ordinary conditions. The interpretation of these findings involves a knowledge of the effects of pressure on the individual component reactions of the overall polymerization process, which has been derived to a considerable extent from studies of the pressure-dependence of similar reactions in non-polymerizing systems. Most of the data available relates to free-radical polymerizations, and this account is concerned mainly (but not exclusively), with reactions of this type.

A brief outline of the chief features of pressure effects in kinetics and equilibria is given in the next section, as a basis for the discussion of polymerization. More detailed information is available in several recent reviews.[1,2,3,4]

6.2 PRINCIPAL EFFECTS OF PRESSURE ON EQUILIBRIUM AND RATE CONSTANTS IN LIQUIDS

6.2.1 Pressure and equilibria

The thermodynamics of chemical equilibria in liquids and solutions show that the equilibrium constant, K, depends on pressure according to equations of the form

$$\left(\frac{\partial\, RT \ln K}{\partial P}\right)_T = -\Delta V \tag{1}$$

The exact significance of ΔV depends on the type of concentration scale used in the definition of K; but, generally speaking, it is the excess of the molar volumes or partial molar volumes of the products over those of the reactants, in the equilibrium mixture. It follows from equation (1) that when the formation of the reaction products involves a decrease in volume (ΔV negative) the equilibrium constant is increased by pressure. At a temperature of 298 K, if the value of ΔV is -10 cm³/mol, the equilibrium constant is increased by factors of 1·5 at 1 kb and 60 at 10 kb; while if ΔV is -20 cm³/mol these factors are 2·2 and 3500. The calculation neglects the pressure-dependence of ΔV but serves to indicate the general magnitude of the effects on equilibria.

Many studies of chemical equilibria between non-ionic substances in liquids have demonstrated pressure effects of this kind. In the self-condensation of two molecules of cyclohexanone at 393 K the equilibrium constant is increased by a factor of 4·5 when the pressure is raised from 2 kb to 5 kb. The change corresponds to an average ΔV of -12 cm³/mol, which is close to the difference between the molar volumes of the pure products and reactants at ordinary pressure. Again, the equilibrium constant for the dimerization of NO_2 (in CCl_4 solution) is increased about four times between 1 bar and 1500 bar, corresponding to a volume change of -23 cm³/mol.

Liquid-phase equilibria between neutral molecules and ions are displaced in the direction of further ionization at higher pressure because of the decrease in volume produced by the strong interactions between ionic charges and the surrounding molecules, ΔV for the ionization of carboxylic acids in water is usually between -11 cm³/mol and -13 cm³/mol, while for amines it is close to -28 cm³/mol. The effect is larger in non-aqueous solvents, for example ΔV for piperidine in water is $-24\cdot3$, but in methanol it is $-49\cdot5$. The ionic product of water itself ($K_w = 1\cdot01 \times 10^{-14}$ mol/kg at 1 bar, 298 K) is increased four times at 2 kb, and there is evidence that it approaches unity at extremely high pressures, (250–300 kb).

The value of ΔV for a particular equilibrium is chiefly determined by such factors as the net change in the number of covalent bonds when the products are formed from the reactants; the formation or cancelling of ionic charges;

and the scission or formation of cyclic structures. These factors are discussed again in later sections.

6.2.2 Pressure and the rate constants of reactions

The dependence of the rate constant, k, of a chemical reaction on the absolute temperature, T, is expressed by the Arrhenius equation

$$k = A\,e^{-E/RT} \tag{2}$$

The equivalent expression in the quasi-thermodynamic transition-state theory of reaction rates is

$$k = \frac{\mathbf{k}T}{h}\,e^{\Delta S^{\ddagger}/R}\,e^{-\Delta H^{\ddagger}/RT} \tag{3}$$

in which $\mathbf{k}$ is Boltzmann's constant, h is Planck's constant, and $\Delta S^{\ddagger}$ and $\Delta H^{\ddagger}$ are the entropy and enthalpy of activation at constant pressure.

The results of some early investigations of the variation of rate constants with pressure were discussed in terms of the pressure-dependence of A and E (or of $\Delta S^{\ddagger}$ and $\Delta H^{\ddagger}$), but more insight is usually obtained if the variation of k is expressed by means of the volume of activation, $\Delta V^{\ddagger}$. In transition-state theory this quantity appears in the equation

$$\left(\frac{\partial\,RT\ln K^{\ddagger}}{\partial P}\right)_{T} = -\Delta V^{\ddagger} \tag{4}$$

where $K^{\ddagger}$ is the equilibrium constant for the formation of the transition state from the reactants, (cf. equation (1)). The theory leads to the equation

$$\left(\frac{\partial\ln k}{\partial P}\right)_{T} = \frac{-\Delta V^{\ddagger}}{RT} \tag{5}$$

where $\Delta V^{\ddagger}$ represents the volume change which accompanies the formation of the transition state from the reactants. (A small but significant extra term appears in equation (5) if the concentrations are expressed in volume units, e.g. on the molar scale.)

According to equation (5) the isothermal variation of k with P is determined by the sign and magnitude of $\Delta V^{\ddagger}$. A value of $-20\ \text{cm}^3/\text{mol}$ for a reaction at 333 K would, from equation (5), correspond to an increase in the rate constant by a factor of 9 at 3 kb, and a factor of 1500 at 10 kb. Usually, however, the plot of $\log k$ against P is curved towards the pressure axis, so that the numerical value of $\Delta V^{\ddagger}$, irrespective of its sign, tends to decrease with increasing pressure. Examples of such curves, for a number of organic reactions, are shown in Figure 6.1.

Of these reactions the most strongly accelerated is the Diels-Alder dimerization of isoprene for which $\Delta V^{\ddagger}$ is, initially, about -26 cm³/mol. This fairly large contraction can be ascribed to the partial formation of the two new covalent bonds in the transition state. The radical-displacement reaction (curve 2), in which the H atom is abstracted from the —SH group, has a $\Delta V^{\ddagger}$ of about -17 in the first part of the pressure range. It also proceeds via a

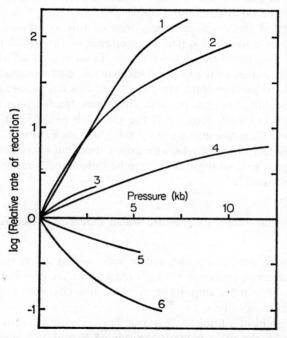

Figure 6.1 Effects of pressure on rates of reactions in liquids (1) Dimerization of isoprene.[5] (2) Abstraction of H atoms from n-hexylthiol by diphenylpicrylhydrazyl radicals.[6] (3) Iodine-catalyzed isomerization of *trans*-1,2-dichloroethylene.[7] (4) Base-catalyzed hydrolysis of monochloroacetate ions.[8] (5) Decomposition of di-t-butyl peroxide in toluene.[9] (6) Base-catalyzed hydrolysis of chloroform[10]

bimolecular transition state, but in this case bond-breaking is synchronous with bond-formation. The volume contraction due to the latter is evidently dominant, in agreement with expectation, (section 6.4.2).

In reaction (3) the rate-determining step is believed to be the addition of an I atom to a C atom of the double bond. $\Delta V^{\ddagger}$ is apparently -12, but the effect of pressure on the rate may be partly obscured by a reduction in the equilibrium concentration of I atoms. The alkaline hydrolysis of sodium monochloroacetate, curve (4), is a bimolecular substitution which is an ionic analogue of the radical displacement reaction (2). There is no net change in the number of ionic charges, although the anionic charge is presumably more dispersed in the transition state.

In two of the reactions the rate constants decrease with increasing pressure. These are the homolytic dissociation of di-t-butyl peroxide (volume-expansion is associated with bond-lengthening in the transition state), and the alkaline hydrolysis of chloroform, for which $\Delta V^{\ddagger}$ is initially $+16$ cm³/mol. This large positive volume of activation is strong evidence for the formation of the free carbene CCl_2 in the rate-determining step.

Measurements of the pressure-dependence of rate constants have been made for about five hundred different reactions[3] in the liquid phase, and there has been considerable progress in the development of correlations between $\Delta V^{\ddagger}$, the nature of the reaction mechanism, and the structure of the reactants and the transition state. (In some cases this has proved a powerful tool for distinguishing between possible alternative mechanisms.) A central problem which is not fully resolved is the variation of $\Delta V^{\ddagger}$ with pressure, which is apparent from the curvature of the plots in Figure 6.1. No exact theoretical analytical function which describes these curves is available, but the pressure-dependence of log k may often be satisfactorily represented by a quadratic polynomial.[11]

6.2.3 Some physical effects of pressure which influence chemical reactivity

Beside the more specifically chemical effects, which are outlined above, pressure may influence reactions through changes in the physical properties of the systems. The most important of these are changes in phase, and increases in liquid viscosity.

The freezing point of a liquid or gas usually rises with increasing pressure. For organic liquids the effect is often between 15 K and 25 K per 1000 bar, and at ordinary temperature few remain unfrozen at 10 kb. Benzene, for example, has a melting point of 393 K at 5·44 kb; while styrene monomer (normal freezing point 242 K) freezes at 333 K at 6·5 kb, and at 373 K at 10·7 kb. Some chemical effects associated with liquid → solid phase transitions under pressure are considered later. In multicomponent fluid systems pressure may also bring about separation into two immiscible fluid phases, or may cause a two-phase system to revert to a single phase. There have been several studies of the conditions under which a polymer-rich phase separates from solutions of polyethylene in highly compressed hydrocarbon gases.

The effect of pressure on the viscosities of organic liquids may be remarkably large. An increase from 1 bar to 10 000 bar may cause the viscosity to increase by a factor of from 10 to 1000, (and sometimes very much higher). The factor depends to some extent on the molecular weight, but is much more sensitive to molecular structure, and is greatest for complex molecular shapes. At sufficiently high viscosities the rate at which reacting molecules diffuse towards each other may be less rapid than their rate of reaction when in contact; and the speed of the reaction becomes diffusion-controlled. The rates of some liquid-phase bimolecular reactions which have very low activation energies are diffusion-controlled at ordinary pressure, and may decrease continuously with increasing pressure. For example, Ewald[12] found that the rate of the bimolecular quenching of the fluorescence of anthracene by CBr_4 is decreased by a factor of 2·6 between 1 bar and 2500 bar, in hexane at 303 K; while the viscosity of hexane increases 3·1 times over the same range. Bimolecular reactions between small molecules with activation energies of 15 kcal/mol or 20 kcal/mol, should only become diffusion-controlled at very high viscosities, but Hamann[13] has demonstrated that this does occur at pressures of 20 kb to 40 kb (with η in the range 10^6–10^7 P). In polymerization the pressure-retardation of the rate of chain termination by bimolecular reactions between radicals appears to be due to diffusion-control.

6.3 GENERAL FEATURES OF RADICAL POLYMERIZATION AT HIGH PRESSURE

The effect of pressure on polymerization has been most extensively investigated for styrene, but the polymerization and copolymerizations of a number of other vinyl monomers show similar general characteristics. If a chemical initiator, such as benzoyl peroxide (BPO) or azo-bis-isobutyronitrile (AIBN), is present, the overall rate of the reaction, R_{pol}, is usually about 7 or 8 times higher at 3 kb than at ordinary pressure, either in undiluted monomer or in a solvent. The curve of log R_{pol} against pressure approximates to a straight line over a considerable range, and the volume of activation, $\Delta V_{pol}^{\ddagger}$, for the overall process, (obtained from the slope), is generally between −17 and −20 cm³/mol. The order of reaction with respect to initiator concentration remains close to 0·5, as at ordinary pressure.

The polymer molecular weight is also higher when the reaction is carried out under pressure. In the case of styrene the average molecular weight of the product rises continuously with reaction pressure over the first 3 kb, and then tends to level off at a value about 3 or 4 times greater than that at 1 bar. Recent studies show that the distribution curves remain similar in form but are displaced towards higher values.[14,15] GPC fractionation of polystyrene produced at 1 bar and at 5 kb gave curves of integral weight distribution against log MW which are parallel over their central portions, although the pressure displacement corresponds to a four-fold increase in molecular weight. Comparable effects of pressure on the molecular weight of the product have been observed in the radical polymerization and copolymerization of other vinyl monomers.

The effect of pressure on the rates of some vinyl polymerizations is shown in Figure 6.2. The thermally-initiated polymerization of styrene, curve (1), is accelerated more strongly than the same reaction with BPO initiator, which follows curve (2). Curve (2) represents not only the polymerization of styrene (with BPO) at both 333 K and 353 K, but also the polymerization of styrene in toluene solution, and (if an approximate correction for the compressibility

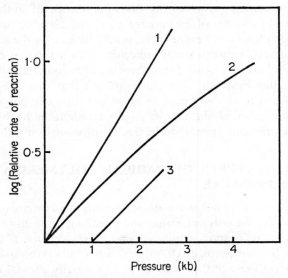

Figure 6.2 The effect of pressure on the rates of some radical polymerizations (1) The 'thermal' polymerization of styrene[16] (without initiator). (2) The polymerization of styrene with initiator;[16,17] and of styrene in toluene solvent;[17] and the copolymerization of styrene with methyl methacrylate.[18] (3) The polymerization of ethylene at 402 K in an ethylene/propane mixture[19]

of the system is made) the rate of copolymerization of styrene with methyl-methacrylate. The curve for the polymerization of ethylene at constant molar concentration in an ethylene/propane mixture (402 K, di-t-butyl peroxide initiator) also has a slope very similar to that of the initial portion of curve (2).

The high pressure radical polymerization of styrene (with an initiator) was first investigated in detail by Norrish and his colleagues.[20,21] They showed that the rate of homolytic dissociation of the initiator is reduced by pressure (see Figure 6.1, curve (5)), but chain propagation (the successive bimolecular additions of monomer molecules to the growing radical) is considerably accelerated. The rate of termination of kinetic chains by bimolecular reactions between pairs of large radicals is decreased because it is diffusion-controlled.

If no other factor were operative the molecular weight of the polymer would increase with pressure to about the same extent as the rate of reaction, but this is not the case. It was therefore concluded that the pressure-acceleration of chain transfer (radical displacement) with the monomer must be comparable to that of chain growth. At ordinary pressure both the kinetic chains and most of the molecular chains are terminated by radical-radical reactions, which largely control the molecular weight; but at high pressures (approx. 3 kb and above in the styrene polymerization) the molecular chains are ended by radical transfer to the monomer, although the kinetic chains still terminate in the usual way. The transfer process does not affect the rate of polymerization, but under pressure the molecular weight is controlled by monomer transfer.

The usual equation for the rate of radical polymerization of monomer M (with initiator I) is

$$R_{pol} = k_{pol}[M][I]^{0.5} = k_p \left(\frac{k_d}{k_t}\right)^{0.5} [M][I]^{0.5} \tag{6}$$

in which k_p, k_d and k_t are the rate constants for chain propagation, initiator dissociation, and chain termination. In combination with equation (5) this yields the expressions

$$\frac{\partial \ln k_{pol}}{\partial P} = \frac{-\Delta V_{pol}^{\ddagger}}{RT} \tag{7}$$

and

$$\Delta V_{pol}^{\ddagger} = \Delta V_p^{\ddagger} + \Delta V_d^{\ddagger}/2 - \Delta V_t^{\ddagger}/2 \tag{8}$$

in which the volumes of activation of the three component reactions are indicated by the appropriate subscripts. From measurements of the pressure-coefficients of these rate constants, the values of the volumes of activation in styrene polymerization are (approximately) $\Delta V_p^{\ddagger} \simeq -13$ cm³/mol; $\Delta V_t^{\ddagger} +$ 10 cm³/mol to $+12$ cm³/mol; and $\Delta V_d^{\ddagger}$ (for BPO) $\simeq +5$ cm³/mol. If these are summed according to equation (8) the overall volume of activation, $\Delta V_{pol}^{\ddagger}$, is about -16 cm³/mol or -17 cm³/mol, (with an uncertainty of several cm³/mol), in agreement with experimental values for this quantity of -17 cm³/mol to -18 cm³/mol.

The similarity of $\Delta V_{pol}^{\ddagger}$ for a number of chemically-initiated vinyl radical polymerizations indicates that the volumes of activation for chain growth and termination are not much affected by changes in substituent groups. In contrast the pressure-acceleration of the thermally-initiated polymerization (see Figure 6.2), is considerably greater, $(\Delta V_{pol}^{\ddagger} = -25.8$ cm³/mol). The difference can be rationalized if it is assumed that thermal initiation involves a bimolecular reaction between two molecules of monomer, which, unlike initiator dissociation, should be pressure-accelerated. The rate of thermal initiation is then expressed by

$$R_{ti} = k_{ti}[M]^2 \tag{9}$$

and kinetic analysis gives an equation of the same form as equation (8), but with $\Delta V_{ti}^{\ddagger}$ (which is negative) substituted for $\Delta V_d^{\ddagger}$, (which is positive). On this basis Guarise[16] has calculated that the volume of activation for thermal initiation of styrene polymerization is $-12 \cdot 6$ cm³/mol. The difference between the pressure-dependence of thermally-initiated and chemically-initiated polymerizations is thus attributed to the difference between the mechanisms by which the primary radicals are generated.

This outline of the main features of radical polymerization under pressure is supported by data from many studies on radical chain growth, radical transfer reactions (which largely control the molecular weights of high pressure polymers), and radical-producing dissociations. A brief review of some of the more interesting results, and their relation to polymerization processes, follows in the next section.

6.4 REACTIONS OF RADICALS WITH MOLECULES, AND RADICAL-PRODUCING REACTIONS, UNDER PRESSURE

6.4.1 Radical addition to vinyl monomers, (chain growth)

The addition of a monomer molecule to a chain radical involves the formation of a covalent bond with one of the C atoms of the double bond. The volume contraction which accompanies this process is mainly due to the large differences between the van der Waals and covalent radii of the atoms. Its magnitude can be approximately estimated if it is assumed that the two atoms between which the bond is formed are initially 4 Å apart (the van der Waals 'touching' distance), and that this then shortens to the C—C bond length of $1 \cdot 54$ Å. At the same time the C=C bond ($1 \cdot 33$ Å) lengthens to become a C—C bond ($1 \cdot 54$ Å). If the net linear contraction of $2 \cdot 25$ Å occurs along the axis of a cylinder which has a radius equal to that of the methylene group (2 Å) the decrease in volume corresponds to -17 cm³/mol. In the transition state the contraction will be rather less because bond-formation is not complete.

In general agreement with this simple model $\Delta V^{\ddagger}$ for the addition of an I atom to a double bond is about -12 cm³/mol (section 6.2.2), and the two measurements (made by different techniques) of $\Delta V_p^{\ddagger}$ in styrene[21,22] give values of $-13 \cdot 4$ cm³/mol and $-11 \cdot 5$ cm³/mol. Unfortunately the results obtained for other pressure polymerizations yield only the ratio $k_p/k_t^{\frac{1}{2}}$, and not the separate rate constants, but some information about $\Delta V_p^{\ddagger}$ can be gained indirectly from studies of copolymerization.

The reactivity ratios in the copolymerization of monomer A with monomer B are defined by

$$r_{\mathrm{A}} = \frac{k_{p\mathrm{AA}}}{k_{p\mathrm{AB}}}, \qquad r_{\mathrm{B}} = \frac{k_{p\mathrm{BB}}}{k_{p\mathrm{BA}}} \qquad (10)$$

in which the k's are the rate constants for the four possible radical-monomer additions. The pressure-dependence of reactivity ratios can be expressed in the form

$$-RT\frac{\partial \ln r_A}{\partial P} = \Delta V^{\ddagger}_{pAA} - \Delta V^{\ddagger}_{pAB} \tag{11}$$

and so be used to derive data on differences between the volumes of activation for the various radical-monomer addition reactions. Measurements of r as a function of pressure have been made for about thirty systems.[4,18,23] In the majority of cases the variation of r with pressure is either within the limits of experimental uncertainty, or is relatively small and indicates differences of only 1 cm³/mol to 2 cm³/mol between the volumes of activation. As many of the systems investigated share common monomers the results are strong evidence that the acceleration of chain growth by pressure is of similar magnitude in a large number of free-radical vinyl polymerizations.

6.4.2 Pressure and chain transfer in vinyl polymerization

Radical transfer or displacement reactions are bimolecular substitutions in which the attacking radical forms a covalent bond with an atom of the substrate molecule, while the rest of the molecule separates as a new free radical. Approximate theoretical calculations for bimolecular substitutions indicate that in the transition state the reduction of the distance between the two atoms which become bonded is more important than the fractional stretching of the bond which breaks, so that a net contraction, and a negative volume of activation, are expected. The strong pressure-acceleration of the reaction between diphenylpicrylhydrazyl and n-hexyl thiol (Figure 6.1) confirms that this is so; and there is much evidence that the similarity of the accelerations of chain growth and chain transfer is a dominant factor in radical polymerizations under pressure.

The effect is particularly clear for polymerizations in which chain transfer to the monomer controls the molecular weight, even at normal pressures. Propylene, unlike ethylene, cannot be polymerized to a solid product of high molecular weight by the radical mechanism, even at very high pressures, because of its readiness to undergo chain transfer, with formation of the allyl radical. Brown and Wall[24] found that the rate of the γ-ray initiated polymerization increases about 75 times between 5 kb and 16 kb at 294 K ($\Delta V^{\ddagger}_{pol}$ is about -10 cm³/mol), but the molecular weight of the polymer is increased over the same pressure range by a factor of only 2 (from 1500 to 3000). Zharov and co-workers,[25] who used AIBN initiator, found the exponent of [I] in the rate equation to be nearly 1 below 3 kb, but to be 0·55 at 6·5 kb. The rate increased with pressure between 3 kb and 8 kb, but the increase in the molecular weight of the polymer did not exceed 30 per cent. Generally similar results have also been reported by Osugi and his colleagues.[26] The near-constancy of the molecular weight, in comparison with the large pressure-dependence of the rate, shows that the transfer reaction with the monomer is accelerated nearly as much as chain propagation by an increase of pressure. Another example of the effect is the polymerization of allyl acetate.[27] The average molecular weight remains at about 2000 for reaction pressures up to 8·5 kb.

Other supporting evidence of the strong pressure-dependence of radical chain-transfer has been derived from work on transfer to solvent in polymerizing systems. The solvent transfer constant, C_s ($= k_s/k_p$, where k_s is the rate constant of the radical displacement reaction) decreases from 98×10^{-4} at ordinary pressure to 84×10^{-4} at 4 kb, for the polymerization of styrene in CCl_4 solvent.[22] This relatively small change shows that k_s increases nearly as much with pressure as does k_p. C_s for styrene polymerizing in tetrachloroethylene is also practically independent of pressure; and although there is some evidence[17,29] that k_s increases rather less rapidly with pressure than k_p when the atom transferred from the solvent is hydrogen, this does not invalidate the main conclusion.

6.4.3 Pressure and radical production: the dissociation of initiators

There is little information available about the effect of pressure on the 'thermal' generation of radicals from the monomer itself (see section 6.3); but a number of studies of homolytic dissociations have been made. Ewald[30] found that the rate constant for the first-order dissociation of pentaphenyl ethane in toluene is halved at 1·5 kb (ΔV about $+13$ cm³/mol), and determined $\Delta V^{\ddagger}$ to be $+3·8$ for the dissociation of AIBN in toluene. Various values for the dissociation of BPO in different solvents range from $+5$ cm³/mol to $+10$ cm³/mol; while for di-t-butyl peroxide the measurements of Walling and Metzger[9] yield values ranging from $+5·4$ cm³/mol in toluene to $+13·3$ cm³/mol in carbon tetrachloride. Simple geometric models of the type discussed earlier could account for such variations only in terms of different degrees of stretching of bonds in the transition states (e.g. between 10 per cent and 30 per cent), but Walling and Metzger suggested that they are associated with the 'cage' effect. The t-butoxy radical may react rapidly with a neighbouring toluene molecule so that in this solvent the reaction is completed within the 'cage' and the observed $\Delta V^{\ddagger}$ relates to the primary dissociation. In CCl_4 the radicals may recombine unless they separate by diffusion, and if diffusion is slower than recombination the observed $\Delta V^{\ddagger}$ will include an important (positive) contribution from the volume of activation of the diffusion process. Neumann and his colleagues[31] have recently made careful studies of the decompositions of several peresters in solvents such as chlorobenzene and cumene, at pressures up to 4 kb. They too find that the apparent volumes of activation vary considerably from one system to another (from a few cm³/mol to about $+13$ cm³/mol), but conclude that the true value of $\Delta V_d^{\ddagger}$ probably does not exceed $+4$ cm³/mol.

6.5 SOME FURTHER EFFECTS OF PRESSURE ON RADICAL AND IONIC POLYMERIZATIONS

6.5.1 Ionic polymerization

A few measurements of the effect of pressure on the rates of ionic polymerizations have been reported and some results are illustrated in Figure 6.3.

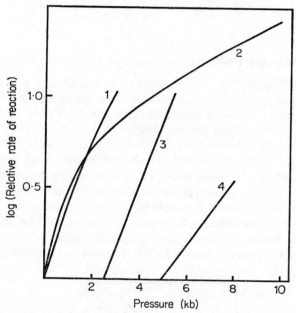

Figure 6.3 The effect of pressure on the rates of some ionic polymerizations (1) styrene[25] (HClO₄ catalyst); (2) 3-methylbutyl vinyl ether[32] (iodine catalyst); (3) tetrahydrofuran[33] (BF₃ diethyletherate); (4) α-methylstyrene[34] (trichloracetic acid)

It may be noted that the reactions were not always investigated in detail, and that control of the purity and dryness of the monomer was not always stringent. The reactions are, however, obviously accelerated by pressure, and comparison with Figure 6.2 shows the general order of magnitude of the effects to be similar to those found for radical polymerizations. The lower part of the curve for the polymerization of styrene with HClO₄ yields a value for $\Delta V_{pol}^{\ddagger}$ of between -20 cm³/mol and -25 cm³/mol.

The pressure-dependence of the separate component reactions of the ionic polymerization processes has not been well established. Pressure must certainly promote the formation of polar and ionic complexes in the initiating steps, as with charge-forming processes in general. In discussing the iodine-catalysed polymerization of the vinyl ether, curve (2), Hamann and Teplitzky[32] point out that the initiating steps are probably equivalent to the rate-determining step in the addition of iodine to olefins in solvents of low dielectric constant. This addition reaction is accelerated about three-fold at 1 kb, that is to about the same extent as the polymerization, so that the increase in polymerization rate might be attributed simply to faster initiation. It seems likely, however, that the bimolecular addition of monomer to the growing ion, or ion pair, will also be accelerated, and so contribute to the total effect.

6.5.2 Polymer-monomer equilibria: pressure and ceiling temperatures

In addition polymerizations the existence of an equilibrium between polymer and monomer produces the characteristic phenomenon of a fairly sharp ceiling temperature (T_c) above which formation of high polymer does not occur (see Chapter 16). If the polymer chains are long $T_c = \Delta H/\Delta S$, where ΔH and ΔS are the enthalpy and entropy changes which occur when a monomer unit is added to the chain. ΔH and ΔS are usually both negative, so that T_c is increased if $-\Delta S$ is decreased, (e.g. by increasing the concentration of monomer in a solvent). The volume change for polymer formation, ΔV, is also usually negative, so that T_c should be increased by increasing pressure. The results illustrated in Figure 6.4 confirm that this is so. The ceiling temperature for polymerization of α-methylstyrene increases by about 17°/kb, and that of tetrahydrofuran by about 20°/kb. The relation between log T_c and

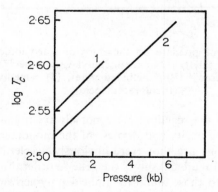

Figure 6.4   The effect of pressure on the polymerization ceiling-temperature of (1) tetrahydrofuran,[35] and (2) α-methylstyrene[34]

P is linear, and may be expressed in the form of the Clapeyron-Clausius equation

$$dT_c/dP = T_c \Delta V/\Delta H \tag{12}$$

and for each substance the line passes through the known value of T_c at ordinary pressure. The polymerization ceiling-temperatures of chloral[36] and of n-butyraldehyde[37] are changed by pressure in a similar way.

If reported values of ΔH (at ordinary pressure) are substituted in equation (12) an approximate estimate of ΔV may be obtained. For α-methylstyrene ΔV is calculated to be -15 cm³/mol, in fairly reasonable agreement with the value obtained earlier from a very simplified model of the addition of a radical to an olefin. A similar calculation from the results for tetrahydrofuran indicates that ΔV is between -9 cm³/mol and -10 cm³/mol. This is of particular interest because there is no net change in the number of covalent bonds; the addition of a monomer molecule to the chain involves the formation of a new bond, but another bond is broken to open the five-membered ring. It is possible that the considerable volume-contraction may come from an ability of chains to pack more densely than ring molecules in the liquid, but it is likely to be due, more specifically, to the elimination of an 'excluded volume' associated with the ring. Evidence has accumulated[38,39] that ring-opening reactions are accompanied by decreases in volume which may exceed 10 cm³/mol, and the contraction which occurs during the liquid-phase polymerization of tetrahydrofuran is probably an example of the same effect.

The pronounced effect of pressure on ceiling temperature can lead to the observation of a 'threshold pressure' for polymerization, followed by a rapid increase of rate with further increase in pressure, when a system is investigated at a fixed temperature and at successively higher pressures. A number of such findings for vinyl monomers, and also for the polymerization of aldehydes to polyoxymethylene chains, are probably to be explained in this way. An obvious corollary is that in studies of the potential polymerizability of a substance under pressure the reaction temperatures should not exceed the corresponding ceiling temperatures, (but should be above the freezing point at each pressure).

6.5.3 Pressure and the micro-structure of polymers

The general trend towards higher molecular weights with increasing reaction pressure may be accompanied by changes in the detailed molecular structure of the polymer. The underlying mechanisms are imperfectly understood and the changes cannot as yet be related satisfactorily to pressure effects in non-polymeric reactions; but a brief summary is included to complete the discussion of homogeneous-phase polymerizations.

A number of investigations in this area have been concerned with the structure of polyethylene. It is generally the case that the number of short side-chains in the polymer decreases as the reaction pressure is increased; possibly because internal transfer reactions, (proceeding via cyclic transition states), which may be the origin of the side-chains, are less accelerated by pressure than is chain growth. For example, Woodbrey and Ehrlich[40] showed that there are only 8 CH_3 groups (side-chain terminals) per 1000 CH_2 groups in polyethylene formed at 3 kb, compared with 35 per 1000 in polymer made at 800 bar. There is also a large reduction in the number of vinylidene $(CH_2{=}C<)$ side-chains over this pressure range.

The polymers formed from dienes at high pressures have also been investigated. Richardson[41] found that polyisoprenes and polybutadienes formed between 5 and 10 kb contained similar proportions of 1,2- and of cis- and trans- 1,4-structures to those in polymers made at ordinary pressure. Jenner and his co-workers confirmed that there is no large effect of pressure (up to 14 kb) on the microstructure of poly-1,3-dienes produced by radical polymerization; but in the anionic polymerization of isoprene (initiated by n-butyl lithium), in organic solvents, they find that increasing pressure causes a large decrease in the proportion of cis-1,4-structures in the polymer produced.[42] Non-conjugated dienes may polymerize to form linear, saturated cyclopolymers, which contain ring structures as repeat units in the molecules. Pressure has been found to accelerate strongly the cyclopolymerization of diallyl cyanamide,[43] and to increase the degree of cyclization in the polymerization of acrylic anhydride; and is also known[44] to promote the cyclopolymerization of n-perfluoropenta-1,4-diene.

The effect of reaction pressure on the stereo-regularity of polymers has been investigated chiefly for methyl methacrylate (see, for example, Walling and Tanner[45]) and it is found that the proportion of isotactic linkages is significantly increased; but Osugi and his co-workers[26] could not detect a similar change in high-pressure polypropylene.

Detailed interpretations of these effects of pressure on polymer microstructure will not be possible until more information has been obtained on the mechanisms of the competing reactions and the geometry of their transition states.

6.5.4 Phase-separations and phase-changes in high pressure polymerization

The separation of polymer, or a polymer-rich phase, during the course of reaction is a well known source of kinetic complications. The few studies of such polymerizations at high pressures show very varied effects which are sometimes quite unlike the normal response to pressure. For example, the rate of the bulk polymerization of acrylonitrile (with AIBN) decreases by about 50 per cent when the pressure is raised to 4 kb, and the molecular weight of the polymer is less, but at 6 kb the reaction is explosively rapid.[46]

The high pressure polymerization of vinyl chloride (which is characterized by phase-separation and autocatalysis) is less unusual. Guarise and his colleagues[47] observed a fairly normal increase in rate ($\Delta V^{\ddagger}_{pol} = -18$ cm³/mol) up to 2·5 kb, and infer from their detailed data that the volume of activation for transfer to the monomer is less negative, (by about 6·5 cm³/mol) than that for chain-growth. Some evidence has also been obtained concerning the effect of freezing the reacting system at high pressure during the polymerization of styrene.[48] When the solidification pressure is reached there is a marked increase in reaction rate. The reaction continues in the solid state as the pressure is increased further, but when a certain excess pressure is attained the rate falls abruptly to a very low value. Pressure (and pressure combined with shear) has significant effects on solid-state polymerizations.[49,50,51] This is a field of growing activity but the results are not reviewed here because their interpretation is still tentative, and no instructive comparison with other types of solid-state reaction is yet possible.

6.6 CONCLUSION

Many of the effects of pressure on polymerizations in liquids (and highly compressed gases) can be explained in terms of changes in the rate constants of the various component reactions, and of displacement by pressure of reaction equilibria. These are most usefully expressed by means of the volumes of activation ($\Delta V^{\ddagger}$) and total volume changes (ΔV); and where these quantities have been evaluated for polymerization processes their sign and general order of magnitude correspond well with those which have been measured for similar types of reaction between small molecules, and with predictions based on simplified models of transition states. Large areas remain to be explored, and much more data on both chemical and physical changes in high pressure systems is needed, but it appears likely that the theoretical framework developed to explain pressure effects in other types of reaction will prove adequate to account for results in the field of polymerization.

6.7 REFERENCES

1. S. D. Hamann, in *High Pressure Physics and Chemistry*, Vol. 2, ed. R. S. Bradley, Academic Press, New York, 1963.
2. E. Whalley, in *Advances in Physical Organic Chemistry*, Vol. 2, ed. V. Gold, Academic Press, New York, 1964.
3. W. J. le Noble, in *Progress in Physical Organic Chemistry*, Vol. 5, eds. A. Streitwieser and R. W. Taft, Interscience, New York, 1967.
4. K. E. Weale, *Chemical Reactions at High Pressures*, Spon, London, 1967.
5. C. Walling and J. Peisach, *J. Amer. Chem. Soc.*, **80**, 5819, (1958).
6. A. H. Ewald, *Trans. Faraday Soc.*, **55**, 792, (1959).
7. A. H. Ewald, S. D. Hamann and J. E. Stutchbury, *Trans. Faraday Soc.*, **53**, 991, (1957).
8. M. W. Perrin, *Trans. Faraday Soc.*, **34**, 144, (1938).
9. C. Walling and G. Metzger, *J. Am. Chem. Soc.*, **81**, 5365, (1959).

10. W. J. le Noble and M. Duffy, *J. Am. Chem. Soc.*, **86**, 4512, (1964).
11. H. S. Golinkin, W. G. Laidlaw and J. B. Hyne, *Can. J. Chem.*, **44**, 2193, (1966).
12. A. H. Ewald, *J. Phys. Chem.*, **67**, 1727, (1963).
13. S. D. Hamann, *Trans. Faraday Soc.*, **54**, 507, (1958).
14. C. Macosko and K. E. Weale, *Am. Chem. Soc.*, *158th Meeting*, *Abstracts of Papers*, 1969.
15. W. F. Dellsperger and K. E. Weale, to be published.
16. G. B. Guarise, *Polymer*, **7**, 497, (1966).
17. A. C. Toohey and K. E. Weale, *Trans. Faraday Soc.*, **58**, 2439, 2446, (1962).
18. J. A. Lamb and K. E. Weale, in *Physics and Chemistry of High Pressures* (*Symp.*), Soc. Chem. Ind., London, 1963.
19. R. O. Symcox and P. Ehrlich, *J. Am. Chem. Soc.*, **84**, 531, (1962).
20. F. M. Merrett and R. G. W. Norrish, *Proc. Roy. Soc.*, **A206**, 309, (1951).
21. A. E. Nicholson and R. G. W. Norrish, *Disc. Faraday Soc.*, **22**, 97, 104, (1956).
22. C. Walling and J. Pellon, *J. Am. Chem. Soc.*, **79**, 4776, (1957).
23. K. Hamanoue, *Rev. Phys. Chem. Japan*, **38**, 120, (1968).
24. D. W. Brown and L. A. Wall, *J. Phys. Chem.*, **67**, 1016, (1963).
25. A. A. Zharov, A. A. Berlin and N. S. Enikolopyan, *J. Polymer Sci.*, *Part C*, **16**, 2313, (1967).
26. J. Osugi, K. Hamanoue and T. Tachibana, *Rev. Phys. Chem. Japan*, **38**, 96, (1968).
27. C. Walling and J. Pellon, *J. Am. Chem. Soc.*, **79**, 4872, (1957).
28. K. Salahuddin and K. E. Weale, work to be published.
29. V. M. Zhulin, M. G. Gonikberg, A. L. Goff and V. N. Zagorbinina, *Polymer Sci. USSR*, **11**, 877, (1969).
30. A. H. Ewald, *Disc. Faraday Soc.*, **22**, 138, (1956).
31. R. C. Neumann Jr. and J. V. Behar, *J. Am. Chem. Soc.*, **91**, 6024, (1969); see also *J. Am. Chem. Soc.*, **92**, 2440, (1970); **93**, 4242, (1971).
32. S. D. Hamann and D. R. Teplitzky, *J. Phys. Chem.*, **65**, 1654, (1961).
33. S. A. Mehdi and K. E. Weale, work to be published.
34. J. G. Kilroe and K. E. Weale, *J. Chem. Soc.*, **1960**, 3849.
35. M. Rahman and K. E. Weale, *Polymer*, **11**, 122, (1970).
36. W. K. Busfield and E. Whalley, *Trans. Faraday Soc.*, **59**, 679, (1963).
37. Y. Ohtsuka and C. Walling, *J. Am. Chem. Soc.*, **88**, 4167, (1966).
38. W. J. le Noble, *J. Am. Chem. Soc.*, **82**, 5253, (1960).
39. A. R. Osborn and E. Whalley, *Trans. Faraday Soc.*, **58**, 2144, (1962).
40. J. C. Woodbrey and P. Ehrlich, *J. Am. Chem. Soc.*, **85**, 1580, (1963).
41. W. S. Richardson, *J. Polymer Sci.*, **13**, 321, (1954).
42. G. Jenner, J. Hitzke and M. Millet, *Bull. Soc. Chim. France*, **1970**, 1183.
43. J. P. J. Higgins and K. E. Weale, *J. Polymer Sci.*, **B7**, 153, (1969); and **A1**, 1705, (1970).
44. D. W. Brown, J. E. Fearn and R. E. Lowry, *J. Polymer Sci.*, **A3**, 1641, (1965).
45. C. Walling and D. D. Tanner, *J. Polymer Sci.*, **A1**, 2271, (1963).
46. V. M. Zhulin, R. I. Baikova and M. G. Gonikberg, *Izvest. Akad. Nauk. SSSR.*, *Otdel. Khim. Nauk.*, **1964**, 1133.
47. A. Crosato-Arnaldi, G. B. Guarise and G. Talamini, *Polymer*, **10**, 385, (1969).
48. G. B. Guarise, G. Palma, E. Siviero and G. Talamini, *Polymer*, **11**, 613, (1970).
49. M. Prince and J. Hornyak, *J. Polymer Sci.*, **5**, 531, (1967).
50. V. M. Kapustyan, A. A. Zharov and N. S. Enikolopyan, *Dok. Akad. Nauk. SSSR.*, **179**, 627, (1968).
51. A. G. Kazakevich, A. A. Zharov, P. A. Yampol'skii, N. S. Enikolopyan and V. I. Gol'danskii, *Dok. Akad. Nauk. SSSR.*, **186**, 1345, (1969).

7

Emulsion Polymerization

W. Cooper

Dunlop Research Centre, Birmingham

7.1 INTRODUCTION

Although the focal points in research on vinyl polymerization in recent years have been concerned with ionic processes, the interest and importance of emulsion systems has not significantly decreased and there is a continuing effort on the details of the polymerization mechanism. The technological basis likewise remains large, the output of emulsion polymers, elastomeric and plastic, being of the order of several million tons per annum.

Most emulsion polymerizations are initiated by free-radicals and show many of the characteristics of other free-radical systems, the main mechanistic differences resulting from the localization of the propagating free-radicals in small volume elements. Polymerization in emulsion not only permits the synthesis of polymers of high molecular weight at high rates, often in cases where bulk polymerization is not possible or is inefficient, but has other important practical advantages. The water absorbs much of the heat of polymerization and facilitates reaction control, and the product is obtained in a relatively non-viscous and easily handled form, to be used as such or readily to be converted into solid polymer after coagulation, washing and drying. In copolymerization, although copolymerization theory is applicable to emulsion systems, differences in solubility of the monomers in the aqueous phase may result in deviations from the behaviour in bulk, with important practical consequences.

The technological origins of emulsion polymerization go back to the 1920's and before the World War II there was a well established industry for the production of synthetic rubbers and plastics by emulsion techniques. During and immediately after the war years, there was considerable progress in the development and application of new, versatile and efficient initiating systems and at about this time the decisive role of the emulsifier on the polymerization was appreciated.

During the past two decades there has been a considerable effort concerned with testing the qualitative and quantitative aspects of the theory of emulsion polymerization. Much of this has been academic in character but a clearer appreciation of the principles involved has assisted in the formulation of practical emulsion systems on rational grounds, rather than on pure empiricism.

7.2 GENERAL FEATURES OF EMULSION POLYMERIZATION

Emulsion polymerizations are open to great variation, from simple systems containing a single monomer, emulsifier, water and a simple initiator to those containing two or three monomers, added at once or in increments, mixed emulsifiers and auxiliary stabilizers, and complex initiator systems including chain transfer agents.

The ratio of monomer to aqueous phase can be varied widely but in technical practice it is generally limited to the range 30/70 to 60/40. The higher levels approach the limit which can be achieved by direct polymerization and owing to the substantial amount of heat to be dissipated can often only be obtained by incremental addition of monomer. A further complication is that the latex increases substantially in viscosity, particularly with the more water-soluble monomers which are miscible with the polymer, and then falls as the reaction is taken to completion. This stage is not infrequently accompanied by an increase in the latex particle size by aggregation and, when surface stabilization is inadequate, by coagulation.

7.2.1 The emulsifier

The emulsifier influences the polymerization rate and the properties of the latex, including particle size, particle size distribution and stability, and its choice in practical systems is important. Emulsifiers are classified into anionic, cationic, and non-ionic types but all are characterized by the presence in the molecule of water soluble and water insoluble portions. They form normal solutions in solvents (aqueous and non-aqueous) at high dilution but abruptly change into aggregated forms, termed micelles, at specific and usually relatively low concentrations. In aqueous solutions the inner portion of the micelle contains the hydrocarbon or other water insoluble component of the molecules, while the hydrophilic portion is located at the surface. The change at the critical micelle concentration (c.m.c) occurs so as to minimize the free-energy of the solution (heat is liberated) and, in general, the molecular weight of the micelle increases as the size of the water insoluble segment increases. In compounds containing linear aliphatic hydrocarbon chains there is little activity below C_{10} while limited solubility restricts the use of emulsifiers containing alkyl groups greater than C_{18}. Micelle formation is accompanied by changes in the properties of the solution the best known of which is the sharp fall in the surface tension of the solution.

Important anionic emulsifiers are fatty acid soaps and salts of linear and branched chain alkane hydrogen sulphates and sulphonic acids, and alkyl benzene and alkyl naphthalene sulphonic acids. The usual counter-ions are sodium, potassium and ammonium. The fatty acids give insoluble salts with most other metals, although the sparingly soluble lithium salts have been used. Certain emulsifiers give other soluble salts, for example of calcium, but this is more in connection with their applications as wetting agents for use with hard water.

Cationic surfactants are almost restricted to quaternary ammonium salts with bromide or chloride counter-ions. They are not widely used commercially and frequently give low rates of polymerization. This arises particularly when the positive charge on the polymer particle is opposite to that of the initiating radicals (e.g. SO_4^-).

There are numerous non-ionic emulsifiers of widely differing structure. They are often complex mixtures and may be of high molecular weight. The more important contain polyethylene oxide chains with one end attached to a hydrophobic grouping (e.g. oleyl, p-nonylphenyl, lauryl or a polypropylene oxide chain) and the other end a hydroxyl group or a methyl group. (It is also possible for the molecule to contain an ionizable grouping such as sulphate or sulphonate). Other hydrophilic groupings are carbohydrate derivatives, e.g. of sorbitol. The onset of micelle formation occurs at comparable concentrations by weight to the ionic surfactants, but the surface tension reduction is not usually so great. Like the ionic emulsifiers it is presumed that the water soluble part of the molecule envelops an inner hydrophobic portion, and in some cases, but not all, there is clear evidence for the formation of (uncharged) micelles. Most of these materials have molecular weights in the range 300 to 8000 but in principle there is no specific limit, as shown by certain water soluble high polymers such as partly acetylated polyvinylalcohol.

These latter types are often used as stabilizers and appear to have properties in between the micelle-forming emulsifiers and the completely water-soluble gums. They reduce the surface tension of water and a break equivalent to a c.m.c. can be detected. The extent to which this occurs depends on the precise structure of the polymer. Polyvinylalcohol itself has little surface activity and the greatest effect is in polymers in which acetate groupings are present in short blocks.[1] (Hydrolysis of an acetate group in polyvinylacetate facilitates attack on an adjacent unit in the polymer chain which tends to give rise to block structures, but the distribution can be controlled during hydrolysis to some extent). It is these which have the best stabilizing effect in emulsion polymerization and they are used particularly in polyvinylacetate emulsions.

Similar considerations probably apply to salts of partly hydrolysed acrylate polymers.[2] A triethylamine salt of a 90/10 butyl acrylate/acrylic acid co-polymer reduced the surface tension of its aqueous solution to a value in the

Table 7.1 Properties of emulsifiers[3]

Emulsifier	c.m.c. mol/1	per cent	Micellar weight		Area/molecule Å²
Potassium laurate	0·0125	0·30	11 900	H₂O	32
			87 000	(1·6 M KBr)	
				(0·1 M K₂CO₃)	
Potassium stearate	0·0005	0·016	—		—
Potassium oleate	0·0012	0·04	—		28
Potassium perfluorononoate	0·0009	0·05	—		—
Potassium abietate	0·012	0·39	—		25–40
Sodium p-dodecyl benzene sulphonate	0·0016	0·055	8 200	(H₂O)	32–40
Sodium lauryl sulphate	0·0087	0·25	17 100	(H₂O)	
			32 200	(0·1 M NaCl)	23
Sodium lauryl sulphonate	0·0095	0·26	14 700	(H₂O)	—
Sodium dioctylsulpho succinate	0·0007	0·03	21 300	(H₂O)	25
Cetyl trimethyl ammonium bromide	0·001	0·036	61 700	(0·013 M KBr)	—
Cetyl pyridinium chloride	0·006	0·21	32 300	(0·0175 M NaCl)	46
Dodecylamine hydrochloride	0·014	0·31	12 300	(H₂O)	26
Octylphenol polyoxyethylene(9)*	0·0002	0·012	66 700	(H₂O)	53
Nonylphenol polyoxyethylene(30)*	0·00025	0·026	—		101
Methylpolyoxyethylenedecyl-ether[12]*	0·0011	0 055	37 100	(H₂O)	—
Sorbitan monolaurate	0·002	0·067	—		36
Block 80/20 ethylene oxide-propyleneoxide MW 8200	6·9 × 10⁻⁶	0·006	8 400	(H₂O)	920

* Figures in parenthesis indicate the number of oxyethylene groups in the molecule.

1. Branched chain surfactants have a similar but rather higher c.m.c. for a given hydrophobic grouping.
2. Molecular weights of ionic micelles are not greatly affected by temperature (a small reduction occurs) but they are greatly increased, to over 10^6 in some cases, by electrolytes.
3. Molecular weights of non-ionic micelles are not affected by salts but rise with an increase in temperature, with precipitation and separation into two phases (the 'cloud point').
4. There is a constitutive relationship between c.m.c. and number of carbon atoms in most types of emulsifier. The polyoxyethylenephenylethers and similar compounds have c.m.c.'s which relate to the molar ratio (R) of ethylene oxide to phenol (in the case of the nonylphenyl ethers in water, $\log$ (c.m.c.) $= 0\cdot056R + 3\cdot87$). Above molecular weights of about 600 the efficiency for micelle formation falls with increasing ethylene oxide content.
5. There are substantial discrepancies in the literature values for micellar weights of certain emulsifiers. Thus, Ludlum[4] reports 8200 and 19 000 for carefully purified para and meta sodium dodecyl benzene sulphonates whereas for an unspecified sodium dodecyl benzene sulphonate Mankowich[5] reports 1000 to 1700 and his value of 7200 for sodium dioctylsulphosuccinate is lower than the value quoted in Table 7.1. Likewise the value of 8400 for the 8200 molecular weight block ethylene oxide-propylene oxide copolymer, which implies that this type of emulsifier does not form micelles is surprising in view of its c.m.c. being at concentrations well below those used for the determination of molecular weight. If the polymer had no micelle forming character the surface tension of its aqueous solution might be expected to fall steadily with concentration, and not to show a well defined break.

region of 35 dyn/cm at 0·5 per cent concentration (corresponding to 3·6 × 10^{-3} mol/l of acrylate salt).

Properties of a number of widely used emulsifiers are given in Table 7.1.

Choice of stabilizer in emulsion systems has always been rather empirical. In some cases it is determined by subsequent processing or use of the polymer, as in the case of synthetic rubbers in which the fatty acid or rearranged rosin acid participate in the vulcanization of the elastomer and the use of simple soaps permits coagulation of the latex by acid-brine. In other cases, latex viscosity, flow properties and mechanical or freeze-thaw stability are the dominating factors, and for these purposes a different choice would be made.

There have been numerous attempts to classify the behaviour of emulsifiers in regard to their effects on the properties of emulsions. In view of the diversity of chemical structures, the differences in ionic charge, etc., there is no obvious property, but a measure of the relative oil and water solubilities of the emulsifier has been shown to have a useful correlation with properties. This has been expressed numerically as HLB numbers*, according to which most surface-active agents have values from about 1 to 40, lower numbers indicating oil solubility and higher numbers, water solubility.[6] Some typical HLB numbers are, polyoxyethylene sorbitan monolaurate 16·7, polyoxyethylene alkylaryl ethers 10 to 18, sodium lauryl sulphate 40, sorbitan monolaurate 8·6.

This somewhat arbitrary numerical characterization has been shown to be related to the spreading coefficients of the emulsifiers, defined in terms of the tensions at the liquid surfaces and interface.[7]

$$S = \gamma_{\text{water}}^{\text{oil or}} - \left\{\gamma_{\text{or oil}}^{\text{water}} + \gamma_{\text{interface}}\right\}$$

The spreading coefficient is thus a measure of the difference between the work of adhesion of the surfaces of the two phases and the work of cohesion of the oil or water phase. It is found that optimum emulsion stability is associated with a small negative spreading coefficient. The reasons for this are not clear but it is possible that the emulsifier retards spreading of small liquid droplets and so tends to retain them as such, and the relationship of this property to the stability of oil in water emulsions is thus fairly clear. Its relevance to emulsion polymerization however is less obvious. The distribution of monomer is important only in so far as its transfer to the growing polymer particle is affected and as a polymer emulsion can be formed from a separate layer of monomer, stabilization of the monomer emulsion is unimportant. Coverage of polymer particles is obviously influenced by emulsifier structure, although the spreading of a highly viscous polymer particle would hardly seem to be of consequence. Rate of polymerization and latex stability are, however, affected by particle-particle interactions and

---

* The contraction HLB stands for hydrophile-lyophile balance.

possibly by interaction between the particle surface and the surrounding aqueous medium. It is conceivable that a parallel relationship between particle aggregation, or ingress or egress of free-radicals, and stabilization of emulsions exists, and if so an HLB correlation could be of value.

So far the HLB concept has only received a limited examination and it is not possible to come to a firm conclusion. With styrene and vinyl acetate maxima in conversion rate, latex stability and inverse of particle diameter were observed at HLB's in the region of 15 to 16 and 16 to 17·5 respectively,[8] but with vinyl chloride no particular significance of HLB number was found.[9]

The nature of the chemical groupings in the emulsifier may be important since side reactions can occur which preclude the use of certain types. For example, in the polymerization of vinyl acetate the pH of the emulsion falls to a value in the region of 5. Acid stable emulsifier must, therefore, be used; fatty acid soaps, which have a pH in the region of 10, would hydrolyse the monomer with the formation of acetaldehyde and loss of latex stability. With this particular monomer cationic emulsifiers give unstable latices, so the choice tends to be alkyl or alkylaryl sulphate or sulphonate types, often in conjunction with non-ionic stabilizers.

The polymerization of monomers which give highly reactive radicals, such as ethylene and vinyl acetate, may be inhibited by emulsifier, and those containing unsaturation in the hydrocarbon portion retard the polymerization quite significantly. With ethylene even potassium myristate retards polymerization to some extent.[23]

Sodium abietate from wood rosin retards the polymerization of butadiene/styrene, due to the presence of conjugated double bonds in the molecule, but the rosin acids (usually as 2/1 blends with fatty acid soaps) confer advantages on the elastomer. To avoid this difficulty the abietic acid is catalytically disproportionated to give a mixture of aromaticized and reduced acids.[11]

Fluoropolymers are made exclusively by emulsion polymerization and here it is necessary to use emulsifiers which are highly fluorinated to avoid incorporation of non-fluorine containing material into the polymer. The CF_2 grouping has a greater surface tension depressant action and aggregation power than the CH_2 grouping in a homologous series and in this instance adequate emulsifier activity is achieved at C_8–C_9.

7.2.2 Auxiliary stabilizers

These materials are added to many emulsion polymerization recipes with the objectives of controlling particle size and distribution and latex stability, and their precise role is by no means clear. They are usually high molecular weight water soluble polymers which are adsorbed on the surface of the polymer particles. In mixtures with other emulsifiers there appears to be competition between auxiliary stabilizers and the micellar type of emulsifier for adsorption on the latex particles.

Many of these substances are creaming agents for latices, the explanation for this being associated with the ability of a single molecule to bring together reversibly a number of latex particles. In the polymerization they probably have a similar effect, and if the surface of the particle is not fully stabilized by the micellar soap the auxiliary stabilizer may cause limited particle aggregation to occur. The adsorbed stabilizer may have portions of the molecules extending into the aqueous phase thus altering the viscosity of the latex, and subsequent addition of other stabilizers can have a marked effect on latex viscosity and the sensitivity to shearing forces[12] when solid particles are present.

Typical auxiliary stabilizers are polyvinylalcohol, sodium polyacrylate, carbohydrates such as gum arabic, polyethyleneoxide, and water soluble cellulose derivatives such as methyl and carboxymethyl cellulose.

7.2.3 Initiators

The most widely used initiators for emulsion polymerization are water soluble persalts, e.g. ammonium, potassium, and sodium persulphates and organic hydroperoxides, such as cumene hydroperoxide. Hydrogen peroxide is used but it is less reproducible in its effect and less efficient than the persulphates. The major reaction of peroxide initiators is homolytic cleavage of the O—O bond, hydrogen peroxide and the persulphates giving the hydroxyl radical and the sulphate ion radical respectively. The peroxides decompose at rates which depend on the medium and some of this may be by non-radical paths. Thus potassium persulphate decomposes at rates varying by a factor of five in the presence of common emulsifiers,[13] and twenty times faster in aqueous vinyl acetate solution compared with water.[14] The decomposition is also pH dependent; increasing rapidly below pH 4, when hydroxyl radicals may be produced by the reaction.

$$SO_4^- + H_3O^+ \rightarrow H_2SO_4 + HO\cdot$$

Ammonium persulphate, which is more soluble than the potassium salt, undergoes side reactions involving oxidation of the ammonium ion.

The persulphates are limited in their use to a temperature range over which they decompose at a reasonable rate, usually in the region of 40–80 °C. Organic peroxides vary widely in stability dependent on structure and it is thus possible to choose a suitable initiator for a given temperature range. Diisopropyldiperoxycarbonate and di-t-butyl peroxide have half-life periods for decomposition of 10 hours at 35 °C and 126 °C respectively. However, most organic peroxides are soluble in monomer and insoluble in water which could cause difficulties by initiating polymerization in the monomer droplets, and the storage of unstable free-radical generators, necessarily at low temperatures, is inconvenient.

For polymerizations at low temperatures 'redox' catalysts are generally used, in which two or more relatively stable compounds interact to produce free-radicals at temperatures as low as −50 °C. Such systems are, moreover, particularly suitable for emulsion polymerization.

Two simple redox systems are potassium persulphate and sodium metabisulphite, and hydrogen peroxide and ferrous sulphate. In these both components are soluble in the aqueous phase. They are effective and will induce rapid polymerization at or below room temperature. There is a tendency however with this type of initiator for rapid and uncontrolled generation of

Table 7.2   Redox systems for emulsion polymerization

Oxidizing agents	Reducing agents
Potassium persulphate	Ferrous sulphate usually with a second
Potassium ferricyanide	reducing agent such as glucose,
Ceric sulphate	fructose, dihydroxyacetone,
Potassium permanganate	tetraethylenepentamine, sodium
t-Butyl hydroperoxide	formaldehyde sulphoxylate or
Cumene hydroperoxide	sodium hydrosulphide
Pinane hydroperoxide	Sodium hyposulphite
Diisopropylbenzene	Sodium metabisulphite
hydroperoxide	Sodium sulphide
	Sodium thiosulphate
	Hydrazine hydrate

free-radicals, which increases greatly with rise in temperature (activation energies are normally in the region of 12–15 kcal/mol) and the polymerization may die out before it is complete. It is possible to make repeated additions of catalyst components but a better solution to this problem has been to have one component of the 'redox' system soluble in the aqueous phase and the other in the monomer, the rate-determining step then being diffusion of the oil soluble component from the monomer into the aqueous phase. Organic peroxides such as benzoyl peroxide can be used for this purpose but they have proved to be less efficient than the organic hydroperoxides which have a better partition coefficient between the monomer and water phases. Examples of suitable oxidants are given in Table 7.2.

The reducing portion of the redox system (usually but not invariably dissolved in the aqueous phase) has also been the subject of much attention. It frequently composes more than one compound and there may be a variable valency transition metal taking part in the cycle.

$$Ox + Fe^{++} \rightarrow R\cdot + Fe^{+++}$$
$$Fe^{+++} + Red \rightarrow Fe^{++} + R\cdot'$$

Free-radicals are produced from both the oxidizing and reducing agents, and in principle both can be intercepted by the monomer to form a growing chain. The actual behaviour will depend on the solubility and reactivity of the radical species involved and marked differences would be expected, for example, between the cumyloxy radical ($PhCMe_2O\cdot$) and the thiosulphate ion radical $S_2O_3\cdot^-$.

The iron-based redox systems have been studied in depth, and it has been found that the availability of the metal can be a determining factor in controlling the rate of polymerization. The iron salt may be precipitated as the sparingly soluble sulphide or sequestered with potassium pyrophosphate. In the latter case the activity of the metal ion is dependent on the structure of the ferrous pyrophosphate complex and is varied greatly by heating the components in aqueous solution prior to forming the catalyst. Other sequestering agents, such as ethylene diamine tetra-acetic acid and its salts, citrates and acetyl acetone, may be used as alternatives to the phosphates. Examples of reducing agents are given in Table 7.2. Variable valency metals other than iron can be used in redox systems, including copper, manganese and cerium, but they are not widely used as such metals tend to be undesirable impurities in polymers.

Very many of the redox systems were developed for the emulsion polymerization of butadiene/styrene, and fast, reproducible high conversion polymerization can be carried out at temperatures in the region of $-20\,°C$ to $-40\,°C$. There is no information on the rates of free-radical production in many of these complex multicomponent redox initiators but in one of the simpler systems a rough estimate can be made. The reaction of ferrous sulphate and potassium persulphate has a rate constant (extrapolated to infinite dilution) of $42\cdot6$ l mol^{-1} s^{-1} at $10\,°C$.[15a]* An exact comparison with potassium persulphate is not easy because of the dependence of rate constants on reagent concentrations and the presence of soaps and other compounds, but the radical generation from an equimolar mixture of ferrous sulphate and potassium persulphate is about 100 times greater at $10\,°C$ than that from potassium persulphate alone at $50\,°C$.

Other free-radical generators, such as azo compounds, have not been used to any great extent in emulsion polymerization, but recently two water soluble azo compounds which are active in the temperature range 40–$80\,°C$ have been described.[16]

$$N_2(CMeCNCH_2SO_3^-Na^+)_2; \quad N_2[CMe_2C(NH_2)_2^+Br^-]_2$$

These are useful since the decomposition of azo initiators is less affected by other compounds than is the case with peroxides, and the latter is of interest

---

* The corresponding reaction of ferrous sulphate with cumene hydroperoxide has about $\frac{1}{8}$ the rate of this.[15b]

in that it gives a cation-radical on homolytic breakdown, instead of the more usual neutral or anion radicals.

7.2.4 Chain transfer agents

Molecular weights of emulsion polymers are controlled by the same means as those of bulk or solution polymers, but control may be more essential since, owing to the reduced rate of chain termination in emulsion systems, polymer molecular weights are much higher. The most important group of chain transfer agents are the aliphatic mercaptans although other materials, such as diisopropyl xanthogen disulphide and halogenated hydrocarbons, can also be employed. They are used particularly with the hydrocarbon monomers where transfer of growing chains with monomer is slow; monomers such as vinyl acetate with a large transfer coefficient do not normally require chain transfer agents. It should be emphasized that for applications of latices, for example as films or in adhesives, etc., high molecular weight is normally a desirable feature and the need for molecular weight control applies more to those polymers which are to be isolated and subsequently subjected to mechanical working or which give rise to undesirable polymer cross-linking in the absence of chain transfer agents. These considerations apply to the polymers and copolymers of butadiene and most of the information on mercaptan modifiers relates to these polymers.

The mercaptans used are C_8–C_{16} straight and branched chain compounds which are soluble in the monomer. The mercaptan is consumed during the polymerization and the rate is determined partly by rate of reaction with polymer free-radicals and partly by diffusion from the monomer droplets to the polymerization loci at a satisfactory rate.

7.2.5 Other additives

The most usual additives in emulsion polymerization are electrolytes such as sodium chloride or potassium sulphate. Their effect is to reduce the number of micelles in the emulsifier solution but they do not have a predictable effect either on the rate of polymerization or the number of particles in the latex. Electrolytes are, however, often included in recipes for the direct polymerization of high concentration (60 per cent) latices. Possibly they affect the stability of the latex particles which then undergo limited aggregation.

Antifreeze agents are necessary in emulsion polymerizations carried out at low temperatures. Methanol and glycerol have been used in 'cold rubber' formulations at temperatures down to $-40\,°C$. High concentrations are required (25–100 per cent on the aqueous phase) but at least with butadiene/styrene monomers there is no inhibition of polymerization and normal latices are obtained.

Other additives used in emulsion polymerization are the sequestering agents which form part of the initiator system by removing or stabilizing heavy metal ions in a form such that polymerization rate is controlled. pH buffers are occasionally added to polymerization recipes but generally are not of significance.

7.2.6 Polymerization

Initially an emulsion of monomer as relatively large droplets above 10 000 Å in diameter is produced. With the systems normally used the emulsion is not particularly stable and on standing a separate layer of monomer forms quickly, leaving a relatively small amount dissolved in the aqueous solution. As polymerization proceeds there is usually, but not in all cases, an accelerating rate after an induction period of variable duration. The induction period is caused by traces of oxygen or other impurities and may be of negligible duration at elevated temperatures. The separate phase of monomer disappears at a conversion which depends on its structure and is usually in the range 15–80 per cent, and with those systems where the polymer is swollen with monomer the emulsion becomes more viscous. The reason for this viscous stage is not entirely clear. There will, of course, be particle-particle interaction which could arise from a tendency of the swollen particles to stick together in reversible aggregates. With inadequately stabilized latices there is a marked tendency for coagulation to occur at this point, and this arises particularly at high polymer concentrations when there will be a large particle surface area. It has also been found that emulsions containing poly-functional cross-linking monomers tend to coagulate at lower concentrations than those from which the cross-linking monomer has been omitted—also implying interparticle interaction.

A qualitative interpretation of the main features of emulsion polymerization was put forward by Harkins in 1946[17] and although there have been additions and modifications of the theory and studies of the quantitative aspects of emulsion polymerization arising from the concept have revealed deviations from what would be expected, the overall view of emulsion polymerization remains widely accepted.

It was considered that the large water-insoluble monomer droplets had a relatively small proportion of the emulsifier adsorbed on their surface. Solubilization of monomer occurred in the micelles of the emulsifier in the aqueous phase in which polymerization was initiated by entry of free-radicals from the aqueous solution. At an early stage in the polymerization a multiplicity of small polymer particles would be formed, isolated from one another by the aqueous phase and by the emulsifier layer adsorbed on their surfaces. Monomer would transfer from the droplets into the growing particles, and during this period the rate would remain approximately constant. In the final stage of the polymerization when all the monomer has been transferred to the particles the rate would diminish as its concentration fell. A free-radical could enter a monomer droplet and initiate polymerization but in view of the relatively small number and hence small surface area (about 10^{11} droplets/cm^3 with less than one tenth the area of emulsifier molecules) this is not important compared with entry into the micelles.

It is apparent that the number of particles will be influenced by the emulsifier concentration and this, together with the behaviour of free-radicals in latex particles, dominates such features as rate of polymerization and molecular weight of polymer. A rough calculation suggests that the number of particles in the emulsion will be very large. As the aqueous system usually contains from 2–5 per cent emulsifier it is readily deduced from the data in Table 7.1 that each cubic centimetre of water will contain in the region of 10^{17}–10^{18} micelles, with a surface area sufficient to provide a monomolecular layer for 10^{15} particles of diameter 500 Å. If there were 10^{15}–10^{16} particles/cm^3 of aqueous phase and each one contained a free-radical then with a 40/60 monomer/water ratio this would be equivalent to a free-radical concentration of 10^{-6}–10^{-5} mol/l in the monomer. This is some 10^2 to 10^3 times higher than in most bulk systems and would result in very high polymerization rates.

7.3 KINETICS OF EMULSION POLYMERIZATION

The first treatment of emulsion polymerization kinetics was by Smith and Ewart.[18] One of the more important assumptions of this theory is that the rate and kinetic chain length are determined by the ingress of radicals into the growing particles and not by the competing rates of propagation and polymer radical termination reactions as in the case of bulk polymerization. An alternative viewpoint was later proposed by Medvedev[19] which differs significantly in that it is suggested that following generation of free-radicals polymerization occurs in an adsorbed layer of monomer at the particle surface. In the Smith-Ewart theory polymerization is considered to be initiated in the micelle and to continue within the monomer-polymer particle. The two treatments give rise to a different dependence of rate on initiator and emulsifier concentrations.

The Smith-Ewart theory has received a greater measure of support, but there are major deviations with all monomers and support for the view that polymerization occurs at the particle surface has come from several later investigations.

In the Smith-Ewart theory, following initiation within the soap micelles, three limiting cases were considered relating to the average number ($\bar{n}$) of free-radicals in a particle: (a) $\bar{n}$ is small compared with unity, due to rapid transfer of radicals out of the active particle, (b) $\bar{n}$ is approximately 0·5, which results from the assumption that the lifetime of two radicals in a particle is small compared with the interval between entry of successive radicals, and a particle will thus contain a single radical or none, (c) $\bar{n}$ is much greater than unity.

The second case has been found to approximate most closely to emulsion polymerization, although examples are known where the average number of free-radicals is significantly less or greater than 0·5. In particles containing a number of free-radicals, the situation approaches a bulk polymerization. Analysis shows this to arise in larger particles, and above a diameter of 1 μ bulk kinetics are followed.

The assumptions which have formed the basis of most calculations are:

(a)   Polymerization occurs in the soap micelles by radicals entering from the aqueous phase.

(b)   The rate of initiation is constant and particles are produced until all the micellar soap has left the solution.

(c)   New particles are not formed nor are they lost by coalescence after nucleation is complete.

(d)   Entry of a second radical into a particle containing a free-radical results in rapid termination.

(e)   The monomer/polymer ratio remains constant while there are monomer droplets present in the system.

(f)   The rate of increase in volume of a particle with time is constant.

(g)   Transfer of radicals out of the particles is inappreciable.

The rate equation for polymerization

$$\frac{-d[M]}{dt} = k_p[M][R\cdot]$$

where $-d[M]/dt$ is the rate in mol l^{-1} s^{-1}, k_p the propagation velocity constant (1 mol^{-1} s^{-1}) and [M] and [R·] the monomer and free-radical concentrations in mol/l respectively, is modified for systems in which there is a variable water to monomer ratio and a separate water phase. To ensure self-consistent units the rate R_p may be expressed as mol/cm³ of water/s. Whence

$$R_p = k_p C_M \frac{N\bar{n}}{N_A}$$

C_M is the equilibrium molar concentration of monomer in the swollen particles, N is the number of particles/cm³, containing on average $\bar{n}$ radicals, and N_A is Avogadro's number. In other investigations the volume fraction of monomer ϕ_M in the swollen particle is used

$$\phi_M = (M_0 C_M/d_m)10^{-3}$$

M_0 is the molecular weight of the monomer and d_m its density. The rate of polymerization may be expressed as R_p' the rate of formation of polymer in cm^3/cm^3 of water/second

$$R'_p = k'_p \phi_M (d_m/d_p) \frac{N\bar{n}}{N_A}$$

k'_p is in units of cm^3 mol^{-1} s^{-1} ($= 10^3 k_p$).

The consequences of particle nucleation in the micelles and rapid termination within the particles is that there would be an accelerating period during which the number of particles increases followed by a region of constant polymerization rate during which the average number of radicals/particle will be 0·5. With the disappearance of the monomer droplets there will be a fall in the concentration of monomer in the particle and a steady fall in polymerization rate.

Since the rate of volume increase of a particle with time is constant, it follows that the integral of the areas of the particles produced will be proportional to the $\frac{5}{3}$ power of the time. Hence the time at which nucleation is complete, that is when the area of the particle surface is equal to the area of emulsifier molecules, will be proportional to the $\frac{3}{5}$ power of the emulsifier concentration (S). The number of particles produced is proportional at constant initiation rate to the nucleation time. The number of particles/gram of emulsifier thus varies with the $-\frac{2}{5}$ power of the emulsifier concentration. Calculations have led to the following relationships.[20]

(i)   The time for nucleation of particles

$$t_{cr} = 0\cdot 365 (S/r)^{0\cdot 6}/(k)^{0\cdot 4}$$

k is a constant for the system corresponding to the growth of a single particle and is related to the monomer concentration, the number of radicals in each particle and the propagation velocity constant. r is the rate of radical production $= 2N_A k_d[\text{I}]$, where I is the concentration of initiator and k_d is its first order decomposition velocity constant.

(ii)   The number of particles/cm^3 of water

$$N = 0\cdot 208 \, S^{0\cdot 6}(r/k)^{0\cdot 4}$$

The mean volume particle diameter (D_v) is related to N by the equation

$$\pi D_v^3 N = (m/w)(d_m/d_p)$$

where m/w is the volume ratio of monomer to water.

(iii)   The conversion at which the nucleation process is complete

$$P_{cr} = 0\cdot 209 S^{1\cdot 2}(k/r)^{0\cdot 2}(1 - \phi_M)$$

(iv)   The molecular weight of the polymer, determined by propagation in the interval between entry into the particle of an initiating and a terminating radical

$$\bar{M}_n = k'_p d_m \phi_M N/\tau$$

An examination of the literature shows that the behaviour of no single monomer is wholly explained, either in terms of these theories or their later refinements, and none of the assumptions hold in entirety. The best agreement with Smith-Ewart kinetics has been observed, as would be expected, with water insoluble monomers such as styrene. In addition to complications resulting from water solubility of monomer and transfer of radicals in and out of growing particles, factors such as the solubility of monomer in polymer, radical trapping, and the onset of diffusion-controlled termination in high viscosity systems, are of major significance. This being so, there seems little merit for the purpose of this review in analysing the kinetics in detail with the objective of testing the theoretical relationships. Rather the main assumptions and the relationships will be considered in the light of the experimental data.

7.3.1 Nucleation

If Smith-Ewart kinetics are followed the rate should accelerate and then remain constant until the monomer concentration in the particle starts to decline. This first stage is frequently observed but in general from rate measurements alone it is not possible to identify it with the nucleation process, as it could equally well be caused by retardation from traces of oxygen or other impurities. The more water soluble monomers in inhibitor-free systems give zero-order rates to very high conversions[21] and even with styrene the acceleration period is not always observed.

At the end of the nucleation process micellar soap should no longer be present in solution. This is generally true, and there is a rise in surface tension of the latex at this point. Foaming, which may occur on stirring monomer emulsions, and is indicative of the presence of micellar soap, has also been found to cease at an early stage in the polymerization of vinylidene chloride.[22] In accord with theory the conversion at which foaming ceased increased with the 1·2 power of the emulsifier concentration. The conversion at completion of nucleation is readily calculated from the kinetic parameters but although calculation and experiment are not infrequently of the same magnitude the data are sufficiently scattered so as not to be a reliable test of the theory.

7.3.2 Particle number

This has been extensively studied and is the most useful in assessing the merits of the kinetic schemes. According to theory the number of particles should rise to a maximum value during nucleation and then remain constant. Above 10 per cent conversion this appears to be true and with adequately stabilized latices there is no change in particle number up to complete conversion. Polymerization in the presence of seed latices takes place in the preformed particles and even with appreciably water soluble monomers seeded latices maintain constancy of particle number.[21b] Evidently unstable particles formed in the aqueous phase aggregate with the preformed particles.

In the early stages of polymerization the more water soluble monomers give very small particles, as small as 30 Å in diameter,[23,61] and at a relatively low conversion aggregation occurs to a constant number of particles. In the case of vinyl acetate a four-fold fall in particle number has been observed at low emulsifier concentrations but, as would be expected, the change at high emulsifier concentrations was smaller.[24]

Krishan and Margaritova[101] interpret coalescence as arising from the requirement of constant particle surface area during polymerization. With methyl methacrylate the particle diameter increases linearly with conversion indicating particle aggregation. However, in this case, the particle number although it falls rapidly in the early stages only changes slightly with a change in polymer concentration from 20–34 g/100 cm³ H₂O, although the corresponding change in particle diameter is from about 500–900 Å; clearly a critical interpretation of more data is required for definite conclusions to be drawn.* There is evidence to show that the particle surface is unsaturated with respect to emulsifier when it is present in low concentrations so that some aggregation is not unexpected. Complete surface coverage is accompanied by a break in the rate against emulsifier concentration curve, which is interpreted as the building of polymolecular layers of emulsifier on the particle surface. This behaviour is not universal, however, and with vinyl chloride the exponent for the dependence of rate on emulsifier increases with increase in its concentration.[38]

Clearly the number of growing centres and their stability later in the polymerization must depend on the monomer, emulsifier and initiating radical and no universal explanation can be given based on any single theory. In fact, it is not essential to have emulsifier present in order to form a fine particle size latex. Methyl methacrylate, for example, in aqueous solution polymerizes to give dispersions containing 10^{14}–10^{15} particle/cm³.[25] The polymer radicals precipitate at very low concentrations (~ 0.03 per cent) to form particles which then absorb monomer from the aqueous solution, and as conversion proceeds partial aggregation occurs since the particle surface is unstabilized. There is no reason to eliminate this type of initiation at higher monomer concentrations in the presence of emulsifiers.

With water soluble monomers this behaviour may not be unexpected, but there are also results on styrene which indicate that micelles do not play an essential role in particle nucleation, notwithstanding the agreement with theory in the presence of emulsifiers.[26] There is virtually no polymerization of styrene with a water soluble initiator in the absence of emulsifier, but potassium laurate at concentrations below the c.m.c. (0.005–0.01 mol/l) has been found to increase the number of particles by up to 100 times. Also recipes based on nonylphenoxypolyethyleneoxyethanol and small concentrations of sodium lauryl sulphate (2–10×10^{-4} mol/l), in which the number of

---

* There is for example evidence that with styrene the number of particles may increase in the steady polymerization rate stage,[88] contrary to theory and to most other observations.

micelles, as judged by light scattering measurements, remained approximately constant in the region of $1\cdot5$–$1\cdot75 \times 10^{16}$/cm³, exhibited great variations in the number of particles formed; from $8\cdot7 \times 10^{13}$ in the absence of sodium lauryl sulphate to $9\cdot5 \times 10^{14}$ at the highest concentration.

It is considered that the primary particles are formed in the aqueous phase on to which the emulsifier is adsorbed irrespective of whether it is present in molecular or micellar form. Aggregation of these primary particles could occur later by coagulation with termination of the propagating chains. Molecularly dissolved emulsifier would be adsorbed in proportion to its concentration and so stabilize the newly formed particles. This would account for the increase in number of particles and, in consequence, the increase in polymerization rate with increase in concentration of emulsifier below the c.m.c.

From these observations it would seem that two distinct processes may be involved in particle nucleation. With styrene, micellar initiation may be the

Table 7.3 Particle diameters in polystyrene latices[27]

Percentage sodium lauryl sulphate	Percentage $K_2S_2O_8$	Percentage Na_2SO_4	D (Å)	
			calc.	obs.
4·8	1·54	0	480	640
0·256	0·029	0	1570	1260–1580
0·152	0·010	0	1980	1730–1900
0·08	0·0042	0·5	2484	2400

$m/w = \frac{40}{60}$.
Temperature 40 °C.

principal locus of polymerization at emulsifier concentrations above the c.m.c. but with aqueous initiation occurring at lower emulsifier concentrations. With methyl methacrylate, methylvinylpyridine and other appreciably water soluble monomers, where the polymerization rate shows no great change either below or above the emulsifier c.m.c., aqueous phase initiation may predominate under all conditions. A relevant observation here is that in the case of vinyl chloride, if allowance is made for the aqueous phase polymerization, the dependence of rate on emulsifier and initiator concentrations agrees with theory.[36]

The dependence of particle number on emulsifier and initiator concentrations has been widely used as a test of the validity of the Smith-Ewart kinetic scheme. With styrene the theoretical exponents of 0·6 and 0·4 have been found and there is excellent agreement between the theoretical and calculated particle diameters for various emulsifier concentrations (Table 7.3). With some monomers, however, low or even zero order in emulsifier concentration has been found, or there is a marked fall in the exponent at low emulsifier concentration. Some values reported in the literature are given in Table 7.4.

Table 7.4 Dependence of polymerization rate (R_p) and particle number (N) on emulsifier and initiator concentrations

			Reaction order				
			Emulsifier		Initiator		
Monomer	Emulsifier	Initiator	R_p	N	R_p	N	Reference
Styrene							
	SF	PP	*	0·6	*	0·4	28
	KL	PP	—	—	0·4	0·4	29
	KOS	PP	0·6	0·6	0·4	0·4	116†
	AH	γ	—	2	—	—	30
	AO	γ	—	—	—	0·22–0·35	31
	SLS	γ	—	—	0·8–1·0	—	104

* R_p proportional to N up to 10^{14} particles/cm³, H_2O exponent $-0{\cdot}83$ for $N \approx 10^{15}$.[82]
† Similar results for methyl and dimethyl styrenes but exponents for N with regard to emulsifier and initiator are 0·72 and 0·18 and 1·0 and 0·09 respectively.

Monomer	Emulsifier	Initiator	Emulsifier R_p	Emulsifier N	Initiator R_p	Initiator N	Reference
Vinyl acetate							
	AVA	PP	0·6	—	0·7	—	21a
	SLS	PP	0	*	0·5	*	32
	SLS	PP	0	—	0·64	—	33
	SLS	PP	0	—	1·0	1·2	111
	Seeded polymerization		0	—	0·8	0·3	
	EO/PO F68 F77	PP	—	3	—	—	24
	—	PP	—	—	—	0·5	34
	SLS PEDDE	PP	0	—	—	—	35
	SLS	γ	0·9–1·0	†	0·7–0·9	†	104

* Rate proportional to $N^{0{\cdot}2}$.
† Rate proportional to $N^{0{\cdot}7}$.

Monomer	Emulsifier	Initiator	Emulsifier R_p	Emulsifier N	Initiator R_p	Initiator N	Reference
Vinyl caproate							
	SLS PEDDE	PP	0·6	—	—	—	35
Vinyl stearate	OPEO	PP	Passes through maximum	No definite relationship	0·8	0	90

Table 7.4 *Continued*

Monomer	Emulsifier	Initiator	Emulsifier R_p	N	Initiator R_p	N	Reference
Vinyl chloride							
	SLS	PP	0·28	—	0·5	—	36
			0·59 (corrected)	—	0·4 (corrected)	—	
	SLS	PP SBS	0	1·0	—	0	37
	AB	PP	0	1·0	—	0	37
	AH AN	PP	0 above c.m.c	>1·0	0·6– 0·7	0	37
	SHOS	PP	0 (below c. 0·4 per cent)	1·2	0·8	—	38
			0·6 (above c. 1 per cent)	0·6			
	SLS SB	PP	~1·0 non-linear	*	0·5	0	39

* Order increases with number of particles 0·05–0·15 ($N = 10^{13}$–10^{16}/cm³ H_2O).

Monomer	Emulsifier	Initiator	Emulsifier R_p	N	Initiator R_p	N	Reference
Methyl methacry- late							
	SLS	AS	~0·5	~3	~0·5	0	44*
	SLS	Ca/H	0·3– 0·4	—	~0·45	—	45
	SLS	Fe/H/O	0·25	—	0·5	—	⎫
	SLS	Mn/H/O	0·18	—	—	—	⎬ 46
	SLS	Mn/H	0·55	—	—	—	⎭
	AO	γ	—	—	0·4	—	31
	SAS	BP	1·0 falling to 0	—	0·5 falling to 0	—	101
			at high concentrations of emulsifier or initiator				
	—	Ag/AS	—	—	0·5	—	47

* Similar relationships for n-butyl methacrylate.

Table 7.4   Continued

Monomer	Emulsifier	Initiator	Emulsifier R_p	N	Initiator R_p	N	Reference
Vinylidene chloride							
	KL	PP	~0·6	—	0·5	—	40
	SLS	AS	0·6 (initial rate) 0 (final rate)	—	0·6 (initial rate) 1·0 (final rate)	—	22
Ethylene	KM	PP	−0·6	—	0·5	—	61
2-Methyl 5-vinyl pyridine	SLS	PP	variable	—	—	—	21b
	SO	PP	0	—	—	—	
Acrylo-nitrile	SLS	SC/SS	0·25	—	0·36–0·42	—	41
Acrolein	PA/SO$_2$	PP	0·2	—	0·4	—	42
Butadiene	KS	PP	~0·6	—	—	—	43
	KS	D/A	R_p proportional to N				117
2,3-Dimethyl butadiene	SPS PEE	AS	1·0	—	0·5	—	102
Chloroprene	KL	PP	—	—	0·4	0·4	29
	SL SAS	PP	1·0 falling to 0	—	0·5 falling to 0	—	

at high concentrations of emulsifier or initiator

Initiators
AS	Ammonium persulphate
Ag/AS	Silver nitrate/ammonium persulphate
BP	Benzoyl peroxide
Cu/H	Cupric ion/hydrazine
D/A	Potassium dichromate/arsenious oxide
Fe/H/O	Ferric ion/hydrazine/oxygen
Mn/H	Manganous ion/hydrazine
Mn/H/O	Manganous ion/hydrazine/oxygen
PP	Potassium persulphate
SBS	Sodium bisulphite
SC/SS	Sodium chlorate/sodium sulphite
γ	Gamma radiation

Emulsifiers
AB	Sodium dibutylsulphosuccinate
AH	Sodium dihexylsulphosuccinate
AN	Sodium dinonylsulphosuccinate
AO	Sodium dioctylsulphosuccinate
AVA	Partly acetylated polyvinylalcohol
EO/PO	Block copolyethyleneoxidepropylene oxide; F68, F77 Pluronics (Wyandotte Chem.)

If propagation occurs exclusively in the micelles the polymerization rate will be proportional to N and the two will have the same dependence on initiator and emulsifier concentrations. This has been observed (Table 7.4) but not in every case. Thus methyl vinyl pyridine polymerized in seed latex gave particles of the expected diameter but rates varied to a much smaller extent.[21b] Similarly, with vinyl chloride the exponent for N and R_p was only the same at high emulsifier concentrations. At low concentrations (below 1 per cent of sodium lauryl sulphate) N varied with an exponent of 1·2 but the rate was independent of concentration.

7.3.3 Free-radicals/polymer particle

In earlier work propagation velocity constants calculated assuming $\bar{n} = 0\cdot5$ were in fair agreement with published values obtained from bulk or solution polymerization. However, this assumption does not hold for all monomers, nor, as more precise calculations have shown, does it hold with any system where large particles are formed.

There is clear evidence for the transfer of radicals out of polymer particles (for example, the smaller after-polymerization with intermittent initiation than would be expected) but provided radical ingress is rapid compared with departure the 0·5 value will be maintained. This condition is broadly true for styrene below 25 per cent conversion but with vinyl chloride estimates of $\bar{n}$ from the rate of polymerization, particle number and propagation velocity constant are much lower than 0·5, in the range 10^{-1} to 5×10^{-4}.[39] This evidently results from quite rapid transfer of radicals out of the latex particles, which is possibly due to the insolubility of the polymer localizing the free-radicals near to the surface. Transfer of radicals out of particles occurs with other monomers including styrene. From the decay in polymerization rate following intermittent initiation with radiation the half-life of a polystyrene

KL	Potassium laurate
KM	Potassium myristate
KOS	Potassium octadecylsulphate
KS	Potassium stearate
MC	Methyl cellulose
PA/SO₂	Polyacrolein/sulphur dioxide
PEDE	Polyethyleneoxidedodecylether
PEE	Esterified polyethyleneoxide
SAS	Sodium alkyl sulphate
SHOS	Sodium-2-hydroxyoctadecylsulphonate
SF	Soap flakes (sodium stearate/palmitate/oleate)
SL	Sodium laurate
SLS	Sodium lauryl sulphate
SO	Sodium oleate
SPS	Sodium pentadecyl sulphate
OPEO	Octylphenoxypolyethyleneoxide

radical at 20 °C is only 400 s.[48] This is sufficient, however, to give rise to multiple peaks in the molecular weight distribution of the polymer obtained by repeated intermittent illumination.[49]

At higher conversions with styrene, methyl methacrylate, vinyl acetate, and probably most monomers which are solvents for the polymer, deviations in the opposite sense occur, and free-radical contents of from 2–10 per particle have been estimated for conversions in the range 50–80 per cent.[50] The explanation for this arises from two causes. Firstly, owing to the high viscosity of the monomer/polymer particles termination velocity constants are low. For styrene k_t has been estimated to be in the range 5–30×10^3 l mol^{-1} s^{-1} compared with over 10^7 l mol^{-1} s^{-1} for low conversion bulk cystems. This augments the second factor tending to increase free-radical sontent, namely the increase in particle size with conversion. The radical concentration has been shown to depend on particle size and rates of initiation and termination by the relationship

$$\bar{n} = \frac{aI_0(\alpha)}{4I_1(\alpha)}$$

provided that transfer of radicals out of particles does not occur. I_0 and I_1 are zero and first order Bessel functions of the first kind of the variable

$$\alpha = \left(\frac{8v}{k_t\tau}\right)^{\frac{1}{2}}$$

v is the volume of the particle and τ the time interval between entry of particles.

This function has been calculated for various values of α. At small values $\bar{n}$ is 0·5 while at large values it becomes equal to $\alpha/4$, the same as the stationary concentration in bulk system. The particle size at which this will occur will depend on k_t and τ (proportional to $1/k_d$) and with the values typical of emulsion polymerizations, bulk kinetics will be established in particles of 10 000 Å diameter, and appreciable deviations from $\bar{n} = 0\cdot5$ will be observed for particle diameters greater than 1000 Å. As $\bar{n}$ increases above 0·5 it is clear that there will be an increasing positive deviation from polymerization rates calculated from Smith-Ewart kinetics. Gardon[50] has shown that the yield of polymer (P) is given by

$$P = At^2 + Bt$$

At^2 and Bt are the terms corresponding to bulk and ideal Smith-Ewart kinetics respectively. This has been shown to hold for a number of monomers (Figure 7.1). The values of A and B for styrene and methyl methacrylate and values for $\bar{n}$ are given in Table 7.5.

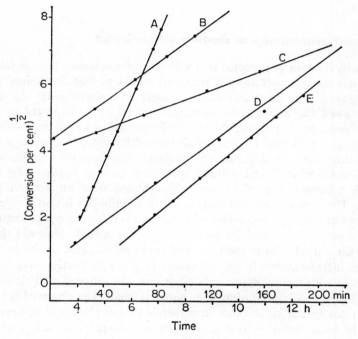

B- Butadiene/styrene) Time in hours
C- Butadiene)

A- Vinyl acetate)
D- Vinylidene chloride) Time in minutes
E- Methyl methacrylate)

Figure 7.1

Table 7.5 Kinetic values for methyl methacrylate and styrene[50]

	Methyl methacrylate at 40 °C ($m/w = 35/65$)		Styrene at 60 °C ($m/w = 40/60$)
Sodium lauryl sulphate (per cent)	0·63	2·58	0·67
Potassium persulphate (per cent)	0·23	0·23	0·24
$B \times 10^5$	5·48	7·55	10·2
$A \times 10^9$	2·29	7·45	10·9
$\bar{n}$ (25 per cent conversion)	0·58	0·63	—
$\bar{n}$ (50 per cent conversion)	—	—	1·3
$\bar{n}$ (60 per cent conversion)	0·96	1·25	—
$\bar{n}$ (80 per cent conversion)	2·90	2·34	—

A further test of the constancy of radical concentration is the effect of added initiator following the nucleation step. If $\bar{n}$ is constant the rate would be unaffected and this has been found to be true for a hydroperoxide-amine initiator with butadiene[51] and for persulphate initiated vinyl chloride polymerization where $\bar{n} \ll 0.5$.[52] Styrene under conditions where $\bar{n}$ would be expected to be greater than 0·5 shows an increase in rate on post-addition of potassium persulphate initiator.[50]

7.3.4 Monomer concentration in emulsion polymerization

Superficially it might be expected that with those monomers which dissolve their polymers the concentration of monomer would be that determined by the conversion at any particular time, whereas with those which do not dissolve or swell the polymer the concentration might be expected to be negligible. Constancy of polymerization rate frequently observed in emulsion polymerization would thus have to result from diffusion of monomer as a rate controlling step. In fact, in an emulsion polymerization monomer diffusion is not limiting[53] and with all systems something approaching an equilibrium concentration of monomer is reached over an appreciable conversion. The reason for establishment of an equilibrium is that the gain in free energy of mixing of monomer with polymer is balanced by the change in surface energy of the particles on swelling.[54] The actual value will thus depend on the particle size of the latex and might be expected to rise as the particle size increases. Significant differences in ϕ_M with particle size have been observed with methyl methacrylate but over the range of particle sizes occurring during a polymerization, the change is small. The observed fall in ϕ_M may have a greater significance than merely that of particle size, since it has recently been found[112a] that even with styrene polymerization when Smith-Ewart kinetics are obeyed conditions can be found where there is a continuous fall in monomer concentration during polymerization (ϕ_M initially 0·8 at low conversion falling to 0·4 at 60 per cent conversion). The fact that the rate and number of particles are constant in this period implies that ϕ_M is constant in the region where polymerization occurs, but that swelling of the particle is not equilibrium, and the centres of the particles are depleted in monomer. The observations of Robb[88] on changes in particle number in the later stages of polymerization need to be considered when making estimates of ϕ_M. Values of ϕ_M which have been reported[27] are

Styrene	0·60	Methyl methacrylate	0·73
Vinyl acetate	0·85	n-Butyl acrylate	0·65
Butadiene	0·56		

(The monomer/polymer composition is often expressed in terms of the weight ratio $W_M = \phi_M/\rho(1 - \phi_M)$ where ρ is the ratio of polymer and monomer densities (d_p/d_m)).

The presence of a completely insoluble polymer, such as polyvinylidene-chloride, also gives rise to a finite concentration of monomer in the particles. It would appear here that adsorption of monomer on to the surface is the explanation; the concentration might be expected to change with growth of the polymer particles in such a case.

7.3.5 Polymer molecular weights

In emulsion systems the polymer molecular weight average should be determined by the growth period between the entry of two successive radicals into the latex particle and is readily calculable from the Smith-Ewart scheme. Provided that the conditions are such that transfer of growing chains with monomer is not limiting (i.e. with styrene and methyl methacrylate the DP is below 10^5) and the radical flux and particle size are such that the polymerization approximates to the ideal case, good agreement has been found between calculation and experiment.[50] Nevertheless, the dependence of molecular weight on conversion is complex—in some cases there is a rise but in others there is not.[16] These changes and the full interpretation of the broad molecular weight distributions[55] await further clarification.

A consideration of all these results, and particularly the data in Table 7.4, shows that Smith-Ewart kinetics are at best followed only in a limited number of cases. It is often found that the experimental conditions chosen can lead to deviations from the ideal but it is equally clear that even under carefully chosen and favourable conditions exact agreement with theory is rare. Nevertheless, there are many features of emulsion polymerization which the theory explains admirably—qualitatively and at least semi-quantitively.

Possibly the most important problem relates to the body of evidence indicating the importance of the particle surface in the polymerization. In only a few examples does the particle number correlate with a high order of the emulsifier concentration, but the localization of hydrophilic radicals at the particle surface, the adsorption of monomer on the surface of insoluble polymer and the rapid transfer of free-radicals out of particles, all suggest that at least in some cases surface reactions may be dominant. Unfortunately, although the reaction orders for dependence of rate on initiator and emulsifier concentration are different when these assumptions are made, the spread of experimental results is too great to distinguish unequivocally between the rival theories.

The relationships derived by Medvedev[19] lead to the following dependence of rate on emulsifier and initiator concentrations.

		Reaction order with respect to	
		Emulsifier	Initiator
1.	Radicals produced in aqueous phase		
	(a) rapid recombination	1·0	0·25
	(b) slow recombination	0·5	0·5
2.	Radical produced at adsorbed emulsifier layer	1·0	0·5
3.	Radicals produced in monomer phase	1·0	0·5

In these, stationary state kinetics are employed. The overall rate is assumed to be first order in emulsifier while the initiator is considered either to generate radicals in the aqueous phase, some of which react with the emulsifier, or to be adsorbed from the aqueous phase into the surface layer of the emulsifier or, if monomer soluble, to react with adsorbed emulsifier.

It is apparent that schemes such as these will give polymerization rates dependent on some power of the emulsifier concentration, which may not be dissimilar from the experimental findings. However, although reactions involving the particle surface are well supported and Smith-Ewart kinetics underestimate their importance, the body of detailed evidence from rate and molecular weight studies to prove these particular schemes is not available.

7.4 SEMI-CONTINUOUS AND CONTINUOUS EMULSION POLYMERIZATION

It is widespread technical practice in emulsion polymerization, particularly with more reactive monomers which have a high equilibrium monomer concentration in the polymer particles, and in the preparation of high concentration latices, to add some of the components in stages during the polymerization. A portion of the monomer, usually about 10 per cent, is used to start the reaction with the whole or part of the other components of the system, and then the remainder of the monomer is added continuously or in increments at such a rate to maintain satisfactory temperature control and latex stability. Incremental addition of emulsifier also affects polymerization rate but the most important effect here concerns particle size since the number of micelles available for nucleation may be varied. Addition of initiator and chain transfer agent incrementally is usually carried out when they are decomposing or reacting rapidly and it is desired to maintain a constant rate of radical production or concentration of chain transfer agent throughout the polymerization.

If all the emulsifier is added to a portion of the monomer charge a large number of particles will be produced in which polymerization of the monomer feed will subsequently occur. Provided the rate of monomer addition is slow the equilibrium concentration is not reached and the rate remains below the steady rate. The rate of polymerization R_p (below the steady state) has been shown to be related to the rate of monomer addition R_a by $1/R_p = 1/K + 1/R_a$, where $K = k_p N\bar{n}/N_A$. In the polymerization of styrene with N in the region 4×10^{14} particles/cm^3, a good reciprocal plot of R_a and R_p has been found[56] but for a styrene polymerization in which $N = 1\cdot4 \times 10^{15}$ and for methyl acrylate, linear plots of R_p against R_a were observed.[57] However, a small value of $1/K$ (i.e. when N is large) makes it difficult, in view of the small amount of data available, to decide with certainty which is the better representation.

A delay between addition of the monomer feed after the initial addition permits completion of the nucleation process but otherwise has a relatively small influence.

Initial addition of only part of the emulsifier will produce a small number of growing particles. The added monomer will grow into these particles and would give a large particle size latex provided no new particles are produced. As the total surface area of the particles increases they will become progressively unstable unless further emulsifier is added, but provided this is at such a rate that it only compensates for the generation of fresh surface no new particles will be produced, and the net effect will be to produce a larger particle diameter latex with a narrow size distribution.[57] The dependence of rate on monomer addition will follow the same pattern as with monomer feed, but the steady polymerization rate will be reached more quickly since the number of growing centres will be smaller.

If the emulsifier addition is sufficiently high so that micelles reappear into the aqueous phase some of the monomer will initiate new particles and in this case the particle size distribution is broadened. This may be the explanation for the broad distributions sometimes observed[58] using monomer/emulsion feed.

In addition to techniques on incremental addition of components to batch reactors, there has been much effort with various types of continuous reactor in emulsion polymerization, in order to achieve better control over reaction rates, heat transfer, etc., and a reduction in the number of manual operations. In principle, polymerization can be carried out in a tubular reactor under conditions which simulate batch polymerization but with only a small mass of material reacting at any given time. This necessitates 'plug flow' in the transport of reactants along the tube with sufficient turbulence such that each volume element of the moving fluid will have the same residence time. A distribution of flow velocities ('laminar flow') along the tube will result in broadening of residence times, with consequent effects on reaction rate.[58a] Laminar flow has also been shown to affect molecular weight distribution in bulk polymerization[58b] but there is no information on its effect in emulsion systems. More usual arrangements for a continuous flow polymerization are continuously stirred tank reactors (a single reactor or several in cascade), sometimes in conjunction with a tubular reactor. These result in greatly broadened distributions of residence times, which in polymerization is reflected in differences in rate, molecular weight, and molecular weight distribution. The differences diminish as the number of reactors in cascade is increased and theoretically with an infinite number tubular flow is achieved. It will be appreciated that as the vessels need not be of the same size, residence times can vary from reactor to reactor. Another conclusion is that for a given output level the volume of a cascade system must be considerably larger than that of a batch reactor.[58a]

In emulsion polymerization where there is concurrent nucleation, particle growth and possibly aggregation, and monomer depletion, the kinetics are complex but have been analysed for a water insoluble monomer such as styrene in terms of the number of polymer particles produced and the concentration of monomer at the steady state.[58c,58d] Calculation shows that in the first reactor the conversion and particle number will pass through a maximum with increase in residence time before reaching a steady value, and with a corresponding fall in the concentration of micellar emulsifier in the aqueous solution. Provided the time is such that all the emulsifier has been adsorbed on the surface of the particles no new particles will be produced in the later stages of the cascade system. It has been calculated, for a somewhat different kinetic model than the Smith-Ewart scheme, that the optimum residence time in the first reactor to give the maximum number of particles, that is maximum rate, is given by $\theta_i = 0.82 \, t_s$ where t_s is the time for disappearance of the emulsifier in the batch system operated under similar conditions. The number of particles and hence the rate under these conditions will be 57.7 per cent of those in the batch.

With the assumption of Smith-Ewart kinetics qualitatively similar but quantitatively different effects are calculated.[58c]

Experimental findings appear to confirm the theoretical calculations. Wall et al.[58e] with a single stirred reactor found the conversion of styrene to polymer to pass through a maximum with increase in residence time before reaching a steady state, and with a system operating at 76 per cent conversion the rate was 53.2 per cent of that of the batch system. More detailed data on styrene, including changes in particle number and soap concentration, have been given by Nomura et al.[58d]

In copolymerization there is some evidence to suggest that the polymers obtained in continuous systems are of more uniform composition than when prepared in batch processes.[58e]

7.5 EMULSION POLYMER LATICES

The properties of aqueous dispersions of fine polymer particles in the main lies outside the scope of this review, but there are two aspects which are relevant here since they intimately concern the polymerization mechanism. These are the particle surface chemistry and the particle size distribution.

7.5.1 Surface chemistry of emulsion polymer particles

The particles in most emulsion polymer latices vary from about 500–2500 Å in average diameter, and would be formed by a fifty-fold variation in emulsifier concentration and a ten-fold variation in initiator concentration. At low emulsifier levels the surface of the particles is incompletely covered, but with the higher levels usually employed a substantial fraction of the polymer surface is covered with a mono-molecular layer of the emulsifier. Thus the area of soap molecules in 1 cm³ of 2·5 per cent potassium laurate solution, approximately 2×10^{21} Å², is equivalent to the surface area of 0·33 g of suspended polymer; $D = 750$ Å ($N \simeq 2 \times 10^{15}$).

As most polymers are hydrophobic the hydrocarbon portions of the emulsifier will be on the particle surface with the ionized groupings in the water, and it is apparent that each polymer particle in the latex will carry a large number of soap molecules ($\sim 3 \times 10^4$ in the example chosen), with a substantial electrical charge. The sign of the charge is dependent on the type of emulsifier and significantly influences the stability of the latex since it is the electrical repulsion which prevents spontaneous coagulation.

It is possible to remove much of the emulsifier by dialysis or by ion exchange, although the ease with which this is accomplished depends on the nature of the emulsifier and the polymer, and is not always possible.[59] The emulsifier-free polystyrene latices have been shown to contain carboxyl groups on the surface when hydrogen peroxide has been used as initiator and carboxyl and sulphate groups with a persulphate initiator. The number of carboxyl groups/particle has been found to be substantial, there being one group for 200–1250 Å² of particle surface and the number of carboxyl groups/polymer molecule ranged from 1 to 6.[60a] Other polystyrene latices have been shown to possess hydroxyl and sulphate end groups, with 45–100 per cent of the sulphate groups on the particle surface; a similar density of charged groupings (one per 200–3100 Å² of particle surface) has been found.[60b] The relative proportions of hydroxyl and sulphate groups in the polymer varied with the pH of the emulsion, hydroxyl content increasing and sulphate decreasing with decreasing pH.

Carboxyl groups have been found in polyethylene obtained by potassium persulphate initiation,[61] and as the extent of reaction increased with the concentration of potassium myristate emulsifier it seems likely that growing chains transfer to the emulsifier by reaction at the particle surface. Reaction of soap with persulphate radicals has been considered a likely possibility,[62] and they could also be involved in an initiation step.

The existance of many ionized groupings on the surface suggests that the entering radical (possibly an oligomer after reaction of the initiator free-radical with a few monomer molecules) has its hydrophilic grouping located on the surface and that monomer diffuses towards the surface for the growth reaction. Termination would therefore occur near the particle surface and radical transfer to the aqueous medium would be facilitated. It should be appreciated that the loss of an uncharged radical from a latex particle may be more easily accomplished than the entry of an ion-radical from the initiator with a charge of the same sign as that of the particle. Conversely charged initiator radicals with the opposite sign could have enhanced entry rates.

Copolymerization of hydrophilic and hydrophobic monomers will clearly have a marked influence on the surface chemistry of the latex particles. Thus in a copolymer from a water soluble monomer such as methacrylic acid with butadiene the combined carboxylate groupings will confer surface activity on the polymer. The surface of the particle will have many ionic groupings thus reducing the amount of emulsifier required. Such latices are of considerable technical importance in foams and in carpet backing materials.

The surface charge may have a significant effect on the stability of the latex, and it is possible that films cast from the latex may have different properties from the bulk polymer owing to the localization of hydrophilic groupings on the boundaries between the particles.

7.5.2 Particle size and particle size distribution

With catalysts and emulsifier systems which give satisfactory polymerization rates and stable latices particle sizes usually lie in the range 600–1000 Å. This imposes a practical upper limit of 50/50 for the monomer/water ratio since there is an approximately linear relationship between the viscosity of the latex and its concentration.[63] A viscosity of more than 10 poise, above which there would be severe restrictions on its technical value, would only correspond to about 40–50 per cent concentration, which is unsatisfactory for many industrial applications. Concentrations of polymer in the region of 60 per cent are normally desired and for this level it is necessary to obtain average diameters in the region of 1500–2000 Å. This can be achieved by the use of low emulsifier concentrations, and electrolytes also appear to be beneficial.[64] As would be expected, polymerization times are long and the latices are relatively unstable. In these systems it has been shown that relatively small particles (400–500 Å at 25 per cent conversion) are initially formed which agglomerate at higher conversions to the 1600–1800 Å range.[65] Auxiliary stabilizers are usually necessary to avoid partial coagulation during polymerization.

These problems have led to alternative techniques for high concentration latex production and there are now several methods by which the particle size of a normal low-solids latex is increased. The latex concentration is then raised by evaporation or centrifugation.[63] Amongst the techniques which have been used for this purpose are the addition of salts or solvents, partial destabilization by freezing and pressure agglomeration.

Particles are initiated continuously during the nucleation period and there will be statistical variations in the growth of individual particles. In consequence a homodisperse latex will not be produced and the dispersity will depend on the growth laws for different sized particles. If the growth of a particle of volume V is related to its diameter by the equation $dV/dt = kD^x$ then for ideal kinetics where $x = 0$, that is the growth rate of a particle is independent of size, $dV/dt = k$ and there will be a normal distribution of particle volumes, the standard deviation of which will be determined by the polymerization kinetics. A further consequence of $x = 0$ is a narrowing of the particle distribution as growth proceeds. (The distribution will narrow or broaden on increase in size dependent on whether x is less or greater than 3.) This type of distribution appears to hold as would be anticipated for fine particle sizes, and is assumed in Smith-Ewart kinetics.

With $x = 2$ or 3 growth would produce a normal or log-normal distribution of particle radii respectively. The latter approximates most closely to the distributions of large particle diameter latices[66] and agrees with the experimentally determined exponent $x = 2.6$, found for large polystyrene particles.[67]

In addition to the distributions, the average values need to be calculated. Some techniques, e.g. electron microscopy permit the calculation of any desired average but others may give only one; soap adsorption measurements, for example, give a volume-area average ($\bar{D}_A$). The other more important averages are number ($\bar{D}_N$), weight ($\bar{D}_W$) and volume ($\bar{D}_V$).

$$\bar{D}_N = \sum N_i D_i / \sum N_i \qquad \bar{D}_W = \sum W_i D_i / \sum W_i = \sum N_i D_i^2 / \sum N_i D_i$$

$$\bar{D}_A = \sum N_i D_i^3 / \sum N_i D_i^2 = 6V/A$$

V is the volume and A the surface area of the particles.

$$\bar{D}_V = (\sum N_i D_i^3 / \sum N_i)^{\frac{1}{3}} = (\bar{D}_N \bar{D}_W \bar{D}_A)^{\frac{1}{3}}$$

The ratios of these averages are convenient measures of the breadth of a distribution.

The production of narrow distribution latices has attracted considerable interest and for this purpose it is necessary to carry out the bulk of the polymerization in a fixed number of particles and to prevent particle aggregation. Examples of the processes which have been employed are the following:

(a) Lithium soaps as emulsifiers. Lithium soaps are sparingly soluble and have c.m.c.'s near their solubility limit. Narrow distribution polystyrene latex ($\bar{D}_W/\bar{D}_N = 1.042$ falling to 1.007) has been obtained by use of mixtures of lithium oleate and stearate.[68]

(b) Low soap concentration with incremental addition. Typically 0·5–2 per cent of sodium lauryl sulphate or potassium laurate is employed, 25 per cent present initially and the remainder incrementally during the course of the reaction. Latices with average diameters in the range 1290–2900 Å and $\bar{D}_W/\bar{D}_N = 1·01–1·02$ have been obtained.[69] In these the expected narrowing of the distributions with conversion is observed.

(c) Seed polymerization. The advantage of this technique is that the seed latex can be examined and characterized before the second step in the polymerization, and by choice of monomer to seed ratio the desired particle size can be obtained. To avoid a separate monomer phase incremental addition of monomer may be employed. Water solubility of monomer might be expected to cause difficulty by forming separate particles in the aqueous phase. Although the rates of polymerization suggest that this occurs any such particles become incorporated into the larger seed particles and particle sizes are as anticipated.

The main difficulty with seed and low emulsifier systems is that the surface of the particle is unsaturated with respect to emulsifier and hence the latex is sensitive to agitation. Work has been described in which stable narrow distribution latices are obtained using small amounts of micellar ionic emulsifiers with a non-ionic emulsifier. In published recipes employing a 1/10 ratio of sodium lauryl sulphate and isooctylphenoxypolyethyleneoxide (9–10 ethylene oxide units in the molecule) as emulsifiers stable polystyrene latices ($m/w = 50/90$) were obtained. These required little stirring to maintain stability and possessed narrow distributions ($\bar{D}_W/\bar{D}_N = 1·03–1·01$).[68]

Recently there have been comprehensive investigations on the preparation of very narrow distribution ($\bar{D}_W/\bar{D}_N$ below 1·01) latices from styrene and styrene-divinyl benzene.[68a] The recipes contained potassium persulphate as initiator and mixed ionic (sodium dodecyl benzene sulphonate), non-ionic emulsifiers. In addition to giving narrow distributions of particle sizes the non-ionic emulsifier increased the latex particle size. The particle size was varied from 0·088 μ to 0·60 μ by changing the concentration of ionic emulsifier and ratio of ionic to non-ionic. Larger particle size latices (up to 1·25 μ, $\bar{D}_W/\bar{D}_N = 1·001–1·0001$) were obtained by use of seed latices. Divinyl benzene reduced the particle size of latices to a small extent but did not affect the narrow particle size distribution. Empirical equations were given for the calculation of amounts of emulsifier, initiator, water and seed latex for particular sized latices.

7.6 INVERSE EMULSION POLYMERIZATION[70]

In this technique a water soluble monomer, usually as an aqueous solution, is dispersed in an organic solvent containing an emulsifier of low HLB number and a peroxide initiator. Sodium p-vinyl benzene sulphonate as a

20 per cent aqueous solution emulsified in xylene exhibited an increased rate of polymerization compared with the rate in solution. The number of particles increased rapidly with emulsifier concentration (exponent initially above 1·0) to a limiting value. Polymerization took place both in the monomer droplets and in the emulsifier micelles. As yet there has been only very limited work in this area but other water soluble monomers such as ethylene sulphonic acid or acrylic acid should be polymerizable by this reverse method. In many cases an aqueous solution of monomer would need to be used, so as to preclude the solution of much of the monomer in the oil phase or in the event of the monomer being a solid.

7.7 POLYMERIZATION IN EMULSION BY NON-RADICAL PROCESSES

In most emulsion systems the initiating species are free-radicals, and variable valency metals which may be present participate only in the free-radical forming reaction and do not significantly influence the propagation reaction. The fact that olefins and diolefins can form relatively stable complexes with certain metals, for example copper and silver of group IB metals and many of the group VIII elements, has led to a study of their effect in polymerization.[71] The only clearly established polymerizations which can be carried out in emulsion with metal salt initiators, where free-radical initiation can be excluded, is the polymerization of butadiene to the *trans*-1,4-form by rhodium containing catalysts. In this system polymerization occurs because the π-allylic rhodium complexes are relatively stable to hydrolysis by water. The major features of interest, namely the initiation and propagation mechanisms, lie outside this review but there are several points of interest relating to the role of the emulsifier. Firstly, although polymerization can be effected in homogeneous medium, emulsifiers have a beneficial effect on the rate of polymerization. With rhodium chloride there is little or no polymerization in the absence of emulsifier and the rate increases with emulsifier concentration—although not with any clearly defined relationship.[72] Secondly, only certain emulsifiers, for example sodium lauryl sulphate and sodium dodecyl benzene sulphonate are effective,[73] and this has led to the suggestion that the active rhodium complex forms part of a micellar complex.

The molecular weights of the polymers are low and not greatly affected by change in reaction variables. This would suggest that initiation and termination processes may involve the emulsifier, possibly accompanied by a valency change in the rhodium since hydrogen donors and reducing agents have a marked accelerating influence on the reaction.[74] There is evidence for the combination of emulsifier in the polymer although this is not entirely conclusive,[75] but end groups from the metal salt and from the water have been demonstrated.[76]

Further work is likely to modify considerably the view as to the reaction mechanism but with this reservation it may be considered that initiation results from the formation of a π-allylic complex by the addition of a grouping (e.g. H, Cl or, possibly, emulsifier) to the diene concurrently with a reduction in the valency state of the metal. The emulsifier may then influence the stability of the propagating chain, possibly by changing the environment of the complex against termination by water molecules.

The cyclic siloxane $[CMe_2SiO]_4$ has been polymerized in the presence of cationic emulsifying agents to give a positively charged emulsion.[77] The initiation is considered to be effected by traces of a quaternary ammonium hydroxide concentrated at the interface between the siloxane and the water.[78] Anionic emulsions of these polymers have been obtained using dodecylbenzene sulphonic acid as a combined initiator-emulsifier. In amounts of 0·12–5 per cent on the monomer, polymer dispersions with particle diameters of 740–70 Å have been reported.[78] These are much smaller than the initial dispersions of the monomer. Molecular weights decrease rapidly with increase in polymerization temperature, from $1·72 \times 10^5$ to $4·9 \times 10^4$ over the range 25–90 °C. The mechanism of polymerization is not known in detail but the cyclic siloxanes do not polymerize by free-radical processes. In the early stages of polymerization low molecular weight silanols are formed by addition of water to the cyclic siloxane and higher polymers form by a rapid addition reaction; the molecular weight in consequence rises with conversion.

Cyclic sulphides, such as ethylene sulphide or propylene sulphide can be polymerized in emulsion by an anionic mechanism using zinc or cadmium salts as initiators. The polymerizations may be carried out in the presence of conventional emulsifiers[79] or the catalyst may be modified to form surface activity during the polymerizations, e.g. with sodium glutamate as co-catalyst.[80] In both cases stable polymer latices are obtained. There is no information on the effect of emulsifier concentrations on the polymerization rate or particle size of the emulsions.

7.8 INDIVIDUAL MONOMERS

7.8.1 Styrene

Industrially, polystyrene is valued for its optical and electrical properties and ease of injection moulding, and for these purposes control of molecular weight and the absence of impurities are essential. As a result most of the polymer is produced by bulk or suspension polymerization, and emulsion polymerization is restricted to areas where the use of a latex as such is required and where there is no disadvantage from the presence of an emulsifier, as in polystyrene latices for the reinforcement of flexible latex foams.

Much of the published work on the emulsion polymerization of styrenes has related to studies of the mechanism of the reaction. Styrene is only sparingly soluble in water (0·029 per cent at 20 °C) but in soap solutions of 2–5 per cent concentration the solubility is greatly increased to a value in the region of 0·5–1·0 per cent. With emulsifier concentrations above the c.m.c. Smith-Ewart kinetics are followed but the presence of methanol in the water or acrylonitrile or methacrylonitrile in the monomer changes the locus of the initiation reaction and there is clear evidence for this occurring in the aqueous phase.[35] Essentially similar conclusions apply to the polymerization in the presence of low emulsifier concentrations.[26] A value of k_p has been calculated for styrene at 60 °C in emulsion. It is 122 $1mol^{-1} s^{-1}$, which is in good agreement with bulk or solution polymerization.[81] Most work on styrene has

Table 7.6 Styrene polymerization[28] (monomer/water = 40/60)

Polymerization Temperature °C	Emulsifier per cent[a]	$K_2S_2O_8$ per cent	Polymerization rate		[η] maximum
			per cent/minute[b]	$10^{19} \times$ g/sec/ particle	
30·5	0·5	0·172	0·09	0·42	17·4
50	0·5	0·172	0·40	0·78	9·72
70	0·5	0·172	1·49	1·44	5·29
70	0·5	0·516	—	1·25	—
70	2·0	0·172	—	1·95	—
90	0·5	0·172	4	2·63	1·43

a. Soap flakes, predominantly sodium stearate/palmitate.
b. Conversion rate calculated from yield of polymer at time of maximum molecular weight.

been carried out with persulphate initiators but normal kinetics appear to be followed with the oil soluble cumene hydroperoxide as initiator.[82] In many of these systems difficulties can arise from uncertainty as to the rates of radical formation from the initiator and the water soluble azo-compounds have advantages in this respect.[16]

Typical polymerization data for styrene are given in Table 7.6.[28]

7.8.2 Vinyl acetate

The main applications of polyvinylacetate are in emulsion paints and adhesives, in which the latex is used directly, usually at 50 per cent concentrations. Emulsion systems are therefore of considerable industrial importance.

Vinyl acetate is appreciably water-soluble (2·5 per cent w/v) and hydrolyses readily, and as there is a fall in pH to 5–6 during polymerization it is clear that fatty acid soaps would not be satisfactory emulsifiers. Moreover, acetaldehyde formed by hydrolysis would interfere with the polymerization.

Also vinyl acetate produces radicals of high reactivity and most unsaturated compounds result in retardation. The generally used emulsifiers are mixtures of anionic and non-ionic types, typically sodium lauryl sulphate and alkyl-phenoxypolyoxyethylenethanol, often with a secondary stabilizer such as partly hydrolysed polyvinyl acetate. The last of these may be used as the sole emulsifier for certain applications, particularly adhesives. Ammonium and potassium persulphates are the preferred initiators and the polymerizations are carried out using gradual monomer addition at reflux temperature to complete conversion.

For many purposes copolymers with glass transition temperatures below room temperature are prepared. This is to give film-forming polymers without the necessity for external plasticizers. The most usual co-monomers are acrylic esters—particularly n-butyl and 2-ethylhexylacrylates and higher branched-chain fatty acid vinyl esters (vinyl versatate).[83,84] The quantitative aspects of the polymerization show marked differences from those observed with styrene. The rate of polymerization is high with or without an emulsi-fier.[35] The number of particles either remains constant following the initial reaction or falls with conversion[24] and the rate remains constant up to 80 per cent conversion. In seed latices the particle size increases with decrease in seed to monomer ratio but polymerization rate is hardly affected.[32] It is becoming appreciated that aqueous phase initiation is the main locus of polymerization with aggregation of the small polymer particles (which are soluble in sodium lauryl sulphate solution at low molecular weights) into larger particles as polymerization proceeds, but a major point not fully established is whether the surface of the latex particles is the place where polymerization mainly occurs. The high observed rate compared with that anticipated from the rate of homogeneous polymerization[33] and the relation-ship between particle surface and conversion is held to be good evidence for the latter view.[85] Vinyl acetate prepared in emulsion with γ-initiation gives very high molecular weight polymer (molecular weight independent of dose rate and emulsifier concentration.[104])

Polymerization data for vinyl acetate are given in Table 7.7.[21a] The high activity of the vinyl acetate radical gives rapid chain transfer and it has been suggested that this reaction occurs with the acetyl group of the monomer to give a water soluble butyrolactonyl radical

$$\mathrm{-\!\!(CH_2)_2\dot{C}HOCO\!\!-}.$$

This can initiate additional polymerization in the aqueous phase or terminate a growing water soluble polymer chain (DP50–300).[111 and 111a] A detailed kinetic analysis shows that the experimental dependences of rate on initiator concentration (given in Table 7.4) in direct and seeded polymerizations are accounted for when these transfer and termination reactions are allowed for.

Table 7.7 Polymerization of vinyl acetate[21a]
temperature 70 °C, monomer 20 g, emulsion 100 g

Emulsifier[a]	$K_2S_2O_8$ per cent	Polymerization rate per cent/minute
2·04	0·02	2·3
1·52	0·02	1·98
1·01	0·02	1·4
0·75	0·02	1·0
1·01	0·04	2·2
1·01	0·012	1·0
1·01	0·008	0·8
1·01	0·004	0·3

a. Partly acetylated polyvinylalcohol.

7.8.3 Vinyl chloride

Polyvinylchloride is currently manufactured by bulk, suspension and emulsion polymerization and although the last is produced on a very large scale, it is probably declining in importance compared with the highly efficient bulk processes developed in recent years. In addition to the homo-polymer, which has few applications in the latex form, copolymers of vinyl chloride with vinyl acetate, alkyl vinyl ethers and olefins such as ethylene and propylene are made by the emulsion process.[86] The reactions are normally carried out at about 50 °C, although lower temperature systems are available, and pressures up to about 100 p.s.i. Technically, polymerizations tend to be rather slow but accelerate rapidly with rise in temperature and with con-version. Water soluble initiators such as hydrogen peroxide or potassium persulphate are widely used, but a room temperature active, monomer soluble catalyst, is diisopropylperoxycarbonate.

Molecular weight control is important. The best mechanical properties, in particular, impact resistance, are obtained with high molecular weights but processability, which is carried out at temperatures near to the decomposition temperature of the polymer (thermal stabilizers are required), is much better with the lower molecular weight polymers and copolymers, particularly when these stabilize the polymer end groups against thermal elimination of hydro-gen chloride.[87]

Polymerization shows continuous acceleration with conversion up to about 60 per cent and the rate does not fall significantly before about 80 per cent conversion provided the initiator is not exhausted. There are few features of the polymerization which obey the Smith-Ewart kinetic scheme. This is largely the result of the high solubility of monomer in water (0·6 per cent w/w) and the low solubility of monomer in polymer. There is a separate phase until about 75 per cent conversion—which compares with about 50 per cent for styrene and 15 per cent for vinyl acetate. Addition of initiator during the polymerization does not increase the rate and likewise additives which reduce the rate of initiation (e.g. a sequestering agent such as citric acid in a copper redox initiator) are without effect.[52]

The behaviour of the emulsifier is complex. At low concentrations of sodium lauryl sulphate the number of particles increases at a fairly high power (1·2) with little effect on rate, but above 10^{-2} mol/l both rate and particle number increase normally (exponent 0·6) with concentration.

A large number of different emulsifiers has been studied, but there appears to be no obvious correlation between particle size, polymerization rate and latex stability.[8] Non-ionic types appear to be somewhat less effective than ionic ones and there is little advantage in the use of mixed emulsifiers. No correlation has been observed between the surface tension of the solution, polymerization rate or polymer molecular weight.[89]

Evidence for initiation in the aqueous phase has come from experiments on seeded polymerizations, where the rate is independent of the number of particles. The increase in rate with conversion could result either from an increase in $\bar{n}$ or from the trapping of radicals in the insoluble polymer. Compared with the ideal conditions $\bar{n}$ is very low due to rapid transfer of free-radicals out of the growing particles. The insolubility of monomer in polymer would suggest that the former is adsorbed at the particle surface, where most of the propagation occurs.

7.8.4 Conjugated dienes: butadiene, isoprene, 2,3-dimethyl butadiene, chloroprene, 1,3-pentadiene, chloroprene and their copolymers

These polymers are of great economic importance, the total manufacture of butadiene/styrene rubbers based on emulsion processes being above 4 000 000 tons in 1971. A relatively small proportion of this is used in the latex form, most being prepared for solid rubbers. With butadiene free-radical initiation emulsion polymerization is the only practical method since, in common with other conjugated dienes such as isoprene and 2,3-dimethyl butadiene, the propagation reaction is slow and the termination reactions are fast. With values of $k_p/k_t^{\frac{1}{2}}$ in the region of 10^{-4} bulk or solution polymerizations are slow and they give polymers of inadequate molecular weight.

Butadiene polymers are subject to chain branching and gel formation by secondary attack of free-radicals on preformed polymer and, in view of the critical importance of molecular weight, molecular weight distribution and polymer linearity on the physical properties of elastomers; these aspects have been studied in depth. The advantages of low temperature polymerization on polymer properties has also stimulated a great deal of effort on initiation systems which give adequate polymerization rates in the region of 0 °C.

The latices are used in many specialized applications, for example in foam rubber, adhesives, and carpet backing, and there are special requirements with regard to viscosity, mechanical and freeze-thaw stability and coagulation characteristics, usually in high concentration (50–60 per cent) emulsions.

1. Butadiene polymers and copolymers

Polymers are currently prepared with two types of system. One employs potassium persulphate and saturated fatty acid soap at 50 °C and the other an oil soluble hydroperoxide and a water soluble reducing agent at 5–10 °C with a rosin based or mixed fatty acid/rosin acid soap. The conversion time curve is characterized by a slow initial accelerating rate followed by a more or less constant region to about 60–70 per cent conversion. The reactions slow down above this but they are in any case terminated because of excessive branching and the onset of gelation. In addition to butadiene polymers, copolymers with styrene, styrene and vinyl pyridine, acrylonitrile and itaconic acid, are in manufacture. There are numerous references in the literature to the preparation of other copolymers, the more important of which are acrylates and methacrylates, methacrylonitrile, 1,3-pentadiene, methyl vinyl and methyl isopropenyl ketones and vinyl ethinylcarbinol. Several of these have been available in semi-commercial production. The broad features of all of these are similar except where the monomer has special reactivity which restricts the type of emulsifier (e.g. methacrylic acid) and they can be prepared using persulphate or redox systems. There are, however, differences in rate and the characteristics of the latex, which may reflect differences in polymerization mechanism.

In spite of the significant water solubility of butadiene (0·13 per cent at 15 °C/793 mm Hg) polymerization rates approximate to Smith-Ewart kinetics. From the rate of polymerization and assuming $\bar{n} = 0·5$, the propagation constant was calculated to be 8·4 l mol^{-1} s^{-1} at 10 °C.[51]

Control of molecular weight in butadiene polymers and copolymers is almost exclusively carried out by use of mercaptan chain-transfer agents, the role of which in emulsion systems is more complex than in bulk or solution polymerization. Primary, secondary and tertiary mercaptans, containing from eight to sixteen carbon atoms, are used and their reactivity is determined by two factors, namely solubilization from the monomer droplets into the growing polymer/monomer particles and the chain transfer activity of the mercaptan with the free-radicals. As solubilization of mercaptan is determined by the longest carbon chain in the molecule, it follows that a primary mercaptan will diffuse less rapidly than a secondary or tertiary mercaptan of the same molecular weight; typical values are 1·32, 2·09 and 2·37 × 10^{-4} g/cm^2, respectively, for the primary, secondary and tertiary C12 mercaptans.[91] Conversely the rate of reaction with a free-radical falls in the order primary > secondary > tertiary and the two opposing trends give optimum performance in practical systems at about C10 to C12. Thus at various conversions to polymer the utilization of mercaptans observed is given in Table 7.8. The reactivity of the mercaptan is often expressed by a coefficient r given by $-d \ln S/dX = rS$, where S is the concentration of mercaptan and X the fractional conversion of monomer to polymer. The optimum regulating activity as judged by the reduction of the molecular weight for a given amount of transfer agent is with $r = 2-4$, which corresponds to 70–90 per cent consumption at 60–70 per cent conversion.[92] The optimum value of r is smaller if higher conversions are intended and its value depends on the polymerization recipe to some extent. Blends of different mercaptans may be used with advantage and it has been shown that there is a gain in efficiency and a narrowing of the $\bar{M}w/\bar{M}n$ ratio of the polymer by carrying out the mercaptan addition incrementally,[93] but the conditions for this are rather critical.

Table 7.8 Utilization of mercaptans in emulsion polymerization of butadiene/styrene[92]

Mercaptan	Utilization per cent	Regulating index r	Mercaptan	Utilization per cent	Regulating index r
p Cl2	90	2·6	t Cl0	100	4·3, 5·7
p Cl4	25	—	t Cl2	90	2·8, 3·7
p Cl6	10	—	t Cl4	60	1·6
			t Cl6	30	0·54

The molecular weight distribution of butadiene polymer changes markedly with conversion.[94] The distribution for a typical SBR polymer at about 25 per cent conversion is markedly skew towards the high molecular weight end with a molecular weight peak in the region of 50 000. As conversion proceeds the high molecular weight tail increases due to depletion of chain transfer agent and to the secondary reactions between polymer and active free-radicals. At high conversions the distribution becomes much broader and binodal distributions may result. The shape of the distribution curve will obviously depend on the structure of the mercaptan modifier employed. At small r values reasonable agreement with theoretical calculation has been observed and relatively narrow ($\bar{M}w/\bar{M}n = 2\cdot75-3$) single peaked distributions have been found. Rapid removal of mercaptan (large r) has given broad distributions ($\bar{M}w/\bar{M}n = 4-16$).

A study of the composition of SBR latex particles shows that the S/B ratio and the total monomer concentration decreases with conversion.[92a] Measurements on cross-linking during polymerization have given values of $k_x/k_p = 7\cdot5 \times 10^{-5}$ at 5 °C, with $E_x - E_p = 4\cdot85$ kcal/mol. The k_x/k_p ratio depends on copolymer composition as well as temperature. The increase in cross-linking at higher temperatures is due also to a decrease in the mercaptan regulating index with increasing temperature.

Owing to the value of butadiene/styrene and other copolymer latices in foam rubber, films and adhesives there has been much work on the development of high concentration latices by direct polymerization. For average particle sizes in the region of 2000 Å about 2 per cent soap would be used with a monomer/water ratio of 100/70. The initial rate is only about 2·5 per cent/h and batch times for 60 per cent conversion may be from 50–60 h. For this reason agglomeration and concentration of conventional emulsions is preferred.

There appears to have been no detailed studies on the mechanism of butadiene/acrylonitrile polymerization in spite of the importance of these polymers as solid rubbers and latices and even less on the other copolymers. Terpolymers of butadiene, styrene and vinyl pyridine are manufactured on a substantial scale for use in tyre cord adhesives, and the carboxylic acid

Table 7.9 Propagation velocity constants for diene polymerization[97]

	A	E	k_p 60 °C
Butadiene	$1·2 \times 10^8$	9·3	100
Isoprene	$1·2 \times 10^8$	9·8	50
2,3-Dimethyl butadiene	$8·9 \times 10^7$	9·0	120

copolymers are increasingly used in adhesives, carpet backing and foam-backed fabrics. The ability of the carboxyl group to react with metal oxides or to co-condense with thermosetting resins leads to many new and interesting applications.

2. Isoprene

The emulsion polymers of isoprene are of little value as elastomers and the polymerization has not attracted much attention. It is broadly similar to butadiene although there are differences in detail.[95] The main feature is the propagation velocity constant compared with butadiene and dimethyl butadiene (Table 7.9).

Isoprene copolymerizes with many monomers including butadiene, styrene and acrylonitrile, but only the latter has achieved importance. This has been obtained as a low viscosity, high solids latex with a 70/30 isoprene/acrylonitrile ratio. It is of interest in that cast films are of high strength without reinforcement with fillers and it has comparable oil swelling resistance to the butadiene acrylonitrile copolymers.[96]

3. 2,3-Dimethylbutadiene

This monomer has been polymerized with persulphate and peroxamine catalysts in the presence of soap, alkylsulphonate and mixed anionic and non-ionic emulsifiers.[97,102] As with other diene monomers branching readily occurs and the polymer viscosity rises to a maximum and then falls with conversion. The rate of branching is less with 2,3-dimethylbutadiene than with isoprene or butadiene and as the energy of activation for branching is almost double that of propagation more linear polymers are obtained at low polymerization temperatures.

4. 1,3-Pentadiene

This monomer calls for little comment. There have been no quantitative measurements made on its polymerization and published information is restricted to the technological evaluation of its copolymers with butadiene or butadiene/styrene.[98] As this monomer exists in separable *cis* and *trans* forms a study of the kinetics of its free-radical initiated polymerization would be of considerable interest.

5. Chloroprene

Chloroprene polymerizes readily in bulk, solution or emulsion to give low melting crystalline or elastomeric polymers of predominantly *trans*-1,4-structure. Owing to its strength, flame, weather and oil resistance, it is of great technical significance and it has some limited latex applications. It is prepared in emulsion with simple fatty acid or rosin acid soaps and persulphate initiators. The product is obtained either as the homopolymer or as a copolymer containing sulphur. The molecular weight of the homopolymers is controlled by the addition of mercaptan chain transfer agents, whereas the high molecular weight polymer containing sulphur (about 0·5 per cent) is converted into a processable material by treatment with certain accelerators for the vulcanization of rubber, in particular dialkyl thiuram disulphides, in alkaline medium. These cleave the polysulphide bonds in the polymer and stabilize the end groups.[99]

Polymerization of chloroprene shows many complexities. The monomer is auto-oxidized readily to form an active initiator and it gives dimers which can participate in the propagation reaction. The polymers undergo cross-linking both in preparation and, unless protected, on storage. The monomer dissolves in water (0·11 per cent) and this is increased by emulsifier (c. 1 per cent in 2 per cent sodium dodecyl sulphate solution).[100] The polymerization rates are not linear and are affected by stirring, even in the presence of micellar emulsifier. The dependence of polymerization rate on initiator concentration appears to be normal[29] at low concentrations but at high concentrations of initiators and emulsifier the rate becomes independent of concentrations.[101]

7.8.5 Vinylidene chloride

Although vinylidene chloride is insoluble in the polymer its solubility, which is 0·6 per cent in water and 1·4 per cent in 0·2 M potassium laurate, increases substantially in the presence of polymer by adsorption at the surface of the polymer particles. The polymerization starts without a significant acceleration stage and this rate is maintained to a high conversion. At low emulsifier concentrations the polymerization slows down at a relatively low conversion which is followed by acceleration to a rate equal to or greater than the initial rate.[22] The first two stages in the polymerization are influenced by stirring in that the early rate is decreased and the second slow stage is increased. This is consistent with improved transport of monomer from the globule to the latex particle, which with gentle agitation or low emulsifier concentration is insufficient to maintain the equilibrium concentration until a later stage in the polymerization. Particle numbers apparently decrease by aggregation but it is difficult to envisage coalescence of hard, high melting point particles, other than as loose agglomerates. The maintenance of high polymerization rates at conversions up to 85 per cent would be anticipated from the low value of C_M (1·0 mol/l) which is lower than with most monomers, as well as by the appreciable water solubility of the monomer.

7.8.6 Ethylene

Most polyethylene is made by high pressure homogeneous polymerization or at low pressures using co-ordination catalysts, but emulsion polymers and copolymers are in relatively small volume production for applications such as polishes. The emulsion polymerization of ethylene requires high pressures (up to 300 atm), as with homogeneous radical polymerization,[10] but copolymers with vinyl acetate, which has a reactivity close to that of ethylene, may be prepared at lower pressures. The pressure employed determines the concentration of ethylene in the aqueous phase, and in consequence the proportion of ethylene in the copolymer.

Ethylene is appreciably soluble in water (0·8–1·7 per cent at 3000–4500 p.s.i. and 85 °C) and the polymerization follows the same pattern as does vinyl acetate; C_M at 85 °C is in the region of 5 mol/l.[23] In the initial stages of the polymerization a large number of particles is produced ($c.$ 10^{18}/ml of diameter $c.$ 30 Å) which aggregate later to a particle size of about 300 Å. Using 0·2–0·3 per cent potassium persulphate as initiator and 1·7–1·9 per cent potassium fatty acid soaps (laurate, myristate and stearate) as emulsifiers, polymerization at 85 °C and 3000 p.s.i. gave 20–30 per cent latex concentrations. Potassium stearate gave larger particle sizes than the laurate or myristate, and the addition of 10–15 per cent of a water-soluble alcohol such as t-butanol reduced the tendency to form small particles in the early stages of polymerization. An interesting observation was the presence of combined potassium myristate in the polymer arising from chain transfer with the growing polymer chains. The polymerization rate and molecular weight of the polymer fell with increasing soap concentration, indicating termination of the growing chains by reaction with the soap at the particle surface.

7.8.7 Methyl methacrylate

Polymethylmethacrylate is produced by casting processes or as beads by suspension polymerization. The emulsion process is applied more to its copolymers with other acrylic esters and styrene, for use in coatings. The monomer dissolves in water to about 1·5 per cent and its polymerization in solution follows homogeneous kinetics until polymer particles are precipitated, when most of the features of emulsion systems with aqueous phase initiation are observed.[45] A number of investigations have been carried out using redox catalysts based on metal salts (Cu^{2+}, Fe^{3+} and Mn^{3+}) with hydrazine[46] and exponents for the dependence of polymerization rate on emulsifier and initiator concentrations are given in Table 7.4. The polymerization rates depend significantly on the stoichiometry of the catalyst and on the monomer concentration in these relatively complex systems.

7.8.8 Other monomers: methyl acrylate, 2-methyl-5-vinylpyridine, acrolein, acrylonitrile

The emulsion polymerization of several other monomers has been studied. These include methyl acrylate, acrylonitrile, 2-methylvinylpyridine, methylvinyl ketone and acrolein. The polymers themselves are not of great significance but they may be present in copolymers of considerable importance.

1. Methyl acrylate

This monomer behaves in a comparable manner to vinyl acetate, (its solubility in water is *c.* 3 per cent) and it has been polymerized in aqueous solution using potassium persulphate as initiator.[103] Without emulsifier the number of particles ($3·9 \times 10^{12}$ particles/cm^3 at 12 per cent conversion) fell by a factor of five due to particle aggregation. The rate was increased by anionic soaps, no doubt resulting from reduced coalescence of particles, whereas cationic emulsifiers decreased the rate. The explanation here is possibly that termination of growing chains is increased by an enhanced rate of entry of the sulphate anion-radical into the oppositely charged polymer particles altering the electrical balance of the emulsion and causing particle aggregation to occur.

2. 2-Methyl-5-vinylpyridine

This monomer (solubility in water 1·04 per cent *w/v* at 60 °C) gives a linear polymerization rate up to 85 per cent conversion, which is not simply related to the emulsifier concentration.[21b] Polymerization in seeded latices gave no significant change in particle number but rates were not proportional to the numbers of particles, being faster than anticipated. This was attributed to polymerization in the aqueous phase followed by adsorption of the polymer particles thus formed in the preformed particles, but it could also result from increased values of $\bar{n}$ during polymerization.

3. Acrolein

This monomer forms a 20 per cent (v/v) solution in water and as the aldehyde group participates in its polymerization cross-linked insoluble polymers are produced. Treatment of the polymers with sulphur dioxide converts them into water soluble film forming materials which are stabilizers for the polymerization. Polymerizations are much faster with these stabilizers and the molecular weights are higher than with conventional emulsifying agents.[42]

4. Acrylonitrile

Acrylonitrile is quite soluble in water (7·8 per cent w/w at 20 °C) but the polymer is insoluble both in water and in monomer and only gives stable emulsions at high emulsifier concentrations. The polymerization from aqueous solution or in emulsion shows the characteristics of a water soluble monomer, resembling in many respects vinyl chloride. The rate remains constant over a wide conversion probably as a result of polymerization at the particle surface.[41]

7.9 EMULSION POLYMERIZATION RECIPES

Apart from the complex low temperature redox systems for butadiene copolymers, of which a vast number have been examined, most emulsion systems are simple. Type and concentration of emulsifier are obviously important but the precise conditions of the experiment (rate of addition of monomer, temperature control) may be equally critical in determining the properties of a given latex, and such information is rarely disclosed in the literature.

The following, therefore, are to be regarded as illustrative of typical, and in one or two instances unusual, emulsion recipes, but are not necessarily those which are employed on a large scale. (In all systems parts are by weight unless stated. Polymerization times are in hours).

Butadiene/styrene In view of the different reactivity ratios a charge ratio of 70-72/28-30 of butadiene/styrene is required to give a polymer containing 23 per cent combined styrene at about 65 per cent conversion. In recipes 1-4 below, 100 parts of total monomer are taken with butadiene/styrene in the ratio 70/30.

1. 'Hot' S.B.R. recipe[105]

Soap flakes	5
Potassium persulphate	0·3
n-Dodecyl mercaptan	0·5
Water	190
Time/temperature	12/50 °C
Yield	65 per cent

2. 'Cold' S.B.R. recipes—Normal and low iron content[105,106,107]

	(i)	(ii)	(iii)	(iv)	(v)	(vi)
Dresinate 214*	4·5	—	4·7	4·7	4·5	2·2
Fatty acid soap	—	5	—	—	—	2·2
Cumene hydroperoxide	—	0·085	0·2	0·1	—	—
p-Menthane hydroperoxide	—	—	—	—	0·10	0·15
Diisopropyl benzene hydroperoxide	0·15	—	—	—	—	—
Sodium sulphate	—	—	—	—	0·79	—
Potassium chloride	—	—	0·5	0·5	—	—
Ferrous sulphate 7H$_2$O	—	0·079	—	0·028	0·05	0·30
Potassium hydroxide	0·2	—	0·09	0·09	—	—
Sodium sulphate	—	0·068	—	—	—	—
Tetraethylene pentamine	0·08	—	0·1	—	—	—
Sodium formaldehyde sulphoxylate	—	—	—	—	0·15	—
Glucose	—	—	—	1·0	—	—
Ethylene diamine tetraacetic acid	0·008	—	—	0·1	0·07	0·02
Potassium pyrophosphate	—	1	—	—	—	0·80
Trisodium phosphate 8H$_2$O	—	—	—	—	—	0·45
Mixed tertiary mercaptans	0·18	0·2	0·24	0·24	0·15	0·16
Daxad 11†	0·05	—	0·1	0·1	0·15	0·15
Water	200	180	180	180	180	180
Time/temperature	15/50 °C	12/0 °C	17/5 °C		17/5 °C	
Conversion (per cent)	60	60	60		60/65	

* Dresinate 214 is the potassium salt of disproportionated rosin acid soap (Hercules Powder Co.).

† Sodium salt of naphthalene sulphonic acid/formaldehyde condensation product (W. R. Grace & Co.).

It is necessary to add a 'short stop' to butadiene/styrene systems. Typically, for a redox system, this would be sodium dimethyldithiocarbamate or sodium polysulphide (0·1–0·2) with a trace of sodium nitrite, usually added with an antioxidant (1·0) such as phenyl-2-naphthylamine.

3. Cationic systems[108]

Dodecylamine hydrochloride	5	—
Cetyl trimethylammonium chloride	—	3
t-Butyl hydroperoxide	0·3	—
Diisopropyl benzene hydroperoxide	—	0·1
Ferrous sulphate 7H$_2$O	—	0·1

Mixed tertiary mercaptans	0·35	0·15
Aluminium chloride	0·36	0·5
Potassium chloride	0·3	0·1
Water	120	180
Time/temperature	22/50 °C	12/5 °C
Conversion (per cent)	85	60

4. Direct 'high solids' latex[64]

Potassium laurate	2
Daxad 11	1·5
Cumene hydroperoxide	0·35
Ferrous sulphate 7H$_2$O	0·4
Potassium pyrophosphate	0·36
Mixed tertiary mercaptans	0·2
Potassium chloride	2
Water	55
Time/temperature	35/10 °C
Conversion (per cent)	60

5. Other systems for butadiene/styrene[109,110]

Butadiene	74
Styrene	26
Pyridine	12
Hydrazine hydrate	0·2
Sodium hydroxide	0·9
Oleic acid	3·3
Copper sulphate 5H$_2$O	3×10^{-4}
Ferrous sulphate 7H$_2$O	5×10^{-4}
Cobalt chloride 6H$_2$O	2×10^{-4}
Tartaric acid	5×10^{-3}
Carbon tetrachloride	1·2
Water	200
Time/temperature	10/21 °C
Yield	86 per cent

Butadiene	51	75
Styrene	49	25
Water	180	180
Sodium dresinate	—	5
Sodium diamyl sulphosuccinate	5	—
Potassium ferricyanide	—	0·3
t C$_{12}$ mercaptan	0·24	—
t C$_8$ mercaptan	—	0·25

p-Methoxyphenyl diazo
(2-naphthyl) ether   0·2   0·2
Time   40/50 °C   6/40 °C
Yield   58 per cent   60 per cent

6. Butadiene redox system[117]

Butadiene   100
Potassium stearate   5
Potassium dichromate   0·3
Arsenious oxide   0·1
Water   180
Temperature   40 °C
Yield   31·8 per cent/h

7. Redox recipe for methyl methacrylate[111]

Methyl methacrylate $6·2 \times 10^{-2}$ M
Potassium permanganate $3·79 \times 10^{-5}$ M
Oxalic acid $0·08 \times 10^{-2}$ M
Sodium cetyl sulphate   0·1 per cent
Time/temperature   1/32 °C
Yield   ~90 per cent

8. Polyacrylate latex[112]

Ethyl acrylate   93
2-Chloroethylvinyl ether   5
p-Divinyl benzene   2
Sodium lauryl sulphate   3
Sodium pyrophosphate   0·7
Potassium persulphate   1
Water   133
Time/temperature   8/60 °C
Yield   ~100 per cent

9. Rhodium chloride catalyzed emulsion polymerization of butadiene[113]

Butadiene   100
Sodium dodecyl benzene sulphonate   5
Rhodium chloride $3H_2O$ 1
Water   200
Temperature   80 °C
Yield   21 per cent/h

10. Chloroprene polymerization[99,114]

Chloroprene	100	100
Wood rosin	4	4
Sulphur	—	0·7
n-Dodecyl mercaptan	0·35	—
Sodium hydroxide	0·95	0·95
Daxad 11	0·6	0·6
Potassium persulphate	0·6	0·6
Water	165	170

Followed by

Sodium lauryl sulphate	0·42⎫
Daxad-11	0·2 ⎬ in water 42 ml
Phenyl-2-naphthylamine	3·5⎫
Phenothiazine	0·2⎬ in benzene 28 ml

Added at 70 per cent conversion.
Polymerization times and temperatures variable.

11. Alkylene sulphide polymerization[79,80]

Propylene sulphide	80 ml	50 ml
Ethylene sulphide	20 ml	—
Allyloxymethylthiirane	20 ml	—
Sodium glutamate	—	1·5
Zinc oxide	4	—
0·880 Ammonia	4 ml	—
Cadmium carbonate	—	2
Sodium dodecylbenzenesulphonate	4	—
Time/temperature	17/20 °C	17/25 °C
Yield	73 per cent	100 per cent

12. Vinyl chloride polymerization[115]

Vinyl chloride	100·0	90
Sodium lauryl sulphate	3·0	0·5
Ammonium persulphate	0·25	—
Sodium hydrosulphite	0·1	—
Ethylenediaminetetra-acetic acid	0·1	—
Vinyl isobutyl ether	—	10
Sodium dihydrogen phosphate	—	0·3
Disodium hydrogen phosphate	—	0·2
Sodium bisulphite	—	0·2
Benzoyl peroxide	—	0·25
Water	296	250
Time/temperature	21/30–40 °C	48/45 °C
Yield	70 per cent	75 per cent

9

13. Narrow distribution styrene copolymer latices[68a]

	(a)	(b)	(c)
Styrene	113·8	92	104·5
Divinyl benzene	1·2	23	0·5
Seed latex*	—	—	88
Water	60	67·5	95
Triton-X100			
(18 per cent w/v)	18	36·2	—
Ultrawet-K			
(10 per cent w/v)	3·75	—	—
Siponate-DS10			
(1 per cent w/v)	—	12·1	5·0
0·1 N Sodium hydroxide	—	7·4	—
Potassium persulphate	25	3·7	6·0
(3 per cent w/v)			
$D(\mu)$	0·250	0·225	0·350
	Polymerization at 65–70 °C for 24 hr		Polymerization at 70 °C. Emulsifier and monomers added dropwise over 3 hr and then polymerization to completion.

* 0·177 μ diameter latex prepared by similar formulation to (a).
Ultrawet-K—Sodium dodecylbenzenesulphonate containing 10 per cent sulphate. (Atlantic Refining Co.).
Siponate-DS10—Purified sodium dodecylbenzenesulphonate. (Alcolac Chem. Corp.).
Triton-X100—Iso-octylphenoxypolyethoxyethanol (Rohm and Haas Co.).

7.10   REFERENCES

1. K. Noro, *British Polymer J.*, **2,** 128, (1970); M. Shiraishi, *British Polymer J.*, **2,** 135, (1970).
2. N. B. Graham and H. W. Holden, *Polymer*, **10,** 198, (1970).
3. K. Shinoda, T. Nakagawa, B. Tamamushi and T. Isemura, *Colloidal Surfactants*, Academic Press, 1963.
4. D. B. Ludlum, *J. Phys. Chem.* **60,** 1240, (1956).
5. A. M. Mankovitch, *Ind. Eng. Chem.*, **47,** 2175, (1955); *J. Phys. Chem.*, **58,** 1027, (1954).
6. W. C. Griffin, *J. Soc. Cosmet. Chem.*, **1,** (5), 311, (1949); **5,** (4), 249, (1954).
7. S. Ross, E. S. Chen, F. Becher and H. J. Ranauto, *J. Phys. Chem.*, **63,** 1681, (1959).
8. G. Greth and J. E. Wilson, *J. Appl. Polymer Sci.*, **5,** 135, (1961).
9. F. Testa and G. Vianello, *J. Polymer Sci.*, **C27,** 69, (1969).
10. A. F. Helin, H. K. Stryker and G. J. Mantell, *J. Appl. Polymer Sci.*, **9,** 1797, (1965).
11. J. L. Azorlosa, *Ind. Eng. Chem.*, **41,** 1626, (1949).
12. J. G. Brodnyan and E. L. Kelley, *J. Polymer Sci.*, **C27,** 263, (1969); R. W. Kreider, *J. Polymer Sci.*, **C27,** 275, (1969).
13. P. A. Vinogradov, P. P. Odintsova and A. A. Shitova, *Vysokomol, Soed.*, **1,** 98, (1962).
14. C. E. M. Morris and A. G. Parts, *Makromol. Chem.*, **119,** 212, (1968).

15. J. W. L. Fordham and H. L. Williams, *J. Amer. Chem. Soc.*, (a) **73**, 4856, (1951), (b) **73**, 1634, (1951).
16. J. W. Breitenbach, K. Kuchner, H. Fritze and H. Tarnawiecki, *British Polymer J.*, **2**, 13, (1970).
17. W. D. Harkins, *J. Chem. Phys.*, **13**, 381, (1945); **14**, 47, (1946); *J. Amer. Chem. Soc.*, **69**, 1428, (1947).
18. W. V. Smith and R. H. Ewart, *J. Chem. Phys.* **16**, 592, (1948).
19. S. S. Medvedev, *Int. Symp. Macromol. Chem. Prague 1957*, p. 174, Pergamon, 1959.
20. J. L. Gardon, *Rubber Chem. and Tech.*, **43**, 74, (1970).
21. (a) J. T. O'Donnell, R. B. Mesrobian and A. E. Woodward, *J. Polymer Sci.*, **28**, 171, (1958); (b) L. Crescentini, G. B. Gechele and M. Pizzoli, *European Polymer J.*, **1**, 293, (1965).
22. P. M. Hay, *et al.*, *J. App. Polymer Sci.*, **5**, 23, 31, 39, (1961).
23. H. K. Stryker, A. F. Helin and G. F. Mantell, *J. App. Polymer Sci.*, **9**, 1807, (1965).
24. D. M. French, *J. Polymer Sci.*, **32**, 395, (1958).
25. R. M. Fitch, M. B. Prenosil, K. J. Sprick, *J. Polymer Sci.*, **C27**, 95, (1969).
26. C. P. Roe, *Ind. Eng. Chem.*, **60**, (9), 20, (1968).
27. J. L. Gardon, *J. Polymer Sci.*, **A1**, **6**, 643, (1968).
28. W. V. Smith, *J. Amer. Chem. Soc.*, **70**, 3695, (1948); **71**, 4077, (1949).
29. Z. Manyasek and A. Rezabek, *J. Polymer Sci.*, **56**, 47, (1962).
30. J. W. Vanderhoff, E. B. Bradford, H. L. Tarkowski and B. W. Wilkinson, *J. Polymer Sci.*, **50**, 265, (1961).
31. G. J. K. Acres and F. L. Dalton, *J. Polymer Sci.*, **A1**, 3009, (1963).
32. R. Patsiga, M. Litt and V. Stannett, *J. Phys. Chem.*, **64**, 801, (1960).
33. A. S. Dunn and P. A. Taylor, *Makromol. Chemi.*, **83**, 207, (1965). A. S. Dunn and L. C.-H. Chong, *British Polymer J.*, **2**, 49, (1970).
34. D. H. Napper and A. G. Parts, *J. Polymer Sci.*, **61**, 113, (1962).
35. S. Okamura and T. Motoyama, *J. Polymer Sci.*, **58**, 221, (1962).
36. G. Vidotto, A. C. Arrialdi and G. Talamini, *Makromol Chem.*, **134**, 41, (1970)
37. E. Peggion, F. Testa and G. Talamini, *Makromol Chem.*, **71**, 173, (1964).
38. H. Gerrens, W. Fink and E. Kohnlein, *J. Polymer Sci.*, **C16**, 2781, (1967).
39. J. Ugelstad, P. C. Mørk, P. Dahl and P. Rangnes, *J. Polymer Sci.*, **C27**, 49, (1969).
40. H. Weiner, *J. Polymer Sci.*, **7**, 1, (1951).
41. W. M. Thomas, E. H. Gleason and G. Mino, *J. Polymer Sci.*, **24**, 43, (1957).
42. H. Chadron, R. C. Schulz and W. Kern, *Makromol. Chem.*, **32**, 197, (1959).
43. M. Morton, P. P. Salatiello and H. Landfield, *J. Polymer Sci.*, **8**, 111, (1952).
44. J. G. Brodnyan, J. A. Cala, T. Konen and E. L. Kelley, *J. Colloid Sci.*, **18**, 73, (1963).
45. J. Bond and P. I. Lee, *J. App. Polymer Sci.*, **13**, 1215, (1969).
46. P. I. Lee and H. M. Longbottom, *J. App. Polymer Sci.*, **14**, 1377, (1970).
47. G. S. Whitby, M. D. Gross, J. R. Miller and A. J. Castanza, *J. Polymer Sci.*, **16**, 549, (1955).
48. J. Romatowski and G. V. Schulz, *Makromol Chem.*, **85**, 227, (1965).
49. G. V. Schulz and J. Romatowski, *Makromol Chem.*, **85**, 195, (1965).
50. J. L. Gardon, *J. Polymer Sci.*, **A1**, **6**, 687, (1968).
51. M. Morton, P. P. Salatiello and H. Landfield, *J. Polymer Sci.*, **8**, 215, (1952).

52. J. Ugelstad and P. C. Mørk, *British Polymer J.*, **2**, 31, (1970).
53. J. L. Gardon, *J. Polymer Sci.*, **A1, 6**, 623, (1968).
54. M. Morton, S. Kaizerman and M. W. Altier, *J. Colloid Sci.*, **9**, 300, (1954).
55. S. Jovanovic, J. Ramatowski and G. V. Schulz, *Makromol. Chem.*, **85**, 187, (1965).
56. R. A. Wessling, *J. Appl. Polymer Sci.*, **12**, 309, (1968).
57. H. Gerrens, *J. Polymer Sci.*, **C27**, 77, (1969).
58. J. J. Krackeler and H. Naidus, *J. Polymer Sci.*, **C27**, 207, (1969).
58a. S. D. Holdworth, *Chem. and Proc. Eng.*, **6**, 312, (1965).
58b. K. C. Denbigh, *Trans Faraday Soc.*, **40**, 352, (1944).
58c. D. B. Gershberg and E. Longfield, *Symp. Polymn. Kinetics and Catalyst Systems*, Pt. 1, A.I.Ch.E., 1961.
58d. M. Nomura, H. Kajima, M. Harada, W. Eguchi and S. Nagata, *J. App. Polymer Sci.*, **15**, 675, (1971).
58e. F. T. Wall, C. J. Delbecq and R. E. Florin, *J. Polymer Sci.*, **9**, 177, (1952).
59. H. Edelhauser, *British Polymer J.*, **2**, 119, (1970).
60. (a) J. Hearn, R. H. Ottewill and J. N. Shaw, *British Polymer J.*, **2**, 116, (1970); (b) H. J. Van den Hul and J. W. Vanderhoff, *British Polymer J.* **2**, 121, (1970).
61. H. K. Stryker, G. J. Mantell and A. F. Helin, *J. App. Polymer Sci.*, **C27**, 35, (1969).
62. I. M. Kolthoff and I. K. Miller, *J. Amer. Chem. Soc.*, **73**, 5118, (1951).
63. L. Talalay, *Proc. 4th Int. Rubber Conf.*, London, 1962.
64. H. S. Smith, H. G. Werner, J. C. Madigan and L. H. Howland, *Ind. Eng. Chem.*, **41**, 1584, (1949).
65. R. W. Brown and L. H. Howland, *Rubber World*, **132**, 471, (1955).
66. R. H. Allen, L. D. Yats and T. Alfrey, Amer. Chem. Soc., Meeting Los Angeles, 1963, (*Org. Coatings and Plastics Chem.* **1**, 183).
67. J. W. Vanderhoff, J. F. Vitkuske, E. B. Bradford and T. Alfrey, *J. Polymer Sci.*, **20**, 225, (1956).
68. D. J. Williams and M. R. Granlio, *J. Polymer Sci.*, **C27**, 139, (1969).
68a. M. E. Woods, J. S. Dodge and I. M. Krieger, *J. Paint Technol.*, **40**, 544, (1968); J. S. Dodge, M. E. Woods and I. M. Krieger, ibid. **42**, 71, (1970). Y. S. Papir, M. E. Woods, and I. M. Krieger, ibid., **42**, 571, (1970).
69. D. J. Williams and E. G. Bobalek, *J. Polymer Sci.*, **A1**, 4, 3065, (1966).
70. J. W. Vanderhoff, E. B. Bradford, H. L. Tarkowski, J. B. Shaffer and R. M. Wiley. *Advances in Chem.* No. **34**, 32, (1962).
71. C. S. Marvel, *et al.*, *J. Polymer Sci.*, **3**, 181, (1948).
72. R. E. Rinehart, H. P. Smith, H. S. Witt and H. Romeyn, *J. Amer. Chem. Soc.*, **84**, 4145, (1962).
73. M. Morton, I. Piirma and B. Das, *Rubber Plastics Age*, **46**, 404, (1965).
74. R. E. Rinehart, *J. Polymer Sci.*, **C27**, 7, (1969).
75. R. Dauby, F. Dawans, Ph. Teyssié, *J. Polymer Sci.*, **C16**, 1989, (1967).
76. V. N. Sokolov, B. D. Babitskii, V. A. Kormer, I. Ya Poddubnyi and N. N. Chesnovkova, *J. Polymer Sci.*, **C16**, 4345, (1969).
77. F. J. Hyde and J. R. Wehrly, *U.S.P.*, **2**, 891, 920, (1959).
78. D. R. Weyenberg, D. E. Findlay, J. Cekada and A. E. Bey, *J. Polymer Sci.*, **C27**, 27, (1969).
79. Dunlop Ltd, British Pat. 1 123 801, (1965).
80. Dunlop Ltd, British Pat. 1 117 307, (1964).
81. B. M. E. Van der Hoff, *J. Polymer Sci.*, **44**, 241, (1960).

82. B. M. E. Van der Hoff, *J. Phys. Chem.* **60**, 1250, (1956).
83. H. A. Oosterhof, *J. Oil Col. Chem. Ass.*, **48**, 256, (1965).
84. A. McIntosh and B. E. L. Reader, *J. Oil Col. Chem. Ass.* **49**, 525, (1966).
85. B. G. Elgood, E. V. Gulbekian and D. Kinsler, *J. Polymer Sci.*, **B2**, 257, (1964).
86. M. J. Cantow, C. W. Cline, C. A. Heiberger, D. Th. A. Huibers and R. Phillips, *Modern Plastics*, **June 1969**, 126.
87. S. Ohtsuka, S. Yoshikawa and Y. Hoshi, *S. P. E. Journal*, **May 1966**, 75.
88. I. D. Robb, *J. Polymer Sci.*, **A1, 7** 417, (1969).
89. H. Hopff and I. Fahla, *British Polymer J.*, **2**, 40, (1970).
90. D. E. Moore, *J. Polymer Sci.*, **A1, 5**, 2665, (1967).
91. R. L. Frank, P. V. Smith, F. E. Woodward, W. B. Reynolds and P. J. Canterino, *J. Polymer Sci.*, **3**, 39, (1948).
92. C. A. Uranek and J. E. Burleigh, *Kaut. Gummi Kunst.*, **19**, 532, (1966); *J. App. Polymer Sci.*, **14**, 267, (1970).
92a. G. M. Burnett, C. G. Cameron and P. L. Thorat, *J. Polymer Sci.*, **A1, 8**, 3435, 3443, (1970).
93. C. A. Uranek and J. E. Burleigh, *J. App. Polymer Sci.*, **9**, 1273, (1965); C. Booth, L. R. Beason and J. T. Bailey, *J. App. Polymer Sci.*, **5, 13**, 116, (1961).
94. E. W. Duck, *British Polymer J.*, **2**, 60, (1970).
95. M. Morton, J. A. Cala and I. Piirma, *J. Polymer Sci.*, **15**, 167, (1955).
96. R. A. Stewart, S. N. Angove, E. S. Graham, G. Hilditch and F. L. White, *Rubber World*, **152, 6**, 52, (1965); *Trans. Inst. Rubber Ind.*, **42**, T1, (1966).
97. M. Morton and W. E. Gibbs, *J. Polymer Sci.*, **A1**, 2679, (1963).
98. I. A. Livshitz, *et al.*, *Soviet Rubber Tech.*, **20, 7**, 2, 20; **8**, 2, (1961).
99. C. E. Hollis, *Chem. & Ind.* (*Lon.*), **1969**, 1030.
100. M. Morton, J. A. Cala and M. W. Altier, *J. Polymer Sci.*, **19**, 547, (1956).
101. T. Krishan and M. Margaritova, *J. Polymer Sci.*, **52**, 139, (1961).
102. R. Sattlemeyer and G. D. Bereznoj, *Makromol. Chem.*, **93**, 280, (1966).
103. C. E. Morris, A. E. Alexander and A. G. Parts, *J. Polymer Sci.*, **A1, 4**, 985, (1966).
104. V. Stannett, J. A. Gervasi, J. J. Kearney and K. Araki, *J. App. Polymer Sci.*, **13**, 1175, (1969).
105. E. W. Duck, *Enc. Polymer Sci. & Tech.*, **5**, 802, (1966).
106. R. Spolsky, H. L. Williams, *Ind. Eng. Chem.*, **42**, 1847, (1950).
107. C. F. Fryling and A. E. Follett, *J. Polymer Sci.*, **6**, 59, (1951).
108. L. H. Howland, V. C. Neklutin, R. W. Brown and H. G. Werner, *Ind. Eng. Chem.*, **44**, 762, (1952).
109. Michelin, French Pat., 1 217 683 (1958).
110. Phillips Pet. Co., British Pats, 644 620, 649 654, 656 727, (1946).
111. S. R. Palit, and R. S. Konar, *J. Polymer Sci.*, **58**, 85, (1962).
111a. M. Litt and V. Stonnett, *J. Polymer Sci.*, **A1, 8**, 3607 (1970).
112. B. F. Goodrich, British Pat. 687 526, (1949).
112a. M. R. Grancio and D. J. Williams, *J. Polymer Sci.*, **A1, 8**, 2617, (1970).
113. R. E. Rinehart, H. P. Smith, H. S. Witt and H. Romeyn, *J. Amer. Chem. Soc.*, **83**, 4864, (1961).
114. W. E. Mochel and J. H. Peterson, *J. Amer. Chem. Soc.*, **71**, 1426, (1949); H. W. Walker and W. E. Mochel, *Proc. 2nd. Rubber Tech. Conf. London 1948*, p. 69.
115. Unpublished recipes.

116. H. Gerrens and E. Kohnlein, *Z. Elektrochem.*, **64**, 1199, (1960).
117. M. Morton, P. P. Salatiello and H. Landfield, *Ind. Eng. Chem.*, **44**, 739, (1952).

7.11 ADDENDUM

Dunn[1] has reviewed the role of the emulsifier, in particular the aspects of number of particles, monomer concentration in the latex particle and the effects of the particle surface on the reactions. It was also reported that, contrary to earlier reports (see section 7.2.3), the decomposition of potassium persulphate is not influenced by emulsifier or dissolved monomer, the higher rates being ascribed to radical induced decomposition of the persulphate.

Brooks has carried out calculations which confirm that diffusion of monomer to the particle and within it is not generally a limiting factor in emulsion systems but that interfacial (coulombic) barriers could be limiting.[2] It has been observed that added electrolyte (in the presence of excess emulsifier), which would reduce repulsion, increases the polymerization rate.[3]

Gardon[4] has given a theoretical treatment of radical termination in the polymer particle during the nucleation stage, which predicts the fall in the average number of radicals from 1 to 0·5 per particle as polymerization proceeds.

Further work on the control of molecular weight of butadiene-styrene emulsion polymers by thiols has been reported.[5] It is shown that for incremental addition of modifier to a given final conversion a higher regulating index is required, as compared with a single addition of modifier.

Further studies have been made on the kinetics of the emulsion polymerization of styrene in a continuous stirred reactor.[6] Reasonable agreement between calculation and experiment for polymerization rate, particle formation and size and average molecular weights was obtained.

The evidence for enhanced stereospecificity in polymers prepared in emulsion systems with conventional free-radical initiators is fragmentary, but it has been shown that polymeric emulsifiers from polyvinyl pyridine quaternized with nonyl bromide increase the syndiotacticity of polymethyl methacrylate—possibly by orienting the monomer as it approaches the reactive site on the particle surface.[7]

7.11.1 References

1. A. S. Dunn, *Chem. & Ind. (Lon.)*, **1971**, 1406.
2. B. W. Brooks, *Brit. Polymer J.*, **3**, 269, (1971).
3. J. L. Mateo and I. Cohen, *J. Polymer Sci.*, **A2**, 711, (1964).
4. J. L. Gardon, *J. Polymer Soc.*, **A1, 9**, 2763, (1971).
5. C. A. Uraneck and J. E. Burleigh, *J. App. Polymer Sci.*, **15**, 1757, (1971).
6. A. W. DeGraff and C. W. Pochlein, *J. Polymer Sci.*, **A2, 9**, 1955, (1971).
7. A. Ya. Chernikova and S. S. Medvedev, *Dok. Akad. Nauk. S.S.R.*, **182**, 1369, (1968); *Chem. Abs.* **70**, 29359, (1969).

8

Carbonium Ions

R. Baker

Department of Chemistry, University of Southampton

8.1 GENERATION AND STABILITY OF CARBONIUM IONS

Carbonium ions can be generated in a number of general ways (Table 8.1). Heterolytic cleavage of a bond so that the carbon atom looses both bonding electrons is the most common method of formation. To obtain stable solutions of carbonium ions it is necessary to use a solvent that is chemically inert towards the ion and formation is favoured by high dielectric constant media and those capable of stabilising the anion. Formation of carbonium ions by addition of a cation to an olefin is the basis of cationic olefin polymerization and it should be apparent that this method cannot normally be used for forming stable carbonium ions. Electron impact can be used for generating carbonium ions from radicals and although not synthetically useful this method has proved of value in studies of the energetics of carbonium ion formation. These ions can also be generated by the oxidation of radicals in solution.

$$Ph_3C^+X^-$$

(1)   (2)   (3)

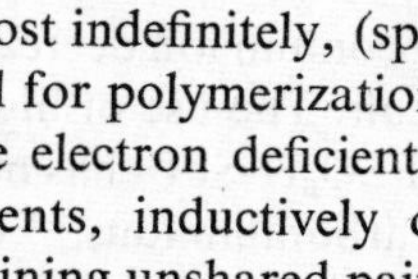

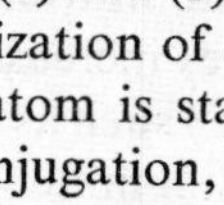

A number of carbonium ions can be formed which are extremely stable and can be kept as salts almost indefinitely, (species (1), (2) and (3)); many of these are particularly useful for polymerizations. Stabilization of an organic cation occurs whenever the electron deficient carbon atom is stabilized by electron-supplying substituents, inductively or by conjugation, with aryl groups or with atoms containing unshared pairs of electrons such as oxygen, nitrogen or sulphur. The cycloheptatrienyl cation, (species (3)), is an example of a 6π-electron system which is predicted by molecular orbital theory to be very stable. Other positive ions may be formed in strong acid solution such as sulphuric acid (equation (4) and characterized by their UV spectra.[1] As the

Table 8.1 Formation of carbonium ions

(a) Heterolytic fission

$RX \longrightarrow R^+ + X^-$ X = halogen, OTs*, OBs* etc.

$R_2S^+R \longrightarrow R_2S + R^+$

$RNH_2 \xrightarrow{HNO_2} RN_2^+ \rightarrow R^+ + N_2$

$ROH \xrightarrow{H_2SO_4} RO^+H_2 \rightarrow R^+ + H_2O$

$3\ Ph_3COEt + 4\ BF_3 \rightarrow 3\ Ph_3C^+ + 3\ BF_4^- + B(OEt)_3$

$t\text{-BuF} + SbF_5 \rightarrow t\text{-Bu}^+ + SbF_6^-$

$Ph_3CH + Cr(VI) \rightarrow Ph_3C^+ + Cr(IV)$

(b) Addition of cation to neutral molecule

$$PhCH{=}CH_2 \xrightarrow{H^+} PhCHCH_3 \atop +$$

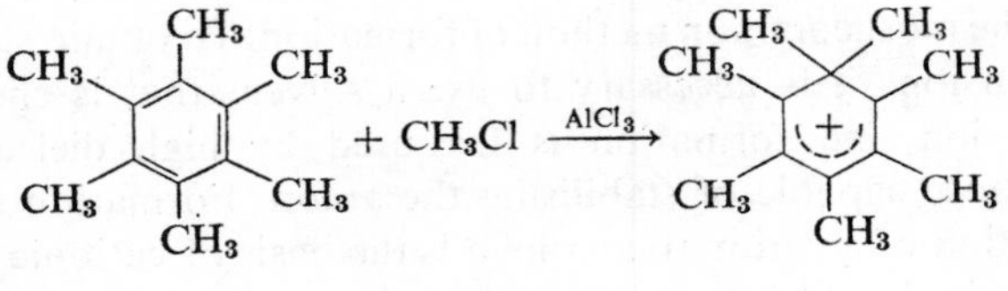

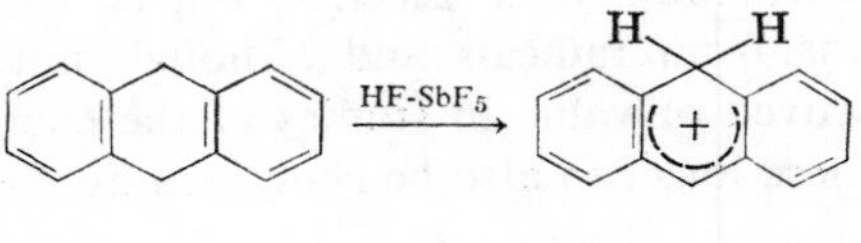

(c) Electron removal from neutral species

$RCO_2^- \rightarrow RCO_2^{\cdot} + e$

$RCO_2^{\cdot} \rightarrow R{\cdot} + CO_2$ Kolbé anodic oxidation

$R{\cdot} \rightarrow R^+ + e$

---

* Toluene-p-sulphonate (OTs) and p-bromobenzenesulphonate (OBs).

degree of stabilization of the carbonium ion decreases, generation in sulphuric acid becomes increasingly difficult. The use of highly acidic solvents (termed 'super acids') such as SbF_5 and FSO_3H-SbF_5 has made possible the generation of an almost infinite variety of carbonium ions.[2,3] Coupled with this approach NMR spectrometry has been developed as a highly discriminating technique for assignment of carbonium ion structure (equations (5) and (6)). Thus the cation generated as shown in equation (6) exists for only a short time before cyclization takes place. Further evidence has been obtained by these techniques for the comparative stability of primary, secondary and tertiary carbonium ions; n-butyl, s-butyl, isobutyl and t-butyl fluorides all lead to the formation of the same t-alkylcarbonium ion in SbF_5.

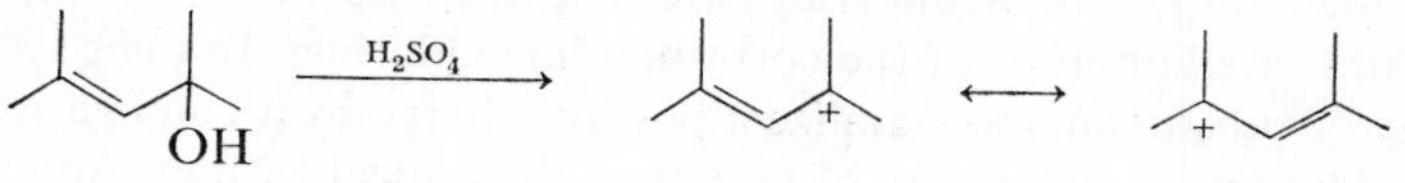

(4)

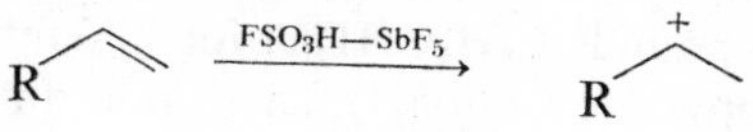

(5)

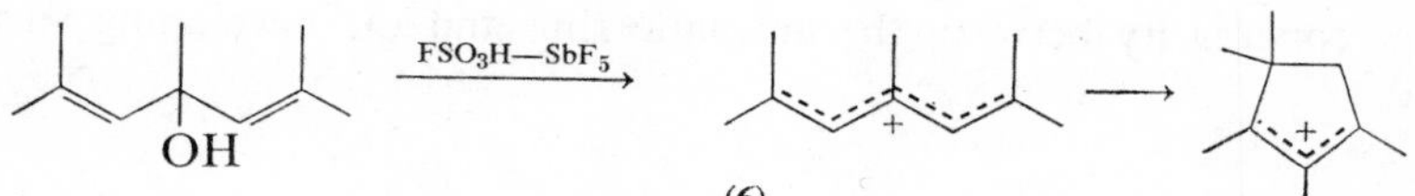

(6)

Numerous methods have been developed for the examination of the factors governing the stability of carbonium ions. Stability is, of course, only a relative term since in chemical systems carbonium ions are always associated with anions and solvent. Intermolecular ion-solvent forces can be discounted if studies are undertaken in the gas-phase and scales of stabilities have been obtained by the measurement of ionization potentials (Table 8.2).

$$R\cdot \rightarrow R^+ + e^-$$

Measurement of the rates of carbonium ion formation in S_N1 reactions have been widely used to examine the extent to which a positive charge can be accommodated in an organic structure. This method depends upon the thesis that the greater the stability of the ion the more rapidly it will be formed. The rate comparison measures the free-energy difference between the parent molecule and the transition state and is a measure only as far as the energies of the transition state reflect the respective energies

Table 8.2 Ionization potentials (kcal mol^{-1}) of radicals derived from electron impact studies

Radical	Ionization potential	Radical	Ionization potential
$\cdot CH_3$	$229\cdot4 \pm 0\cdot7$	$(C_6H_5)_2CH$	$168\cdot8 \pm 2\cdot3$
$CH_3\cdot CH_2$	$202\cdot5 \pm 1\cdot2$	$CH_2{=}CH$	$217\cdot9 \pm 1\cdot2$
$CH_3\cdot CH_2\cdot CH_2$	$200\cdot4 \pm 1\cdot2$	$CH_2{=}CH\cdot CH_2$	$188\cdot2 \pm 0\cdot7$
$(CH_3)_2CH$	$182\cdot2 \pm 1\cdot2$	$CH_3\cdot CH{=}CH\cdot CH_2$	$177\cdot8 \pm 1\cdot2$
$(CH_3)_3C$	$171\cdot8 \pm 1\cdot2$	$HC{\equiv}C\cdot CH_2$	$190\cdot2 \pm 1\cdot8$
$C_6H_5\cdot CH_2$	$178\cdot9 \pm 1\cdot8$		
$p\text{-}CNC_6H_5\cdot CH_2$	$197\cdot9 \pm 2\cdot3$		
$p\text{-}MeOC_6H_5\cdot CH_2$	$157\cdot7 \pm 2\cdot3$		

of carbonium ions. Support for this is found in the observation that the heats of formation, ΔH_f (obtained by solution calorimetry of a number of ions from their respective alcohols in HSO_3F-SbF_5) bear a close relationship to the rates of ethanolysis of the corresponding chlorides. In some systems, the ability of substituents to stabilize a positive charge by a conjugative effect is dependent upon coplanarity of groups to the central carbon. An example of this effect is found in the comparison of the relative rates of solvolysis in 80 per cent ethanol of the compounds C_6H_5CHRCl for different R· groups, methyl, 540; ethyl 125; isopropyl 27; t-butyl 1. These rate differences have been ascribed, in part, to steric repulsions between the alkyl group and the ortho-hydrogens of the ring, which increase with size of alkyl group, and prevent coplanarity between the aromatic ring and the developing vacant orbital.

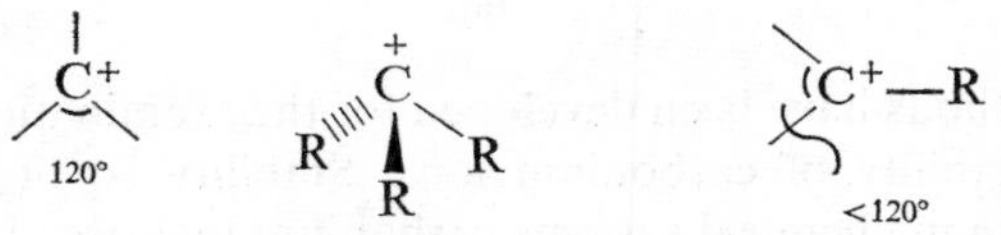

Out-of plane strain   In-plane strain

Figure 8.1 Strain in carbonium ions

The ideal structure for a carbonium ion has the three sp^2 orbitals at angles of 120°. Any change in geometry will be strongly resisted by a carbonium ion since the hybridization and the repulsion energies are increased. Two types of distortion are possible; deformations in which the carbonium ion carbon and the three attached atoms remain in the same plane, and distortions leading to non-planar structures (Figure 8.1). This latter type of distortion is the geometry required for carbonium ion formation in some bridged ring systems in which planar ions cannot be formed for structural reasons.

'In-plane strain' (Figure 8.1) occurring when one of the angles of a carbonium is required to have an angle < 120 ° has been termed 'I'-strain. Thus cyclopropane derivatives are extremely unreactive in solvolysis reactions since one of the angles of the carbonium ion would be required to be 60° (the abnormal reactivity of cyclobutane compounds is discussed later). Cyclopentane derivatives show a greater reactivity than acyclic derivatives since, although a small constraint is imposed from angle considerations, the rehybridization of a sp^3 carbon to the sp^2 carbonium ion is accompanied by a decrease of four axial carbon-hydrogen non-bonded interactions. Relief of non-bonded interactions are, in general, an important factor in governing the rate of formation of carbonium ions. Thus, if in a carbonium ion precursor, R_3C-X, the R groups are bulky with substantial non-bonded interaction, strain (termed 'B' strain) is relieved on ionization. Since the R·

groups are farther apart in the transition state than in the ground state, enhanced reactivity results. This factor contributes to the reactivity of t-butyl chloride in solvolysis reactions and is even greater as the size of R increases.

8.2 NEIGHBOURING GROUP PARTICIPATION[5]

One of the major advances in the study of carbonium ion reactions was the recognition of neighbouring group participation. The requirement for this effect is that a substituent G, possessing lone pairs is located in the molecule in such a position to become bonded or partially bonded to the reaction centre. A stabilization of a carbonium ion in this way results in an increased rate of formation of this ion; when this is observed the substituent is said to provide anchimeric assistance. The anchimerically assisted reaction rate constant k_Δ, is frequently accompanied by the component of the reaction proceeding without participation, k_c (Figure 8.2). The rate of

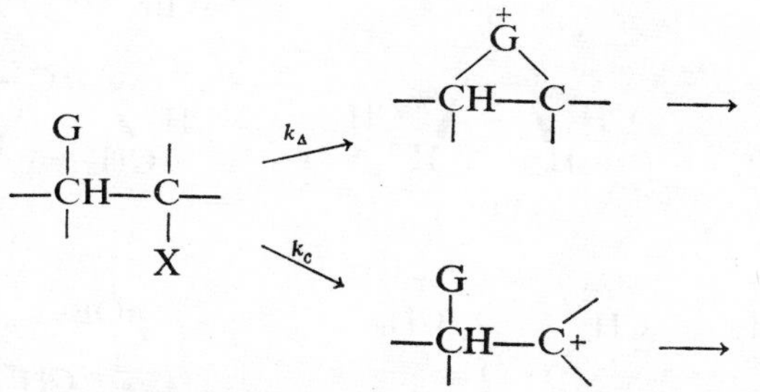

Figure 8.2 Neighbouring group participation

acetolysis of *trans*-2-acetoxycyclohexyl-p-toluenesulphonate, formula (7), is 2000 times faster than that of *cis*-isomer. Confirmation that the symmetrical ion is the intermediate in the reaction is the formation of racemic product when optically active material is used. Further indications of a reaction involving a neighbouring group is that a product is produced which differs from that which would be expected in the absence of participation. In some cases the participation occurs after the rate determining ionization so that, although the structure of the product is affected, no anchimeric assistance is observed. This may be a product with retained configuration, a ring-closed product, or one in which the participating group has migrated. Due to the incursion of bromonium ions in the reaction of hydrobromic acid with *dl-erythro-* and *dl-threo*-3-bromobutanol species (8) and (9), products are formed with retention of configuration; *meso-* and *dl*-2,3-dibromobutane are produced, respectively. For halogens generally, nucleophilic participation increases in the order Cl < Br < I. Participation by methoxyl groups is

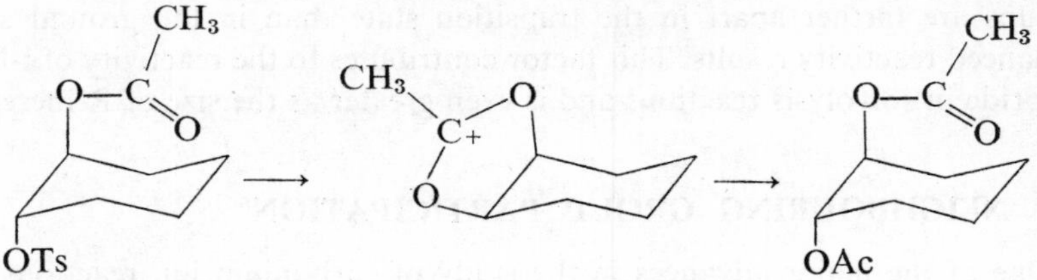

(7)

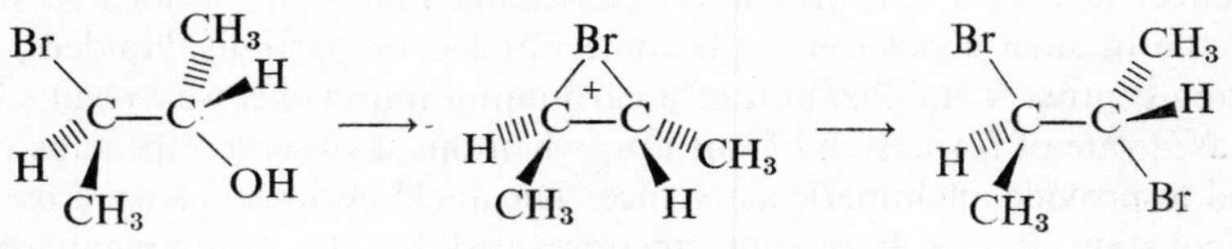

(8)

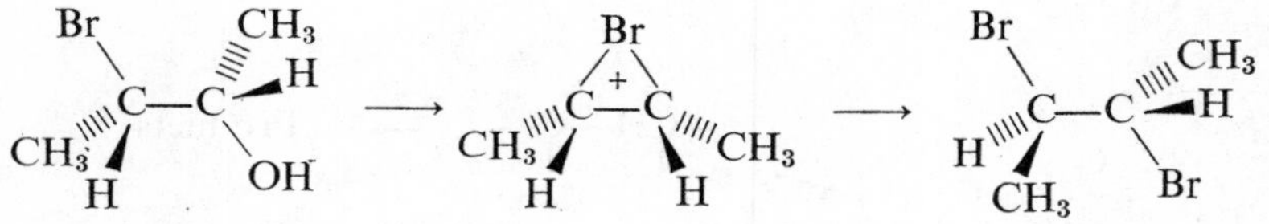

(9)

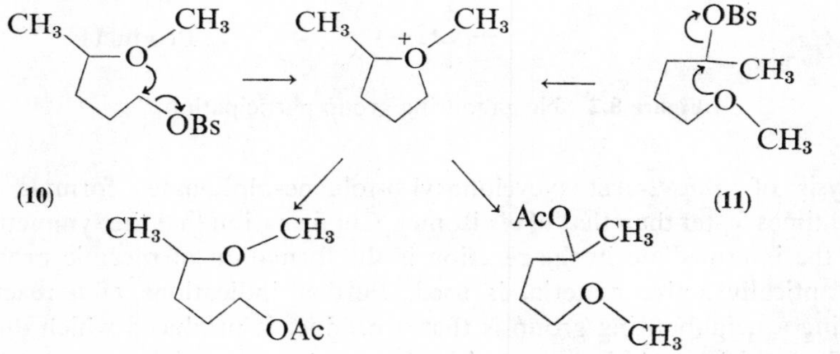

(10)   (11)

common; the acetolysis products of 4-methoxypentyl- species (10) and 4-methoxy-1-methylbutyl-p-bromobenzene sulphonate species (11) are identical, indicating the common intermediate. Participation has been observed with a wide range of other substituents such as hydroxyl, amines, carboxyl, thioethers, esters, amides and aldehydes.

The probability of neighbouring group participation is dependent upon the size of ring formed in the transition state. Increases in the number of degrees of freedom with the length of carbon chain causes a consequent larger decrease in entropy on going from the initial state to the transition state for the cyclization reaction. Conversely, the strain energy decreases with the size of the ring formed. The balance of strain energy and entropy factors favours neighbouring group effects for the formation of three-, five- and six-membered rings.

Double bonds have been found to be particularly effective as neighbouring groups. Acetolysis of 4-methylpent-3-enyl-p-toluenesulphonate species (12) is 1200 times faster than that of ethyl p-toluenesulphonate and yields 2-cyclopropylpropene and 4-methylpent-3-enyl acetate. The requirement for effective participation of the double bond is an ideal stereochemical arrangement in relation to the p-orbital of the carbonium ion. The additional steric constants found in cyclic systems compared to acyclic compounds has produced examples where participation is particularly favourable; the reverse is also found. One of the most striking examples of double bond participation is found in the acetolysis of *anti*-norborn-2-en-7-yl tosylate

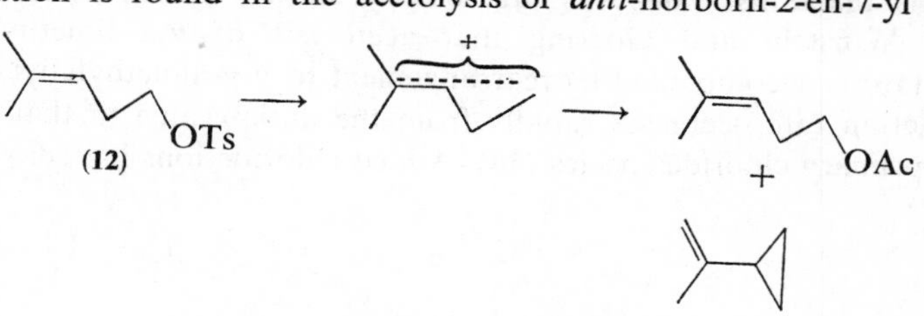

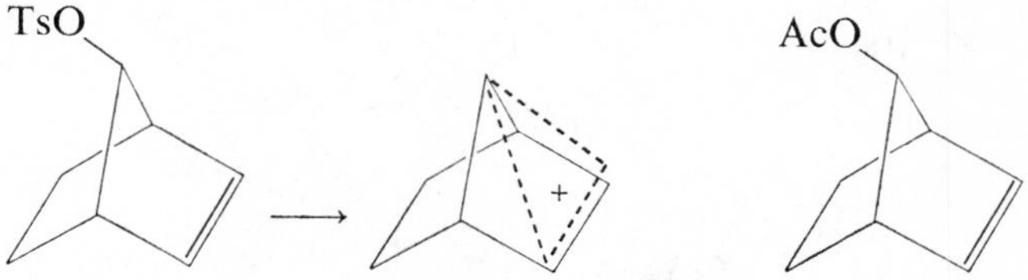

species (13) which proceeds 10^{11} faster than that of the analogous saturated compound, norbornyl-7-yl tosylate. The π-electrons of the double bond are particularly well placed to interact with the developing carbonium ion at the 7-position to yield the ion (13); indicative of the intervention of this intermediate is the reaction product, anti-norborn-2-en-7-yl acetate species (14) obtained with retention of configuration. In some cases phenyl groups and σ-bonds can also act as neighbouring groups but these are discussed later in the section on reactions of carbonium ions.

8.3 ION PAIRS AND SOLVENT EFFECTS[6]

Carbonium ions in solution derive much of their stability from solvation effects and are formed more easily in more polar media. Numerous scales of solvent polarity have been considered and the simplest is that based upon the di-electric constant of the medium. As a measure of 'ionizing power' of a medium this has been of only limited success since, together with the electrostatic effect which a solvent exerts on the stability of the positive ion, the nucleophilicity of the medium is also important. Among other scales of solvent polarity, Y, ε_T and Z-values, the latter two based on spectroscopic measurements have proved the most useful although still far from ideal.

Carbonium ions can also gain considerable stabilization from other ions in solution. From simple electrostatic theory it is predicted that the presence of added salts in solution will increase the rate of carbonium ion formation due to the increase in ionic strength of the medium. It is also predicted that ions of opposite charge can co-exist in solution as ion pairs, kept together by electrostatic attraction. The detection of ion pairs from conductivity data was followed by evidence for the occurrence of ion pairs as intermediates in solvolysis reactions. This was first clearly shown by the observation of Young, Winstein and Goering that acetolysis of α,α-dimethylchloride species (16) is accompanied by rearrangement to γ,γ-dimethylallyl chloride. The reaction rate decreases rapidly from the initial value to that expected for the primary chloride species (18). Added chloride ions have no effect on

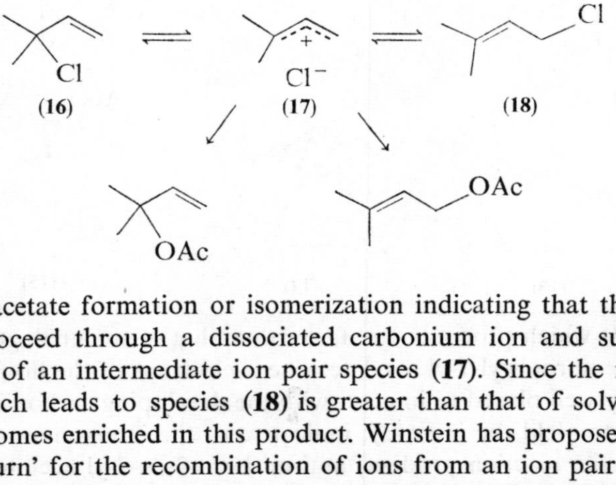

the rate of acetate formation or isomerization indicating that the reaction does not proceed through a dissociated carbonium ion and suggests the intervention of an intermediate ion pair species (17). Since the rate of the reaction which leads to species (18) is greater than that of solvolysis, the solution becomes enriched in this product. Winstein has proposed the term 'internal return' for the recombination of ions from an ion pair to form a covalent bond in contrast to 'external return' which signifies the recombination of a carbonium ion and an anion from the solution. Isomerization of α,α-dimethylallyl chloride does not occur in ethanol which has a higher dielectric constant than acetic acid since in the former, dissociation of ions is more favoured.

Ion pairs have also been detected in S_N1 reactions by the loss of optical activity of the substrate prior to solvolysis. If k_α is the rate of racemization and k_t the overall rate of the reaction (measured by titration), k_α is frequently larger than k_t. This was suggested to be consistent with ion pair intermediates which can either return to the starting material (and then produce racemization) or react further with solvent to yield product. Studies of salt effects have proved extremely successful in the investigation of ion pairs. A 'special salt effect' discovered by Winstein and co-workers in the acetolysis of a number of alkyl-arenesulphonates led to the proposal that at least two types of ion pairs were present in many solvolyses.[6] This effect, illustrated in Figure 8.3

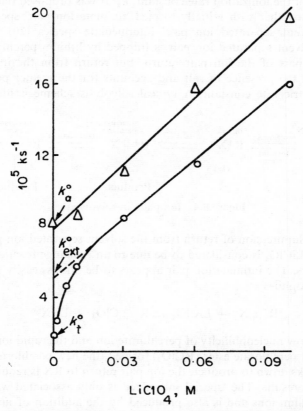

Figure 8.3 The effect of lithium perchlorate on the acetolysis of *threo*-3-p-anisyl-2-butyl bromobenzenesulphonate at 25°

for *threo*-3-p-anisyl-2-butyl-p-bromobenzenesulphonate is a strikingly rapid non-linear rise in the titrimetric rate for low lithium perchlorate concentrations, followed by a linear increase in rate constant with the concentration of added salt. The linear rate increase of k_t and a similar effect on the polarimetric rate constant, k_α, is described by the equations:

$$k_\alpha = k_\alpha^0(1 + b[\text{LiClO}_4])$$
$$k_t = k_t^0(1 + b[\text{LiClO}_4])$$

where b is per cent change in k_t for 0·01 M change in $LiClO_4$ and k_α^0 and k_t^0 are the rate constants at zero salt concentrations; for secondary arenesulphonates the constant b varies from 10 to 40. It is clear that the special salt effect is concerned with the reduction of ion pair return and k_{ext}^0, the intercept produced by extrapolation of the linear portion, is the rate constant predicted in the absence of this return. The striking fact appears that k_{ext}^0 still falls quite short of the ionization rate constant, k_α^0. It was proposed that for some solvolyses, together with initially formed 'intimate ion pair' species (19) a second 'solvent separated ion pair' intermediate species (20) is present. Only the solvent separated ion pair is trapped by lithium perchlorate, thus eliminating part of the ion pair return, but return from the intimate ion continues in the presence of salt and accounts for the greater polarimetric than titrimetric rate constant. A typical solvolysis scheme is illustrated in

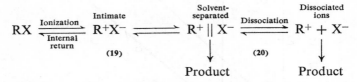

Figure 8.4 Ion pairs in solvolysis

Figure 8.4. Suppression of return from the solvent separated ion pair by the addition of $LiClO_4$ is considered to be due to an exchange reaction between two ion pairs. The intimate ion pair appears to be far less easily trapped by added nucleophiles

$$R^+ \parallel X^- + LiClO_4^- \rightleftharpoons R^+ \parallel ClO_4^- + Li^+X^-$$

Due to the low nucleophilicity of perchlorate ion and the rapid ionization of any covalent perchlorate ester $(RClO_4)$ formed, the new perchlorate ion pair rapidly breaks down to product, the ion pair return to RX is reduced and an enhanced k_t results. The 'special salt effect' is only associated with longer-lived carbonium ions and is also produced by the addition of acetates and bromides. Evidence has been discussed for the intervention of two kinds of ion pairs in the solvolysis and the question must be asked concerning the relative importance compared to that of free ions in reactions. The extent to which two types of ion pair and the free carbonium ion take part in solvolytic reactions depends upon the structure of the carbonium ion and the solvent. The longer the lifetime of the carbonium ion intermediate the greater the likelihood of formation of the solvent-separated and dissociated ion. In general the higher the dielectric constant of the medium the smaller the extent of ion-pairing of carbonium ion and negative species. The importance of information of ion-pairing in solutions of cations can be illustrated by considering the use of cations in the initiation of polymerization. In these reactions the reactivity of the free ions is much larger than that of ion pairs.[7]

8.4 REACTIONS OF CARBONIUM IONS

Since carbonium ions are highly energetic intermediates a number of reaction paths are possible (Figure 8.5). Competition between substitution and

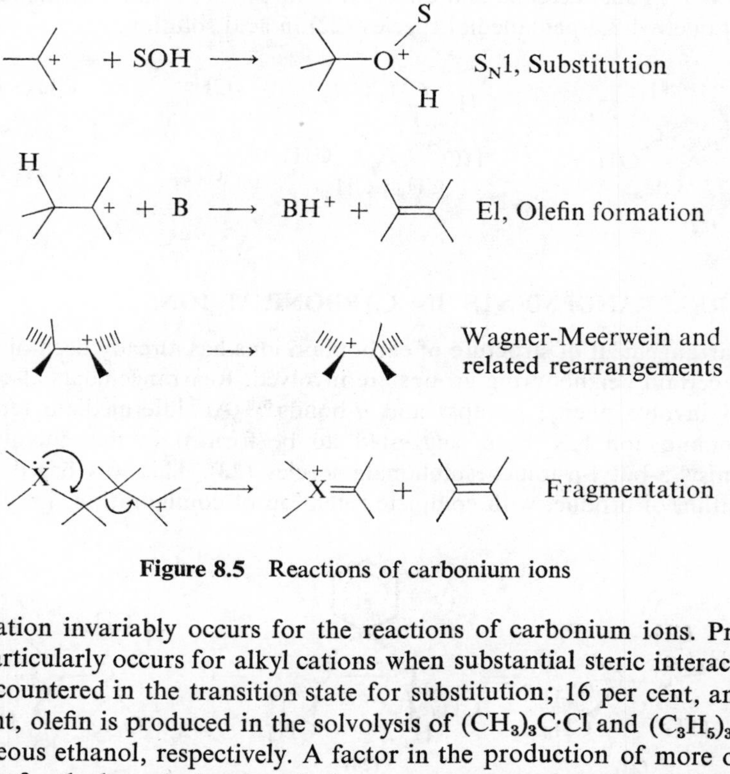

Figure 8.5 Reactions of carbonium ions

elimination invariably occurs for the reactions of carbonium ions. Proton loss particularly occurs for alkyl cations when substantial steric interactions are encountered in the transition state for substitution; 16 per cent, and 19 per cent, olefin is produced in the solvolysis of $(CH_3)_3C\cdot Cl$ and $(C_3H_5)_3C\cdot Cl$ in aqueous ethanol, respectively. A factor in the production of more olefin product for the branched cations is the greater hyperconjugative stabilization of the double bond by alkyl substituents species (**21**). This concept provides a satisfactory explanation for the production of the most highly substituted

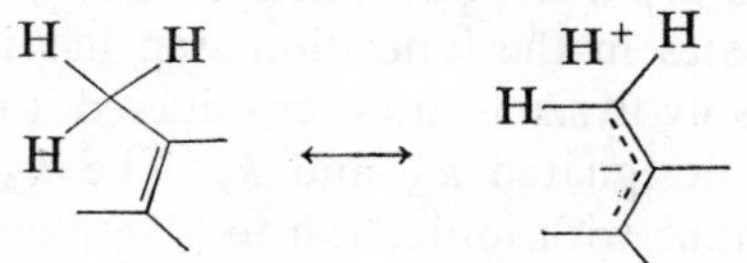

(21)

olefin (Saytzeff orientation) from carbonium ions. The stability of the products is thus reflected in the transition states leading to their formation. Steric factors must also play a part in determining the stabilities of the products and in the olefin production from $(CH_3)_2CH \cdot CH_2CH(OBs)CH_3$ in acetic acid, the *trans*-2-ene/*cis*-2-ene ratio is 2.0. Thus the eclipsing interactions in the *cis*-olefin are reflected in the product forming transition state.

Together with the loss of a proton from carbonium ions to yield olefins in some cases a fragmentation involving the loss of a larger part of the molecule can occur.[8] Thus, acetone and dimethyl-2-butene are formed by the reaction of tetramethyl-2,4-pentanediol species (**22**) in acid solution.

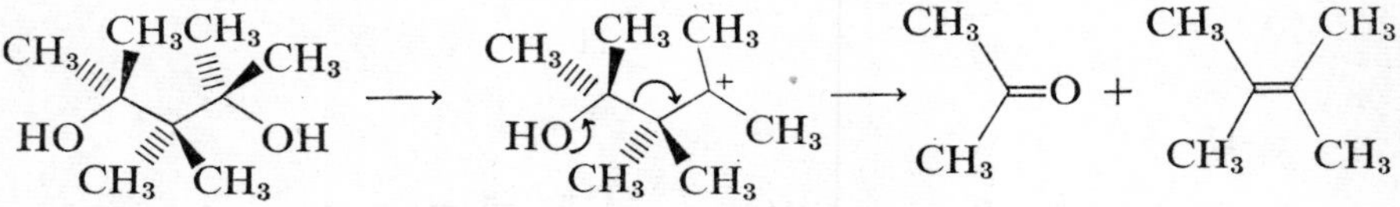

8.5 REARRANGEMENTS IN CARBONIUM IONS

Rearrangement of structure of carbonium ions has already been discussed when certain neighbouring groups are involved. Rearrangements also occur which involve phenyl groups[5] and σ-bonds.[5,9] An intermediate termed a phenonium ion has been suggested to be formed in the solvolysis of 2-p-anisyl-3-butyl-p-toluenesulphonate species (**23**). This was based on the formation of product with complete retention of configuration, i.e. *threo*-2-

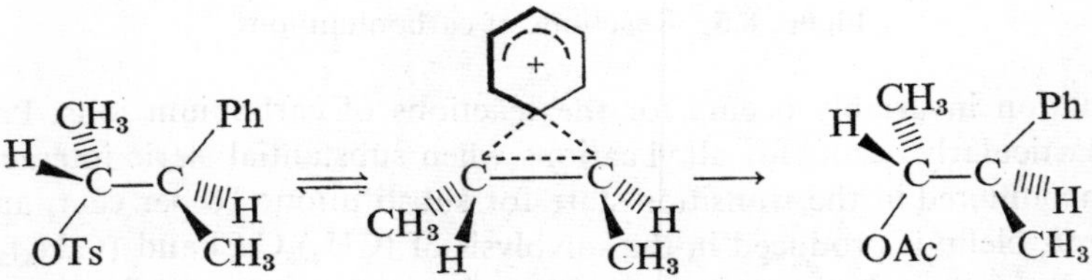

p-anisyl-2-butyl acetate. The formation of a completely racemized product from optically active material is also indicative of an intermediate such as species (**24**) which has a plane of symmetry. The question of whether the phenyl group participates in the ionization step has involved considerable controversy but the solvolysis is now considered to occur through two competing pathways, designated k_Δ and k_s. The k_Δ path, involves aryl participation concomitant with ionization to give phenonium ions, and the k_s pathway involves solvent in the normal way for substitution (Figure 8.6).

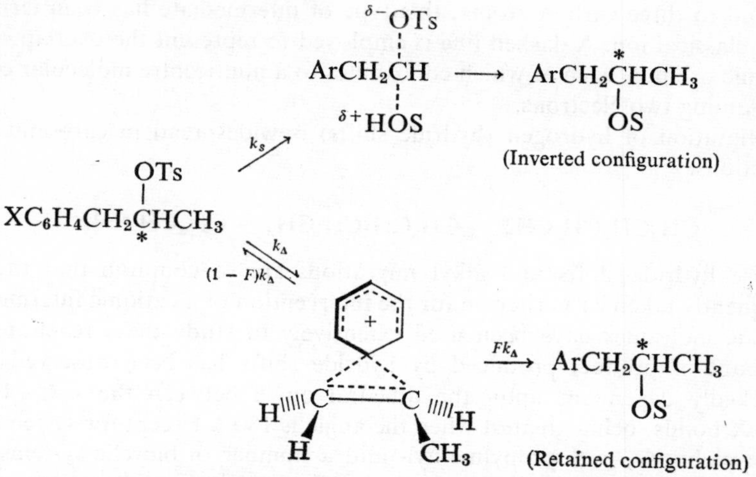

$$\underset{\delta+ HOS}{\underset{|}{ArCH_2\overset{\delta- OTs}{\underset{}{\overset{|}{\underset{}{C}}}H}}} \longrightarrow \underset{OS}{\underset{|}{ArCH_2\overset{*}{C}HCH_3}}$$

(Inverted configuration)

$$\underset{(25)}{\underset{*}{XC_6H_4CH_2\overset{|}{\underset{}{C}}HCH_3}} \quad \overset{k_s}{\nearrow}$$

$$\overset{k_\Delta}{\underset{(1-F)k_\Delta}{\rightleftarrows}}$$

$$\overset{Fk_\Delta}{\longrightarrow} \underset{OS}{\underset{|}{ArCH_2\overset{*}{C}HCH_3}}$$

(Retained configuration)

Figure 8.6 Mechanism of solvolysis involving phenonium ions

A large number of cases where alkyl migrations are known such as in the solvolysis of neopentyl chloride species (25), the nitrous acid deamination of neopentylamine species (27) and the conversion of pinacol to pinacolone. The migration may take place synchronously with the ionization process or after the formation of the carbonium ion. Evidence exists for both of these alternatives but, in either event, rearranged product results. It is important to consider the intermediate that is required for a bridged carbonium ion involving a σ-bond species (26). The bonding electron pair cannot be fully

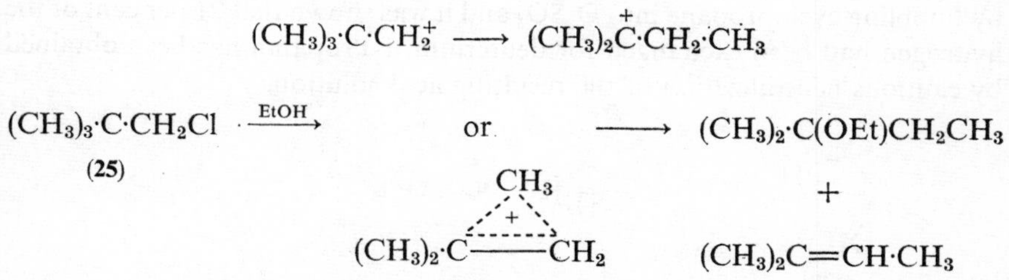

$$(CH_3)_3 \cdot C \cdot CH_2^+ \longrightarrow (CH_3)_2 \overset{+}{C} \cdot CH_2 \cdot CH_3$$

$$\underset{(25)}{(CH_3)_3 \cdot C \cdot CH_2Cl} \xrightarrow{EtOH} \quad \text{or} \quad \longrightarrow (CH_3)_2 \cdot C(OEt)CH_2CH_3$$

$$\underset{(26)}{(CH_3)_2 \cdot \overset{CH_3}{\overset{\cdot}{\underset{+}{C}}} CH_2} \qquad +$$

$$(CH_3)_2C{=}CH \cdot CH_3$$

$$\underset{(27)}{(CH_3)_3 \cdot C \cdot CH_2NH_2} \xrightarrow{HNO_2} (CH_3)_3C \cdot CH_2\overset{+}{N}_2 \longrightarrow (CH_3)_3C \cdot \overset{+}{C}H_2 + N_2$$

$$\downarrow$$

$$(CH_3)_2C{=}CHCH_3 + \underset{OH}{\underset{|}{(CH_3)_2\overset{}{C} \cdot CH_2CH_3}} \longleftarrow (CH_3)_2\underset{+}{C} \cdot CH_2CH_3$$

bound to three carbon atoms; this type of intermediate has been termed a non-classical ion. A dashed line is employed to represent the overlap of two atomic orbitals, each of which contributes to a multicentre molecular orbital containing two electrons.

Migration of hydrogen (hydride shifts) is widespread in carbonium ion reactions.

$$CH_3CH_2CH_2CH_2^+ \rightarrow CH_3CH_2\overset{+}{C}H\cdot CH_3 \rightarrow CH_3\overset{+}{C}H\cdot CH_2CH_3$$

These hydride shifts and alkyl migrations are so common that they are frequently taken as a criterion for the intervention of a cationic intermediate. Cyclic molecules have been used extensively to study these reactions; the amount of product produced by hydride shifts has been observed to be markedly dependent upon the dihedral angle between the —C—H and C—X bonds, being greatest when the angle is 180 °. Except for systems such as cyclobutyl-, cyclopropylmethyl- and a number of bicyclic systems, little evidence is available that σ-bond participation is important in the rate-determining step for fragmentation of carbonium ions.

A number of carbonium ion rearrangements of simple acyclic systems have been demonstrated to occur through protonated cyclopropane species (28).[10] Thus, the deamination of 1-aminopropane labelled at the 1-position with ^{14}C yields 1-propanol with 8 per cent of the label at the 3-position. Although this result could also be interpreted as the result of 1,3-hydride shifts or methyl migration, later studies and isotopic labelling experiments confirmed the existence of protonated cyclopropane intermediates. For example, cis- and trans-dimethylcyclopropanes were isolated from the deamination of 2-amino-3-methylbutane. Protonated cyclopropanes have also been generated by bubbling cyclopropane into D_2SO_4 and it was shown that 21 per cent of the hydrogen had been exchanged for deuterium. 1-Propanol has been obtained by cautious neutralization of the resulting acid solution.

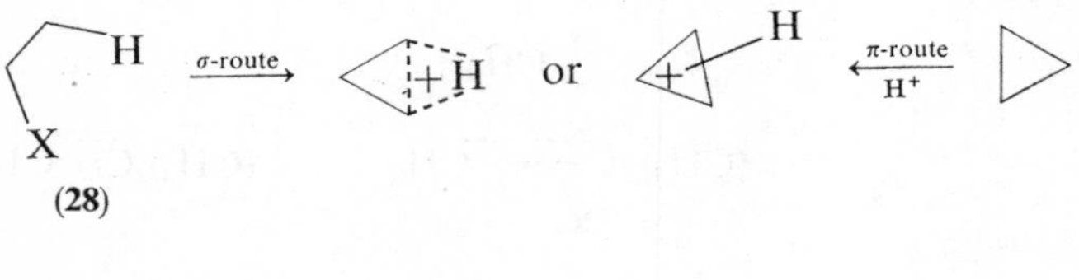

(28)

8.6 OTHER REACTIONS

Emphasis has been placed so far on the reactions of carbonium ions which involve no other reagent (i.e. elimination; fragmentation and rearrangement). Substitution reactions differ in that reaction occurs with a nucleophile and similar processes involving attack on other electron rich species (olefins and aromatic compounds) are extremely important in carbonium chemistry.

Association of carbonium ions with olefins is the fundamental process associated with cationic polymerization and is discussed in detail in the next chapter.

The reaction of carbonium ions with aromatic π-electrons is also commonly observed in aromatic alkylations. This is the Friedel-Crafts process and is carried out using alkyl halides, alcohols, olefins and acetylenes in the presence of acidic catalysts such as $AlCl_3$, BF_3, equation (**29**). Kinetic studies of a wide

$$RCl + AlCl_3 \longrightarrow R^+ + AlCl_4^-$$

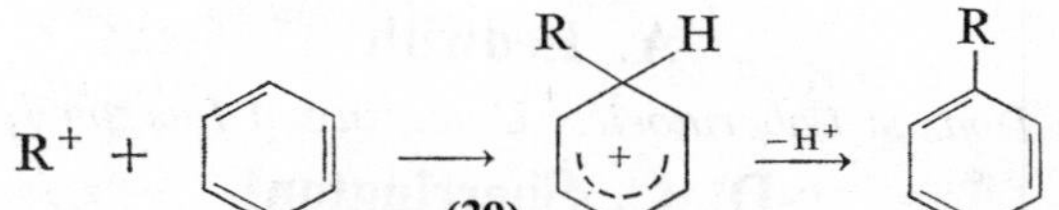

range of aromatic alkylations indicate that the formation of the σ-complex is rate-determining and that the reaction rate is given by:

$$\text{rate} = k[\text{alkylating agent}][\text{aromatic cpd}]\phi$$

where ϕ is a function of the catalyst concentration. The attack of carbonium ions on —C—X multiple bonds, is typified by N-alkylation of nitriles.

$$R^+ + R'C{\equiv}N \rightarrow R'C^+{=}NR \rightarrow R'CONHR$$

Hydride transfer reactions from a molecule to carbonium ions represents a further general class of reactions and this can be visualized as cationic attack on a σ-bond. Thus hydride abstraction from isopentane by the t-butyl cation occurs very rapidly at room temperature.

$$(CH_3)_3C^+ + (CH_3)_2CH{\cdot}CH_2CH_3 \rightarrow (CH_3)_3CH + (CH_3)_2\overset{+}{C}{\cdot}CH_2CH_3$$

8.7   REFERENCES

1. G. A. Olah and C. W. Pittman, *Adv. Phys. Org. Chem*, **4**, 305, (1966).
2. G. A. Olah and P von R Schleyer (eds), *Carbonium Ions*, Vol. 1, Interscience, Wiley, New York, 1968.
3. D. Bethel and V. Gold, *Carbonium Ions*, Academic Press, London, 1967.
4. F. P. Lossing, *Mass Spectroscopy*, C. A. McDowell (ed.) p. 433, McGraw-Hill, New York, 1963.
5. B. Capon, *Quart. Rev.*, **18**, 45, (1964).
6. S. Winstein, B. Appel, R. Baker and A. Diaz, *Organic Reaction Mechanisms*, Chem. Soc. Spec. Publ., No. 19, 109, 1965.
7. A. Ledwith, Adv. in Chemistry Series, No. 91, p. 317, American Chemical Society, 1969.
8. C. A. Grob, *Angew. Chem. Intern. Ed.*, **8**, 535, (1969).
9. G. D. Sargent, *Quart. Rev.*, **20**, 301, (1966).
10. C. J. Collins, *Chem. Rev.*, **69**, 543, (1969).

9
Reactivity and Mechanism in Cationic Polymerization

A. Ledwith

Donnan Laboratories, University of Liverpool,

D. C. Sherrington

Department of Pure and Applied Chemistry,
University of Strathclyde, Glasgow

9.1 INTRODUCTION

Cationic polymerization is the conversion of low molecular weight monomeric molecules into high molecular weight polymeric ones, via a mechanism involving stepwise growth of a carbonium (R_3C^+), carboxonium ($R\overset{+}{O}=CR_2$), oxonium (R_3O^+), sulphonium (R_3S^+) or immonium ($R_2\overset{+}{N}=CR_2$) ion. The types of monomer susceptible to polymerization by such intermediates have already been discussed (Chapter 1) where it was pointed out that both vinyl

$$n\ CH_2{=}CHR \longrightarrow \mathord{\sim\!\!\sim\!\!\sim} C^{\oplus} \longrightarrow \mathord{\sim\!\!\sim}(CH_2{-}CHR)_{\overline{n}}\mathord{\sim\!\!\sim}$$

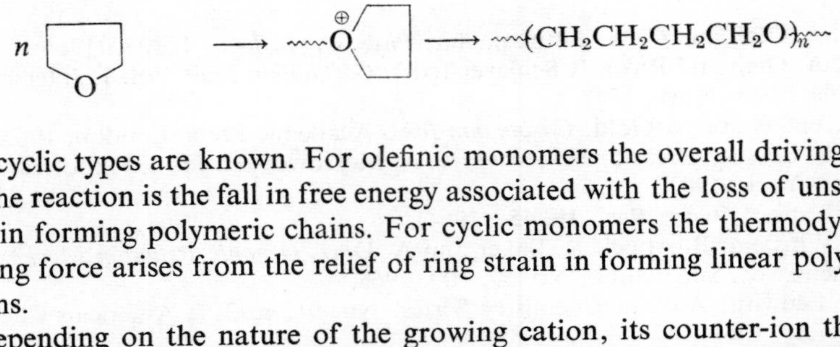

$$n\ \underset{O}{\bigsqcup} \longrightarrow \mathord{\sim\!\!\sim}\overset{\oplus}{O}{\bigsqcup} \longrightarrow \mathord{\sim\!\!\sim}(CH_2CH_2CH_2CH_2O)_{\overline{n}}\mathord{\sim\!\!\sim}$$

and cyclic types are known. For olefinic monomers the overall driving force for the reaction is the fall in free energy associated with the loss of unsaturation in forming polymeric chains. For cyclic monomers the thermodynamic driving force arises from the relief of ring strain in forming linear polymeric chains.

Depending on the nature of the growing cation, its counter-ion the solvating and dissociating ability of any solvent employed, and the temperature, both free cationic and various ion pair (or higher aggregated) species may at some time contribute to the reaction, since, under favourable conditions, these different ionic forms may all be present in equilibrium, for example.

$$\sim\!C^\oplus X^\ominus \quad \rightleftharpoons \quad \sim\!C^\oplus//X^\ominus \quad \rightleftharpoons \sim\!C^\oplus + X^\ominus$$

Intimate (or contact)   Solvent separated   Free solvated
ion pairs   ion pairs   ions

Classical initiation, propagation, transfer and termination reactions have all been characterized for particular types of cationic polymerization, namely

$$Y^\oplus + CH_2\!\!=\!\!CHR \longrightarrow YCH_2\!\!-\!\!\overset{\oplus}{C}HR \quad \text{(Initiation)}$$

$$\sim\!CH_2\!\!-\!\!\overset{\oplus}{C}HR + CH_2\!\!=\!\!CHR \longrightarrow \sim\!CH_2\!\!-\!\!CHR\!\!-\!\!CH_2\!\!-\!\!\overset{\oplus}{C}HR$$
$$\text{(Propagation)}$$

$$\sim\!CH_2\!\!-\!\!\overset{\oplus}{C}HR + CH_2\!\!=\!\!CHR \longrightarrow \sim\!CH\!\!=\!\!CHR + CH_3\!\!-\!\!\overset{\oplus}{C}HR$$
$$\text{(Transfer to monomer)}$$

$$\sim\!CH_2\!\!-\!\!\overset{\oplus}{C}HR + X^\ominus \longrightarrow \sim\!CH_2\!\!-\!\!\overset{\displaystyle X}{\overset{|}{C}}HR \quad \text{(Termination)}$$

In contrast to other types of polymerization, however, most cationic systems, particularly those involving vinyl monomers, are typified by complex initiation equilibria, rapid propagation reactions, and chain breaking dominated by transfer rather than true termination processes. It is the purpose of this chapter to discuss these observations in more detail bearing in mind the accepted principles which have been derived from more conventional organic reactions.

Literature on all aspects of cationic polymerization up to 1963 has been comprehensively reviewed in a book edited by P. H. Plesch;[1] other leading workers in the field have contributed additional surveys of particular aspects.[2-9]

9.2 THE NATURE OF THE ACTIVE INTERMEDIATE IN CATIONIC POLYMERIZATION

In spite of the great volume of literature on cationic polymerization there is in fact very little direct evidence that carbonium ions or related cations are present during vinyl polymerizations. Likewise, results from ring-opening systems have tended to assume, rather than prove, the presence of cationic intermediates e.g. oxonium ions in the polymerization of tetrahydrofuran. Carbonium ions are much more reactive than corresponding radicals and it is not surprising that they have eluded detection (Chapter 8). In addition to recombining easily with attendant counter-ions, they may undergo isomerization to form more stable structures, so that the resulting polymer can possess a structure differing from that of the monomer unit from which it was formed.[10] The difficulty of forming carbonium ions coupled with the ease of their destruction or diversion often precludes establishment of a stationary state concentration of active centres. Even when such a state is achieved equilibrium concentrations are often very small and this again makes the active intermediate difficult to identify. A very similar situation arises in free-radical polymerization, where, until quite recently, exceptional circumstances were required before the propagating free-radical could be detected by ESR techniques. Cationic systems are further complicated by difficult experimental conditions and the sensitivity of the processes to trace amounts of polar impurities. Indeed, this is one of the reasons why it has proved impractical to set up a useful general reaction scheme comprising a sequence of elementary reactions as can be done for free-radical polymerizations. The majority of cationic systems have something unique associated with them, more often than not arising in the initiation mechanisms. Many of these as we shall see (section 9.3) can be complex and are still by no means fully understood.

9.2.1 Spectroscopic detection of cationic intermediates

Many olefins when dissolved in concentrated acids yield broad ultra-violet (UV)/visible absorption bands presumably characteristic of the carbonium ions formed by protonation and subsequent isomerizations or condensations. Evans and Hamann[11] moved nearer to detecting carbonium ions in polymerizations when they showed that the spectrum from the system 1,1-diphenylethylene/BF_3/H_2O in benzene was essentially that of the 1,1-diphenylethylene carbonium ion. BF_3/H_2O is a useful cationic initiator system (see section 9.3) and evidence of simple protonation of the olefin by this combination was significant in the development of theories of cationic polymerization. When the system styrene/$AlCl_3$ (or $SnCl_4$) in 1,2-dichloroethane containing traces of water was examined[12] it was believed that an absorption maximum corresponding to a growing polymeric cation had at last been obtained. However, supporting studies[13] have shown that the principle originator of this spectrum is in fact a stable indanyl cation generated in a termination reaction. Indeed, more recently,[14] related species have been detected during the cationic polymerization of indene itself, for example

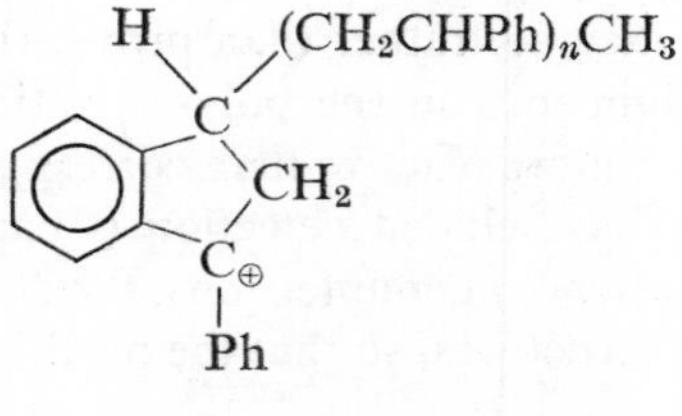

To date, therefore, there is still no direct spectroscopic evidence for the presence of carbonium ions in polymerizing systems, unlike the corresponding situation in anionic polymerization. In this case characteristic absorption bands with high extinction coefficients of growing polymeric carbanions have not only been recorded, but have proved invaluable in the study of the kinetics of anionic polymerization.[15] A number of attempts to identify low molecular weight model carbonium ions have been successful, however. Olah and his coworkers[143] have recorded the low temperature ultra-violet and nuclear magnetic resonance (NMR) spectra of the p-methoxystyryl cation, while more recently[144] the NMR spectrum of the styryl carbonium ion itself has been obtained.

All the other useful methods for detecting carbonium ions are indirect and often qualitative in their approach (see Chapter 8). The evidence is therefore circumstantial—but the volume of information is so large that the conclusions reached are essentially indisputable.

9.2.2 Effects of additives and impurities on cationic polymerization

The deliberate use of basic or nucleophilic additives has been invaluable in providing evidence for the occurrence of cationic species in these systems. Almost invariably typical radical inhibitors, for example quinones, oxygen and α,α'-diphenyl-β-picrylhydrazyl (DPPH), have no inhibiting or retarding effects on polymerization rates. Polar additives have much more dramatic effects. Trace quantities of water or other protic agents are often used as cocatalysts in combination with, say, Lewis acids such as BF_3, $AlCl_3$, $TiCl_4$ (see section 9.3.2), but in some systems water, particularly in excess, may act as an inhibitor. The mechanism of inhibition is not clear since in theory the destruction of a cationic centre by protic nucleophiles should produce equivalent numbers of protons, i.e. chain initiators. In practice the situation is governed by kinetic factors and each carbonium ion generated by, say,

$$H^{\oplus} + CH_2{=}CHR \longrightarrow CH_3{-}\overset{\oplus}{C}HR$$

$$\overset{H_2O}{\diagup}{}_{k_i} \qquad {}_{k_p}\overset{CH_2{=}CHR}{\diagdown}$$

$$\begin{array}{c} CH_3{-}CHR \\ \diagdown \\ OH + H^{\oplus} \end{array} \qquad \text{Polymer}$$

protonation of an olefin can react either with water (k_i) or with monomer (k_p). While $k_i \gg k_p$ and water is present in excess then excessive solvolytic transfer-type reactions will always lead to apparent termination or formation of ill-defined oligomeric products containing hydroxyl end-groups. Water has a particularly destructive effect on the radiation induced polymerization of styrene (section 9.4) where water concentrations $\geqslant 10^{-3}$ M suppress cationic propagation in favour of a radical addition process.[17]

Other types of basic additive may also be effective terminators of cationic polymerization (e.g. amines, ethers, sulphides) and this follows from the well-known greater thermodynamic stability of ammonium, oxonium, and sulphonium ions over related carbonium ion species. Indeed this effect is highlighted by the failure of olefins (and cyclic compounds) containing heteroatoms (O, N, S) to copolymerize cationically with simple alkenes such as isobutene and styrene. It needs to be stressed that molecules containing such heteroatoms need not be deliberately added but may frequently be encountered as adventitious impurities in solvents, monomers, catalysts and the walls of reaction vessels, with resulting irreproducibilities in reaction rates and polymer molecular weights.

The concept of destroying activity by the introduction of a nucleophilic species was originally directed towards showing, qualitatively, the presence of cationic intermediates, but has now been extended to provide a method of quantitative assay of the number of active propagating centres. Thus a basic reagent B, added to a polymerization mixture, combines with the growing centre (P_n^+) to form an enol-group ($P_n B^+$) on the dead polymer and permits assay of the number of active centres, given the existence of a suitable monitoring technique for $P_n B^+$. Though the method has not yet been successfully applied in vinyl systems, Saegusa[18] has determined the propagating oxonium ion concentrations in bulk tetrahydrofuran polymerizations by termination with PhONa, and spectrophotometric determination of the PhO-groups in the isolated polymer, for example

$$P_n + \text{\textasciitilde\textasciitilde\textasciitilde}\overset{\oplus}{O}\!\diagup\!\!\!\square + PhO^{\ominus} \longrightarrow \overset{\cdot}{P}_n{-}O(CH_2)_4{-}OPh$$

9.2.3 Copolymerization, conductance and solvent effects

Copolymerization studies have also been used to discover the nature of a polymerization, especially in the case where a particular monomer is polymerizable by more than one type of intermediate. In principle a second monomer, known to polymerize via one intermediate only, is added, the formation or otherwise of a copolymer proving or disproving the presence of such intermediates. The method is qualitative and fraught with difficulties, because the large differences in reactivity of monomers susceptible to cationic polymerization means that copolymer formation by such intermediates is often impossible. For example, p-methoxystyrene can be homopolymerized in styrene solvent providing the initiator reacts only with the former monomer.

As far as cationic systems are concerned, therefore, the technique tends to be a negative one, and is used to show the absence of radical intermediates, rather than the definite presence of cationic ones. Nevertheless, in combination with the effect of added impurities, copolymerization studies have been found useful, particularly in the studies of N-vinyl-carbazole polymerizations.[19,20]

The detection of ionic intermediates by conductance experiments has also been attempted, but here again the method is qualitative and often misleading. Before charge carriers can be detected via conductance methods, the species must be essentially unassociated, i.e. devoid of attendant counter-ions (see section 9.2.4). The proportion of free species present is a function of temperature, dielectric constant, etc. In addition the mobility of such ions is dependent on the viscosity of the reaction medium. Even in homogeneous solution polymerizations the viscosity will be continuously increasing with percentage conversion of monomer and, in the case of bulk polymerizations, can become extremely large. It would be expected that in using a more polar solvent for a given polymerization, an increase in conductance would result, and this has been observed.[21] The sharp rise in conductance found at the end of some polymerizations[21] is, however, more difficult to interpret and illustrates the care required in drawing conclusions from complex systems.

The effect of varying the solvent on the rate of a reaction is a technique used generally in studies of organic reaction mechanisms, but its use as a test for the presence of an ionic mechanism in polymerizations has been rather indiscriminate. Thus the often used and over-simplified proposition, that solvents of high polarity (as measured, say, by dielectric constant) should accelerate ionic polymerization, is quite misleading. The effect of the polarity of a solvent on a polymerization reaction involving ionic species is a complex phenomenon because such a process is composed of many elementary reactions each of which has its own solvent dependence. For example, most initiation reactions involve development of charge and will, in general, be favoured by increasing the solvent polarity or dielectric constant. On the other hand many propagation reactions, especially those involving free cations, will occur via transition states in which there is considerable charge delocalization and which would generally be *destabilized* by solvents of increasing dielectric constant, although, in favourable cases, increasing solvent *polarizability* could have a markedly favourable effect on transition state formation, and consequently on rates of polymerization.[22] A further complication arises because many organic ionic reactions involve more than one charged species, usually in a fixed equilibrium, each with its own reactivity, and variation of the solvent automatically alters the relative abundance of each (see below). The effects of solvent variation are, therefore, discussed in more depth in the individual sections dealing with initiation, propagation etc.

9.2.4 Free ions and ion pairs

Winstein[23] has shown conclusively that many organic ionic materials $(A^{\oplus}B^{\ominus})$ in a suitable solvent can exist in at least three forms, a contact (or intimate) ion pair, a solvent separated ion pair, and free solvated ions, for example

$$A^{\oplus}B^{\ominus} \;\rightleftharpoons\; A^{\oplus}//B^{\ominus} \;\overset{K_d}{\rightleftharpoons}\; A^{\oplus}B^{\ominus}$$

Intimate or	Solvent separated	Free solvated
contact ion pair	ion pair	ions

The value of K_d varies according to the nature of $A^{\oplus}$ and $B^{\ominus}$, the solvating and dissociating ability of the solvent employed and the temperature. Dennison and Ramsey[24] have treated quantitatively the effect of solvent and have shown the relationship below to hold.

$$-\ln K_d = z^2/aDkT$$

where $z =$ charge on ions, $a =$ sum of van der Waals' ionic radii, $D =$ dielectric constant of the solvent, $k =$ Boltzmann constant, and $T =$ temperature. The treatment, however, is over-simplified to some extent in that it requires ions to be present only as very close pairs or completely free. In addition, Winstein was able to show that the rates at which these various species react with a given substrate can be substantially different. In general, a free cation would be expected to react more rapidly than any of its corresponding ion pairs because of its higher effective charge density. The fundamental source of reactivity of a cation can be regarded as its ability to polarize susceptible molecules, and clearly the larger the charge density, the greater the polarizing power. Thus in ionic polymerizations as well as the difficulty of deciding the chemical nature of the species involved in initiation, propagation etc., there is also the problem of the physical nature of each in relationship to its counter-ion.

Because of the overwhelming difficulties in determining merely the organic chemistry of most initiation reactions in cationic systems, the additional problem of free ion and ion pair participation has received little attention, and indeed is unlikely to do so for some time. Relatively slow equilibria are involved in initiation reactions and so the dramatic effects found in the more straightforward propagation reactions are unlikely to be parallelled. These will be discussed in detail in the section dealing with propagation and for the purpose of simplification the problem of free ion equilibria in initiation phenomena will largely be overlooked.

Fortunately there is available a simple conductance technique for monitoring ion pair free ion equilibria, which whilst giving no information about different ion pairs, readily produces data on the proportion of free ions to all ion pairs. The equilibrium is treated as approximating to the dissociation of a weak electrolyte, e.g. acetic acid, in water

$$A^{\oplus}B^{\ominus} \overset{K_d}{\rightleftharpoons} A^{\oplus} + B^{\ominus}$$

$$CH_3COOH \overset{K}{\rightleftharpoons} CH_3COO^{\ominus} + H^{\oplus}$$

For such systems Ostwald's Dilution Law represents fairly accurately the behaviour of free ions and ion pairs.

$$\alpha = \frac{\lambda}{\lambda_0}$$

and

$$\frac{1}{\lambda} = \frac{1}{\lambda_0} + \frac{c\lambda}{K(\lambda_0)^2}$$

where λ = equivalent conductance of a solution concentration, c, λ_0 = equivalent conductance at infinite dilution, and α = degree of dissociation. Fuoss[25] has shown, however, that for ionic concentrations as low as 10^{-4} M, Ostwald's equation is not exact because of its neglect of long range interionic attraction upon the conductance and activities of ions. Maintaining the concept of the 'sphere in continuum' model, in which ions are regarded as hard spheres immersed in a continuous medium, he corrected the equation from first principles and derived the relationship below.

$$\frac{F(z)}{\lambda} = \frac{1}{\lambda_0} + \frac{[cf_{\pm}^2 F(z)]}{K(\lambda_0)^2}$$

where, $F(z) = \frac{4}{3}\cos^2\frac{1}{3}\cos^{-1}(-3\sqrt{3}z/2)$, z = constant $\sqrt{\lambda}c(\lambda_0)^{-\frac{3}{2}}$, and $f_{\pm}$ = mean activity coefficient (= constant αC). From measurements of λ at various values of c in a given solvent and at given temperature, and application of this equation, it is possible to obtain data on K, that is in a static system, it is possible to determine the proportions of free ions and ion pairs.

Specific examples of the effects of ion pair dissociation equilibria on reaction rates, for both carbonium ion and carbanion producing systems, are discussed in Chapters 8 and 10 respectively.

9.2.5 Pseudo-cationic systems

The polymerization of styrene by sulphuric[26,27] and perchloric acids[8,28–35] has been studied in various solvents by a number of groups of workers. The most thoroughly studied system is styrene/perchloric acid in dichloromethane and 1,2-dichloroethane over the temperature range −90 °C to +30 °C. In order to explain some of the experimental observations Plesch[8] proposed a pseudo-cationic propagating species which is neither a free ion nor an ion pair, but in fact a relatively unreactive oligostyryl perchlorate covalent ester, $CH_3(CHPhCH_2)_n CHPh\cdot O\cdot ClO_3$. Pepper and collaborators[30,31] take the view that while this species is definitely present in the system, it does not itself propagate, but can be regarded as a dormant store of potentially reactive intermediates.

$$\sim St^{\oplus} + ClO_4^{\ominus} \rightleftharpoons \sim St^{\oplus}ClO_4^{\ominus} \rightleftharpoons \sim St—O—ClO_3$$

    Free ions   Ion pairs   Covalent ester

There appears to be no disagreement concerning the process of initiation and the question will be dealt with in detail in the section on propagation.

9.3 INITIATION REACTIONS

The fundamental process involved in any initiation reaction is the generation of an active site, in this case a cation, on some part of a monomer molecule, for example

$$XY + CH_2{=}CHR \longrightarrow XCH_2{—}\overset{\oplus}{C}HR\ Y^{\ominus}$$

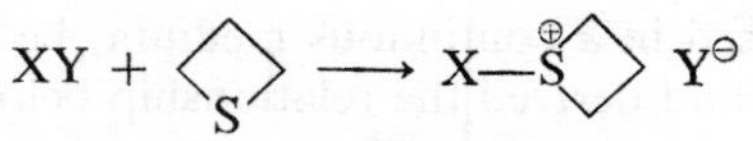

$$XY + \left\langle \underset{S}{\bigcirc} \right\rangle \longrightarrow X{—}\overset{\oplus}{\underset{S}{S}}\bigcirc\ Y^{\ominus}$$

More often than not in cationic systems this process can be broken down into a pre-initiation equilibrium (or equilibria), involving the formation of a reactive charged species along with its counter-ion, followed by addition of the charged pair to the monomer. The charged moieties are formed either in a bimolecular reaction or via the apparent heterolytic splitting of a single molecule. For example,

$$AlR_3' + RCl \qquad R^{\oplus}[AlR_3'Cl]^{\ominus}$$
$$I_2 \underset{-S}{\overset{S}{\rightleftharpoons}} (SI)^{\oplus}I^{\ominus} \quad \text{or} \quad I^{\oplus}(SI)^{\ominus}$$

S = solvent, monomer, I_2 etc.

The initial equilibrium is usually a relatively slow process while the subsequent attack on monomer (true initiation) is often rapid and sometimes comparable to the rate of propagation.

A wide variety of systems have now been shown to be effective initiators of cationic polymerization, and any list is bound to be less than comprehensive. Some catalysts will polymerize virtually all cationically active monomers, vinyl and cyclic types, and among the more well-known ones are high energy radiation and complex acids derived from BF_3, $AlCl_3$, AlR_3, $SnCl_4$, $TiCl_4$, $SbCl_5$. Other catalyst systems find application with only specific vinyl or cyclic monomers (generally the more reactive ones): Ziegler-Natta catalysts, I_2, protic acids, organic electron acceptors (e.g. tetracyanoethylene), stable carbonium ion (e.g. $Ph_3C^+SbCl_6^-$), and aminium ion salts (e.g. $(p\text{-}BrPh)_3N^{\oplus}$ $SbCl_6^-$). Others are useful for the polymerization of cyclic monomers only: oxonium ion salts (e.g. $Et_3O^{\oplus}SbCl_6^{\ominus}$), carboxonium ion salts (e.g.

$$(CH_3\overset{\oplus}{O}{=}CH_2)AlCl_4^-),$$

acylium salts (e.g. $CH_3CO^{\oplus}AlCl_4^{\ominus}$), and aryl diazonium salts

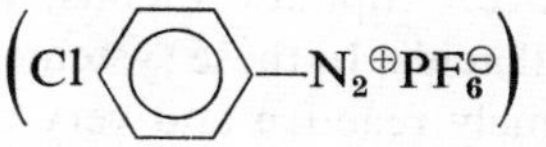

$$\left(Cl\langle\bigcirc\rangle{-}N_2^{\oplus}PF_6^{\ominus}\right)$$

When the catalyst is a preformed salt it is usually effective in its own right but on occasions an accompanying cocatalyst is required. The phenomenon of cocatalysts has probably produced more difficulties and disagreements concerning the chemistry of cationic systems than most other factors combined, and even today the exact situation of some systems is still a little confused. Furthermore, because of the high reactivity and the problems caused by the mere handling of some of these materials the position may not change for some time, although an elegant rationalization of some systems has now been achieved.[36]

9.3.1 Mechanism of the initiation of polymerization of vinyl monomers

It is the purpose of this section to give a mechanistic outline of some of the more important catalyst systems known, but to avoid gross experimental detail and results, except where necessary for the understanding of a fundamental principle. In general both cation and counter-ion will be shown in equations, but, except where specifically mentioned, no deliberate attempt has been made to show these species as paired or free.

(i) *Protonic acids* (Brönsted acids)

In 1932 Whitmore[37] proposed a mechanism for the protonic-acid initiation of polymerization in which a proton is added directly to the vinyl double bond.

$$M^{\oplus}B^{\ominus} + CH_2{=}CHR \rightarrow CH_3{-}\overset{\oplus}{C}HR + B^{\ominus}$$

The reaction is bimolecular, the proton choosing the site of maximum electron density for attack while also producing the thermodynamically most stable carbonium ion. This, however, is by no means the whole explanation since not all protonic acids initiate vinyl polymerization. Clearly the anion is of utmost importance; if this is a strong nucleophile, then subsequent reaction of the initially formed carbonium ion with its own counter-ion will take place more rapidly than the reaction with monomer, i.e. propagation. As a result the overall reaction will be the addition of the acid across the vinyl double bond of the monomer.

$$H^{\oplus}B^{\ominus} + CH_2{=}CHR \longrightarrow CH_3{-}\underset{\underset{B}{|}}{C}HR$$

It is not surprising to find therefore that acids such as perchloric and sulphuric, whose anions are fairly non-nucleophilic are in general better initiators[38] than, say, simple hydrogen halides. The actual monomer involved is also of great importance. Aliphatic olefins, even isobutene, polymerize only with difficulty with all acids. In these systems the carbonium ions formed by protonation are extremely reactive and very often the overall reaction is again the addition of the acid across the double bond. Styrene type monomers, however, display a higher tendency to polymerize under the action of protonic acids, α-methylstyrene and p-methoxystyrene being particularly reactive. There is little doubt that the underlying reason for this is the greater stability of the carbonium ion produced by protonation, and a diminished tendency to react with the counter-ion. The reactivity of the latter can be strongly suppressed by solvation, and the possibility of the system undergoing polymerization thereby enhanced, explaining why the catalytic activity of these acids depends markedly on the polarity of the medium, and other reaction conditions.

By far the most important system studied is the anhydrous perchloric acid/styrene combination. Pepper and co-workers[28–31] regard the initiation as proton addition with the production of growing free ions and ion pairs. Some reaction with the counter-ion does take place, however, and an equilibrium is believed to exist which involves some kinetically dormant covalent ester. Plesch and collaborators[8] on the other hand, while supporting the initiation mechanism, believe the ester to be capable of propagation (pseudocationic).

The polymerization of vinylethers by the action of protonic acids has not been studied in very great detail. Shostakovskÿ[39] has shown, however, that sulphuric acid can induce the polymerization of alkyl but not arylvinylethers, in keeping with the expected reactivity of these two groups of monomers. In the case of the aromatic ethers a relatively stable onium complex is formed, which slowly isomerizes and dissociates to form a propagating species.

Various heterogeneous acid catalysts have been known for many years and some have found application as cracking and isomerization catalysts in the petroleum industry. These normally consist of sulphuric acid/metal sulphate complexes[40] and provide very efficient heterogeneous systems for the polymerization of alkyl and aryl olefins and of alkylvinylethers. Many molecules with free electron pairs, including olefinic π-electron clouds, can be adsorbed on to the surface of these catalysts. If the adsorbed molecules

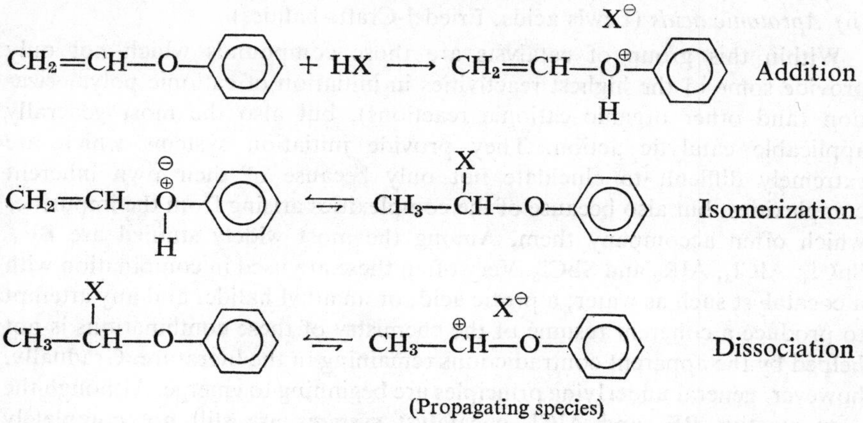

$$CH_2=CH-O-\bigcirc \;+\; HX \;\longrightarrow\; CH_2=CH-\overset{\oplus}{\underset{H}{O}}-\bigcirc \quad X^{\ominus} \qquad \text{Addition}$$

$$CH_2=CH-\overset{\oplus}{\underset{H}{O}}-\bigcirc \;\rightleftharpoons\; CH_3-\overset{X}{\underset{}{CH}}-O-\bigcirc \qquad \text{Isomerization}$$

$$CH_3-\overset{X}{\underset{}{CH}}-O-\bigcirc \;\rightleftharpoons\; CH_3-\overset{\oplus}{CH}-O-\bigcirc \quad X^{\ominus} \qquad \text{Dissociation}$$

(Propagating species)

can react with a proton, then the surface acts as a very strong protonic acid, with a counter-ion (the surface) which is a very poor nucleophile. Lal and McGrath[41] have proposed such a co-ordinated cationic mechanism to explain why polymers resulting from these catalysts are often stereo-regular, for example

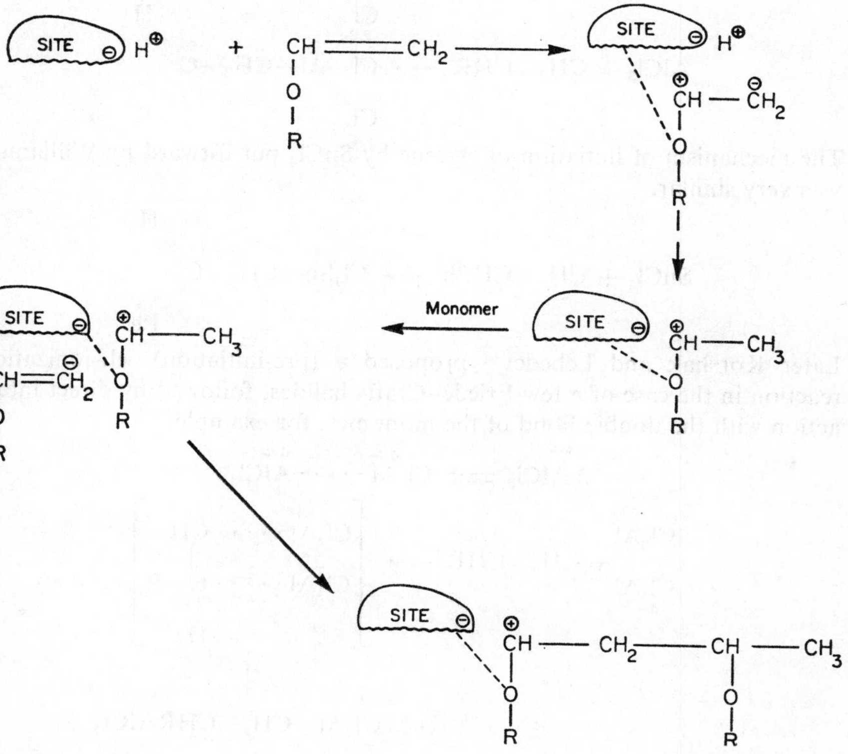

10

(ii) *Aprotonic acids* (Lewis acids, Friedel-Crafts halides)

Within this group of catalysts are those compounds which not only provide some of the highest reactivities in initiation of cationic polymerization (and other organic cationic reactions), but also the most generally applicable catalytic action. They provide initiation systems which are extremely difficult to elucidate not only because of their own inherent complexities, but also because of the complexities arising from the impurities which often accompany them. Among the most widely studied are BF_3, $SnCl_4$, $AlCl_3$, AlR_3 and $SbCl_5$. Very often these are used in combination with a cocatalyst such as water, a protic acid, or an alkyl halide, and any attempt to produce a coherent résumé of the chemistry of these combinations is not helped by the apparent contradictions remaining in the literature. Gradually, however, general underlying principles are beginning to emerge. Although the very reactive BF_3 and $AlCl_3$ cocatalyst systems are still not completely understood, the elegant work of Kennedy[36] on AlR_3/RCl systems with styrene has gone a long way towards elucidating the mechanism of alkyl halide cocatalysis.

Hunter and Yohe[42] were the first to formulate a mechanism of initiation involving direct interaction of an aprotonic acid with monomer. For example,

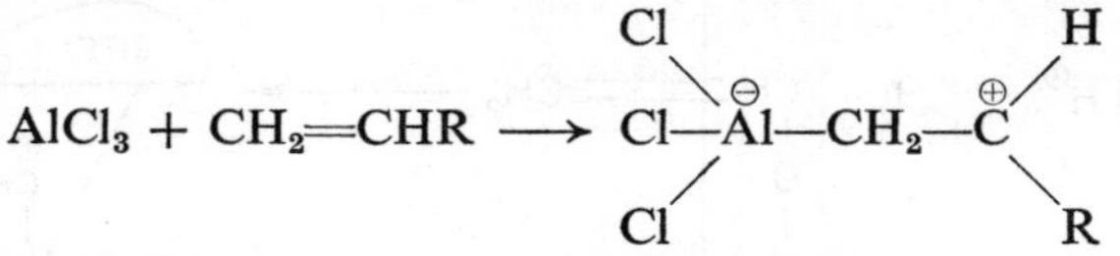

$$AlCl_3 + CH_2{=}CHR \longrightarrow \overset{Cl}{\underset{Cl}{\overset{|}{Cl{-}\overset{\ominus}{Al}}}}{-}CH_2{-}\overset{H}{\underset{R}{\overset{\oplus}{C}}}$$

The mechanism of initiation of styrene by $SnCl_4$ put forward by Williams[43] was very similar.

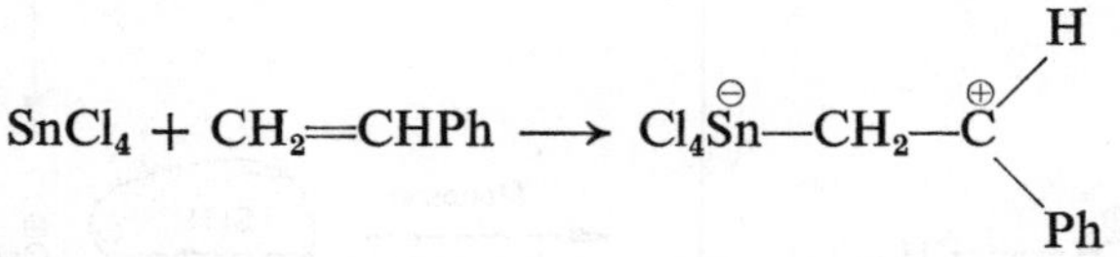

$$SnCl_4 + CH_2{=}CHPh \longrightarrow Cl_4\overset{\ominus}{Sn}{-}CH_2{-}\overset{H}{\underset{Ph}{\overset{\oplus}{C}}}$$

Later Korshak and Lebedev[44] proposed a (pre-initiation) self-ionization reaction in the case of a few Friedel-Crafts halides, followed by direct interaction with the double bond of the monomer, for example

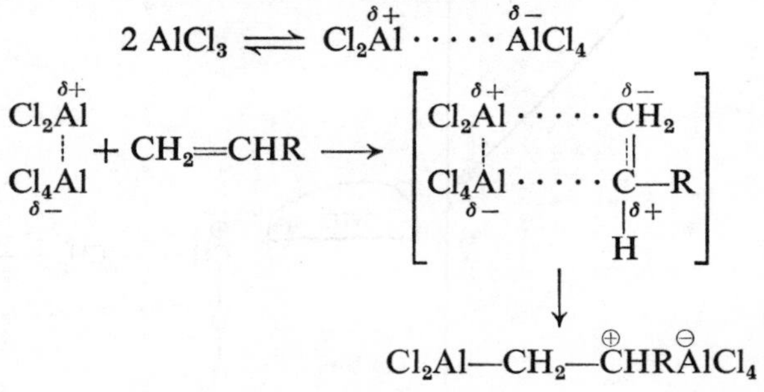

$$2\,AlCl_3 \rightleftharpoons \overset{\delta+}{Cl_2Al}\cdots\cdots\overset{\delta-}{AlCl_4}$$

$$\underset{\underset{\delta-}{Cl_4Al}}{\overset{\overset{\delta+}{Cl_2Al}}{\vdots}} + CH_2{=}CHR \longrightarrow \left[\begin{array}{c}\overset{\delta+}{Cl_2Al}\cdots\cdots\overset{\delta-}{CH_2} \\ \vdots \qquad\qquad \| \\ \underset{\delta-}{Cl_4Al}\cdots\cdots\underset{\overset{|}{\underset{H}{\overset{\delta+}{}}}}{C{-}R}\end{array}\right]$$

$$\downarrow$$

$$Cl_2Al{-}CH_2{-}\overset{\oplus}{C}HR\overset{\ominus}{AlCl_4}$$

Much more recently Chmelíř *et al.*[45] have invoked a similar auto-ionization process for the system $AlBr_3$/isobutene, with again the formation of a propagating ion pair,

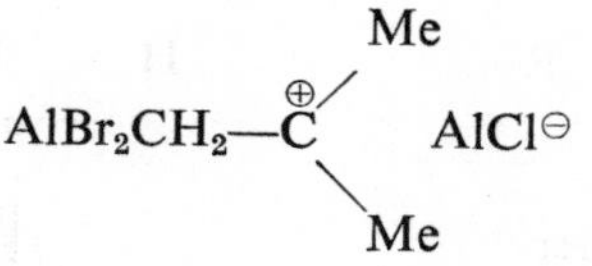

$$AlBr_2CH_2-\overset{\overset{\displaystyle Me}{\diagup}}{\underset{\diagdown}{\overset{\oplus}{C}}}\quad AlCl^{\ominus}$$
$$\qquad\qquad\qquad\; Me$$

Mechanisms such as these cannot account completely for the cocatalytic effect of water or solvent, whose presence it now appears is virtually indispensable for initiation to proceed. Almost without exception claims of catalysis without an accompanying cocatalyst have been disproved by more exact work, though the report by Chmelíř[45] that Al_2Br_6 initiates isobutene polymerization without the aid of water still remains unchallenged. Initiation of certain vinylethers also appears to date to require no cocatalysts.[46]

Lewis acids of the Friedel-Crafts type are characterized by having an electron deficient central atom, which on interaction with olefinic or aromatic π-electrons, generally give rise to weak π-complexes[47] ($\Delta H_{formation} \leqslant 2$ kcal mol^{-1}.[48]). To promote initiation, the presence of another potential ionizable molecule is necessary. For the π-complex to initiate directly it would have to rearrange into an isomeric carbonium ion. For example,

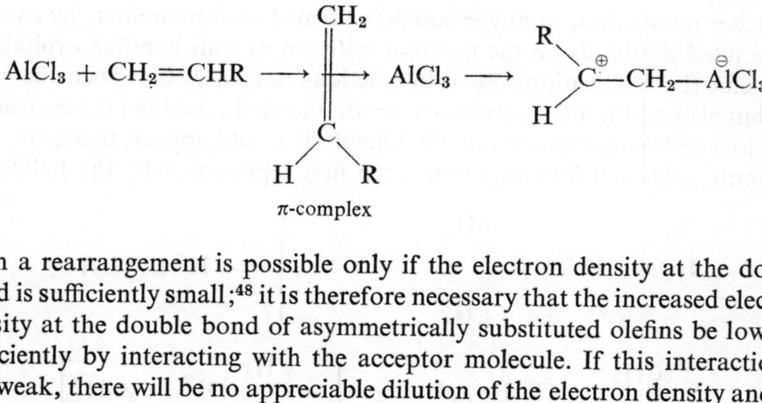

$$AlCl_3 + CH_2{=}CHR \longrightarrow \underset{\underset{\displaystyle \underset{H\quad R}{\diagup\diagdown}}{C}}{\overset{\overset{\displaystyle CH_2}{\|}}{\cdot}}\ AlCl_3 \longrightarrow \underset{H}{\overset{R}{\diagdown}}\overset{\oplus}{C}{-}CH_2{-}\overset{\ominus}{AlCl_3}$$

$$\pi\text{-complex}$$

Such a rearrangement is possible only if the electron density at the double bond is sufficiently small;[48] it is therefore necessary that the increased electron density at the double bond of asymmetrically substituted olefins be lowered sufficiently by interacting with the acceptor molecule. If this interaction is too weak, there will be no appreciable dilution of the electron density and the π-complex will not rearrange. The catalytic activity of Lewis acids therefore depends to a large extent on the nature of the cocatalyst (H_2O, HCl etc.), which yields the initiating particle (usually H^+ or R^+) whose empty orbitals are more suitable for co-ordination with the π-electrons than the vacant orbitals in the molecule of the Lewis acid. Whether the electron rearrangement is a concerted process as indicated or whether the initiating particle is generated first as a separate entity, as shown below, probably depends upon reaction conditions. When the cocatalyst is the last component to be added,

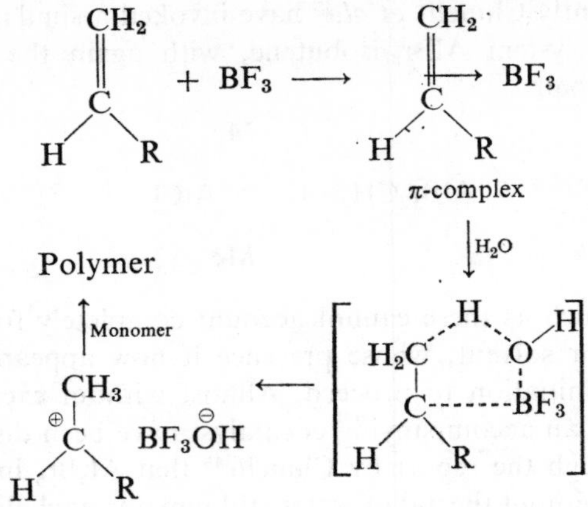

$$
\underset{\substack{\text{H} \quad \text{R}}}{\overset{\substack{CH_2 \\ \parallel \\ C}}{}} + BF_3 \longrightarrow \underset{\substack{\text{H} \quad \text{R}}}{\overset{\substack{CH_2 \\ \parallel \\ C}}{}} \!\!\!\!\! \cdot BF_3
$$

π-complex

$$\Big\downarrow H_2O$$

Polymer

$\uparrow$ Monomer

$$
\underset{\substack{\text{H} \quad \text{R}}}{\overset{\substack{CH_3 \\ \oplus \\ C}}{}} \quad BF_3\overset{\ominus}{O}H \quad \longleftarrow \quad
\left[\underset{\substack{\text{H} \quad \text{R}}}{\overset{\substack{H \cdot \cdots H \\ H_2C \qquad O \\ C\text{------}BF_3}}{}} \right]
$$

then presumably the olefin acceptor complexation is complete and initiation involves the attack of cocatalyst upon this unit. On the other hand where catalyst and cocatalyst are premixed, then an effective protonic acid, as shown below, is available to attack uncomplexed monomer. The transition state for both reactions would be very similar. In a situation where the Lewis acid is in excess then there is at least the possibility of protonic acid attacking monomer π-complexed with aprotonic acid.

With aprotonic acids other than BF_3 and AlX_3 there arises the possibility of further interaction, in any π-complex formed with monomer, by overlap of the filled d-orbitals on the acceptor with empty anti-bonding orbitals on the olefin (back donation). Such interactions not only strengthen the complex but also add to the electron density on the olefin, making the rearrangement to a carbonium ion even more difficult. It would appear, therefore, that aprotonic acids fall into two types: the first represented by the halides of

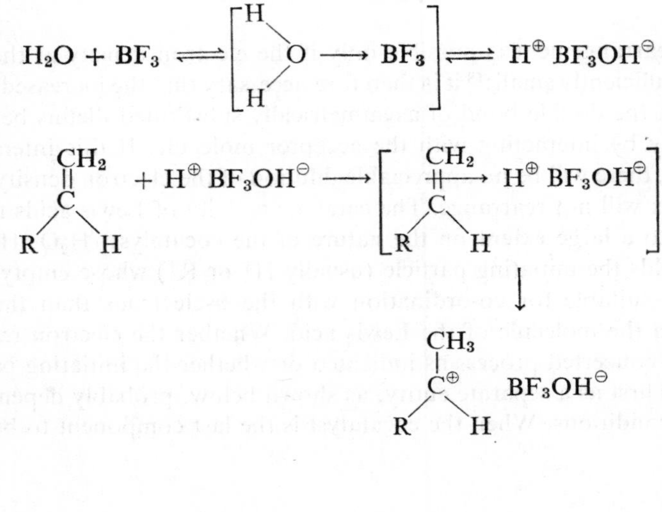

$$
H_2O + BF_3 \rightleftharpoons \left[\overset{\substack{H \\ | }}{\underset{\substack{| \\ H}}{O}} \longrightarrow BF_3 \right] \rightleftharpoons H^{\oplus} BF_3OH^{\ominus}
$$

$$
\underset{\substack{\text{R} \quad \text{H}}}{\overset{\substack{CH_2 \\ \parallel \\ C}}{}} + H^{\oplus} BF_3OH^{\ominus} \longrightarrow \left[\underset{\substack{\text{R} \quad \text{H}}}{\overset{\substack{CH_2 \\ \parallel \\ C}}{}} \!\!\! \longrightarrow H^{\oplus} BF_3OH^{\ominus} \right]
$$

$$\Big\downarrow$$

$$
\underset{\substack{\text{R} \quad \text{H}}}{\overset{\substack{CH_3 \\ \oplus \\ C}}{}} \quad BF_3OH^{\ominus}
$$

boron and aluminium (and gallium) having no d-electrons in the valence sphere of the central atom, while the second, which includes all the compounds based on larger atoms, has d-electrons available for back donation. Interaction of both types with olefins and other bases may well be different as indicated, depending on the extent to which the d-orbitals of the central atom play a part in the structure of the π-complex. In practice, BF_3 and $AlCl_3$ are generally found to initiate polymerization more rapidly than, say, $TiCl_4$ or $SnCl_4$,[49] and for this reason there is a shortage of kinetic data on systems employing these 'fast' catalysts compared with those using the 'slow' class. In addition this differential provides some indication that pre-formed protonic acid is not the active catalyst since $H^{\oplus}BF_3OH^{\ominus}$ and $H^{\oplus}SnCl_4OH^{\ominus}$ for example, might be expected to have similar reactivities.

Cocatalyst 'poisoning' is yet another complication in these systems. Experimentally it is found that there is a critical concentration of cocatalyst at which the rate of polymerization is a maximum. This is found especially in the case of cocatalysis by water.[50] The actual concentration at which maximum catalytic efficiency is observed varies, as might be expected, with the nature of the catalyst, cocatalyst, monomer, solvent and other reaction conditions. An explanation normally forwarded[51] is that excess of cocatalyst, a more basic molecule than the monomer, competes successfully for the complex protonic acid formed in the initial equilibrium, for example

$$BF_3 + H_2O \rightleftharpoons H^{\oplus}[BF_3OH]^{\ominus} \xrightarrow{H_2O} [H_3O]^{\oplus}[BF_3OH]^{\ominus} \quad \text{etc.}$$

In the case of water this yields an oxonium ion salt whose reactivity towards the monomer is substantially lower than the complex protonic acid. No data on separate rates of initiation (rather than overall rates of polymerization) are available and hence this scheme cannot be questioned, although it is known that cocatalyst concentrations also effect the molecular weights of polymers obtained (i.e. by transfer reactions), and it is possible that the 'poisoning' effect of excess cocatalyst may be due to reactions subsequent to the initiation step and not prior to it. Once the propagating carbonium ion is formed then in the presence of, say, excess water there will be competition between the propagation reaction with monomer and a chain breaking process with the cocatalyst,

$$\sim C^{\oplus}X^{\ominus} + M \xrightarrow{\text{propagation}} \sim CM^{\oplus}X^{\ominus}$$

$$\sim C^{\oplus}X^{\ominus} + H_2O \xrightarrow{\text{chain breaking}} \sim C\text{—OH} + HX$$

Dominance by the latter reaction would reduce overall rates of polymerization and in the extreme, produce an apparent complete inhibition of polymer formation.

The mechanism of alkyl halide cocatalysis is more fully understood, due largely to the work of J. P. Kennedy[36,52] and his collaborators. Combinations of aluminium chloride, alkyl aluminium chlorides and trialkyl aluminiums with alkyl halides are widely used in the production of polyisobutene and butyl rubbers, (polyisobutene/isoprene copolymers). Their mode of action, however, has been best elucidated in the polymerization of styrene, where the intermediate propagating carbonium ion has greater stability. The actual catalyst systems investigated were $RCl/AlEt_2Cl$, where R represents numerous alkyl and aryl groups. Kennedy has proposed the comprehensive initiation scheme shown below.

Mechanism of the initiation of styrene polymerization by the catalyst system $RCl/AlEt_2Cl$

$$RCl + AlEt_2Cl \rightleftharpoons R^{\oplus}AlEt_2Cl_2^{\ominus} \overset{St}{\rightleftharpoons} RSt^{\oplus}AlEt_2Cl_2^{\ominus} \xrightarrow{Monomer} Polymer$$

$$\Big\downarrow \text{Initiation}$$
$$\rightleftharpoons \quad REt + AlEtCl_2$$

$$RCl + AlEtCl_2 \rightleftharpoons R^{\oplus}AlEtCl_3^{\ominus} \overset{St}{\rightleftharpoons} RSt^{\oplus}AlEtCl_3^{\ominus} \xrightarrow{Monomer} Polymer$$

$$\Big\downarrow \text{Initiation}$$
$$\rightleftharpoons \quad REt + AlCl_3$$

$$RCl + AlCl_3 \rightleftharpoons R^{\oplus}AlCl_4^{\ominus} \overset{St}{\rightleftharpoons} RSt^{\oplus}AlCl_4^{\ominus} \xrightarrow{Monomer} Polymer$$
$$\text{Initiation}$$

In the first set of equations RCl cocatalyst and $AlEt_2Cl$ catalyst interact to produce the carbonium ion $R^{\oplus}$ and the attendant gegen-ion $AlEt_2Cl^{\ominus}$. At this point $R^{\oplus}$ can either initiate polymerization of styrene (St) or can extract an ethyl group from the counter-ion to produce a saturated hydrocarbon, REt, and a new Lewis acid, $AlEtCl_2$. In the absence of styrene monomer only the latter reaction is possible and this has been demonstrated experimentally. The new Lewis acid in combination with RCl can initiate polymerization of further styrene molecules by the processes indicated. Proof that the initiation mechanism is the addition of a carbonium ion, derived from the cocatalyst, to a monomer molecule was obtained by labelling techniques, radio-labelled carbon in the alkyl (or aryl) halide being readily detected as polymer end-groups.

The efficiency of the initiation reaction, that is

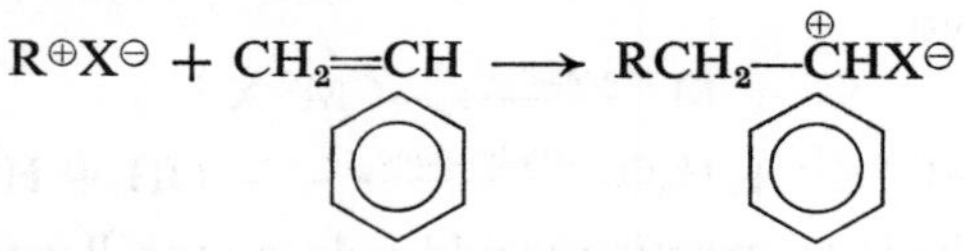

$$R^{\oplus}X^{\ominus} + CH_2{=}CH \longrightarrow RCH_2{-}\overset{\oplus}{C}HX^{\ominus}$$

is determined by the carbonium ion stability and, or, carbonium ion availability. Ionization according to the equation

$$RCl + AlEt_2Cl \rightleftharpoons R^{\oplus}[AlEt_2Cl_2]^{\ominus}$$

being facilitated by low bond dissociation energies of the alkyl (or aryl) halide and low ionization potentials of the alkyl group, and importantly, by solvation of the ions produced. With alkyl chlorides yielding primary or secondary carbonium ions (e.g. n-butyl or iso-propyl chloride) the forward reaction will not occur significantly. If these very reactive ions could be formed they would immediately initiate polymerization of styrene producing the relatively more stable styryl cation,

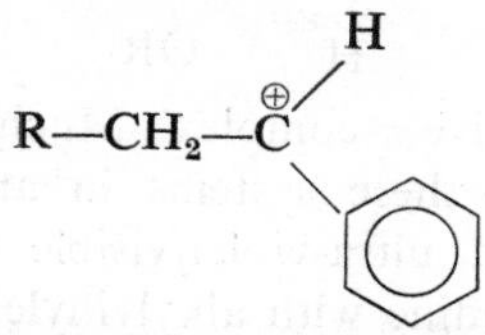

$$R-CH_2-\overset{\oplus}{C}\overset{H}{\diagup} \text{(with phenyl ring)}$$

Tert-butyl chloride yields a tertiary carbonium ion of higher stability but still thermodynamically unstable relative to the styryl cation, so that rapid initiation results. On the other hand, triphenylmethyl chloride produces the very stable trityl cation which is in fact too unreactive to catalyse styrene polymerization. In general, therefore, a carbonium ion more unstable than, in this case, the styryl cation is required, but not so unstable that its production from the interaction of alkyl halide with aprotonic acid becomes impossible.

(iii) Iodine

Iodine is an efficient initiator of the polymerization of some of the more reactive vinyl monomers, e.g. p-methoxy styrene,[53] alkylvinylethers,[22,54] N-vinylcarbazole and to a lesser extent styrene and acenaphthylene.[55] Alkylvinylethers represent the most studied systems, and as early as 1878 Wislicenus[56] obtained polyalkylvinylethers using molecular iodine. Eley and Richards[57] demonstrated that ions, not free-radicals, as proposed earlier,[58] were the active intermediates and suggested that a number of equilibria were involved prior to polymerization.

Mechanism of initiation of polymerization of vinylethers by molecular iodine[57]

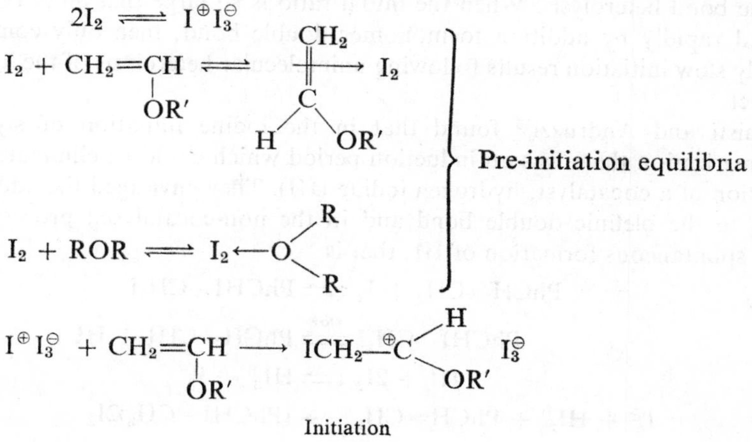

$$2I_2 \rightleftharpoons I^{\oplus}I_3^{\ominus}$$

$$I_2 + CH_2{=}CH{\underset{OR'}{|}} \rightleftharpoons \begin{matrix} CH_2 \\ \| \\ C \\ H \diagdown OR' \end{matrix} \longrightarrow I_2$$

$$I_2 + ROR \rightleftharpoons I_2{\leftarrow}O{\overset{R}{\underset{R}{\diagup}}}$$

Pre-initiation equilibria

$$I^{\oplus}I_3^{\ominus} + CH_2{=}CH{\underset{OR'}{|}} \longrightarrow ICH_2{-}\overset{\oplus}{C}{\overset{H}{\underset{OR'}{\diagup}}} I_3^{\ominus}$$

Initiation

ROR is the ether solvent and

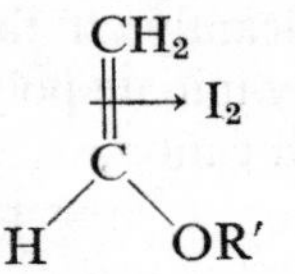

$$\begin{array}{c} CH_2 \\ \| \ \longrightarrow \ I_2 \\ C \\ H \quad OR' \end{array}$$

represents a so-called 'inactive π-complex'. Much more recently Ledwith and Sherrington[54] re-examined these systems in an attempt to clarify some anomalies in the reported ultra-violet/visible spectra,[21] and found that interaction of molecular iodine with alkylvinylethers in methylene chloride leads initially to equilibrium formation of appropriately substituted 1,2-diiodoethanes

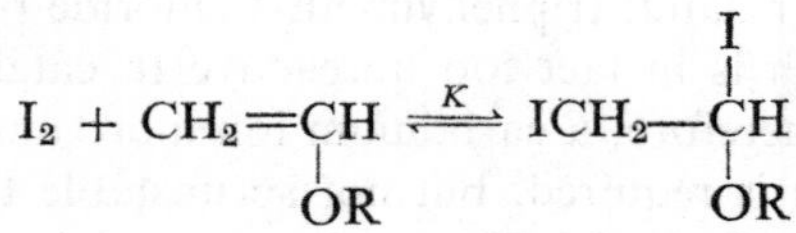

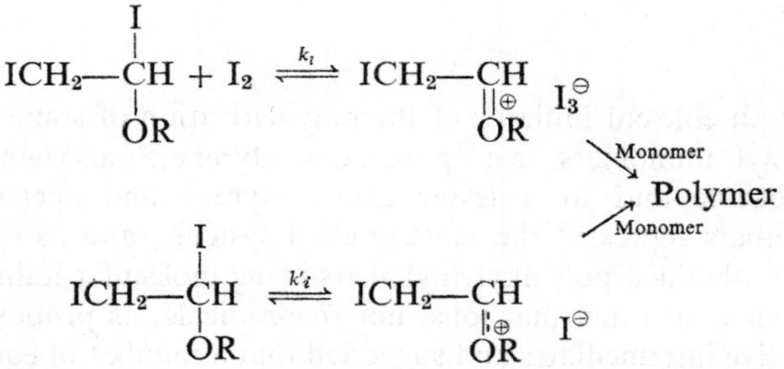

$$I_2 + CH_2{=}CH \underset{OR}{\overset{K}{\rightleftharpoons}} ICH_2{-}\underset{OR}{\overset{I}{CH}}$$

$$ICH_2{-}\underset{OR}{\overset{I}{CH}} + I_2 \overset{k_i}{\rightleftharpoons} ICH_2{-}\underset{OR}{\overset{I}{\underset{\oplus}{CH}}} \ I_3^{\ominus}$$

$$\searrow \text{Monomer}$$
$$\text{Polymer}$$
$$\nearrow \text{Monomer}$$

$$ICH_2{-}\underset{OR}{\overset{I}{CH}} \overset{k'_i}{\rightleftharpoons} ICH_2{-}\underset{OR}{\overset{I}{\underset{\oplus}{CH}}} \ I^{\ominus}$$

When the initial ratio [Monomer]/[I_2] is sufficiently small for appreciable quantities of free I_2 to remain, initiation occurs by I_2 catalysis of carbon-iodine bond heterolysis. When the initial ratio is so large that all I_2 is consumed rapidly by addition to monomer double bond, then only comparatively slow initiation results following unimolecular heterolysis of the iodine adduct.

Guisti and Andruzzi[55] found that in the iodine initiation of styrene polymerization there was an induction period which could be eliminated by addition of a cocatalyst, hydrogen iodide (HI). They envisaged the addition of I_2 to the olefinic double bond and in the non-cocatalysed process the slow spontaneous formation of HI, that is

$$PhCH{=}CH_2 + I_2 \rightleftharpoons PhCHI{-}CH_2I$$

$$PhCHI{-}CH_2I \overset{slow}{\rightleftharpoons} PhCH{=}CHI + HI$$

$$HI + 2I_2 \rightleftharpoons HI_2^{\oplus} + I_3^{\ominus}$$

$$I_3^{\ominus} + HI_2^{\oplus} + PhCH{=}CH_2 \longrightarrow (PhCHI{-}CH_3)2I_2$$

Participation of HI as shown in the scheme results in the formation of PhCHI—CH$_3$, ionization of which yields an active cationic centre. Deliberate addition of HI accelerated establishment of these equilibria, and hence formation of active centres. It is interesting to note the similarity with the case of vinylethers, and it is possible here also, that excess I$_2$ may produce propagating species by catalysis of the heterolytic fission of carbon–iodine bonds in

$$\text{PhCH—CH}_2\text{I} \quad \text{or} \quad \text{PhCH—CH}_3$$
$$\mid \qquad\qquad\qquad \mid$$
$$\text{I} \qquad\qquad\qquad\quad \text{I}$$

(iv) Stable carbonium ion salts

A number of years ago triphenylmethyl cation, Ph$_3$C$^\oplus$, formed *in situ* by dissociation of trityl chloride, was shown[59] to initiate polymerization of 2-ethylhexylvinylether in m-cresol solvent. The complexities of ionization-dissociation equilibria, however, precluded absolute evaluation of kinetic data. More recently certain stable carbonium ion salts, notably hexachloro-antimonate (SbCl$_6^\ominus$) salts of cycloheptatrienyl (C$_7$H$_7^\oplus$) and triphenylmethyl (Ph$_3$C$^\oplus$) cations, have been shown[60,61] to be very efficient initiators of the polymerization of certain olefins. The very stability of these salts means that only reactive monomers can be initiated and this has been confirmed by the polymerization of alkylvinylethers,[22] p-methoxystyrene, indene[14] and N-vinylcarbazole[62] and, as we shall see later, certain cyclic ethers such as tetrahydrofuran.[63] In fact such stable salts are merely a special case of systems such as those studied by Kennedy[36,52]

$$\text{RCl} + \text{AlCl}_3 \rightleftharpoons \text{R}^\oplus[\text{AlCl}_4]^\ominus$$

or by Olah and his coworkers,[64] where the carbonium ion, R$^\oplus$, is sufficiently

$$\text{RF} + \text{BF}_3 \rightleftharpoons \text{R}^\oplus\text{BF}_4^\ominus$$

stable in the presence of a large, diffuse, non-nucleophilic anion to be isolable as a crystalline salt.

Since the discovery of the effectiveness of the SbCl$_6^\ominus$ salts, trityl salts with different anions have also been used[65,66] as well as other stable cation types such as xanthylium[14] and pyrylium derivatives.[67]

The advantages of using these catalysts are that the initiator systems are fairly simple, quantitative in their mode of action, and usually free from pre-initiation equilibria. In the case of the polymerization of N-vinylcarba-zole using C$_7$H$_7^\oplus$SbCl$_6^\ominus$, it appears that initiation involves the direct addition of C$_7$H$_7^\oplus$ to monomer and takes place virtually instantaneously. In studying this polymerization therefore, it is, in effect, propagation only that is observed. This situation has been used to full advantage to investigate propagation reactions in detail (see later) and isolation of a simple adduct from the reaction carried out in methanolic solvents confirms the initiation mechanism shown below.

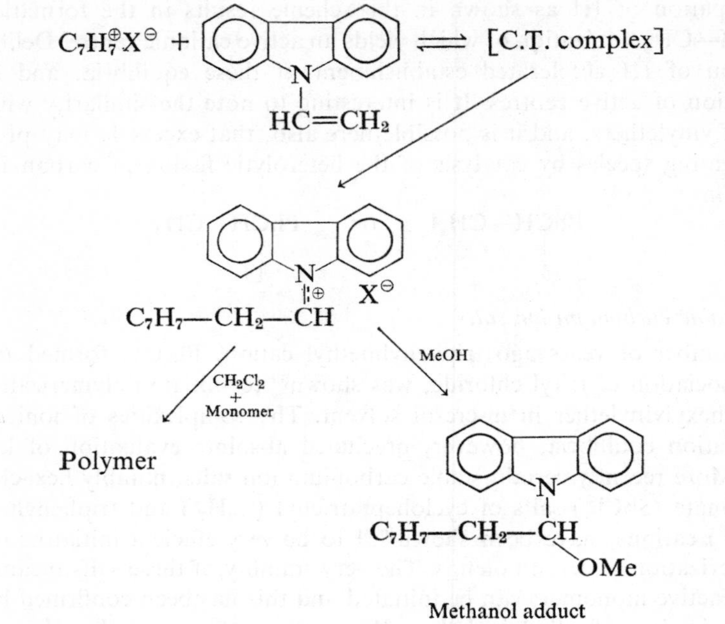

$$C_7H_7^{\oplus}X^{\ominus} + \text{[carbazole-N-vinyl]} \rightleftharpoons [\text{C.T. complex}]$$

$$C_7H_7\text{—}CH_2\text{—}CH \quad X^{\ominus}$$

Polymer ← $\begin{array}{c} CH_2Cl_2 \\ + \\ \text{Monomer} \end{array}$ $\xrightarrow{\text{MeOH}}$ $C_7H_7\text{—}CH_2\text{—}CH$ OMe

Methanol adduct

In the case of alkylvinylethers, initiation appears to be a little slower than the subsequent propagation reaction, but is nevertheless quantitative. At least three different mechanisms are possible for initiation using, say, $Ph_3C^{\oplus}$ cation.

(1) Direct addition

$$Ph_3C^{\oplus} + CH_2\text{=}CHOCH_2R \rightarrow Ph_3C\text{—}CH_2\text{—}CH\overset{\oplus}{=}OCH_2R$$

(2) Hydride ion abstraction from the alkoxy group

$$Ph_3C^{\oplus} + CH_2\text{=}CHOCH_2R \rightarrow Ph_3CH + CH_2\text{=}CH\text{—}\overset{\oplus}{O}\text{=}CHR$$

(3) Electron transfer from olefin to cation

$$Ph_3C^{\oplus} + CH_2\text{=}CHOCH_2R \rightarrow Ph_3C^{\circ} + [CH_2\text{=}CHOCH_2R]^{\oplus}$$

These have been discussed in detail[22] and bearing in mind independent evidence from Russian workers,[68,69] it seems reasonable to conclude that initiation involves primary addition of the cation to the vinylether as in the case of N-vinylcarbazole.

A further extension of this type of catalyst involves use of stable aminium salts,[70] for example

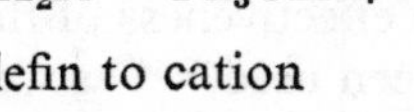

$$(Br\text{-phenyl})_3N^{\oplus}$$

$SbCl_6^{\ominus}$ and $ClO_4^{\ominus}$. Since triarylamines are non-basic and cannot form quaternary ammonium salts, these stable radical cations salts can react only by an electron transfer process. Indeed, they have been shown to be powerful one-electron oxidizing agents,[71] and more recently, useful catalysts for the polymerization of N-vinylcarbazole[72] via intermediate formation of the radical cation of N-vinylcarbazole. Such species are likely to dimerize producing a dication (see below) analogous to the double-ended styryl carbanions produced in the anionic polymerization of styrene initiated by radical anions; the dication may be scavenged by methanol or other protic agents, or may yield polymers in aprotic media.

Mechanism proposed for the initiation of N-vinylcarbazole polymerization by aminium salts

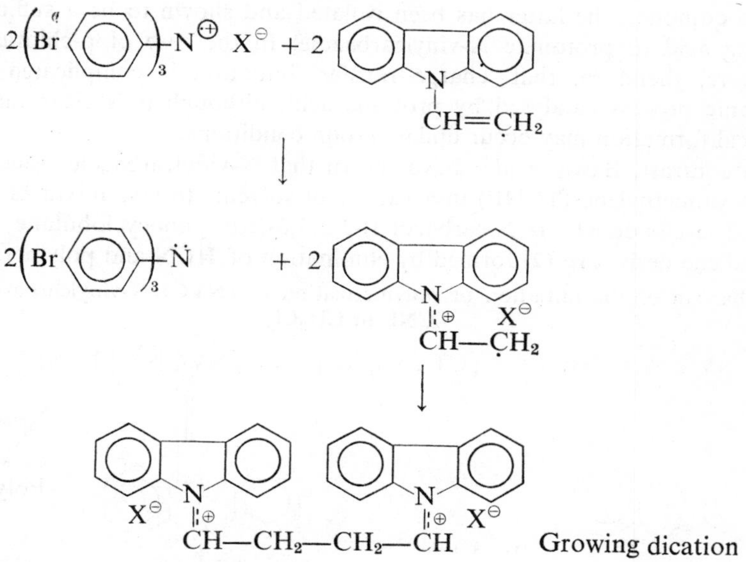

(v) *Charge-transfer initiation*

There are, in the literature, many reports of so-called 'charge-transfer' initiation of vinyl polymerization. In many cases the monomer involved is N-vinylcarbazole, and initiators employed include compounds such as tetranitromethane, p-chloranil, 1,3,5-trinitrobenzene, tetracyanoethylene, trichloroethylene and maleic anhydride.[73–75] Furthermore, other initiators such as nitro compounds, carbon tetrachloride and carbon tetrabromide may react by similar mechanisms.

The term 'charge-transfer polymerization' was first used in this context by Ellinger[73] since only 'partial electron transfer' from N-vinylcarbazole to the various acceptor molecules appeared to take place although the fact that the early experiments were carried out mainly in molten monomer (m.p. 64 °C) would certainly have helped to disguise the effects of additives such as water.

Independent work by Scott et al.[75] supported the view that the organic compounds act as one electron acceptors producing the radical cation of the monomer and this has been borne out by more recent studies, that is

$$Monomer + Acceptor \rightleftharpoons [CT\ Complex] \rightleftharpoons Monomer^{\oplus} + Acceptor^{\ominus}$$

$\downarrow$ Monomer

Polymer

The p-chloranil and tetracyanoethylene systems are the two which subsequently have been most thoroughly investigated. Shirota et al.[76] have reported that with pure p-chloranil almost no polymerization occurs, and that the real initiator is an acidic impurity, 3,5,6-trichloro-2-hydroxy-1,4-benzoquinone. The latter has been isolated and shown to be a sufficiently strong acid to protonate N-vinylcarbazole. In the case of p-chloranil, it appears, therefore, that 'charge-transfer' initiation is complicated by a cationic process catalysed by protonic acid, although it is clear that ion radical formation may occur under certain conditions.

In contrast, Bawn et al.[77] have shown that N-vinylcarbazole reacts with tetracyanoethylene (TCNE) in a variety of solvents to give mixtures of the 2 + 2 cycloadduct, 1-(N-carbazolyl)-2,2,3,3-tetracyanocyclobutane (1), a butadiene derivative (2) formed by elimination of HCN and polymer.

Mechanism of the initiation of polymerization of (NVC) N-vinylcarbazole by TCNE in CH$_2$Cl$_2$

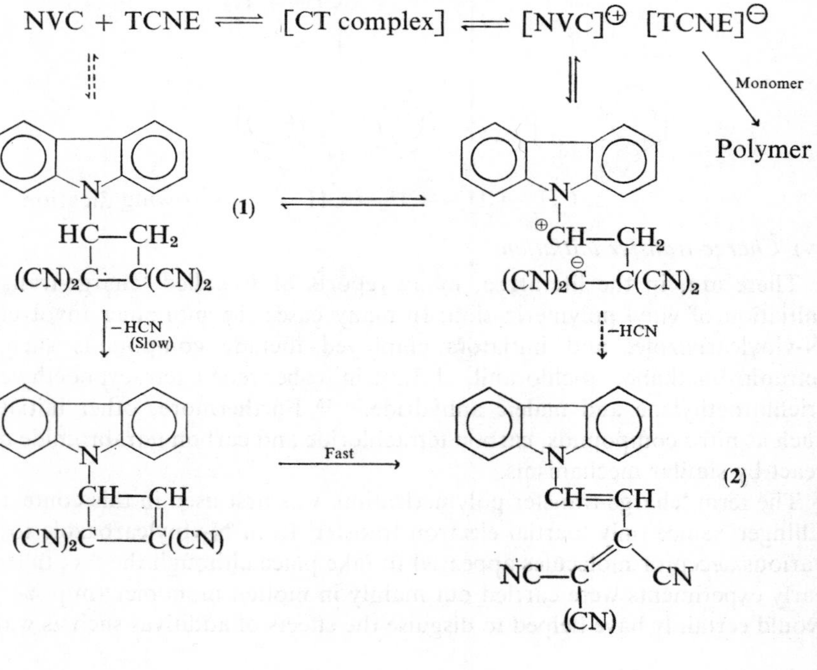

$$NVC + TCNE \rightleftharpoons [CT\ complex] \rightleftharpoons [NVC]^{\oplus}\ [TCNE]^{\ominus}$$

Initiation of NVC polymerization can be achieved by TCNE directly or by the cyclobutane, indicating the latter to be in equilibrium with its components. ESR experiments have shown TCNE to be the counter-ion during polymerization. The work of Nakamura et al.[78] largely substantiates these results.

(vi) *High energy radiation*

In 1957 isobutene was successfully polymerized at $-78\,°C$ by high energy radiation,[79] and shown to be a cationic process by the use of additives. Since then many monomers have been polymerized by this method, and the work of Williams[80] has been of particular interest.

On irradiation with γ-rays from a ^{60}Co source an electron is ejected from suitable liquid monomers,[81] and so presumably electrons become solvated by the latter. The monomeric radical cation formed is then free to propagate

$$CH_2{=}CHR \xrightarrow{\;\gamma\text{-ray}\;} \dot{C}H_2{-}\overset{\oplus}{C}HR + e^{\ominus}$$

via a radical mechanism at one end and a cationic one at the other. Many cationic propagation steps will take place for each radical addition, because of the difference in reactivity of the two species. Thus even if the radical ends are never destroyed by, say, dimerization their contribution to polymerization will be undetectably small. Using this technique, extremely low concentrations of active centres can be achieved and the cations formed have no effective counter-ion. Problems involving initiation equilibria and resulting initiator fragments, together with ion pair contributions do not arise. The method therefore allows the propagation of free cations to be investigated, and has proved invaluable as a yardstick for more conventional chemically initiated systems.

9.3.2 Mechanism of the initiation of ring opening polymerization (cyclic ethers)

Most of the initiating systems described here are useful for all cyclic ethers although their relative efficiencies vary in a non-predictable way from one ring system to another. Most of the detailed kinetic studies have been carried out using tetrahydrofuran (THF) and whilst, the important mechanisms in this section are largely based on this monomer, they are generally applicable to other cyclic ethers and sulphides.[82,83]

THF was the first cyclic ether to be polymerized to a high polymer by cationic processes;[84] the early work was due to Meerwein and his collaborators and many of the factors governing catalyst reactivity were recognized and characterized by these workers. A much wider range of effective initiators is now available and the most important of these are discussed under the appropriate sub-heading.

(i) Oxonium Salts

Initiation of cyclic ether polymerization requires reaction between the ether oxygen and an ion pair comprising a highly electrophilic cation (X^+) and a counter-ion of low nucleophilicity (Y^-), for example

$$X^{\oplus}Y^{\ominus} + O\!\!\diagdown\!\!\square \;\rightleftharpoons\; X\!\!-\!\!\overset{\oplus}{O}\!\!\diagdown\!\!\square \;\; Y^{\ominus}$$

Propagation follows when the initially formed cyclic oxonium ion undergoes ring opening (by reaction with more monomer) more rapidly than competing side reactions such as inter- or intramolecular hydride ion abstractions, for example

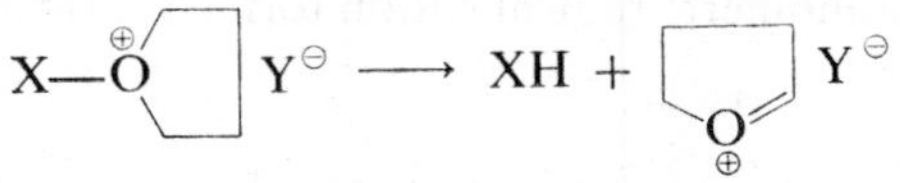

$$X\!\!-\!\!\overset{\oplus}{O}\!\!\diagdown\!\!\square\; Y^{\ominus} \;\longrightarrow\; XH + \square\!\!\diagup_{\underset{\oplus}{O}}\; Y^{\ominus}$$

One of the simplest types of cation that can be used is a trialkyl oxonium ion,[85,86] or possibly a very reactive quaternary ammonium ion.[87] The former can be regarded as specifically (ether) solvated alkyl cations ($R_2O + R^{\oplus} \rightleftharpoons R_3O^{\oplus}$) and consequently they are very powerful alkylating agents. They readily act as efficient initiators for cyclic ether polymerization but the actual initiation process involves regeneration of the parent dialkylether which may act subsequently as a transfer agent, for example

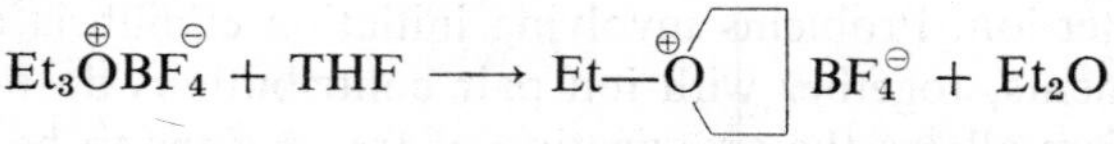

$$Et_3\overset{\oplus}{O}BF_4^{\ominus} + THF \;\longrightarrow\; Et\!\!-\!\!\overset{\oplus}{O}\!\!\diagdown\!\!\square\; BF_4^{\ominus} + Et_2O$$

Furthermore, although this particular reaction would seem to be the most direct method for forming the cyclic oxonium ion necessary for propagation, kinetic studies have shown that the initiation reaction (ether exchange) is still significantly slower than the ensuing propagation. Trialkyl oxonium salts are sufficiently stable to be used as preformed initiators, but it is often more convenient to generate them in situ by reactions between a Friedel-Crafts halide and a reactive epoxide. Epichlorohydrin is particularly useful for this purpose, for example

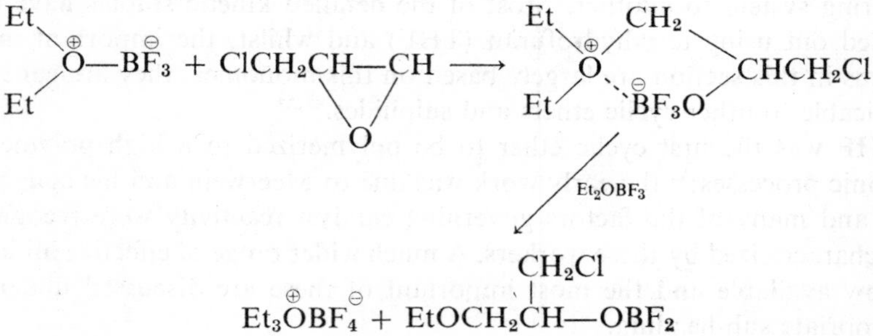

$$Et_3\overset{\oplus}{O}BF_4^{\ominus} + EtOCH_2\overset{\underset{|}{CH_2Cl}}{CH}\!\!-\!\!OBF_2$$

(ii) *Protonic acids*

Initiation of cyclic ether polymerization might be expected as a result of protonation of the oxygen atom by strong acids. As with vinyl monomers this only occurs if the acid anion is of low nucleophilicity. Acids such as $HFeCl_4$, HBF_4, $HSbCl_6$ and $HClO_4$ fall into this category and form a secondary oxonium ion with THF capable of propagating. Sulphuric acid on the other

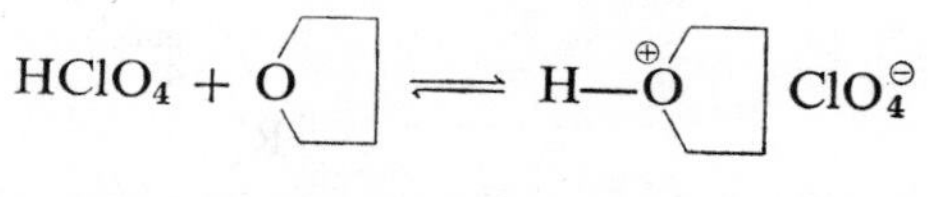

$$HClO_4 + O \quad \rightleftharpoons \quad H—\overset{\oplus}{O} \quad ClO_4^{\ominus}$$

hand is not an initiator because the bisulphate anion is a stronger nucleophile than the monomer, and competes successfully for the oxonium ion. How-

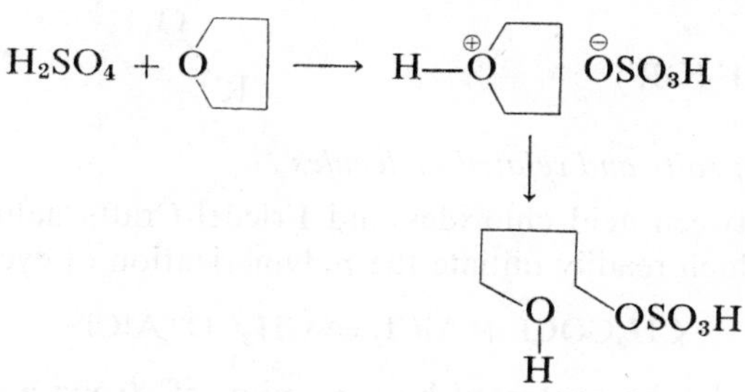

$$H_2SO_4 + O \quad \longrightarrow \quad H—\overset{\oplus}{O} \quad \overset{\ominus}{O}SO_3H$$

$$\downarrow$$

$$\underset{H}{O} \quad OSO_3H$$

ever, fluorosulphuric acid, HSO_3F and chlorosulphuric acid, HSO_3Cl, are quite effective initiators.

Anions of low nucleophilicity are most readily formed by complexing Lewis acid halides with a suitable proton donor, for example $BX_3 + HX \rightleftharpoons H^{\oplus}BX^{\ominus}$, and provide a wide range of initiator systems, most of which are best prepared *in situ*. Such systems however, are ill-defined and, as with vinyl monomers, adventitious impurities very often play a large role in the initiation phenomena.

(iii) *Carboxonium ion salts*

These initiators are readily formed by interaction between organic molecules having labile halogen atoms and suitable Lewis acids such as $AlCl_3$, BF_3 etc.[88] Particularly useful are the alpha-chloro-ethers, for example

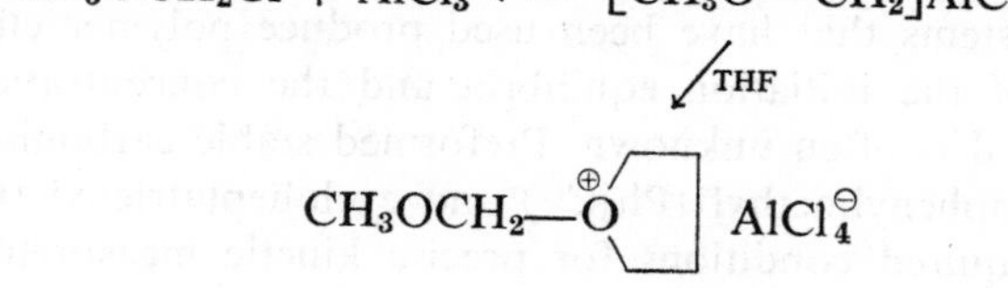

$$CH_3OCH_2Cl + AlCl_3 \rightleftharpoons [CH_3\overset{\oplus}{O}=CH_2]AlCl_4^{\ominus}$$

$$\swarrow THF$$

$$CH_3OCH_2—\overset{\oplus}{O} \quad AlCl_4^{\ominus}$$

Alternatively, carboxonium salts can be prepared by the reaction of Friedel-Crafts halides with ortho-esters or acetals, for example

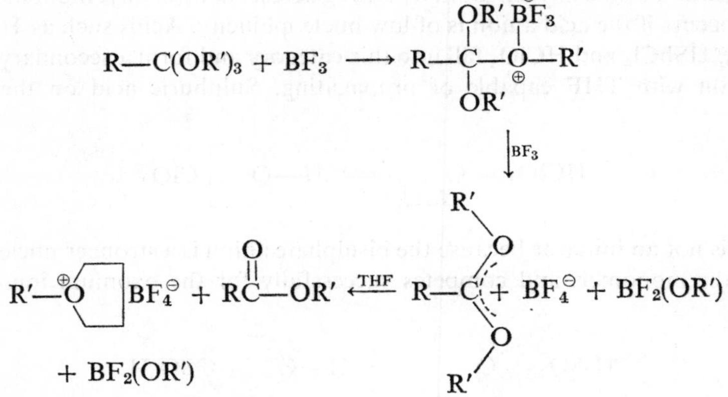

$$R—C(OR')_3 + BF_3 \longrightarrow R—\overset{\overset{\displaystyle OR'}{|}}{\underset{\underset{\displaystyle OR'}{|}}{C}}—\overset{\ominus}{O}\overset{\oplus}{—}R' \quad \overset{\ominus}{BF_3}$$

$$\downarrow BF_3$$

$$R'—\overset{\oplus}{O}\boxed{}\;BF_4^{\ominus} + R\overset{\overset{\displaystyle O}{\|}}{C}—OR' \quad\overset{THF}{\rightleftharpoons}\quad R—C\!\!\overset{\displaystyle O\diagdown R'}{\underset{\displaystyle O\diagup R'}{\big\langle}}\; + BF_4^{\ominus} + BF_2(OR')$$

$$+\ BF_2(OR')$$

(iv) *Acylium ion salts and related molecules*

Reactions between acid chlorides and Friedel-Crafts halides give rise to acylium salts which readily initiate the polymerization of cyclic ethers[88]

$$CH_3COCl + AlCl_3 \rightleftharpoons CH_3CO^{\oplus}AlCl^{\ominus}$$

and which can also be generated by the action of strong acid (non-nucleophilic anion) on carboxylic acid anhydrides. If aromatic acid chlorides are

$$(CH_3CO)_2O + HClO_4 \rightleftharpoons CH_3CO^{\oplus}ClO_4^{\ominus} + CH_3COOH$$

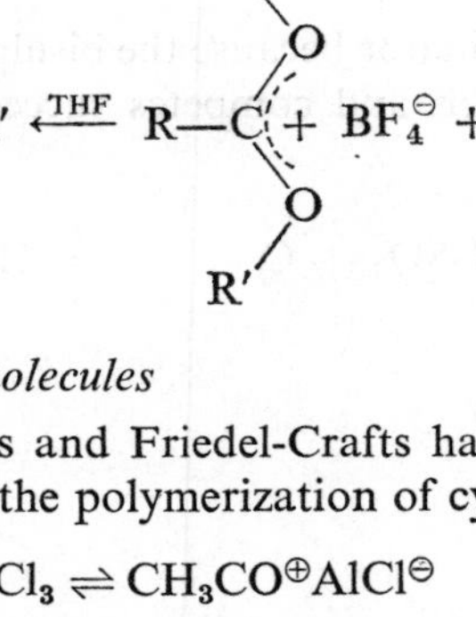

$$\downarrow THF$$

$$CH_3CO—\overset{\oplus}{O}\boxed{}\;ClO_4^{\ominus}$$

used instead of aliphatic derivatives then the acylium ion may be isolated as stable salts, for example $PhCO^{\oplus}SbCl_6^{\ominus}$, permitting better control of the initiation reactions.

(v) *Stable carbonium ion salts*

In order to study the polymerization of cyclic ethers quantitatively an easily characterizable and reproducible initiation mechanism is required. While most systems that have been used produce polymer efficiently, the exact nature of the initiation equilibria and the concentration of active species produced is often unknown. Preformed stable carbonium ion salts, in particular triphenylmethyl ($Ph_3C^{\oplus}$) and cycloheptatrienyl ($C_7H_7^{\oplus}$) salts, provide the required conditions for precise kinetic measurements on the polymerizations.[63] Triphenylmethylhexachloroantimonate and the corresponding hexafluorophosphate have proved especially useful. Initiation using cycloheptatrienyl salts seems to be much slower. Trityl salts have the added advantage that they are highly coloured and their concentrations in solution are easily monitored by spectrophotometric techniques.

When solutions of these salts are added to cyclic ethers there is rapid decolouration of the catalyst followed by polymerization of the ether. The initial decolouration is due to oxonium ion formation, followed by a rapid reaction producing triphenylmethane[89] before the onset of polymerization.

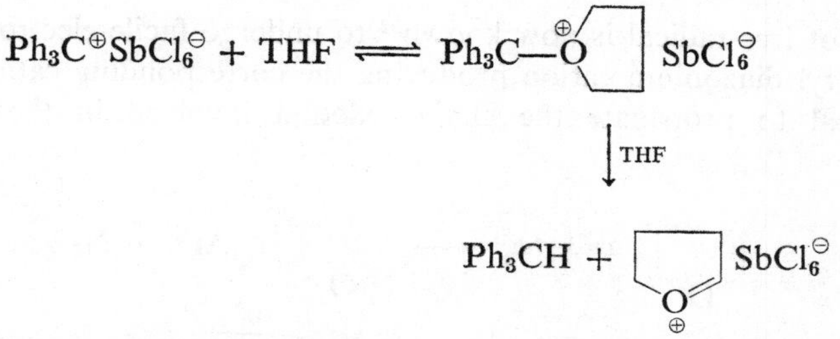

$$Ph_3C^{\oplus}SbCl_6^{\ominus} + THF \rightleftharpoons Ph_3C\!-\!\overset{\oplus}{O}\!\!\diagdown \quad SbCl_6^{\ominus}$$

$$\Big\downarrow THF$$

$$Ph_3CH + \quad \underset{\oplus}{\overset{}{O}}\!\!\diagdown \quad SbCl_6^{\ominus}$$

The hydride ion abstraction process shown here is in marked contrast to the reaction of these salts with active olefins where direct addition of the carbonium ion has been demonstrated to be the primary mechanism of initiation.[22,62]

(vi) *Aryl diazonium salts*

Aryl diazonium salts decompose thermally and photochemically by both free-radical and ionic mechanisms, and the reactions can be utilized to effect initiation of polymerization of many cyclic ethers though the precise mechanisms of all the initiation processes have not yet been satisfactorily explained. Dreyfuss and Dreyfuss[90,91] first showed that p-chlorophenyl diazonium hexafluorophosphate was a good initiator for polymerization of tetrahydrofuran and other cyclic ethers and, moreover, gave good evidence of the absence of termination and transfer processes. Whatever the undoubtedly complex nature of the diazonium ion decomposition processes, initiation of polymerization was thought to occur by overall hydride ion abstraction—one possibility is indicated below

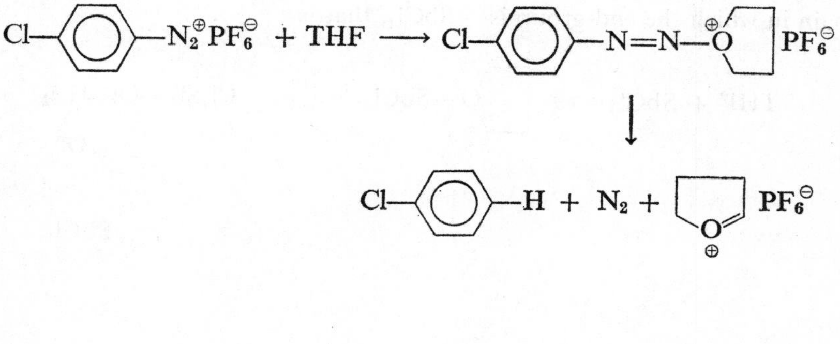

$$Cl\!-\!\!\diagup\!\!\bigcirc\!\!\diagdown\!\!-\!N_2^{\oplus}PF_6^{\ominus} + THF \longrightarrow Cl\!-\!\!\diagup\!\!\bigcirc\!\!\diagdown\!\!-\!N\!=\!N\!-\!\overset{\oplus}{O}\!\!\diagdown \quad PF_6^{\ominus}$$

$$\Big\downarrow$$

$$Cl\!-\!\!\diagup\!\!\bigcirc\!\!\diagdown\!\!-\!H + N_2 + \quad \underset{\oplus}{\overset{}{O}}\!\!\diagdown \quad PF_6^{\ominus}$$

However, it is equally likely that the primary mode of decomposition of the aryl diazonium salt is homolytic yielding, ultimately, α-alkoxy radicals from the cyclic ether.

$$ArN_2^+ + (X) \longrightarrow Ar\cdot + N_2$$

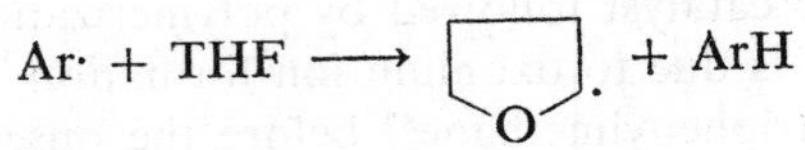

$$Ar\cdot + THF \longrightarrow \underset{O}{\boxed{}}\cdot + ArH$$

This type of free-radical is now known[92] to undergo facile electron transfer with the aryl diazonium cation producing the corresponding cation and an aryl radical to propagate the chain reaction involved in the initiation process.

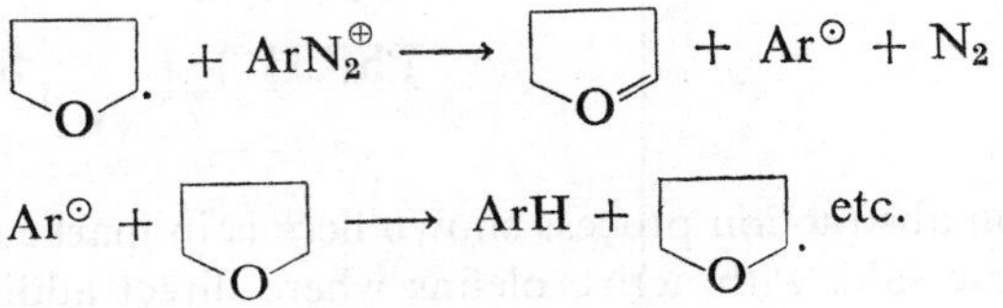

$$\underset{O}{\boxed{}}\cdot + ArN_2^\oplus \longrightarrow \underset{O}{\boxed{}}^\oplus + Ar^\odot + N_2$$

$$Ar^\odot + \underset{O}{\boxed{}} \longrightarrow ArH + \underset{O}{\boxed{}}\cdot \quad \text{etc.}$$

(vii) *Friedel-Crafts halides*

These compounds have been used widely with many cocatalysts to produce various reactive cations for THF polymerizations as already discussed. There are however, many reports[93] that several of these halides will initiate cyclic ether polymerization without a cocatalyst. The situation is analogous to that found in the literature on initiation of olefin polymerization by these reagents and once again it seems likely that in perfectly dry systems Friedel-Crafts halides in general are not effective initiators without assistance from adventitious impurities, particularly water. One possible exception is SbCl₅[94] which differs from say BF₃ and AlCl₃ in that antimony possesses a stable lower valence state. In addition it is a well-known chlorinating agent,[95] and could function by first chlorinating the ether, after which the resulting α-chloroether would function as cocatalyst as indicated earlier. Alternatively the SbCl₅–THF complex could react with more SbCl₅ to form a growing polymer chain in which the end-group is —SbCl₄, that is

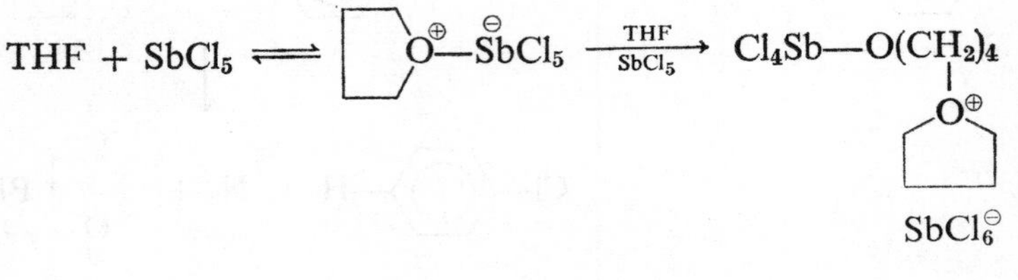

$$THF + SbCl_5 \rightleftharpoons \underset{O^\oplus - \overset{\ominus}{Sb}Cl_5}{\boxed{}} \xrightarrow[SbCl_5]{THF} Cl_4Sb\!-\!O(CH_2)_4$$

$$\underset{SbCl_6^\ominus}{\underset{O^\oplus}{\boxed{}}}$$

Clearly whichever mechanism is correct only one polymer chain results from an initiation reaction involving two molecules of $SbCl_5$, one of which is effectively reduced in the process. Similar behaviour might also be anticipated for Lewis acids having other metals in high oxidation states.

9.3.3 Effect of solvent and temperature on the initiation reaction

There is very little reliable kinetic data available on rates of initiation as opposed to overall rates of polymerization. A discussion of the effect of temperature on such reactions must therefore be rather speculative. A change in the reaction temperature is likely to shift the position of any chemical initiation equilibria, and the physical equilibrium involving free ions, ion pairs and any other species. In addition the rate at which each species reacts with monomer will also be altered.

In general the latter would be expected to follow a simple Arrhenius dependence, i.e. a decrease in temperature would result in a decrease in reactivity. In the system isobutene/methylene chloride/$AlCl_3$ the rate of polymerization has such a dependence,[96] and indeed, with higher α-olefins, polymerization can be completely stopped by lowering the temperature.[97] Presumably the initiation reaction, whatever it might be, is the rate controlling process, and these systems are merely showing simple Arrhenius behaviour. A definite example of this situation is the system $C_7H_7^+SbCl_6^-$ N-vinylcarbazole in CH_2Cl_2. Here the initiation is known to involve a direct addition reaction,[62] and lowering the temperature slowly reduces the rate of this process, until at $\sim -50\,°C$ no net reaction takes place. Instead a stable pink-coloured charge-transfer complex can be observed.

With many systems, however, decreasing the temperature increases the overall rate of polymerization,[98,99] and presumably, therefore, the rate of initiation. In these cases either a pre-initiation equilibrium is disturbed in such a manner as to produce an increase in the charged species subsequently involved in the actual initiation reaction, or the physico-chemical equilibrium involving free ions and ion pairs is altered in favour of the more reactive moieties (probably free ions), or a combination of both operates. A more precise analysis than this is impossible in the absence of data such as pre-initiation equilibrium constants and ion pair dissociation constants. It seems predictable that the correct choice of system could produce a maximum or minimum in a plot of rate of reaction against temperature.

The effect of solvent on a given catalyst system is very closely related to the effect of temperature, since alteration of the latter changes the dielectric constant of a given solvent which, in effect, is equivalent to changing the solvent itself. It is not surprising to find, therefore, that the solvent dependences of initiation reactions are also complex. Where reaction conditions are such that both free ions and ion pairs are present then increasing the solvation power of the solvent would be expected to increase the proportion of free ions. In general, free ions are likely to have a higher reactivity so that the rate of initiation should increase. If data for ion pair and free ion contributions could be obtained separately then, and only then, could the well-established principles of physical organic chemistry, governing solvent dependence, be applied.

Consider a simple two stage initiation scheme involving only free ions (or only ion pairs).

$$RX + MX_n \rightleftharpoons R^{\oplus} + MX_{n+1}^{\ominus}$$

$$R^{\oplus} + CH_2 = CH \xrightarrow{\text{Initiation}} RCH_2 - \overset{\oplus}{CH}$$
$$\qquad\qquad\quad | \qquad\qquad\qquad\qquad\quad |$$
$$\qquad\qquad\quad R \qquad\qquad\qquad\qquad\quad R$$

Inert solvents of high polarity will lower the energy of the polar products of the pre-equilibrium more than the neutral reactants and, since this equilibrium is often rate controlling for the whole polymerization reaction, increasing the solvent polarity often enhances the overall rate of polymerization.[100] On the other hand the process of addition of $R^{\oplus}$ to the monomer, that is the true initiation reaction, occurs via a transition state involving charge dispersal, for example

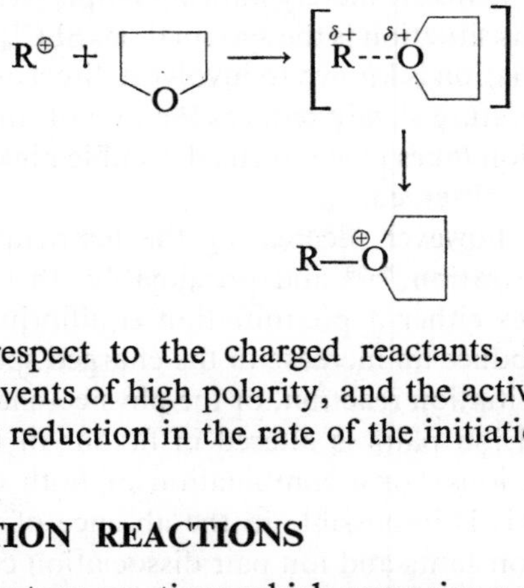

Therefore, with respect to the charged reactants, the transition state is destabilized by solvents of high polarity, and the activation energy increased with a consequent reduction in the rate of the initiation reaction.

9.4 PROPAGATION REACTIONS

Of all the elementary reactions which comprise a polymerization process the propagation reaction is probably the most fundamental, since it is the repetitive act which produces polymeric species. A single propagation step involves the addition of only one monomer molecule to the active centre, in this case a cation.

$$\text{\textasciitilde\textasciitilde\textasciitilde}M^{\oplus} + M \rightarrow \text{\textasciitilde\textasciitilde\textasciitilde}M - M^{\oplus}$$

However, in an idealized model this same reaction is repeated over and over again, because, rather uniquely, the growing polymer chain as 'seen' by one incoming monomeric unit appears identical to a second such unit after the first has undergone addition. In this respect chain growth polymerization is unique among organic reaction mechanisms.

In cationic systems the active centre is a powerful electrophile. A propagating carbonium ion in a vinyl polymerization can be regarded as having one *p*-orbital, formally localized on the terminal carbon atom, completely devoid of electrons. In practice conjugative effects usually intervene, so that the electron deficiency of the relevant atom is somewhat reduced, e.g. in alkylvinylether polymerizations the positive charge will be delocalized over both

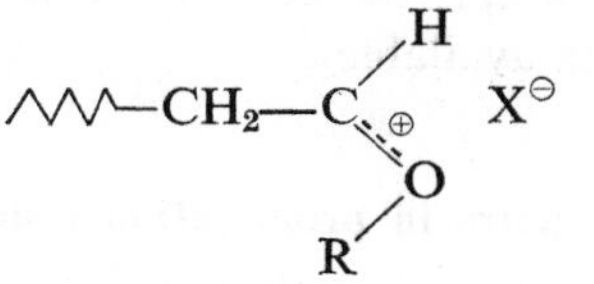

$$\text{/\!\!\sqrt{\!}\!\!\sqrt{}}-CH_2-\overset{\displaystyle H}{\underset{\displaystyle O}{\overset{\displaystyle |}{\underset{\displaystyle |}{C_{\oplus}}}}}\qquad X^{\ominus}$$
$$R$$

the oxygen atom and the terminal carbon atom and similar situations arise in the case of olefins having other types of electron-rich substituents.

The propagating species in cyclic ether polymerization is an oxonium ion and propagation involves attack of one or other of the α-carbon atoms of the ring by the lone pair of electrons of an incoming cyclic ether molecule, that is

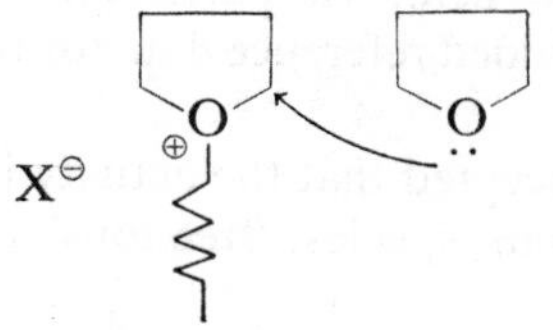

A qualitative understanding of this can be obtained by assuming some contribution to the structure of the active centre from such species as

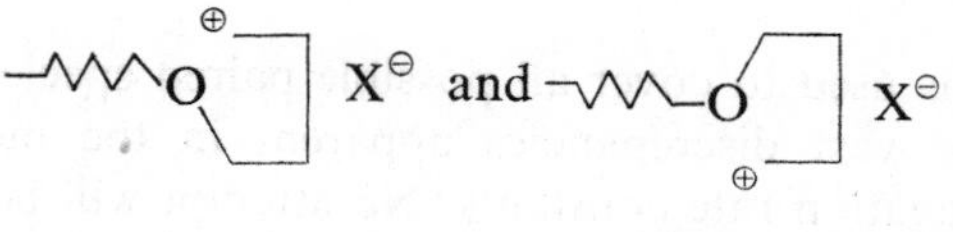

Each of the α-carbon atoms has a reduced electron density relative to a saturated atom in an alkane chain and hence some carbonium ion character.

Propagation in both examples involves the overlap of filled orbitals on a monomer molecule with empty, or partially empty orbitals of similar energy and symmetry on the propagating cation. In olefin polymerization the overlap involves π-electrons of the monomer, whereas in ring opening polymerization it is an unshared pair of electrons on the incoming ether molecule. This is in marked contrast to anionic polymerization where vinyl propagation is via filled orbitals on the active centre (usually a simple carbanion) overlapping with π^*-antibonding orbitals in the olefin. Antibonding orbitals are somewhat higher in energy than the π-bonding orbitals involved in cationic propagation and it would be expected that in corresponding cationic and anionic propagation reactions the latter would be a slightly more activated process. It follows therefore that k_p anionic is less than k_p cationic. Such a comparison, however, is possible only if the physical nature of the two propagating species with respect to their counter-ions and solvent is the same. The only circumstances under which this condition applies is if both active centres are completely dissociated and under the same reaction conditions. We shall see that these simple theoretical predictions are borne out by the limited experimental data available.

9.4.1 Free ions and ion pairs in propagation reactions

We have already noted that it was Winstein, studying solvolysis reactions, who first demonstrated that free ions and various ion pairs can have different reactivities. However, it is in the field of ionic polymerization, in particular in ionic propagation reactions, that his ideas have been substantiated in a most dramatic way. The most elegant results have been obtained from work on anionic propagations (see Chapter 11), but parallel results are slowly emerging in the cationic field. In particular radiation induced cationic polymerizations have provided reference data for those working on chemically initiated systems.

It is now universally accepted that the activity in cationic propagation can involve at least two different species, 'free ions' and 'ion pairs'. That is

$$\text{\small www}M^{\oplus} + M \rightarrow [\text{\small www}M^{\overset{\delta+}{}}\text{------}\overset{\delta+}{M}] \rightarrow \text{\small www}M_2^{\oplus}$$

$$\text{\small www}M^{\oplus}X^{\ominus} + M \rightarrow [\text{\small www}M^{\overset{\delta+}{}}\text{------}\overset{\delta+}{M}] \rightarrow \text{\small www}M_2^{\oplus}X^{\ominus}$$
$$X^{\ominus}$$

The latter term is used to cover all possible paired species and presumably accounts for the vast discrepancies apparent in the published literature relating to propagation rate constants. No attempt will be made, therefore, to present a comprehensive review of kinetic data. Instead the section will be limited to some of the more fundamental work which has appeared in the last few years.

(i) *Vinyl monomers*

For simplicity the rate of propagation, R_p, can be regarded as being composed of a free ion and ion pair contribution. The overall or observed rate constant, $k_{p\,\text{obs.}}$, will then be given by

$$k_{p\,\text{obs.}} = \alpha k_p^{\oplus} + (1 - \alpha)k_p^{\oplus}$$

where α = degree of dissociation of ion pairs into free ions, $k_p^{\ominus}$ = propagation constant of free ions, and $k_p^{\oplus}$ = propagation constant of ion pairs. In any kinetic experiment which is to be of value, first of all data on the rate of propagation alone (as opposed to overall rate of polymerization) must be ascertained. Secondly, the proportions of free ions and paired species (i.e. α) must be known. Finally, in order to determine the two unknowns $k_p^{\oplus}$ and $k_p^{\ominus}$, either their approximate relative values must be known, or the proportion of each species present must be altered, a new $k_{p\,\text{obs.}}$ determined, and hence two simultaneous equations established and solved, that is

$$k'_{p\,\text{obs.}} = \alpha' k_p^{\oplus} + (1 - \alpha')k_p^{\oplus}$$
$$k''_{p\,\text{obs.}} = \alpha'' k_p^{\oplus} + (1 - \alpha'')k_p^{\oplus}$$

An analysis similar to this has been achieved with anionic systems using a 'living polymer' technique[101] which eliminates the initiation reaction. No counterpart to these systems has so far been characterized for cationic polymerization of vinyl monomers, although it appears possible for cyclic ether polymerizations. Usually, excessive transfer and termination reactions mean that either a completely dead system remains after polymerization, or the number of active centres is not known accurately. Furthermore, any activity that does remain normally resides on small molecular species as a result of transfer when the monomer is largely consumed, so that conductance studies on polymerization reactions cannot give information on the true ion pair dissociation constants for the propagating species. An approximate value for these can be obtained only by means such as extrapolation from data on stable carbonium ion salts. Cationic polymerizations must always involve an initiation mechanism, and almost without exception catalyst systems are quantitatively (and often qualitatively) ill-defined (see section 9.3). High dielectric constant ($\varepsilon = 10$–30) solvents are frequently required in order to obtain homogeneity, so that relatively large concentrations of free ions may be involved, with consequent high overall rates of polymerization, and adding to the existing experimental difficulties.

A number of attempts[28,53,102] have been made to characterize cationic polymerizations in a manner similar to that used in radical polymerizations, that is a kinetic scheme involving initiation, propagation, transfer and termination reactions is suggested, and an overall (or observed) velocity constant related theoretically to the velocity constants of the elementary reactions. Almost without exception no fundamentally useful results have been obtained in this way, and there is little hope that such a general method for the study of cationic polymerizations will ever be achieved, since almost every cationic system possess some unique feature.

Methods which have been partially successful involve reagents and special circumstances which simplify the expression relating the observed velocity constants to individual steps in the polymerizations. Usually, conditions have been chosen such that only free ions are present, and $k_{p\,obs.} = k_p^{\oplus}$. Radiation induced polymerizations avoid complications of initiation (see section 9.3) and enable propagation alone to be investigated[17,79,80,103,104] Concentrations of ions produced are so small that even in the low dielectric constant media employed (bulk monomer), contributions from ion pair species can be neglected and propagation data obtained are for free ions. Rate coefficients obtained in this way are shown in Table 9.1 and discussed later.

Table 9.1 Propagation rate constants ($M^{-1}\,s^{-1}$) for the cationic polymerization of olefins

Monomer	Solvent	$T\,(°C)$	Initiator	k_p	Reference
Cyclopentadiene	None	−78	Radiation	6×10^8	79
Styrene	None	15	Radiation	$3·5 \times 10^6$	80
α-Methylstyrene	None	0	Radiation	4×10^6	80
Isobutylvinylether	None	30	Radiation	3×10^5	80
Isobutylvinylether	None	42·5	Pulse radiation	1×10^6	103
Isobutylvinylether	CH_2Cl_2	0	$C_7H_7^{\oplus}\ SbCl_6^{\ominus}$	5×10^3	22
Tert-butylvinylether	CH_2Cl_2	0	$C_7H_7^{\oplus}\ SbCl_6^{\ominus}$	$3·5 \times 10^3$	123
Methylvinylether	CH_2Cl_2	0	$C_7H_7^{\oplus}\ SbCl_6^{\ominus}$	$1·4 \times 10^2$	123
2-Chloroethylvinylether	CH_2Cl_2	0	$C_7H_7^{\oplus}\ SbCl_6^{\ominus}$	2×10^2	123

An alternative approach used by Ledwith and his collaborators[22,62] utilizes initiator systems based on stable carbonium ion salts (e.g. $Ph_3C^{\oplus}$ and $C_7H_7^{\oplus}SbCl_6^{\ominus}$). For reactive monomers, such as alkylvinylethers and N-vinylcarbazole, initiation is essentially instantaneous on mixing reagents, and the concentration of active centres generated approximates very closely to the concentration of initiator salt added. Here again, termination does not occur during the kinetic life-times and measurement of the overall rate of polymerization gives an estimate of the rate coefficient for propagation. The polymerizations have been studied in methylene chloride at 0 °C and −25 °C, and at these temperatures conductance experiments show that the catalyst salts are essentially completely dissociated into free ions. A study of a large variety of stable salts of cations of widely differing structure with the same anion ($SbCl_6^{\ominus}$) showed that values for the ion pair dissociation constants did not vary by more than one order of magnitude at a given temperature in methylene chloride, and thus the dissociation constants for the propagating polymeric salts would not be expected to differ substantially from those of the initiators. Kinetic data obtained therefore refer to the reactions of free propagating cation with monomer, $k_p^{\oplus}$ (Table 9.1). All values of $k_p^{\oplus}$ reported so far lie in the range 10^{+3}–10^{+9} $M^{-1}s^{-1}$, and this breadth of results is not surprising in view of the wide variety of monomers studied, the different reaction conditions, and the different techniques employed. Isobutylvinyl-ether has been studied by both techniques and enables some comparison to be made. Williams[80] estimates that $k_p^\oplus = 3 \times 10^{+5}$ M^{-1}s^{-1} for the bulk mono-mer at 30 °C with an enthalpy of activation of 6·6 kcal mol^{-1}, whereas Ledwith[22] reports $k_p^\oplus = 3\cdot5 \times 10^3$ M^{-1}s^{-1} at 0 °C in methylene chloride solvent with an activation enthalpy of 6 kcal mol^{-1}. Extrapolation of this latter data to 30 °C gives a value of 2×10^4 M^{-1}s^{-1} in CH$_2$Cl$_2$, apparently one order of magnitude less than that from the radiation induced polymerization studies. However, if proper account is taken of the effect of solvent, then this apparent discrepancy is minimized and the two values are in quite good agreement considering the very high reactivities and the vastly different techniques used. By analogy with results in the anionic field, it now appears likely that the much lower values of k_p previously reported for some of these mono-mers[20,26,28,105] represent either ion pair propagation data, or result from incorrect analysis of the number of active centres. The reader is referred to the review by Plesch[8] for further discussion of this point.

It is worth noting that Aso[106] has succeeded in showing qualitatively the presence of free ions in the polymerization of o-divinylbenzene catalysed by Ph$_3$C$^\oplus$BF$_4^\ominus$, by studying the effects on polymer structures induced by addition of tetra-alkylammonium fluoroborates, and this new technique might also be used to give quantitative rate data.

Studies of the polymerization of styrene permit a useful comparison between the reactivity of a free cation with its monomer and a corresponding free anion with the same monomer. As already pointed out, simple orbital theory would predict the cationic process to be more facile, i.e. $k_p^\oplus$ should be greater than $k_p^\ominus$ (free anionic propagation rate constant) under identical reaction conditions. Swzarc's data[15] for the living styryl anion in tetra-hydrofuran (THF) at 25 °C ($k_p^\ominus = 6\cdot5 \times 10^4$ M^{-1}s^{-1}), and that of Williams[80] for the styryl cation in bulk monomer at 15 °C ($k_p^\oplus = 3\cdot5 \times 10^6$ M^{-1}s^{-1}), appear to confirm the simple predictions. This follows since bulk styrene and THF have similar dielectric constants, minimizing solvation effects, and the free anionic reaction appears to have an activation enthalpy of only ~ 2 kcal mol^{-1}, while that of the corresponding free cation appears to be even smaller, < 1 kcal mol^{-1}, minimizing temperature effects.

(ii) Cyclic monomers

This group of monomers has received considerably less attention than the olefinic type as far as investigations of absolute reactivity are concerned, and the data are correspondingly sparse.[8,107]

In general, the physical organic solution chemistry of cyclic monomers would be expected to parallel qualitatively that of olefin monomers, and, under suitable conditions, therefore, both free ion and ion pair species would be expected to contribute. However, in oxonium ion bulk polymerization of cyclic ethers (e.g. THF) it is not the reactivity of a paired ion with that of a 'bare' free ion, that is to be compared, but with that of a free ion specifically solvated by a monomer possessing high co-ordination power (but nevertheless, a low dielectric constant). Such solvation results in a marked reduction in the charge density of the ion, and hence its reactivity. The further reduction in charge density brought about by pairing of the solvated cation with an anion, may be relatively insignificant when compared to the effect which pairing has on the reactivity of a corresponding 'bare' free ion in vinyl polymerization. Many cyclic monomer propagation reactions may therefore appear to involve only one type of active centre. Indeed, under certain circumstances, it appears that the reactivity of an ion pair may exceed that of the free ion. Ledwith,[61] for example, has shown that the hydride ion transfer between THF and triphenylmethyl cation occurs more rapidly with $Ph_3C^{\oplus}$-$SbCl_6^{\ominus}$ ion pairs than with the free cation. Such a reversal of reactivity in the initiation reaction could be repeated in the propagation reaction. Table 9.2 summarizes some of the results obtained for cyclic monomers.

Table 9.2 Propagation rate constants ($M^{-1} s^{-1}$) for cationic polymerization of cyclic monomers

Monomer	Solvent	T °C	Initiator	k_p	Reference
BCMO[a]	PhCl	70	$(i\text{-}Bu)_3Al + H_2O$	8·5	8
THF	None	0	$Et_3Al + H_2O + P^b$	$6·4 \times 10^{-3}$	18
THF	$ClCH_2CH_2Cl$	0	$Et_3O^{\oplus}BF_4^{\ominus}$	$4·8 \times 10^{-3}$	86
THF	CH_2Cl_2	−0·5	$Et_3O^{\oplus}BF_4^{\ominus}$	$1·4 \times 10^{-3c}$	107
THF	None	50	$Ph_3C^{\oplus}SbCl_6^{\ominus}$	$1·4 \times 10^{-2}$	61
1,3-Dioxolan	CH_2Cl_2	0	$HClO_4$	10	8
1,3-Dioxepan	CH_2Cl_2	0	$HClO_4$	3×10^3	8
3,3-Dimethyl Thietan	CH_2Cl_2	20	$Et_3O^{\oplus}BF^{\ominus}$	$6·5 \times 10^{-3}$	8

a. Bis-3,3-chloromethyloxetane.
b. P = epichlorohydrin, propene oxide, or β-propiolactone.
c. Ion pair reactivity.

The most recent information on the polymerization of THF has been obtained by Worsfold and Sangster.[107] They report separate values for the free ion and ion pair (BF_4^-) rate constants for propagation of $1·0 \times 10^{-2}$ and $1·4 \times 10^{-3} M^{-1}s^{-1}$ respectively at −0·5 °C in methylene chloride solution. Such data are most rewarding since they confirm the expected lower reactivity of oxonium ions in cyclic ether polymerizations relative to carbonium ions in vinyl monomer systems. In addition the difference in reactivity between the free oxonium ion and its corresponding ion pair is less than an order of magnitude, providing qualitative evidence for the specific solvation of the free oxonium ion by monomer molecules, as indicated above. Parallel results using the monomer 3,3-dimethylthietan have been obtained almost simultaneously by Goethals and his coworkers.[145] The free ion derived from this monomer propagates with a rate constant of $6{\cdot}7 \times 10^{-2}$ $M^{-1}s^{-1}$ in methylene chloride at 20 °C, while its ion pair (BF_4^-) reacts more slowly by a factor of only ~40. Once again this seems to indicate a considerable degree of specific solvation of the free sulphonium ion by monomer.

9.4.2 Pseudo-cationic polymerization

Several groups of workers have studied the polymerization of styrene in methylene chloride initiated by perchloric acid.[28–35] This is the most important system in which pseudo-cationic behaviour is believed to occur, though more recently[55] similar phenomena have been reported in the system styrene/iodine/methylene chloride. The expression pseudo-cationic species was first introduced by Plesch to explain an apparent absence of *charged* active intermediates in the polymerization of styrene by perchloric acid. The experimental observations which lead to the conclusion that propagating species other than ion pairs and free ions at present are numerous and complex, but have recently been reviewed by Plesch.[8] There appears to be agreement among the various groups of workers that covalent ester species participate in equilibrium such as

$$\text{\textormSt}^{\oplus} + ClO_4^{\ominus} \rightleftharpoons \text{wwSt}^{\oplus}ClO_4^{\ominus} \rightleftharpoons \text{wwStOClO}_3$$
Free ions   Ion pairs   Covalent ester

The propagation reaction is complex and recent detailed kinetic studies by Pepper and his collaborators, who pioneered the work in this field, have established the existence of three distinct stages. Stage 1 occurs at low temperatures and consists of an extremely rapid reaction for which conductance experiments have confirmed the presence of ionic species, possibly including free oligostyryl cations. Stage 1 diminishes in value with increasing temperature and is followed by a much slower stage 2. After some time at stage 2, the hitherto colourless solution suddenly turns yellow, there is a simultaneous increase in conductance, and the residual monomer is consumed extremely rapidly, constituting stage 3. The yellow colour has been shown to be due to the 3-aralkyl-1-phenylindanyl cation, and stage 3 is generally regarded as involving propagation by a very small concentration of oligo-

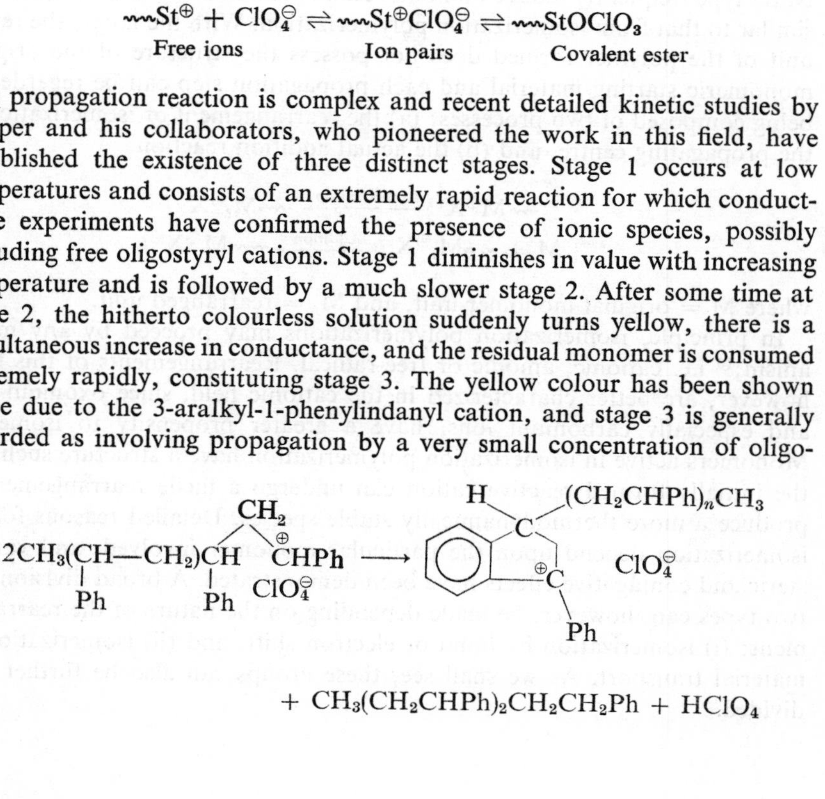

$$+ CH_3(CH_2CHPh)_2CH_2CH_2Ph + HClO_4$$

styryl cations in equilibrium with the indanyl species as indicated. Disagreement arises over stage 2. Does propagation take place via a small concentration of free ions or a larger concentration of some much less reactive species? If species of low reactivity are involved, are these ion-pairs or covalent ester groups? Plesch currently believes that stage 2 involves a free ion contribution together with participation of ester groups in the growth reaction (pseudo-cationic propagation); Pepper and his coworkers[30,31] take the view that ion pairs are responsible, and that the fraction of oligostyryl species present as ester units are dormant and do not themselves propagate. The present authors incline to the second view but believe that further evidence is required to remove the remaining ambiguities.

9.4.3 Propagation with concurrent isomerization

So-called isomerization polymerizations involve propagation reactions in which the active species rearrange to an energetically preferred structure, prior to each addition of monomer. This situation is not to be confused with polymerizations where the monomer is first isomerized, in the presence of the catalyst, before propagation;[108] heterogeneous catalysts of the Ziegler-Natta type frequently induce monomer isomerization and the overall result is similar to that from isomerization polymerization. With the latter, the repeat unit of the polymer formed does not possess the structure of the original monomeric starting material and each propagation step can be regarded as being composed of two processes: (a) the rearrangement or isomerization of the propagating centre, and (b) the actual addition reaction.

$$\text{\Large\sim}\text{\small\sim}M^{\oplus}X^{\ominus} \xrightarrow{\text{Isomerization}} \text{\Large\sim}\text{\small\sim}M_t^{\oplus}X^{\ominus}$$

$$M + \text{\Large\sim}\text{\small\sim}M_t^{\oplus}X^{\ominus} \xrightarrow{\text{Addition}} \text{\Large\sim}\text{\small\sim} M^{\oplus}X^{\ominus}$$

where M = original monomer unit, and M_t = rearranged unit.

In principle, isomerization polymerizations may proceed by any mechanism;[10] i.e. cationic, anionic or free-radical. Rearrangements of this type, however, are better characterized in the cationic field, since oxonium ions and especially carbonium ions, have a greater propensity to isomerize, Monomers active in isomerization polymerization have a structure such that the initially formed reactive cation can undergo a facile rearrangement to produce a more thermodynamically stable species. Detailed reasons for the isomerization depend upon the particular monomer involved, and various steric and conjugative effects have been demonstrated. A broad division into two types can, however, be made depending on the nature of the rearrangement: (i) isomerization by bond or electron shift, and (ii) isomerization by material transport. As we shall see, these groups can also be further subdivided.

(i) *Bond and electron rearrangements*

(a) *Intra-intermolecular polymerizations.* This type of reaction is also referred to as cyclopolymerization since the intermolecular propagation step is often preceded by an intramolecular cyclization reaction. It follows, therefore, that each monomer molecule must possess at least two positions of unsaturation or reactivity. Polymerization of non-conjugated aliphatic and some alicyclic dienes, such as 2,6-diphenyl-1,6-heptadiene, fall into this group, for example

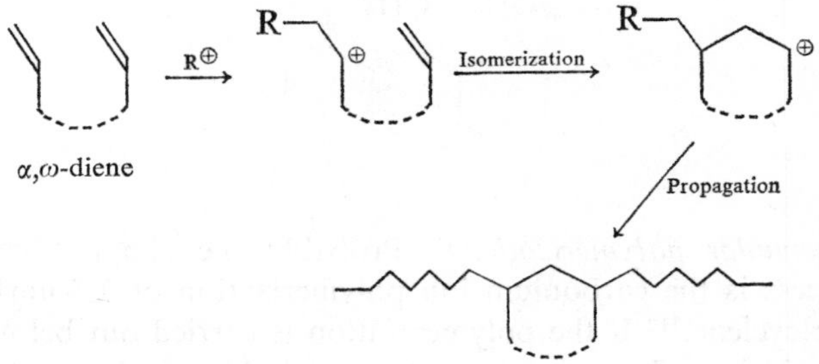

α,ω-diene

In this case, intramolecular attack of the initially formed electrophilic site on another part of the same molecule results in the formation of a different carbonium ion. If the monomer is symmetrically substituted, then the iso-merization process does not produce a carbonium ion of particularly increased stability. However, the overall propagation produces a polymer in which one area of unsaturation in the monomer has disappeared in forming a chain molecule, while the other has been consumed in the formation of an unstrained alicyclic ring and the overall lower thermodynamic state of the polymer contributes substantially to the driving force of the isomerization reaction. A similar reaction has been demonstrated with dialdehydes.[110]

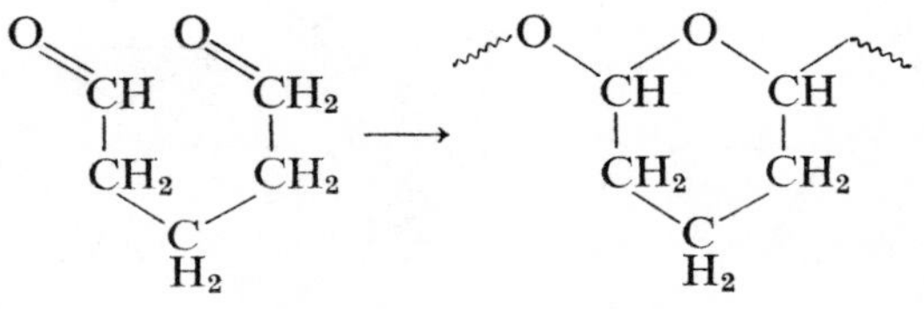

Polymerization of 9-vinylanthracene induced by various Lewis acids has been viewed as an isomerization polymerization of a conjugated triene.[111] The product can be converted, probably by protonation-deprotonation to a

polymer of 9,10-dimethylanthracene

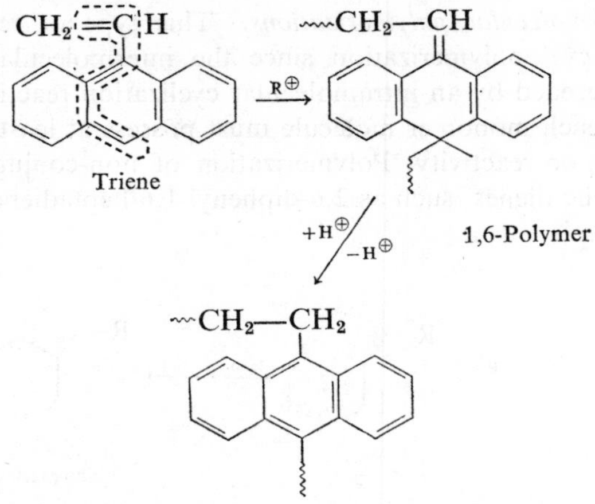

Triene

1,6-Polymer

(b) *Transannular polymerizations.* Probably the simplest case of this type of process is the carbonium ion polymerization of 2,5-norbornadiene to polynortricyclene.[112] If the polymerization is carried out below −100 °C to eliminate side reactions, for example cross-linking, a pure white toluene-soluble polymer is obtained and X-ray studies indicate that the tricyclic units are probably randomly stacked in the chain structure. The carbonium ion rearrangements involved are a common feature of the organic chemistry of norbornadiene, for example

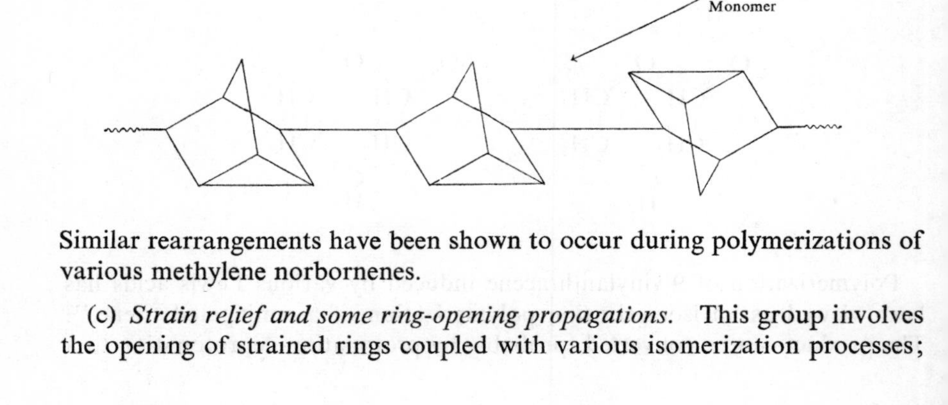

Monomer

Similar rearrangements have been shown to occur during polymerizations of various methylene norbornenes.

(c) *Strain relief and some ring-opening propagations.* This group involves the opening of strained rings coupled with various isomerization processes;

it does not include simple ring opening polymerizations of cyclic monomers such as tetrahydrofuran. Cyclopropane is the simplest strained cyclic monomer and can be polymerized by $AlBr_3$ to low molecular weight, ill-defined oils.[113] Probably the primary carbonium ion generated from this monomer isomerizes virtually instantaneously to a mixture of more stable species, leading eventually to a complexity of low-molecular weight branched products. In contrast the 1,1-dimethyl derivative[114] gives a uniform polymer.

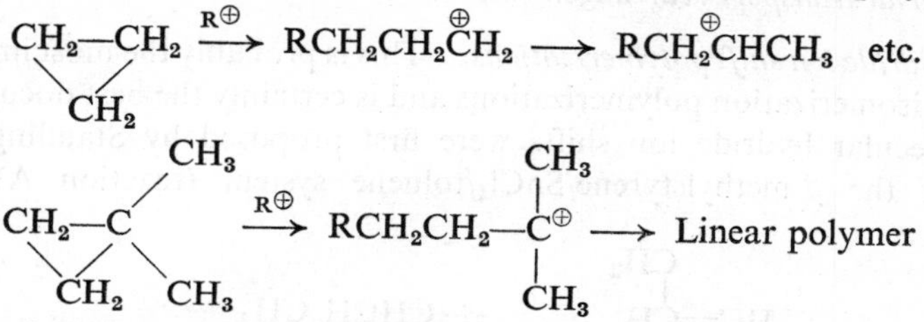

$$CH_2\!-\!CH_2 \xrightarrow{R\oplus} RCH_2CH_2\overset{\oplus}{C}H_2 \longrightarrow RCH_2\overset{\oplus}{C}HCH_3 \quad \text{etc.}$$
$$\diagdown\!CH_2\!\diagup$$

$$CH_2\!-\!C \xrightarrow{R\oplus} RCH_2CH_2\!-\!C^{\oplus} \longrightarrow \text{Linear polymer}$$

Terpene derivatives[10] undergo polymerization with isomerization when treated with Friedel-Crafts halides as exemplified by β-pinene. The related,

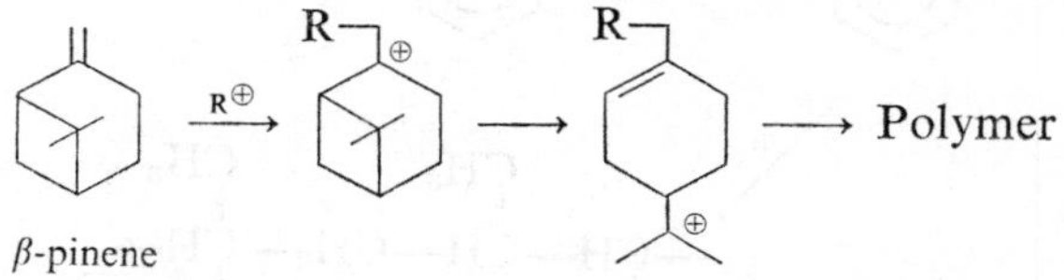

β-pinene

but less strained, exo-methylene cyclohexane cannot be polymerized under the same circumstances, although myrcene yields the same polymer as from β-pinene via an intra-intermolecular propagation mechanism. Isomerization

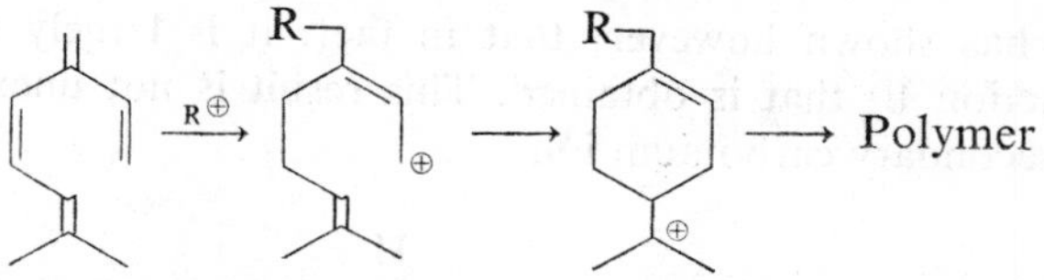

during cationic polymerization of vinylcyclopropane yields a polymer structure apparently arising by a 1,5 propagation mode.

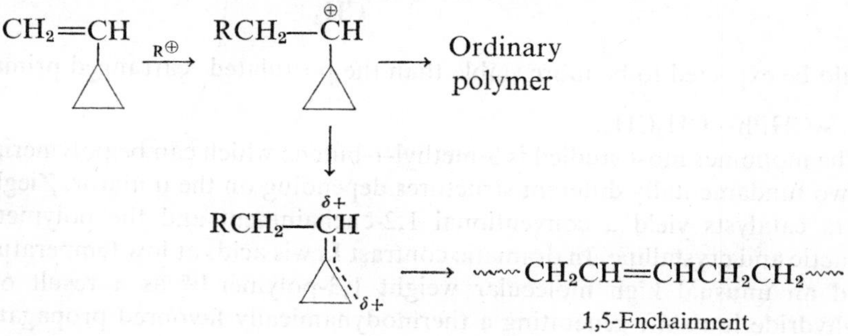

1,5-Enchainment

Structurally related vinylepisulphide also appears to undergo a ring-opening rearrangement polymerization.

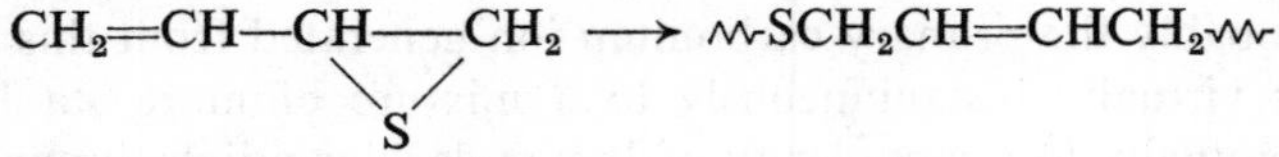

$$CH_2{=}CH{-}CH{-}{-}CH_2 \longrightarrow \text{\small\textit{w}}{-}SCH_2CH{=}CHCH_2{-}\text{\small\textit{w}}$$
$$\diagdown \; S \; \diagup$$

(ii) *Material transport rearrangements*

(a) *Hydride ion shift polymerizations.* This is probably the most important group of isomerization polymerizations and is certainly the best documented. Intramolecular hydride ion shifts were first proposed by Staudinger[115] in 1935 for the β-methylstyrene/SnCl$_4$/toluene system (reaction A). More

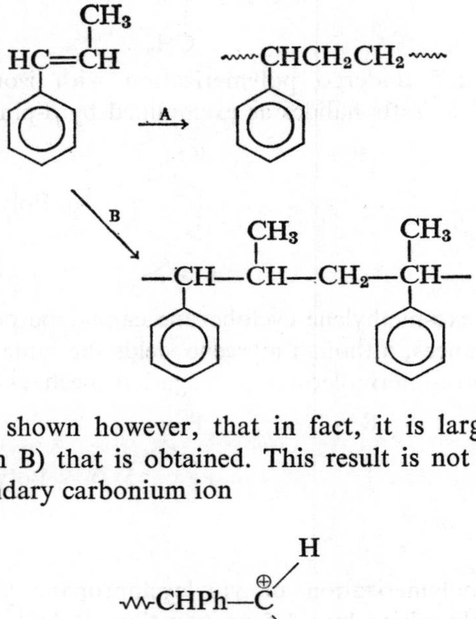

recent work has shown however, that in fact, it is largely the ordinary polymer (reaction B) that is obtained. This result is not unexpected since the original secondary carbonium ion

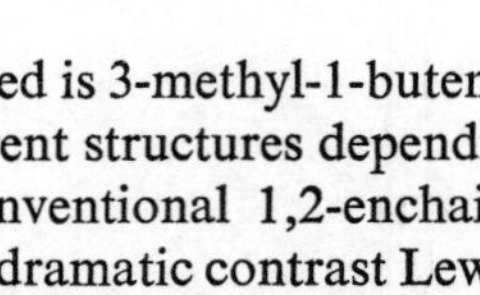

would be expected to be more stable than the postulated rearranged primary one, $\text{\small\textit{w}}CHPh{-}CH_2\overset{\oplus}{C}H_2$.

The monomer most studied is 3-methyl-1-butene which can be polymerized to two fundamentally different structures depending on the initiator. Ziegler-Natta catalysts yield a conventional 1,2-enchainment and the polymer is isotactic and crystalline. In dramatic contrast Lewis acids at low temperatures yield an unusual high molecular weight 1,3-polymer,[116] as a result of a 1,2-hydride ion shift generating a thermodynamically favoured propagating

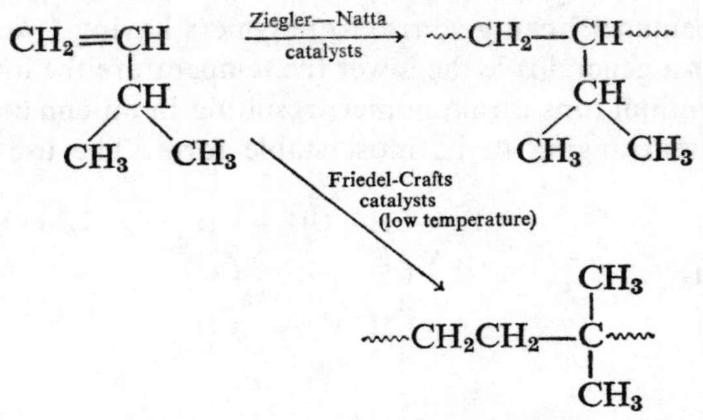

$$CH_2{=}CH \xrightarrow[\text{catalysts}]{\text{Ziegler—Natta}} \text{wwCH}_2{-}CH\text{ww}$$

(with $CH(CH_3)CH_3$ substituent on the vinyl carbon, and the same substituent on the polymer repeat unit)

$$\xrightarrow[\substack{\text{catalysts} \\ \text{(low temperature)}}]{\text{Friedel-Crafts}}$$

$$\text{wwCH}_2CH_2{-}\underset{\underset{\displaystyle CH_3}{|}}{\overset{\overset{\displaystyle CH_3}{|}}{C}}\text{ww}$$

tertiary cation. The 1,3-polymer obtained at $-130\,^{\circ}\text{C}$ is highly crystalline owing to the high chain symmetry, but polymerization at temperatures above

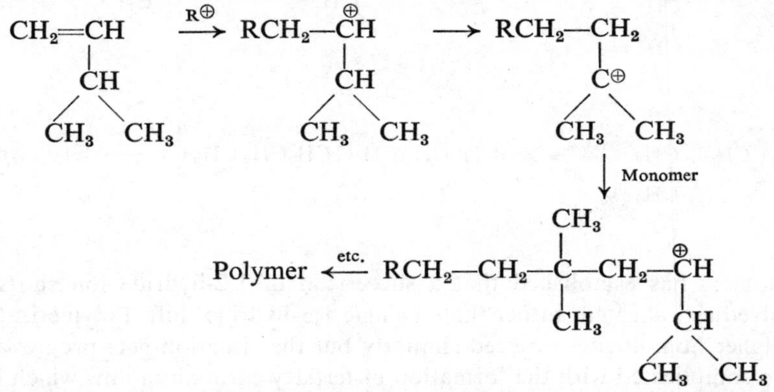

$$CH_2{=}CH \xrightarrow{R^{\oplus}} RCH_2{-}\overset{\oplus}{CH} \longrightarrow RCH_2{-}CH_2$$

(each with $CH(CH_3)CH_3$ group; third species has $\overset{\oplus}{C}(CH_3)CH_3$)

$$\Big\downarrow \text{Monomer}$$

$$\text{Polymer} \xleftarrow{\text{etc.}} RCH_2{-}CH_2{-}\underset{\underset{\displaystyle CH_3}{|}}{\overset{\overset{\displaystyle CH_3}{|}}{C}}{-}CH_2{-}\overset{\oplus}{CH}$$

(with $CH(CH_3)CH_3$ on the terminal $\overset{\oplus}{CH}$)

$-100\,^{\circ}\text{C}$ yields a rubbery product thought to be a random copolymer of 1,3- and 1,2-units.[10]

Vinylcyclohexane is structurally related to the 3-methyl-1-butene, and polymerizes in a similar manner

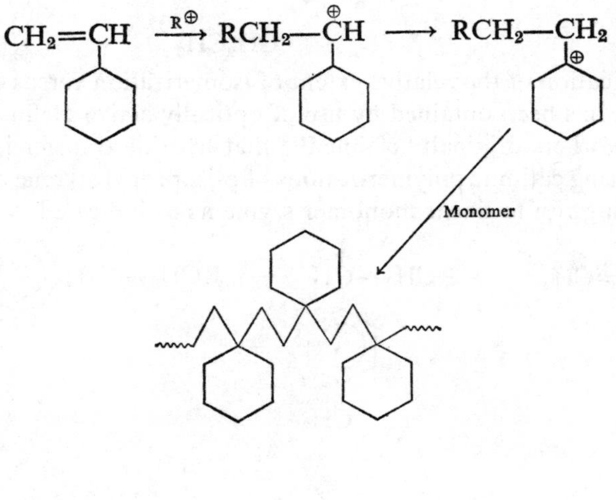

$$CH_2{=}CH \xrightarrow{R^{\oplus}} RCH_2{-}\overset{\oplus}{CH} \longrightarrow RCH_2{-}CH_2$$

(each bearing a cyclohexyl group; third species has a cyclohexyl cation)

$$\Big\downarrow \text{Monomer}$$

4-methyl-1-pentene[117] can give rise to polymers having 1,4-, 1,3- and 1,2-repeat-units; as a general rule the lower the temperature the lower the rate of attack of carbonium ions on monomer, resulting in an enhanced possibility of the cation rearranging to its most stable form. The use of deuterated

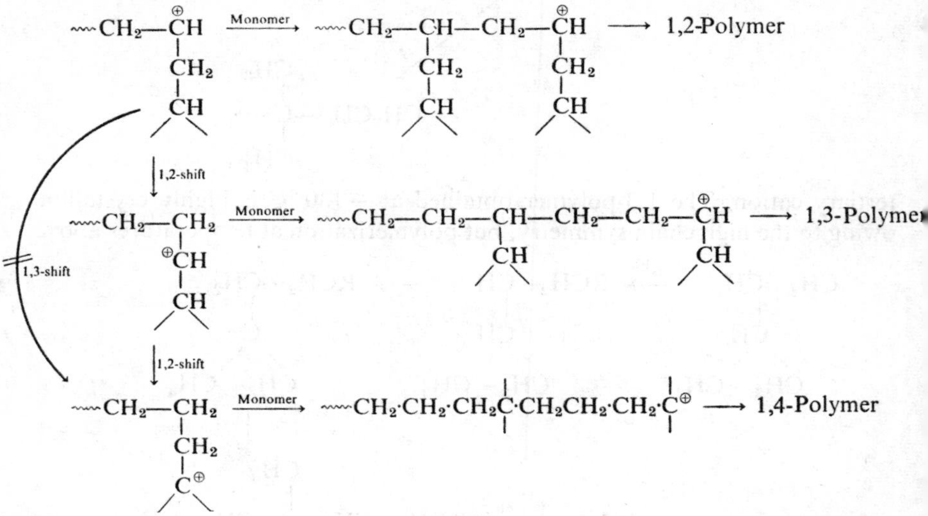

monomers has established that a succession of 1,2-hydride ion shifts are involved as indicated, rather than a single 1,3-hydride shift. Polymerizations of higher homologues proceed similarly but the situation gets progressively more complicated with the formation of tertiary carbonium ions which have difficulty in reacting with monomer because of steric hindrance ('buried' carbonium ions). For example

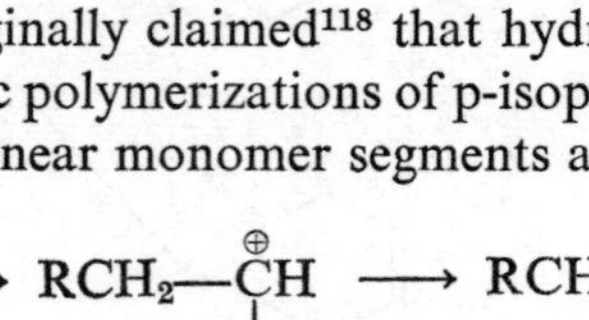

Some indication of the relative extent of isomerization versus conventional propagation has been obtained by use of optically active olefins.[10]

Russian workers originally claimed[118] that hydride transfer isomerization occurred during cationic polymerizations of p-isopropyl styrene and p-methyl styrene, giving rise to linear monomer segments as indicated

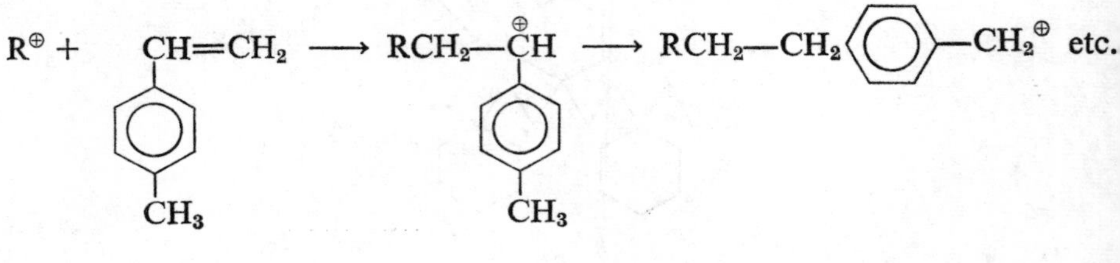

However, more recent, careful work,[119] has shown the earlier claims to be erroneous. All the evidence suggests that alkyl styrenes, even ortho-iso-propylstyrene, polymerize by conventional 1,2-vinyl additions. There have been other claims for hydride shifts in cationic polymerizations, including that of isopropylvinylether, however, the structural evidence produced to support the isomerized units leaves room for ambiguity.[120] For a critical survey of this field, the reader is referred to the review by Kennedy.[108]

(b) *Group migration polymerizations.* The only accepted case of a group migration reaction occurring during propagation is that which occurs in the cationic polymerization of 3,3'-dimethyl-1-butene.[121] At $\sim -130\,^\circ$C a high molecular weight polymer can be obtained but at higher temperatures only oligomers can be isolated. The failure to produce a high molecular weight product above $\sim -130\,^\circ$C is attributed to the slow propagation of a severely sterically hindered active centre

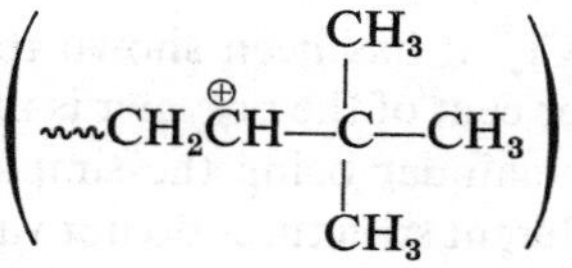

$$\left(\text{www}CH_2\overset{\oplus}{C}H-\underset{\underset{CH_3}{|}}{\overset{\overset{CH_3}{|}}{C}}-CH_3 \right)$$

Molecular models indicate that the unrearranged cation cannot propagate beyond about the trimer stage, because of the highly hindered carbonium ion produced. Under these conditions the cation stabilizes itself by proton ejection, producing neutral oligomers. At very low temperatures, however, this secondary ion is stabilized enough (i.e. has a sufficient life-time) for a methide shift to occur before proton expulsion, producing the energetically favoured tertiary ion. The latter is much less hindered and can propagate with reasonable efficiency.

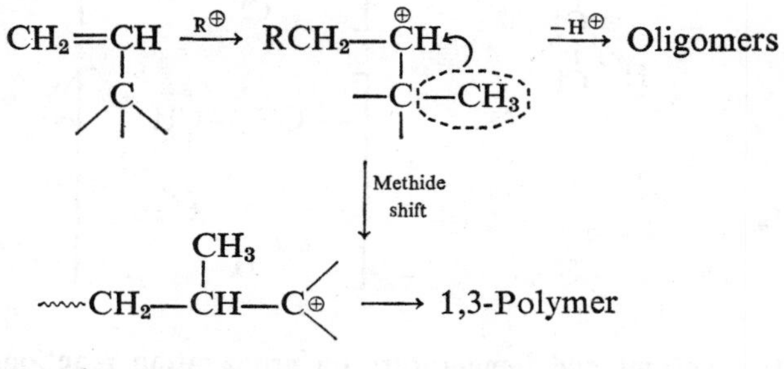

$$CH_2{=}CH \xrightarrow{R^\oplus} RCH_2{-}\overset{\oplus}{C}H \xrightarrow{-H^\oplus} \text{Oligomers}$$

with C and $-C-CH_3$ (methide shift)

$$\text{www}CH_2{-}\underset{}{\overset{\overset{CH_3}{|}}{C}H}{-}\overset{\oplus}{C} \longrightarrow \text{1,3-Polymer}$$

(d) *Halide ion migration polymerization.* Theoretically, 3-chloro-3-methyl-1-butene could yield three different repeat units in the polymer chain when polymerized by a cationic mechanism

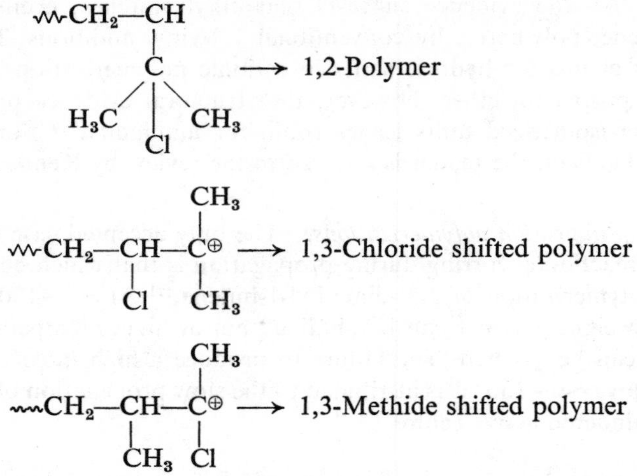

$$\text{w-CH}_2\text{—CH} \quad \underset{\overset{|}{\underset{\displaystyle H_3C \overset{|}{\underset{\displaystyle Cl}{}} CH_3}{C}}}{\quad} \longrightarrow \text{1,2-Polymer}$$

$$\text{w-CH}_2\text{—CH—}\overset{\overset{\displaystyle CH_3}{|}}{\underset{\underset{\displaystyle CH_3}{|}}{C}}\!\!^{\oplus} \longrightarrow \text{1,3-Chloride shifted polymer}$$
$$\qquad\quad \underset{\displaystyle Cl}{|}$$

$$\text{mwCH}_2\text{—CH—}\overset{\overset{\displaystyle CH_3}{|}}{\underset{\underset{\displaystyle Cl}{|}}{C}}\!\!^{\oplus} \longrightarrow \text{1,3-Methide shifted polymer}$$
$$\qquad\quad \underset{\displaystyle CH_3}{|}$$

Experimentally, using $AlCl_3$, it has been shown that methide ion shift does not occur, but about 50 per cent of the polymer is made up of units formed by chloride ion shifts, the remainder being the simple 1,2-addition product.[122] The proportions of the different structures do not vary much with temperature and the slight effect observed is the reverse of that found with pure hydrocarbon monomers (e.g. 3-methyl-1-butene) that is increasing the temperature favoured the chloride migration slightly.

Kennedy[10] has suggested that the difference may reside in the relative stabilities of the corresponding transition states, for example

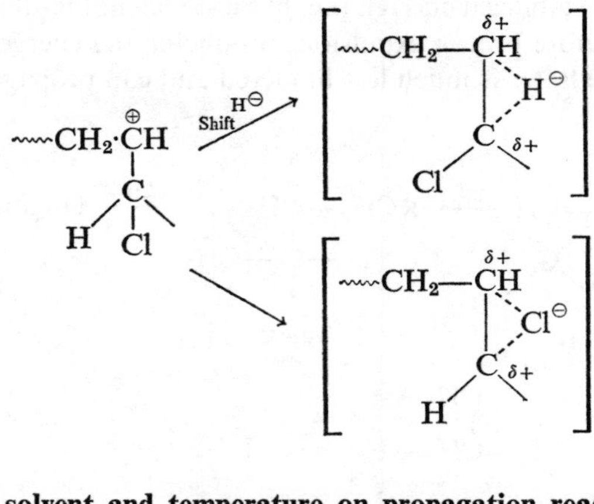

9.4.4 Effect of solvent and temperature on propagation reactions

The effect of both solvent and temperature on propagation reactions have long been misinterpreted largely, as with initiation phenomena, because of the failure to isolate the process of propagation from the other elementary reactions which comprise polymerization. In fact, relevant information on propagation is so limited that a discussion of these effects must be largely speculative.

The effect of temperature on some special systems has already been mentioned, for example propagation depropagation equilibria, isomerization propagations etc., but the majority of vinyl monomers do not display these special properties and change in temperature can be regarded as having two concurrent effects. The first is an alteration in the proportions of free ion and ion pair species present, and the second is the change in rate at which each of these react. Considering the latter point, a simple Arrhenius dependence would be expected for the rate constant of propagation of free ions and ion pairs, and in the case of free ions this has been demonstrated.[22,62] Enthalpies of activation estimated in general are low ($\sim$6 kcal mol^{-1} for isobutylvinylether[22] and N-vinylcarbazole[62] and 2 kcal mol^{-1} for styrene[80] and tert-butylvinylether.[123] Corresponding data for ion pair propagations are not yet available but temperature effects will be more complex because of the competing ion pair dissociation equilibria. Lowering the temperature would be expected to increase the proportion of free ions because decrease of temperature increases the dielectric constant of the solvent sufficiently to overcome the simpler thermal considerations of the equilibrium. (Almost invariably ion pair dissociation constants are found to increase with temperature decrease.) An increase in the proportion of the more reactive species present, therefore, produces an increase in the value of the composite rate

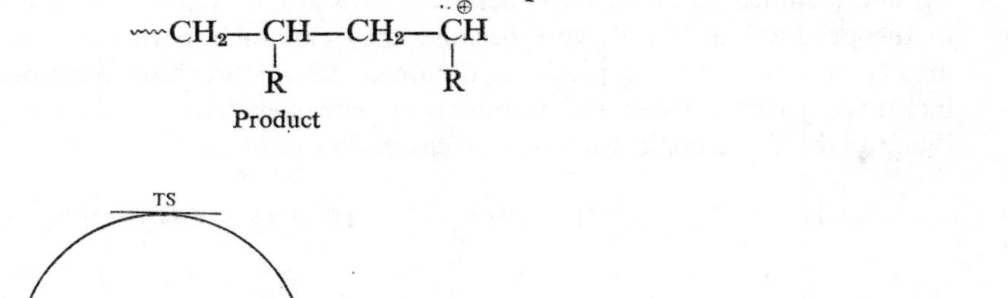

Reactants

Transition state (TS)

Product

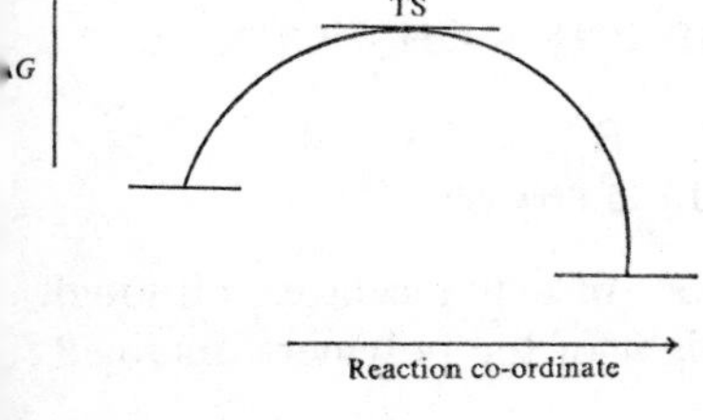

ΔG

TS

Reaction co-ordinate

constant. Apparently negative activation enthalpies may be anticipated and have been observed in the field of anionic polymerization.[101]

Clearly, the effect of solvent and temperature are closely related. When a more polar solvent is used for systems in which both ion pairs and free ions are present then the effect will be similar to that of reducing the temperature. If the polarity change is dramatic then a correspondingly large change in the composite propagation constant will be anticipated.

Overall the free energy change is negative by virtue of the loss of unsaturation in going from a $\ce{C=C}$ double bond to a $\ce{-C-C-}$ single bond. Analysis of the diagram shows that a charge dispersal process is required for conversion of reactants to transition state. In the absence of specific solvation phenomena, the latter will therefore be destabilized relative to the ground state by solvents of high dielectric constant. Increasing the ionic solvation power of the polymerization medium will therefore increase the activation energy for propagation and decrease the rate constant for the reaction as observed experimentally for many small molecule systems. Such a correlation may explain the apparent difference in free ion propagation rate constants for isobutylvinylether,[22,80] where a change from bulk monomer to methylene chloride solvent (dielectric constant $4 \rightarrow 10$) causes an apparent reduction in $k_p^{\oplus}$ of $3 \times 10^5 \ M^{-1}s^{-1}$ to $2 \times 10^4 \ M^{-1}s^{-1}$ at 30 °C.

9.5 CHAIN BREAKING REACTIONS

A chain breaking reaction is one which interrupts the repetitive action of propagation, exemplified for the extreme case, termination

$$\ce{\sim CH2-\overset{\oplus}{C}HX^{\ominus}} \underset{R}{\overset{|}{}} \longrightarrow \text{Dead polymer}$$

From a technical point of view these reactions are an inherent disadvantage in the production of polymers because they can limit both the yield and molecular weight of the products obtained. Chain breaking reactions fall into two types: transfer and termination processes. The former terminate the growth of one chain with concurrent initiation of another.

$$\ce{\sim CH2-\overset{\oplus}{C}HX^{\ominus}} + \ce{CH2=CH} \longrightarrow \ce{\sim CH=CH} + \ce{CH3-\overset{\oplus}{C}HX^{\ominus}}$$

Dead Polymer

There is, therefore, no reduction in the number of active centres, although such reactions limit the overall molecular weight of the polymers obtained.

On the other hand, a termination reaction stops the growth of a propagating chain and removes or stabilizes the source of activity so that the re-initiation process cannot take place.

$$\sim CH_2\overset{\oplus}{-}\underset{\underset{R}{|}}{CHX}^{\ominus} + B^{\ominus} \longrightarrow CH_2\underset{\underset{R}{|}}{-}CHB + X^{\ominus}$$

The number of active centres is therefore reduced and the molecular weight and yield of polymer obtained also limited.

A clear distinction between the two types of reaction is not always apparent. Transfer can be visualized as a chain cutting reaction

～～～～～～～～～～～～～～～～～～～～～～～～～ (a) No transfer

～～～～～ ～～～～～ ～～～～～ ～～～～～ (b) Transfer operative

When transfer becomes excessive, however, no high molecular weight units are formed, products with mere oligomeric dimensions being obtained. In these circumstances the transfer reaction becomes, in effect, a termination reaction with respect to formation of high polymers. In addition a situation can arise where a dead polymer is formed from an active centre with concurrent generation of some reactive species whose reactivity differs somewhat from the reactivity of the original catalytic site. If the difference in reactivity is large, then effective termination is achieved; however, if the difference is quite small, then the reaction can be regarded as a transfer process. A proper distinction between the two is therefore governed by kinetic factors.

Both types of reaction may be spontaneous and an inherent part of particular systems, especially for monomer transfer reactions. The ease with which a propagating vinyl monomer transfers its activity to a new monomeric unit in cationic systems often parallels the ease of the propagation reaction itself. No doubt this arises because of the rather similar energy considerations and orbital overlap involved in these processes.

In some types of polymerization, especially anionic polymerization (Chapter 11), both chain transfer and termination processes are absent, but cationic polymerizations tend to be dominated by transfer reactions except for the polymerization of tetrahydrofuran with PF_6^- counter-ions. As a general rule a reduction in the polymerization temperature reduces the rate of chain breaking reactions more than propagation processes, so that higher molecular weights and polymer yields are obtained at lower temperatures.

Quantitative data on chain breaking processes in cationic systems are sparse, and difficult to interpret anyway, because of possible contributions from impurities. The emphasis in this section will therefore be on mechanism, rather than reactivity, and some of the more important systems will be dealt with in detail.

9.5.1 Transfer reactions

There are a wide variety of possible transfer reactions which can occur in cationic systems. Some of them are general and apply to any monomer susceptible to attack from electrophilic reagents, while others are specific to certain monomers only. In all cases however, these reactions are only special examples, in the polymer field, of processes well documented for low molecular weight organic cations.

One of the simplest transfer reactions possible is the spontaneous regeneration of protonic catalyst species. The ejected proton is a powerful electrophile

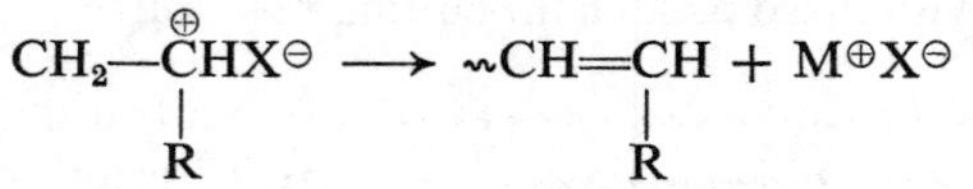

$$\underset{\underset{R}{|}}{CH_2{-}\overset{\oplus}{C}HX^{\ominus}} \longrightarrow \underset{\underset{R}{|}}{\sim CH{=}CH} + M^{\oplus}X^{\ominus}$$

and can readily add to another monomer molecule, initiating a new polymer chain; the dead polymer so formed possesses terminal unsaturation and is potentially active.

Alternatively, transfer can be a simple bimolecular process involving the growing chain and monomer, and indeed proton transfer to monomer is the most often assumed molecular weight controlling reaction in cationic

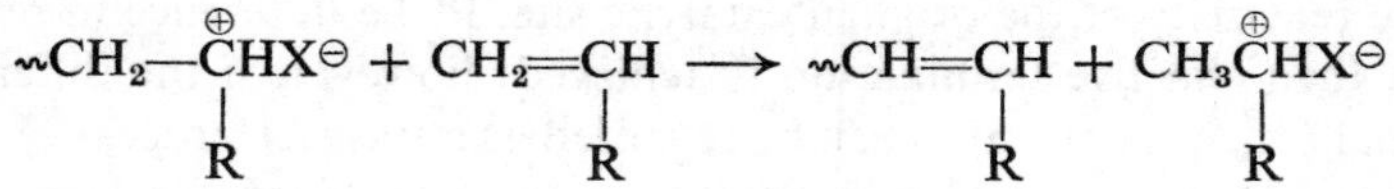

$$\underset{\underset{R}{|}}{\sim CH_2{-}\overset{\oplus}{C}HX^{\ominus}} + \underset{\underset{R}{|}}{CH_2{=}CH} \longrightarrow \underset{\underset{R}{|}}{\sim CH{=}CH} + \underset{\underset{R}{|}}{CH_3\overset{\oplus}{C}HX^{\ominus}}$$

polymerizations.[5,124,125] Frequently proton transfer to monomer is so facile that some specific and favourable mechanism would seem to be operative.

In theory proton transfer to some deliberately added reagent could constitute a transfer process; however, to produce such a reaction would require a material with a high basicity. Such compounds are more likely to produce complete termination rather than a transfer mechanism. (If $H^{\oplus}B$

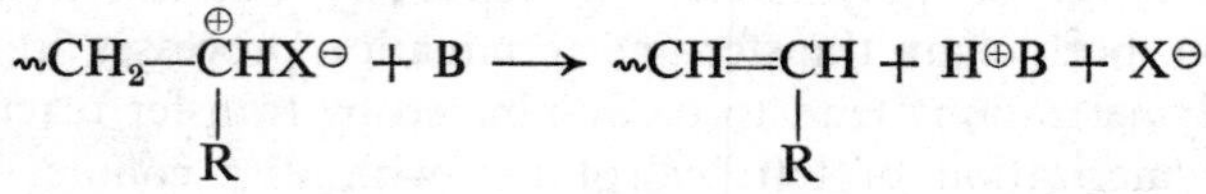

$$\underset{\underset{R}{|}}{\sim CH_2{-}\overset{\oplus}{C}HX^{\ominus}} + B \longrightarrow \underset{\underset{R}{|}}{\sim CH{=}CH} + H^{\oplus}B + X^{\ominus}$$

can re-initiate polymerization then the overall reaction is a transfer process; on the other hand, if $H^{+}B$ is a stable entity, effective termination is achieved.) Some oxygen compounds do appear to act as transfer agents, either by a substitutive transfer, for example

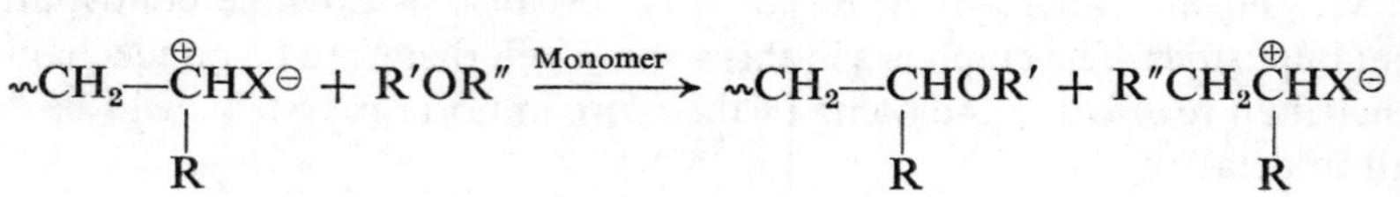

$$\underset{\underset{R}{|}}{\sim CH_2{-}\overset{\oplus}{C}HX^{\ominus}} + R'OR'' \xrightarrow{\text{Monomer}} \underset{\underset{R}{|}}{\sim CH_2{-}CHOR'} + \underset{\underset{R}{|}}{R''CH_2\overset{\oplus}{C}HX^{\ominus}}$$

or a proton exchange reaction proceeding via a secondary oxonium ion, for example

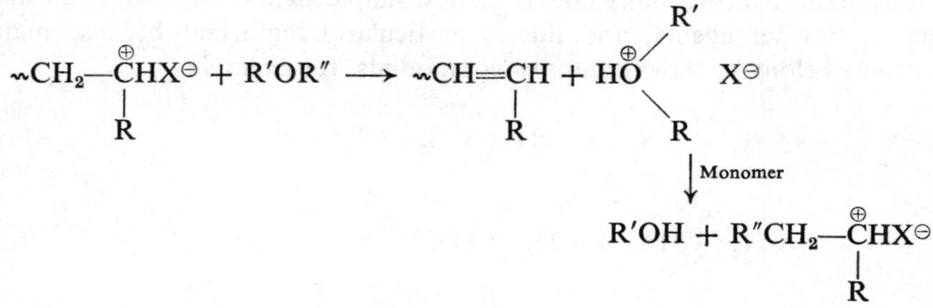

$$\text{\textasciitilde CH}_2\text{—}\overset{\oplus}{\text{CHX}}{}^{\ominus} + \text{R}'\text{OR}'' \longrightarrow \text{\textasciitilde CH}=\text{CH} + \text{H}\overset{\oplus}{\text{O}}\underset{\text{R}}{\overset{\text{R}'}{<}} \quad \text{X}^{\ominus}$$

$$\underset{\text{R}}{|} \qquad\qquad \underset{\text{R}}{|}$$

(Monomer)

$$\longrightarrow \text{R}'\text{OH} + \text{R}''\text{CH}_2\text{—}\overset{\oplus}{\text{CHX}}{}^{\ominus}$$

$$\underset{\text{R}}{|}$$

If growing or dead polymer is involved in transfer, an active centre is generated on the polymer backbone, leading to formation of a branch site, for example

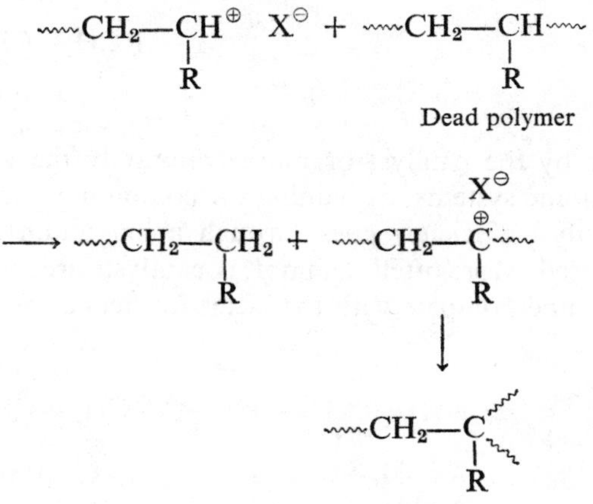

$$\text{\textasciitilde CH}_2\text{—}\overset{\oplus}{\text{CH}} \; \text{X}^{\ominus} + \text{\textasciitilde CH}_2\text{—}\text{CH}\text{\textasciitilde}$$

$$\underset{\text{R}}{|} \qquad\qquad \underset{\text{R}}{|}$$

Dead polymer

$$\longrightarrow \text{\textasciitilde CH}_2\text{—}\text{CH}_2 + \text{\textasciitilde CH}_2\text{—}\overset{\oplus}{\underset{\text{R}}{\text{C}}}\text{\textasciitilde}$$

$$\underset{\text{R}}{|} \qquad\qquad \text{X}^{\ominus}$$

$$\longrightarrow \text{\textasciitilde CH}_2\text{—}\underset{\text{R}}{\overset{\text{\textasciitilde}}{\text{C}}}$$

Branched polymer

A special case of this transfer process occurs when a growing polymer uses its own backbone as a transfer site in a so-called 'back biting' reaction. Usually only short side chains are produced by this reaction since 'back biting' takes place at conformationally controlled sites a few units back along the chain from the active centre, for example

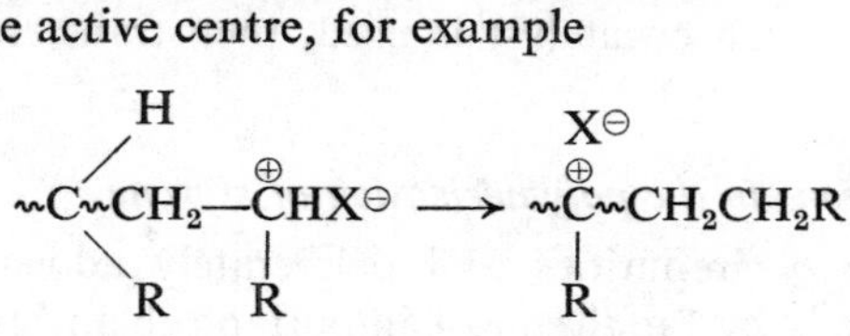

$$\underset{\text{R}\quad\text{R}}{\text{\textasciitilde}\overset{\text{H}}{\underset{}{\text{C}}}\text{\textasciitilde CH}_2\text{—}\overset{\oplus}{\text{CHX}}{}^{\ominus}} \longrightarrow \underset{\text{R}}{\text{\textasciitilde}\overset{\oplus}{\text{C}}\text{\textasciitilde CH}_2\text{CH}_2\text{R}} \quad \text{X}^{\ominus}$$

A rather similar reaction is thought to take place in the high temperature, high pressure, radical polymerization of ethylene, and is responsible for the production of the short chain branches, which reduce the melting point and crystallinity relative to linear high density polyethylene.

Aromatic hydrocarbons and relatively simple halide molecules can also act as transfer agents, and this is particularly important because many solvents belong to these groups of compounds, for example

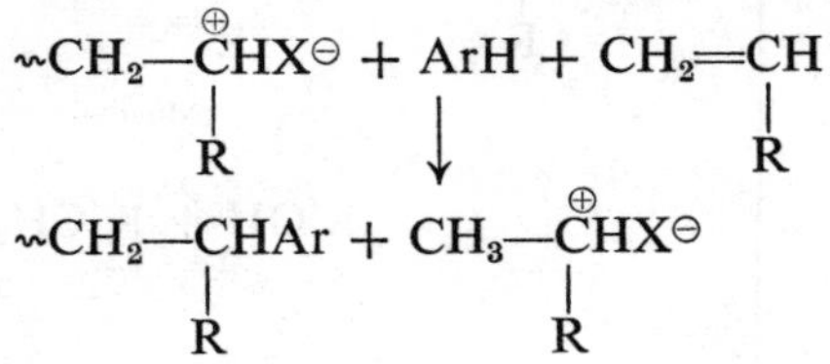

$$\sim CH_2\overset{\oplus}{-}\overset{\oplus}{C}HX^{\ominus} + ArH + CH_2{=}CH$$

$$\sim CH_2{-}CHAr + CH_3{-}\overset{\oplus}{C}HX^{\ominus}$$

and

$$\sim CH_2{-}\overset{\oplus}{C}HX^{\ominus} + R'Y \longrightarrow \sim CH_2{-}CHY + \overset{\oplus}{R'}X^{\ominus}$$

$$\qquad\qquad\qquad\text{(Halide)}\qquad\qquad\qquad\qquad\downarrow\text{ Monomer}$$

$$RCH_2{-}\overset{\oplus}{C}HX^{\ominus}$$

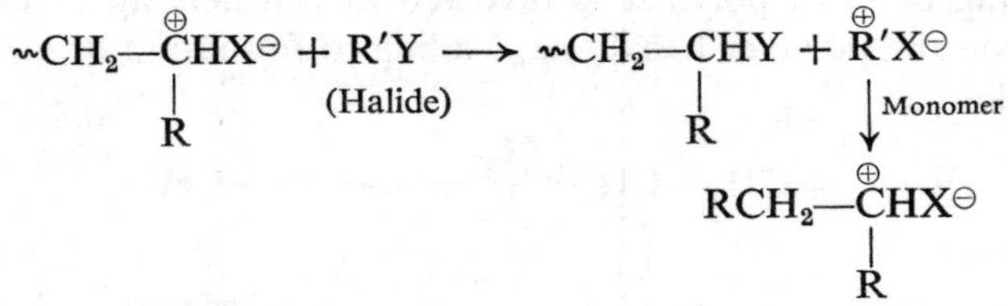

Finally transfer by the catalyst or more commonly the cocatalyst is very prevalent in cationic systems, explaining the common experimental observation that cocatalyst efficiency goes through a maximum as its ratio with catalyst is increased. More often than not, cocatalysts are more basic than the monomer itself, and compete with the latter for active centres, for example water

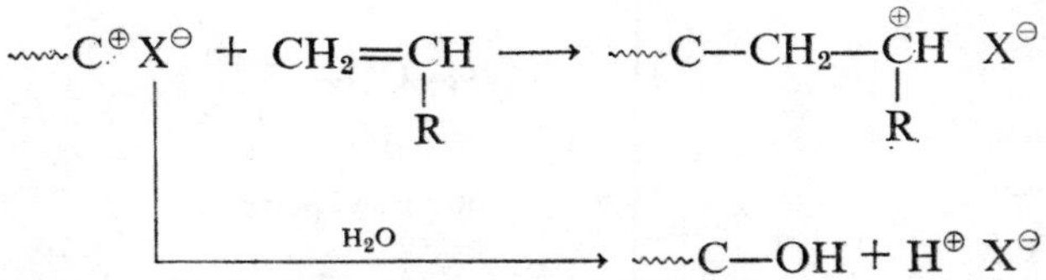

$$\sim\sim C^{\oplus}X^{\ominus} + CH_2{=}CH \longrightarrow \sim\sim C{-}CH_2{-}\overset{\oplus}{C}H \; X^{\ominus}$$

$$\overset{\text{H}_2\text{O}}{\longrightarrow} \sim\sim C{-}OH + H^{\oplus} X^{\ominus}$$

While the concentration of monomer overwhelms that of cocatalyst, the transfer effect of the latter will be small. When the proportions of unused cocatalyst become significant, for example towards the end of a polymerization or when too much cocatalyst is used, then transfer reactions become prevalent.

(i) *Transfer reactions in the polymerization of styrene*

In the absence of impurities and deliberately added reagents, simple monomer transfer is by far the most important chain breaking reaction in styrene polymerization. The rate of monomer transfer appears to vary somewhat with the catalyst used and the polarity of the medium.[126] This is difficult to explain and could arise merely from an incorrect assessment of the degree of transfer by this mechanism. On the other hand, if ion pairs rather than free ions, play a role in transfer, then some dependence on reaction medium would be expected. Spontaneous transfer has been demonstrated with sulphuric acid as the catalyst, and aromatic reagents and oxygen containing compounds are also known to provide effective transfer sites.[126]

Probably the most characteristic feature of styrene transfer reactions, however, is the occurrence of terminal indane skeletons in the dead polymer molecules produced. Intramolecular electrophilic aromatic substitution, by a 'back biting' mechanism, gives rise to an indane skeleton, with effective proton transfer to monomer

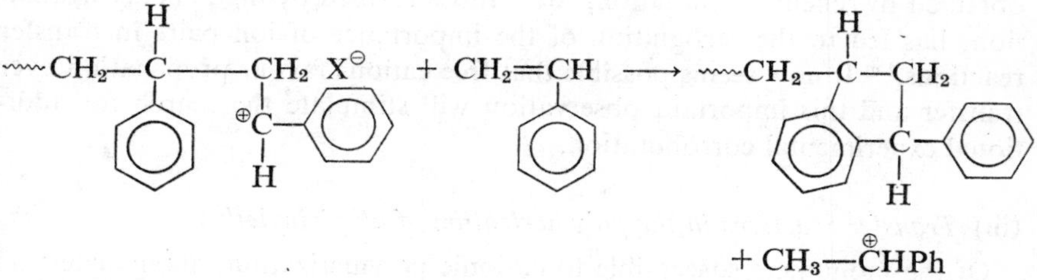

In fact, any transfer reaction for styrene can be envisaged as leading to either a terminal double bond in the polymer, or an indane residue; the structures are isomeric and almost certainly arise simultaneously.

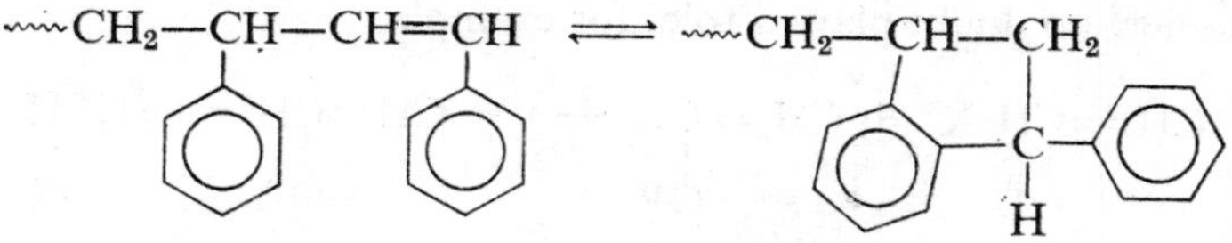

(ii) *Transfer reactions in isobutene polymerization*

The propagating cation in isobutene polymerization is a highly reactive, substituted tert-butyl cation and as such is highly prone to proton elimination. Low molecular weight products are usually formed, unless the temperature is drastically reduced.[127] The predominant terminal groups are olefinic linkages formed by proton elimination, or, e.g. hydroxyl groups formed by combination with co-catalyst (BF_3/H_2O system). When alkyl halides are used as the solvent then transfer, with incorporation of a halogen atom into the polymer may take place,[128] for example

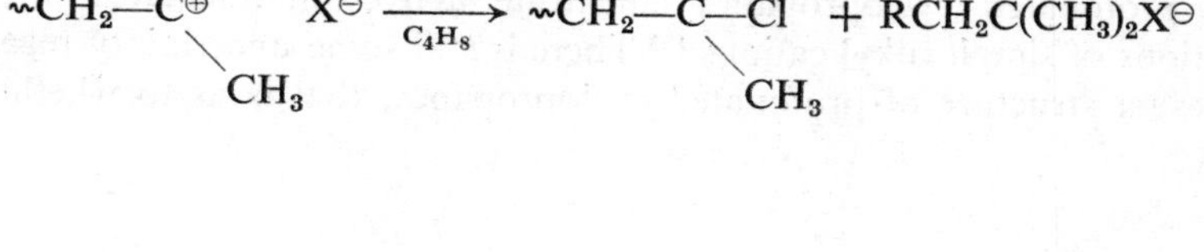

Generally speaking, low temperatures are employed in the polymerization of isobutene and this makes the system much less sensitive to side reactions. The most widely used catalysts are $AlCl_3$/alkyl halide or AlR_3/alkyl halide combinations. These are used commercially for the production of rubbers based on polyisobutene, and Kennedy[129] has reviewed the effect of added impurities. In some cases purely a transfer phenomenon is observed, but in most a poisoning (i.e. termination) effect is also operative. Olefins with α-branches prove to be strong molecular weight depressors but not terminators, with the transfer activity probably due to the fact that such olefins may be incorporated into the polymer chains to yield sterically hindered tertiary carbonium ions. Comparison of molecular weights for polyisobutenes obtained by 'chemical initiation,' with those formed by high energy irradiation, has led to the recognition of the importance of ion pairs in transfer reactions.[130] It now seems possible that free cations favour propagation over transfer and this important observation will stimulate the search for additional experimental corroboration.

(iii) *Transfer reactions in the polymerization of alkylvinylethers*

Of the monomers susceptible to cationic polymerization, alkylvinylethers probably show the greatest tendency to monomer transfer. Ledwith and his coworkers[22] have shown that the propagation reaction in the polymerization of isobutylvinylether is only 30–50 times faster than transfer, with both processes having rather similar activation energies. Direct proton transfer to monomer, no doubt plays a role, for example

$$\text{\sim\sim}CH_2{=}\underset{\underset{OR}{|\oplus}}{CH}\ X^{\ominus} + CH_2{=}\underset{\underset{OR}{|}}{CH} \longrightarrow \text{\sim\sim}CH{=}\underset{\underset{OR}{|}}{CH} + CH_3\underset{\underset{OR}{|\oplus}}{CH}\ X^{\ominus}$$

but the tremendous efficiency of monomer transfer in this case suggests that other forms of monomer transfer may be operative, and provides an opportunity for speculation.

In essence the proton transfer equilibrium shown above represents the relative basicities of a vinylalkylether monomer and the corresponding alkenylalkylether, formed as a terminal unit. Studies on the mechanism of acid catalysed hydrolysis of unsaturated ethers[131,132] indicate that propenylalkylethers may be very much more reactive (i.e. more basic) than corresponding vinylalkylethers, arguing against the proposed proton transfer, except at very high monomer concentrations. There are at least two alternative reactions which may explain the ease of monomer transfer.

Recent studies of carbonium ion behaviour have established beyond doubt that protonated cyclopropanes constitute active intermediates in many reactions of simple alkyl cations.[133] There is still some uncertainty regarding the exact structure of protonated cyclopropanes, that is as to whether the ring is 'face protonated', 'edge protonated' or 'corner protonated', but a primary alkyl cation may be represented as an equilibrium between classical and non-classical structures. For a simple alkane structure the order of

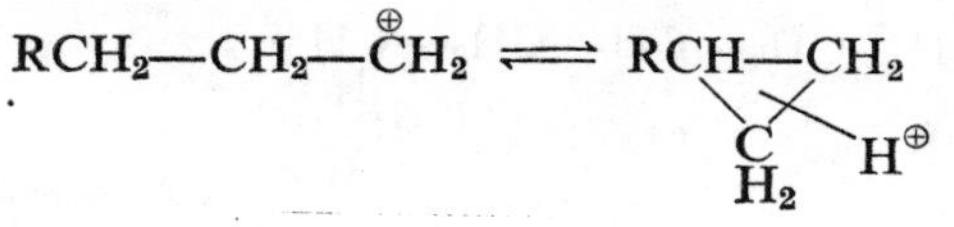

$$RCH_2-CH_2-\overset{\oplus}{C}H_2 \rightleftharpoons RCH-CH_2$$

carbonium ion stability falls in the sequence tert $>$ sec $>$ protonated cyclopropyl $>$ primary. However, for substituted derivatives, the free energy differences between protonated cyclopropyl and secondary or tertiary classical structures are not very great.[134] It is plausible therefore, to suggest that the propagating classical cation in vinylether polymerization is in equilibrium with a very small concentration of the nonclassical protonated cyclopropyl derivative, for example

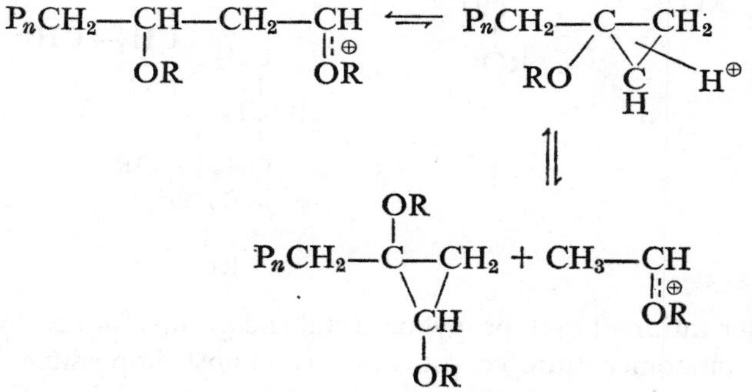

$$P_nCH_2-CH-CH_2-CH \rightleftharpoons P_nCH_2-C-CH_2$$

$$P_nCH_2-C-CH_2 + CH_3-CH$$

For the particular example considered, the presence of stabilizing alkoxy groups would overwhelmingly favour the open chain classical structure, but cyclopropanes are much less basic than similarly substituted olefins, providing a simple explanation for the observed facile monomer transfer reaction.

A second alternative monomer transfer process would be possible if the propagating carboxonium ion derived stabilization by neighbouring group effects of monomer segments in the same chain. The cyclic oxonium ion formed by intramolecular stabilization would have electrophilic reactivity at positions 1, 2 and 3, according to the nature of the substituents, but here again reactivity at position 1 would be heavily favoured by the alkoxy substituent. Reaction at position 1 leads to propagation and would be expected to dominate; reaction at position 2 constitutes a monomer transfer reaction in which an alkyl group has been transferred; and reaction at position 3 results in propagation with branching. A similar cyclic oxonium ion was proposed some years ago[135] to explain the stereoregular nature of vinylether polymerizations with heterogeneous catalysts.

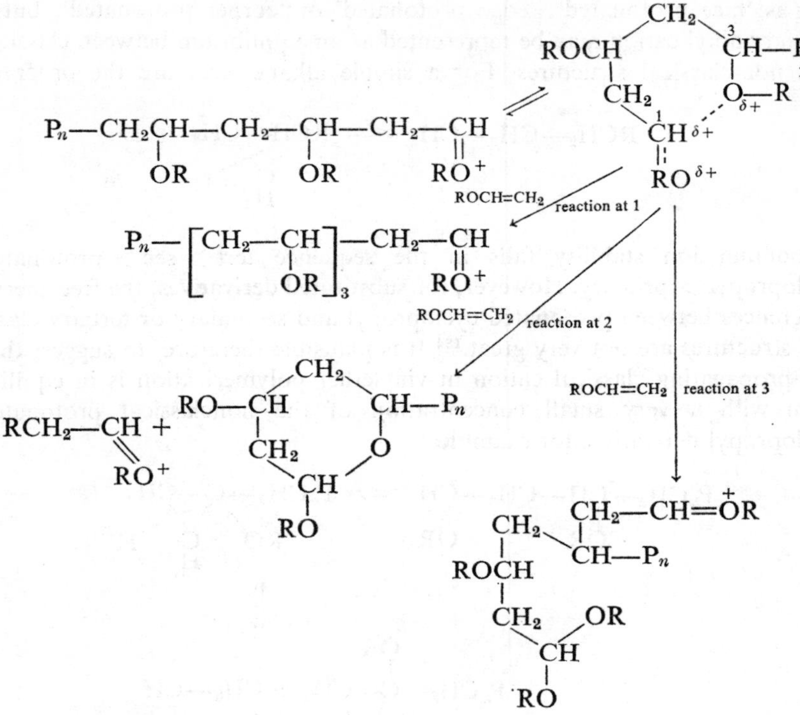

$$
P_n-CH_2CH-CH_2-CH-CH_2-CH \quad \rightleftharpoons \quad
\begin{array}{c}
ROCH \\ | \\ CH_2
\end{array}
\begin{array}{c}
\overset{3}{CH_2} \\ \overset{2}{CH}-P_n \\ O{-}R_{\delta+} \\ \overset{1}{CH}_{\delta+} \\ RO_{\delta+}
\end{array}
$$

with the lower chain bearing OR, OR, RO^+ substituents

$$ROCH{=}CH_2 \quad \text{reaction at 1}$$

$$
P_n-\left[CH_2-CH \atop OR \right]_3 -CH_2-CH \atop RO^+
$$

$$ROCH{=}CH_2 \quad \text{reaction at 2}$$

$$ROCH{=}CH_2 \quad \text{reaction at 3}$$

$$
RCH_2-CH \atop RO^+ \quad + \quad
\begin{array}{c}
CH_2 \\ RO{-}CH \quad CH{-}P_n \\ CH_2 \quad O \\ CH \\ RO
\end{array}
$$

$$
\begin{array}{c}
CH_2 \quad CH_2-CH{=}\overset{+}{O}R \\ ROCH \quad CH-P_n \\ CH_2 \quad OR \\ CH \\ RO
\end{array}
$$

Characterization of cyclopropyl or acetal end groups formed by any of the suggested monomer transfer processes is almost impossible because of anticipated reactivity towards electrophilic species remaining at the completion of polymerization. It is perhaps worth noting, however, that the semi-liquid nature of homogeneously polymerized polyalkylvinylethers may indicate a degree of branching, as well as low molecular weights, both of which would be expected following internal cyclic oxonium ion formation.

(iv) *Transfer reactions in the polymerization of N-vinylcarbazole*

The molecular weight data obtained on polymers of N-vinylcarbazole prepared using certain stable carbonium ion salts ($C_7H_7^+SbCl_6^-$ and $C_7H_7^+ClO_4^-$)[62] indicate that transfer reactions are considerably less efficient than in many other cationic systems. Interestingly, significantly lower molecular weight products are obtained from the perchlorate salt and the temperature dependence of molecular weight is also smaller. These results would appear to indicate a definite molecular weight dependence on counter-ion. As already noted such a phenomenon has been tentatively proposed for isobutene polymerization[141] and, if genuine, is of extreme importance. In the case of tropylium ion initiated polymerization of N-vinylcarbazole, propagation occurs largely by free ions,[62] but it could be that the small equilibrium concentration of ion pairs is largely responsible for the transfer reaction.

If the transfer reaction is simply a monomer transfer process involving the migration of a proton, then participation of certain counter-ions might lower the enthalpy of activation (and hence the rate) of the reaction, relative to the situation involving a free ion as indicated

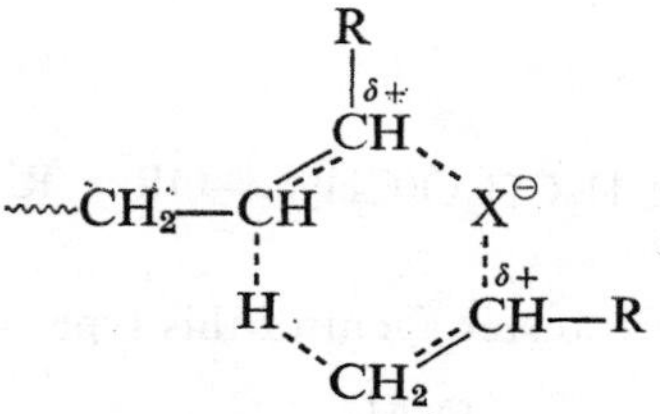

Related work by Tazuke[136] provides evidence for the effect of ion pairs containing perchlorate ion, in facilitating chain transfer by methanol, during the polymerization of NVC initiated by protonic acids. However, these systems are not so well characterized and it may be that the apparent effect of perchlorate ion arises from quite different phenomena. This is especially likely in the case of perchlorate ion because of its well-known hydrogen bonding characteristics towards protic agents. It is interesting to note that a preferential effect of certain counter-ions on proton transfer between carbanions has been noted in several systems,[137,138] and the structurally related hydride ion transfer between tetrahydrofuran and triphenylmethyl cation occurs more rapidly with $Ph_3C^+SbCl_6^-$ ion pairs than with the free cation.[61]

(v) Transfer reactions in ring-opening polymerizations

Polymerizations of cyclic monomers such as tetrahydrofuran (THF) exhibit many transfer reactions without the facility for simple proton transfer to monomer, although excess protic reagents do induce a transfer reaction, for example excess cocatalyst, HX, when used with BF_3. Similarly, protonic

acids such as HSO_3F give rise to excessive transfer reactions and only low molecular weight polymers result.

Acyclic ethers have been shown to be very efficient transfer agents in THF polymerization,[91] and also in the polymerization of trioxane.[139] The reaction in both cases amounts to an ether exchange process as shown below

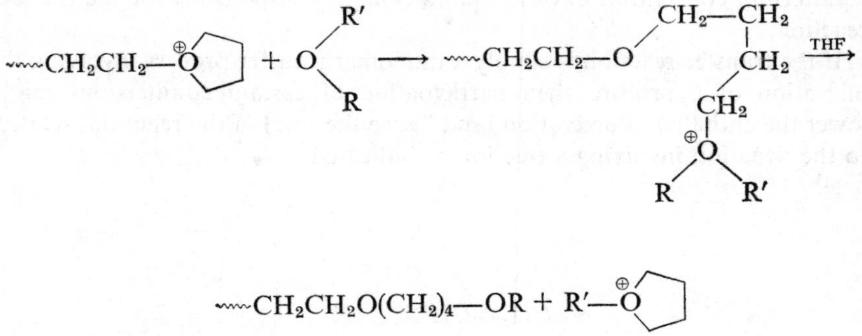

$$\sim\sim CH_2CH_2-\overset{\oplus}{O}\rceil + O\overset{R'}{\underset{R}{\diagup}} \longrightarrow \sim\sim CH_2CH_2-O\diagup\begin{matrix}CH_2-CH_2\\ |\\ CH_2\\ |\\ CH_2\\ |\\ \overset{\oplus}{O}\\ \diagup\diagdown\\ R \quad R'\end{matrix} \xrightarrow{\text{THF}}$$

$$\sim\sim CH_2CH_2O(CH_2)_4-OR + R'-\overset{\oplus}{O}\rceil$$

The most efficient chain transfer agent of this type is trimethylorthoformate.

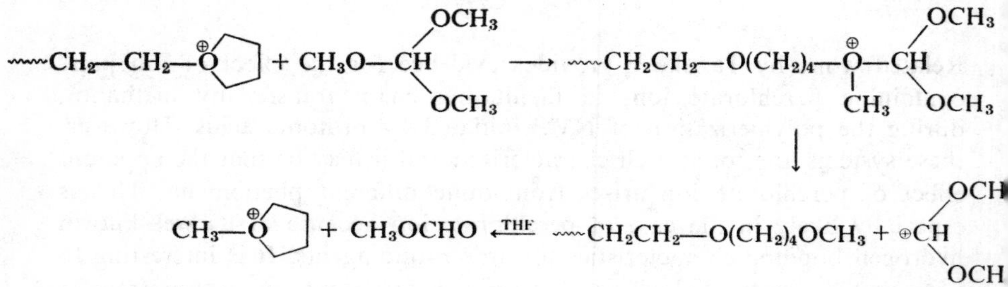

$$\sim\sim CH_2-CH_2-\overset{\oplus}{O}\rceil + CH_3O-\overset{OCH_3}{\underset{OCH_3}{\diagup}}CH \longrightarrow \sim\sim CH_2CH_2-O(CH_2)_4-\overset{\oplus}{\underset{CH_3}{O}}-\overset{OCH_3}{\underset{OCH_3}{\diagup}}CH$$

$$\downarrow$$

$$CH_3-\overset{\oplus}{O}\rceil + CH_3OCHO \xleftarrow{\text{THF}} \sim\sim CH_2CH_2-O(CH_2)_4OCH_3 + \oplus\overset{OCH}{\underset{OCH}{\diagup}}CH$$

The reaction can be used to control the equilibrium molecular weight of a given polymerizing system, and the polymers obtained have predominantly methyl end groups.

9.5.2 Termination reactions

In very pure systems true termination reactions are scarce in cationic polymerization, however, because of the high reactivity of the intermediates involved, only extremely low levels of impurities are required for a visible effect.

Probably the largest source of spontaneous termination in cationic systems is reaction of the active centre with all or part of the counter-ion. In the polymerization of styrene initiated by perchloric acid, some covalent per-chlorate species are formed, and these either propagate further with great

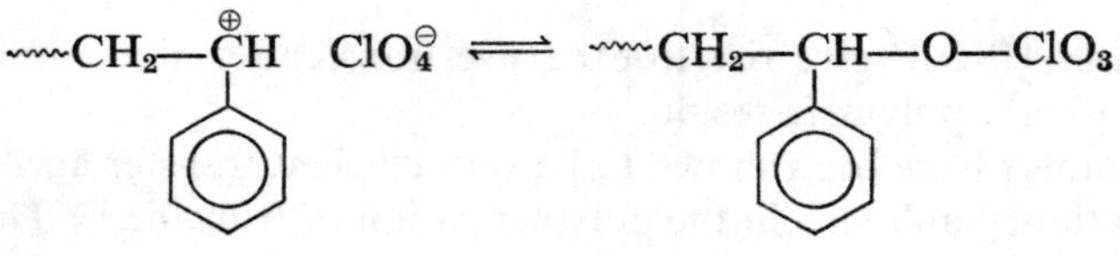

$$\sim\sim CH_2-\overset{\oplus}{CH} \quad ClO_4^{\ominus} \rightleftharpoons \sim\sim CH_2-CH-O-ClO_3$$

sluggishness or are completely dormant.[8,30,31] Perchlorate anions in the polymerization of N-vinylcarbazole do not appear to give rise to a termination reaction.[62] Tetrachloroaluminium ($AlCl_4^-$) and hexachloroantimonate ($SbCl_6^-$) ions on the other hand do provide a source of termination process by incipient dissociation to chloride ions. The $AlCl_4^-$ ion is used widely as the counter-ion in the polymerization of isobutene and chloride incorporation into the polymer appears to be the only spontaneous termination possible.[140]

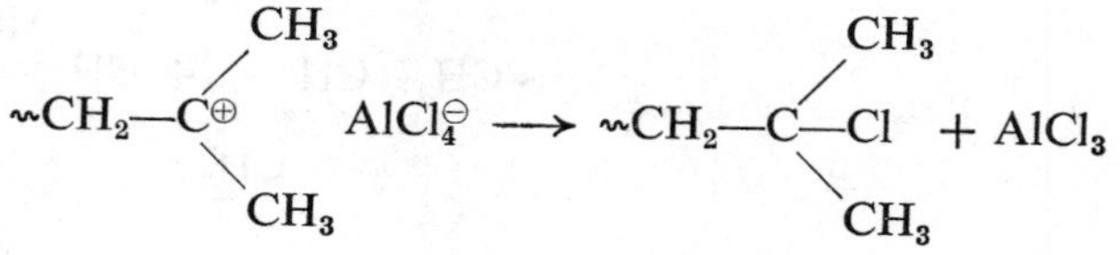

$$\sim CH_2-\overset{\displaystyle CH_3}{\underset{\displaystyle CH_3}{\overset{\oplus}{C}}} \quad AlCl_4^{\ominus} \longrightarrow \sim CH_2-\overset{\displaystyle CH_3}{\underset{\displaystyle CH_3}{C}}-Cl \ + AlCl_3$$

A similar reaction is envisaged in the polymerization of vinylethers by hexachloroantimonate salts.[22] Antimony pentachloride is itself a good

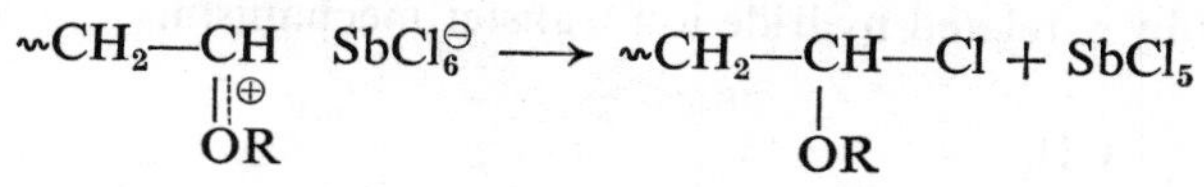

$$\sim CH_2-\underset{\displaystyle OR}{\overset{\displaystyle\oplus}{CH}} \quad SbCl_6^{\ominus} \longrightarrow \sim CH_2-\underset{\displaystyle OR}{CH}-Cl + SbCl_5$$

initiator for the polymerization of alkylvinylethers,[125] however, in this reaction there is a net loss of $SbCl_5$ species consequent upon generation of a catalyst site. Thus for every two active centres lost, only one is regenerated—

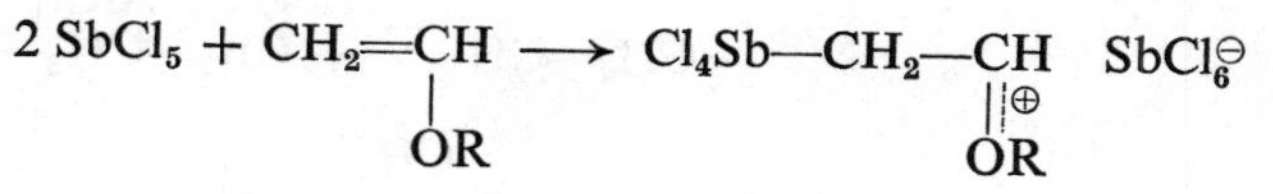

$$2\ SbCl_5 + CH_2=\underset{\displaystyle OR}{CH} \longrightarrow Cl_4Sb-CH_2-\underset{\displaystyle OR}{\overset{\displaystyle\oplus}{CH}} \quad SbCl_6^{\ominus}$$

hence this is a termination reaction. Further evidence regarding this complex process is afforded by the isolation of polymers which have apparently undergone an elimination reaction on the backbone in the presence of $SbCl_5$.[22]

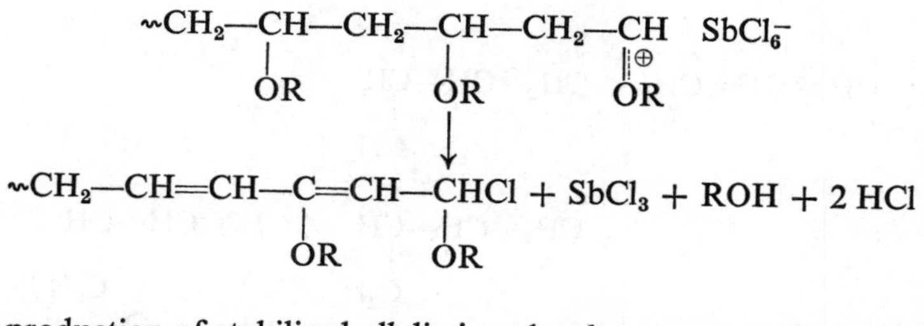

$$\sim CH_2-\underset{\displaystyle OR}{CH}-CH_2-\underset{\displaystyle OR}{CH}-CH_2-\underset{\displaystyle OR}{\overset{\displaystyle\oplus}{CH}} \quad SbCl_6^-$$

$$\downarrow$$

$$\sim CH_2-CH=CH-\underset{\displaystyle OR}{C}=CH-\underset{\displaystyle OR}{CHCl} + SbCl_3 + ROH + 2\ HCl$$

The production of stabilized allylic ions has been suggested as a termination reaction in the polymerization of propylene by the catalyst complex $AlCl_3/HBr$ and a similar hydride ion transfer was thought to be a possible termination reaction in the polymerization of isobutene. However, extended

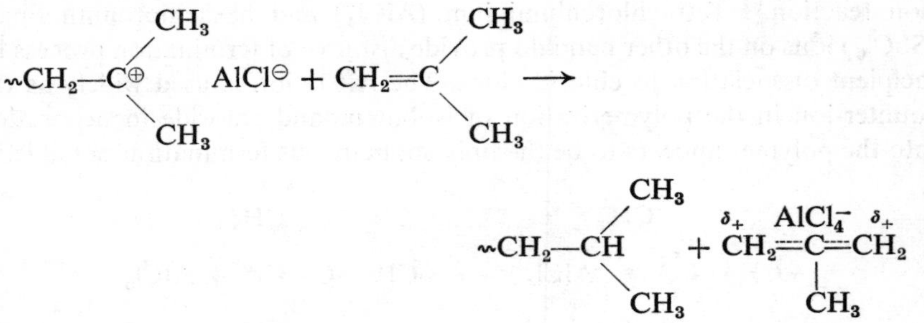

$$\text{~CH}_2\text{—}\overset{\text{CH}_3}{\underset{\text{CH}_3}{\overset{|}{\underset{|}{C}}}}{}^{\oplus} \quad \text{AlCl}^{\ominus} + \text{CH}_2\text{=}\overset{\text{CH}_3}{\underset{\text{CH}_3}{\overset{|}{\underset{|}{C}}}} \longrightarrow$$

$$\text{~CH}_2\text{—}\overset{\text{CH}_3}{\underset{\text{CH}_3}{\overset{|}{\underset{|}{C}}}}\text{H} \quad + \overset{\delta_+}{\text{CH}_2}\text{=}\overset{\text{AlCl}_4^-}{\overset{}{C}}\text{=}\overset{\delta_+}{\text{CH}_2}$$

$$\underset{\text{CH}_3}{\overset{|}{}}$$

studies by Kennedy[141] have shown that this reaction is probably not very important in homopolymerization but addition of simple n-alkenes (e.g. propene, but-1-ene, pent-1-ene and hex-1-ene) does produce termination, presumably by a related hydride ion transfer mechanism.

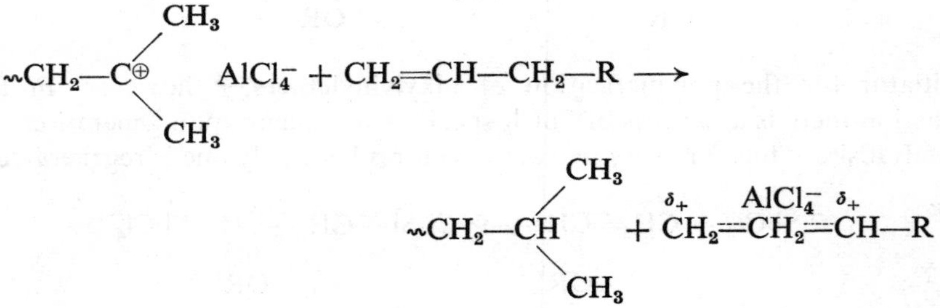

$$\text{~CH}_2\text{—}\overset{\text{CH}_3}{\underset{\text{CH}_3}{\overset{|}{\underset{|}{C}}}}{}^{\oplus} \quad \text{AlCl}_4^- + \text{CH}_2\text{=CH—CH}_2\text{—R} \longrightarrow$$

$$\text{~CH}_2\text{—}\overset{\text{CH}_3}{\underset{\text{CH}_3}{\overset{|}{\underset{|}{C}}}}\text{H} \quad + \overset{\delta_+}{\text{CH}_2}\text{=}\overset{\text{AlCl}_4^-}{\text{CH}_2}\text{=}\overset{\delta_+}{\text{CH}}\text{—R}$$

Increasing termination efficiency with increasing molecular weights from propylene to hex-1-ene might be due to the enhanced electron donating character of the larger alkyl groups, which in turn will augment the stability of the allylic ions:

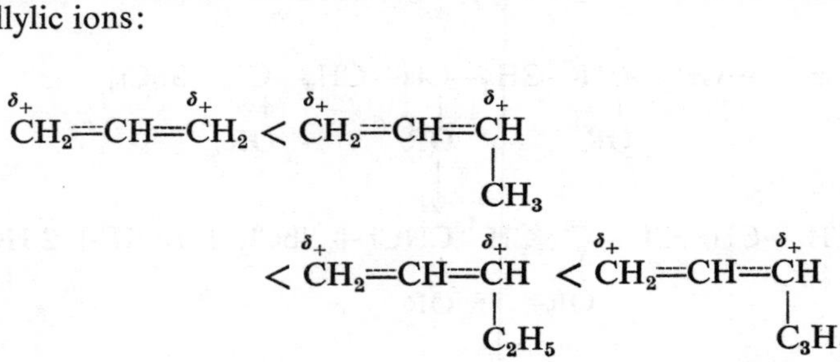

$$\overset{\delta_+}{\text{CH}_2}\text{=}\overset{\delta_+}{\text{CH}}\text{=}\overset{\delta_+}{\text{CH}_2} < \overset{\delta_+}{\text{CH}_2}\text{=}\overset{\delta_+}{\text{CH}}\text{=}\overset{\delta_+}{\underset{\underset{\text{CH}_3}{|}}{\text{CH}}}$$

$$< \overset{\delta_+}{\text{CH}_2}\text{=}\overset{\delta_+}{\text{CH}}\text{=}\overset{\delta_+}{\underset{\underset{\text{C}_2\text{H}_5}{|}}{\text{CH}}} < \overset{\delta_+}{\text{CH}_2}\text{=}\overset{\delta_+}{\text{CH}}\text{=}\overset{\delta_+}{\underset{\underset{\text{C}_3\text{H}_7}{|}}{\text{CH}}}$$

Conjugated dienes are found to have a similar effect and here again this is probably due to the formation of an allylic cation, though in this case the latter is incorporated on the end of a growing chain. This macrocation may

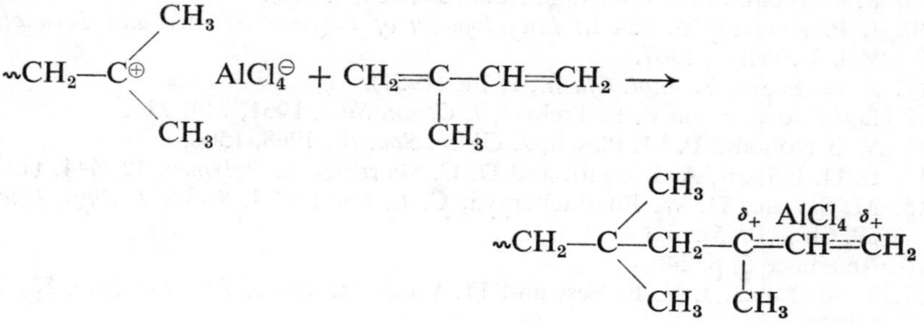

$$\text{wCH}_2\text{—}\overset{\displaystyle \overset{CH_3}{\diagup}}{\underset{\diagdown}{C^{\oplus}}}_{CH_3} \quad AlCl_4^{\ominus} + CH_2\text{=}\overset{\displaystyle |}{\underset{CH_3}{C}}\text{—CH=CH}_2 \longrightarrow$$

$$\text{wCH}_2\text{—}\overset{\displaystyle \overset{CH_3}{\diagup}}{\underset{\diagdown}{C}}_{CH_3}\text{—CH}_2\text{—}\overset{\displaystyle |}{\underset{CH_3}{C}}\text{=}\overset{\delta_+ \ AlCl_4 \ \delta_+}{CH\text{=}CH_2}$$

further stabilize itself by proton ejection in which case an overall transfer phenomenon results. Such terminal conjugation has been found experi-

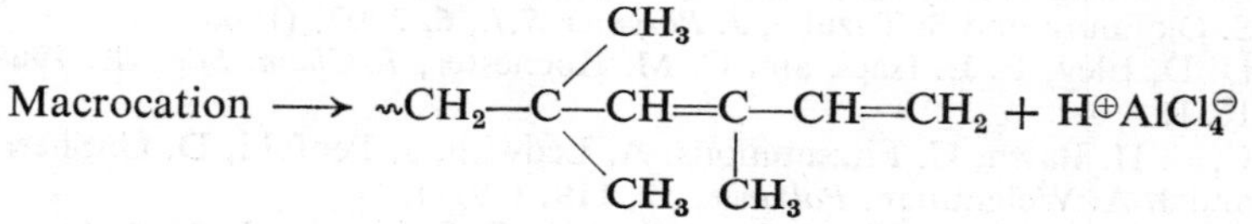

$$\text{Macrocation} \longrightarrow \text{wCH}_2\text{—}\overset{\displaystyle \overset{CH_3}{\diagup}}{\underset{\diagdown}{C}}_{CH_3}\text{—CH=}\overset{\displaystyle |}{\underset{CH_3}{C}}\text{—CH=CH}_2 + H^{\oplus}AlCl_4^{\ominus}$$

mentally[142] and these reactions with conjugated dienes are of technical importance, because incorporation of diene units into polyisobutene chains is a crucial step in the synthesis of butyl rubbers.

The polymerization of aromatic monomers by cation intermediates provides the possibility of the involvement of delocalized nuclei in termination reactions. It has already been pointed out that intramolecular attack of the growing styryl cation on one of the aromatic groups on its own backbone can give rise to transfer effects via formation of indane units. These appear to participate in hydride transfer equilibria with the growing cations to give stable indanyl cations and hence overall termination.

9.6 REFERENCES

1. *The Chemistry of Cationic Polymerization* ed. P. M. Plesch, Pergamon Press, New York, 1963.
2. D. C. Pepper, p. 1314. in *Friedel-Crafts and Related Reactions* ed. G. A. Olah, Wiley (Interscience), New York, 1964.
3. D. C. Pepper, *European Polymer J.*, **1**, 11, (1965).
4. J. P. Kennedy and A. W. Langer, *Fortschr. Hochpolymer. Forsch.*, **3**, 508, (1965).
5. Z. Zlamal, p. 231 in *Kinetics and Mechanism of Polymerization, Vinyl Polymerization* Vol. I, pt. II ed. G. E. Ham, Marcel Dekker, New York, 1969.
6. *Kinetics and Mechanism of Polymerization* Vol. 2., Ring-Opening Polymerization, ed. K. C. Frisch and S. L. Reegen, Marcel Dekker, New York, 1969.
7. A. Ledwith and C. Fitzsimmonds, p. 377 in *Polymer Chemistry of Synthetic Elastomers* Pt. 1, ed. J. P. Kennedy and E. Tornquist, Wiley (Interscience), New York, 1968.

8. P. M. Plesch, *Fortschr. Hochpolymer. Forsch.*, **8**, 137, (1971).
9. F. C. Whitmore, *Chem. Eng. News*, **26**, 688, (1948).
10. J. P. Kennedy p. 754 in *Encyclopedia of Polymer Science and Technology* Vol. 7., Wiley, 1967.
11. A. G. Evans, *J. Appl. Chem.*, **1**, 24, (1951).
12. D. O. Jordan and F. E. Treloar, *J. Chem. Soc.*, **1961**, 729, 734.
13. V. Bertoli and P. M. Plesch, *J. Chem. Soc. B.*, **1968**, 1500.
14. E. D. Eckard, A. Ledwith and D. C. Sherrington, *Polymer*, **12**, 444, (1971).
15. M. Szwarc, D. W. Bhattacharyya, C. L. Lee and J. Smid, *J. Phys. Chem.*, **69**, 612, (1965).
16. Reference 5. p. 240.
17. T. M. Bates, J. V. F. Best and Ff. Williams, *Trans. Faraday Soc.*, **58**, 192, (1962).
18. T. Saegusa, S. Matsumoto and Y. Hashimoto. *Polymer J. (Japan)*, **1**, 31, (1970).
19. L. P. Ellinger, *Polymer*, **5**, 559 (1964); **6**, 549, (1965).
20. S. Okamura and S. Tazuke, *J. Polymer Sci.*, **6**, 2907, (1968).
21. D. D. Eley, F. L. Isack and C. M. Rochester, *J. Chem. Soc. A.*, **1968**, 872, 1651.
22. C. E. H. Bawn, C. Fitzsimmons, A. Ledwith, J. Penfold, D. C. Sherrington and J. A. Weightman. *Polymer*, **12**, 119, (1971).
23. S. Winstein, P. E. Klinedinst Jr., and G. C. Robinson, *J. Amer. Chem. Soc.*, **83**, 885, (1961).
24. J. T. Dennison and J. B. Ramsey, *J. Amer. Chem. Soc.*, **77**, 2615, (1955).
25. R. M. Fuoss and F. Accascina *Electrolyte Conductance* Interscience, New York 1959.
26. D. C. Pepper *et al.*, *Proc. Roy. Soc. A.*, **263**, 58, 63, 75, 82, (1961).
27. K. Ikeda, T. Higashimura and S. Okamura, *Chem. High Polymers (Japan)*, **26**, 364, (1969).
28. D. C. Pepper and P. J. Reilly, *J. Polymer Sci.*, **58**, 639, (1962).
29. D. C. Pepper and P. J. Reilly, *Proc. Roy. Soc.*, **291**, 41, (1966).
30. L. E. Darcy, W. P. Millrine and D. C. Pepper, *Chem. Comm.*, **1968**, 1441.
31. B. MacCarthy, W. P. Millrine and D. C. Pepper, *Chem. Comm.*, **1968**, 1442.
32. A. Gandini and P. M. Plesch, *Proc. Chem. Soc.*, **1964**, 240.
33. A. Gandini and P. M. Plesch, *J. Polymer Sci. B.*, **3**, 127, (1965).
34. A. Gandini and P. M. Plesch, *J. Chem. Soc.*, **1965**, 4826.
35. A. Gandini and P. M. Plesch, *Europ. Polymer J.*, **1968**, 4, 55.
36. J. P. Kennedy, *J. Macromol. Sci. Chem. A3(5)*. **1969**, 861.
37. F. Whitmore, *J. Amer. Chem. Soc.*, **54**, 3274, (1932).
38. Reference 5, p. 252.
39. M. F. Shostakovsky and M. F. Burmistrova, *J. Appl. Chem. Russ.*, **15**, 249, (1942).
40. S. Okamura, T. Higashimura and T. Watanabe, *Die Makromol Chemie*, **50**, 137, (1961).
41. J. Lal and J. E. McGrath, *J. Polymer Sci. A-2.*, **1964**, 3369.
42. W. H. Hunter and R. V. Yohe, *J. Amer. Chem. Soc.*, **55**, 1248, (1933).
43. H. Williams, *J. Chem. Soc.*, **1940**, 775.
44. V. V. Korshak and N. N. Lebedev, *Dokl. Akad. Nauk. SSSR.*, **57**, 263, (1947).
45. M. Chmelir, M. Marek and O. Wichterle, *I.U.P.A.C. Symposium on Macromolecular Chem., Prague. 1965*, preprints p. 110.
46. Reference 1 p. 383.

47. R. Nakane and T. Watanabe, *J. Chem. Soc. (Japan)*; **35**, 1747, (1962).
48. M. J. S. Dewar, *Bull. Soc. Chim. France*, **1951**, C71.
49. Reference 5 p. 260–264.
50. C. G. Overberger, R. J. Ehring and R. A. Marcus, *J. Amer. Chem. Soc.*, **80**, 2456, (1958).
51. Reference 5 p. 271.
52. J. P. Kennedy, *J. Macromol. Sci. Chem. A3(5)*, **1969**, 885.
53. S. Okamura, N. Kanoh and T. Higashimura, *Die Makromol. Chem.*, **47**, 19, (1961).
54. A. Ledwith and D. C. Sherrington, *Polymer*, **12**, 344, (1971).
55. D. Giusti and F. Andruzzi, *Symposium of Macromol. Chem., Prague, 1965*, Paper P 492.
56. J. Wislicenus, *Ann.*, **92**, 106, (1878).
57. D. D. Eley and A. W. Richards, *Trans. Faraday Soc.*, **45**, 425, (1949).
58. W. Chalmers, *Can. J. Res.*, **7**, 464, 472, (1936).
59. D. D. Eley and A. W. Richards, *Trans. Faraday Soc.*, **45**, 436, (1949).
60. A. Ledwith, *J. Appl. Chem.*, **17**, 344, (1967).
61. A. Ledwith, *A.C.S. Advan. in Chem. Series*, **1969**, No. 91, 317.
62. P. M. Bowyer, A. Ledwith and D. C. Sherrington, *Polymer*, **12**, 509, (1971).
63. C. E. H. Bawn, R. M. Bell, C. Fitzsimmonds and A. Ledwith, *Polymer*, **6**, 661, (1965).
64. G. A. Olah, H. W. Guinn and S. J. Kuhn, *J. Amer. Chem. Soc.*, **82**, 426, (1960).
65. T. Kunitake, Y. Matsuguma and C. Aso, *Polymer J.*, **2**, 345, (1971).
66. Y. Matsuguma and T. Kunitake, *Polymer J.*, **2**, 353, (1971).
67. C. E. H. Bawn, R. A. Carruthers and A. Ledwith, *Chem. Comm.*, **1965**, 522.
68. D. N. Kursanov, M. E. Vol'pin and I. S. Akhren, *Dokl. Akad. Nauk. USSR.*, **120**, 531, (1958).
69. Ibid., **30**, 159, (1960).
70. A. Ledwith, *Accounts Chem. Res.*, **5**, 133, (1972).
71. C. E. H. Bawn, F. A. Bell and A. Ledwith, *Chem. Comm.*, **1968**, 599.
72. D. C. Sherrington, Ph. D. Thesis, Liverpool, 1969.
73. L. P. Ellinger, *Chem. and Ind.*, **51**, 1982, (1963).
74. L. P. Ellinger, *Polymer*, **6**, 549, (1965).
75. H. Scott, A. G. Miller and M. M. Labes, *Tetrahedron Lett.*, **17**, 1073, (1963).
76. T. Natsuume, Y. Shirota, M. Hirota, S. Kusabayashi and H. Mikawa, *Chem. Comm.*, **1969**, 289.
77. C. E. H. Bawn, A. Ledwith and M. Sambhi, *Polymer*, **12**, 209, (1971).
78. T. Nakamura, M. Soma, T. Onishi and T. Tamura, *Die Makromol. Chem.* **135**, 341, (1970).
79. W. H. Davison, S. H. Pinner and R. Worrall, *Chem. and Ind.*, **1957**, 1274.
80. M. D. Bonin, M. L. Calvert, W. L. Miller and Ff. Williams, *J. Polymer Sci. B-2*, **1964**, 143; K. Ueno, Ff. Williams, K. Hayashi and S. Okamura, *Trans., Faraday Soc.*, **63**, 1478, (1967).
81. Y. Tabata, p. 305 in *Kinetics and Mechanism of Polymerization* Vol. 1. Vinyl Polymerization pt. II. ed. G. E. Ham, Marcel Dekker, New York, 1969.
82. Y. Ishii and S. Sakai in reference 6, p. 13.
83. P. Dreyfuss and M. P. Dreyfuss in reference 6. p. 111.
84. H. Meerwein, E. Battenberg, H. Gold, E. Pfeil and G. Willfang, *J. Prakt. Chem.*, **134**, 83, (1939).
85. B. A. Rozenberg, E. B. Lyudvig, A. R. Gantmakhers, and S. S. Medvedev, *Vysokomoleckul. Soedin.*, **1965**, 7, 188.

86. D. Vofsi and A. V. Tobolsky, *J. Polymer Sci.*, *A-3*, **1965**, 3261.
87. W. Ziegenbeim and K. M. Hornung, German Patent 1 159 651, 1965.
88. H. Meerwein, D. Delfs and H. Morsal, *Ang. Chem.*,**72**, 927, (1960).
89. I. Kuntz, *J. Polymer Sci. B*-4, **1966**, 427.
90. M. P. Dreyfuss and P. Dreyfuss, *Polymer*, **6**, 93, (1965).
91. M. P. Dreyfuss and P. Dreyfuss, *J. Polymer Sci.*, *A-4*, **1966**, 2179.
92. J. I. G. Cadogan, R. M. Paton and C. Thomson, *Chem. Comm.*, **1970**, 1229; V. Habmann, C. Rüchardt and C. C. Tan, *Tetrahedron Letters*, **1971**, 3885.
93. Reference 7 p. 394.
94. F. B. Lyudvig, B. A. Rozenberg, T. M. Zvereva, A. R. Gantmakhers and S. S. Medvedev, *Vysokomolekul. Soedin.*, **7**, 269, (1965).
95. J. Holmes and R. Petit, *J. Org. Chem.*, **28**, 1695, (1963).
96. J. M. Beard, P. H. Plesch and P. P. Rutherford, *J. Chem. Soc.*, **1964**, 2566.
97. A. D. Ketley, *J. Polymer Sci.*, *B-2*, **1964**, 827.
98. A. G. Evans, *J. Appl. Chem.*, **1**, 240, (1951).
99. W. R. Longworth and P. H. Plesch, IUPAC Symposium, Wiesbaden, 1959, Preprints III-A-11.
100. N. Kanoh, A. Gotoh and T. Higashimura, *Die Makromol Chem.*, **63**, 115, (1963).
101. For a comprehensive review see M. Szwarc, *Carbanions, living polymers and electron transfer processes*, Interscience, New York., 1968.
102. S. Okamura, N. Kanoh and T. Higashimura, *Die Makromol. Chem.*, **47**, 35, (1961).
103. K. Ueno, K. Hayashi and S. Okamura, *J. Macromol. Sci. A2*, **1968**, 209.
104. J. Kohler and V. Stannett, *Polymer Reprints*, **12**, No. 1, 98, (1971).
105. T. Higashimura, p. 313 in *Structure and Mechanism in Vinyl Polymerization* eds T. Tsuruta and K. F. O'Driscoll, Dekker, New York, 1969.
106. C. Aso, T. Kunitake, V. Matsuguma, and Y. Imaizumi, *J. Polymer. Sci.*, *A-1*, **6**, 3049, (1968).
107. J. M. Sangster and D. J. Worsfold, *Polymer Preprints*, **13**, No. 1, 72, (1972).
108. J. P. Kennedy and T. Otsu, *Adv. Polymer Sci.*, **7**, 369, (1970).
109. J. P. Kennedy and A. W. Langer, *Adv. Polymer Sci.*, **3**, 508, (1964).
110. C. G. Overberger, S. Ishida and H. Ringdorf, *J. Polymer Sci.*, **62**, 51, (1962).
111. R. M. Michel, *J. Polymer Sci.*, *A-2*, **2**, 2533, (1964).
112. J. P. Kennedy and J. A. Hinlicky, *Polymer*, **6**, 133, (1965).
113. C. F. H. Tipper and D. A. Walker, *J. Chem. Soc.*, **1959**, 1352.
114. A. D. Ketley, *Polymer Letters*, **1**, 313, (1963).
115. H. Staudinger and E. Dreher, *Ann.*, **517**, 73, (1935).
116. J. P. Kennedy, L. S. Minckler Jr., G. C. Wanless and R. M. Thomas, *J. Polymer Sci. A2*, **1964**, 1441, 2093.
117. G. G. Wanless and J. P. Kennedy, *Polymer*, **6**, 111, (1965).
118. N. D. Prischepa, Y. Y. Goldfarb and B. A. Krentsel, *Vysokomolekul. Soedin*, **8**, 1658, (1966).
119. J. P. Kennedy, P. L. Magagnini, and P. H. Plesch, *J. Polymer Sci.*, *A-1*, **9**, 1635, 16, (1971).
120. C. E. Schildknecht, *Polymer Eng. Sci.*, **6**, 1, (1966).
121. J. P. Kennedy, J. J. Elliott and B. E. Hudson Jr., *Die Makromol. Chem.*, **79**, 109, (1964).
122. J. P. Kennedy, P. Borzel, W. Naegele and R. G. Squires, *Die Makromol. Chemie*, **93**, 191, (1966).
123. E. C. Lockett, Ph.D. Thesis, University of Liverpool, 1972.
124. P. H. Plesch, *Progress in High Polymers*, **2**, 137, (1968).

125. D. D. Eley in reference 1 p. 375.
126. Reference 1 p. 285–298.
127. F. S. Dainton and G. B. B. M. Sutherland, *J. Polymer Sci.*, **4**, 37, (1949).
128. J. P. Kennedy and R. M. Thomas, *J. Polymer Sci. A-1*, **1963**, 331.
129. J. P. Kennedy p. 291 in *Polymer Chemistry of Synthetic Elastomers* Pt. 1, ed. J. P. Kennedy and E. Tornquist, Wiley (Interscience), New York, 1968.
130. J. P. Kennedy, A. Shinkawa, and Ff. Williams, *J. Polym. Sci.*, *A-1*, **9**, 1551, (1971).
131. A. Ledwith and H. J. Woods, *J. Chem. Soc. B.*, **1966**, 753.
132. D. M. Jones and N. F. Wood, *J. Chem. Soc. B.*, **1964**, 5400.
133. C. C. Lee, *Prog. in Phys. Org. Chem.*, **7**, 129, (1970).
134. G. J. Karabatsos, N. Msi and S. J. Keyerson, *J. Amer. Chem. Soc.*, **92**, 621, (1970).
135. C. E. H. Bawn and A. Ledwith, *Quart. Rev.*, **16**, 362, (1962).
136. S. Tazuke, *Chem. Comm.*, **1970**, 1277.
137. T. E. Hogan-Esch and J. J. Smid, *J. Amer. Chem. Soc.*, **98**, 2764, (1967).
138. D. H. Hunter and Y. T. Lin, *J. Amer. Chem. Soc.*, **90**, 5921, (1968).
139. V. Jaacks, H. Baader and W. Kern, *Die Makromol. Chem.*, **83**, 56, (1965).
140. Reference 129, p. 309.
141. J. P. Kennedy and R. G. Squires, *J. Makromol. Sci.* (*Chem.*) *A1.*, **1967**, 805.
142. Z. Zlamal and A. Zazda, *J. Polymer Sci.*, *A-1*, **1963**, 3199.
143. G. A. Olah, M. B. Comisarow, E. Namanworth and B. Ramsey, *J. Amer. Chem. Soc.*, 1967, **89**, 5259.
144. G. A. Olah, R. D. Porter and D. P. Kelly. *J. Amer. Chem. Soc.*, 1971, **93**, 464.
145. W. Drijvers and E. J. Goethals. Paper presented at the IUPAC Symposium on Macromolecular Chemistry, Boston, July 1971, p. 663.

10

Carbanions

E. C. Dart

Imperial Chemical Industries Limited, Corporate Laboratory, Runcorn

10.1 INTRODUCTION

One of the major influences on structure reactivity relationships in carbanion chemistry has been the development of a pK_a scale for carbon acids. In addition to facilitating assessment of the ease with which specific protons may be abstracted from a given substrate, these give a quantitative picture of the extent to which structural and electronic effects in carbanions affect their stability. For such reasons a considerable portion of the following discussion is devoted to defining a pK_a scale and relating structural effects to pK_a.

Similarly, studies on the effect of solvent on pK_a have provided an insight into the factors controlling anion solvation. The work of Cram[3] has shown that hydrogen bonding solvents exert a significant stabilizing influence on anions, although in this respect Ritchie[1] argues also for the importance of solute-solvent dispersion interactions. In relatively non-polar aprotic solvents Smid[41] has shown that the anion is essentially unsolvated and yet the nature of its ion-pairing with cation and the solvation of the latter can influence reactivity to a remarkable degree.

However, reactivity deals with the kinetic rather than the thermodynamic approach and the necessity for defining the precise nature of the reactive intermediate and if possible the position of the transition state on the reaction co-ordinate cannot be over-stressed. Studies like that of Bordwell[2] and coworkers on the rates of base catalysed hydrogen isotope exchange of various nitrocycloalkanes demonstrate this admirably. He assembles evidence to show that the transition state for proton removal resembles un-ionized nitroalkane reactant more than its anionic intermediate. The result is a poor correlation between nitroalkane exchange rate and stability of the intermediate anion as measured by the pK_a of its conjugate acid.

Sections on the configurational stability of carbanions, the stereochemistry of their reactions, and the reactivity of delocalized anions were thought to be worthy of inclusion because of their particular relevance to diene polymerizations. From a practical standpoint a summary of the more general methods of carbanion preparation is included together with their more important reactions.

10.2 pK_a, AN INDEX OF CARBANION STABILITY

The stability of a carbanion R^- and to some extent its reactivity is related to the degree of dissociation of the carbon acid RH in the equilibrium:

$$RH \overset{K_a}{\rightleftharpoons} R^- + H^+ \tag{1}$$

In the gas phase the equilibrium constant K_a represents an absolute measure of acidity and is dependent only upon the concentrations of the various species involved.

$$K_a^G = \frac{[R^-][H^+]}{[RH]} \qquad pK_a^G = -\log_{10} K_a^G \tag{2}$$

But, although this serves as an important theoretical reference, acidities are more readily measureable in solution where solute-solute and solute-solvent interactions may modify K_a^G. The general solution acidity constant K_a^S is defined in terms of activities rather than concentrations, where a_X denotes the activity of a species X, and f_X its activity coefficient.

$$K_a^S = \frac{a_{R^-} \cdot a_{H^+}}{a_{RH}} = K_a^G \cdot \frac{f_{R^-} \cdot f_{R^+}}{f_{RH}} = K_a^G F^S \tag{3}$$

The factor F^S represents the extent to which the interaction of carbon acids and their conjugate bases with solvent may modify the gas phase acidity constant K_a^G. Thus variations in solvent polarity, in mutual solute-solvent polarizability, and in the ability of the solvent to act as a hydrogen bond donor or proton acceptor have been found to have a significant influence on the magnitude of K_a^S. Since one or more of these factors may dominate for a particular solute-solvent combination absolute pK_a values can vary in a relatively unpredictable manner from solvent to solvent.

Hydrogen bonding solvents generally have a stabilizing effect on carbanions. Thus on going from water to the dipolar aprotic solvent dimethyl sulphoxide (DMSO) acetylacetone, benzoylacetone, and nitromethane show increases in pK_a of 4·4, 2·5 and 5·7 respectively.[118] This effect has also been noted in comparative acidities in t-butanol and DMSO, the pK_a of t-BuOH in the latter is estimated to be 13 units higher than in t-BuOH making the t-butoxide ion a much less stable species but conversely a more powerful base in the aprotic solvent.[3]

In direct contrast the pK_a's of 9-carbomethoxyfluorene, 9-cyanofluorene, and fluoradene decrease on going from methanol to DMSO each by c. 6 pK_a units.[1] Here, stabilization by hydrogen bonding is overridden by what Ritchie believes to be a combination of two factors; the greater solvent basicity of DMSO, that is the greater stability of HDMSO⁺ over MeOH₂⁺ in the equilibria (4) and (5), and the ability of DMSO to stabilize the anions of these particular carbon acids by

$$RH + DMSO \rightleftarrows [HDMSO]^+ + R^- \tag{4}$$

$$RH + MeOH \rightleftarrows Me\overset{+}{O}H_2 + R^- \tag{5}$$

dispersion interactions. These arise from induced dipole-dipole forces between solute and solvent and are dependent upon their mutual polarizabilities.

In spite of this seemingly complex situation relative pK_a values between structurally related carbon acids do not vary dramatically with solvent. In the range benzene, cyclohexylamine (CHA) and DMSO the relative pK_a values of the hydrocarbon acids of Table 10.1 vary surprisingly little.

Table 10.1 pK_a Values of structurally related hydrocarbon acids measured in different solvents

Compound	Cyclohexylamine[8]	Benzene[7]	DMSO[9]
9-Phenylfluorene	18·5	21	18·6
Indene	20·0	21	18·2
Fluorene	23·1	20·5	22·1
p-Biphenylyldiphenyl-methane	30·2	31	25·3
Triphenylmethane	31·5	33	27·2
Diphenylmethane	33·1	35	28·6

Remembering that the values quoted are referred to water as the standard state (vide supra) those transferred from benzene and cyclohexylamine show good agreement. The DMSO values are somewhat more compressed but this may reflect the fact that in the latter solvent we are dealing with dissociated ions rather than ion pairs.[11]

For acids whose anions have differing solvation requirements, the ΔpK_a^S values are not constant. Picric acid, for instance, which is able to self disperse its negative charge in its anionic form by delocalization, is a stronger acid than HBr in DMF, while in protic type solvents like water the halogen acid is much the stronger.[4] Evidently, in H-bonding solvents, anions of the more localized variety, derive a great deal of their stabilization from hydrogen bonding interactions with solvent.

10.2.1   Measurement of pK_a

In order that chemical structure and carbanion reactivity can be correlated with pK_a it is desirable that the pK_a's of all carbon acids be referred to a common solvent. Water was chosen for this purpose presumably because of the vast body of dissociation data already accumulated in the medium. The majority of carbon acids, however, have little solubility in water and, of those that dissolve, the majority are sufficiently weak so as to be negligibly dissociated. These difficulties are to some extent overcome by measuring pK_a values in an appropriate solvent and then transferring them to water via some compound whose pK_a is measurable in both solvents. The success of this approach depends upon how closely this 'anchor' compound resembles the group of acids whose pK_a values are being transferred and current choices are perhaps still unsatisfactory in this respect.[5]

Various direct and indirect methods are available for pK_a determination but it is outside the scope of this work to discuss each of these in any degree of detail. Perhaps the most instructive for the non-specialist involve indirect methods using equilibrium or kinetic techniques.

Equilibrium methods

ΔpK_a values may be measured from an equilibrium such as that shown in equation (6).

$$RH + R_1^-, M^+ \rightleftarrows R^-, M^+ + R_1H \qquad (6)$$

The position of equilibrium can either be estimated optically or by carbonation providing that the rates of carbonation of the respective anions greatly exceed their equilibration rates. Thus given a suitable starting point such as a compound of sufficient acidity and solubility to be measured in aqueous solution a scale of pK_a^W values can be developed. The equilibrium approach was pioneered by Conant and Wheland[6] and was later expanded by McEwen.[7] More recently Streitwieser and coworkers[8] and Steiner and Gilbert[9] have extended McEwen's study in benzene and ether solvents to cyclohexylamine and dimethylsulphoxide respectively. Again the pK_a^S values of the hydrocarbons common to each study are strikingly similar.

Kinetic methods

Measurements of pK_a based on spectroscopic and equilibrium studies are only useful in the pK_a range where anions are coloured and where solvents of the requisite ion solvating power are of sufficiently low acidity that they do not interfere with the equilibria involved. These factors become more of a problem with increasing pK_a so that with weaker carbon acids like benzene and toluene alternative methods have to be sought.

One approach, first applied to carbanions by Shatenstein,[10] involves the use of the Brönsted Catalysis Law, equation (7)

$$\log k = \alpha p K_a + \beta; \qquad 0 < \alpha < 1 \qquad (7)$$

α is a structurally sensitive parameter and β is a constant. Applied to carbanions this equation relates the rate of hydrogen isotope exchange of a suitably labelled carbon acid in a non-labelled solvent with its pK_a as determined by the equilibrium method. Having established a correlation, the pK_a values of unknown acids should be obtainable from a knowledge of their catalytic exchange rates.

$$RD + B^- \underset{k_{-1}}{\overset{k_1}{\rightleftharpoons}} R^- \cdots DB \underset{BH}{\overset{k_2}{\longrightarrow}} RH + B^- \qquad (8)$$

$$k_e = \frac{k_1 k_2}{k_{-1} + k_2} \qquad \text{for} \quad k_2 \gg k_{-1}; \quad k_e = k_1 \qquad (9)$$

This approach is particularly useful for the weaker carbon acids since only a small steady state concentration of anion is required to promote measureable exchange. However for the correlation to apply it is important that steric effects remain constant and that the rate determining step in the kinetic exchange process resembles closely the thermodynamic ionization of the acid.[11] Over a wide pK_a range a Brönsted plot may not remain linear. For small pK_a differences between catalyst and substrate the transition state for the rate process may lie close to products giving a slope that tends to unity. For larger differences the transition state may resemble reactants giving an α that tends to zero. The Brönsted correlation developed by Streitwieser and coworkers[12] compares the rate of H/D and H/T exchange in MeOH/MeONa with the pK_a obtained by the equilibrium method in CsCHA/CHA. In accord with expectations the plot was found to show curvature with an α of c. 0·4 around fluorene (pK_a 22) to c. 0·9 around triphenylmethane (pK_a 33). To extrapolate further into the weak acid region it is useful to adopt the assumption that $\alpha \rightarrow 1$. Then each factor of ten change in catalytic exchange rate corresponds to a ΔpK_a of one unit. This is in fact how the pK_a values in the >35 range have been estimated in Table 10.3. However, since the extrapolation is made from a correlation involving delocalized structures to carbon acids whose anions are mainly localized and range in hybridization from sp^2 to sp^3 its validity is open to question. Nevertheless kinetic methods do provide a reasonable guideline to relative acidities in the weak acid region.

10.2.2 The relation of carbanion structure to pK_a

The range of estimated dissociation constants for carbon acids covers some sixty powers of ten ranging from pentacyanocyclopentadiene ($pK_a^W = -11$)

with an acidity greater than perchloric acid to cyclohexane which is currently believed to have a pK_a^W of 49.

Structurally, the factors responsible for this are related to the ability of the conjugate base to accommodate negative charge. This can be best achieved in three ways.

1. Delocalization. This incorporates hydrocarbon and heteroatom carbon acids and effects due to aromaticity.
2. Hybridization of the orbital carrying the negative charge.
3. Stabilization by inductive and field effects.

Anions derived from acetone and nitromethane acquire most of their stabilization by delocalization of the negative charge into the more electronegative oxygen atoms.

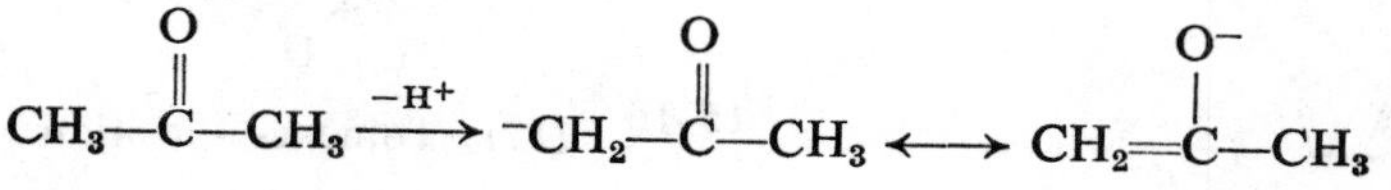

$$CH_3-\overset{\overset{\displaystyle O}{\|}}{C}-CH_3 \xrightarrow{-H^+} {}^-CH_2-\overset{\overset{\displaystyle O}{\|}}{C}-CH_3 \longleftrightarrow CH_2{=}\overset{\overset{\displaystyle O^-}{|}}{C}-CH_3 \qquad (10)$$

$$CH_3-\overset{\overset{\displaystyle O}{\diagup\!\!\!\!\diagdown}}{\underset{\underset{\displaystyle {}^-O}{+}}{N}} \xrightarrow{-H^+} {}^-CH_2-\overset{\overset{\displaystyle O}{\diagup\!\!\!\!\diagdown}}{\underset{\underset{\displaystyle O^-}{+}}{N}} \longleftrightarrow CH_2{=}\overset{\overset{\displaystyle O^-}{\diagup}}{\underset{\underset{\displaystyle O^-}{+}}{N}} \qquad (11)$$

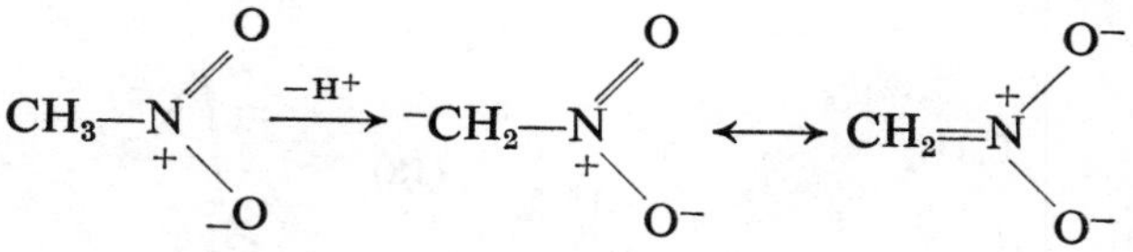

The stabilization afforded by a second keto- or nitro-substituent is fairly additive but steric hindrance to coplanarity reduces the effect of a third (Table 10.2). In the case of the sulphonyl series, where stabilization by $d\text{-}p$ orbital overlap is possible, or in the linear cyanomethane series this problem does not arise.

Table 10.2 pK_a^W Values of heteroatom carbon acids[15]

Compound[a]	pK_a	Compound	pK_a
CH_3COCH_3	20	$CH_3SO_2CH_3$	23
$CH_2(COCH_3)_2$	9	$CH_2(SO_2CH_3)_2$	14
$CH(COCH_3)_3$	6	$CH(SO_2CH_3)_3$	0
CH_3NO_2	11	CH_3CN	25
$CH_2(NO_2)_2$	4	$CH_2(CN)_2$	12
$CH(NO_2)_3$	0	$CH(CN)_3$	-15^{16}
		CN CN ![ring] CN CN H CN	-11^{17}
CF_3H	28^{18}	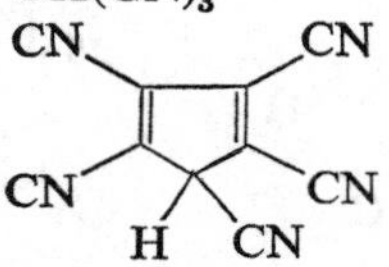	
$(CF_3)_3CH$	11^{18}		

a. All future pK_a^W values recorded in the text will be abbreviated to pK_a.

Table 10.3 pK_a Values of some hydrocarbons[a]

No.	Compound	pK_a	No.	Compound	pK_a
(4)	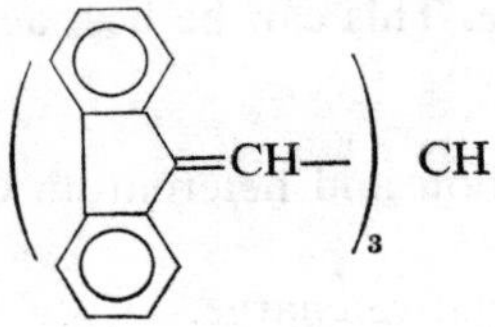 $=CH-$ CH (subscript 3)	9·4	(13)	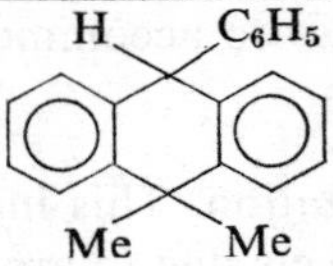	28·0[b]
			(14)	$(C_6H_5)_3CH$	31·5
			(15)	$(C_6H_5)_2CH_2$	33·1
(5)	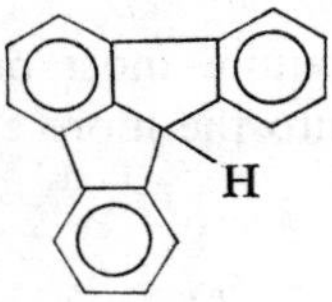	13·9	(16)	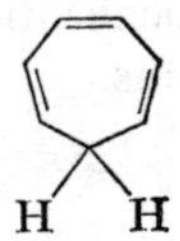	36·0
(6)		15–16	(17)	Toluene-α-position	39·0
(7)	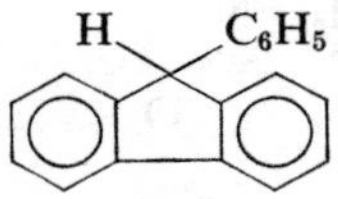	18·5	(18)	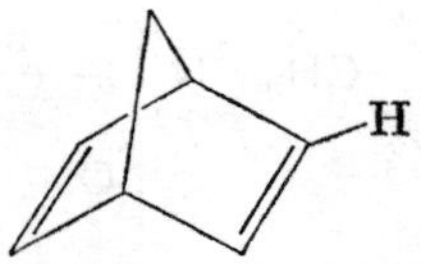	40·0[c]
(8)		19·6	(19)	1-Hexene-3-position	40·0[c]
	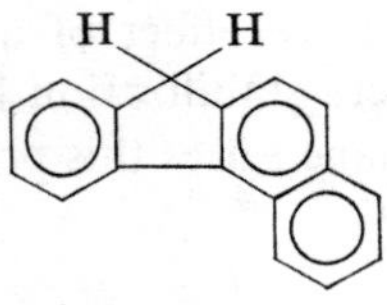		(20)	Ethylbenzene-α-position	40·0
(9)		20·1	(21)	Cumene-α-position	41·0
			(22)	Benzene	41·0
(10)		22·4	(23)	Ethylene	42·0
(11)	$HC \equiv CH$	25·0	(24)		42·0
(12)		26·5			

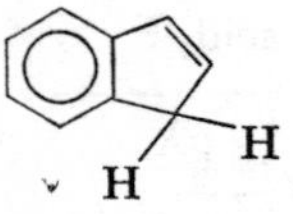

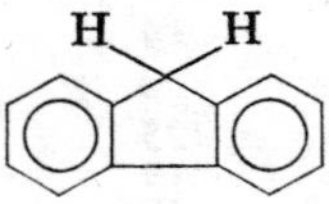

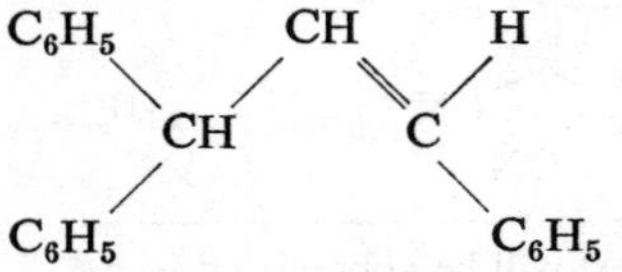

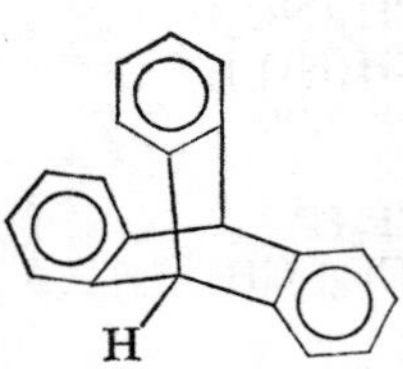

Table 10.3 *(Continued)*

No.	Compound	pK_a	No.	Compound	pK_a
(25)	Propene-2-position	43·0	(29)	Ethane	48·0
(26)	Cyclopropane	44·0			
(27)	Cyclobutane	47·0	(30)	Neopentane	48·0
			(31)	Cyclopentane	48·0
			(32)	Propane-2-position	49·0
(28)	Methane[d]	47·0	(33)	Cyclohexane	49·0

a. Taken from references 21 & 22 unless otherwise stated.
b. Reference 20.
c. Calculated from relative exchange rates in references 20 & 87.
d. Note reference 74.

A further illustration of the different orbital overlap requirements for sulphones and ketones is shown in the series (**1–3**) below.

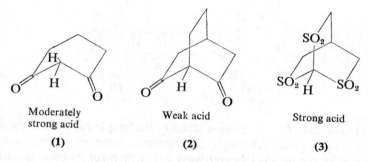

Moderately strong acid	Weak acid	Strong acid
(1)	**(2)**	**(3)**

The bridgehead electron pair in the anion derived from (**2**) is rigidly held in an sp^3 orbital. Its geometrical disposition relative to the ketone π-system makes mutual overlap minimal, and as a result, bicyclic diketone (**2**) is a weaker acid than monocyclic ketone (**1**).[13] Overlap between the carbanion electron pair and the less geometrically demanding sulphur d-orbitals, on the other hand, makes the trisulphone (**3**) a relatively strong acid.[14]

Apart from the less rigid directional requirements of d-orbitals, elements that have them available appear to be far more carbanion stabilizing than those that do not. A good example of this is shown in the base catalysed detritiation of $C_6H_5CT(SC_6H_5)_2$ which proceeds at a rate 10^6 times greater than that of its oxygen analogue $C_6H_5CT(OC_6H_5)_2$.[19]

A list of relevant pK_a values for hydrocarbon acids is given in Table 10.3. A typical illustration of the effect of delocalization is shown in a comparison of pK_a's in the series methane (pK_a 47) toluene (pK_a 39) diphenylmethane

(pK_a 33) and triphenyl methane (pK_a 31·5) Substitution of a phenyl group for hydrogen in methane results in an acidity increase of 8 pK_a units. The ΔpK_a between toluene and diphenylmethane is similar to that between methane and toluene indicating that the introduction of a second phenyl group produces an approximately additive effect. The fact that triphenylmethane does not conform to this trend is in line with the expectation that the triphenylmethyl anion for steric reasons is unable to attain complete coplanarity. If two of the rings of triphenylmethane are forced into coplanarity by replacing two of the ortho-hydrogen atoms with a C—C bond as in 9-phenylfluorene, (**7**), a 10^{13} increase in acidity results. Streitwieser[20] has estimated that c. 10^4 of this is due to the effect of increased coplanarity, the remainder being due to the anion stabilizing effect of the cyclopentadiene ring. This conclusion was based on the observation that 9,9-dimethyl-10-phenyldihydroanthracene, (**13**), a triphenylmethane analogue with a pK_a some 4 units lower, confers two ring coplanarity on the latter without the benefit of cyclopentadiene stabilization.

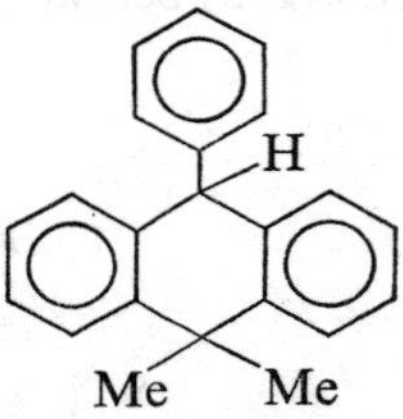

(**13**)	(**7**)
pK_a28	pK_a18·5

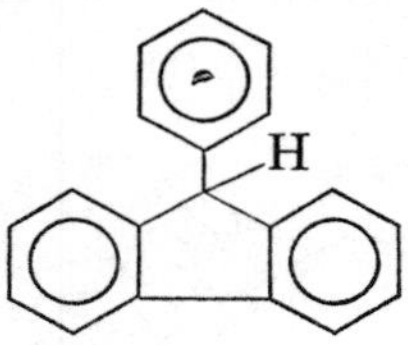

For predictions of relative anion stability the simple reasonance picture is sometimes inadequate. Although, for example, more resonance structures can be written for the cycloheptatrienyl anion than for its cyclopentadienyl counterpart the latter is more stable by 20 pK_a units. In molecular orbital terms the prediction of this sort of phenomenon is straightforward. In fact one of the successes of the simple Hückel MO theory has been in predicting that planar cyclic structures containing conjugated $(4n + 2)$ π-electrons ($n = 0, 1, 2, 3 \cdots$) should benefit from a substantial delocalization energy.[23] Other so-called aromatic anions that fulfil this condition are the cyclobutadiene dianion, the cyclooctatetraene dianion[24] and the cyclononatetraene anion.[25] Such anions are particularly stable because they have completely filled bonding molecular orbitals in contrast to the cycloheptatrienyl anion which has to accommodate two of its electrons in antibonding orbitals. Breslow has termed $4n$ π-electron systems such as this antiaromatic[26] and suggests that they are destabilized relative to localized analogues. Thus the base catalysed hydrogen deuterium exchange of the cyclopropene (**34**), involving an anti-aromatic cyclopropenyl anion intermediate was found to be four powers of ten slower than its cyclopropane analogue (35). Whereas anions like cyclopentadienyl and cyclononatetraenyl can be prepared

(34)   (35)

by proton abstraction in the presence of a suitable base, the cyclooctatetraenyl anion, having no suitable proton, cannot be obtained in this way. However, by treatment of cyclooctatetraene with alkali metals in ether solvents or in liquid ammonia two electrons are taken up and the tub-shaped molecule is converted into its planar aromatic dianion.[27]

Indene and fluorene, whose anions are isoelectronic with naphthalene and anthracene, can be regarded as cyclopentadienyl anions fused respectively to one and two benzo groups. That the introduction of these should result in a less rather than a more stable species is not evident from resonance theory but is in accord with MO calculations.[5]

Delocalization effects are generally of such magnitude that they tend to overshadow the lesser contribution from inductive or field effects. Inductive effects describe electrostatic interactions operating through bonds and are usually negligible if more than three bonds separate a substituent from the reactive centre. Field effects on the other hand encompass similar interactions operating directly through space or through the polarizable solvent medium. In a three-dimensional structure both of these effects will be operative and it is generally difficult to assess the relative contributions of each.

Such effects are important in comparing the stabilities of the trypticyl with the methyl anion. Trypticene (24) Table 10.3, a triphenylmethane analogue, is 5 pK_a units more acidic than methane. This difference cannot be accounted for by invoking the resonance explanation since, in the bicyclic compound, the aromatic π-system and the orbital containing the negative charge are orthogonal. Evidently the trypticenyl anion is stabilized electrostatically by a combination of field and inductive effects to the extent of $\sim 2 pK_a$ units per phenyl group.[28]

The nature of inductive stabilization in localized aromatic anions has been examined in some detail by Streitwieser and Lawler.[29] Using labelling techniques the kinetic acidities of a series of polycyclic hydrocarbons with structures ranging from benzene to anthracene were determined. As in the case of the trypticenyl anion the orbital containing the unshared electron pair is orthogonal to the aromatic π-system and the negative charge cannot be delocalized by resonance. In view of this the variation in isotopic exchange rates is quite striking (Table 10.4). The relative rates were found to be best

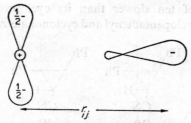

Figure 10.1 Classical model for the interaction of a carbanion lone pair with a quadrupole. Field effect function $F_i = \sum_j 1/r_{ij}$; r_{ij} is the distance between the lone-pair-containing carbon i and the aromatic carbon j

correlated with a simple electrostatic field model (Figure 10.1) in which the interaction is that of a negative point charge with a series of quadrupoles, each quadrupole being an sp^2 carbon bearing a positive charge and a p orbital containing one electron shared between its upper and lower lobes. The interaction is summed for each carbon atom in the molecule. Since the net effect of each is stabilizing, arene anion stability is expected to increase with carbon number.

Since methyl groups are electron donating in character their effect on anions should be destabilizing. This is borne out in a comparison of the acidities of toluene, ethylbenzene and cumene which show an increase of 1 pK_a unit per added methyl group.

An interesting effect has been noted by Streitwieser and Mares[30] for the fluorine substituent. Whereas in trifluoromethane it is clearly anion stabilizing (Table 10.2), in the 9-position of fluorene it leads to a reduction in acidity. This has been attributed to an unfavourable electrostatic interaction between the p-orbital lone pairs on fluorine and the aromatic π-system.

Like their arene analogues anions derived from acetylene, ethylene and ethane have little outlet for delocalization. Nonetheless, the pK_a values of their conjugate acids range through 25 units. This difference has been attributed to the varying degree of electronegativity of sp, sp^2 and sp^3 hybrid orbitals.[31]

Table 10.4 Relative H/T exchange rates for labelled arenes in LiCHA/CHA at 49·9 °C[28]

Hydrocarbon	Position of label	Relative exchange rate
Benzene	—	1·0
Biphenyl	3	4·3
	4	2·7
Naphthalene	1	7·3
	2	4·7
Phenanthrene	9	21·0
Anthracene	1	15·0
	9	75·0

Electrons in s-orbitals are, on average, much closer to the carbon nucleus than p-electrons and benefit, as a consequence, from the more favourable electrostatic interaction. In hybrid orbitals containing varying amounts of s- and p-character the stability of an electron pair is related to the per cent of s-orbital character. Thus in the series ethane, ethylene, acetylene the increased acidity is a consequence of the amount of s-orbital used in the corresponding C—H bond. It is interesting to note that the 50 per cent s-character of the charge bearing orbital in acetylene confers upon its anion a stability equivalent to those with substantial delocalization energy.

Increased s-character in a C—H bond also results whenever the carbon skeleton is highly strained. To accommodate the strain in small ring systems carbon atoms tend to rehybridize using a greater amount of more flexible p-orbitals within the carbon framework. This leaves proportionately more s-character to be accommodated in the exocyclic C—H bonds. In the cyclo-alkane series the kinetic acidities have been shown to parallel the per cent of s-orbital character and are in the order cyclopropane > cyclobutane > cyclopentane > cyclohexane.[32]

The introduction of strain into olefins modifies the hybridization of the sp^2 carbons and exerts a significant influence on acidity. This is reflected in the estimated pK_a values of propylene (**24**), norbornene (**36**), norbornadiene (**18**) and 2,3,3-trimethyl-1-cyclopropene (**37**).[31,32]

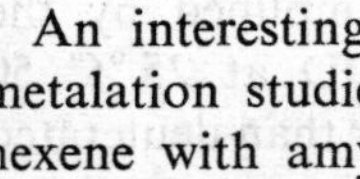

CH_3—$\overset{\underset{\textstyle H}{\mid}}{C}$=$CH_2$			
pK_a 43	pK_a 41·5	pK_a 40	pK_a ~29
(**24**)	(**36**)	(**18**)	(**37**)

An interesting blend of delocalization and strain factors comes from metalation studies on cyclohexene and cyclopentene. Treatment of cyclohexene with amyl sodium followed by carbonation gives a product acid derived from allylic proton abstraction. With cyclopentene strain factors predominate and vinyl proton abstraction is favoured.[33]

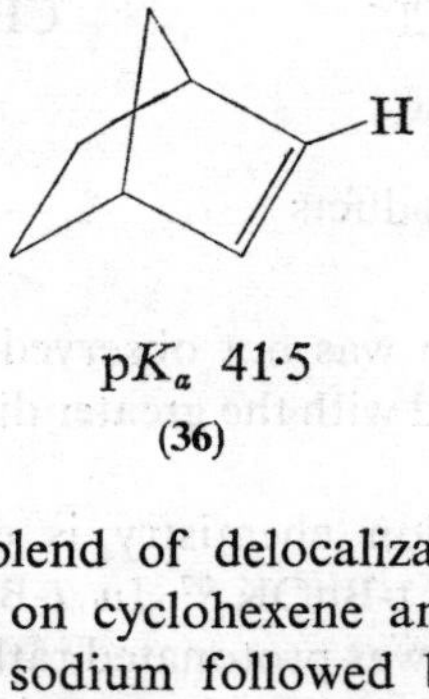

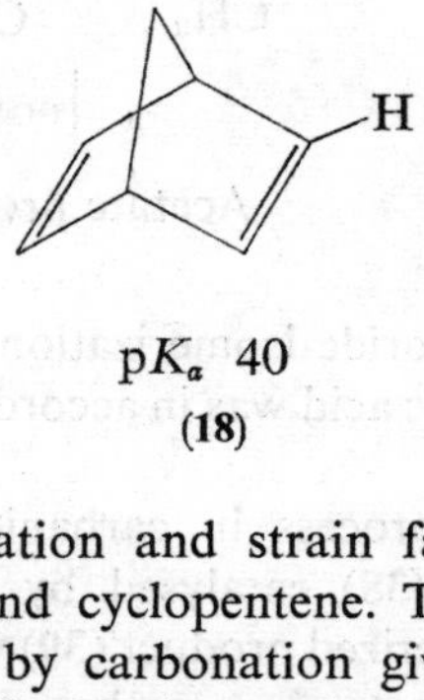

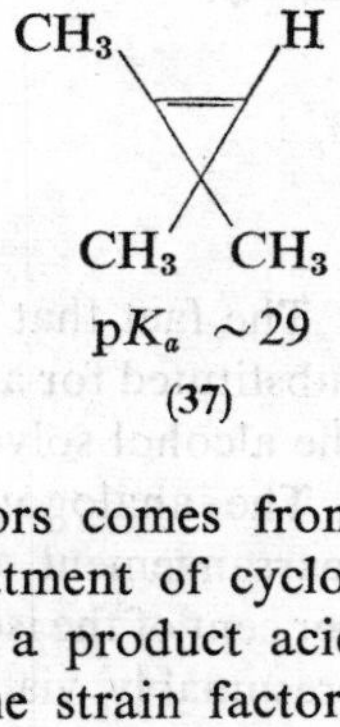

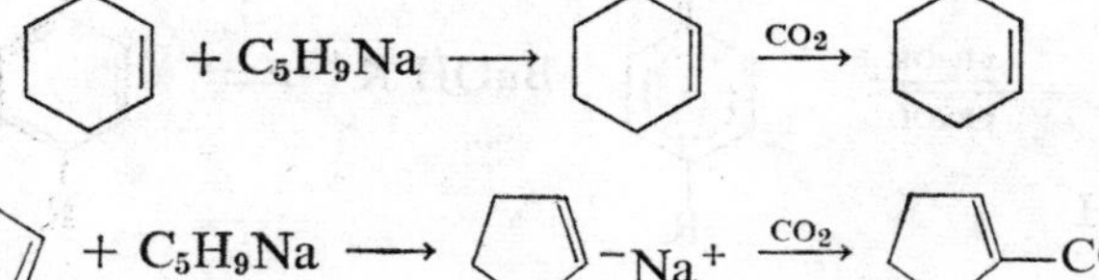

$$\bigcirc\!\!| + C_5H_9Na \longrightarrow \bigcirc^{-Na^+} \xrightarrow{CO_2} \bigcirc^{CO_2H} \qquad (12)$$

$$\bigcirc\!\!| + C_5H_9Na \longrightarrow \bigcirc^{-Na^+} \xrightarrow{CO_2} \bigcirc\!-CO_2H \qquad (13)$$

10.3 ION PAIRING AND REACTIVITY

Whereas in hexamethylphosphoramide and similarly highly solvating media free ions have been shown to exist,[34] in less polar solvents like THF and dioxane ions are generally paired or, as is the case in hydrocarbon solvents, even more highly associated.[35] Ion pairing considerations may not affect estimates of relative thermodynamic carbanion stability[36] but as we shall see later the extent of ion pairing and its nature have a profound effect on reactivity.

Chemical evidence for the intermediacy of ion pairs in organic reactions first came from the solvolytic studies of Young, Winstein and Goering.[36] In the acetolysis of α,α-dimethylallyl chloride isomerization of the starting chloride to its γ,γ-isomer was found to proceed at a comparable rate to product acetate formation. Since the pure chloride in the absence of solvent did not rearrange and since added chloride ion had no effect on the rate of acetate formation, isomerization was postulated to proceed through ion pair intermediates.

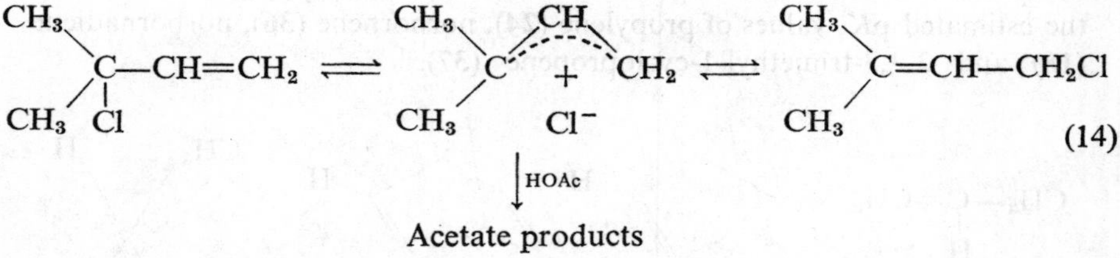

$$\begin{array}{ccc} \underset{CH_3}{\overset{CH_3}{\underset{|}{C}}}-CH=CH_2 & \rightleftharpoons & \underset{CH_3}{\overset{CH_3}{C}} + CH_2 \quad \rightleftharpoons \quad \underset{CH_3}{\overset{CH_3}{C}}=CH-CH_2Cl \\ \end{array}$$
(14)

Acetate products

The fact that chloride isomerization was not observed when ethanol was substituted for acetic acid was in accord with the greater dissociating power of the alcohol solvent.

The analogous process in carbanion chemistry is exemplified by the rearrangement of (38) catalysed by t-BuOK.[37] In t-BuOD at 25 °C 50 per cent of the isomerized product (39) was protonated rather than deuterated presumably via a potassium carbanide ion pair containing an associated molecule of t-BuOH.

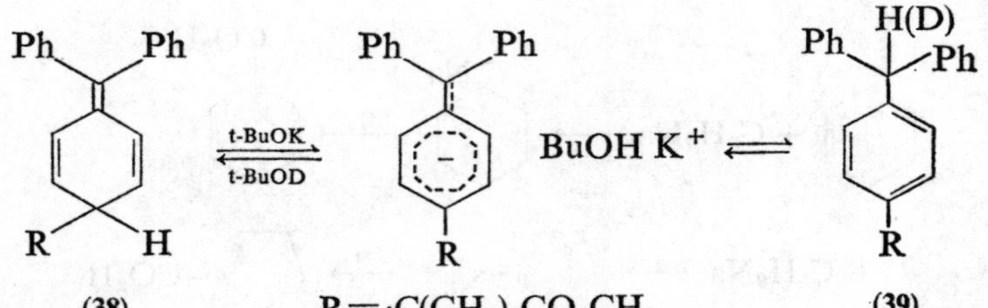

$$R = \cdot C(CH_3)_2CO_2CH_3$$

(38)   (39)

The role of ion pairs in determining the steric course of electrophilic substitution has been extensively investigated by Cram[38] using a technique borrowed from carbonium ion chemistry—that of comparing changes in optical rotation (k_α) with rate of loss of leaving group—in this case isotopic label (k_e).

In solvents of low dissociating power, for example THF, t-BuOH, 1 M in ammonia as base, the optically active fluorene derivative (40) exchanges D for H with predominantly retention of configuration $k_e/k_\alpha > 1$. Exchange is believed to occur by rotation of the NH_3D^+ ion within an ion pair. When the highly polar solvent dimethyl sulphoxide is used, dissociation takes

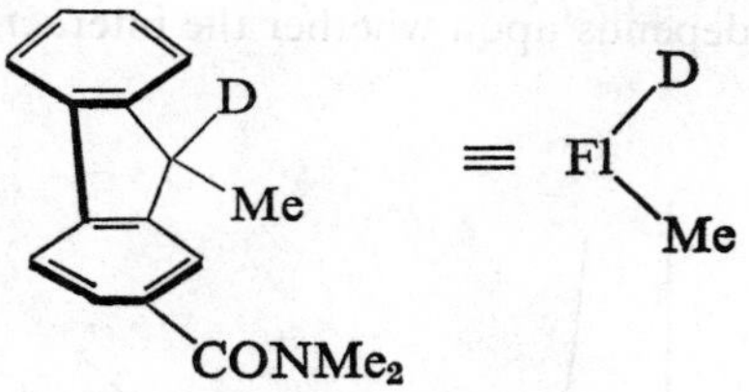

(40)

precedence over ion pair rotation, the label is lost in the bulk solvent, and the now planar fluorenylide ion is protonated with equal probability at either face to give a product that is both racemized and exchanged ($k_e/k_\alpha = 1$).

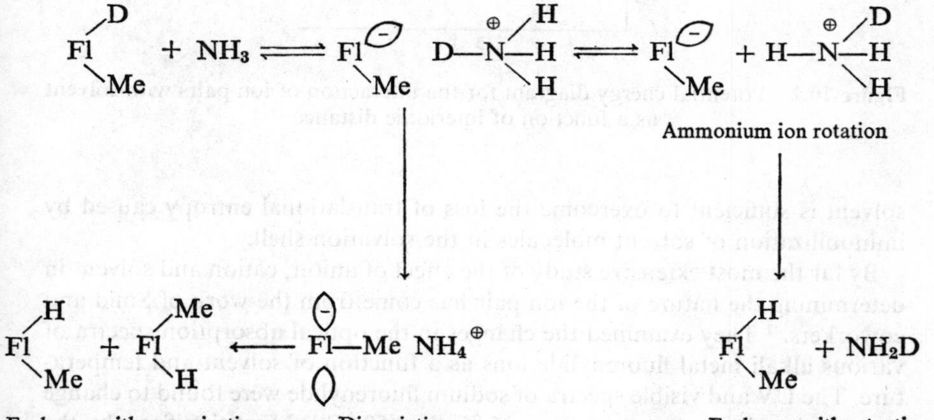

Ammonium ion rotation

Exchange with racemization   Dissociation   Exchange with retention

More recently the application of physical techniques to the study of ion pair equilibria has provided evidence that supports Winstein's[39] earlier proposals, again based on solvolytic studies, that ion pairs may exist in at least two forms, as contact or intimate ion pairs (R^-Li^+), and as solvent separated ion pairs ($R^+//Li^+$) and that these by their very nature may have differing reactivities. The energetics involved in their interconversion are

$$RLi \rightleftharpoons R^-Li^+ \rightleftharpoons R^-//Li^+ \rightleftharpoons R^- + Li^+ \qquad (16)$$

summarized in the potential energy diagram in Figure 10.2.[40]

The energy minimum corresponding to the smaller interionic distance represents the contact pair. As the ions are drawn apart their potential energy increases until the distance between them, represented by the second minimum becomes sufficient to accommodate a solvent molecule. Whether the latter situation is favoured depends upon whether the interaction between ions and

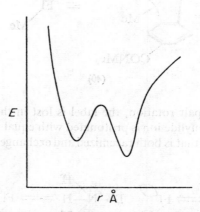

Figure 10.2   Potential energy diagram for the interaction of ion pairs with solvent as a function of interionic distance

solvent is sufficient to overcome the loss of translational entropy caused by immobilization of solvent molecules in the solvation shell.

By far the most extensive study of the effect of anion, cation and solvent in determining the nature of the ion pair has come from the work of Smid and coworkers.[41] They examined the changes in the optical absorption spectra of various alkali metal fluorenylide ions as a function of solvent and temperature. The UV and visible spectra of sodium fluorenylide were found to change on reducing the temperature from 25 °C to −50 °C.[42] Most significantly, the band at 356 nm was replaced by one at 373 nm (Figure 10.3). The involvement of free fluorenylide ions in the equilibrium was excluded on the basis of conductance measurements[43] and the failure of the relative band heights to respond to changes in carbanion concentration and the addition of

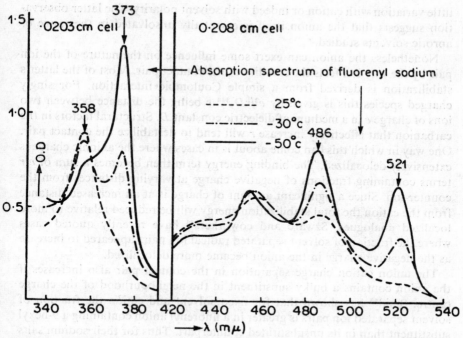

Figure 10.3 Absorption spectrum of fluorenyl sodium in THF as a function of temperature. Reprinted with the permission of the author and the Journal of the American Chemical Society

ionizable common salt (NaBPh$_4$). Smid therefore concluded that the spectral change was the result of a switch in equilibrium from contact to solvent separated ion pair.

The effect of the size of the cation on the position of the absorption maximum was found to be significant only in the case of the contact pair. (Table 10.5). The absorption maximum of the solvent separated pair showed

Table 10.5 Variation of the position of the absorption maximum with cation for fluorenylide ion pairs in THF at 25 °C

Cation	Cation radius Å	λ_{max} nm
Li$^+$	0·60	349
Na$^+$	0·96	356
K$^+$	1·33	362
Cs$^+$	1·66	364
F$^-$//M$^+$	~4·5	373
F$^-$	∞	374

little variation with cation or indeed with solvent polarity. The latter observation suggests that the anion itself is essentially unsolvated in the range of aprotic solvents studied.*

Nonetheless the anion can exert some influence on the nature of the ion pair by the way it affects the stability of the contact pair. Most of the latter's stabilization is derived from a simple Coulombic interaction. For singly charged species this is given by e^2/rD,[34a] r being the distance between two ions of charge e in a medium of dielectric constant D. Structural factors in the carbanion that effectively increase r will tend to destabilize the contact pair. One way in which this can come about is in cases where the anionic charge is extensively delocalized. The binding energy term then becomes a sum of e/r terms containing fractions of negative charge at varying distances from the counter-ion. Since a significant amount of charge is at an increased distance from the cation the total stabilization energy will be reduced relative to more localized analogues. Szwarc and coworkers[44] have recently quoted cases where the fraction of solvent separated radical ion pairs appeared to increase as the negative charge in the anion became more delocalized.

The anion cation charge separation in the contact pair also increases if the anion contains a bulky substituent in the neighbourhood of the charge centre. Smid[45] has shown that for fixed solvent and cation the fraction of solvent separated ion pairs is greater in a fluorenyl anion containing a 9-hexyl substituent than in its unsubstituted counterpart. Thus for their sodium salts in THF at 25 °C the ratios of solvent separated to contact ion pairs are >50 and 4·6 respectively.

The contact-solvent separated ion pair equilibrium is particularly sensitive to changes in solvent and cation. Preference for the solvent separated state is a function of the intensity of the cation solvent interaction and is greatest for small highly polarizing cations (Li > Na > K > Cs) and basic solvents (THF > 2,5-DHF > F) (Table 10.6). It is worth noting that even down to −70 °C in THF the large Cs⁺ prefers the contact state. At this temperature the lithium and sodium salts are solvent separated and the potassium salt exists as a mixture of the two forms.

For the solvent separated state to prevail the energy gained by specific cation solvation must more than compensate for the loss of translational entropy of the solvent molecules comprising the co-ordination shell. This entropy loss is greater in THF where four molecules of the solvent are probably required to solvate the lithium cation than in DME where only two molecules are needed to achieve the same degree of co-ordination. Thus as is shown in Table 10.6 the ratio of sodium fluorenylide solvent separated to contact pairs at 25 °C in THF and DME is 4·6 and >50 respectively.

---

* This is to be contrasted with the behaviour of anionic species in hydrogen bonding type solvents.

Extending this argument it now becomes clear why the polyethyleneglycol-dimethyl ethers $CH_3O(CH_2CH_2O)_x CH_3$[46] and the cyclic crown ethers[47,48] for example compound (41), are able, even when added in relatively small amounts, to convert relatively poorly solvating media into solvents that match DMSO and HMPA in properties. Since only one of these large

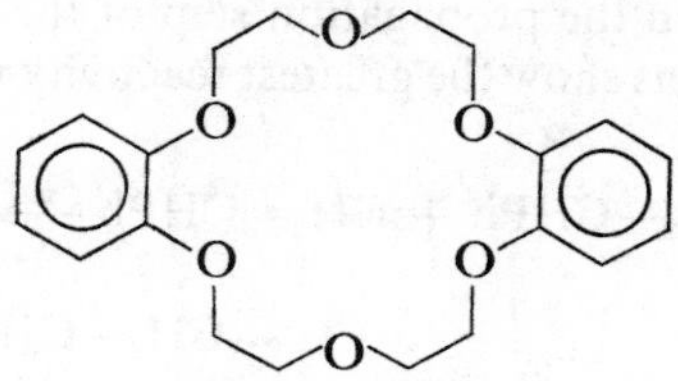

(41)

multidentate molecules is needed to form a solvent cage around the cation the unfavourable decrease in entropy of solvation is kept to a minimum.

Crown ethers such as the one shown above significantly increase the solubility of potassium hydroxide in hydrocarbon solvents. In benzene, the

Table 10.6 Ratios of contact to solvent separated alkali metal fluorenylide ion pairs as a function of solvent at 25 °C[45,42]

Cation	THF	2,5-DHF	F	1,2-DME
Li	4·6	1·1	0·01	>50
Na	0·05	—	—	4·0
K	<0·05	—	—	<0·05
Cs	<0·01	—	—	<0·01

THF = Tetrahydrofuran, DHF = Dihydrofuran, DME = Dimethoxyethane, F = Furan

relatively unsolvated hydroxide ion is a much more reactive species than its aqueous counterpart and promotes hydrolyses of hindered esters of 2,4,6-trimethylbenzoic acid that proceed only with difficulty in aqueous solution.[48]

Similarly explained is the effect of additives like N,N,N′,N′-tetramethyl-ethylenediamine (TMEDA) on the reactivity of butyllithium in hydrocarbon media[47] see below. Evidently this bidentate amine breaks up the highly associated organometallic into its monomeric form creating a more potent species in the form of a TMEDA-separated ion pair.

A striking example of the role played by solvent in ion pair equilibria is to be found in the interconversion of anions (42) and (43).[50] In what must be a

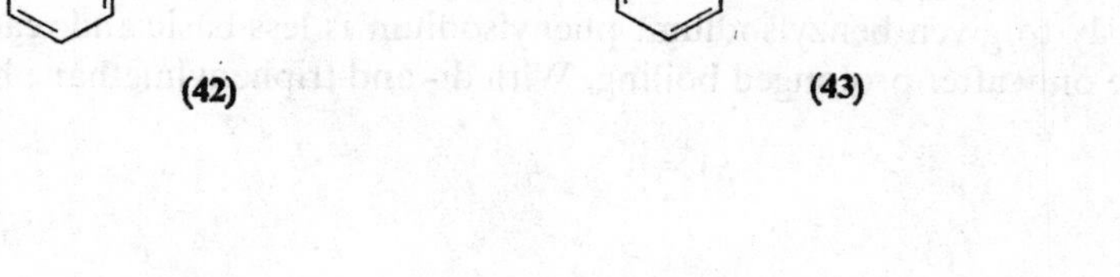

(42)   (43)

solvent separated pair the diphenylcyclopropylmethyl anion is delocalized and its lithium cation well solvated by THF. Rearrangement to the localized isomer is the energetic consequence of favouring the contact pair by changing to a poorer cation-solvating solvent.

One reaction where ion pairs and free ions have been shown to have different reactivities is in the propagation step of the anionic polymerization of styrene. Here, free ions show the greatest reactivity with the more accessible

$$
\text{~(CH}_2\text{—CHPh)}_n\overset{\overset{\text{M}^+}{|}}{\text{CH}_2\text{—CHPh}} + \text{CH}_2\text{=CHPh} \xrightarrow{k_p}
$$

$$
\text{~(CH}_2\text{—CHPh)}_{n+1}\text{—CH—}\overset{\overset{\text{M}^+}{|}}{\text{CHPh}}
$$

solvent separated ion pair being more reactive than the contact pair. The propagation rate constants for polystyryl alkali metal ion pairs at 25 °C in dioxane, a solvent that favours the contact pair for all of the alkali metal cations, vary from $0.9 \text{ M}^{-1}\text{s}^{-1}$ for Li^+ to $24.5 \text{ M}^{-1}\text{s}^{-1}$ for Cs^+. In THF, a solvent which favours a solvent separated pair for Li^+ and a contact pair for Cs^+, the respective rate constants are $160 \text{ M}^{-1}\text{s}^{-1}$ and $22 \text{ M}^{-1}\text{s}^{-1}$.[51] The k_p for the free polystyryl ion under these conditions has been calculated to exceed that of the solvent separated ion pair by a factor of 400. Its reactivity is essentially independent of solvent in the range DME, THF, THP, THF/dioxan, THF/benzene, further evidence that the anion is relatively unsolvated in these media.[53] This order of reactivity may not be general and depends upon the transition state requirements of the reaction under scrutiny. In some electron and proton transfer reactions the solvent separated pair has been found to react faster than the free ion with the contact pair being the least reactive.[34a,54]

10.4 PREPARATION OF CARBANIONS

10.4.1 From other carbanions by metalation[56,57]

Alkali metal salts of stronger carbon acids may be prepared from the salts of their weaker analogues by the equivalent of a simple acid base reaction; the extent to which exchange is achieved depends upon the pK_a difference between the acids involved and the nature of solvent and counter-ion.

$$
\text{RNa} + \text{R'H} \rightleftharpoons \text{RH} + \text{R'Na} \tag{18}
$$

In order to displace the equilibrium to the right the preferred metalating agents are those of extremely high basicity (high pK_a) like the alkyl and phenyl alkali metal salts. Whereas amyl sodium metalates toluene instantaneously to given benzylsodium, phenylsodium is less basic and reacts with toluene only after prolonged boiling. With di- and triphenylmethane however metalation occurs in the cold.[58] Phenyllithium is less reactive than its corresponding sodium and potassium derivatives and is a poor metalating agent even for triphenylmethane.[59] For weak acids pK_a 43–44 appears to be the limit beyond which amyl sodium is unable to metalate on an acceptable time scale. This is evidenced by the fact that cyclopropane and nortricyclane are compounds with the highest recorded pK_a's that have been metalated.[60]

Polar solvents have been reported to have a beneficial effect on metalation rates which increase steadily in the order heptane, benzene, THF, DME, DMF, DMSO, HMPA.[56,61] However it is important to remember that such a generalization makes no allowance for the specific effects of solvent and cation on ion pair equilibria and reactivity. Each case therefore should be judged on its merits. The lower reactivity of the alkyllithiums in hydrocarbon solvents is probably a result of their polymeric nature since the addition of strong cation complexing agents like TMEDA break up the aggregates and create a highly reactive TMEDA-separated monomeric ion pair. Thus whereas toluene is very slowly metalated by butyllithium it reacts rapidly and quantitatively with butyllithium-TMEDA.[49,62]

In such reactions the proton abstracted need not necessarily be that associated with the most stable carbanion. In cases where two protons of comparable acidity are in the same molecule the proton removed may be the result of kinetic rather than thermodynamic control. After three hours of treatment with amylsodium cumene is metalated 42 per cent α-, 39 per cent m-, 19 per cent p-. After twenty hours the mixture equilibrates to afford 100 per cent of the thermodynamically more stable α-product.[63] Similar competitive abstractions have been observed between vinyl and allylic protons.[64] At short reaction times 1-dodecene is metalated by amylsodium predominantly at the terminal vinyl position. Under equilibrating conditions the main products are those due to metalation at the allylic position.

Generation of anions of the stronger carbon acids ($pK_a < 30$) does not require bases of such high reactivity, and milder non-hydrocarbon bases (e.g. $NaNH_2/NH_3$) may be used. The choice is again dependent upon the relative pK_a's of the two acids involved. Some representative values of non-carbon acids are shown in Table 10.7.

Table 10.7 pK_a values for some amines and alcohols[6,7]

Compound	pK_a
NH_3	36
Cyclo-$C_6H_{11}NH_2$	34
$C_6H_5NH_2$	27
$(C_6H_5)_2NH$	23
$(CH_3)_3COH$	19
$(CH_3)_2CHOH$	18
CH_3OH	16

10.4.2 Direct metalation

The direct replacement of hydrogen by metal constitutes an alternative route to organoalkali compounds. In the case of fluorene the reaction has been

$$RH + M \rightarrow RM + \tfrac{1}{2}H_2 \qquad (19)$$

shown to proceed via the radical anion which rapidly decomposes to the carbanion.[61] Compounds up to the acidity of diphenylmethane[66] have been prepared by this technique.

10.4.3 Metal/metal exchange

Organoalkali metal compounds were first isolated from interaction of organomercury derivatives with the corresponding alkali metal.[67] To avoid

$$2Li + R_2Hg \rightleftharpoons 2RLi + Hg \qquad (20)$$

contamination of product by soluble organomercury compound a large excess of alkali metal is generally used to drive the equilibrium to the right. This approach may be applied to other organometallic compounds provided that the displacing metal is more electropositive than the metal it displaces. Alkyl potassium compounds are thus obtained from the interaction of potassium metal and lithium or sodium alkyls.

$$RLi + K \rightarrow RK + Li \qquad (21)$$

A recent method particularly useful for making primary alkyl sodium derivatives from their more readily available lithium analogues involves treatment of the alkyllithium with sodium t-butoxide.[71]

$$RLi + Me_3CONa \xrightarrow{\text{Heptane, 0°C}} RNa + Me_3COLi$$

10.4.4 Halogen-metal interchange

The reaction of a metal with an organic halide also constitutes a useful way of generating carbanionic derivatives. It is especially useful for the preparation of group I and II organometallics.

$$n\text{-BuCl} + 2Li \rightarrow BuLi + LiCl \qquad (23)$$

$$CH_3Br + Mg \rightarrow CH_3MgBr \qquad (24)$$

In the past such secondary processes as transmetalation, equation (25), elimination, equation (27), and substitution reactions, equation (26), rendered the preparation of pure organoalkali compounds by this technique impractical. Now, by judicious control of temperature, the use of finely divided metal dispersions and high speed stirring numerous organoalkali

$$RCH_2CH_2Na + SH \rightarrow SNa + RCH_2CH_3 \qquad (25)$$

$$RCH_2CH_2CH_2CH_2R + NaCl \qquad (26)$$

$$RCH_2CH_2Na + RCH_2CH_2Cl \nearrow$$

$$\searrow$$

$$RCH_2CH_3 + RCH{=}CH_2 + NaCl \qquad (27)$$

compounds can be prepared in good yield.[69] The method is of general applicability but is especially useful for the preparation of alkyl derivatives which are difficult to prepare by alternative methods.

Carbanion reactivity can be modified by changing the counter-ion. This is readily achieved with the alkali metal carbanions by treating with other metal halides. For example

$$RLi + HgCl_2 \rightarrow RHgCl + LiCl \qquad (28)$$

If an excess of the alkyl lithium is used the dialkyl mercurial is obtained.[72]

$$RHgCl + RLi \rightarrow RHgR + LiCl \qquad (29)$$

The direction of displacement here is such that the most electropositive metal finishes up as the inorganic salt.

10.4.5 Halogen-metal interconversion

This method can often be profitably used to obtain organolithium derivatives that are not readily prepared by reaction of organic halides with the metal. Aryl and vinyl halides are particularly unreactive in this respect. The reaction is an equilibrium in which the favoured product is that with the

$$RBr + R'Li \rightleftharpoons RLi + R'Br \qquad (30)$$

most stable anion. The relative amounts of RLi and R'Li will depend therefore on the pK_a's of RH and R'H.[73]

10.4.6 Ether cleavage

A method that bears some resemblance to halogen metal interchange is that of ether cleavage.[57,70] With sodium/potassium alloy, which is more reactive than sodium metal alone, and has the advantage of being a liquid

$$C_6H_5OCH_3 + 2K \rightarrow C_6H_5K + KOCH_3 \qquad (31)$$

at room temperature, cleavage is generally quantitative. Preparations are limited to anions whose carbon acids have a $pK_a <$ benzene or propene. Purely aliphatic ethers are stable towards metals up to 200 °C.

Alkyllithiums however are less stable in ether solvents. When a solution of isopropyllithium prepared in diethyl ether at $-60\,^\circ$C was warmed to room temperature products arising from the decomposition of the solvent were obtained.[52]

$$(CH_3)_2CHLi + C_2H_5OC_2H_5 \rightarrow CH_2{=}CH_2 + C_2H_5OLi + (CH_3)_2CH_2 \quad (32)$$

$$CH_2{=}CH_2 + (CH_3)_2CHLi \rightarrow (CH_3)_2CHCH_2CH_2Li \quad (33)$$

10.5 STEREOCHEMISTRY AND CONFIGURATIONAL STABILITY

Since saturated carbanions are isoelectronic with amines it has long been considered that they too may be pyramidal and similarly rapidly inverting. However extensive ion pairing and some degree of covalent character in

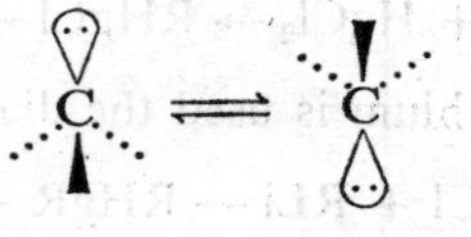

the metal-carbon bond, especially in hydrocarbon solvents, generally slows down the inversion process.

Curtin has demonstrated that optically active sec-butyllithium derived from interchange of sec-butylmercury and 2-octyllithium in pentane is capable of retaining its configuration for long periods at low temperatures. Carbonation at $-40\,^\circ$C gave a product with 83 per cent retained configuration.[75] Similarly, ^{1}H NMR data indicate that menthyllithium (**44**) is conformationally rigid in hydrocarbon solvents at $50\,^\circ$C and in ether solvents below $0\,^\circ$C.[76]

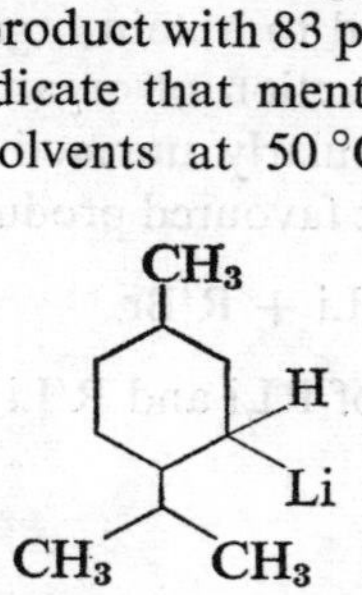

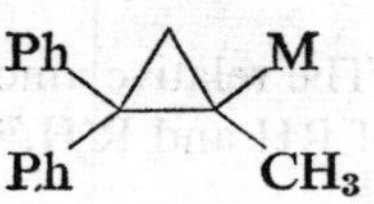

(44)   (45)

Cyclopropyllithiums appear somewhat more stable. The lithium compound derived from optically active 1-bromo-1-methyl-2,2-diphenylcyclopropane by reaction with butyllithium (**45**) retained its optical activity and was stable up to thirty minutes in both hydrocarbon and ether solvents from $-8\,^\circ$C to $35\,^\circ$C. Carbonation of the sodium salt generated in pentane solution from the same bromide and pentyl sodium demonstrated that it too had retained its configuration during its preparation.[77]

Using NMR Roberts and coworkers examined the effect of metal on the configurational stabilities of some neohexylmetallic compounds

$$(Me_3CCH_2CH_2)_nM$$

in diethyl ether. The rates of loss of configuration paralleled the per cent of ionic character of the C-metal bond and were in the order $RLi > R_2Mg > R_2Zn$ with the predominantly covalent R_3Al and R_3Hg compounds showing no spectral change up to 150 °C.[78]

An interesting example of configurational stability and stereochemistry of substitution has been quoted for the norbornyl Grignard reagent.[79] Treatment of either exo- or endo-norbornyl bromide or chloride (46) with magnesium in ether gave an identical mixture of norbornyl magnesium halides as judged by NMR. The pure endo-Grignard, isolated in the mixture

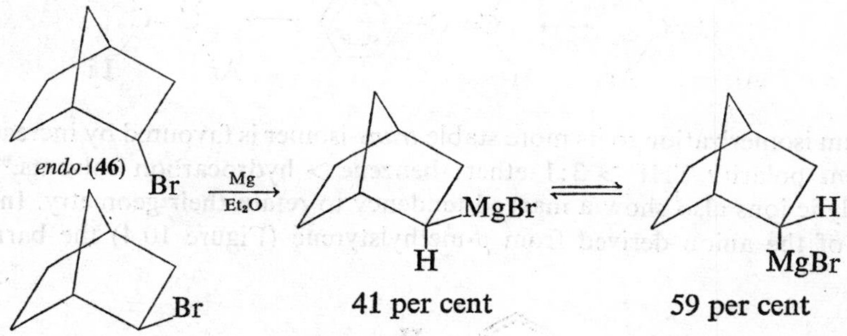

by the addition of benzophenone in sufficient amount to remove the more reactive exo-isomer, reverted to the equilibrium mixture over a period of one day at 37 °C. Carbonation or mercuration of the pure endo-reagent gave products with retained configuration. The lack of stereo-specificity in the preparation of the Grignard reagent is interesting especially in view of the fact that it is configurationally stable on the time scale of the preparation. Evidently, metal halogen exchange reactions pass through radical intermediates which may lose their configuration before being reduced to carbanions. To account for this Walborsky has suggested the following scheme.[80]

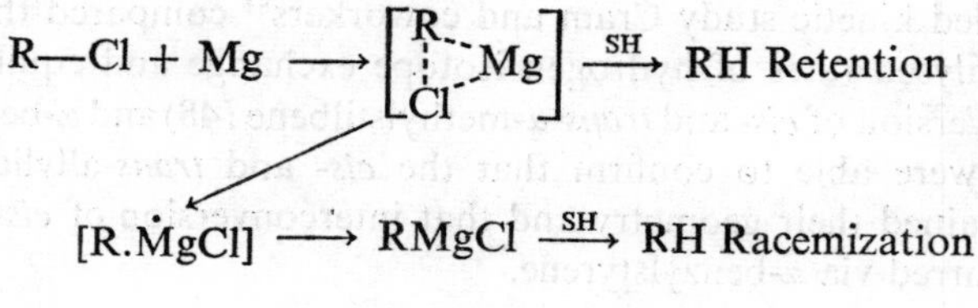

Vinyl carbanions are considerably more configurationally stable than their saturated analogues. Both *cis*- and *trans*-propenyl lithium (**47**) retain their stereochemical integrity for periods of up to one hour in refluxing ether.[81] Vinyllithiums with α-aryl substituents are somewhat less stable. Isomerization

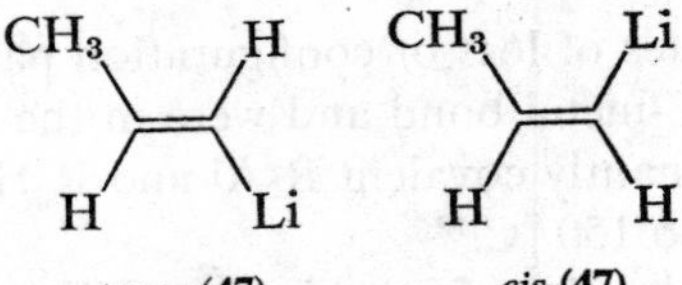

trans-(**47**)   *cis*-(**47**)

is facilitated by the availability of a linear allenic type transition state in which the charge is delocalized into the aromatic rings. In the case of cis-stilbenyl

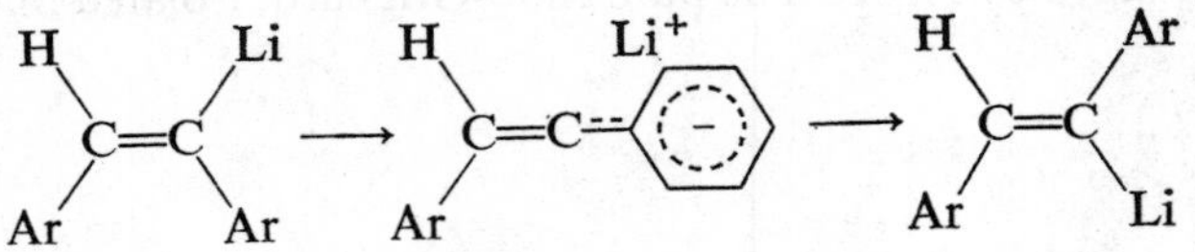

lithium isomerization to its more stable *trans*-isomer is favoured by increasing solvent polarity THF > 3:1 ether: benzene > hydrocarbon solvents.[82]

Allylic ions also show a marked tendency to retain their geometry. In the case of the anion derived from β-methylstyrene (Figure 10.4) the barriers

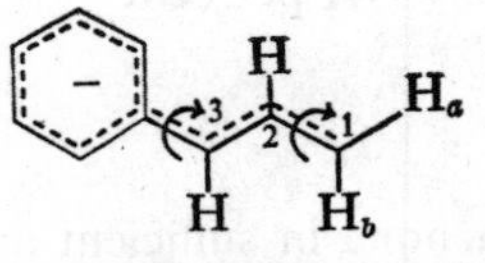

Figure 10.4 The barriers to rotation of the lithium-β-methylstyrenyl anion in THF solution

to rotation about the C_1—C_2 and C_3— aryl bond have been measured by NMR.[83] Rotation about the C_1—C_2 bond was manifested by the collapse of the multiplet due to H_2 coupled to H_a and H_b at 90 °C. Rotation of the C_3— aryl bond is a more facile process which was frozen out at −58 °C when the o-phenyl protons became inequivalent.

In a detailed kinetic study Cram and coworkers[84] compared the t-BuOK/t-BuOH catalysed rates of hydrogen isotope exchange and equilibration in the interconversion of *cis*- and *trans*-α-methylstilbene (**48**) and α-benzylstyrene (**49**). They were able to confirm that the *cis*- and *trans*-allylic ion intermediates retained their geometry and that interconversion of *cis*- and *trans*-stilbene occurred via α-benzylstyrene.

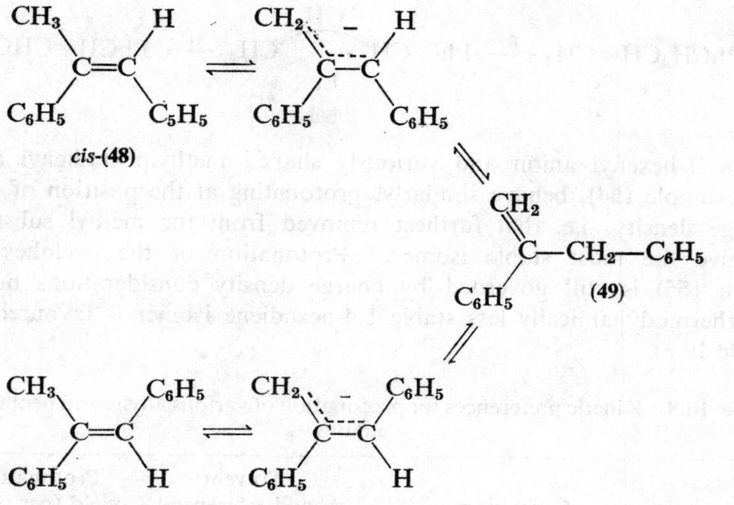

cis-(48)

(49)

trans-(48)

From a similar study Bank and Schriesheim[85] have suggested that the preference for *cis*-isomers in the base catalysed isomerization of terminal alkenes under conditions of kinetic control is not due to ground state conformational factors but to the greater stability of the *cis*- over the *trans*-allylic ion caused by a favourable ion-dipole interaction.

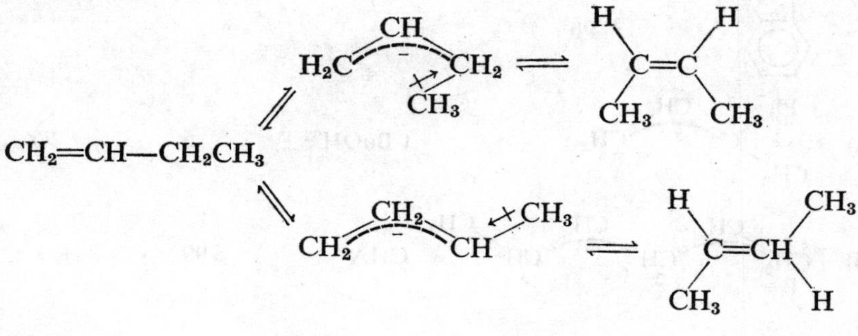

10.6 REACTIVITY OF DELOCALIZED IONS

Delocalized substituted allyl and pentadienyl anions can be regarded as multifunctional in their reactions with electrophilic reagents. Protonation of the cinnamyl anion (**50**), for example, under non-isomerizing conditions yields a 1:1 mixture of allylbenzene and β-methylstyrene, the direction of protonation correlating well with that expected on the basis of Hückel π-electron charge distribution.[86]

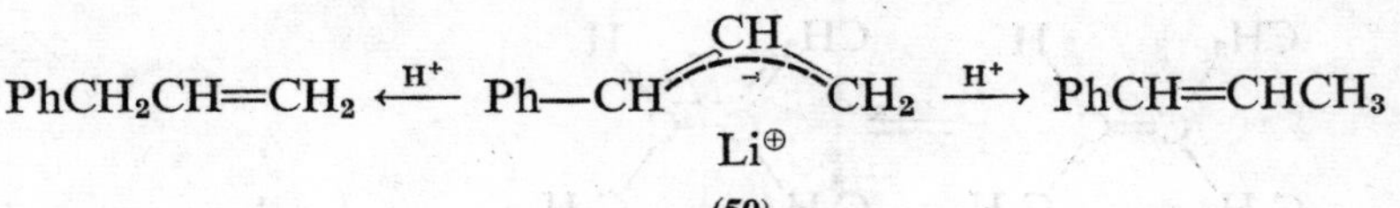

$$PhCH_2CH{=}CH_2 \xleftarrow{H^+} \underset{Li^{\oplus}}{Ph{-}CH \overset{CH}{\underset{}{{-}}} CH_2} \xrightarrow{H^+} PhCH{=}CHCH_3$$

(50)

The 1-hexenyl anion and variously shaped methylpentadienyl anions, for example (54), behave similarly, protonating at the position of highest charge density, i.e. that furthest removed from the methyl substituent, to give the most stable isomer.[87] Protonation of the cyclohexadienyl anion (55) is still governed by charge density considerations but now the thermodynamically less stable 1,4-hexadiene isomer is favoured[88] (see Table 10.8).

Table 10.8   Kinetic preferences for protonation of various allylic and pentadienylic anions

No.	Carbanion	Solvent and reference	Protonation yield (per cent)	
			1-position	3-position
(51)	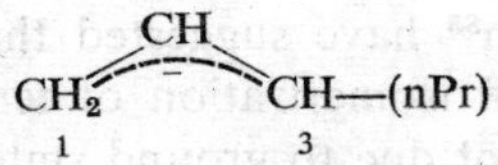 $CH_2{=}CH{-}(nPr)$ (allylic)	CHA[87]	98	2
(52)	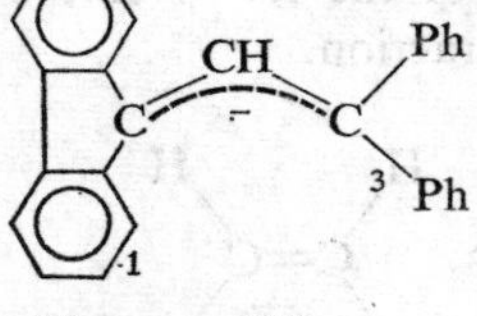 fluorenyl–$C{=}CH{-}CPh_2$	DMSO[89]	93	7
(53)	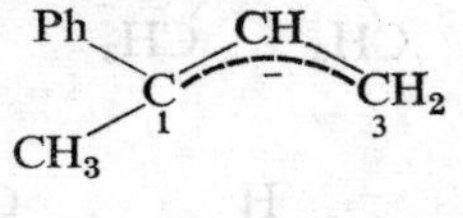 $Ph(CH_3)C{=}CH{-}CH_2$	t-BuOH[92]	5	95
(54)	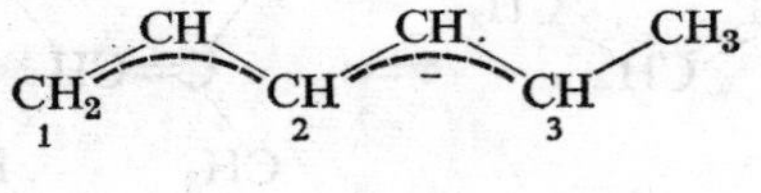 $CH_2{=}CH{-}CH{=}CH{-}CH_3$	CHA[87]	>99	$2+3 < 1$
(55)	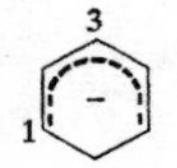 cyclohexadienyl	t-AmOH[88]	11	89
(56)	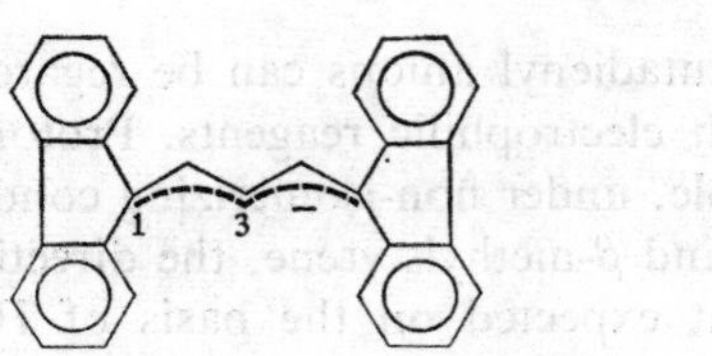 bis(fluorenyl)	DMSO[89]	76	24

In another study the ratios of isomers formed on protonation and methylation of various ambient anions flanked by highly delocalizing aromatic groups such as fluorenyl and phenyl (for example (52) and (56)) have been measured. These also show a qualitative correlation with charge distribution.[89]

In the presence of protic solvents aromatic radical ions or dianions afford dihydroaromatics, whose formation, via anionic intermediates, can be rationalized on a charge density distribution basis. Reduction of naphthalene for example favours 1,4-dihydronaphthalene over its 1,3-isomer.[86]

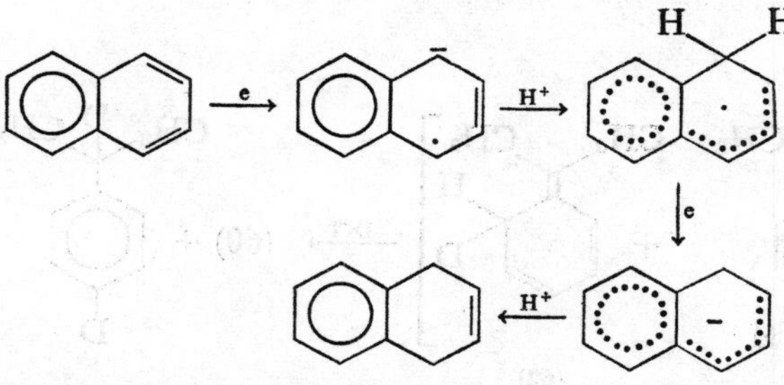

Heteroatom carbanions like phenylnitromethyl anion (57) carry most of their charge on the highly anion stabilizing nitro group and prefer to protonate, as a result, to give the less stable aci-nitro-isomer (58).[90] The qualitative agreement between charge density correlations and products of

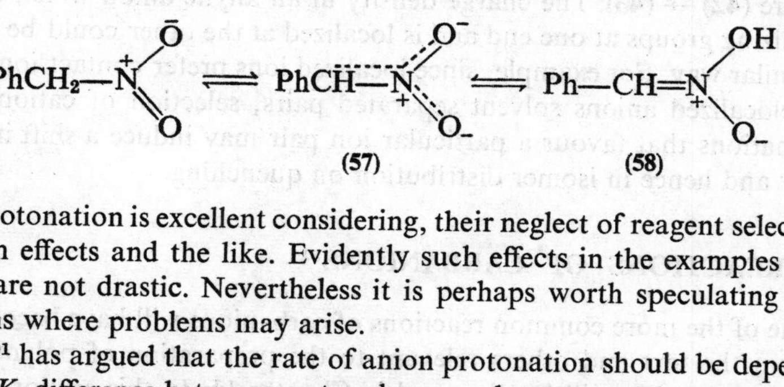

(57)   (58)

anion protonation is excellent considering, their neglect of reagent selectivity, solvation effects and the like. Evidently such effects in the examples so far studied are not drastic. Nevertheless it is perhaps worth speculating about situations where problems may arise.

Eigen[91] has argued that the rate of anion protonation should be dependent on the pK_a difference between proton donor and acceptor. Thus it is possible to conceive of two tautomers A + B which may interconvert via anion C. A, by virtue of its greater ground state stability, has a higher pK_a than B and, on Eigen's hypothesis, should be favoured when C is protonated. Hence, other factors being equal there will always be a tendency to protonate to favour the more stable isomer.

One instance where donor pK_a has been increased with interesting results is in the reaction of α,α-dimethylbenzylpotassium (59) with D_2O and DCl in ether.[93] With deuterated water α-deuterocumene (60) is the sole product, but with DCl, 15 per cent of ring attack at the p-position and 5 per cent at the o-position were observed giving ring-deuterated cumenes that were produced via the isopropylidenecyclohexadiene intermediates, (61) and (62).

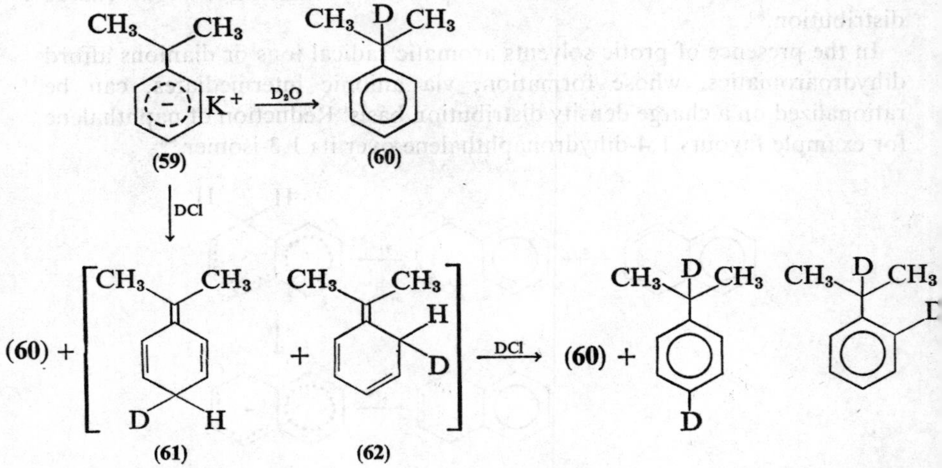

CH₃ CH₃   CH₃ D CH₃

(59)   (60)

(60) + (61) + (62) → (60) + products

(61)   (62)

We noted earlier how variations of solvent, cation and temperature affected ion pair equilibria, and in one example we saw how a variation in these parameters promoted isomerization of a delocalized to a localized structure (42) → (43). The charge density in an allylic anion which contains delocalizing groups at one end and is localized at the other could be affected in a similar way. For example, since localized ions prefer contact ion-pairing and delocalized anions solvent separated pairs, selection of cation-solvent combinations that favour a particular ion pair may induce a shift in charge density and hence in isomer distribution on quenching.

10.7 REACTIONS OF CARBANIONS[94]

Some of the more common reactions of carbanions will have become evident from the text and others relevant to the preparation of polymers with specific end-groups will be discussed in Chapter 11. In this section we shall give some of these a brief consideration but concentrate upon the reaction with alkyl halides and oxygen.

With acids of lower pK_a carbanions are converted to the corresponding hydrocarbon, the rate generally correlating with the pK_a difference between the respective acids.[91] Quenching a carbanion with deuterated or tritium-enriched water constitutes a useful method of introducing a specific isotopic label. For example

$$RLi + D_2O \rightarrow RD + LiOD \qquad (38)$$

Carbanions add to carbonyl compounds in what are generally very fast reactions to produce the corresponding alcohols. With substituted benzophenones in toluene-ether solution the rates of organolithium addition are p-tolyl > phenyl > ethyl > isopropyl.[95] Presumably the order reflects the ease of breakdown of unreactive organolithium aggregates into a more labile monomeric form rather than relative carbanion reactivity.

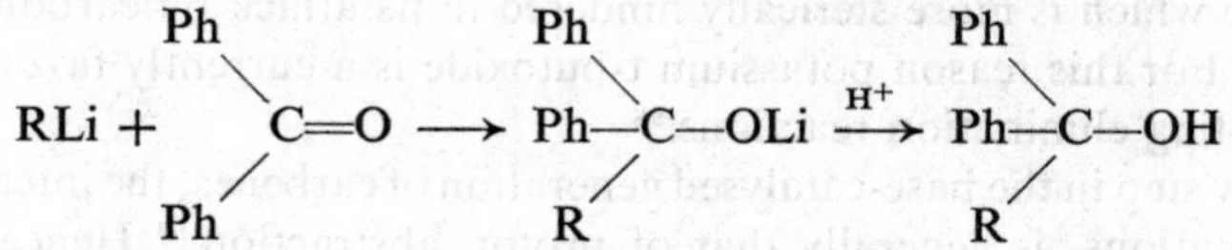

$$RLi + \begin{matrix} Ph \\ \diagdown \\ C=O \\ \diagup \\ Ph \end{matrix} \longrightarrow Ph-\underset{\underset{R}{|}}{\overset{\overset{Ph}{|}}{C}}-OLi \xrightarrow{H^+} Ph-\underset{\underset{R}{|}}{\overset{\overset{Ph}{|}}{C}}-OH \qquad (39)$$

Addition to double bonds generally requires some driving force such as attachment of a suitably activating substituent, equation (40),[96] or release of strain, equation (41),[55] to proceed with ease.

$$RLi + C_6H_5CH{=}CHCOOC_6H_5 \rightarrow C_6H_5CHRCHLiCOOC_6H_5 \qquad (40)$$

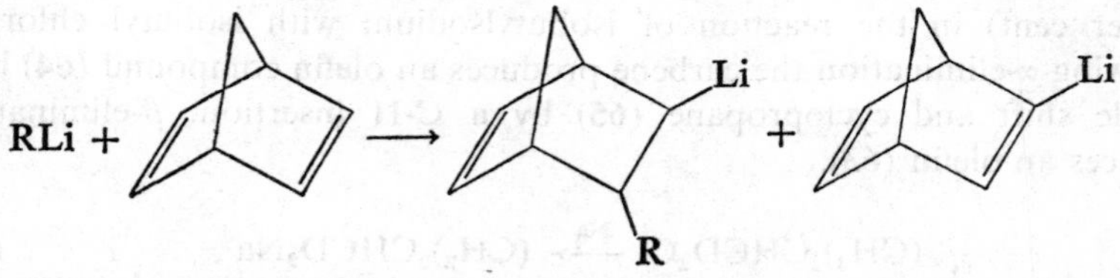

$$RLi + \quad\longrightarrow\quad + \qquad (41)$$

Reaction with organic halides is potentially quite complex. With alkyl halides there are four major possibilities.

$$R_1CH_2CH_2X + R^- \longrightarrow R_1CH_2CH_2R + X^- \qquad \text{Substitution} \qquad (42)$$

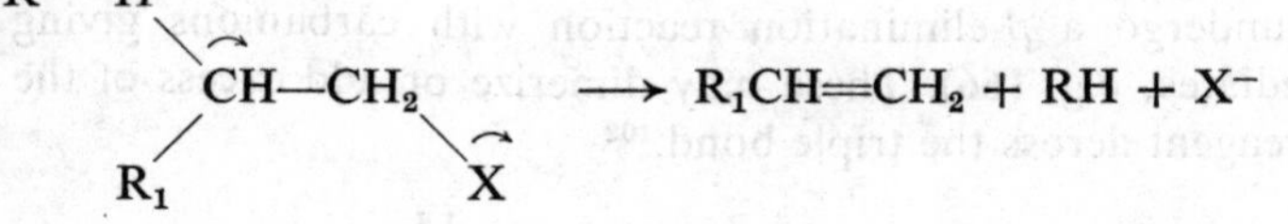

$$\begin{matrix} R^-\frown H \\ \diagdown \\ CH-CH_2 \\ \diagup \quad\quad\diagdown \\ R_1 \quad\quad\quad X\frown \end{matrix} \longrightarrow R_1CH{=}CH_2 + RH + X^- \quad \beta\text{-Elimination}$$

$$(43)$$

$$R^- + R_1CH_2CH_2X \longrightarrow RH + R_1CH_2\overset{-}{C}HX \qquad \alpha\text{-Elimination}$$

$$R_1CH_2\overset{-}{C}HX \longrightarrow [R_1CH_2\overset{..}{C}H] + X^- \longrightarrow R_1CH{=}CH_2 \qquad (44)$$
$$\text{A carbene} \qquad \text{and other products}$$

$$R^- + R_1CH_2CH_2X \longrightarrow R_1CH_2CH_2^- + RX \qquad \begin{matrix}\text{Metal-halogen} \\ \text{interchange}\end{matrix}$$

$$(45)$$

A detailed discussion of the relative preferences for each of these is beyond the scope of this work. However some generalizations are worthy of note.[97] Competition between substitution and elimination depends on a variety of factors; temperature, the nature of the attacking base, the substrate, its leaving group X and the solvent. In general elimination reactions have higher activation energies, making substitutions more favoured at lower temperatures. Also there is evidence that the fraction of substitution increases with the more efficient leaving groups, p-toluenesulphonates being preferred over halides. Elimination may be favoured by higher temperatures or the use of a bulky base which is more sterically hindered in its attack on carbon than on hydrogen. For this reason potassium t-butoxide is a currently favoured base for promoting elimination reactions.[98]

The slow step in the base-catalysed generation of carbenes, the intermediates in α-eliminations, is generally that of proton abstraction.[97] Hence it is the acidity of the protons α to the halide that governs the extent of this reaction. Haloforms are sufficiently acidic to be deprotonated by alkoxides[99] but dihalomethanes and alkyl halides generally require the strongest organometallic bases.[100] Using labelling techniques Doering and Kirmse[101] were able to assess the relative amounts of α- (67 per cent) to β-elimination (33 per cent) in the reaction of isobutylsodium with isobutyl chloride. Following α-elimination the carbene produces an olefin compound (64) by a hydride shift and cyclopropane (65) by a C–H insertion. β-elimination produces an olefin (63).

$$(CH_3)_2CHCD_2Cl \xrightarrow{Na} (CH_3)_2CHCD_2Na \qquad (46)$$

$$(CH_3)_2CHCD_2Cl + (CH_3)_2CHCD_2Na \longrightarrow$$

$$\underset{(63)}{(CH_3)_2C{=}CD_2} + \underset{(64)}{(CH_3)_2C{=}CHD} + \underset{\substack{\diagdown\,CH_2\,\diagup \\ (65)}}{CH_3CH{-\!\!-\!\!-}CHD} \qquad (47)$$

Aryl halides undergo a β-elimination reaction with carbanions giving benzyne intermediates, e.g. (66). These may dimerize or add excess of the organometallic reagent across the triple bond.[102]

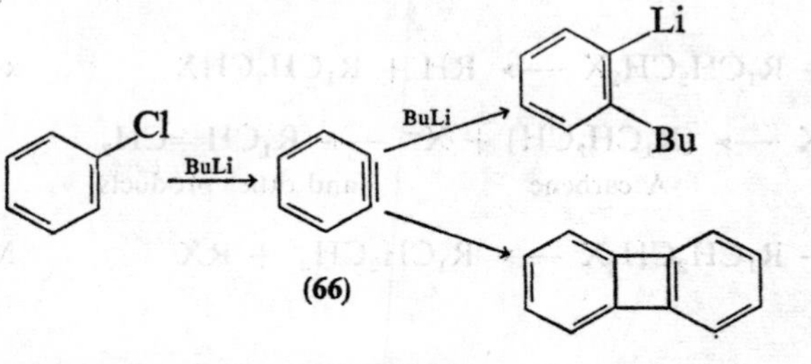

$$(48)$$

(66)

Metal halogen interconversions have already been discussed (section 10.4.5). They are only of importance when the pK_a of the acid $R_1CH_2CH_3$ is greater than that of RH.

In basic solution hydrocarbons of the requisite acidity may be oxidized by molecular oxygen to alcohols or peroxides. Russell and coworkers[103] favoured a radical chain mechanism for the t-BuOK-catalysed oxidation of triphenylmethane and fluorene.

$$R_3C^- + O_2 \rightarrow R_3C\cdot + O_2^{\overline{\cdot}} \tag{49}$$

$$R_3C\cdot + O_2 \rightarrow R_3COO\cdot \tag{50}$$

$$R_3COO\cdot + R_3C^- \rightarrow R_3COO^- + R_3C\cdot \tag{51}$$

$$R_3COO\cdot + O_2^{\overline{\cdot}} \rightarrow R_3COO^- + O_2 \tag{52}$$

$$R_3COO\cdot + R_3C\cdot \rightarrow R_3COOCR_3 \tag{53}$$

$$R_3C\cdot + O_2^{\overline{\cdot}} \rightarrow R_3COO^- \tag{54}$$

This allows for the formation of dimers and peroxides not accounted for by the simple cage-type recombination process.

$$R_3C^- + {}^1O\text{—}O^1 \rightarrow [R_3C^1, {}^1O\text{—}O^- \rightarrow R_3C^1, {}^1O\text{—}O^-] \rightarrow R_3COO^- \tag{55}$$

In a variety of protic and aprotic solvents other than DMSO triphenylmethyl hydroperoxide (R = Ph) is the major product. In DMSO the peroxy-anion is reduced by the solvent to alkoxide anion.

$$R_3COO^- + CH_3SOCH_3 \rightarrow R_3CO^- + CH_3SO_2CH_3 \tag{56}$$

Alkoxy-anions also result when alkyl peroxide anions react with excess of the organometal.

$$ROOMgX + RMgX \rightarrow 2ROMgX \tag{57}$$

Garst and coworkers[104] found that 5-hexenylmagnesium bromide in the presence of oxygen yielded a mixture of cyclic and acyclic alcohols (**68**) and (**67**). In its absence quenching with carbon dioxide gave 95 per cent of open chain acid (**69**). Here we have a striking example of the differing behaviour

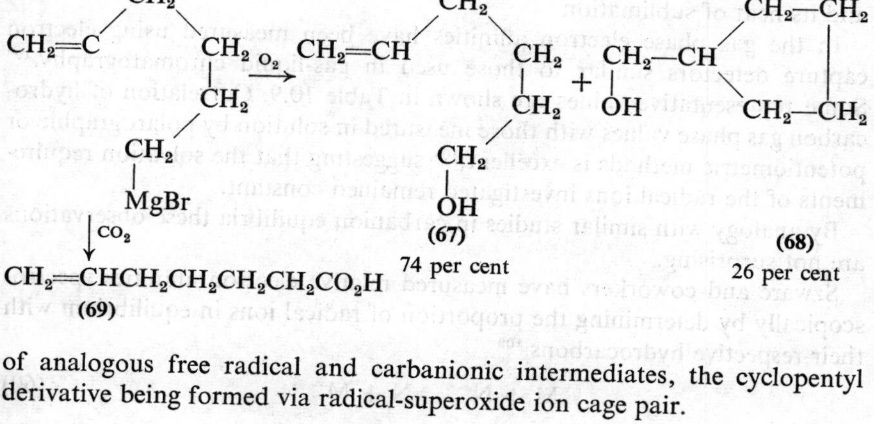

of analogous free radical and carbanionic intermediates, the cyclopentyl derivative being formed via radical-superoxide ion cage pair.

A measure of the sensitivity of carbanions to oxygen should be provided by the ionization potential of the anion.[89] In MO terms this is given by the energy of the highest occupied molecular orbital. For electrochemical oxidation of phenolic anions to their radicals such a correlation is observed.[105]

10.8 RADICAL ANIONS[107]

10.8.1 Generation

While metal salts of carbanions are formally derived from hydrocarbons by displacement of hydrogen by metal, radical anions are formed by addition of an electron to neutral molecules containing suitably low lying vacant molecular orbitals. Naphthalene, for example, is reduced by metallic sodium to sodium naphthalenide and benzophenone to sodium diphenylketyl.

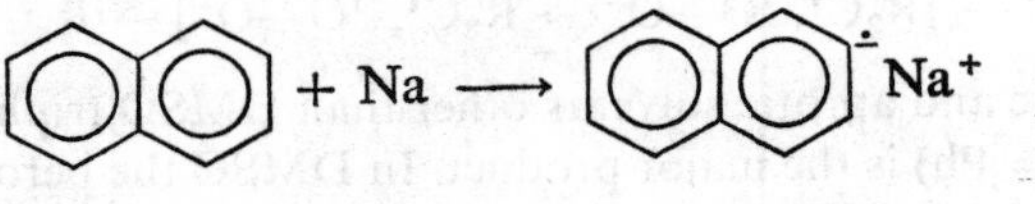

$$\text{(naphthalene)} + Na \longrightarrow \text{(naphthalenide)}^{\cdot -} Na^+ \qquad (58)$$

$$Ph_2CO + Na \longrightarrow Ph_2\dot{C}—O^-Na^+ \qquad (59)$$

Since the molecular orbitals involved are generally antibonding, the energy of the system is raised relative to the neutral molecule. Nonetheless, favourable electron transfer reactions can occur in cases where the antibonding levels involved are of lower energy than the ionization potential of the product radical ion.

In practice radical ions are most frequently generated in aprotic solvents using alkali or alkaline earth metals as reducing agents. The efficiency of the reduction process is dependent upon the electron affinity of the acceptor, the solvation requirements of the cation, the ionization potential of the metal, and its heat of sublimation.

In the gas phase electron affinities have been measured using electron capture detectors similar to those used in gas-liquid chromatography.[108] Some representative values are shown in Table 10.9. Correlation of hydrocarbon gas phase values with those measured in solution by polarographic or potentiometric methods is excellent,[107] suggesting that the solvation requirements of the radical ions investigated remained constant.

By analogy with similar studies in carbanion equilibria these observations are not surprising.

Szwarc and coworkers have measured relative electron affinities spectroscopically by determining the proportion of radical ions in equilibrium with their respective hydrocarbons.[109]

$$M + N^{\cdot -} \rightleftharpoons N + M^{\cdot -} \qquad (60)$$

Table 10.9 The electron affinities of some aromatic hydrocarbons and carbonyl compounds in the gas phase[108]

Compound	Electron affinity, eV
Naphthalene	0.152 ± 0.016
Triphenylene	0.284 ± 0.020
Phenanthrene	0.308 ± 0.024
Anthracene	0.552 ± 0.061
Pyrene	0.579 ± 0.064
Acetophenone	0.334 ± 0.004
Benzaldehyde	0.42 ± 0.010
1-Naphthaldehyde	0.62 ± 0.030
2-Naphthaldehyde	0.745 ± 0.070
9-Phenanthraldehyde	0.655 ± 0.14

The values obtained were found to be in good agreement with those determined by other methods. The advantage of this technique over alternatives lies in its accuracy in measuring small differences in electron affinity.

The solvent can have a significant effect on the efficiency of the metal reducing agent. This is because the solution equilibrium between alkali metals and hydrocarbons generally involves ion pairs which show the same sensitivity to changes in solvent, temperature and cation as the alkali metal fluorenylide ion pairs of Smid and coworkers. Thus in spite of the fact that the heat of sublimation and ionization potential are greater for lithium than for sodium metal, biphenyl is reduced to a greater extent by lithium in THP and DEE.* This switch in reactivity in solution must be accounted for by the differences in the solvation energy of the respective ion pairs. In dioxan, a poor cation solvator, known to favour contact pair formation, the situation is reversed and sodium is the more efficient reducing agent.[110]

Ion pair equilibria in the case of radical anions may be followed by ESR. In the contact pair there is a significant coupling of spins between the alkali metal counter-ion (^7Li, ^{23}Na, ^{39}K and ^{133}Cs are each endowed with a nuclear spin) and the unpaired electron. This is a function of the electrostatic interaction between them. In contact sodium naphthalenide in 2-MeTHF at 50 °C the sodium splitting is 1·55 gauss. On conversion to the solvent separated pair by reducing the temperature to −90 °C the coupling constant is reduced to 0·50 gauss.[111] Thus, by studying changes in line shape with temperature Hirota was able to derive thermodynamic data for the ion pair equilibria of various alkali-metal radical anion salts. His results nicely complement those of Smid[41] described earlier.

* THP = Tetrahydropyran; DEE = Diethoxyethane.

10.8.2 Disproportionation

In solution certain radical anions, for example those derived from anthracene and phenanthrene, can acquire a second electron either from one of their own kind (disproportionation) or from an alternative source to form dianions.

$$ArH^{\cdot -} + ArH^{\cdot -} \rightarrow ArH^= + ArH \tag{61}$$

This process is virtually unknown in the gas phase since the addition of the second electron results in a dinegatively charged species that lacks the overriding benefit of solvation and, as a result, is severely destabilized by Coulombic repulsion.

Initial attempts to understand the factors controlling disproportionation focussed on the effect of solvent on anion.[112] These met with limited success presumably because the ionic species involved were considered to be fully dissociated. In reality the radical ion and dianion are paired with their respective counter-ions and show the now familiar sensitivity to changes in temperature, cation and solvent.

Perhaps this can best be exemplified by considering one of the equilibria involved in the disproportionation of the tetraphenylethylene (TPE) radical anion.[107]

$$2TPE^{\cdot -}, Na^+ \overset{K_1}{\rightleftharpoons} TPE^=, 2Na^+ + TPE \tag{62}$$

Disproportionation is hindered in THF at moderately low temperatures (Table 10.10). Here, the sodium salt of the TPE radical anion exists as a

Table 10.10 Temperature dependence and thermodynamic parameters for the disproportionation of the TPE radical anion in THF[107,117]

Temperature °C	K_1	Activation parameters
20	400	$\Delta H_1 = 19 \pm 2$ kcal/mol
0	39	$\Delta S_1 = 75$ e.u.
−20	2·6	
−37	0·44	at 20 °C

solvent separated ion pair while its dianion is in the contact state. As a result, displacement of the equilibrium towards the radical ion is favoured by cation solvent combinations that show the greatest exothermicity of solvation; that is by small cations and highly basic solvents. The driving force for the production of the dianion in this situation arises in the main from a substantial gain in entropy caused by desolvation of the cation in the solvent separated ion pair.

When both radical anion and dianion are contact pairs, that is when Cs^+ is the cation or when the solvent is dioxan or ether, both known to be poor cation solvators, disproportionation of $TPE^{\pm}$ is quantitative. Now Coulombic stabilization is the governing factor, and that afforded the dianion by two contact cations is almost twice the value for the corresponding radical ion.

10.8.3 Dimerization

Loss of resonance stabilization precludes any tendency for aromatic radical ions to dimerize. Solutions of sodium naphthalenide, for example, can be stored without change for relatively long periods. However, olefins like 1,1-diphenylethylene, styrene and α-methylstyrene dimerize readily and from carbonation studies show a preference for the more thermodynamically acceptable tail to tail dimers.[113]

$$2 \quad \underset{Ph}{\overset{Ph}{\diagdown}} CH{=}CH_2 \rightleftharpoons 2 \quad \underset{Ph}{\overset{Ph}{\diagdown}} \overset{-}{C}{-}CH_2 \rightleftharpoons \underset{Ph}{\overset{Ph}{\diagdown}} \overset{-}{C}{-}CH_2CH_2{-}\overset{-}{C}\underset{Ph}{\overset{Ph}{\diagup}}$$

(63)

Olefins like tetraphenylethylene are presumably prevented from dimerizing on steric grounds and, for the same reason, the diphenylethylene dimer will not add to another molecule of its monomer.[114] It will however initiate polymerization of less bulky monomers like methyl methacrylate.[115]

Dimerization of α-methylstyrene can be limited by the presence of excess of the monomer so that the addition of the latter to the initially formed radical ion is favoured. This species may then dimerize to what is considered to be the head to head, tail to tail tetramer—one of the more commonly used initiators for two-ended polymerization reactions[116] (see Chapter 11).

It is interesting to note that the radical recombination rate constant for the free TPE radical ion in HMPA $(2 \times 10^4 \text{ M}^{-1}\text{s}^{-1})$[106,107] is less than for the ion pair in THF $(1 \times 10^6 \text{ M}^{-1}\text{s}^{-1})$[12] and much slower than the diffusion-controlled recombination rates of simple radicals $(c.\ 10^{10}\text{ M}^{-1}\text{s}^{-1})$. Presumably, the differences are related to the energy required to bring two similarly charged species together.

10.8.4 Radical ions as reducing agents

Organic halides are efficiently reduced to carbanions when treated with solutions of anion radicals. In the presence of protic reagents hydrocarbons are produced. Thus 5-hexenyl bromide[65] is reduced to 1-hexene on the addition of sodium naphthenilide. Radical cyclization and dimerization

$$\text{CH}_2=\text{CHCH}_2\text{CH(Br)CH}_3 + \text{Nap}^{\overset{\cdot}{-}} \text{Na}^+ \longrightarrow \quad + \text{NaBr} + \text{Nap}$$

$$\Big\downarrow \text{Nap}^{\overset{\cdot}{-}}\text{Na}^+$$

$$\xleftarrow{\text{H}^+}$$

reactions are unimportant indicating that reduction of the intermediate radical is a rapid step.

10.9 REFERENCES

1. C. D. Ritchie, *Solute Solvent Interactions*, ed. J. F. Coetzee and C. D. Ritchie, 219 pp, Marcel Dekker, New York 1969.
2. F. G. Bordwell, W. J. Boyle, Jun. and K. C. Yee, *J. Am. Chem. Soc.*, **92**, 5927, (1970).
3. D. J. Cram, *Fundamentals of Carbanion Chemistry*, pp. 42 and 180. Academic Press, New York, 1965.
4. A. J. Parker, *Quart. Revs.*, **16**, 174, (1962).
5. H. Fischer and D. Rewicki, *Progress in Organic Chemistry*, Vol. 7, p. 135, Butterworth, London, 1968.
6. J. B. Conant and G. W. Wheland, *J. Am. Chem. Soc.*, **54**, 1212, (1932).
7. W. K. McEwen, *J. Am. Chem. Soc.*, **58**, 1124, (1936).
8. A. Streitwieser, Jr., J. I. Brauman, J. H. Hammons and A. H. Pudjaatmaka, *J. Am. Chem. Soc.*, **87**, 384, (1965).
9. E. C. Steiner and J. M. Gilbert, *J. Am. Chem. Soc.*, **87**, 382, (1965).
10. (a) A. I. Shatenstein, *Dokl. Akad. Nauk SSSR.*, **60**, 1029, (1949). (b) A. I. Shatenstein, *Advan. Phys. Org. Chem.*, **1**, 153–201, (1963).
11. (a) A. Streitwieser Jun. and J. H. Hammons, *Progress in Phys. Org. Chem.* Vol. 3, p. 56, Wiley, New York, 1965.
12. (a) A. Streitwieser Jun., quoted in C. D. Ritchie and R. E. Uschold, *J. Am. Chem. Soc.*, **90**, 3415, (1968). (b) C. D. Ritchie and R. E. Uschold, *J. Am. Chem. Soc.*, **90**, 2821, (1968).
13. P. D. Bartlett and G. F. Woods, *J. Am. Chem. Soc.*, **62**, 2933, (1940).
14. W. von E. Doering and L. K. Levy, *J. Am. Chem. Soc.*, **62**, 2933, (1940).
15. R. G. Pearson and R. L. Dillon, *J. Am. Chem. Soc.*, **75**, 2439, (1953).
16. R. H. Boyd, *J. Phys. Chem.*, **67**, 737, (1963).
17. O. W. Webster, *J. Am. Chem. Soc.*, **88**, 3046, (1966).
18. D. J. Cram, *Fundamentals of Carbanion Chemistry*, p. 70, Academic Press, New York, 1965.
19. S. Oae, W. Tagaki and A. Ohno, *Tetrahedron*, **20**, 417, 427, (1964).
20. A. Streitwieser Jun., W. M. Padgett II and I. Schwager, *J. Phys. Chem.*, **68**, 2922, (1964).
21. H. Fischer and D. Rewicki, *Progress in Organic Chemistry*, Vol. 7, p. 138, Butterworth, London, 1968.
22. E. M. Kosower, *An Introduction to Physical Organic Chemistry*, p. 27, Wiley, New York, 1968.
23. A. Streitwieser, Jun., *Molecular Orbital Theory for Organic Chemists*, p. 256, Wiley, New York, 1961.
24. T. J. Katz, *J. Am. Chem. Soc.*, **82**, 3784, (1960).

25. (a) T. J. Katz and P. J. Garratt, *J. Am. Chem. Soc.*, **85**, 2852, (1963); (b) E. A. LaLancette and R. E. Benson, *J. Am. Chem. Soc.*, **85**, 2853, (1963).

26. R. Breslow and M. Douek, *J. Am. Chem. Soc.*, **90**, 2698, (1968).

27. (a) W. Reppe, O. Schlichting, K. Klager and T. Toepel, *Ann.* **560**, 1, (1948); (b) A. C. Cope and F. A. Hochstein, *J. Am. Chem. Soc.*, **72**, 2515, (1950).

28. A. Streitwieser, Jun., R. A. Caldwell and M. R. Granger, *J. Am. Chem. Soc.*, **86**, 3578, (1964).

29. A. Streitwieser, Jun., and R. G. Lawler, *J. Am. Chem. Soc.*, **85**, 2854, (1963).

30. A. Streitwieser, Jun., and F. Mares, *J. Am. Chem. Soc.*, **90**, 2444, (1968).

31. D. J. Cram. *Fundamentals of Carbanion Chemistry*, p. 48, Academic Press, New York, 1965.

32. A. Streitwieser, Jun. and W. R. Young, *J. Am. Chem. Soc.*, **91**, 527, 529, (1969).

33. C. D. Broaddus and D. L. Muck, *J. Am. Chem. Soc.*, **89**, 6533, (1967).

34. (a) M. Szwarc, 'Ions and Ion Pairs', *Accts. Chem. Res.*, **2**, 87, (1969). (b) S. N. Bhadani, J. Jagur-Grodzinski and M. Szwarc, *J. Am. Chem. Soc.* **90**, 6421, (1968).

35. (a) D. Margerison and J. P. Newport, *Trans. Faraday Soc.*, **59**, 2058, (1963). (b) T. L. Brown, R. L. Gerters, D. A. Bafus and J. A. Ladd, *J. Am. Chem. Soc.*, **86**, 2134, (1964).

36. W. G. Young, S. Winstein and H. L. Goering, *J. Am. Chem. Soc.*, **73**, 1958, (1951).

37. D. J. Cram, F. Willey, H. P. Fischer and D. A. Scott, *J. Am. Chem. Soc.*, **86**, 5370, (1964).

38. (a) D. J. Cram and L. Gosser, *J. Am. Chem. Soc.*, **85**, 3890, (1963); (b) D. J. Cram and L. Gosser, *J. Am. Chem. Soc.*, **86**, 5445, (1964); see also reference 31.

39. S. Winstein, E. Clippinger, A. H. Fainberg and G. C. Robinson, *J. Am. Chem. Soc.*, **76**, 2597, (1954).

40. E. Grunwald, *Anal. Chem.*, **26**, 1696, (1954).

41. J. Smid, *Structure and Mechanism in Vinyl Polymerization*, eds. T. Tsuruta and K. F. O'Driscoll, 345 pp., M. Dekker, New York, 1969.

42. T. E. Hogen-Esch and J. Smid, *J. Am. Chem. Soc.*, **88**, 307, (1966); **87**, 669, (1965).

43. T. E. Hogen-Esch and J. Smid, *J. Am. Chem. Soc.*, **88**, 318, (1966).

44. (a) P. Chang, R. V. Slates and M. Szwarc, *J. Phys. Chem.* **70**, 3180, (1966). (b) R. V. Slates and M. Szwarc, *J. Phys. Chem.*, **69**, 612, (1965).

45. L. Chan and J. Smid, *J. Am. Chem. Soc.*, **90**, 4654, (1968).

46. L. Chan and J. Smid, *J. Am. Chem. Soc.*, **89**, 4547, (1967).

47. K. H. Wong, G. Konizer and J. Smid, *J. Am. Chem. Soc.*, **92**, 667, (1970).

48. C. J. Pederson, *J. Am. Chem. Soc.*, **89**, 7017, (1967); *J. Am. Chem. Soc.*, **92**, 386, 351, (1970).

49. A. W. Langer, Jun., *Trans. New York, Acad. Sci.*, **27**, 741, (1965).

50. A. Maercker and J. D. Roberts, *J. Am. Chem. Soc.*, **88**, 1742, (1966).

51. D. N. Bhattacharyya, J. Smid, M. Szwarc, *J. Phys. Chem.*, **69**, 624, (1965).

52. P. D. Bartlett, S. Friedman and M. Stiles, *J. Am. Chem. Soc.*, **75**, 1771, (1953).

53. T. Shimonura, J. Smid and M. Szwarc, *J. Am. Chem. Soc.*, **89**, 5743, (1967).

54. T. E. Hogen-Esch and J. Smid, *J. Am. Chem. Soc.*, **89**, 2764, (1967).

55. A. Streitwieser, Jun. and R. A. Caldwell, *J. Org. Chem.*, **27**, 3360, (1962).

56. J. M. Mallan and R. L. Bebb, *Chem. Rev.*, **69**, 693, (1969).

57. M. Schlosser, *Angew. Chem. Internat. Edit.*, **3**, 362, (1964).

58. A. A. Morton and I. Hechenbleikner, *J. Am. Chem. Soc.*, **58**, 2599, (1936).

59. H. J. Gilman and B. J. Gaj, *J. Org. Chem.*, **28**, 1725, (1963).
60. R. A. Finnegan and R. S. McNees, *J. Org. Chem.*, **29**, 3234, (1964).
61. N. L. Bauld and J. H. Zoeller Jun., *Tetrahedron Letters*, **1967**, 885.
62. A. W. Langer Jun., *Polymer Preprints*, Vol. 7, No. 1, p. 132, American Chemical Society Division of Polymer Chemistry, Jan. 1966.
63. R. A. Benkeser and T. V. Liston, *J. Am. Chem. Soc.*, **82**, 3221, (1960).
64. C. D. Broaddus, T. J. Logan and T. J. Flautt, *J. Org. Chem.* **28**, 1174, (1963).
65. J. F. Garst, P. W. Ayres and R. C. Lamb, *J. Am. Chem. Soc.*, **88**, 4261, (1966).
66. H. O. House and V. Kramar, *J. Org. Chem.*, **27**, 4146, (1962).
67. (a) F. S. Acree, *Amer. Chem. J.*, **29**, 588, (1903). (b) G. Buckton, *Liebigs Ann. Chem.*, **1859**, 112, 222.
68. D. Bryce-Smith, *J. Chem. Soc.* (*London*), **1954**, 1079.
69. (a) A. A. Morton, G. M. Richardson and A. T. Hallowell, *J. Am. Chem. Soc.*, **63**, 327 (1941). (b) A. A. Morton, J. B. Davidson and H. A. Newey, *J. Am. Chem. Soc.*, **64**, 2240, (1942). (c) A. A. Morton and H. A. Newey, *J. Am. Chem. Soc.*, **64**, 2248, (1942). (d) A. A. Morton, J. B. Davidson, and R. J. Best, *J. Am. Chem. Soc.*, **64**, 2239 (1942).
70. K. Ziegler, F. Crössman, H. Kleiner and O. Schäfer, *Ann.*, **473**, 1 (1929).
71. L. Lochmann, J. Pospíšil and D. Lim, *Tetrahedron Letters*, **1966**, 257.
72. H. Gilman and R. E. Brown, *J. Am. Chem. Soc.*, **52**, 3314, (1930).
73. R. G. Jones and H. Gilman, *Organic Reactions*, Vol. IV, p. 339, Wiley, New York, 1951.
74. A. Streitwieser Jun. and D. R. Taylor, *Chem. Comm.*, **1970**, 1248.
75. (a) D. Y. Curtin and W. J. Koehl Jun., *Chem. and Ind.* (*London*) **1960**, 262. (b) D. Y. Curtin and W. J. Koehl Jun., *J. Am. Chem. Soc.*, **84**, 1967 (1962).
76. W. H. Glaze and C. M. Selman, *J. Org. Chem.*, **35**, 1987, (1968).
77. (a) H. M. Walborsky, F. J. Impastato and A. E. Young, *J. Am. Chem. Soc.*, **86**, 3283, (1964). (b) J. B. Pierce and H. M. Walborsky, *J. Org. Chem.*, **33**, 1962, (1968).
78. M. Witanowski and J. D. Roberts, *J. Am. Chem. Soc.*, **88**, 737, (1966).
79. (a) N. G. Krieghoff and D. O. Cowan, *J. Am. Chem. Soc.*, **88**, 1322, (1966). (b) F. R. Jensen and K. L. Nakamaye, *J. Am. Chem. Soc.*, **88**, 3437, (1966).
80. H. M. Walborsky and A. E. Young, *J. Am. Chem. Soc.*, **86**, 3288, (1964).
81. D. Y. Curtin and J. W. Crump, *J. Am. Chem. Soc.*, **80**, 1922, (1958).
82. (a) A. N. Nesmeyanov, A. E. Borisov and N. A. Volkenau, *Izv. Akad. Nauk. SSSR, Otdel. Khim. Nauk*, **1954**, 992. (b) A. N. Nesmeyanov and A. E. Borisov, *Tetrahedron*, **1**, 158, (1957).
83. V. R. Sandel, S. V. McKinley and H. H. Freedman, *J. Am. Chem. Soc.*, **90**, 495, (1968).
84. D. H. Hunter and D. J. Cram, *J. Am. Chem. Soc.*, **86**, 5478, (1964).
85. (a) A. Schriesheim and C. A. Rowe Jun., *Tetrahedron Letters*, **10**, 405, (1962). (b) S. Bank, A. Schriesheim and C. A. Rowe, Jun., *J. Am. Chem. Soc.*, **87**, 3244, (1965).
86. A. Streitwieser Jun. *Molecular Orbital Theory for Organic Chemists*, pp. 418 and 428, Wiley, New York, 1961.
87. S. Winstein, E. C. Dart and C. E. Watts, unpublished results.
88. R. B. Bates, R. H. Carnighan and C. E. Stables, *J. Am. Chem. Soc.*, **85**, 3030, (1963).
89. (a) H. Fischer and D. Rewicki, *Progress in Organic Chemistry*, Vol. 7, p. 154, Butterworth, London, 1968. (b) R. Kuhn and D. Rewicki, *Tetrahedron Letters*, **21**, 261, (1965).

90. C. K. Ingold, *Structure and Mechanism in Organic Chemistry*, p. 565, Cornell University Press, Ithaca, New York, 1953.

91. M. Eigen, *Angew. Chem. Internat. Edit.*, **3**, 1, (1964).

92. D. J. Cram. *Fundamentals of Carbanion Chemistry*, p. 207, Academic Press, New York, 1965.

93. G. A. Russell, *J. Am. Chem. Soc.*, **81**, 2017, (1959).

94. (a) J. E. Leffler, *The Reactive Intermediates of Organic Chemistry*, Interscience, New York, 1956. (b) D. C. Ayres, *Carbanions in Synthesis*, Oldbourne, London, 1966.

95. C. G. Swain and L. Kent, *J. Am. Chem. Soc.*, **72**, 518, (1950).

96. W. I. O'Sullivan, F. W. Swamer, W. J. Humphlett and C. R. Hauser, *J. Org. Chem.*, **26**, 2306, (1961).

97. J. Hine, *Physical Organic Chemistry*, McGraw-Hill, New York, 1962.

98. H. C. Brown and R. L. Klimisch, *J. Am. Chem. Soc.*, **88**, 1430, (1966).

99. J. Horiuti and Y. Sakamoto, *Bull. Chem. Soc. Japan*, **11**, 627, (1936).

100. G. L. Closs and L. E. Closs, *J. Am. Chem. Soc.*, **81**, 4996, (1959).

101. W. Kirmse and W. von E. Doering, *Tetrahedron*, **11**, 266, (1960).

102. (a) R. Huisgen and J. Sauer, *Chem. Ber.*, **92**, 192, (1959); (b) R. Huisgen, W. Mack, K. Herbig, N. Ott and E. Anneser, *Chem. Ber.*, **93**, 412, (1960).

103. G. A. Russell and A. G. Bemis, *J. Am. Chem. Soc.*, **88**, 5492, (1966).

104. R. C. Lamb, P. W. Ayres, M. K. Toney and J. F. Garst, *J. Am. Chem. Soc.*, **88**, 4262, (1966).

105. A. Streitwieser, Jun., *Molecular Orbital Theory for Organic Chemists*, p. 186, Wiley, New York, 1961.

106. M. Matsuda, J. Jagur-Grodzinski and M. Szwarc, *Proc. Roy. Soc. (London)*, Ser. A, **288**, 212, (1965).

107. (a) M. Szwarc, *Carbanions, Living Polymers and Electron Transfer Processes*, 297 pp., Interscience, New York, 1968. (b) M. Szwarc, *Progress in Phys. Org. Chem.*, Vol. 6, 323 pp. Wiley, 1968.

108. (a) W. E. Wentworth, E. Chen. and J. E. Lovelock, *J. Phys. Chem.*, **70**, 445, (1966). (b) R. S. Becker and E. Chen, *J. Chem. Phys.*, **45**, 2403, (1966).

109. J. Chandhur, J. Jagur-Grodzinski and M. Szwarc, *J. Phys. Chem.*, **71**, 3063, (1967).

110. (a) A. E. Shatenstein, E. S. Petrov and M. I. Belousova, *Organic Reactivity*, (Tartus State University, Estonia, U.S.S.R.). p **1**, 191 (1964). (b) A. E. Shatenstein, E. S. Petrov and E. A. Yakoleva, International Symposium on Macromolecular Chemistry, Prague, 1965, *J. Polymer Sci. C*, **16**, Part 3, 1729, (1967).

111. (a) N. Hirota, *J. Phys. Chem.*, **71**, 127, (1967). (b) N. Atherton and S. I. Weissman, *J. Am. Chem. Soc.*, **83**, 1330, (1961).

112. N. S. Hush and J. Blackledge, *J. Chem. Phys.*, **23**, 514, (1955).

113. (a) J. L. R. Williams, T. M. Laasko and W. J. Dulmage, *J. Org. Chem.*, **23**, 638, (1958). (b) W. Schlenk, J. Appenrodt, A. Michael and A. Thal, *Chem. Ber.*, **47**, 473, (1914).

114. M. Szwarc, *Proc. Roy Soc.*, **A279**, 260 (1964).

115. D. Freyss, P. Rempp and H. Benoit, *J. Polymer Sci.*, **B2**, 217 (1964).

116. C. L. Lee, J. Smid and M. Szward, *J. Phys. Chem.*, **66**, 904, (1962).

117. R. C. Roberts and M. Szwarc, *J. Am. Chem. Soc.*, **87**, 5542 (1965).

11

Anionic Polymerization

A. Parry

Imperial Chemical Industries Limited, Corporate Laboratory, Runcorn

11.1 INTRODUCTION

Anionic polymerization is concerned with those polymerization reactions in which the reactive species present at the end of a growing polymer chain is an anion. This strict definition may stand in need of modification when considering reactions that verge upon co-ordination and cationic polymerization,[1] but it will suffice at present.

The anionic polymerization of a monomer to a polymer involves the three main processes of initiation, propagation and termination found in all polymerization reactions. These three processes form convenient divisions in which to describe anionic polymerization. A practical example will serve to illustrate these three processes.

A flask fitted with a rubber septum is flushed with dry nitrogen. A 10 per cent solution of pure styrene in dry tetrahydrofuran is injected into the flask by syringe followed by a small quantity of n-butyllithium solution in hexane. A deep red colouration immediately forms in the flask and the reaction mixture becomes warm and increases in viscosity indicating that polymerization is taking place. The polymerization is complete in a few minutes but the red colour will persist for several hours or days depending on the purity of the system. When the polymerization is complete a small quantity of methanol is injected and the red colour is immediately discharged. If the now colourless solution is poured into an excess of methanol a white precipitate of polystyrene forms.

The following processes have taken place:

(a) Initiation

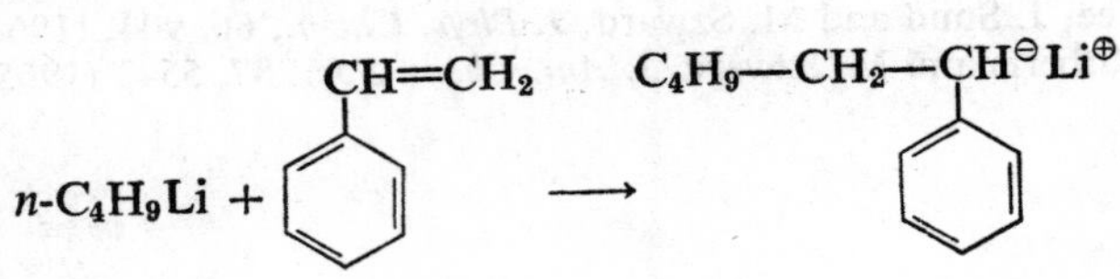

$$n\text{-}C_4H_9Li + \underset{\text{(styrene)}}{\overset{CH=CH_2}{\bigcirc}} \longrightarrow C_4H_9-CH_2-\overset{\ominus}{CH}Li^{\oplus}$$

(b) Propagation

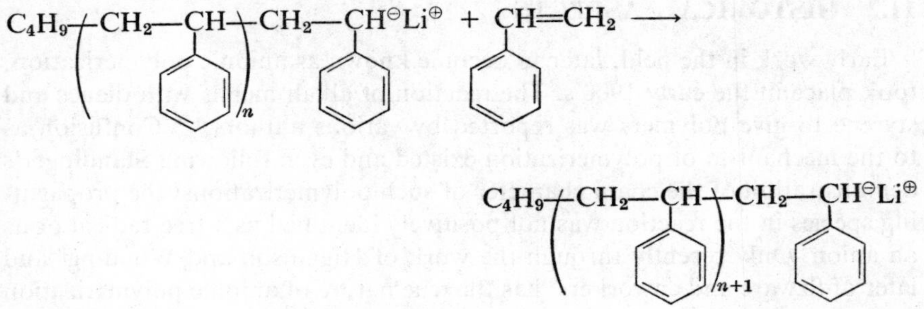

$$C_4H_9\left[CH_2-CH\right]_n CH_2-CH^\ominus Li^\oplus + CH=CH_2 \longrightarrow$$

$$C_4H_9\left[CH_2-CH\right]_{n+1} CH_2-CH^\ominus Li^\oplus$$

(c) Termination

$$C_4H_9\left[CH_2-CH\right]_n CH_2\overset{\ominus}{C}HLi^\oplus + CH_3OH \longrightarrow$$

$$C_4H_9\left[CH_2-CH\right]_n CH_2-CH_2 \quad + CH_3OLi$$

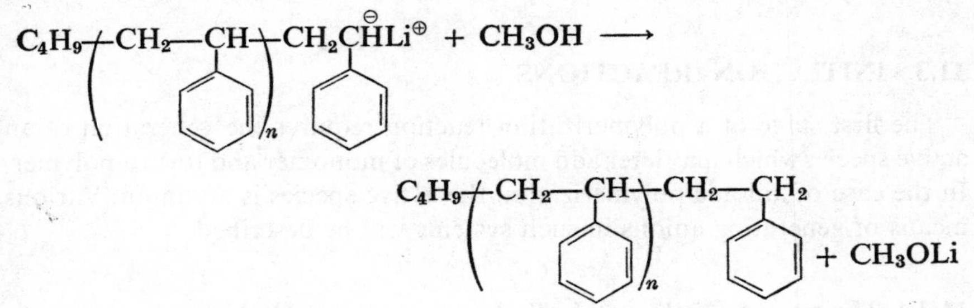

The most striking feature of this polymerization is that, in the absence of a terminating agent, the polymer chain will continue to grow in length until all the monomer has been consumed, and even then the active anion is still present on the end of the polymer chain ready to continue polymerization if more monomer becomes available. This concept has lead Szwarc to coin the name 'living polymers' for these systems.[2] This freedom from termination reactions has made the study of these systems extremely rewarding from a mechanistic point of view, and several important practical considerations also stem from this phenomenon:

(i) The virtual absence of termination reactions and the relative rapidity of the initiation reaction compared with the propagation reaction enables polymers with very narrow molecular weight distribution to be prepared.

(ii) The addition of a different monomer to a living system after the complete polymerization of the first monomer enables block copolymers to be prepared.

(iii) A suitably chosen terminating agent deliberately added to a living polymer system can form polymers with specific end groups such as —COOH, —COCl, —NH₂, —OH etc.

These factors have lead to a tremendous study of anionic polymerization over the past twenty years and recently, industrial processes using anionic polymerization have come into operation using techniques that a few years ago were confined to the laboratory vacuum line.

13

11.2   HISTORICAL ASPECTS

Early work in the field, later to become known as anionic polymerization, took place in the early 1900's. The reaction of alkali metals with dienes and styrene to give polymers was reported by various authors.[3-6] Confusion as to the mechanism of polymerization existed and even following Staudinger's demonstration of the chain character of such polymerizations[7] the propagating species in the reaction was not positively identified as a free-radical or as an anion. Only recently through the work of Higginson and Wooding[8] and later of Szwarc and coworkers[9] has the true nature of anionic polymerization been revealed. Professor Szwarc's pioneering work on living polymers has opened up the field of anionic polymerization as we know it today.

11.3   INITIATION REACTIONS

The first stage of a polymerization reaction requires the generation of an active species which may later add molecules of monomer and form a polymer. In the case of anionic polymerization the active species is an anion. Various means of generating anions in such systems will be described.

11.3.1   The use of alkali metal alkyls

Initiation by metal alkyls is achieved by adding across the double bond of the monomer in what amounts to a Michael addition, thus:

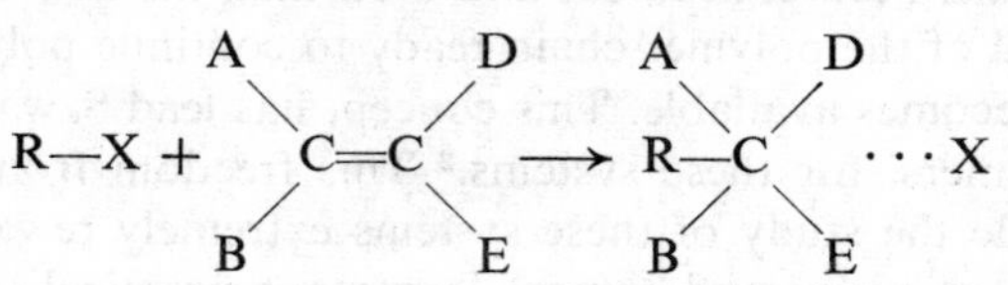

$$
R\text{---}X + \overset{\displaystyle A \qquad D}{\underset{\displaystyle B \qquad E}{C{=}C}} \longrightarrow \overset{\displaystyle A \qquad D}{\underset{\displaystyle B \qquad E}{R\text{---}C \quad \cdots X}}
$$

The $C \cdots X$ bond may have varying character from partially covalent with lithium initiators in non-polar solvents to completely ionic with other alkali metals in polar solvents. The most commonly used metal alkyl initiator is n-butyllithium although others such as benzyl potassium are in common use also.

n-Butyllithium, prepared by the reaction of lithium metal dispersion on an n-butyl halide in an aprotic solvent, is usually obtained commercially as a 10 per cent weight solution in hexane. The pure compound is a viscous liquid which is very reactive and is quickly destroyed on exposure to air. Butyllithium solution may be conveniently dispensed by hypodermic syringe, the operation being conducted in a dry box. The C—Li bond in butyllithium is believed to be partially ionic and partially covalent[10] in contrast to butylsodium where the bond is ionic. In hydrocarbon solution, where the C—Li bond is largely covalent, butyllithium exists as a hexamer in equilibrium with a small amount of the free monomer.[11] It has been found that the active species in most reactions is the monomeric form of n-butyllithium and if the equilibrium is forced towards the monomeric state by complexing with N,N,N′,N′-tetramethylethylenediamine then butyllithium becomes far more reactive than normal and readily carries out metallation reactions.[12] Similarly with initiation reactions it seems that the active species in initiation is actually the monomeric form of n-butyllithium. Secondary and tertiary butyllithiums, less associated in solution than n-butyllithium because of their bulkier side chains, are more active polymerization initiators. The use of n-butyllithium to initiate polymerization in isoprene and styrene is illustrated below.

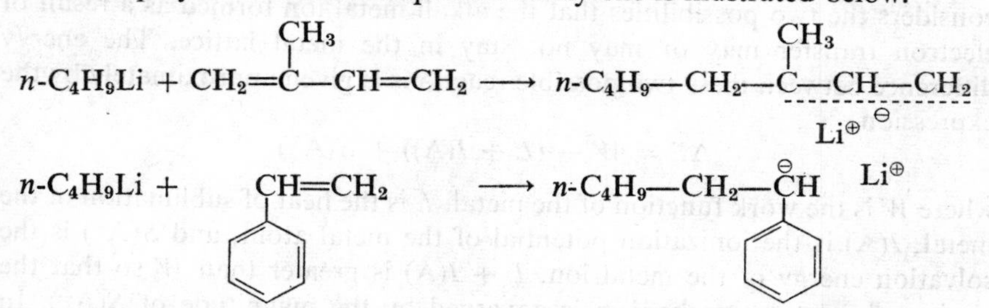

$$n\text{-}C_4H_9Li + CH_2{=}\underset{\underset{CH_3}{|}}{C}{-}CH{=}CH_2 \longrightarrow n\text{-}C_4H_9{-}CH_2{-}\underset{\underset{CH_3}{|}}{C}{-}CH{-}CH_2 \quad \overset{\ominus}{Li^{\oplus}}$$

$$n\text{-}C_4H_9Li + \underset{\bigcirc}{CH{=}CH_2} \longrightarrow n\text{-}C_4H_9{-}CH_2{-}\overset{\ominus}{\underset{\bigcirc}{CH}} \quad Li^{\oplus}$$

The carbanions formed are then ready to attack another molecule of monomer and so propagate the polymerization. The ease with which the above reactions occur depends on the basicity of the initiator with respect to the electron affinity of the particular monomer in the solvent system chosen. Styrene for example is initiated relatively slowly over several minutes in pure hydrocarbon media, but the addition of a small amount of tetrahydrofuran (on an equimolar basis to the amount of initiator used) speeds up the reaction many times.[11] This is believed to involve the formation of a monoetherate with butyllithium and thus effectively bring the initiator into the reactive monomeric state.

It should be noted that monolithium alkyls such as n-butyllithium produce a living polymer species that adds monomer only at one end, the other end of the chain being capped with a butyl group. This is in contrast to most other methods of initiation, see below, that produce living polymer species capable of adding monomer at both ends of the polymer chain. The production of polymer using one ended growth has important consequences both in kinetic studies[13] and in block copolymer preparation.[14] Monofunctional initiators with alkali metals other than lithium are easily prepared by an ether cleavage reaction of the alkali metal with cumylmethylether.[15]

11.3.2 Initiation with alkali metals

The direct reaction of a monomer with an alkali metal is a heterogeneous process and to speed up the reaction the alkali metal is used in a form that presents a large surface area to the monomer such as a mirror or a fine dispersion. In this case a direct one-electron transfer from the alkali metal to the monomer to form a radical anion can take place, for example

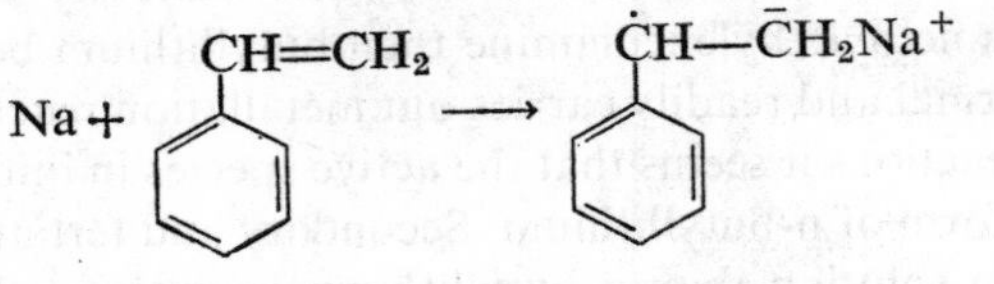

$$Na + \underset{\text{(styrene, } CH{=}CH_2)}{} \longrightarrow \underset{\text{(}\dot{C}H{-}\bar{C}H_2Na^+\text{)}}{}$$

A better understanding of this process is achieved by viewing the total reaction system and not merely the isolated reaction shown above. Szwarc[16] considers the two possibilities that the alkali metal ion formed as a result of electron transfer may or may not stay in the metal lattice. The energy difference between these two possible reactions is given approximately by the expression

$$\Delta E = W - (L + I(A)) + S(A^+)$$

where W is the work function of the metal, L is the heat of sublimation of the metal, $I(A)$ is the ionization potential of the metal atom, and $S(A^+)$ is the solvation energy of the metal ion. $L + I(A)$ is greater than W so that the choice of reaction mechanism is governed by the magnitude of $S(A^+)$. In solvents of poor solvating power for the alkali metal ion such as hydrocarbons $S(A^+)$ is small so that the first reaction is favoured and the metal ion remains in the lattice. The growing polymer chain is thus strongly bound to the metal surface by Coulombic attraction between its anion and the positively charged metal surface. Eventually the growing chain reaches a critical length where favourable entropy considerations cause transfer of a hydride ion to the metal surface and release the 'dead' (non-growing) polymer chain into solution. Initiation by alkali metals in hydrocarbon solvents is normally very slow even if possible at all.

In solvents of good solvating power for alkali metal cations such as ethers the $S(A^+)$ term becomes large enabling the second reaction to become energetically favourable and allow anion/cation pairs to pass into solution as living polymers.

The type of alkali metal used as initiator also has a large effect on the rate of the initiation reaction. The photoelectric work function, or the energy required to completely remove an electron from a solid metal, decreases along the series $Li > Na > K > Rb > Cs$. It would seem at first that this series should thus increase in reactivity from lithium to caesium. However the ionic size also increases in the same order.

Li	Na	K	Rb	Cs
0·68	0·97	1·33	1·47	1·67 Å

This causes the energy of solvation to decrease from lithium to caesium. These two effects compete and the choice of alkali metal giving the fastest initiation reaction may vary with solvent, temperature and electron affinity of the monomer in question.[13]

Transfer of one electron from alkali metal to monomer leads, as we have seen, to the formation of a radical anion. The possibility of transferring a second electron to the monomer to form a dianion depends largely on the monomer. In the case of *trans*-stilbene this reaction is possible.[17] In the case of

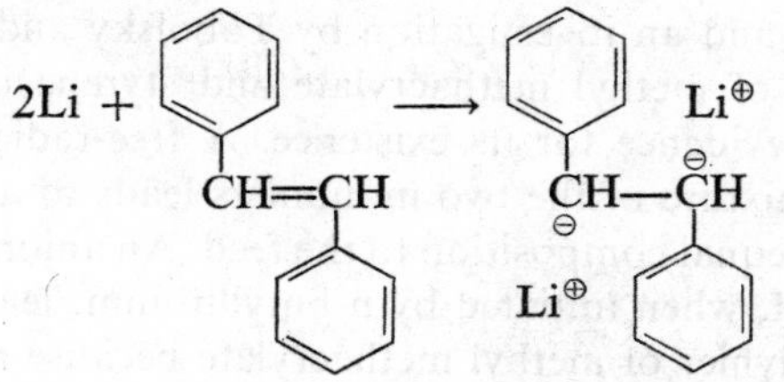

$$2\text{Li} + \overset{\displaystyle \text{C}_6\text{H}_5}{\underset{\displaystyle \text{C}_6\text{H}_5}{\text{CH=CH}}} \longrightarrow \overset{\displaystyle \text{C}_6\text{H}_5}{\underset{\displaystyle \text{C}_6\text{H}_5}{\overset{\ominus}{\text{CH}}-\overset{\ominus}{\text{CH}}}} \ \ \text{Li}^{\oplus} \ \ \text{Li}^{\oplus}$$

styrene, dianion formation is not possible because of the difficulty of stabilizing a second negative charge on the molecule. The styrene radical ion may react in two ways:

(a)   It may dimerize to give a dianion

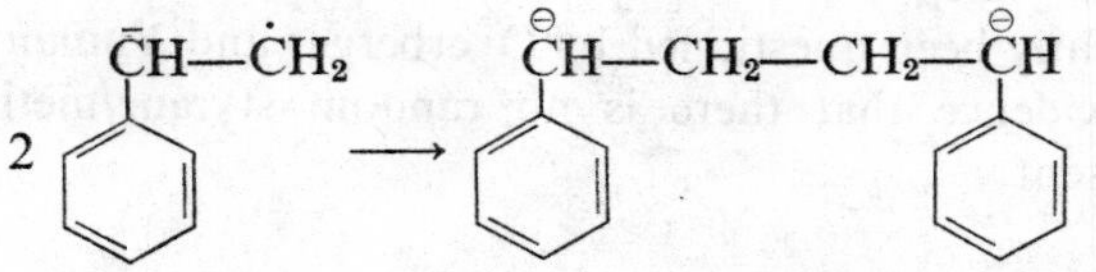

$$2 \ \ \overset{\displaystyle \text{C}_6\text{H}_5}{\overset{|}{\text{CH}}-\dot{\text{CH}}_2} \ \longrightarrow \ \overset{\ominus}{\text{CH}}-\text{CH}_2-\text{CH}_2-\overset{\ominus}{\text{CH}}$$

(b)   It may react with another molecule of styrene to produce a species having both an anion and a free-radical

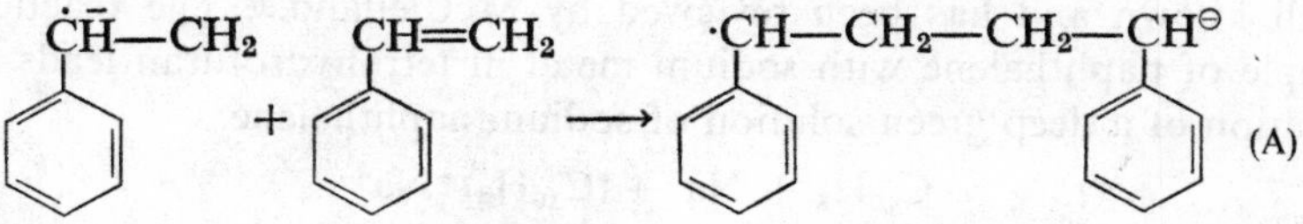

$$\dot{\overset{\displaystyle \text{C}_6\text{H}_5}{\text{CH}}}-\text{CH}_2 \ + \ \overset{\displaystyle \text{C}_6\text{H}_5}{\text{CH=CH}_2} \ \longrightarrow \ \cdot\text{CH}-\text{CH}_2-\text{CH}_2-\text{CH}^{\ominus} \quad \text{(A)}$$

This dimeric radical anion may add further monomer via free-radical or anionic addition, a further electron transfer may occur to give a dimeric dianion or the dimer may itself dimerize to give a tetrameric dianion

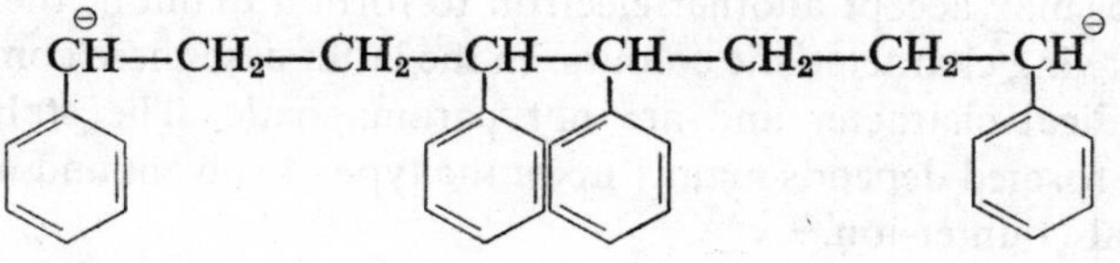

$$\overset{\ominus}{\text{CH}}-\text{CH}_2-\text{CH}_2-\text{CH}-\text{CH}-\text{CH}_2-\text{CH}_2-\overset{\ominus}{\text{CH}}$$

The eventual result of all these processes is to form an initiating species of low molecular weight capable of propagating anionic polymerization at each end of the chain. The speed of the propagation reaction with respect to the relatively slow heterogeneous initiation reaction means that it is very difficult to prepare low molecular weight polymers by this form of initiation, unless very refined techniques are used.[18] In addition with this type of initiation new growing chains are being formed throughout the reaction and the resulting polymer has a broad molecular weight distribution.

The intermediate species (A) having both a free-radical and an anionic end is interesting and an investigation by Tobolsky and coworkers[19] of the copolymerization of methyl methacrylate and styrene initiated by metallic lithium provides evidence for its existence. A free-radical initiated copolymerization of a mixture of the two monomers leads to a random copolymer of approximately equal composition to the feed. An anionic copolymerization on the other hand, when initiated by n-butyllithium, leads to an almost 100 per cent homopolymer of methyl methacrylate because methyl methacrylate anion is less electronegative than styrene and cannot initiate styrene polymerization. When, however, a mixture of methyl methacrylate and styrene was initiated by metallic lithium the polymer was found to contain about 10 per cent polystyrene. Tobolsky's conclusion was that species (A) was the active initiator growing pure polymethylmethacrylate at the anionic end and a methylmethacrylate/styrene copolymer at the free-radical end. This interpretation has recently been questioned by Overberger and Yamamoto[20] however, on NMR evidence that there is no random styrene/methylmethacrylate polymer present.

11.3.3 Initiation by alkali metal complexes

The formation of highly coloured complexes from the reaction of alkali metals with polynuclear aromatic hydrocarbons and ketones in ether solvents is well known and has been reviewed by McClelland.[21] The reaction for example of naphthalene with sodium metal in tetrahydrofuran leads to the formation of a deep green solution of sodium naphthalene.

$$C_{10}H_8 + Na \rightarrow (C_{10}H_8)^{\overline{\cdot}} Na^+$$

An anionic free-radical has been formed by the transfer of one electron from the alkali metal into a low energy unoccupied molecular orbital of the naphthalene molecule. The anionic free-radical is closely associated with a sodium cation to preserve electrical neutrality. Some substances, for example anthracene, may accept another electron to form a dianion, the mono- and dianions having characteristic colours. In the latter case these complexes have no free-radical character and are not paramagnetic. The stability of the complexes formed depends mainly upon the type of solvent and nature of the alkali metal counter-ion.[22]

Alkali metal complexes act readily as initiators of anionic polymerization provided that the relative electron affinities of initiator and monomer are suitable, for example the addition of styrene monomer to a green solution of sodium naphthalene in tetrahydrofuran causes the colour to change instantly to the red colour of living polystyrene anion, and the styrene to polymerize. No further colour change is observed until the living polystyrene is killed, whereupon the solution becomes colourless. A simple one electron transfer has taken place from the naphthalene radical anion to the lowest unoccupied antibonding π-orbital of the vinyl group of styrene to form a new radical anion

$$[\text{Naphthalene}]^{\overline{\cdot}}\,\text{Na}^+ + \text{Styrene} \leftrightharpoons \text{Naphthalene} + [\text{Styrene}]^{\overline{\cdot}}$$

This reaction is very rapid with the equilibrium far to the right at ambient temperatures. The styryl radical anion may then dimerize to a dianion and, as in the case of alkali metal initiation, form a polymer chain capable of growing at both ends.

Potassium benzophenone radical anion was found to be incapable of initiating styrene polymerization although capable of initiating methyl methacrylate polymerization.[22] Potassium benzophenone in THF is capable of accepting a second electron to form a dianion which will initiate styrene polymerization.

In general these hydrocarbon radical anions are only stable in good solvating solvents such as tetrahydrofuran, and an exchange of solvent from ether to benzene causes the equilibrium to shift in favour of free naphthalene and alkali metal powder is precipitated.

An alternative method of initiation to electron transfer from initiator to monomer in these systems, is bond formation between the two species.[21] Dilithiostilbene initiates the polymerization of isoprene in this way, polymer formation occurring at each end of the initiator.

The two modes of initiation, electron transfer or bond formation may be distinguished by the fact that only in the latter are initiator molecules incorporated in the polymer.

11.3.4 Initiation by living polymers

A living polymer may itself initiate the polymerization of a different monomer provided that the energetics of the reaction are favourable. The reaction

$$P^{\ominus}X^{\oplus} + M \rightarrow PM^{\ominus}X^{\oplus}$$

is governed by the relative basicities of the two species, anion donor and monomer acceptor in the particular counter-ion/solvent system chosen. A monomer reactivity table may be constructed for some of the more common monomers.[14]

Table 11.1 Monomer reactivity table

Relative reactivity group	Examples of monomer types	Typical propagating ends	pK_a range*
1	Styrenes, dienes	$-CH^{\ominus}$  (phenyl)	40–42
2	Acrylic esters	$\begin{array}{c} CH_3 \\ \| \\ -C^{\ominus} \\ \| \\ COOCH_3 \end{array}$	24
3	Cyclic oxides	$-CH_2CH_2O^{\ominus}$	15
4	Acrylonitriles, isocyanates	$\begin{array}{c} -CH^{\ominus} \\ \| \\ CN \end{array}$	(25)
5	Nitro alkenes	$\begin{array}{c} -CH^{\ominus} \\ \| \\ NO_2 \end{array}$	11

* Typical pK_a values of parent carbon acids. The absolute values of pK_a will obviously change with solvent but the relative positions will, in general, remain the same.

The positions of monomers in Table 11.1 may be correlated with the pK_a values of their anions as discussed earlier (Chapter 10).

The anions from monomers in a particular group will initiate polymerization of any member of a group lower in the table than themselves but not for a member of a higher group, e.g. polystyryl anion will initiate the polymerization of methyl methacrylate but the reverse will not occur. Members of a group will initiate polymerization in other members of the same group but there is usually a preferred direction e.g. polystyryl anion readily initiates the polymerization of isoprene, but the reverse reaction takes place more slowly. These phenomena have an important bearing on the preparation of block copolymers and will be discussed at a later stage.

11.3.5 Ether cleavage reactions

It may be desired to produce living polymers growing at only one end of the chain and yet having a counter-ion other than lithium. The higher alkali metal analogues of butyllithium are not suitable initiators for anionic

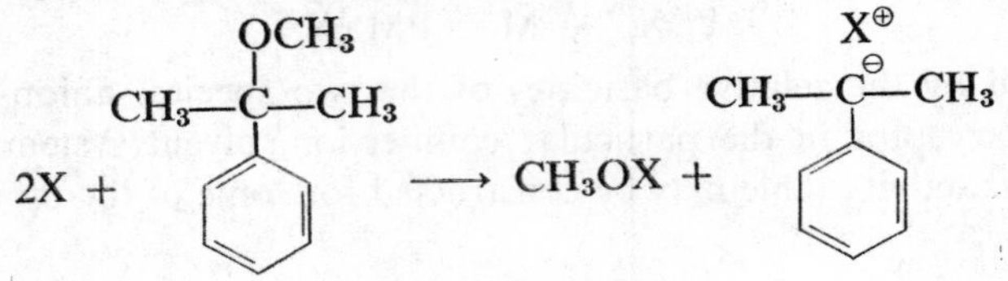

$$2X + \underset{\text{phenyl}}{\overset{\overset{\displaystyle OCH_3}{\underset{\displaystyle |}{CH_3-C-CH_3}}}{}} \longrightarrow CH_3OX + \underset{\text{phenyl}}{\overset{\overset{\displaystyle X^{\oplus}}{\underset{\displaystyle |}{CH_3-\overset{\ominus}{C}-CH_3}}}{}}$$

polymerization as they are largely ionic and insoluble in hydrocarbon media.[10] A convenient preparation of one-ended initiators from sodium, potassium, rubidium and caesium is by the ether cleavage reaction of the alkali metal upon cumylmethylether.[23] The cumylmethylether must be freshly prepared as it tends to decompose to α-methyl styrene on standing.[1]

11.3.6 Electrochemical initiation

In order to overcome the practical difficulties of using alkali metals and their compounds as initiators of anionic polymerization, some investigators have studied the direct electrochemical reduction of monomers to their anion. This method, though potentially very attractive, suffers from many practical difficulties. Funt and coworkers[24] have generated living polystyrene by the electrolysis of solutions of styrene in tetrahydrofuran in the presence of $NaAl(C_2H_5)_4$ or $NaB(C_6H_5)_4$ as supporting electrolyte. They found that it was necessary to use divided cells for the electrolyses as the generation of styryl anions in the cathode compartment was accompanied by the formation of an equimolar quantity of terminating species in the anode compartment. In the divided cell, polystyrene was isolated from the cathode compartment, whereas with an undivided cell no polymerization occurred. The terminating species formed at the anode were believed to be $Al(C_2H_5)_3$ or $B(C_6H_5)_3$ respectively.

Anderson[25] has pointed out that a competing mode of initiation to the direct electron transfer initiation proposed by Funt, would be the reduction of the sodium cation of the electrolyte to sodium metal which then causes conventional initiation. He has demonstrated a less ambiguous method of electrochemical initiation by the use of tetramethylammonium tetraphenylboride as supporting electrolyte for the electrolysis of styrene in hexamethylphosphoramide. Other workers[26] have demonstrated the electro-initiation of α-methyl styrene polymerization.

These systems, although interesting, have many restraints placed upon them. The solvent must be capable of dissolving the supporting electrolyte, monomer and polymer, and also allowing sufficient ionization of the electrolyte, while being inert to attack by reactive carbanions. The supporting electrolyte must also be inert to attack by carbanions and its cation must have a higher reduction potential than that of the monomer in use. These severe restrictions confine electrochemical initiation to the sphere of laboratory research at the present time.

11.4 PROPAGATION OF ANIONIC POLYMERIZATION

Following the initiation reaction where a stabilized carbanion is generated on a molecule of monomer a further molecule of monomer may be attacked to form a new carbanion.

$$R—X + M \rightarrow RM^-X^+$$
$$RM^-X^+ + M \rightarrow RMM^-X^+ \quad \text{etc.}$$

It may be seen that after the first few additions of monomer there is virtually no difference in reactivity between the two species:

$$R—(M)_n M^-X^+$$

and

$$R—(M)_{n+1}M^-X^+$$

and they will both attack new molecules of monomer at the same rate. In addition to this, termination reactions are virtually absent in many of the monomer systems in common use. Consequently, all the chains that have been initiated grow at the same rate until all the available monomer has been consumed. The polymer molecules at this stage still carry their reactive anion, and should more monomer become available the molecular weight will increase by further propagation. Only when the anion is removed (by, e.g. protonation) does the molecule become incapable of further growth.

In practice there is always a finite concentration of free monomer in equilibrium with the living chains but in most systems the equilibrium is almost totally towards polymer formation at ambient temperatures.

One notable exception to the above rule is found in the case of α-methyl styrene.

$$R—(M)_n—M^- + M \rightleftharpoons R—(M)_{n+1}M^-$$

The temperature at which the rates of the forward and reverse reactions in this equilibrium become equal is known as the ceiling temperature[27] and above this temperature no polymerization will occur. The ceiling temperature for a monomer depends upon free monomer concentration, e.g. α-methyl styrene will not polymerize at room temperature at concentrations below 1 molar although high polymer may be formed by cooling to $-40\ ^\circ$C. If the living polymer is not terminated at this temperature and the system is warmed to room temperature then depolymerization will occur to restore equilibrium. If on the other hand the polymerization is terminated at $-40\ ^\circ$C the polymer isolated from the reaction mixture is stable. Depropagation will not occur even at high temperatures unless the polymer is activated by, for example free-radical formation. Higher concentrations of monomer than 1 molar will form polymer at room temperature, polymerization occurring until the free monomer concentration has fallen to a level of 1 molar. In the case of α-methyl styrene initiated by electron transfer, polymerization to low molecular weight may take place at temperatures and concentrations normally forbidden by ceiling temperature considerations. This has been interpreted by Szwarc[28] in thermodynamic terms as a head to head monomer addition to the radical anion followed by a dimerization of the free-radical ends to give a tetramer of structure (1).

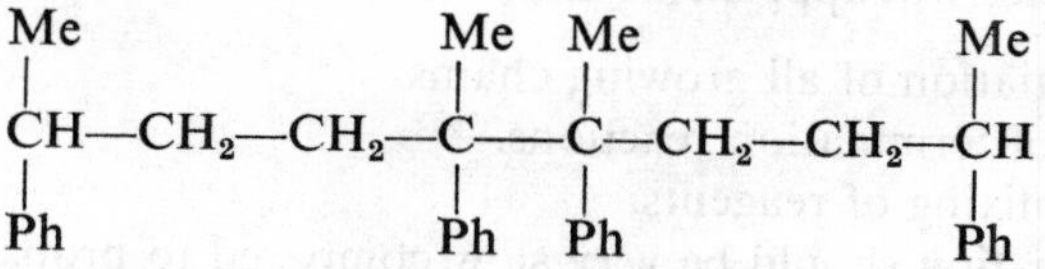

$$
\underset{\text{Ph}}{\overset{\text{Me}}{\mid}}{\text{CH}}-\text{CH}_2-\text{CH}_2-\underset{\text{Ph}}{\overset{\text{Me}}{\underset{\mid}{\overset{\mid}{\text{C}}}}}-\underset{\text{Ph}}{\overset{\text{Me}}{\underset{\mid}{\overset{\mid}{\text{C}}}}}-\text{CH}_2-\text{CH}_2-\underset{\text{Ph}}{\overset{\text{Me}}{\mid}}{\text{CH}}
$$

(1)

The ceiling temperature effects apply only to the conventional head to tail addition process which does not occur in this case. The structure (1) of the α-methyl styrene tetramer has recently been questioned by Richards[29] from NMR and mass spectral evidence. Richards favours structure (2)

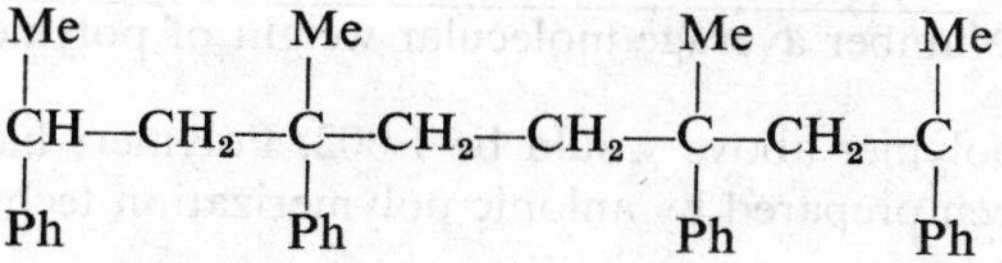

$$
\underset{\text{Ph}}{\overset{\text{Me}}{\mid}}{\text{CH}}-\text{CH}_2-\underset{\text{Ph}}{\overset{\text{Me}}{\underset{\mid}{\overset{\mid}{\text{C}}}}}-\text{CH}_2-\text{CH}_2-\underset{\text{Ph}}{\overset{\text{Me}}{\underset{\mid}{\overset{\mid}{\text{C}}}}}-\text{CH}_2-\underset{\text{Ph}}{\overset{\text{Me}}{\underset{\mid}{\overset{\mid}{\text{C}}}}}
$$

(2)

Whatever the structure of the α-methyl styrene tetramer, its dianion has proved to be a useful initiating species since it is highly coloured and has a low molecular weight compared with low molecular weight living polystyrene. Less sterically strained polymers than poly α-methyl sytrene have higher ceiling temperatures, e.g. living polystyrene has an equilibrium monomer concentration of 9.08×10^{-4} M at 150 °C in benzene.[30]

11.4.1 Molecular weight distribution

Anionic polymerization has been used to prepare polymers with very sharp molecular weight distributions and is at present the best method available for preparing these polymers. Anionically prepared polymers are used as standards for gel permeation chromatography (GPC) a technique for fractionating polymers mainly on a size basis.[31] GPC is commonly used in polymer chemistry for measuring the distribution of individual molecular weights found in a sample of polymer.

Polymers prepared by condensation, free-radical and other methods commonly contain a very large spread of molecular weights of individual molecules from monomers up to very high molecular weight polymer. This is a consequence partly of the many types of occurrence such as termination and forms of transfer that may befall a growing species, and partly of slow initiation reactions in these systems.

In the case of some anionic systems the ideal reaction conditions shown below may be almost achieved and the spread of molecular weights in the resultant polymer will approach a Poisson distribution.[32]

(a)   Rapid initiation of all growing chains.
(b)   Absence of termination reactions.
(c)   Efficient mixing of reagents.
(d)   Depropagation should be very slow compared to propagation.

The Poisson distribution function is very narrow, e.g. a polymer of average molecular weight 500 000 having a Poisson distribution will have 95 per cent of all chains with molecular weights within 10 per cent of the average. The sharpness of a molecular weight distribution is usually represented as the value

$$\frac{\text{Weight average molecular weight of polymer}}{\text{Number average molecular weight of polymer}}$$

which for the polymer above would be 1·002. Polymers having such distributions have been prepared by anionic polymerization techniques.[33]

11.4.2 Kinetics of the propagation reaction

The propagation reaction of anionic polymerization has been extensively investigated from a kinetic view-point. The special nature of the reaction, that is fast initiation rates compared to propagation rates and the virtual absence of termination reactions, has enabled the propagation reaction to be studied in isolation from competing reactions. In other forms of polymerization, initiation, propagation, transfer and termination reactions, all occur simultaneously making the study of one of these processes alone very difficult. The anionic propagation reaction has thus been studied in great detail and many interesting effects have come to light.

11.4.3 Ionic species

The active species in ionic polymerization, that is the polymeric ion, is not an independent agent, as is the free-radical, but is associated in some way with a gegenion to preserve electrical neutrality in the system. Studies of solutions of salts have shown that various species may exist in solution. The three main forms which exist in equilibrium with one another in solution are shown below.[34]

$$A^+B^- \rightleftharpoons A^+//B^- \rightleftharpoons A^+ + B^-$$

Ion pair   Solvent   Free ions
             separated
             ion pair

In addition, at high concentrations of salts, triple ions such as $(A^+B^-A^+)B^-$ and higher aggregates occur. The types of ionic species present in solution depend largely on the nature of the solvent and particularly on its dielectric constant. Strong electrolytes in solvents of high dielectric constant such as water are highly dissociated to free ions and only form ion pairs at high concentrations. On the other hand, in solvents of low dielectric constant, ion pairs predominate and at high concentrations many higher agglomerates exist.[35]

Anionic polymerizations are usually conducted in solvents of low dielectric constant such as hydrocarbons and ethers and hence are greatly influenced by ionic agglomeration. Kinetic investigations have shown the presence of various ionic species and demonstrated their influence on the course of the

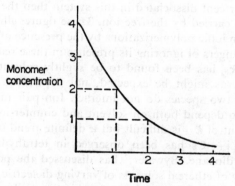

Figure 11.1 First order monomer consumption curve

propagation reaction. The polymerization of styrene in tetrahydrofuran may be used to illustrate this point. The propagation reaction is strictly first order in monomer, that is the half life of the reaction is constant over many half lives, as illustrated in Figure 11.1. The order with respect to initiator and hence concentration of growing chains (assuming complete use of initiator) is, however, complex. The rate constant calculated for the propagation reaction is very dependent on the concentration of growing chains at which it is measured. In the absence of termination reactions the concentration of living chain ends [LE] is constant throughout the reaction and $-d[M]/dt = k_p[LE][M]$ where [M] is monomer concentration. However k_p was found to vary inversely with [LE] and this has now been explained by Szwarc,[36] in terms of the reaction being carried by two types of ionic species, the free ion and the ion pair

Rate constant

Ion pair propagation $(M)_nM^-X^+ + M \rightarrow (M)_{n+1}M^-X^+$ $\qquad k_{ip}$

Rate constant

Free ion propagation $(M)_nM^- + M \rightarrow (M)_{n+1}M^-$ k_{fi}

Dissociation   Ion pair $\overset{K_d}{\rightleftharpoons}$ Free ions

It may be shown that the observed rate constant k_p is related to k_{ip}, k_{fi}, K_d, for small dissociation to free ions, by the expression

$$k_p = k_{ip} + (k_{fi} - k_{ip})K_d^{\frac{1}{2}}[LE]^{-\frac{1}{2}}$$

Thus from this expression if k_p is plotted against $[LE]^{-\frac{1}{2}}$ a straight line will result giving a value of k_{ip} at the intercept. If K_d is then estimated using conductivity measurements then k_{fi} may be found from the slope of the graph. These measurements have been made for the polystyrene/tetrahydrofuran system,[36] and it was found that $k_{fi} \simeq 64\,000$ and $k_{ip} \simeq 80$. This tremendous difference in the two propagation rates means that if the ion pairs are only 1 per cent dissociated in this system then the majority of the propagation is still carried by the free ion. These figures illustrate the great influence exerted on ionic polymerizations by the presence of the counter-ion and indicate the dangers of ignoring its presence in these reactions. The free ion rate constant k_{fi} has been found to be slightly solvent dependent[13] in these systems but, as might be expected, not dependent on the type of counter-ion as the two species do not interact. Ion pair rate constants k_{ip} have been found to depend both on solvent and counter-ion.[13] The experimental measurement of k_{ip} is difficult but a definite trend of decreasing k_{ip} along the series Li $\rightarrow$ Cs has been observed in tetrahydrofuran and an opposite trend in dioxane. Bywater[37] has discussed the polymerization of styrene in a variety of ethereal solvents of varying dielectric constant. In the series of dimethoxyethane, tetrahydrofuran, tetrahydropyran, oxepane (hexamethylene oxide) and dioxane there is a gradual decrease in the concentrations of free ions and solvent separated ion pairs as the solvent dielectric constant decreases and the contribution of ion pairs thus becomes more important.

From a practical view-point the changes in rate in going from a non-polar to a polar solvent are very marked. For example the anionic polymerization of styrene in benzene initiated by n-butyllithium is a slow reaction. On mixing the reagents no colour is apparent at first, and then very slowly the pale red colour of living polystyrene appears. The initiation reaction is very slow lasting several hours and the propagation reaction is no more rapid. If, on the other hand, a small quantity of tetrahydrofuran (equimolar to the amount of n-butyllithium used) is added, then almost instant initiation occurs and the propagation reaction takes place so rapidly that it may become dangerously exothermic.[38]

Temperature effects on anionic polymerizations are mostly conventional, an increase in temperature leading to an increase in propagation rate with activation energies of a few kilocalories per mole. This apparent simplicity disguises in most systems rather complex effects. It must be remembered that in these reactions many different ionic species are present capable of propagating the polymerization at very different rates. Changing the temperature causes a change in the relative proportions of the ionic species present as well as having a different effect on the polymerization rate of each species. Decreasing the reaction temperature in general reduces the ion pair and free ion rates, but increases the number of the reactive free ions present so that temperature effects are not so pronounced as in many reactions. An interesting 'negative activation energy' has in fact been noted for the ion pair rate constant for the polymerization of styrene in tetrahydrofuran using sodium counter-ion.[39] This was explained in terms of contributions from both contact and solvent separated ion pairs (having different reactivities), the relative concentrations of which changed with temperature.

The factor of 10^3 between the free ion and ion pair propagation rate constants would at first appear to invalidate one of the requirements for Poisson molecular weight distribution, namely that all the chains should grow at the same rate. However, the equilibrium between the various ionic species is rapid, and sharp molecular weight distributions are still produced. Schulz and coworkers[40] have measured a broadening of molecular weight distribution due to this effect, and shown that it can be overcome by adding an inert common-ion salt, such as sodium tetraphenylboride, to the polymerization medium to suppress the dissociation of ion pairs to free ions.

The presence of impurities in the reaction medium may have effects both on the rate of propagation and on the stereo-regularity of the polymer produced. Such impurities may be, for example alkali metal alkoxides, formed by reaction of the initiator or growing polymer with traces of water adsorbed on the walls of the reaction vessel. As we have seen, the course of the polymerization may be greatly influenced by the presence of trace amounts of some compounds present in concentrations comparable with the concentration of initiator. This may be of the order of 10^{-5} M for high molecular weight polymer formation. (Due to the living nature of the reaction the polymerization of one mole of monomer to a degree of polymerization of 10^3 requires 10^{-3} mole of initiator.)

Alkali metal alkoxides may often reduce polymerization rates. The alkoxide, which is usually formed from initiator destruction, has of course the same alkali metal as the gegenion. The presence of alkoxide may thus displace the equilibrium dissociation of free ions/ion pairs in favour of the far less reactive ion pairs. In methyl methacrylate polymerization, on the other hand, Wiles and Bywater[41] have shown that the presence of alkali metal alkoxides increases the polymerization rate, by complexing with the ion pair.

The influence of impurities has been shown in the anionic polymerization of styrene. Braun and coworkers[42] reported that the anionic polymerization of styrene with n-butyllithium initiation in toluene, benzene or hexane at

−30 °C formed some isotactic polymer. Bywater[43] reinvestigating this system found that under very pure conditions no isotactic polymer was formed but if small amounts of water were deliberately added, which would react with part of the initiator to form lithium hydroxide, then some of the polymer formed was indeed isotactic.

Studying the propagation reaction of sodium polystyryl in tetrahydropyran Szwarc[44] has shown that the presence of trace amounts of glyme solvents increase the propagation rate almost 200 times. This has been explained partly by an increased formation of free polystyryl ions due to solvation of the Na^+ cation and partly to the formation of reactive solvent-separated ion pairs.

These examples show how sensitive is anionic polymerization towards small changes in reaction conditions and indicate the dangers of extrapolating data measured under very pure conditions to more practical preparations with less rigorous purity control.

11.4.4 Polymer structure

(i) Polymerization of dienes

The anionic polymerization of isoprene and butadiene is of interest from both academic and commercial view-points. The ionic polymerization of dienes proceeds through only one of the two double bonds in the molecule and can form polymers of differing structures and properties. The structure of isoprene is

$$
\begin{array}{c}
CH_3 \\
| \\
CH_2{=}C{-}CH{=}CH_2 \\
\;1\quad\;2\quad\;3\quad\;\;4
\end{array}
$$

A unit of isoprene when incorporated into a polymer chain may have one of four structures as shown on facing page. Natural rubber is *cis*-1,4-polyisoprene and this is usually the most desired form of synthetic polyisoprene as it has the lowest glass transition temperature of the four forms and exhibits good rubber properties over a wide temperature range. Butadiene exhibits only three types of addition since the 1,2- and 3,4-modes are indistinguishable. The type of polymer formed in an anionic polymerization of these monomers is governed by the temperature, solvent and counter-ion employed in the reaction. *Cis*-1,4-polymer production necessitates the use of lithium counter-ions and non-polar solvents such as cyclohexane. The use of other counter-ions or of lithium counter-ion in polar solvents produces polymer of mixed microstructure in which the 1,2-, 3,4- and *trans*-1,4-forms predominate, see Table 11.2.

The type of microstructure of polydienes is generally accepted to be connected with the degree of ionization of the carbon-alkali metal bond, but a

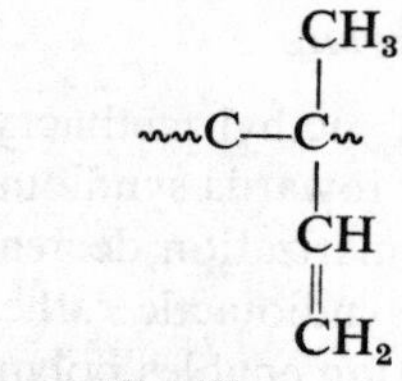

$$\text{wwwC—Cww} \quad \begin{array}{c} CH_3 \\ | \\ CH \\ \| \\ CH_2 \end{array} \qquad \text{1,2-Addition}$$

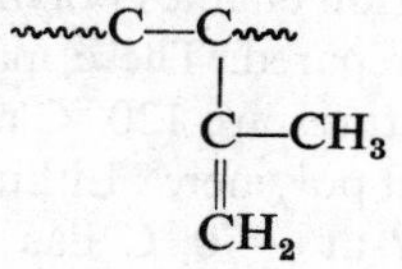

$$\text{wwwC—Cww} \quad \begin{array}{c} C—CH_3 \\ \| \\ CH_2 \end{array} \qquad \text{3,4-Addition}$$

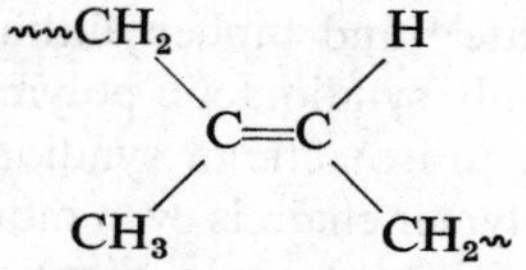

$$\begin{array}{c} \text{wwCH}_2 \quad H \\ C=C \\ CH_3 \quad CH_2\text{w} \end{array} \qquad \textit{Trans}\text{-1,4-addition}$$

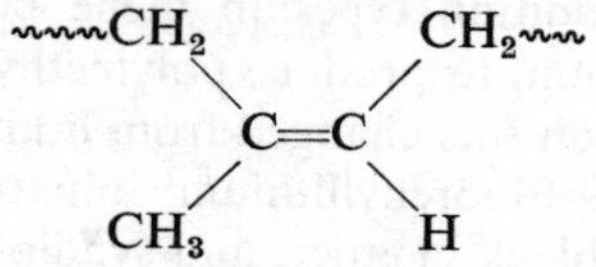

$$\begin{array}{c} \text{wwwCH}_2 \quad CH_2\text{www} \\ C=C \\ CH_3 \quad H \end{array} \qquad \textit{Cis}\text{-1,4-addition}$$

Table 11.2 Microstructure of polydienes prepared by anionic polymerization

Monomer	Solvent	Counter-ion	cis-1,4 per cent	trans-1,4 per cent	3,4 per cent	1,2 per cent	Reference
Isoprene	Hydrocarbon	Li	91	—	9	—	45
	Cyclohexane/THF[a]	Li	—	26	66	9	46
	Cyclohexane	Na	29	29	42	—	46
	THF[a]	Na	—	—	82	18	46
Butadiene	Hexane	Li	43	50	7	—	47
	THF[a]	Li	0	9	91	—	47
	Benzene	Na	23	45	32	—	48
	THF[a]	Na	5	15	80	—	49

a. THF = tetrahydrofuran.

general theory that fits all the experimental data is yet to be evolved. Cyclic transition complexes have been postulated[50] to account for the *cis*-1,4-mode of addition in the lithium catalysed polymerization of isoprene in non-polar solvents.

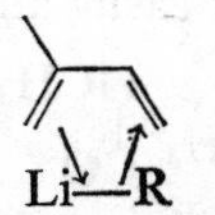

$$\text{Li}\underset{}{\overset{}{\longleftarrow}}R$$

The difficulty of formulating a general theory is increased by the various types of ion pair, association complex and counter-ion that are present in the reaction. For a review of current theories the reader's attention is directed to reference 51.

(ii) *Polymerization of methyl methacrylate*

The free-radical polymerization of methyl methacrylate leads to polymer of random structure with a tendency towards syndiotactic placements which increases as the temperature of polymerization decreases.[52] This is due to a slightly lower activation energy for syndiotactic rather than isotactic placement. The use of anionic polymerization enables polymethylmethacrylates of high syndio- or isotacticity to be prepared. These materials are interesting in that their glass transition temperatures are 120 °C and 60 °C respectively, compared with 100 °C for free-radical polymer.[53] Lithium aluminium hydride initiator in toluene or diethyl ether at −70 °C has been used to prepare highly isotactic polymethylmethacrylate[54] and biphenyllithium in glyme at −70 °C has been used to prepare highly syndiotactic polymethylmethacrylate.[55] The detailed mechanism leading to isotactic or syndiotactic placement of a monomer adding to a growing polymer chain is even more complex than the situation leading to various addition types in diene polymerization. Alkali metal ketyl initiators were found to produce polymethylmethacrylate of differing tacticity as the counter-ion was changed from lithium to sodium in a tetrahydrofuran medium.[54] 9-Fluorenyllithium initiation produced isotactic polymer in toluene, stereo-block in ether, and syndiotactic polymer in tetrahydrofuran.[56] Methyl methacrylate polymerization is complicated when compared with diene polymerization, for not only have various ionic forms to be taken into account, but also the possibility that some of the initiators may be heterogeneous in action and that alkoxides formed by destruction of initiator may complex with the growing species.[57]

Stereo-order in the growing chain has been found to affect the rate of propagation in some monomers. In the case of methyl methacrylate[58] it has been postulated that slow polymerization occurs immediately following initiation until by chance a small sequence of isotactic polymer is formed. This may assume a helical configuration to which further addition of monomer, in an isotactic manner, may occur rapidly. When eventually a 'mistake' is made in the addition, and a syndiotactic placement takes place, then the propagation rate falls drastically until further random additions have again built up a helical system and fast polymerization is resumed. The slow propagation of non-helical polymer is due to the formation by 'back biting' of a cyclic trimer which does not easily add monomer.

$$\begin{array}{c} OCH_3 \\ | \\ CH_3 \;\; CO \;\;\;\; CH_3 \\ | \;\; | \qquad | \\ \sim\sim CH_2-C \;\; Li^+ \cdots C \\ | \qquad\qquad | \\ CH_2 \qquad CH_2 \;\; COOCH_3 \\ \qquad\quad C \\ | \qquad\qquad | \\ CH_3 \qquad\qquad COOCH_3 \end{array}$$

When the polymer chain has assumed a helical configuration the relative positions of the carbanion and carbonyl group are too distant to form the six-membered ring complex.

Many such interesting effects have been noted in methyl methacrylate polymerization. A recent paper[59] describes a replication process reminiscent of DNA synthesis in which helical molecules of syndiotactic polymethylmethacrylate will act as a template for the production of isotactic polymethylmethacrylate grown in its presence. Such studies perhaps herald the advent of a second generation of polymerization reactions with high degrees of control over the polymers produced.

11.5 TERMINATION REACTIONS

The termination of an anionic polymerization reaction may be carried out deliberately so as to put a specific end group on the polymer chain at the end of the reaction, or it may occur through reaction with impurities in the system, or by attack upon its own monomer or polymer. The anionic polymerization of styrene constitutes perhaps the best example of a 'living' system in which termination reactions, in the absence of impurities, are virtually absent. Other monomers do not behave in such an ideal manner. Methods of termination of anionic polymerization will be discussed below.

11.5.1 Reaction with proton donors

When deliberately terminating a polymerization after all the monomer has been consumed it is often desirable to have a known end group on the chain. Termination with a proton donor such as an alcohol or an acid is often carried out to produce polymers with saturated hydrocarbon end groups.

$$PCR_2^-X^+ + CH_3OH \rightarrow PCR_2H + MeOX$$

Termination by admitting air to the apparatus may produce polymer chains with occasional —OH or —COOH end groups formed by termination with O_2 or CO_2. These groups may modify the properties of the polymer in an undesirable manner particularly where low molecular weight polymer is involved, as the ratio of end group to bulk polymer is high in this case.

11.5.2 Transfer reactions

A growing polymer chain may react with another species in such a way that its anion is transferred to the new species and the growing chain is terminated.

$$PCR_2^-X^+ + RH \rightarrow PCR_2H + R^-X^+$$

Two different possibilities then arise:

(a) R⁻ may be too stable to initiate further polymerization of monomer, in which case RH is acting as an inhibitor for the reaction. Small quantities of cyclopentadiene reacting in this way will inhibit isoprene polymerization due to formation of the stable cyclopentadienyl anion ($pK_a \sim 15$) which will not reinitiate isoprene ($pK_a \sim 38$) polymerization.

(b) If R⁻ can initiate further polymerization of the monomer then it is acting as a chain transfer agent. Toluene has been found to be a chain transfer agent in some systems,[60] and in this case benzyl anions formed from toluene are reactive enough to reinitiate polymerization. Toluene is acting to modify the polymerization in the same way as mercaptans are used as chain transfer agents in free-radical polymerization.[61]

The presence of the anionic chain transfer agent leads to a poor molecular weight distribution since transfer is a random process. One advantage of the transfer agent however is that complete conversion to low molecular weight polymer may be achieved with the use of only a small initiator concentration. Normally the production of a low molecular weight polymer needs large quantities of initiator so that the use of a chain transfer agent may be economically advantageous.

Transfer reactions to monomer or to polymer may also occur. Acrylonitrile is particularly susceptible in this respect,[62] and anionic polymerization leads to lower than theoretical molecular weights and production of branched polymer caused by transfer to monomer or polymer respectively.

11.5.3 Isomerization of the polymeric carbanion

Solutions of living polystyrene in tetrahydrofuran suffer a gradual change of absorption spectrum with corresponding loss of activity caused by isomerization of the carbanion to a more stable form.[63] Two consecutive reactions were found to occur in this process of which the first was rate determining:

$$\text{CH}_2\text{—CHPh—CH}_2\text{—}\overline{\text{CH}}\text{PhNa}^+ \rightarrow \text{CH}_2\text{—CHPh—CH}\text{=}\text{CHPh} + \text{NaH}$$

$$\text{CH}_2\text{—}\overline{\text{CH}}\text{PhNa}^+ + \text{CH}_2\text{—CHPh—CH}\text{=}\text{CHPh} \rightarrow \text{CH}_2\text{—CH}_2\text{Ph}$$

$$+ \text{CH}_2\underset{\underset{\text{Na}^+}{|}}{\text{—}\overline{\text{C}}\text{Ph}}\text{—CH}\text{=}\text{CHPh}$$

The 1,3-diphenylallyl anion formed is stable and will not initiate further styrene polymerization.

Styryl anion is relatively stable compared with the anion from methyl methacrylate which only forms true living systems at low temperatures (around $-78\,°C$).[64] At ambient temperatures attack on monomer or polymer ester groups can occur.[65]

11.5.4 Reaction with impurities

Anionic polymerizations are preferably carried out in all-glass reaction vessels under high vacuum or in an inert atmosphere such as argon. Kinetic studies of such systems require very rigorous cleanliness both of reagents and glassware for, as we have seen, trace impurities have a large influence on the course of the reaction. Glassware, for example, has a layer of water adsorbed on its surface which is very hard to remove but which can react readily with carbanions. Flaming the vessel under vacuum or washing with a living polymer solution are the usual methods of removing this layer.[66]

Living polymer solutions are quickly killed on exposure to air as they react with water vapour, oxygen, carbon dioxide etc. very rapidly. Oxygen is particularly troublesome as it reacts relatively slowly with carbanions and so is more difficult to remove by titration with initiator—the preferred method for removing impurities from solvents and monomers.

The reaction of living polymers with active compounds deliberately introduced to the system is of great importance in preparing polymers with specific reactive end groups. These materials may be used in further reactions, e.g. to produce block copolymers by coupling together reactive end groups. Two examples will be considered:

(a)   —COOH ended polystyrene[67]

$$\sim CH_2-CH^-Na^+ + CO_2 \longrightarrow$$
$$\underset{Ph}{|}$$

$$\sim CH_2-CHCOO^-Na^+ \xrightarrow{HCl} \sim CH_2-CHCOOH + NaCl$$
$$\underset{Ph}{|} \qquad\qquad\qquad \underset{Ph}{|}$$

The reaction with CO_2 must employ a high degree of mixing as the —COO$^-$ groups produced associate together very strongly and the resultant high viscosity of the reaction mixture prevents complete reaction of all the anions with CO_2.

(b)   —OH ended polyisoprene[68]

$$\sim Li^+ + \overset{O}{\overset{/\backslash}{CH_2-CH_2}} \longrightarrow$$

$$\sim CH_2-CH_2-O^-Na^+ \xrightarrow{CH_3OH} \sim CH_2-CH_2-OH + LiOMe$$

In this system the ethylene oxide may polymerize as well as adding one unit on to the end of the living polyisoprene. Rapid mixing of the reagents will overcome this tendency.

It should be noted that, although anionic polymerizations are very sensitive to impurities, the reactions become easier to carry out as their scale is increased because such factors as surface to volume ratio are decreasing. This has enabled anionic polymerizations to be carried out on a large scale in industrial processes, such as synthetic rubber production, with good control of polymer properties.

11.6   DESIGN FEATURES FOR ANIONIC POLYMERIZATION SYSTEMS

The question often arises as to which monomers are capable of being polymerized by an anionic mechanism. What factors will indicate whether or not a new monomer will polymerize in this way?

Attempts have been made[69] to correlate monomer anionic reactivity with the Q and e values for monomers originally proposed for the free-radical polymerization of vinyl monomers by Alfrey and Price.[70] In the Q-e scheme Q, the resonance factor, is related to the electron donating or accepting character of the vinyl bond; and e, the polarity factor, is related to the degree of polarization of the vinyl bond. Tables of Q and e values derived experimentally for a range of monomers have been published[70] and a plot of Q against e shows that monomers susceptible to free-radical, anionic or cationic polymerization mechanisms fall roughly into three areas of the graph. Such a plot was used by Konishi[71] to predict that vinylidene chloride would polymerize by an anionic mechanism. This was found to be so, but the polymerization was not clean as considerable elimination of LiCl had occurred.

The reactivity of monomers towards anionic polymerization has also been examined on the basis of LCAO molecular orbital theory.[72] Some limited success has been achieved[73] in relating the π-energy loss in the reaction to the rate of polymerization.

$$\text{Polymer anion} + \text{Monomer} \rightarrow \text{Polymer anion}$$

In practice several factors must be considered when assessing the suitability of a monomer for anionic polymerization:

(a)   Can the monomer form a stable anion either by electron transfer or by attack from a carbanion?

(b)   Can the anion derived from the monomer attack another molecule of the same monomer to form a new anion? This may be prevented by the anion having too great a stability, as is the case with cyclopentadiene or by the effect of steric hindrance preventing polymer formation as in the case of 1,1-diphenylethylene.[74]

(c)   Does the monomer contain reactive groups that could be attacked by the carbanion instead of the conventional attack on the monomer double bond? This can take place in the polymerization of methyl methacrylate if the reaction is carried out at elevated temperatures. Attack by carbanions on the ester group of monomer or polymer can lead to termination of the chain and elimination of alkoxide. Halogen substituted monomers are frequently difficult to polymerize because of elimination of alkali metal halides.

(d)   Can the anion isomerize to an unreactive form or undergo proton transfer from monomer to polymer? Proton transfer was found to occur in the anionic polymerization of acrylamide[75] leading to chain branching and irregular monomer addition.

Although this chapter is mainly concerned with anionic vinyl polymerization it must be noted that other monomers are capable of anionic polymerization. Cyclic monomers such as ethylene oxide, propylene sulphide, cyclic siloxanes, ε-caprolactam etc. may polymerize by a ring opening mechanism. In these cases however the reactive species is not usually a carbanion but —O⁻, —S⁻ etc. An interesting series of polymerizations of cyclic sulphides has recently been described by Morton.[76] It was shown that while the polymerization of ⟨S⟩ is carried by a —S⁻ ion the polymerization of ⟨S⟩ is carried by a carbanion. The carbanionic species was found to be far more reactive than the —S⁻ species, in fact the former was capable of initiating styrene polymerization while the latter species was not.

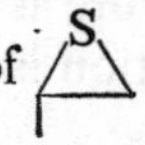

11.7   ANIONIC COPOLYMERIZATION REACTIONS

11.7.1 Random copolymerization

In some respects anionic copolymerization is simpler than free-radical copolymerizations and in other respects more complex.

The simple situation arises in, for example, the addition of an anionic initiator to a mixture of styrene and methyl methacrylate. The red colour of living polystyrene is never observed and the polymer isolated at the end of the reaction is almost pure polymethylmethacrylate containing at most one or two units of styrene. The reason for this is that while styryl anion will very easily initiate the polymerization of methyl methacrylate, methyl methacrylate anion is not sufficiently nucleophilic to initiate the polymerization of styrene. Hence all the initiator is quickly converted to growing polymethylmethacrylate chains. Likewise if methyl methacrylate is added to living polystyrene then a block copolymer of the two monomers will be produced. This is a general phenomena for monomers in different relative reactivity groups (section 11.3.4).

Random copolymerization occurs more readily for monomers within the same relative reactivity group where each anion may initiate either monomer. The copolymerization of styrene and dienes has been particularly well studied.

The familiar Mayo copolymerization equation[77] originally developed for free-radical systems has been used to investigate anionic copolymerizations.[78]

$$\frac{d[M_1]}{d[M_2]} = \frac{[M_1]}{[M_2]} \times \frac{r_1[M_1] + [M_2]}{r_2[M_2] + [M_1]}$$

where $r_1 = k_{11}/k_{12}$ and $r_2 = k_{22}/k_{21}$. The reactivity ratios r_1 and r_2 have the same meaning as in the free-radical system but in this case for two possible anions reacting with two possible monomers. The situation can now become quite complex however, as well as the various rates of propagation for the various free ions, ion pairs etc. found in the anionic polymerization of both single monomers the two systems may interact. Cross-association between ion pairs, solvation by the other monomer of an ion pair, salt effects on the dissociation of an ion pair and penultimate unit effects may all influence the reaction. In fact greatly differing behaviour is observed in these copolymerizations on changing the solvent, counter-ion or temperature as may be expected.

A useful feature of the diene/styrene copolymerization is that all four rate constants may be measured directly, namely

$$\text{p styrene}^- + \text{styrene} \rightarrow \text{p styrene}^- \quad k_{11}$$
$$\text{p diene}^- + \text{diene} \rightarrow \text{p diene}^- \quad k_{22}$$
$$\text{p styrene}^- + \text{diene} \rightarrow \text{p diene}^- \quad k_{12}$$
$$\text{p diene}^- + \text{styrene} \rightarrow \text{p styrene}^- \quad k_{21}$$

k_{11} and k_{22} are known, they are the two propagation rate constants for homopolymerization of the two monomers. k_{12} and k_{21} may be directly measured by spectrophotometry, e.g. by observing the rate of decrease of the styryl anion absorption at 340 mμ as a diene is added to living polystyrene.[79] The reactivity ratios may thus be calculated for these systems and used to predict copolymer composition using the integrated form of the copolymerization equation.

The system styrene/butadiene initiated by n-butyllithium in benzene was found[80] to proceed slowly at first, the solution being pale yellow in colour, and at a later stage to polymerize more rapidly and show the characteristic red colour of living polystyrene. The reason for this was that the cross-over reaction from polystyryl anion to polybutadienyl anion was very rapid compared to the cross-over reaction from polybutadienyl anion to polystyryl anion. Hence there was at first a high concentration of polybutadienyl anions (and no polystyryl absorption spectrum) and the copolymerization was virtually a homopolymerization of butadiene. At a later stage when most of the free butadiene had been polymerized the reaction became almost a homopolymerization of styrene. The polymers produced in this reaction were thus almost of an AB block type. Similar results were obtained for the styrene/isoprene system. The sensitivity of copolymer composition to changes in counter-ion or solvent is shown in the two examples below:

(a)   The copolymerization of styrene and isoprene in benzene using sodium counter-ion was shown to form polymer initially rich in styrene, the reverse of the case with lithium initiation.[82]

(b)   The copolymerization of styrene and isoprene in tetrahydrofuran with caesium counter-ion formed polymer initially rich in styrene.[83]

Reactions may occur in copolymerizations which are not normally encountered in a homopolymerization. The copolymerization of 1-vinyl naphthalene with styrene described by Szwarc is a case in point.[84] It was found that the cross-over reaction from polyvinylnaphthalene anion to styrene monomer was unusual and the following mechanism was proposed:

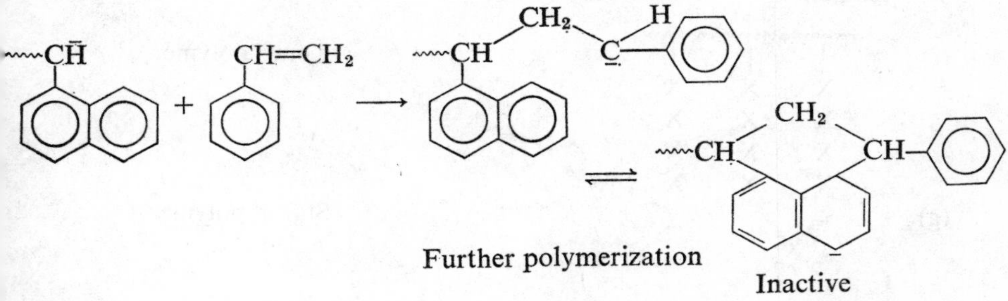

Further polymerization   Inactive

The amount of styryl anion available for further polymerization is small but its removal by polymerization to higher molecular weight allows the equilibrium to shift away from the inactive species until eventually all the inactive ends are used up. No similar phenomena have been noted in the homopolymerizations of these two monomers.

In the preparation of diene/styrene copolymers it is sometimes advantageous to prevent block formation of either polystyrene or polybutadiene. Small quantities of randomizing agents such as potassium t-butoxide or dimethoxyethane are added to the reaction mixture to give the required control over block formation.[85]

11.7.2 Block copolymerization

Polymer chemists have in the past few years become increasingly interested in block copolymers. Anionic polymerization has provided a means of preparing well-defined block copolymers and is a frequently used synthetic technique. A block copolymer contains sequences of homopolymers chemically linked together through covalent bonds. It may have quite different properties to a material of the same constitution but having the various monomers randomly incorporated into the molecule. Block copolymers may be of several different types:

(a) $(XXXX \cdots)_n—(YYYY \cdots)_m$ AB block copolymer

(b) $(YYYY \cdots)_n—(XXXX \cdots)_m—(YYYY \cdots)_n$
                                                                      ABA block copolymer

(c) $(XXXX \cdots)_n—(YYYY \cdots)_m—(ZZZZ \cdots)_o$ ABC block copolymer

(d) $(XXXX \cdots)_a—(YYYY \cdots)_b—(XXXX \cdots)_a—(YYYY \cdots)_b$ etc.
                                                                      Multi-block copolymer

(e) $(XXXX \cdots)—(XXXX \cdots)$ Stereo-block copolymer
      Isotactic   Syndiotactic

(f)   Comb polymer

(g)   Star copolymer

Anionic polymerization techniques enable block copolymers such as these
to be synthesized in the following ways:

(i) *Sequential Monomer Addition*

(a) *Monofunctional initiators.*   Monomer A is initiated by a monofunctional initiator such as n-butyllithium and is allowed to react to completion.
Monomer B is then added and its polymerization is initiated by the polymeric
carbanion of polymer A. When monomer B has completely reacted the
reaction is terminated and worked up to produce an AB block copolymer.
The length of each segment is determined by the amount of each monomer
added. This system is applicable to AB, ABA and higher block numbers
provided that each monomer block is capable of initiating the polymerization
of each following block. An example of this system is the AB block copolymerization of styrene and isoprene initiated by n-butyllithium.[86]

(b) *Difunctional initiators.*   Monomer B is initiated by a difunctional
initiator such as sodium naphthalene and allowed to react to completion.
Monomer A is then added to the still-living system and is polymerized on
both ends of the B chain to form an ABA block copolymer. This system is
applicable to ABA, ABCBA and higher block numbers provided that
relative monomer reactivity is observed as before. An example of this system
is the ABA block copolymerization of styrene/isoprene/styrene.[87]

(ii) *Coupling Reactions*

A living AB block copolymer prepared by the use of a monofunctional initiator as described above is reacted with an excess of a difunctional coupling agent, such as 1,6-dibromohexane, and the AB blocks are coupled together to form an ABBA block copolymer:

$$2(AAA \cdots)_n—(BBB \cdots)_m B^- Li^+ + Br(CH_2)_6Br \rightarrow$$

$$(AAA \cdots)_n—(BBB \cdots)_{m+1}—(CH_2)_6—(BBB \cdots)_{m+1}—(AAA \cdots)_n + 2\ LiBr$$

These coupling reactions are not always 100 per cent efficient and, for instance, a lithium/halogen exchange may occur instead of the desired elimination.[14] Nevertheless the coupling reaction may afford a block copolymer not accessible by other methods of synthesis. An example of such a coupling reaction has been described by Morton for the preparation of a multi-block copolymer of styrene and octamethylcyclotetrasiloxane, using dimethyldichlorosilane as coupling agent.[88]

Interesting multi-block copolymers have been prepared by Richards[89] by coupling difunctional living polymers with difunctional coupling agents. Monomers such as styrene, methyl methacrylate and dienes have been initiated by alkali metals in the presence of difunctional coupling agents such as $Br(CH_2)_nBr$. The polymers only grow to low molecular weight (dimers etc.) before coupling reactions occur. Multi-block copolymers up to 10 000 molecular weight have been prepared by this method.

Star copolymers have been prepared by reacting monofunctional living polymers with multifunctional coupling agents. The number of arms of the star correspond to the functionality of the coupling agent. Star polystyrene has been prepared using 1,2,4,5-tetrachloromethylbenzene as coupling agent.[90]

(iii) *Grafting reactions*

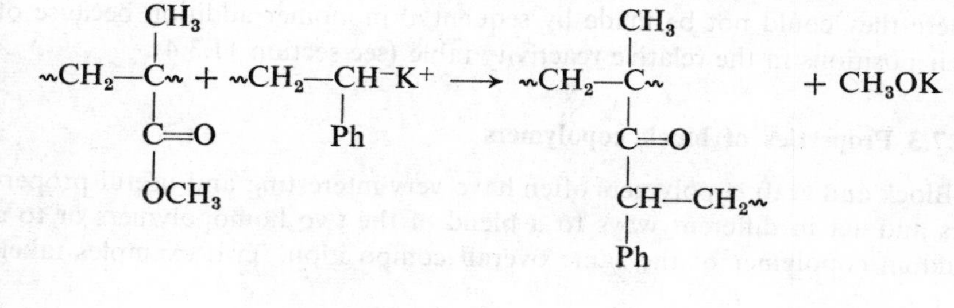

Comb copolymers have been prepared by the reaction of monofunctional living polymers on the polymer backbone of a preformed polymer. Preformed polymethylmethacrylate is attacked by living polystyrene to produce a comb copolymer in this way.[91] The attack may take place at any ester group on the polymethylmethacrylate backbone. The control of grafting sites may be achieved more precisely by building active sites into the backbone polymer, e.g. by incorporating small amounts of p-chlorostyrene into a polystyrene backbone.[92]

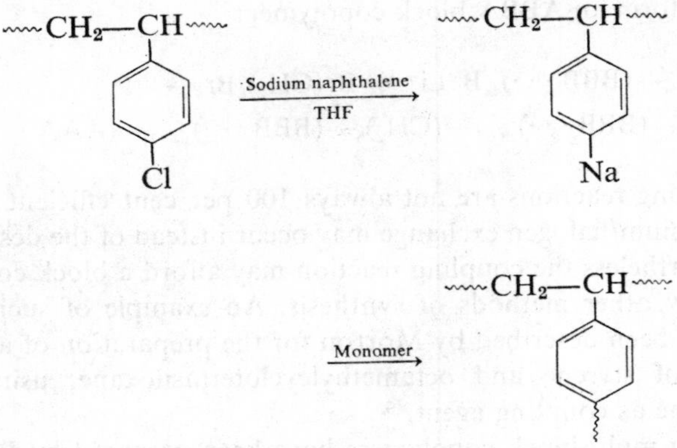

$$\sim\sim CH_2—CH \sim\sim \xrightarrow[\text{THF}]{\text{Sodium naphthalene}} \sim\sim CH_2—CH \sim\sim$$

$$\xrightarrow{\text{Monomer}} \sim\sim CH_2—CH \sim\sim$$

(iv) Termination with reactive groups

Polymers may be produced with reactive end groups by terminating an anionic polymerization with an appropriate killing agent. The reactive group may then serve as a site for the initiation of polymerization of a different monomer by anionic or free-radical mechanism, or may serve as a reactive species in a condensation polymerization. Gerber[93] has reported the use of

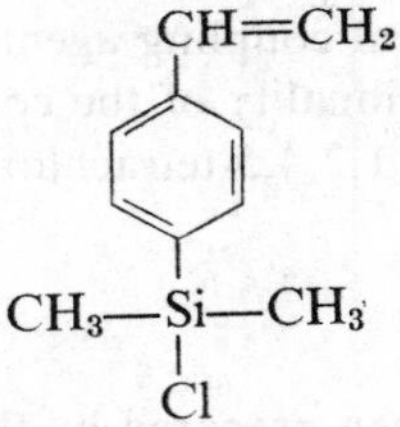

$$CH=CH_2$$

$$CH_3—Si—CH_3$$
$$|$$
$$Cl$$

and similar species as killing agents for various anionic polymerization systems. The reaction of polymer molecules having such functional end groups with alkali metals form new living species capable of block copolymer formation with a new monomer. This enables block copolymers to be formed where they could not be made by sequential monomer addition because of their positions in the relative reactivity table (see section 11.3.4).

11.7.3 Properties of block copolymers

Block and graft copolymers often have very interesting and useful properties and act in different ways to a blend of the two homopolymers or to a random copolymer of the same overall composition. Two examples taken from the large number of block copolymers which have been prepared will illustrate the possibilities of these materials.

(i) *Isoprene/methyl methacrylate AB block copolymers*

These materials exhibit interesting solution properties. The block copolymers are soluble in both acetone and linear saturated hydrocarbon solvents despite the fact that one of the components of the block is normally insoluble in both cases. A film cast from acetone solution, in which the polymethylmethacrylate segment is preferentially solvated, is a glassy film containing rubber particles. On the other hand a film cast from hexane solution, in which the polyisoprene is preferentially solvated, is a rubbery film containing glassy polymethylmethacrylate particles. Thus the same chemical compound can exist in two physically different forms.

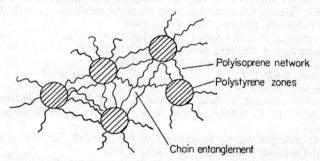

Polyisoprene network

Polystyrene zones

Chain entanglement

Figure 11.2   Structure of styrene/butadiene/styrene thermoplastic elastomer

(ii) *Styrene/butadiene/styrene ABA block copolymers*

These materials, manufactured by the Shell Chemical Company,[94] are thermoplastic elastomers; that is they are capable of being moulded and remoulded at elevated temperatures while behaving as cross-linked rubbers at ambient temperatures. A normal rubber has covalent cross-links built into the polymer during the process of vulcanization and cannot be reprocessed. The physical form of the thermoplastic elastomers is believed to be as shown in Figure 11.2. The polystyrene segments of the block copolymers aggregate to form glassy domains at ambient temperatures which effectively cross-link all the polyisoprene rubbery segments, and in addition act as reinforcing filler particles in the rubber. On heating the material above the glass transition temperature of the polystyrene (90–100 °C) the glassy polystyrene zones soften and the material can flow under stress. On cooling the glassy zones reform.

Such materials indicate the potential of the study of block copolymers and this field will undoubtedly bring further advances in the future.

11.8 REFERENCES

1. M. Szwarc, *Carbanions, Living Polymers and Electron-Transfer Processes*, Interscience, 1968.
2. M. Szwarc, *Nature*, **178**, 1169, (1956).
3. F. E. Matthews and E. H. Strange, British Patent, 24 790, (1910).
4. C. Harries, *Ann.*, **383**, 213, (1911).
5. K. Ziegler, H. Colonius and O. Schäfer, *Ann.*, **473**, 36, (1929).
6. W. Schlenk and E. Bergmann, *Ann.*, **479**, 42, (1929).
7. H. Staudinger, *Berichte*, **53**, 1073, (1920).
8. W. C. E. Higginson and N. S. Wooding, *J. Chem. Soc.*, **1952**, 760.
9. M. Szwarc, M. Levy and R. Milkovich, *J. Am. Chem. Soc.*, **78**, 2656, (1956).
10. G. E. Coates, *Organometallic Compounds*, 2nd Edition, Methuen, London, 1960.
11. S. Bywater, *Fortsch. Hoch Polym. Forsch.*, **4**, 66, (1965).
12. A. W. Langer Jun., *A.C.S. Polymer Preprints, Phoenix*, **7**, 132, (1966).
13. J. Smid in *Structure and Mechanism in Vinyl Polymerization*, eds. T. Tsuruta and K. F. O'Driscoll, Dekker, New York, 1969.
14. L. J. Fetters, *J. Polymer Sci.*, **C26**, 1, (1969).
15. K. Ziegler and H. Dislich, *Berichte*, **90**, 1107, (1957).
16. M. Szwarc, *Makromol. Chem.*, **35**, 132, (1960).
17. US Patent, 3 078 254.
18. Reference 1, page 159.
19. K. F. O'Driscoll, R. J. Boudreau and A. V. Tobolsky, *J. Polymer Sci.*, **31**, 115, (1958); K. F. O'Driscoll and A. V. Tobolsky, *J. Polymer Sci.*, **31**, 123, (1958).
20. C. G. Overberger and N. Yamamoto, *J. Polymer Sci.*, **B3**, 569, (1965).
21. B. J. McClelland, *Chem. Revs.*, **64**, 301, (1964).
22. Reference 13, page 377; S. Inoue, T. Tsuruta and J. Furukawa, *Makromol. Chem.*, **42**, 12, (1960).
23. K. Ziegler and H. Dislich, *Berichte.*, **90**, 1107, (1957).
24. B. L. Funt, D. Richardson and S. N. Bhadani, *Can. J. Chem.*, **44**, 711, (1966).
25. J. D. Anderson, *J. Polymer Sci.*, **A6**, 3185, (1968).
26. N. Yamazaki, S. Nakahama and I. Tanaka. Paper presented at the 17th meeting of the CITCE, Tokyo, Sept. 1966, Abstracts B307.
27. F. S. Dainton and K. J. Ivin, *Quart. Rev.*, **12**, 61, (1958).
28. M. Szwarc, *Trans. Faraday Soc.*, **58**, 2036, (1962).
29. R. L. Williams and D. H. Richards, *Chem. Comm.*, 414, (1967).
30. S. Bywater and D. J. Worsfold, *J. Polymer Sci.*, **58**, 571, (1962).
31. J. C. Moore, *J. Polymer Sci.*, **A2**, 835, (1964).
32. P. J. Flory, *J. Am. Chem. Soc.*, **62**, 1561, (1940).
33. F. M. Brower and H. W. McCormick, *J. Polymer Sci.*, **A1**, 1749, (1963). G. M. Guzman and A. Bello, *Makromol. Chem.*, **107**, 46, (1967).
34. S. Winstein and G. C. Robinson, *J. Am. Chem. Soc.*, **80**, 169, (1958).
35. C. A. Kraus, *J. Phys. Chem.*, **58**, 673, (1954); *J. Phys. Chem.*, **60**, 129, (1956).
36. D. N. Bhattacharyya, C. L. Lee, J. Smid and M. Szwarc, *J. Phys. Chem.*, **69**, 612, (1965).
37. G. Löhr and S. Bywater, *Can. J. Chem.*, **48**, 2031, (1970).
38. D. J. Worsfold and S. Bywater, *Can. J. Chem.*, **40**, 1564, (1962).
39. T. Shimomura, K. J. Tölle, J. Smid and M. Szwarc, *J. Am. Chem. Soc.*, **89**, 796, (1967).

40. R. V. Figini, G. Löhr and G. V. Schulz, *J. Polymer Sci.*, **B3**, 985, (1965).
41. D. M. Wiles and S. Bywater, *J. Phys. Chem.*, **68**, 1983, (1964).
42. D. Braun, W. Betz and W. Kern, *Makromol. Chem.*, **42**, 89, (1960).
43. D. J. Worsfold and S. Bywater, *Makromol. Chem.*, **65**, 245, (1963).
44. M. Shinohara, J. Smid and M. Szwarc, *J. Am. Chem. Soc.*, **90**, 2175, (1968).
45. H. Morita and A. V. Tobolsky, *J. Am. Chem. Soc.*, **79**, 5853, (1957).
46. D. J. Worsfold and S. Bywater, *Can. J. Chem.*, **42**, 2884, (1964); *Rubber Chem. Technol.*, **38**, 627, (1965).
47. R. V. Basova, A. A. Arest-Yakubovich, D. A. Solovykh, N. V. Desyatova, A. R. Gantmakher and S. S. Medvedev, *Dokl. Akad. Nauk, SSSR*, **149**, 1067, (1963), from *Chem. Abstr.*, **59**, 5343, (1963).
48. *Idem, Proc. Acad. Sci.*, *USSR*, **149**, 312, (1963).
49. A. A. Arest-Yakubovich and S. S. Medvedev, *Proc. Acad. Sci.*, *USSR*, **159**, 1305, (1964).
50. R. S. Sterns and L. E. Forman, *J. Polymer Sci.*, **41**, 381, (1959).
51. *Polymer Chemistry of Synthetic Elastomers*, Part II, eds. J. P. Kennedy and E. G. M. Törnquist, Interscience, 1969.
52. F. A. Bovey, *J. Polymer Sci.*, **46**, 59, (1960).
53. *Polymer Handbook*, eds. J. Brandrup and E. H. Immergut, Interscience, 1966.
54. T. Tsuruta, T. Makimoto and Y. Nakayama, *Makromol. Chem.*, **90**, 12, (1966).
55. A. Roig, J. E. Figueruelo and E. Llano, *J. Polymer Sci.*, **B3**, 171, (1965).
56. W. E. Goode, F. H. Owens, R. P. Fellmann, W. H. Snyder and J. E. Moore *J. Polymer Sci.*, **46**, 317, (1960).
57. L. Reich and A. Schindler, 'Polymerization by Organometallic Compounds', *Polymer Reviews*, Vol. 12, Interscience, 1966.
58. D. L. Glusker, E. Stiles and B. Yoncoskie, *J. Polymer Sci.*, **49**, 297, (1961).
59. T. Miyamoto and H. Inagaki, *Polymer Journal*, **1**, 46, (1970).
60. B. W. Brooks, *Chem. Comm.*, **2**, 68, (1967).
61. C. H. Bamford, W. G. Barb, A. D. Jenkins and P. F. Onyon, *The Kinetics of Vinyl Polymerization by Radical Mechanisms*, Butterworth, London, 1958.
62. A. Ottolenghi and A. Zilkha, *J. Polymer Sci.*, **A1**, 687, (1963).
63. G. Spach, M. Levy and M. Szwarc, *J. Chem. Soc.*, **1962**, 355.
64. R. K. Graham, D. L. Dunkelberger and E. S. Cohn, *J. Polymer Sci.*, **42**, 501, (1960).
65. H. Schreiber, *Makromol. Chem.*, **36**, 86, (1959); R. Rempp, V. I. Volkov, J. Parrod and C. Sadron, *Bull. Soc. Chim. France*, **1960**, 1919; P. Rempp and V. I. Volkov, *Compt. Rend.*, **250**, 1055, (1960).
66. D. J. Worsfold and S. Bywater, *Can. J. Chem.*, **38**, 1891, (1960).
67. T. A. Altares Jun., D. P. Wyman, V. R. Allen and K. Meyersen, *J. Polymer Sci.*, **A3**, 4131, (1965).
68. E. J. Goldberg, US Patent 3 055 952, (1962).
69. N. Kawabata, T. Tsuruta and J. Furukawa, *Makromol. Chem.*, **51**, 70, (1962).
70. T. Alfrey Jun. and C. C. Price, *J. Polymer Sci.*, **2**, 101, (1947).
71. A. Konishi, *Bull. Chem. Soc.*, *Japan*, **35**, 193, (1962).
72. K. Higasi, H. Baba and A. Rembaum, *Quantum Organic Chemistry*, Interscience, 1965.
73. A. Eisenberg and A. Rembaum, *J. Polymer Sci.*, **B2**, 157, (1964); J. Moacanin and A. Rembaum, *J. Polymer Sci.*, **B2**, 979, (1964).
74. D. Freyss, P. Rempp and H. Benoit, *J. Polymer Sci.*, **B2**, 217 (1964).
75. D. S. Breslow, G. E. Hulse and A. S. Matlack, *J. Am. Chem. Soc.*, **79**, 3760, (1957).
76. M. Morton and R. F. Kammereck, *J. Am. Chem. Soc.*, **92**, 3217, (1970).

77. F. R. Mayo and F. M. Lewis, *J. Am. Chem. Soc.*, **66**, 1954, (1944).
78. *Progress in Reaction Kinetics*, Vol. 3, ed. G. Porter, Pergamon, New York, 1965.
79. A. F. Johnson and D. J. Worsfold, *Makromol. Chem.*, **85**, 273, (1965).
80. M. Morton and F. R. Ellis, *J. Polymer Sci.*, **61**, 25, (1962).
81. Y. L. Spirin, D. K. Polyakov, A. R. Gantmakher and S. S. Medvedev, *J. Polymer Sci.*, **53**, 233, (1961); C. E. H. Bawn, *Rubber Plastics Age*, **42**, 267, (1961).
82. A. V. Tobolsky and C. E. Rogers, *J. Polymer Sci.*, **38**, 205, (1959).
83. A. Rembaum, F. R. Ellis, R. C. Morrow and A. V. Tobolsky, *J. Polymer Sci.*, **61**, 155, (1962).
84. F. Bahsteter, J. Smid and M. Szwarc, *J. Am. Chem. Soc.*, **85**, 3909, (1963).
85. H. L. Hsieh and W. H. Glaze, *Rubber Chem. Technol.*, **43**, 1, (1970).
86. M. Baer, *J. Polymer Sci.*, **A2**, 417, (1964).
87. S. Schlick and M. Levy, *J. Phys. Chem.*, **64**, 883, (1960).
88. M. Morton, A. A. Rembaum and E. E. Bostick, *J. Appl. Polymer Sci.*, **8**, 2707, (1964).
89. D. H. Richards, N. F. Scilly and Miss F. J. Williams, *Chem. Comm.*, **1968**, 1285.
90. S-P. S. Yen, *Makromol. Chem.*, **81**, 152, (1965).
91. G. Finaz, Y. Gallot, J. Parrod and P. Rempp, *J. Polymer Sci.*, **58**, 1363, (1962).
92. G. Greber and G. Egle, *Makromol. Chem.*, **53**, 208, (1962).
93. G. Gerber, *Makromol. Chem.*, **101**, 104, (1967).
94. Shell Chemical Company, British Patent, 1 000 090.

12

Reactivity and Mechanism in Polymerization by Complex Organometallic Derivatives

A. Ledwith

Donnan Laboratories, University of Liverpool

D. C. Sherrington,

Department of Pure and Applied Chemistry,
University of Strathclyde, Glasgow

A wide and ever growing variety of both simple and complex organometallic derivatives have now been shown to act as effective catalysts of the polymerization of many monomers, olefinic and otherwise. Furthermore, many of these possess powerful stereo-regulating properties and produce polymers with well-defined stereochemical patterns, particularly when used at low temperatures. Classification into catalyst groups with common physico-chemical characteristics is difficult and, in many cases, of dubious significance. There are, however, a few groups about which large funds of both experimental and theoretical information have accumulated in the literature over the past fifteen years. Two of the most important and well definable of these are the Ziegler-Natta catalysts[1,2] and those which we shall refer to as cobalt[3,4] and π-allyl type catalysts.[5] It is no coincidence that both of these are of fundamental importance in the commercial production of some of the more common plastics and rubbers encountered today. The purpose of this chapter is to discuss reactivity and mechanisms in polymerization by these two types of catalyst, with emphasis on the initiation and propagation reactions and their stereochemical implications. The polymerization of ethylene and α-olefins by Ziegler-Natta catalysts will be presented in the first section while the second will deal with the polymerization of conjugated diolefins. The latter section is itself divided into two parts, the first covering Ziegler-Natta systems and the second the newer catalysts based largely on cobalt and nickel complexes.

12.1 POLYMERIZATION OF ETHYLENE AND α-OLEFINS BY ZIEGLER-NATTA CATALYSTS

Ziegler-Natta type[1,2] catalysts are formed by the exposure of a transition metal salt to a metal alkyl or hydride, or alkylated metal halide, usually in an inert atmosphere. Metal alkyls of groups I–III and transition metal salts of groups IV–VIII are usually cited within the definition, though in practice metal alkyls are usually limited to those of aluminium, and the transition metal component to ionic species containing between zero and three d-electrons, and more specifically to titanium and vanadium derivatives. Catalysts based on transition metals further along the first period, especially cobalt and nickel, form homogeneous initiator systems which are particularly useful for the polymerization of conjugated dienes and shall be dealt with separately.

Most vinyl monomers have now been successfully polymerized by Ziegler-Natta catalysts. By far the most important application, however, is in the polymerization of non-polar olefins such as ethylene and propylene, and the conjugated di-olefins, butadiene and isoprene. Ethylene is easily polymerized[1] at low pressures ($\sim$ 1–10 atm) to a highly linear polymer.

$$CH_2 = CH_2 \rightarrow \sim\!CH_2\!-\!CH_2\!-\!CH_2\!-\!CH_2\!\sim$$

The absence of branches, which occur in polyethylenes produced by older methods involving extremes of temperature and pressure, allows the product to crystallize more easily, resulting in a higher melting point and, in general, better physical properties. Propylene can be polymerized exclusively to either the isotactic[2] or the syndiotactic[6] polymer. The isotactic form has a sufficiently high melting point (176 °C) and low production costs for the polymer to find wide domestic and industrial application. On the other hand, the syndiotactic form, which has equally good physical properties, is more costly to produce and is mainly of academic interest.

Research into and speculation concerning the mechanism of α-olefin polymerization has been widespread and many excellent reviews of the topic are already available.[7–12] Mechanistic details of diene polymerization are much more scarce. One reason for this, no doubt, is the wider variety of stereoregular polymers which can be produced (see below).

In any analysis of experimental observations on Ziegler-Natta type catalysts with a view to formulating a mechanism, it is important to realize that both components are often capable of individually initiating polymerization of particular vinyl monomers, and in these cases conventional cationic, anionic or free-radical mechanisms are operative. Such polymerizations are not of the Ziegler type, e.g. $AlEtCl_2$ or $TiCl_3$ will polymerize vinylethers via a cationic mechanism, and isoprene can be anionically polymerized by lithium butyl alone from the Ziegler-Natta system $LiBu/TiCl_4$. Without doubt the possibility of reaction involving only one component has contributed to many of the apparent experimental anomalies from studies of Ziegler-Natta catalysts.

12.1.1 Mechanism of initiation and propagation

The detailed chemical nature of the active centre produced when the two components of a Ziegler-Natta catalyst interact has been debated since the first discoveries were made; many mechanisms have been proposed and much experimental data have been amassed. Although today it is generally accepted that the growth step takes place at a transition metal-carbon bond in both heterogeneous and homogeneous systems, a consideration of earlier theories is useful to show how the present position has evolved.

(i) *Early Theories*

In 1956, Nenitzescu and co-workers suggested that Ziegler polymerization takes place by a free-radical mechanism. They envisaged the partial alkylation of the transition metal halide by the metal alkyl followed by reduction of the transition metal ion and the liberation of an organic radical, which then initiated conventional radical polymerization, for example

$$AlR_3 + TiCl_4 \rightarrow RTiCl_3 + AlRCl_2$$
$$RTiCl_3 \rightarrow R\cdot + TiCl_3$$
$$R\cdot + \text{Monomer} \rightarrow \text{Polymer}$$

The reaction between organometallic compounds and derivatives of titanium had been studied previously, in order to obtain titanium compounds containing Ti—C bonds.[14] Little success was achieved in the isolation of such compounds but reduction of Ti(IV) to Ti(III) was shown to take place and the formation of higher alkanes, for example R_2 derived from $R\cdot$, showed the alkyl group to be the active reducing agent.[15] Topchiev *et al.*[16] proposed a similar mechanism but, in view of the fact that $TiCl_3$ is insoluble in hydrocarbon solvents, the normal polymerization medium, he saw the process as a heterogeneous one in which both radical and reacting ethylene molecules are bound to the $TiCl_3$ surface. A similar heterogeneous reaction was described by Friedlander[17] though in this case initiation was seen as direct electron transfer from the transition metal to the olefin. Both Duck[18] and Van Helden *et al.*[19] supported a radical mechanism but incorporated various aluminium complexes to act as radical stabilizers. Initiation and propagation involving a bound anion was postulated by Gilchrist.[20] Transfer of a simple carbanion from an adsorbed metal alkyl to an adsorbed olefin constituted the initiation step; while addition of the anionic end of the last added olefin to another adsorbed olefin resulted in chain propagation. The anionic end was, however, not fixed at a metal centre, although the growing anion as a whole was bound to the TiCl₃ lattice.

The relatively low reactivity of α-olefins compared with more polar monomers, and the fact that highly stereo-regular polymers were obtained with these catalysts, prompted the idea that a process more complex than that for simple polymerizations must be involved. One such mechanism involves a so-called bimetallic active site, and bimetallic mechanisms have been widely discussed.[21-27] In some theories two transition metal atoms are involved,[21] though more usually two different metals are envisaged. Natta[22] himself favoured this type of intermediate and as generally assumed, the structure of the bimetallic site is

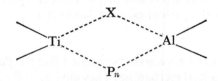

where X = halogen or alkyl, and P_n = growing polymer chain. A monomer molecule becomes co-ordinated to the titanium centre, while simultaneously the Ti-polymer partial bond is broken. Natta suggested that the olefin becomes polarized and inserted into the Al-carbon bond (Figure 12.1). He and his group were able to isolate various stable electron deficient bridged bimetallic complexes of trivalent titanium and aluminium in support of their polymerization scheme. However, these invariably were shown either to have halogen bridging groups, or, in the complexes where no halogen

was present, titanium-cyclopentadienyl $\left(\text{Ti---} \right)$ bonds, and not

simple Ti—C σ-bonds. Furthermore, they exhibited only low catalytic activity for ethylene polymerization and essentially zero reactivity towards propylene. A bimetallic species was also proposed as the intermediate by Boor,[28] who made the important suggestion that chlorine vacancies at the surface of the transition metal halide play a significant role in the stereoregulating properties of Ziegler-Natta catalysts (Figure 12.2).

Arguments for the so-called monometallic mechanisms have been forwarded as strongly as all others. Nenitzescu[13] was probably the first to propose this type of active centre, though in detail it is quite different to the picture largely accepted today. At the time, he considered it a much poorer proposition to the free-radical mechanism which he and his co-workers had also suggested. The metal alkyl component was seen as alkylating the titanium trichloride, and the olefin insertion took place at a tetravalent titanium centre

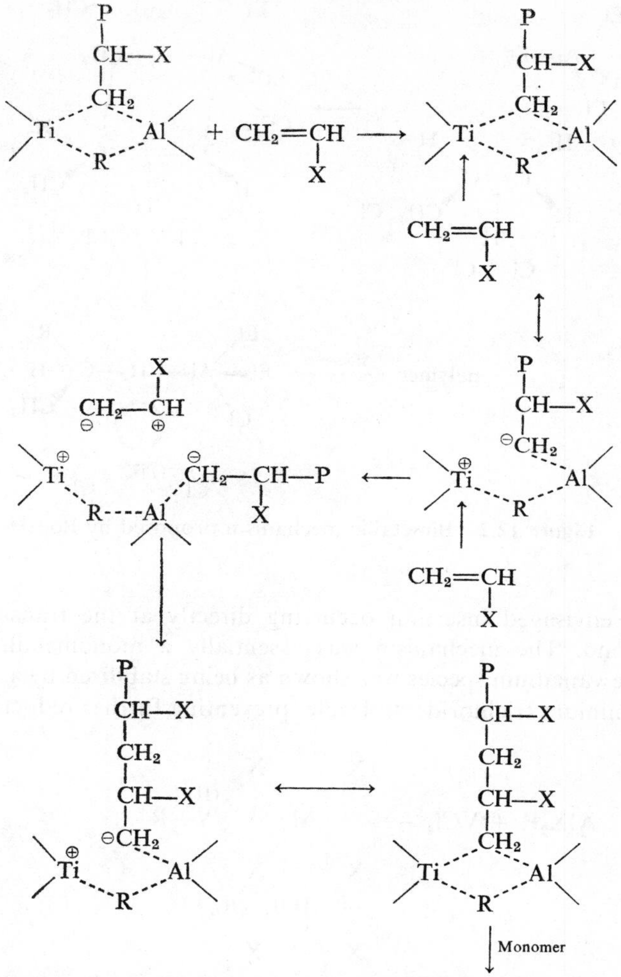

Figure 12.1 Bimetallic mechanism suggested by G. Natta[22]

as indicated

$$R-TiX_3 \longrightarrow R{:}]^- TiX_3]^+$$

$$\downarrow {\scriptstyle CH_2=CH_2}$$

$$R-CH_2-CH_2{:}]^- TiX_3]^+$$

Ludlum[29] proposed a similar insertion reaction but concluded that substantial reduction of the titanium valence state must take place initially, yielding active species such as RTiCl. A similar lower valence state of the alkylated transition metal, in this case vanadium, RVCl, was proposed by Carrick,[30]

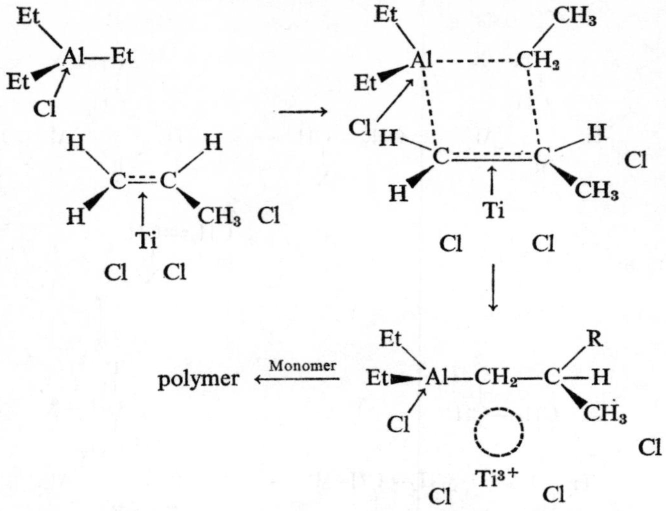

Figure 12.2   Bimetallic mechanism proposed by Boor[28]

who also envisaged insertion occurring directly at the transition metal–carbon bond. The mechanism was essentially a monometallic one even though the vanadium species was shown as being stabilized by a metal alkyl, or an aluminium trichloride molecule, preventing further reduction.

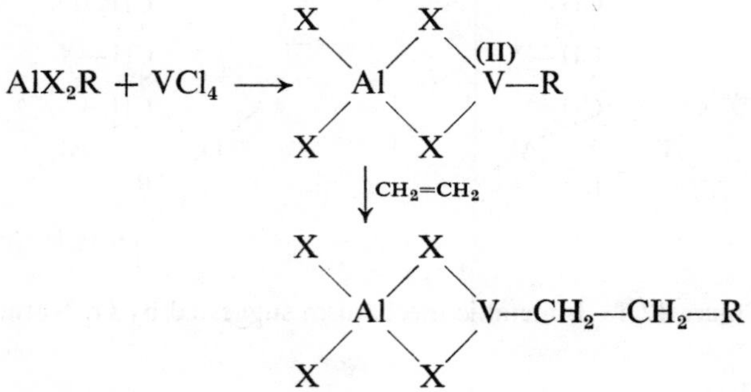

(ii) *The Monometallic Mechanism of Cossee*

The proposal by Cossee[31] that the active site was a transition metal having an overall octahedral configuration in which one position is vacant due to a missing ligand, provides the basis for modern concepts of the mechanisms describing Ziegler-Natta polymerizations. It seems rather remarkable that it was five years from the discovery of transition metal catalysis to the concept of an octahedral reactive centre, in view of the dominance of octahedral configuration and bonding in transition metal chemistry. A diagrammatic representation of the active centre is shown below.

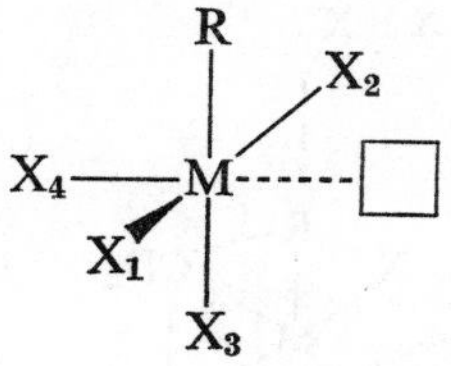

M = transition metal
R = alkyl group or growing polymer chain
X_1–X_4 are anions, normally Cl^-
□ = vacancy

For the catalyst system $TiCl_3/AlEt_3$ the catalytic centre might be formed by the following sequence of reactions:

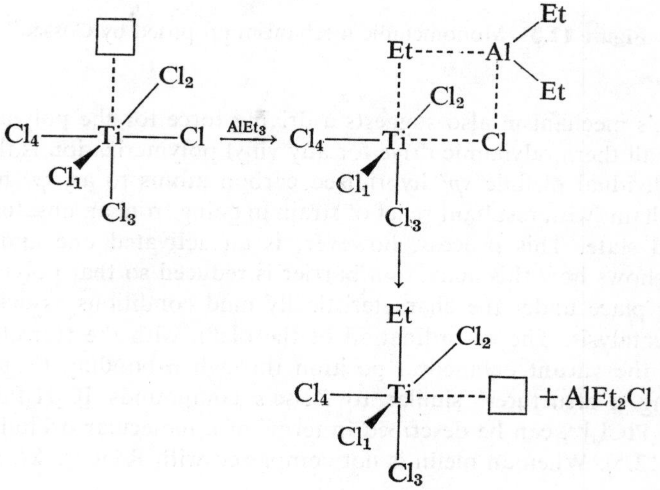

This model provides a transition metal–alkyl σ-bond and the facility for co-ordination of a monomer molecule with the transition element. Propagation of polymerization is assumed to be the interposition of the co-ordinated olefin molecule between the transition metal and the alkyl group via a four-membered ring transition state (Figure 12.3).

One of the main advantages of the Cossee mechanism is its ability, with slight modification, to explain both heterogeneous and homogeneous catalyst systems. With a solid phase catalyst, for example α-$TiCl_3/AlEt_3$, such a situation is most easily visualized at the surface of halogenides with layer structures. Then X_1–X_4 are anions of the lattice of the solid compound.[32] In solution, the active site consists of individual molecules possibly stabilized with other metallic centres via bridging groups. Much more is known today about homogeneous systems and this point will be taken up in more detail later.

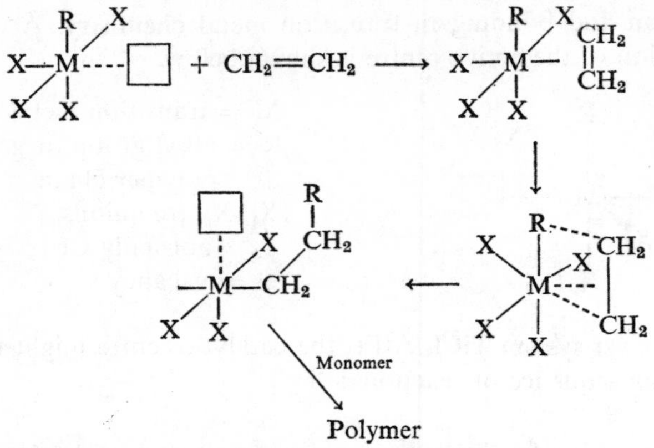

Figure 12.3   Monometallic mechanism proposed by Cossee[31]

Cossee's mechanism also suggests a driving force for the polymerization. The overall thermodynamic drive for any vinyl polymerization is the change from individual olefinic sp^2 hybridized carbon atoms to an sp^3 hybridized carbon chain, with resultant relief of strain in going from an unsaturated to a saturated state. This process, however, is an activated one and Cossee's scheme shows how this activation barrier is reduced so that polymerization can take place under the characteristically mild conditions associated with Ziegler catalysis. The co-ordination of the olefin with the transition metal atom at the vacant octahedral position through π-bonding (Figure 12.4), producing a structure[33] similar to Zeise's compounds $[C_2H_4PtCl_2]_2$ and $K^+[C_2H_4PtCl_3]^-$, can be described in terms of a molecular orbital diagram. (Figure 12.5). When an olefin is not complexed with $RTiCl_4$, ΔE represents

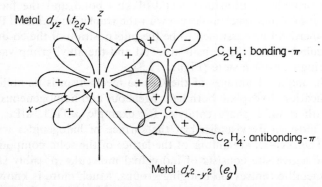

Figure 12.4   Schematic diagram showing the relevant orbitals in a π-bonded complex between a transition metal and ethylene

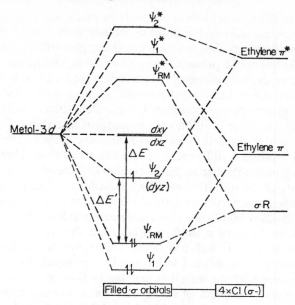

Figure 12.5 Tentative molecular orbital diagram for the octahedral complex $RTiCl_4(C_2H_4)$. For simplicity, the $4s$ and $4p$ orbitals are not included, and the Ti—Cl bond is assumed 100 per cent ionic

the energy that an electron in the Ti—C bond must acquire in order to weaken that bond. Since the catalyst can be stored for long periods without decomposition, ΔE is large enough to maintain the Ti—C bond intact in the absence of olefin. However, if an olefin is co-ordinated to the complex, a new energy level, $\psi_2(d_{yz})$, is formed by mixing metal d-orbitals and ethylene π^*-antibonding orbitals ('back donation'). This new orbital is below the energy level of the original metal $3d$-orbitals, and an electron from the metal-carbon bond (ψ_{RM}) can be excited into it much more readily. The Ti—C bond will therefore be more susceptible to homolysis. This concept, which relates the strength of metal–carbon bonds with the energy gap between the highest filled and the lowest empty or half filled orbital, was originally proposed by Chatt and Shaw[34] in order to explain the inherent instability of alkyls of transition metals. If $\Delta E'$ is smaller than the critical energy gap in the Chatt and Shaw model, the alkyl group will be expelled as a radical which then attaches itself in a concerted process to the nearest carbon atom of the olefin while, at the same time, the other carbon of the olefin bonds to the metal. It is wothwhile pointing out that for simplicity the insertion step can be envisaged as involving the formal breaking of the metal alkyl bond. However, in practice there is no strict requirement for this, and in fact, Cossee has shown that the path of minimal activation proceeds by an alkyl migration path which involves at least partial overlap of alkyl orbitals with orbitals on the metal (d_{yz}). After this first step, the transition metal complex still has its alkyl group now the first segment of a prospective polymer chain, and its vacant position; these two, however, have changed places. When the reaction is repeated, the alkyl group returns to its original position. If the positions of the vacancy and the alkyl group are stereochemically equivalent then co-ordination of an olefin molecule at either position with subsequent insertion will produce the same result, but if there is some non-equivalence, then each position of reaction will produce a different stereochemical addition. These considerations will be pursued further in the discussion of stereo-regulation.

The stability of the transition metal–alkyl bond is in fact the crucial factor in deciding whether a particular system will function as a catalyst or not. The isolated metal–alkyl bonds must be sufficiently stable in the absence of co-ordinated olefin to have a finite lifetime, and this requires a metal ion electronegativity that is not too high. This same bond, however, must become sufficiently destabilized when an olefin molecule is co-ordinated in the vacant position for the insertion reaction to take place. A more quantitative, theoretical, treatment of energy levels and bonding between titanium and ethylene in the growth reaction for soluble catalysts has been made recently.[34a]

Cossee's model predicts that only certain transition elements will provide the correct conditions. According to his molecular orbital diagram, the catalytic activity of the transition metal compound does not require the presence of an electron in each of the d_{yz}-, d_{xz}- and d_{zy}-orbitals. However, no two electrons are allowed to be present simultaneously in the $\psi_2(d_{yz})$-orbital; this means that only transition metal ions having from zero up to three unpaired d-electrons are suitable. It also follows from the diagram that the specific action of the transition metal ion is possible only when the energy of the metal $3d$-level is between the bonding and antibonding energy levels of the olefin. Cossee[31] has made some rough computations on the relative positions of some d-electron energy levels and has shown that those in $TiCl_3$ are very close to the π-electron energy levels of ethylene; whereas, for example, $CrCl_3$ does not have this correlation and is known to be a very poor catalyst. The d-orbital energy level in chromium can be raised to correspond more closely to that of the π-orbitals of ethylene by co-ordination with oxygen instead of chlorine, as in the Phillips-type[35,36] catalyst for the polymerization of ethylene (Chapter 13).

Rodriguez and van Looy[37] advocate a similar mechanism with a transition metal–carbon bond as the actual growth centre, but in addition, they require a complexed base metal alkyl to be an integral part of the site (see below). In the polymerization of propylene to isotactic polymer they argue that the orientation of the methyl group is dictated by the requirement of minimal

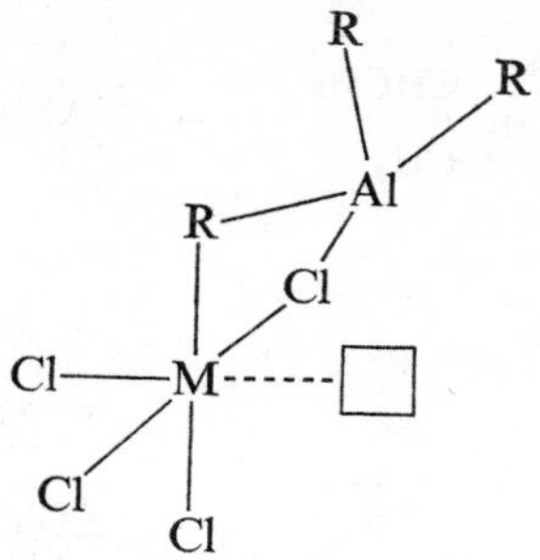

$$
\begin{array}{c}
R \\
\diagdown \quad R \\
\diagup \\
R-\!\!-Al \\
\diagup \\
R \quad \big| \\
\quad Cl \\
\big| \\
Cl-\!\!-M\cdots\cdots\square \\
\diagup \quad \big| \\
Cl \quad Cl
\end{array}
$$

steric interaction with the external groups carried by the aluminium atom. This point will appear again in the discussion of stereoregulation.

(iii) More recent mechanistic developments

Although the mechanism put forward by Cossee is very attractive, there is no direct experimental evidence that complexing occurs prior to the olefin insertion step. Furthermore, as we shall see, the Arlman-Cossee[38] model for the mechanism of isotactic α-olefin polymerization requires the growing polymer chain and the vacancy to exchange positions at each growth step. Between each of these steps the polymer chain must migrate back to its former position since, in the several crystalline modifications of $TiCl_3$ normally employed, the two positions are not stereochemically equivalent (see below). An alternative model, originating with H. G. Tennett and put forward formally by Boor,[39] is a cis-four-centre addition without prior co-ordination. These authors argue that if reaction proceeds via co-ordination of monomer at the active site then the activation energy should be increased not decreased. They suggest that co-ordination stabilizes the ground state more than the transition state, and the rate of insertion is therefore lower than it would be if no prior complex formation took place. The Boor model requires that the metal–carbon bond be highly polarizable and hence metal centres which have small radii and bear a high positive charge are preferred. The driving force consists of the transfer of the electrons as indicated in structure (c) (Figure 12.6), leading to the thermodynamically favoured conversion of the C=C double bond into two single C—C bonds, with retention of the metal–carbon σ-bond. As the olefin approaches the metal centre, the polarized M—C bond becomes longer (structure (b)) and the olefin is inserted directly into it. The advantage of this mechanism is that the polymer chain maintains its position in the octahedral complex, that is structures (a) and (d) are equivalent, and therefore eliminates the need for the polymer migration apparently required for the production of isotactic polymers from α-olefins. This type of concerted reaction, with its accompanying driving force, is also attractive in the light of the discovery of a number of catalysts[40] for ethylene and propylene polymerization, which appear to

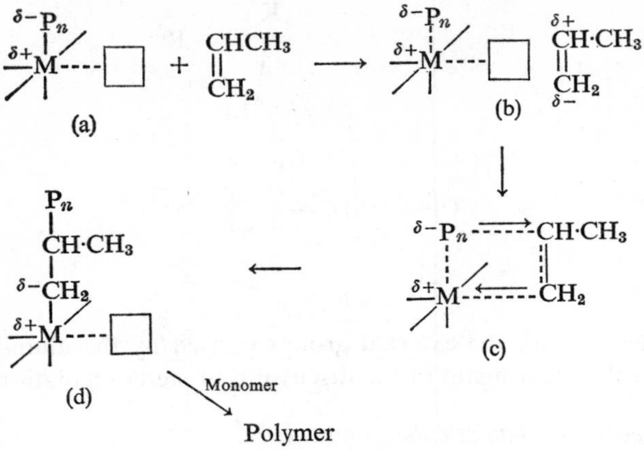

Figure 12.6   Propagation mechanism involving no prior co-ordination of monomer, as proposed by Boor[39]

violate the Cossee model requirement, that only sites in which the transition metal ion has zero to three d-electrons can be active for mono-olefin polymerization. These catalysts, apparently d^4- and d^5-electronic systems, produce linear polyethylene and isotactic polypropylene. The yields are invariably low, however, and very much more experimental evidence is needed before the Cossee model becomes redundant. It might well be that the active sites in these catalysts are in fact higher oxidation states of the metals involved, i.e. $d^0 \rightarrow d^3$ systems, which represent only a tiny fraction of the total positive ionic species present. ESR data,[36] for example seems to indicate that in the Phillips chromium oxide catalyst the species $[Cr(V)O]^{3+}$ is present.

Much more recently Buls and Higgins[41] have proposed a new heterogeneous mechanism, which, while retaining many of the basic ideas of the Cossee model for heterogeneous catalysts, differs substantially in that all of the titanium content of a completely activated catalyst is suggested to be involved in the catalytic process. The latter is therefore not solely a surface phenomenon, and the system studied, $TiCl_3/AlEt_2Cl$, not a typical heterogeneous catalyst. From experimental data on different catalyst/co-catalyst ratios, and from the number of polymer molecules produced per activating $AlEt_2Cl$ used, they have concluded that the basic structure in the unactivated catalyst is a molecule, Ti_8Cl_{24}, with Al_2Cl_6 present in solid solution. (*Note* that commercial $TiCl_3$ is produced by reduction of $TiCl_4$ by aluminium; Al_2Cl_6 is a by-product not completely removed from the system. Such catalysts are then activated by the addition of a co-catalyst—an alkylating agent—in this case $AlEt_2Cl$.) The crystal structures of the modifications of $TiCl_3$ which possess stereo-regulating properties are well known[42] (see later). From this, these authors suggest that each group of eight titanium atoms is arranged in a planar *p*-xylene-like skeleton (Figure 12.7), sandwiched between two layers of ligands each consisting of 12 chlorine atoms. The unit contains seven complete octahedral positions, six of which contain titanium atoms, and the seventh (the central one) is empty. These six titanium atoms are completely co-ordinated, but the remaining two in the unit, those protruding from each end of the hexagon, are in contact with only two chlorine atoms. Al_2Cl_6 has approximately the same crystal parameters as $TiCl_3$, and these molecules could be inserted between the Ti_8Cl_{24} units, particularly in a catalyst sample which has been subjected to ball-milling. Such units would then be stabilized and the two exposed titanium atoms would each be in contact with three chlorine atoms. A complete layer of such alternating units would contain regularly spaced cavities, sufficiently large to allow aluminium alkyls and monomer to diffuse through them. Alkylation of the exposed titanium atoms would

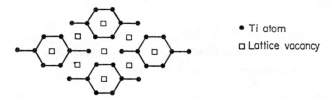

● Ti atom

□ Lattice vacancy

Figure 12.7   Titanium layer in γ-$TiCl_3$ showing Ti_8 units

produce active centres, each with two vacancies, not unlike those proposed by Cossee. Furthermore, the configuration of the chlorine vacancies on each side of an alkylated Ti_8Cl_{24} unit is such that they would be mirror images of each other. The mechanism of polymerization, that is monomer co-ordination followed by insertion into the Ti-alkyl bond, would then be essentially as already outlined, and all the quantum chemical considerations put forward by Cossee would still be valid. To what extent this 'Uniform Site Theory of Ziegler Catalysis' is generally applicable does not appear to have been discussed independently. Its experimental basis seems sound and in the context of the system studied, that is the $TiCl_3$/$AlEt_2Cl$, ball milled catalyst, the mechanism proposed is realistic. Photographic evidence from other systems in favour of polymer growth from crystal edges and defects,[43] however, would appear to be contrary to the 'Uniform Site Theory' as a general explanation of heterogeneous Ziegler-Natta catalysis.

In the field of homogeneous catalysis the elegant work of Olivé and Olivé[44,45] on the system $(C_5H_5)_2TiCl_2$/$AlEtCl_2$ has gone far towards confirming many of the accepted concepts in Ziegler-Natta polymerization of ethylene. This catalyst system yields a high molecular weight linear polymer from ethylene, though is ineffective for the polymerization of α-olefins in general. It has been clearly demonstrated that the equilibria shown in Figure 12.8 are operative. In fact, the intermediate $Cp_2TiEtCl$ is a stable tetrahedral compound which can be isolated. The Ti—C σ-bond presumably is strong because of the presence of two cyclopentadienyl ligands. The pure compound is inactive as a catalyst, but in the presence of AlEtCl$_2$ polymer is readily obtained from ethylene. In combination with AlCl$_3$, however, only low catalytic activity is obtained, showing the active species to be complex A (Figure 12.8). Thus in this instance, a Ti—C bond which is too stable to be a

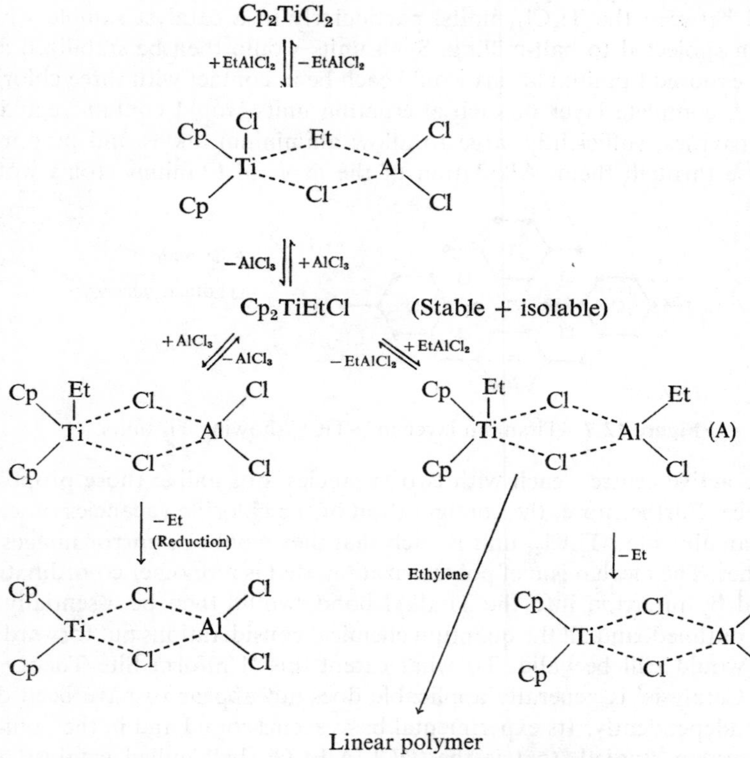

Figure 12.8 Initiation equilibria proposed by Olivé and Olivé[44] in the catalyst system Cp$_2$TiCl$_2$/AlEtCl$_2$

catalytic centre is sufficiently destabilized by co-ordination with an aluminium alkyl to become active. Olivé and Olivé argue that the π-Et bond comes under the *trans*-influence of the bridged aluminium species, and presumably suffers

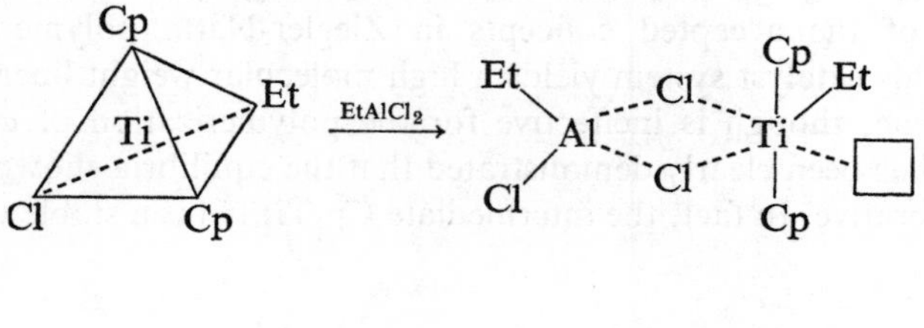

weakening. ESR evidence showed 'electron pressure' in the direction Et →
Al → Cl → Ti, but no such pressure for the system Cl—Al · · · Cl · · · Ti,
so that the corresponding compound with four chlorine atoms co-ordinated
to Al would be expected to be inactive, as indeed it is. The latter point is very
important since it discounts the idea that $Cp_2TiEtCl$ is inactive solely on
stereochemical grounds, i.e. because there is no vacancy at which an ethylene
molecule can approach the Ti centre.

It is now generally accepted that growth takes place at a transition metal–
carbon σ-bond in the Ziegler-Natta catalysis of ethylene and α-olefin polymerization. The only outstanding question concerning the structure of the
active site is whether or not the latter is stabilized by the present of a base
metal derivative bonded via halogen bridges. The two possible structures
are shown in Figure 12.9.

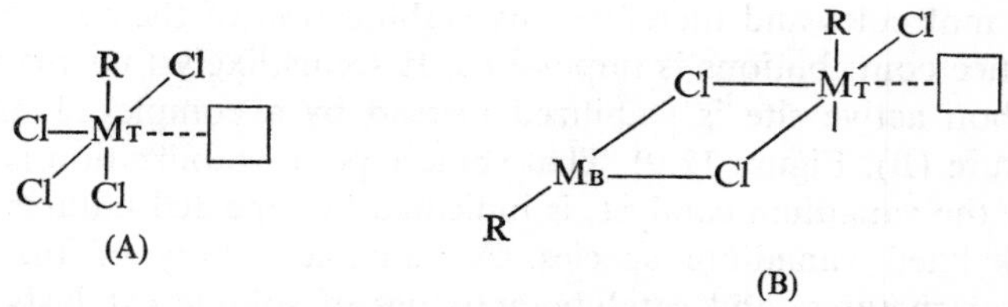

Figure 12.9 Possible structures of the active site in Ziegler-Natta catalysts:
□ = vacancy; R = alkyl group, M_T = transition metal (Ti or V); M_B = base
metal (normally Al)

Propagation is then visualized as a series of growth steps whereby olefin is
inserted into the M_T—R bond. It seems likely that each olefin molecule is
initially complexed to the central metal ion prior to insertion, though this
point is not entirely settled. Whether the actual point of attachment of the
growing polymer chain to the active octahedral site remains fixed or not is
also in contention. A summary of the mechanistic pictures for heterogeneous
and homogeneous catalysts is given below.

(a) *Heterogeneous catalysts.* A typical heterogeneous catalyst is one
prepared by mixing $TiCl_3$, which is insoluble in hydrocarbon solvents, with
$AlEt_3$, which is soluble. A solid residue is formed and alkylation of the
surface resulting in the formation of Ti—C σ-bonds undoubtedly takes place.
Metal alkyl molecules also appear to become complexed on the surface of a
$TiCl_3$ crystal, but in this case it seems unlikely that they become an integral
part of the active centre. Probably, the crystal lattice provides sufficient
stabilization of Ti—C bonds to make complex formation with a base metal
alkyl unnecessary for Ti—C bonds to survive or become active. Recent
molecular orbital calculations by Begley and Pennella[46] have shown that in
a $RTiCl_4$ structure (Figure 12.9) of the form (A) (based on a $TiCl_3$ lattice)
there is a barrier against decomposition of 1·07 eV, which they suggest is

still high enough to maintain the Ti—C bond intact. They support the argument of Cossee[47] that only by further weakening of the Ti—C bond through complexing of the transition metal with the olefin can polymerization take place. Barriers of $3 \cdot 02$ eV, $1 \cdot 07$ eV and $0 \cdot 7$ eV were calculated for excitation of electrons from the highest filled orbital to the lowest empty or partly filled orbital, when the Ti—C bond is present in the structures CH_3TiCl_3 (an isolated molecule), $RTiCl_4$ and $RTiCl_4^-$ olefin complex, respectively. The latter pair were taken to be part of a crystal lattice.

(b) *Homogeneous catalysts.* These are much fewer in number and have some very different characteristics from heterogeneous catalysts. The two most important examples are $AlEt_2Cl/VCl_4$ which polymerizes propylene to a syndiotactic polymer and $(C_5H_5)_2TiCl_2/AlEtCl_2$ which polymerizes ethylene to linear high density polythene. In solution, the active complex exists as individual molecules and therefore any stabilization of their structure from crystal lattice contributions is impossible. It seems likely that the transition metal–carbon active site is stabilized instead by a complexed metal alkyl (i.e. structure (B), Figure 12.9). That some type of stabilization is required, at least by the vanadium catalyst, is indicated by repeated failures to isolate simple alkylated vanadium species. A valuable survey of the magnetic properties, structures, and catalytic activities of soluble catalysts based on vanadium and chromium has been given by Henrici-Olivé and Olivé.[44a]

Very recent work[47a] has demonstrated that some fairly simple covalent alkyl derivatives of transition metals may be efficient homogeneous catalysts for ethylene polymerization. Catalysts have the general structure $M^{n+}(CH_2Y)_n$ where M is a transition metal (Ti, Zr) and Y can be C_6H_5, $Si(CH_3)_3$, $C_{10}H_7$ etc. Here of course, there is no ambiguity as to the requirement for a non-transition metal component, a co-ordinated-olefin-Cossee-type mechanism seems likely, and there is no doubt that rapid development of both theories and practical applications will result from these early disclosures.

12.1.2 Experimental evidence for existing mechanisms

The experimental data available up to 1966 has been admirably reviewed by Boor,[10] who more recently has up-dated this publication.[12] Only a few references, which contain crucial information relevant to the formulation of the mechanisms already discussed, will therefore be presented here.

There is overwhelming evidence that the primary reaction occurring when Ziegler catalyst components are mixed is the alkylation of the transition metal compound by the metal alkyl. For example, Rodriguez, Van Looy and Gabant[37,48] investigated in detail, in the absence of solvents, the reactions of $TiCl_3$ and aluminium alkyls ($AlMe_3$, $AlMe_2Cl$ and $AlEt_2Cl$). From the stoichiometry data, CH_3/CD_3 exchange, and infrared analysis combined with electron microscopy, rapid exchange of alkyl and halogen was shown to take place. When the alkyl substituent was labelled with carbon-14, the radioactive carbon was found at the end of the polymer chain showing insertion to have taken place.[49] Polymerizations terminated by addition of tritiated alcohols yielded polymers containing tritium,[50] again showing growth at a positive metallic centre, i.e. so-called 'co-ordinated anionic' polymerization. Hydrogen was shown to be an effective transfer agent,[50] reducing polymer molecular weights without affecting catalyst activity. The most reasonable explanation for this again requires growth to be at a metal–carbon bond.

$$\underset{/\overset{|}{\diagdown}}{-\overset{\overset{\displaystyle P_n}{\diagup}}{M}}----\square \;+\; H_2 \;\xrightarrow{\text{Transfer}}\; \underset{/\overset{|}{\diagdown}}{-\overset{\overset{\displaystyle H}{\diagup}}{M}}----\square \;+\; H—P_n$$

Further evidence that propagation takes place at alkylated transition metal centres rather than at base metal derivatives is three-fold. Firstly, there are now many catalysts known[51] which are free of base metal alkyl components and, although they are usually of lower activity, high molecular weight stereoregular polymers of α-olefins have been obtained. Secondly, experiments in which the transition metal component is varied tend to produce large variations in polymer yield and stereoregularity, whereas variation of the metal alkyl in general produces relatively minor effects.[52] Finally, a simple but fundamentally important observation is that there are no specific catalysts for ethylene or α-olefin polymerization known that are based only on the metal alkyl component, there is always the requirement for a transition element to be present.

12.1.3 Oxidation state of the transition metal at the active site

Since the valence state of the central transition metal ion significantly influences the structure of the active centre, much experimental effort has been directed towards measuring this property. In the case of heterogeneous catalysts the number of active sites present is small relative to the total number of transition metal ions present, and hence it is difficult to relate unequivocally an experimental oxidation state value to the activity of the catalyst. In principle, homogeneous systems should provide a more straightforward situation, but even with this type of catalyst, since two components are involved, there can be no guarantee that complete homogeneity with regard to the oxidation state of the species present is achieved.

The earliest theories on the mechanism of Ziegler polymerizations using $TiCl_4$ required the reduction[13] of the transition element to the 3^+ or even 2^+ oxidation states. With a steady improvement in experimental techniques and better interpretation of the data obtained, oxidation state values that have been determined have gradually risen. Because of the variety of the systems studied, it is difficult to extrapolate a specific result into a general observation. Carrick and co-workers, however, have pointed out that there are two opposing factors which come into play when the oxidation state of a transition metal is changed. Firstly, there is a decrease in electronegativity of the metal centre as its valence state decreases, e.g. TiIV$(1 \cdot 6) \rightarrow$TiII$(1 \cdot 1)$. The lower electronegativity makes the transition metal–carbon bond more polar so that any co-ordinated, and therefore polarized, monomer is more readily incorporated into the growing polymer chain. Secondly, with the progressive filling of transition metal orbitals with decrease of oxidation state, the electron acceptor character of the metal decreases. This has the effect of reducing the possibility of effective complexing of the olefin at the metal centre. One factor therefore encourages propagation directly while the other opposes indirectly by reducing co-ordination power prior to monomer insertion. Thus, as far as oxidation state is concerned, there exists a delicate balance between efficient catalysis and inactivity. In some cases, depending on the particular transition metal, the ligands attached to it, and the nature of the metal alkyl only a single valence state of the transition metal component will be active, while in others several valence states may well give rise to effective catalytic sites.

12.1.4 Origin of stereo-regulation

The ability of Ziegler-Natta catalysts to polymerize ethylene and α-olefins (non-polar monomers), under relatively mild conditions, to high molecular weight polymers is in itself remarkable. Even more incredible though, are their stereo-regulating properties which give rise to polymers of high steric purity. All of the early catalysts were heterogeneous and the importance of the crystal surface in the stereo-regular polymerization of α-olefins was quickly recognized. To date, all effective isotactic polymerizations of propylene require a solid transition metal salt to be present, while the corresponding syndiotactic form has now been prepared from both homogeneous and heterogeneous sites. Although the specific stereo-regulating capacity and activity of a catalyst depend on the particular combination of components used, the general property is a function of the transition metal salt alone. This is confirmed by the observation that stereo-regular polymers can be obtained from catalysts free of metal alkyls, though often with reduced efficiency.

Experiments with deuterium labelled propylene have shown that the insertion of the α-olefin into the metal–carbon bond during propagation takes place by a '*cis*-opening' of the double bond. The olefin adds in a head to tail fashion by 1,2-addition and the unsubstituted carbon of the olefin, invariably adds to the transition metal component of the metal–carbon growth site (Figure 12.10). It should be noted however that these conclusions assume that '*trans*-opening', followed by bond rotation, does not take place.

A number of proposals have been made to explain stereoregular placement in α-olefin polymerization. Natta believed that isotactic propagation at the heterogeneous site takes place because such presumed bimetallic centres have asymmetric character, i.e. each site can complex an α-olefin molecule in one of two different ways. Opening of the double bond when the olefin is co-ordinated in one mode gives a '*d*' configuration, while propagation from the other mode produces an '*l*' configuration. Polymerization at one site would therefore produce *d*-isotactic polymer and *l* isotactic polymer would result from the other. Since it is now generally accepted that the active site is an octahedral monometallic centre, Natta's model as a whole must be rejected. However, the concept of the monomer complexing in two differing modes

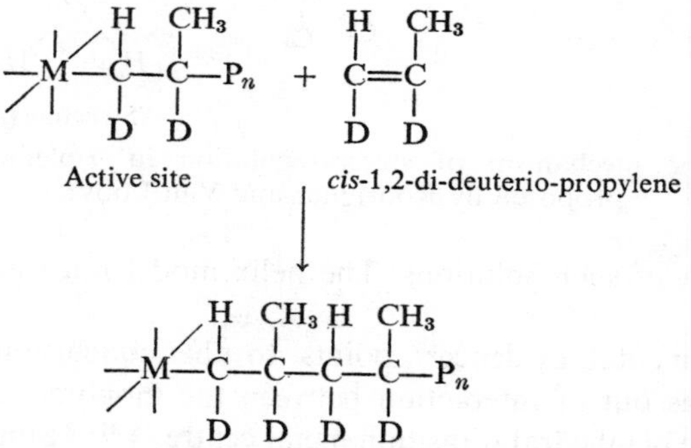

Active site · · · · · · · · · · *cis*-1,2-di-deuterio-propylene

Figure 12.10 Polymerization of deuterium labelled propylene (*cis*-opening of the double bond in 1,2-addition)

remains a fundamentally important one. Rodriguez and van Looy[37] have up-dated Natta's mechanism by proposing an active site incorporating a base metal alkyl with an alkylated octahedral transition metal atom. The latter are located on the lateral faces and defects of a transition metal halide crystal and contain two chlorine vacancies. Isotactic placement is thought to occur because the α-olefin can complex with the exposed titanium in only one mode, i.e. with its alkyl substituent directed away from the alkyl groups on the Al atom (structure (A) Figure 12.11). Once again, however, this model suffers from the evidence of catalysis in systems free from base metal alkyls.

Coover[56] suggested that the growing polymer is a helix and that the rate of placement preserving the helical symmetry is faster than the rate of any other placement, hence producing isotactic polymer from α-olefins. Hoeg and Liebman[57] have shown that solution polymerizations of propylene at high temperatures (above the melting point of the polymer) yielded an isotactic product, and that bonds due to highly ordered helices are absent in the

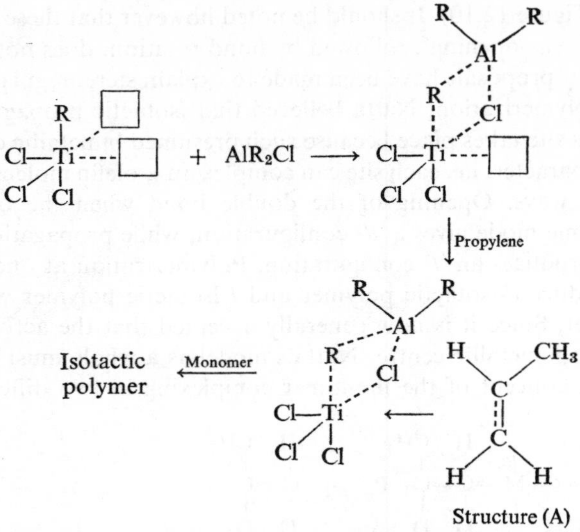

Structure (A)

Figure 12.11 The mechanism of stereo-regulation in α-olefin polymerization proposed by Rodriguez and Van Looy[37]

infrared spectra of such solutions. The helix model must also therefore be rejected.

Most experimental evidence[58] points to the conclusion that isotactic placement arises out of interaction between the monomer and the ligands surrounding the octahedral transition metal centre.[28,31,38] Similar interactions are thought to be responsible for syndiotactic placement at heterogeneous sites,[47] though syndiotactic placement with homogeneous catalysts is thought to arise from the interaction between the last added monomer unit and the monomer being complexed at the same transition metal centre.[50]

(i) Cossee-Arlman Heterogeneous Model

The most comprehensive and satisfactory explanation of the stereo-regulating properties of heterogeneous Ziegler-Natta catalysts has been proposed by Cossee and Arlman. The catalyst system $TiCl_3/AlEt_3$ is described in detail, although the results can be taken as general for all systems involving transition metal halide crystals with layer lattice structures. The diagrammatic representation of the octahedral active site and the schematic chemical mechanism have already been discussed, along with the simple underlying quantum chemical principles (section 12.1.1(ii)). The detailed stereochemical course of the reaction will now be considered.

$TiCl_3$ occurs in a number of crystalline modifications, α, β and δ, depending on the method of preparation and isolation. The degree of stereoregularity in the polymers produced when these are used as catalysts varies considerably.

The highest steric purity is obtained when well-crystallized α-TiCl₃ is used. This modification consists of stacks of elementary crystal sheets each containing two chloride layers with one titanium layer in between. Thus, along the principal axis of the crystal, two chloride layers alternate with one titanium layer. In all cases the Ti³⁺ ions are located in octahedral interstices of the Cl lattice, the Cl⁻ ions themselves are hexagonally close-packed. In a TiCl₃ sheet, only two thirds of the octahedral interstices contain Ti³⁺; the Ti³⁺ are arranged in regular hexagons with an empty Ti site at the centre (Figure 12.7) It can be shown that, for a stoichiometric crystal of TiCl₃ to be electroneutral, it is necessary that the crystal should contain a number of Cl⁻ vacancies on the surface. In fact, in any co-ordination lattice, boundary problems exist. The regular structure of the interior is interrupted, which means that anomalous co-ordination numbers have to be adopted for the surface ions. Surface structures are not known in atomic detail, but Arlman has computed the number of likely vacancies, ignoring the other possibilities for maintaining electroneutrality, e.g. interstitial Ti³⁺ ions or substitution of Ti⁴⁺ for a limited number of Ti³⁺ ions. In addition, based on calculations of the Coulomb energies needed to remove a Cl⁻ ion from crystallographically different sites of a piece of crystal, Arlman[32] concluded that the vacancies should be found on the edges of the elementary sheets and not in basal planes of the crystal. This theoretical prediction confirmed earlier experimental observations that polymers appeared to grow from crystal edges.[59] More recent work[43] on the polymerization of propylene in the gas phase, on dry single crystals of α-TiCl₃ activated with Al(Me)₃ vapour, has also shown that growth takes place on lateral edges and dislocation lines of the catalyst crystal.

For a full description of the crystallography of the TiCl₃ lattice and for the way in which lattice defects orientate monomer co-ordination and chain growth in a stereoregular manner, the reader is referred to a definitive review by Cossee.[47] Basically the Cossee-Arlman model assumes co-ordination of monomer with titanium at a vacancy in the TiCl₃ lattice; migration of an alkyl (polymeric) group from titanium to the substituted end of the co-ordinated α-olefin with formation of a new —CH₂—Ti bond has the effect of filling the initial lattice vacancy and simultaneously creating a new one in close proximity. Two possibilities now arise: either the alkyl group stays where it is and a second insertion takes place, or it flips back to the original position before a second monomer unit is incorporated. In the first case the polymer produced will be syndiotactic, in the second, isotactic. The two positions between which the growing chain can oscillate are non-equivalent, hence, one of them should be preferred for steric reasons.

Apparently in normal polymerization procedures carried out between 50 °C and 100 °C, the growing polymeric alkyl group shifts to the most favourable position before the next monomer molecule is inserted.

Kinetic analysis has shown that, on average, surface sites are empty, so that there will be ample opportunity for the growing alkyl group to shift back to the favoured position following each insertion. This shift, which requires a circular movement with the Ti^{3+} ion in the centre, will, in the same way as the migration in the insertion step, be aided by the presence of the d_{yz}-orbital of the metal. The formation of isotactic or syndiotactic polymer is thus governed by the ratio of the rates of the alkyl shift and the monomer insertion. Decreasing temperature will reduce the rate of both alkyl shift and alkyl migration in the insertion step. However, the rate at which monomer complexes with the vacant site in the crystal may well be increased at lower temperatures, so that the overall rate of insertion may be enhanced. Syndiotactic placement would therefore be expected to be favoured by lowering the temperature. In practice, syndiotactic polypropylene has been obtained only in low yields with heterogeneous catalysts, the best results being obtained at low temperatures ($\sim -80\,^{\circ}C$). Invariably, larger proportions of isotactic polymer are produced at the same time. Recently a modification of the Cossee-Arlman mechanism has been proposed[49a] in which the active titanium site is required to be protruding above the two neighbouring layers in the crystal lattice. In this way monomer complexation is possible with both titanium vacancies (pointing outside the crystal surface) eliminating the requirement for alkyl shift of the Ti—R bond.

(ii) *Boor-Youngman Homogeneous Model*

These authors[50] have examined the catalyst system $AlEt_2Cl/VCl_4$ in detail. The solvents employed were heptane and toluene, which gave complete homogeneity at the temperature used for polymerizations ($-78\,^{\circ}C$). They considered the active centre to be an octahedral complex containing an alkyl group and an open co-ordination site or vacancy through which an olefin can become complexed. In contrast to the heterogeneous case, each centre is separate in solution, and stabilization is achieved most probably by complex formation with one or more aluminium alkyl molecules via halogen bridges.

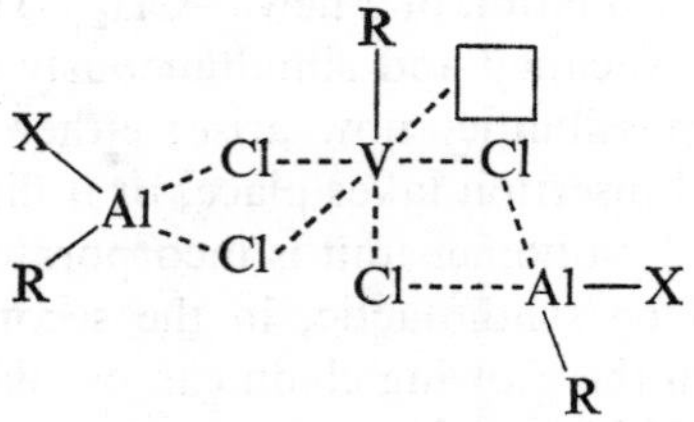

Boor and Youngman[50] have shown that, while 1-butene and ethylene copolymerize rapidly with this catalyst, 1-butene homopolymerizes only slowly. This suggests that steric interaction between the substituents on the

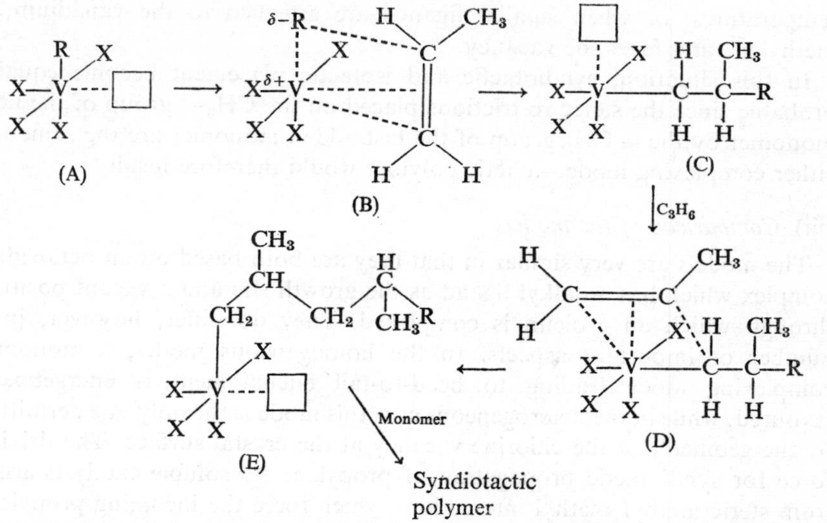

Figure 12.12   Polymerization of propylene leading to syndiotactic polymer, after Boor and Youngman[50]

last added and new monomer units dominate the transition state for propagation. The sequence of steps envisaged which lead to syndiotactic placement of propylene is shown in Figure 12.12. The important features of the mechanism are:

1.   The monomer co-ordination mode shown in Figure 12.13, structure (B), which leads to head to tail enchainment is energetically favoured.
2.   Rotation about the R—V bond is hindered as it is difficult for the CH$_3$-group of the last added monomer unit to pass over the adjacent ligand.
3.   One orientation of the methyl group on the incoming monomer leading to syndiotactic placement is more favourable than the other, due to methyl-methyl interactions.
4.   Whenever the rotation barrier [see 2.] is overcome, for example at higher

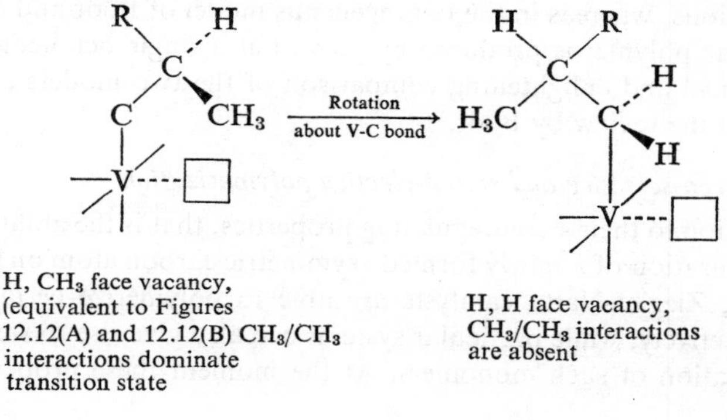

H, CH$_3$ face vacancy, (equivalent to Figures 12.12(A) and 12.12(B) CH$_3$/CH$_3$ interactions dominate transition state

H, H face vacancy, CH$_3$/CH$_3$ interactions are absent

temperatures, or when smaller ligands are attached to the vanadium, a methylene unit faces the vacancy.

In this situation, syndiotactic and isotactic placement become equally probable since the steric restrictions placed on the CH_3- group of the new monomer by the $-CH_2$ group of the last added monomer are the same for either complexing mode—atactic polymer would therefore result.

(iii) Comparison of the models

The models are very similar in that they are both based on an octahedral complex which has an alkyl ligand as the growth site and a vacant position through which an α-olefin is complexed. They do differ, however, in a number of important aspects. In the homogeneous model, a monomer complexing mode leading to head-to-tail enchainment is energetically favoured, while in the heterogeneous case this mode is the only one permitted by the geometry of the chlorine vacancy at the crystal surface. The driving force for syndiotactic propagation of propylene by soluble catalysts arises from steric methyl–methyl interactions which force the incoming propylene always to have a configuration which is opposite to that of the last added propylene molecule. The ligand environment about the metal centre either prevents or minimizes rotation of the M—C growth bond, but otherwise does not influence orientation. The driving force for isotactic propagation of α-olefins on heterogeneous catalysts originates from steric interactions between the α-olefin (specifically the α-substituent) and the ligands which form the environment of the transition metal at the surface of the crystal. This interaction permits the α-olefin to approach the M—C bond in only one configuration, thereby forming long sequences of units in one or the other configuration. Monomer-monomer interactions are minimized by rotations about the M—C growth bond, which move the α-substituent of the last added monomer away from the vacancy; thus the syndiotactic driving force is absent. Finally, in the Arlman-Cossee model, isotactic placement requires the R— group to return to its original position between growth steps, and syndiotactic placement requires consecutive insertions to take place with the site in alternate configurations, whereas in the homogeneous model of Boor and Youngman, syndiotactic polymer is produced by growth at a single octahedral position. For a critical and enlightening comparison of the two models the reader is referred to the review by Boor.[10]

12.1.5 Stereo-selective and stereo-elective polymerization

In addition to their stereoregulating properties, that is the ability to control the configuration of a newly formed asymmetric carbon atom on the polymer backbone, Ziegler-Natta catalysts are able to polymerize certain α-olefins stereo-selectively, while particular systems can also give rise to stereo-elective polymerization of such monomers. At the moment these properties, which are not confined to Ziegler-Natta catalysts,[60] are not as technically important as the straightforward stereo-regulating properties, but from a theoretical point of view they are of equal, if not more, significance. The process of stereo-selection involves an optically *inactive* catalyst and an α-olefin with an optically active carbon atom in its side chain, while stereo-election requires an optically *active* catalyst and an α-olefin with an asymmetric carbon atom specifically in the α-position of the side chain. These observations were first made by Pino and his collaborators[61,62] and the properties and configurations of polymers obtained are discussed in detail in Chapter 15. In essence it appears that Ziegler-Natta catalysts are able to choose predominantly one or other enantiomer from a racemic monomer mixture, with the result that there occurs a prevalence of one of the enantiomeric forms in each single polymer chain, namely stereo-selectivity, for example

$$
\begin{array}{c}
H_2C{=}CH \\
| \\
CH_2 \\
| \\
{*}CHCH_3 \\
| \\
CH_2 \\
| \\
CH_3
\end{array}
\quad
\xrightarrow[\text{TiCl}_4]{\text{Al(i-Bu)}_3}
\quad
\xrightarrow[\substack{\text{separation} \\ \text{on optically} \\ \text{active column}}]{\substack{\text{Adsorption} \\ \text{chromatographic}}}
\quad
\begin{array}{l}
\text{Optically active} \\
\text{polymer fractions} \\
\text{having opposite} \\
\text{signs of rotation}
\end{array}
$$

(R, S) 4-Me-1-hexene   * = asymmetric C atom

When catalysts containing an optically active metal alkyl were used in the polymerization of 4-Me-1-hexene similar results were obtained.[61,62] However, when α-olefins having an asymmetric carbon in the α-position with respect to the double bond (e.g. 3-Me-1-pentene) were polymerized with bis-[(S)-2-Me-butyl]-Zn/TiCl$_4$ catalyst, the polymer obtained and the monomer left unreacted were both optically active.[63] Both racemic (R,S)3-Me-1-pentene and (R,S)-3,7-diMe-1-octene were examined.

$$
\begin{array}{c}
CH_2{=}CH \\
| \\
{*}CHCH_3 \\
| \\
CH_2 \\
| \\
R'
\end{array}
\quad
\xrightarrow{\text{ZnX}_2^S/\text{TiCl}_4}
\quad
\begin{array}{l}
\text{Optically active polymer} \\
+ \text{optically active} \\
\text{unreacted monomer}
\end{array}
$$

* = asymmetric centre, (R,S) monomer R' = —CH$_3$ for 3-Me-1-pentene, or —CH$_2$CH$_2$CH(CH$_3$)$_2$ for 3,7- diMe-1-octene, X^S = (S)-2-methyl-butyl

The sign of rotation of the recovered monomer indicated that the monomeric antipode having the same absolute structure as the optically active (S)-2-methyl-butyl groups present in the catalyst mixture was preferentially polymerized. Pino estimated from the optical activity of the recovered mono-mers and from the conversion of monomer to polymer that the ratio of the polymerization rates of the two antipodes was between 1·2 and 1·7. The polymers obtained were optically active, and the sign of rotation corre-sponded to that of the monomeric enantiomer which appeared to be pref-erentially polymerized. In the case of poly-3,7-diMe-1-octene, the optical activity increased as molecular weight and stereoregularity of the fraction increased. This type of polymerization was termed stereo-elective by Pino. Apparently, if the asymmetric centre on the alkyl group of the catalyst is too far away from the active centre, the formation of optically active polymers does not take place;[64] it is as if there is no asymmetry at all in the catalyst alkyl group, and stereo-selectivity with a racemic monomer may occur but not stereo-electivity. The latter requires both the asymmetric carbon atom on the catalyst to be close to the catalytic site, and the side chain asymmetric carbon atom of the monomer to be adjacent to its double bond.

A number of mechanisms for the stereo-selective and elective processes have been proposed and have been reviewed by Boor[10] who has also suggested the most plausible explanation.[12] Boor[12] pointed out that since isotactic polymerizations are involved, the active site probably has the form proposed by Cossee, and that all Ziegler–Natta catalysts are potentially stereo-selective. Using the octahedral representation of Cossee, these sites can be visualized as mirror images. For an heterogeneous catalyst these octahedral sites are

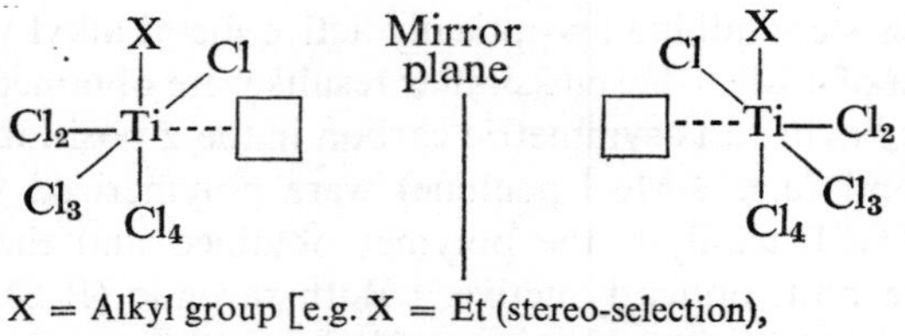

X = Alkyl group [e.g. X = Et (stereo-selection),
or X = (S)-2-Me-1-butyl (stereo-election)]

part of a crystal lattice, and since the environments of the ligands Cl_3 and Cl_4 are not equivalent, the sites are true mirror images. According to this model, the ligand environment at the metal centre effects the preferential adsorption of one or the other enantiomer independently of the last added unit. The mir-ror image sites may be designated M^R—X and M^S—X.

Stereo-selection of, say, racemic (RS)-4-Me-1-hexene takes place because M^R—X sites complex and preferentially polymerize the R enantiomer, while an equal number of M^S—X sites preferentially complex and polymerize, at an equal rate, the S enantiomer. The ability of M^R—X and M^S—X sites to selectively polymerize R and S monomeric enantiomers, respectively, is attributed to steric interactions between the α-substituent of the monomer, and the ligand environment around the transition metal–carbon bond. This view is supported by the experimental evidence of Pino that as the isotacticity of poly-4-Me-1-hexene increased, so did the optical activity of the polymer fraction. Since both the isotactic placement and stereo-selective growth are of steric origin, the less exposed, more isotactic-stereo-regulating sites have a greater capacity to complex and preferentially polymerize one or the other enantiomer. An alternative explanation considered was that only one type of site exists on the TiCl$_3$ surface, but that one half acquire R character, and the other S character, upon the equally probable addition of either the R or S enantiomer. The configuration of the last added monomer unit therefore determines a preference for the enantiomer having the same configuration, i.e. optical activity of the catalyst alkyl group after the first addition determines which monomeric enantiomer is subsequently complexed and polymerized. This idea was rejected since optically active polymers were not formed from 4-Me-1-hexene when optically active metal alkyls were used as cocatalysts.

Stereo-election takes place when both the catalyst and the monomer side chain C atom α- to the double bond are optically active. The active sites involved are now designated M^R—X^S and M^S—X^S (for an optically active catalyst alkyl group, X^S). It is proposed that the stereo-elective process then takes place because the first addition of the R enantiomer to a M^R—X^S bond is much more difficult than the first addition of S enantiomer to an M^S—X^S bond (where X^S and S enantiomer have the same configuration). In other words, when the enantiomer and the X^S substituent have the same configuration, the first addition is facile, but when they have opposite configurations it is difficult. This is true only when the asymmetric carbon atom of the X group is close to the metal centre. However, once, an R enantiomer is added to the M^R—X^S site, then subsequent additions presumably occur as easily as in the case of S enantiomer addition to M^R—X^S, and its subsequent addition.

12.2 POLYMERIZATION OF CONJUGATED DI-OLEFINS (DIENES) BY ZIEGLER-NATTA CATALYSTS AND RELATED TRANSITION METAL COMPLEXES

Conjugated di-olefins can be polymerized to give both 1,4- and 1,2-polymers (or 3,4- if the di-olefin is not symmetric). In the case of 1,4-polymerization, there exists the possibility of obtaining two geometrically isomeric polymers, 1,4-*cis* and 1,4-*trans*. Since in the 1,2-case an asymmetric carbon atom is introduced into the saturated polymer chain, the possibility of isotactic and syndiotactic placement arises, just as with simple vinyl polymerization. When the conjugated diene possesses at least one internal double

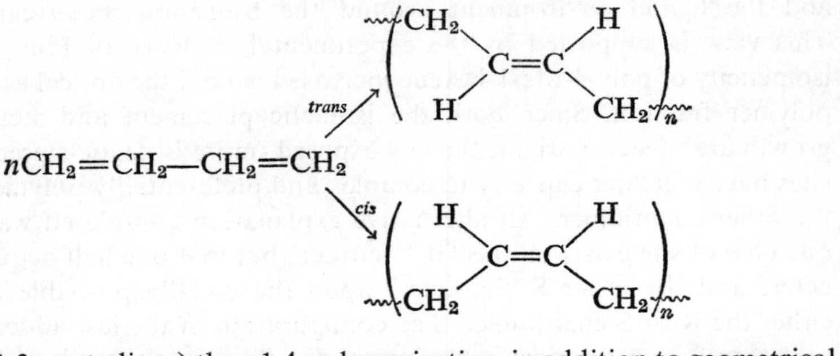

$$nCH_2=CH_2-CH_2=CH_2 \quad \xrightarrow{trans} \quad \left(\begin{array}{c} \sim CH_2 \\ \diagdown \\ C=C \\ \diagup \quad \diagdown \\ H \quad\quad CH_2 \sim \end{array} \right)_n$$

$$\xrightarrow{cis} \quad \left(\begin{array}{c} H \quad\quad H \\ \diagdown \quad\quad \diagup \\ C=C \\ \diagup \quad\quad \diagdown \\ \sim CH_2 \quad CH_2 \sim \end{array} \right)_n$$

bond (e.g. 1,3-pentadiene) then 1,4-polymerization, in addition to geometrical isomerism also brings about optical isomerism, with the further possibility of

$$nCH_2=CH_2-CH=CH_2 \longrightarrow +CH_2-\overset{*}{C}H+_n$$
$$\mid$$
$$CH=CH_2$$

the isotactic and syndiotactic forms of each. In the case of two simpler conjugated di-olefins, all the four possible polymer structures from butadiene,

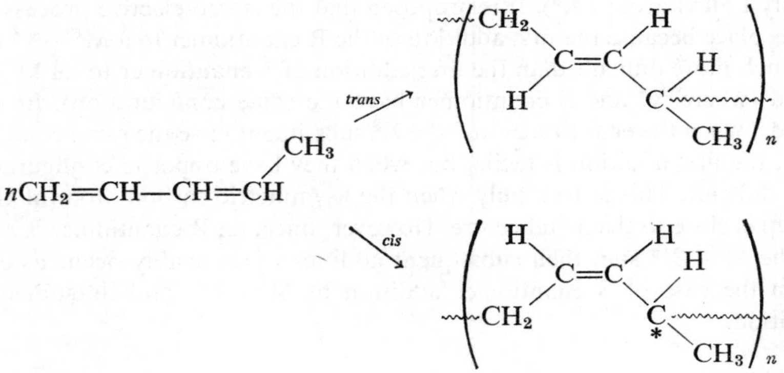

$$nCH_2=CH-CH=CH \quad \xrightarrow{trans} \quad \left(\begin{array}{c} \sim CH_2 \quad\quad H \\ \diagdown \quad\quad\quad \diagup \\ C=C \\ \diagup \quad\quad\quad \diagdown \\ H \quad\quad\quad C \sim \\ \quad\quad\quad CH_3 \end{array} \right)_n$$

$$\xrightarrow{cis} \quad \left(\begin{array}{c} H \quad\quad\quad H \\ \diagdown \quad\quad\quad \diagup \\ C=C \\ \diagup \quad\quad\quad \diagdown \\ \sim CH_2 \quad\quad \overset{*}{C} \sim \\ \quad\quad\quad CH_3 \end{array} \right)_n$$

and some of the six possible forms from isoprene, have been isolated in a very high degree of steric purity.[65]

Among the many combinations of organometallic compounds with complexes of transition metals which polymerize conjugated dienes, of particular note are those based on organoaluminium compounds with complexes of cobalt, nickel, titanium and vanadium.[66] There are also numerous examples of stereo-regular polymers obtained from catalysts containing a transition element alone, e.g. π-allyl nickel halides.[67] Furthermore, it is not essential for an organic grouping to be present in the catalyst at the outset, the organometallic derivative of the transition metal can be formed *in situ* by interaction with the monomer or the solvent, e.g. $Ni(CO)_4$ and $Co_2(CO)_8$.[68]

The distinction between conventional Ziegler-Natta catalysts and the newer systems based predominantly on cobalt and nickel may in the light of further experimental developments be marginal. Many of the practical results and mechanistic interpretations from experiments using mono-olefins may be regarded as valid in the case of diene polymerization by Ziegler-Natta catalysts; however, to apply these findings indiscriminately to the case of cobalt and nickel based catalysts, which are ineffective for α-olefin polymerization, would be too presumptuous. Because of the greater complexity of the problem—a larger number of possible polymerization modes and a wider selection of effective catalysts—it is not yet possible to treat the polymerization of di-olefins in the same detail as mono-olefins. Some important conclusions have been drawn, however, and these will now be discussed. A division will be made between the two classes of catalyst, and emphasis will be placed on theoretical interpretation of experimental facts, rather than their detailed presentation, though where necessary, relevant practical information will be outlined.

12.2.1 Ziegler-Natta catalysis of the polymerization of conjugated dienes

(i) *Nature of the catalysts*

Those Ziegler-Natta catalysts which polymerize conjugated di-olefins to polymers of high steric purity are largely heterogeneous. The nature of the transition metal, its valency, the crystalline form of the compound, the type of substituents bonded to it, and the possible presence of complexing agents exert a large influence on the catalysis. When aluminium alkyls are used as cocatalysts, the 1,4-*cis*-structure is largely obtained with compounds of metals of group IV (and VIII) of the periodic table. The 1,4-*trans*-structure is predominantly obtained with group IV and group V metal halides, whereas group VI metals favour a vinyl 1,2-polymerization.

The influence exerted by the crystalline form of the transition metal compound is most evident in the case of $TiCl_3$. Of the four modifications which have been described,[42] the α, γ and δ forms, having a layer structure, favour the formation of 1,4-*trans*-polydiolefins, while the β form, having a fibre-like structure, preferably yields the 1,4-*cis*-structure. The type of substituent bonded to the transition metal plays a decisive role not only on the type of catalysis (a few substituents do give rise to homogeneous systems), but also on the structure of the resulting polymer. Whereas the halides tend to direct the polymerization 1,4-, the use of polar substituents such as alcoholates, amides or acetylacetonates gives rise to vinyl polymerization (1,2-polybutadienes and 3,4-polyisoprenes). In the case of the titanium halides, furthermore, the type of halogen greatly influences the polymer structure. The stereo-regular polymerization of butadiene with catalysts prepared from

R_3Al and TiX_4 has been reported for $X = F^-$, Cl^-, Br^- and I^-. Of these, the iodide catalyst gives a polymer with much the higher 1,4-*cis*-content.[69] The nature of the metal in the organometallic compound exerts a remarkable influence on the catalysis, more marked than in the case of α-olefin polymerization. While the group I metal alkyls tend to give vinyl structures (sodium and potassium alkyls), with group II and III metal derivatives it is generally possible to obtain both isomeric 1,4-structures.

All of these generalizations are based on numerous experimental results and without exception there are examples known which contradict the conclusions drawn. In view of the multicomponent nature of Ziegler-Natta systems, and their susceptibility to contamination by small traces of impurities, it would be remarkable if such deviations were unknown. In addition, it is likely that specific ratios of particular components, especially in the presence of deliberately added electron donor or acceptor impurities, could give rise to active sites with abnormal properties.

(ii) *Mechanism of polymerization*

Many of the systems used for the stereo-regular polymerization of α-olefins also yield polymers of high steric purity from dienes. In the case of those catalysts with a definite lattice structure, for example TiX_3, it seems reasonable to extrapolate the arguments concerning the nature of the active site in mono-olefin polymerization to the case of diolefins. Halide vacancies on the crystal surface of such catalysts must therefore be regarded as the source of catalytic activity. The active site envisaged consists, as before, of an octahedral transition metal ion with a metal–alkyl bond. At least one of the co-ordination sites is vacant to allow monomer molecules to be complexed before insertion into the metal–polymer bond. Since diene molecules are structurally quite different to mono-olefins, the nature of the interaction of monomer with the active site may vary somewhat. Conjugated di-olefins can exist in two conformations, a single s-*cis*- and a single s-*trans*-form. This arises from the required coplanarity of the double bonds to allow π-electron

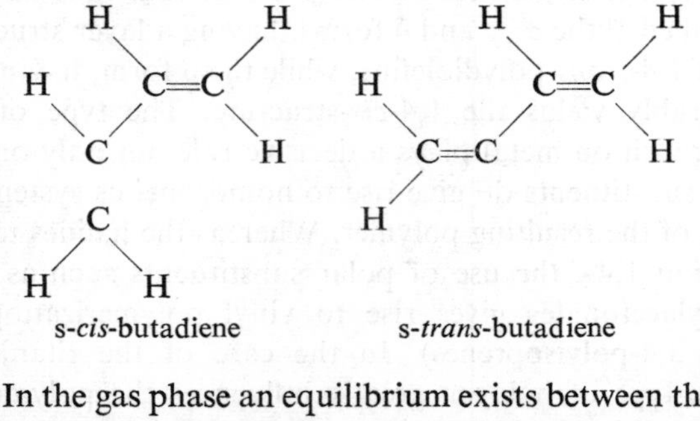

s-*cis*-butadiene   s-*trans*-butadiene

conjugation. In the gas phase an equilibrium exists between the two, although the s-*trans*-form predominates. For example, in butadiene under normal conditions the energy difference is 2·3 kcal mol^{-1} in favour of the s-*trans*-conformation and this means that the conformational isomerization reaction is achieved very easily, especially under the influence of catalysts. The presence of two double bonds in these monomers means that at the active centre there is the possibility of co-ordination with one or both double bonds, depending on the space available around the metal ion. Simple models shown that at an active site of type (A) (Figure 12.13) with one

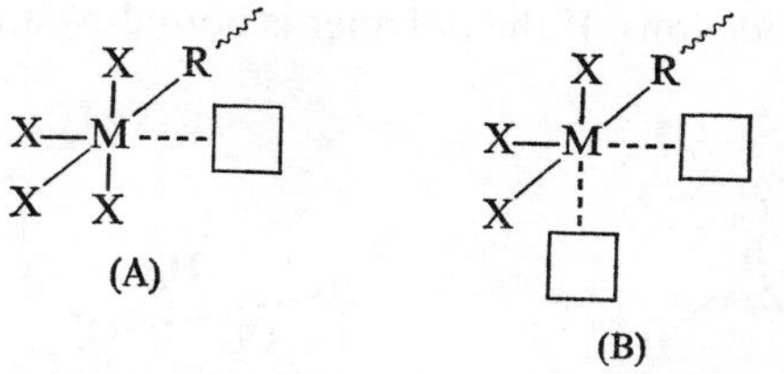

(A)

(B)

Figure 12.13 Possible structures of the active site in diene polymerization by Ziegler-Natta catalysts

vacancy, the only possibility of complex formation is via the use of one double bond with the molecule in the s-*trans*-form.[70] However, on a lattice such as β-TiCl$_3$, where there are also sites with two vacancies (type (B)) there are, in principle, two possibilities: co-ordination of the molecule in the s-*trans*-form via one double bond, or in the s-*cis*-form using two double bonds. In addition, Cooper has pointed out that it is at least possible that two adjacent transition metal atoms may be involved in monomer co-ordination, most likely with the molecule in its s-*trans*-form.[71]

The fact that a double bond is present in a position β to the metal-bonded carbon atom of the polymer chain (that is the terminal group in the polymer chain is a substituted allyl unit) permits the possibility of π-bonding of the growing chain to the metal atom.[72] Dynamic equilibria between σ-bonded and π-bonded allyl metal compounds are well-characterized and known to be affected by donor ligands, for example

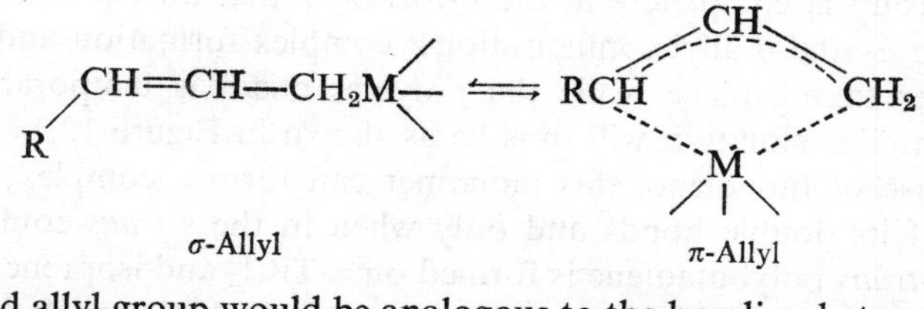

σ-Allyl π-Allyl

A σ-bonded allyl group would be analogous to the bonding between growing polymer and metal atom in α-olefin polymerization but σ-π-allyl type equilibration could occur during propagation. Molecular orbital considerations and qualitative kinetic data[74] suggest that increasing the electron density in the complex (decreasing the ionization potential of the metal orbitals) results in a shift of the equilibrium towards the σ-form. It seems, therefore, that for elements like titanium, with a rather low ionization potential, the σ-form will be more stable than the π-form, but for metals further along the period, the reverse may well be true.

(iii) *Origin of stereo-regulation*

With α-TiCl$_3$ based catalysts the active sites have only one vacant position and four fixed chloride ions. If the polymer is bound by a π-allyl linkage then

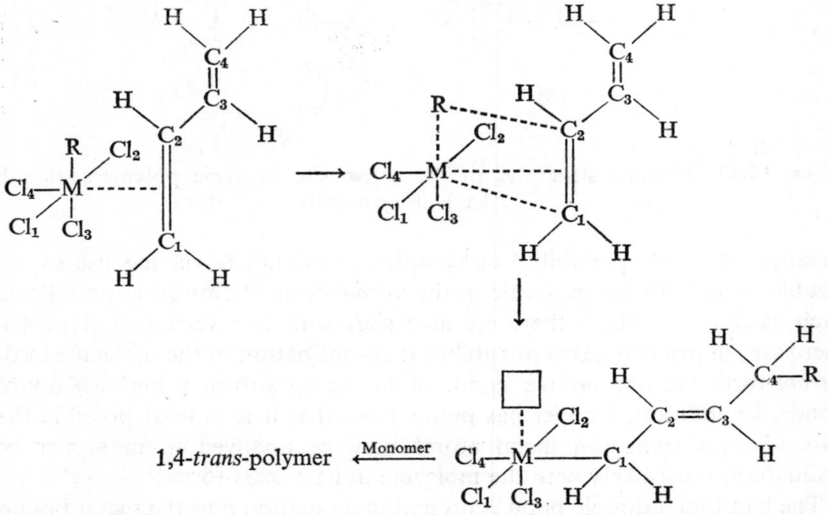

Figure 12.14   Polymerization of butadiene to 1,4-*trans*-polymer by α-TiCl$_3$

the polymeric alkyl group would effectively occupy two co-ordination positions, not one, and there would be no opportunity for monomer co-ordination to occur and hence no propagation. It must therefore be assumed that either the alkyl group is completely in the σ-form, or that an equilibrium exists between the π- and σ-allyl configurations; complex formation and insertion of monomer taking place while the polymer chain is temporarily in the σ-allyl form. The situation will thus be as shown in Figure 12.14.

In the case of butadiene, this monomer can form a complex, involving only one of its double bonds and only when in the s-*trans*-conformation. Indeed 1,4-*trans*-polybutadiene is formed on α-TiCl$_3$ and isoprene also gives the *trans*-polymer but the reaction rate is very much slower.[76] β-TiCl$_3$ has two types of active site with one and two vacancies respectively. Those with only one vacancy will not differ in their behaviour from the sites on α-TiCl$_3$, so these would be expected to produce 1,4-*trans*-polymer. The sites having two vacancies may, if the polymer chain is σ-bonded, form a complex with both double bonds of the monomer in the s-*cis*-conformation. Clearly, on these sites 1,4-*cis*-polybutadiene is to be expected. In practice it has been found[77] that β-TiCl$_3$ does give two pure polymers from butadiene, separable by extraction, namely 1,4-*cis*- and 1,4-*trans*-polybutadienes in comparable quantities. With isoprene and β-TiCl$_3$ a similar result is predictable. However,

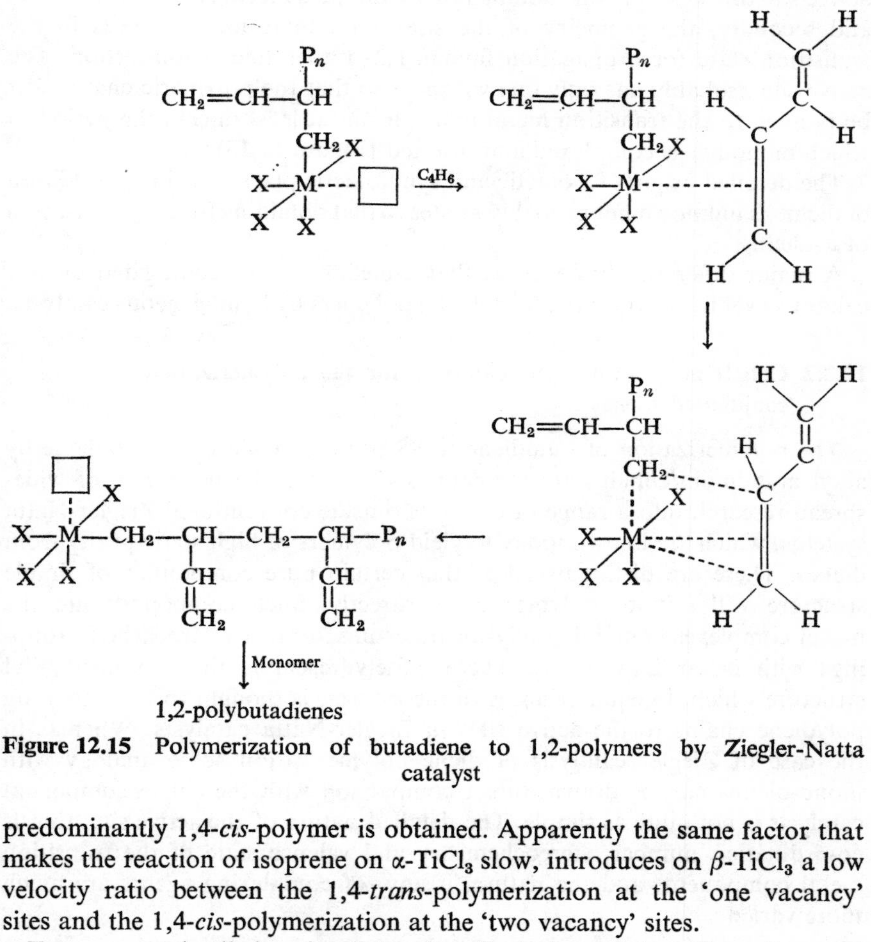

Figure 12.15 Polymerization of butadiene to 1,2-polymers by Ziegler-Natta catalyst

predominantly 1,4-*cis*-polymer is obtained. Apparently the same factor that makes the reaction of isoprene on α-TiCl$_3$ slow, introduces on β-TiCl$_3$ a low velocity ratio between the 1,4-*trans*-polymerization at the 'one vacancy' sites and the 1,4-*cis*-polymerization at the 'two vacancy' sites.

For a more detailed and enlightening discussion of the effect of crystal structure on mechanism of polymerization the reader is referred to the definitive review by Cossee.[47]

The first catalyst system[78] reported to yield 1,2-polybutadiene was the homogeneous combination of vanadium triacetylacetonate with trialkyl-aluminium in aromatic solvents. The polymer is partially crystalline and

15

syndiotactic. Both isotactic and syndiotactic placement can be obtained with a number of chromium based catalysts all of which are homogeneous.[79] Isotactic 1,2-polybutadiene can be obtained using aluminium trialkyls with chromium triacetylacetonate while other systems yield polymers whose tacticity depends on the Al/Cr ratio. It seems likely that 1,2- as opposed to 1,4-stereo-regulation arises because firstly, the monomer is complexed at the active site through only one double bond with the molecule in its s-*trans*-form and secondly, the geometry of the site and interatomic distances in the transition state for propagation favour 1,2- rather than 1,4-insertion. The active site probably has only one vacancy, so that the polymeric chain must be bonded to the transition metal in a σ-form, at least during the period in which monomer is complexed and inserted (Figure 12.15).

The detailed origin of isotactic and syndiotactic placement in 1,2-addition of dienes is unknown but probably relates to that obtaining for polymerization of α-olefins.

A major difference however, is that α-olefins, unlike conjugated dienes, cannot as yet be polymerized to isotactic polymers by homogeneous catalysts.

12.2.2 Cobalt and π-allyl type catalysts for the polymerization of conjugated dienes

The polymerization of butadiene to 98 per cent 1,4-*cis*-polybutadiene by alkyl aluminium/cobalt chloride catalysts[3] marked the beginning of widespread research into a range of catalysts outside conventional Ziegler-Natta systems, which are known today to yield polymers of high steric purity from dienes. These are distinguished in that certain pure compounds of simple structure will initiate polymerization directly. Such compounds are the nickel complexes containing allyl or other unsaturated hydrocarbon groupings with delocalized bonds. These closely resemble the π-bonded allyl structure which, in equilibrium with the σ-form, is thought to bind growing polydiene chains to the active sites in Ziegler-Natta catalysts. Whereas in the case of Ziegler catalysis of diene polymerization some analogy with mono-olefins can be drawn, direct comparison with the single component catalysts is not quite as simple. The detailed nature of the active site, that is co-ordination number, stereochemistry and valence state of the transition metal component, and the nature, if any, of complexing agents are much more varied.

(i) Catalyst structure

Most reviews on Ziegler-Natta catalysts have made some reference to these newer systems, particularly the cobalt and nickel based ones. However, it is only recently that some sort of comprehensive picture has emerged.[71] The most well defined of the nickel initiators are the π-allyl nickel halides.[67,80]

These are dimeric in non-polar solvents and as the overall rate of polymerization of butadiene is half-order in catalyst concentration, the monomeric species is presumably responsible for initiation.[81] It is conceivable that dissociation is promoted by complex formation with monomer in the s-*cis*-form where both double bonds are involved, that is

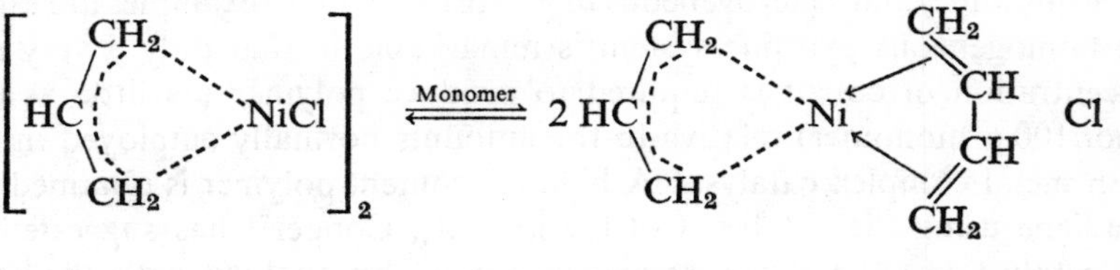

Polymerizations are slow but increase as the halide is changed from $Cl \rightarrow Br \rightarrow I$, which probably parallels their extent of dissociation. The systems are homogeneous but the exact nature of the active complex and certainly its stereochemistry, co-ordination number etc., are unknown. It does seem likely, however, that an equilibrium exists between a covalent and an ionic form. Much faster polymerizations are achieved by the addition of Lewis acids such as $AlCl_3$, VCl_4 or $TiCl_4$. Undoubtedly, these promote the formation of full ionic compounds and in aromatic solvents, arene complexes have been isolated as indicated

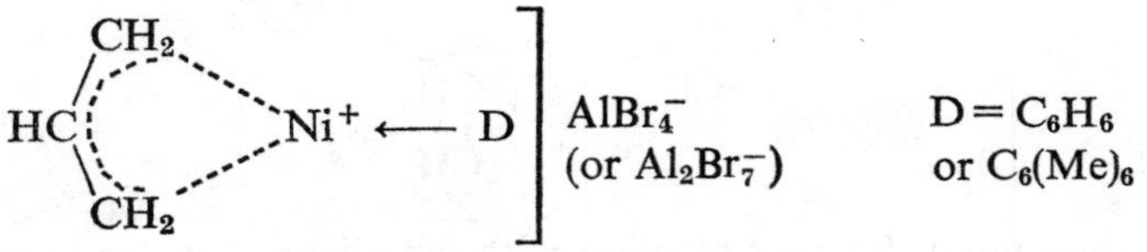

By analogy, the ionic complex involving a polymerizable diene may be of the form

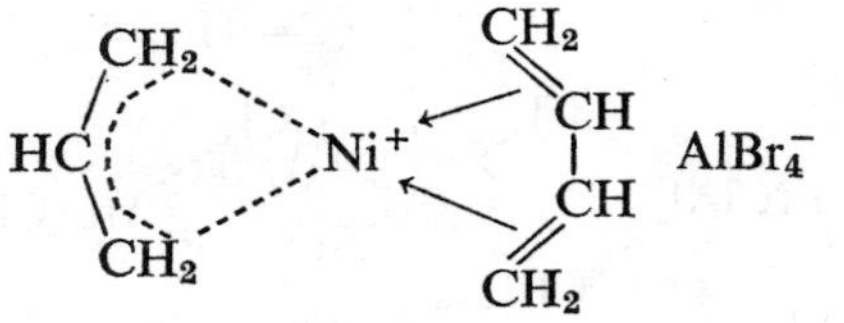

High reactivity may be the consequence of the ionic structure, and by the same token the low activity of the π-allyl nickel halides may be due to a relatively low degree of ionization. Additional evidence supports the view that increasing positive charge on the nickel increases catalytic activity.[82] Many other nickel compounds[71] have now been shown to be effective catalysts for diene polymerization, and in general the presence of Lewis acids, especially aluminium halides, increases their effectiveness. Almost certainly the active centres consist of complexes similar to those shown above.

Cobalt complexes, generally stabilized by carbonyl ligands, have been shown to function in the same way as the nickel compounds although, in general, they appear to be less stable. Of far more importance, though, are those cobalt catalysts based on simple cobalt(II) chloride.[3] Among the many systems which have been characterized, both homogeneous (e.g. AlEt$_2$Cl/ CoCl$_2$/pyridine) and heterogeneous (e.g. AlEt$_2$Cl/CoCl$_2$) examples are known. The homogeneous pyridine system is remarkable in that only a very small concentration of cobalt is required to produce polymer (as little as 0·006 mmol/100 g monomer) relative to the amounts normally employed in transition metal complex catalysis. A high *cis*-content polymer is obtained from butadiene using the catalyst CoCl$_2$/Al$_2$Cl$_3$Et$_3$. Cooper[71] has suggested that this catalyst has the structure shown below, by analogy with the square

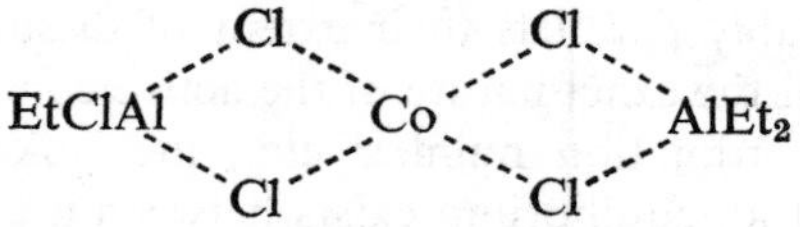

$$\text{EtClAl} \overset{\displaystyle Cl}{\underset{\displaystyle Cl}{<\!\!>}} \text{Co} \overset{\displaystyle Cl}{\underset{\displaystyle Cl}{<\!\!>}} \text{AlEt}_2$$

planar configuration of Co(AlCl$_4$)$_2$.[83] Alkylation of the cobalt is then readily achieved by a simple rearrangement, producing a structure with a transition

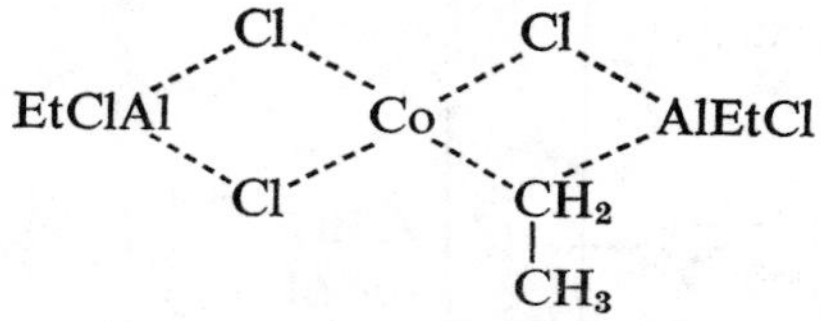

$$\text{EtClAl} \overset{\displaystyle Cl}{\underset{\displaystyle Cl}{<\!\!>}} \text{Co} \overset{\displaystyle Cl}{\underset{\displaystyle \underset{\displaystyle CH_3}{\overset{\displaystyle |}{CH_2}}}{<\!\!>}} \text{AlEtCl}$$

metal–carbon bond. Further reaction with butadiene, say, could proceed as indicated

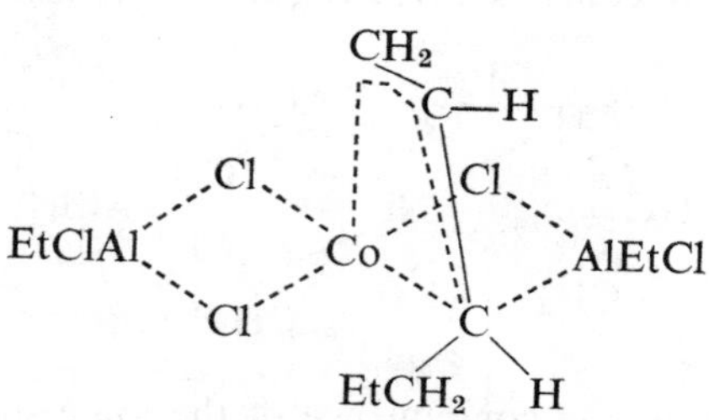

$$\text{EtClAl} \overset{\displaystyle Cl}{\underset{\displaystyle Cl}{<\!\!>}} \text{Co} \overset{\displaystyle Cl}{\underset{\displaystyle \underset{\displaystyle EtCH_2\quad H}{C}}{\overset{\displaystyle \overset{CH_2}{C-H}}{}}}{<\!\!>}} \text{AlEtCl}$$

However, the allylic grouping formed now occupies two co-ordination sites, leaving only one position available for π-complexing with a second monomer molecule, if we assume a maximum co-ordination number of six for cobalt. For the production of 1,4-*cis*-polymer it seems likely that the incoming monomer must be pre-complexed before insertion into the polymer chain, and furthermore, complexed in the s-*cis*-conformation which requires two co-ordination positions. In view of the structure of the nickel complexes previously discussed, a more probable structure for the catalyst would be an ionic complex, for example

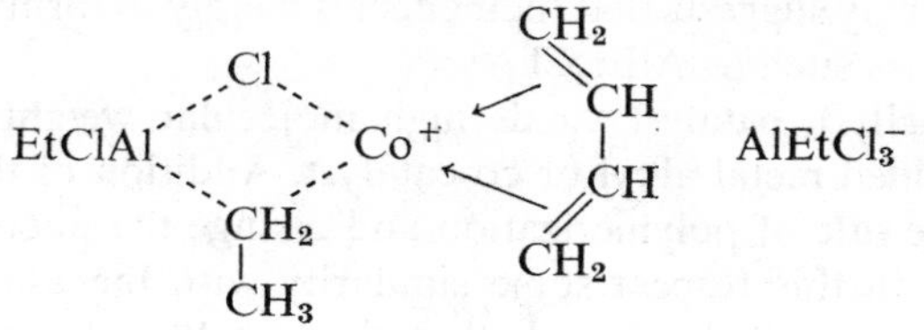

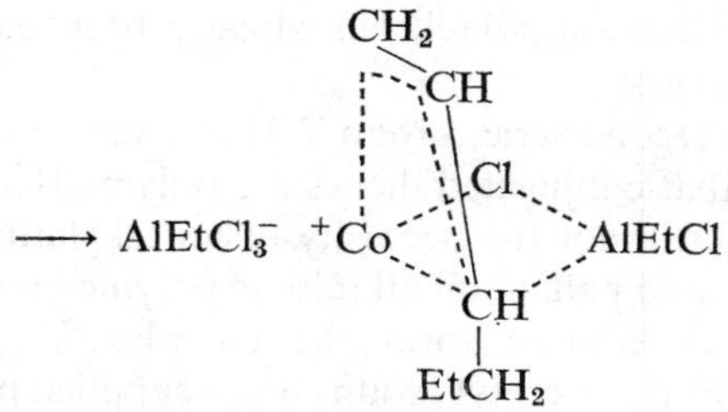

The positively charged active centre has sufficient co-ordination sites available for complex formation with subsequent butadiene molecules in the s-*cis*-form. Cooper[71] has also suggested an alternative structure in which stabilization with the aluminium alkyl via a halide bridge is redundant once propagation starts; such a structure is identical to those proposed in the case of π-allyl nickel catalysts for which there is substantial experimental evidence.

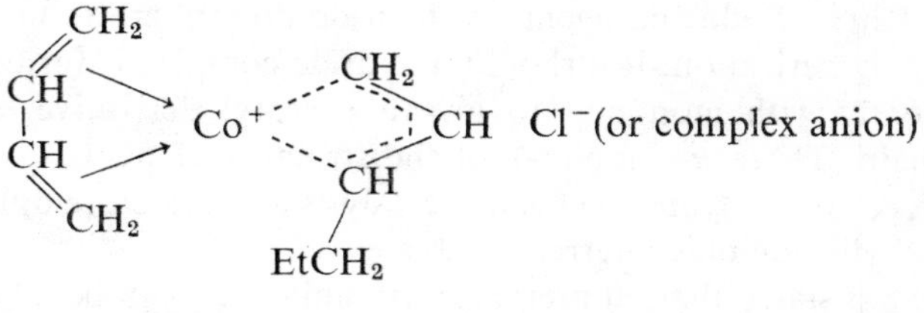

The role of the aluminium alkyl is therefore reduced to that of an alkylating agent, as in the case of monometallic mechanisms for the polymerization of α-olefins by Ziegler-Natta catalysts. In addition though, such alkyls may encourage the formation of ionic species by forming stable diffuse anions.

A complication in deducing the structure of the active species in the $CoX_2/AlEt_2Cl$ catalysts is the requirement for a cocatalyst. Water, oxygen and methanol have all been shown to be effective, and early work in this field led to widely differing hypotheses.[84] Sinn et al.,[85] for example, pointed out that the effect of water was analogous to its co-catalytic function in Friedel-Crafts polymerizations. They concluded therefore, that the 1,4-*cis*-polymerization of butadiene by cobalt catalysts involved a 'cationic co-ordinated' mechanism. In this the monomer was first orientated by the cobalt, then released as the cationic end of the chain attacked it. Although by no means all recorded observations have been reconciled even today, the work of Gippin[86] strongly suggests that their effect is merely to form a stronger Lewis acid from species such as $AlEt_2Cl$.

The soluble $Cr(\pi\text{-allyl})_3$ catalyst yields high molecular weight 1,2-polybutadiene without added metal alkyl or co-catalyst. Addition of these does, however, increase the rate of polymerization and change the microstructure of the polymer, and in this respect some similarity with the π-allyl nickel halides emerges. It seems likely though that this catalyst is more closely related to $Cr(acac)_3$ which has already been mentioned in the section on Ziegler catalysis.

With the second series group VIII transition elements, the main feature of interest is that conjugated dienes are polymerized directly by simple salts or complexes, without the necessity of an alkylating agent. Salts of rhodium, ruthenium, and palladium all initiate polymerization. Rhodium chloride and butadiene in ethanol form the complex,[87] $[(\pi\text{-}C_4H_7)_2RhCl_2]_2C_4H_6$, the hydrogen for the π-crotyl groups being supplied by the ethanol. This does not represent the catalytic complex which occurs in polymerization systems since this molecule is itself inactive. In aqueous solution rhodium chloride requires certain sulphate or sulphonate emulsifiers to be present before polymer is obtained,[88] and the emulsifier becomes incorporated into the polymer chain. In constrast, rhodium nitrate in water or ethanol will polymerize monomer without emulsifier. Experimental evidence seems to show that, in the formation of the active catalyst, the rhodium salt is first reduced to a lower oxidation state, Rh(I). Reducing agents or hydride donors are found to greatly accelerate polymerization. If a rhodium hydride complex is formed it would react efficiently with monomer to give a π-crotyl derivative, a potential polymer chain. Those π-complexes of the structure $(RhClL_2)_2$ where $L_2 = C_4H_6$, C_8H_{12} or $(C_2H_4)_2$ are all effective catalysts and therefore differ from the inactive π-allylic complex referred to above.

Cooper[71] has stated that, at present, no definite view can be taken as to the precise structure of the propagating species in rhodium catalysts, but probably the most realistic interpretation is that initiation forms a low valence state π-allylic complex, which could be a dimer or an ionic species derived

from it. The presence of emulsifying ligands may stabilize unbridged species, such ions forming part of a micelle. At the present time catalysts based on rhodium are of mainly academic interest.

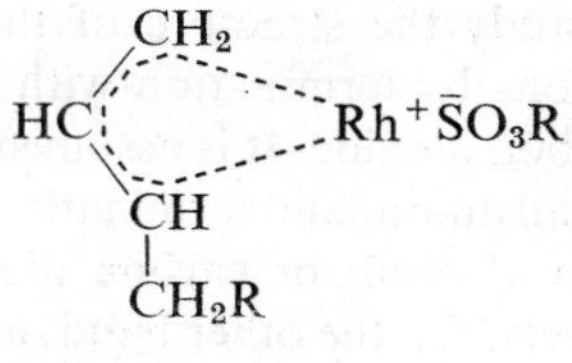

$$HC\underset{CH}{\overset{CH_2}{\lessgtr}}Rh^+\bar{S}O_3R \qquad R = Cl \text{ or } H$$

$$CH_2R$$

(iii) *Mechanism of polymerization*

With only a tentative knowledge of catalyst structure available, a detailed interpretation of polymerization mechanism is impossible. One of the earliest suggestions[89] was that growth of the polymer chain occurs on the base metal, usually aluminium, with co-ordination of the monomer at the transition metal atom, possibly in the form of a positive ion. This argument has been re-iterated in a slightly different form by Gippin[86] who has suggested that, in the case of the system cobalt octoate/aluminium chloride, butadiene co-ordinates in the s-*cis*-form to cobalt, then adds 1,4- to an aluminium alkyl ion pair (Figure 12.16). Scott *et al.*[83,90] investigated the catalyst system,

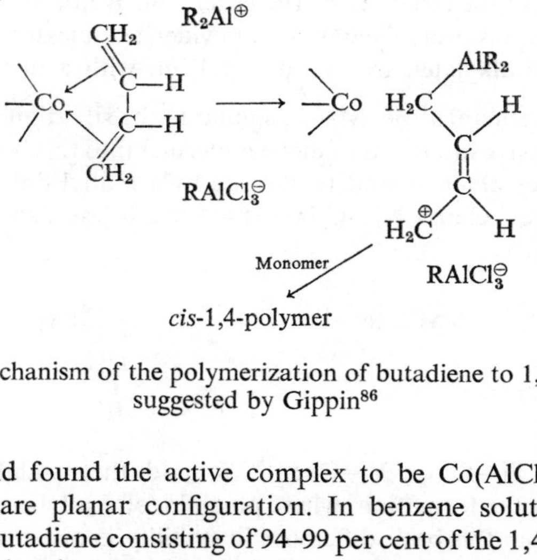

Figure 12.16 Mechanism of the polymerization of butadiene to 1,4-*cis*-polymer as suggested by Gippin[86]

$CoCl_2/AlCl_3$, and found the active complex to be $Co(AlCl_4)_2$, containing cobalt in a square planar configuration. In benzene solution containing thiophene, polybutadiene consisting of 94–99 per cent of the 1,4-*cis*-configuration was obtained and each polymer chain was shown to have one thiophene end group. More recent work on this system[91] has shown that the presence of thiophene is not necessary to obtain stereo-regular polymers, but that its inclusion is useful because it inhibits cationic polymerization due to excess $AlCl_3$, furthermore the complex resulting from the reaction of $CoCl_2$ and $AlCl_3$ is thought to be octahedral, with propagation involving chain growth from the cobalt centre, following prior co-ordination of each butadiene molecule.

Attempts have been made to study the structure of the growing chain polymer in cobalt catalysed reactions by termination with reagents such as isotopically labelled alcohol or carbon dioxide. It is assumed that with dienes the allylic chain end will behave as an anion, and termination will incorporate the replaceable hydrogen from an alcohol, or carbon dioxide to form an alkane or carboxylate ion respectively. On the other hand, a growing carbonium ion would incorporate alkoxyl from the alcohol on termination, e.g.

$$\text{\textasciitilde}CH_2\overset{-}{X}{}^+ + ROH^* \rightarrow \text{\textasciitilde}CH_2\text{---}H^* + ROX$$

$$\text{\textasciitilde}CH_2\overset{-}{X}{}^+ + {}^*CO_2 \rightarrow \text{\textasciitilde}CH_2\overset{*}{C}OO^-X^+$$

$$\text{\textasciitilde}CH_2\overset{+}{X}{}^- + \overset{*}{R}OH \rightarrow \text{\textasciitilde}CH_2OR^* + HX$$

A large volume of data[66] on these newer catalysts is available and whereas for Ziegler-Natta catalysts (group IV and V transition metals) the evidence points clearly to a 'co-ordinated anionic' mechanism, with group VIII metals no clear cut view emerges. It could be that a duality of mechanisms exists depending on the exact composition of the catalysts, and, in view of the possible variety in the structure of the latter, this is not unreasonable.

On the whole, however, the balance of evidence indicates the active centre to consist of an alkylated transition metal ion with a metal–carbon bond polarized in the general sense, $\overset{\delta+}{M}\text{---}\overset{\delta-}{C}$, similar to the situation in Ziegler-Natta catalysts. Successive insertions of diene molecules into this bond gives rise to a growing polymer chain bound to the site by a π-allyl linkage. It is known that simple σ-Co—C and Ni—C bonds are much less stable than, say, the

corresponding Ti—C or V—C entities, and that stability is increased substantially by the formation, where possible, of π-allyl compounds. There is little doubt that this basic difference in stability is one reason why cobalt- and nickel-based catalysts fail to polymerize α-olefins, monomers which demand the growing polymer to be attached by a simple M—C σ-bond. Whether the growing polydiene chain is always attached by a π-allyl linkage is not certain, but the σ-form in any dynamic equilibrium will be shorter lived than in the case of diene polymerization by titanium and vanadium species.

The fact that sterically pure polymers are obtained is an indication that each diene molecule approaches the active centre in some regular manner, and the simplest way in which this could be achieved is for monomer to be complexed at the site. There are no instances of monomer π-complexes being identified in polymerization systems, but two complexes of related structure are the rhodium chloride-butadiene complex referred to already, and a cobalt complex $Co(\pi\text{-}C_8H_{13})(\pi\text{-}C_4H_6)$. Here, the butadiene is in the s-*cis*-form and the Co—C bond lengths are not very different from those in the π-allyl group.[92]

A general scheme for the mechanism of polymerization of butadiene is shown in Figure 12.17. No attempt is made to specify the co-ordination number or the stereochemistry of the active site since this undoubtedly varies with different systems. Both cobalt and nickel complexes occur as six co-ordinate octahedral, four co-ordinate tetrahedral and square planar structures, all of which are potentially suitable as active sites. Furthermore, the possibility of co-ordination number and stereochemistry changing during elementary steps cannot be dismissed. In most cases the active site consists of a positive ion accompanied by a complex counter-anion (although a few systems approximate more closely to covalent bimetallic units, i.e. transition metal centres stabilized by complex formation with metal alkyls). Where the catalyst is heterogeneous the physical state will resemble an ionic crystal lattice. More often though, catalysts are soluble in the polymerization medium and, depending on the two ions, the solvating and dissociation abilities of the solvent and the temperature, these will exist as ion pairs (or higher agglomerates) or will be fully solvated and essentially independent of one another. The solvents which are normally employed have very low dielectric constants and it would be expected, therefore, that ionized catalyst complexes would exist as intimate ion pairs, and thus permit the possibility that counter-ions may be involved in the stereo-regulating action of these catalysts. The primary requirement for the positive ion to be active is a transition metal–carbon bond of such a stability that step-wise insertion of monomer molecules can take place. In most cases this situation already exists in that either the transition metal component possesses an alkyl group at the outset, or it readily acquires one by alkylation from a second catalytic component in the system. One or two catalysts, however, react with monomer or solvent to form the required linkage. Propagation at this bond is then achieved by monomer insertion facilitated by prior co-ordination at the same transition metal centre (Figure 12.17). Whether the growing chain is bound by a π- or σ-allyl linkage is not clear and, although the π-form is the more stable, the possibility of a continuous $\sigma \leftrightarrow \pi$ isomerization is very real. On purely stereochemical grounds a π-allyl configuration is most likely during the time no monomer is complexed at the centre, and the act of complex formation may encourage the π-allylic structure to pass into an σ-alkyl one prior to insertion of the monomer into the chain. It is possible to obtain all four sterically pure polymers from butadiene and therefore, as in the case of diene polymerization by Ziegler systems, a number of questions arise. Firstly, what factors favour 1,2-vinyl polymerization against the 1,4-mode of addition? Secondly, with 1,2-reaction what factors govern whether isotactic or syndiotactic placement is achieved, while with 1,4-polymerization what factors control the formation of *cis*- and *trans*-polymers?

The suggestion most frequently made is that the stereo-regularity of polymers is dependent on the conformation of the monomer complexed at

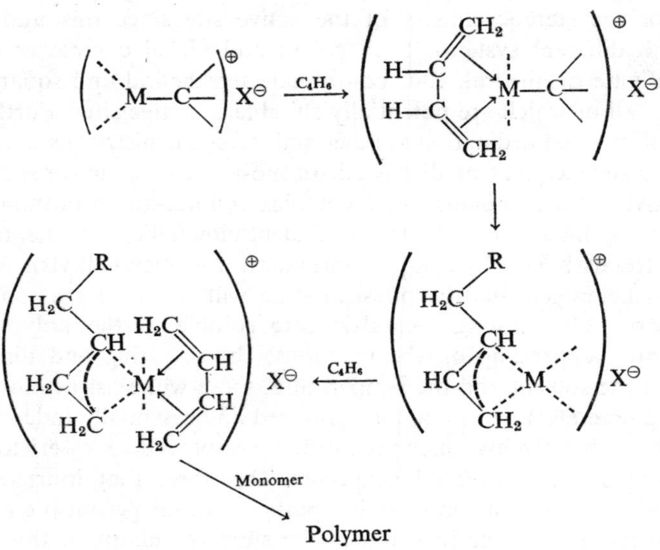

Figure 12.17 Mechanism of polymerization of dienes by transition metal compounds; for simplicity, monomer co-ordination in the s-*cis*-mode only is shown

the active site prior to insertion. However, it is possible to see how 1,2-, as opposed to 1,4-addition may arise without postulating prior co-ordination of monomer (Figure 12.18), although the explanation of the formation of 1,4-*cis* as opposed to 1,4-*trans*-polymer is more difficult without this concept.

Much more widely accepted though, is the view that s-*cis*-co-ordinated monomer (two double bonds) gives rise to 1,4-*cis*-polymer, whereas the s-*trans*-co-ordination of monomer (one double bond only) is the source of 1,4-*trans*- and 1,2-placement. The fact that mixtures of 1,2- and 1,4-*trans*-polymers from a single catalyst are more common that any other combination, and that 1,4-*cis*-polymer is more often produced alone, tends to support this argument. Indirect evidence for these ideas, is that donor molecules tend to give high 1,4-*trans*-content polymers when added to catalysts which normally give a 1,4-*cis*-structure (e.g. THF with π-C_3H_5NiCl[67] and NEt_3

with $CoCl_2/AlEt_2Cl/H_2O$).[93] These function presumably by reducing the number of co-ordination sites available to complex monomer.

Dimerization of butadiene to 3-methyl-1,4,6-heptatriene is brought about by the cobalt catalyst $CoC_{12}H_{19}$ and thus raises doubts as to the generality of the above proposals. It is believed that s-*cis*-co-ordinated butadiene adds to the substituted end of the initially formed π-crotyl grouping.[92] In this

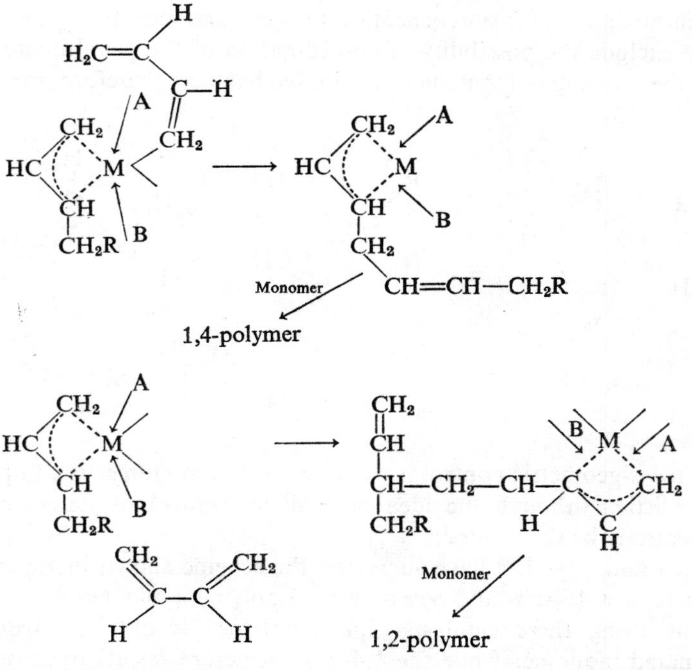

Figure 12.18 Possible mechanism for the formation of 1,2-, as opposed to 1,4-polymers of conjugated dienes. Diene attack at A forms 1,4-polymer while diene attack at B produces 1,2-polymer.

particular case the dimer is eliminated as a stable molecule, but with a different complex the reaction could continue in stepwise manner with formation of a 1,2-vinyl polymer, providing a possible case where s-*cis*-co-ordinated monomer yielded 1,2-polymer.

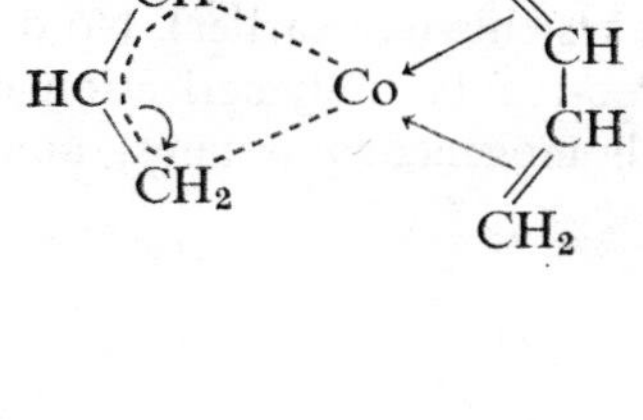

Another tenable possibility is that geometric isomerism is controlled by the π-allylic polymer–metal bond with the syn-form favouring *trans*-polymerization and the anti-form *cis*-addition. Syn/anti-orientations would be dependent on the co-ordinating ability of the double bond nearest the π-allylic polymer end, i.e. the penultimate double bond. In simpler but related compounds NMR studies have demonstrated the ready interconversion of such structures[73] as indicated, although in the case of polymer systems such interconversions may be more difficult for stereochemical reasons, and recent NMR evidence[95] seems to exclude the possibility of co-ordination of the penultimate double bond of the growing polybutadiene chain. On balance, therefore, the concept

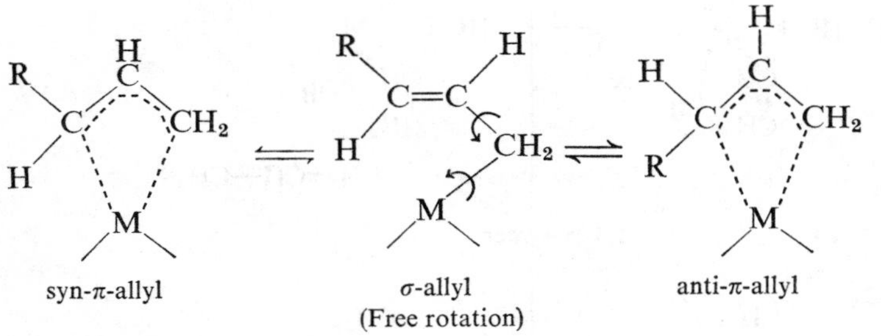

syn-π-allyl σ-allyl   anti-π-allyl
                          (Free rotation)

that *cis/trans*-geometric control is exercised as the monomer is complexed is more realistic, although the idea of π-allylic control of stereo-regularity cannot entirely be discounted.

Dawans and Teyssié[95] have suggested the scheme shown in Figure 12.19 to indicate how 1,4-*cis*- and *trans*- and 1,2-polymers may be obtained from butadiene using these catalysts. 1,4-*cis*-polymer is obtained from s-*cis*-co-ordinated monomer while the 1,4-*trans*-structure results from monomer co-ordination involving only one double bond with the molecule in its s-*trans*-form. The formation of 1,2-units is shown as taking place when complex formation also involves only one double bond, when either the building of a π-allylic polymer linkage is impossible, or the interatomic distances happen to favour 1,2-insertion.

The factors governing isotactic and syndiotactic placement in 1,2-polymerization by these catalysts are unknown. Relatively little effort, theoretical or experimental, has been turned in this direction probably because of the lack of commercial importance attached to the resultant polymers. In this respect a very close analogy exists with the 1,2-polymerization of conjugated dienes by Ziegler-Natta systems (see earlier). No doubt many of the factors which give rise to stereo-regular polymerization of α-olefins, particularly those associated with homogeneous systems, are also operative in these systems.

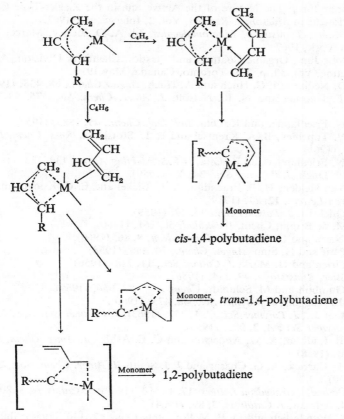

Figure 12.19 Mechanism of 1,4 (*cis-* and *trans-*) and 1,2-polymerization of butadiene proposed by Dawans and Teyssié[95]

12.3 REFERENCES

1. K. Ziegler, E. Holzkamp, H. Breil and H. Martin, *Angew Chem.*, **67**, 541, (1955).
2. G. Natta, *J. Polymer Sci.*, **16**, 143, (1955).
3. Montecatini, Belgian Patent, 575 507 (1959).
4. M. Gippin, *Ind. Eng. Chem., Prod. Res. Develop.*, **1**, 32, (1962).
5. G. Wilke (Studiengesellschaft Kohle), French Patent, 1 410 430. (1964).
6. A. Zambelli, G. Natta and I. Pasquon, *J. Polymer Sci. C.* **4**, 411, (1964).
7. C. E. H. Bawn and A. Ledwith, *Quarterly Reviews, Vol. XVI No. 4*, 361, (1962).
8. M. H. Lehr, *Survey of Progress in Chemistry*, Vol. 3, p. 183, Academic Press, Inc., New York, (1966).
9. E. A. Youngman and J. Boor Jun., 'Syndiotactic Polypropylene,' pp. 33–69 in *Macromol. Reviews.* Vol. 2, Interscience, 1967.

10. J. Boor Jun., 'The Nature of the Active Site in the Ziegler-Type Catalyst', p. 115–261 in *Macromol. Reviews*, Vol. 2, Interscience, 1967.
11. *The Stereochemistry of Macromolecules*, ed. A. D. Ketley, Marcel Dekker, New York, 1967.
12. J. Boor Jun., Organic Coatings and Plastics, Chemistry Division, ACS/CIC Meeting, Vol. 30, p. 208, Toronto, Canada, May 1970.
13. C. D. Nenitzescu, C. Huch and A. Huch, *Angew Chem.*, **68**, 438, (1956).
14. D. F. Herman and N. K. Nelson, *J. Amer. Chem. Soc.*, **75**, 3877, 3882, (1953).
15. H. N. Friedlander and K. Oita, *Ind. Eng. Chem.*, **49**, 1885, (1957).
16. A. V. Topchiev, B. A. Krentsel and L. L. Stotskaya, *Russ. Chem. Rev.*, **30**, 192, (1961).
17. H. N. Friedlander and K. Oita, *A.C.S. Meeting Abstr.*, **130**, 138, (1956).
18. E. W. Duck, *J. Polymer Sci.*, **34**, 86, (1959).
19. R. Van Helden, H. P. Braendlin, A. F. Bickel and E. C. Kooymann, *Tetrahedron Letters*, **12**, 24, (1959).
20. A. Gilchrist, *J. Polymer Sci.*, **34**, 49, (1959).
21. P. M. de Bruijn, Chem. *Weekblad*, **56**, 161, (1960).
22. G. Natta and G. Mazzanti, *Tetrahedron*, **8**, 86, (1960).
23. F. Patit and H. Sinn, *Angew Chem.*, **70**, 496, (1958).
24. F. Eirich and H. Mark, *J. Colloid Sci.*, **11**, 748, (1956).
25. G. Bier, *Kunststoffe*, **48**, 354, (1958).
26. A. Gumbolt and M. Schmidt, *Chem. Ztg.*, **83**, 656, (1959).
27. M. L. Huggins, *J. Polymer Sci.*, **48**, 473, (1960).
28. J. Boor Jr., *J. Polymer Sci, C*, **1**, 257, (1963); *J. Polymer Sci, B*, **3**, 7, (1965); *J. Polymer Sci., A*, **3**, 995, (1965).
29. D. B. Ludlum, A. W. Anderson and C. E. Ashby, *J. Amer. Chem. Soc.*, **80**, 1380, (1958).
30. W. L. Carrick, A. G. Chasar and J. J. Smith, *J. Amer. Chem. Soc.*, **82**, 5319, (1960).
31. P. Cossee, *Tetrahedron Letters*, **17**, 12, 17, (1960); *J. Catalysis*, **3**, 80, (1964).
32. E. J. Arlman, *J. Catalysis*, **3**, 89, (1964).
33. J. A. Wunderlich and D. P. Mellor, *Acta Cryst.*, **7**, 130, (1954); ibid., **8**, 57, (1955).
34. J. Chatt and B. L. Shaw, *J. Chem. Soc.*, **1959**, 705.
34a. D. R. Armstrong, P. G. Perkins and J. J. P. Stewart, *J. Chem. Soc. (Dalton)* **1972**, 1972.
35. P. Cossee and L. L. v. Reijen, in *Actes du Ziene Congrès Intern. de Catalyse, Paris 1960*, p. 1697, Éditions Technip, Paris, 1961.
36. L. L. v. Reijen, P. Cossee and H. J. v. Haren, *J. Chem. Phys.*, **38**, 572, (1963).
37. L. A. M. Rodriguez and H. M. van Looy, *J. Polymer Sci. A1*, **4**, 1951, 1971, (1966).
38. E. J. Arlman and P. Cossee, *J. Catalysis*, **3**, 99, (1964).
39. Reference 10 p. 195.
40. A. S. Matlack and D. S. Breslow, *J. Polymer Sci.*, **3**, 2853, (1965).
41. V. W. Buls and T. L. Higgins, *J. Polymer Sci. A1*, **8**, 1025, (1970).
42. G. Natta, P. Corradini and G. Allegra, *J. Polymer Sci.*, **35**, 559, (1959).
43. J. Y. Guttman and J. E. Guillet, Organic Coatings and Plastics, Chemistry Division, ACS/CIC Meeting, Vol. 30, p. 177, Toronto, Canad, May 1970.
44. G. H. Olivé and S. Olivé, *Ang. Chem. Int. Edtn.*, **6**, 790, (1967).
44a. Ibid., **10**, 776, (1971).

45. G. H. Olivé and S. Olivé 'Co-ordination Polymerization with Soluble Transition Metal Catalysts', *Advances in Polymer Science-Fortschritte der Hochpolymeren-Forschung*, **6,** 421, (1969).
46. J. W. Begley and F. Pennella, *J. Catalysis*, **8,** 203, (1967).
47. P. Cossee, 'The Mechanisms of Ziegler-Natta Polymerizations. II. Quantum Chemical and Crystal Chemical Aspects', Reference 11, Chapter 3.
47a. U. Giannini and U. Zucchini, *Chem. Comm.*, **1968,** 940; D. G. H. Ballard, p. 236 in Abstracts XXIII IUPAC Congress, Boston, 1971.
48. L. A. Rodriguez, H. M. v. Looy and J. A. Gabant, *J. Polymer Sci. A1*, **4,** 1905, 1917, 1927, (1966).
49. A. Zambelli, G. Natta and I. Pasquon, *J. Polymer Sci.*, *C*, **4,** 411, (1963).
49a. G. Allegra, *Die Makromol. Chem.*, **145,** 235, (1971).
50. J. Boor, Jun. and E. A. Youngman, *J. Polymer Sci.*, *A1*, **4,** 1861, (1966).
51. Reference 10, p. 169.
52. Reference 11, Chapter 1, D. O. Jordan 'Ziegler-Natta Polymerization: Catalyst, Monomers and Polymerization Procedures.'
53. W. L. Carrick, A. G. Chasar and J. J. Smith, *J. Amer. Chem. Soc.*, **82,** 5319, (1960).
54. T. Miyazawa and Y. Ideguchi, *Makromol. Chem.*, **79,** 89, (1964).
55. G. Natta, *J. Inorg. Nucl. Chem.*, **8,** 589, (1958).
56. H. W. Coover, Jun., *J. Polymer Sci*, *C*, **4,** 1511, (1963).
57. D. F. Hoeg and S. Liebman, *Ind. Eng. Chem. Proc. Des. Develop.*, **1,** 120, (1962).
58. Reference 10, p. 186.
59. B. Hargitay, L. Rodriguez and M. Mialto, *J. Polymer Sci.*, **35,** 559, (1959).
60. M. Sepulchre, N. Spassky and P. Sigwalt, I.U.P.A.C. XXIII International Congress of Pure and Applied Chemistry, Macromolecular Preprint Vol. II. p. 657, Boston, U.S.A., July 1971.
61. P. Pino, F. Ciardelli, G. P. Lorenzi and G. Natta *J. Amer. Chem. Soc.*, **84,** 1487, (1962).
62. P. Pino, F. Ciardelli and G. P. Lorenzi, *J. Polymer Sci*, *C*, **4,** 21, (1963).
63. P. Pino, F. Ciardelli and G. P. Lorenzi, *Makromol Chem.*, **70,** 182, (1964).
64. M. Goodman, K. J. Clarke, M. A. Stoke and A. Abe, *Makromol. Chem.*, **72,** 131, (1964).
65. M. M. Lehr, pp. 247–257 in *Survey of Progress in Chemistry* Vol. 3, Academic Press, New York, 1966.
66. W. Cooper, Organic Coatings and Plastics, Chemistry Division, ACS/CIC Meeting, Vol. 30, p. 191, Toronto, Canada, May 1970.
67. L. Porri, G. Natta and M. C. Gallazzi, *J. Polymer Sci*, *C16*, **1967,** 2525.
68. S. Otsuka and M. Kawabami, *Ang. Chem. Int. Ed. Engl.*, **2,** 618, (1963).
69. P. H. Moyer and M. L. Lehr, *J. Polymer Sci. A.*, **3,** 217, (1965).
70. British Patent 910 216 (1961).
71. W. Cooper, *Ind. and Engl. Chem. Prod. Res. Develop.*, **9,** 457, (1970).
72. G. Natta, *J. Polymer Sci.*, **34,** 21, (1959).
73. G. E. Coates, M. L. H. Green and K. Wade, *Organometallic Compounds* Vol. 2, pp. 46–48, Methuen, London, 1968.
74. K. Vrieze, P. Cossee, C. MacLean and C. W. Hilbers, *J. Organometallic Chem.*, **6,** 672, (1966).
75. G. L. Statton and K. C. Ramey, *J. Amer. Chem. Soc.*, **88,** 1327, (1966).
76. E. J. Arlman, *J. Catalysis*, **5,** 178, (1966).
77. G. Natta, *Actes Congr. Intern. Catalyse 2*, *Paris*, **1960,** 67.

78. G. Natta, L. Porri, G. Zanini and L. Fiore, *Chim. Ind.* (*Milan*), **41**, 526, (1959).
79. G. Natta, L. Porri, G. Zanini and A. Palvarini, *Chim. Ind.* (*Milan*), **41**, 1163, (1959).
80. V. A. Kormer, B. D. Babitskii, M. I. Loback and N. N. Chesnokova, *J. Polymer Sci. C16*, **1969**, 4351.
81. J. F. Harrod and L. R. Wallace, *Macromolecules*, **2**, 449, (1969).
82. J. P. Durand, F. Dawans and Ph. Teyssié, *J. Polymer Sci. B.*, **6**, 757, (1968); *J. Polymer Sci. A1*, **8**, 979, (1970).
83. D. E. O'Reilly, C. P. Poole Jun., F. Belt and H. Scott, *J. Polymer Sci A2*, **1964**, 3257.
84. Reference 60. pp. 248–250.
85. H. J. Sinn, H. Winter and W. von Tirpitz, *Makromol Chem.* **48**, 59, (1961).
86. M. Gippin, *Ind. Eng. Chem. Prod. Res. Develop.*, **4**, 160, (1965).
87. J. Powell and B. L. Shaw, *Chem. Communications*, **11**, 323, (1966).
88. M. Morton, I. Pürma and B. Das, *Rubber Plastics Age*, **46**, 404, (1965).
89. H. Uelzmann, *J. Polymer Sci.*, **32**, 457, (1958).
90. H. Scott, R. E. Frost, R. F. Belt and D. E. O'Reilly, *J. Polymer Sci.*, **42**, 3233, (1964).
91. J. G. Balas, H. E. De la Mare and D. C. Schissler, *J. Polymer Sci*, *A3*, **1965**, 2243.
92. G. Allegra, F. Lo Giudice, G. Natta, U. Giannini, G. Fagherazzi and P. Pino, *Chem. Communications*, (*London*), **1967**, 1263.
93. W. Cooper, G. Degler, D. E. Eaves, R. Hank and G. Vaughan, Amer. Chem. Soc. Advan. Chem. Ser. No. 52, pp. 46, 1966.
94. T. Matsumoko and J. Furukawa, *J. Polymer Sci*, *B5*, **1967**, 935.
95. F. Dawans and Ph. Teyssié, Organic Coatings and Plastics, Chemistry Division, ACS/CIC Meeting, Vol. 30, p. 208, Toronto, Canada, May 1970.

13

Reactivity and Mechanism with Chromium Oxide Polymerization Catalysts

D. R. Witt

Phillips Petroleum Co., Bartlesville, Oklahoma

13.1 INTRODUCTION

A unique process for the polymerization of ethylene and α-olefins was discovered and developed by Phillips Petroleum Company.[1] The heart of this process is the catalyst which consists of chromium oxide supported on silica or silica-alumina. Other support materials such as titania, zirconia, thoria or alumina may also be used but show no advantage over silica or silica-alumina. This supported chromium oxide catalyst made it possible to obtain polyethylene resins at only a few atmospheres of pressure in contrast to the extremely high pressures used in processes which grew out of the discoveries in England in the 1930's.[2] The polymers made by the Phillips process are distinctly different from the commerically available polyethylenes made in the high-pressure process. Having a linear, polymethylene structure, the polymers are highly crystalline, high density resins which exhibit high tensile strength, toughness and rigidity. These polymers are referred to as low-pressure or high density polyethylenes to distinguish them from the 'conventional' low density polymers. Copolymers, containing some short-chain branching, are also readily made by the Phillips process and extends the spectrum of high density resins available for many new applications.

Other α-olefins which have no branching closer to the double bond than the 4-carbon position can also be polymerized with the supported chromium oxide catalyst. Butene-1, pentene-1 and hexene-1 gave polymers ranging from tacky semi-solids to viscous liquids. A hard, brittle polymer was obtained from 4-methyl-1-pentene.

13.1.1 Description of catalyst and catalyst activation

Chromium oxide polymerization catalysts can be prepared by dry-mixing or non-aqueous impregnation but are readily made by impregnation of silica or silica-alumina with an aqueous solution of a soluble chromium compound such as chromium trioxide. Following drying the finely divided catalyst is normally activated by fluidizing in dry air at temperatures ranging from 400 °C to 1000 °C for several hours. It is then stored in dry air or nitrogen. The chromium compound, when in the preferred range of 0·5–5 per cent of the catalyst, is converted mostly to CrO_3 at these elevated temperatures. Evidently the support is not just an inert diluent, for chromium oxide by itself is stable mostly as Cr_2O_3 when heated to 400 °C or higher. At higher loadings a large percentage of the chromium is converted to Cr_2O_3 suggesting agglomeration of the chromium on the support.

13.1.2 Description of polymerization process

The polymerization of ethylene may be carried out in solution or in a slurry. In the solution process an inert hydrocarbon such as cyclohexane is used to dissolve the polymer as it is formed. The finely divided catalyst is thus held in suspension by agitation of a hydrocarbon solution containing dissolved monomer and polymer. Reaction temperatures range from 125 °C to 185 °C and reaction pressures from 20–30 atmospheres. Catalyst removal from the polymer is accomplished by filtration or centrifugation. Polymerization may also be conducted in a slurry process using a paraffinic hydrocarbon which is a poor solvent for the polymer. Reaction temperature is held below 110 °C to prevent solution of the polymer. As the reaction proceeds the catalyst particles increase in size and then divide by spalling and splitting to expose fresh catalyst surfaces to the reacting monomer. The granular polymer that is formed in this process is continuously removed from the reactor to a flash chamber where diluent and unreacted olefin are removed from the polymer. Because of the high productivities obtained it is not necessary to remove the catalyst for most purposes.

A third polymerization process has been described[3] which polymerizes ethylene in the gaseous phase. Catalyst removal from the granular polymer may not be necessary if polymer yields are high enough.

13.1.3 Reaction of CrO_3 with the support

As mentioned the catalyst consists of chromium oxide stabilized on support materials such as silica and silica-alumina. Chromium oxides by themselves will not polymerize ethylene to solid polymers. Nor will chromium oxides mixed with pre-dried silica or silica-alumina support polymerizations unless heated in air at elevated temperatures. If a mixture of chromium trioxide and support material is heated in air above the melting point of the chromium trioxide (approximately 197 °C) all of the chromium distributes itself on the support, giving an orange-coloured catalyst. As this newly formed catalyst is heated above 200 °C it develops polymerization activity when tested at polymerization conditions. Activation at temperatures above 400 °C influences a more rapid increase in the polymerization activity of the catalyst as additional water is removed.

The reactions of CrO_3 with silica results in the formation of a Cr—O—Si bond in surface chromate structure well above 800 °C.[4] It also accounts for the easy recovery of CrO_3 from the catalyst by leaching with water. This conclusion was verified in two ways. Silica was wetted with water and dried to constant weight at 150 °C. It was then wetted with a solution containing a known amount of CrO_3 and again dried to constant weight. Repeated experiments were made and showed that for each mole of CrO_3 present on the silica, slightly more than one mole of bound water was released.

Thermogravimetric analyses were made of silica and identical silica containing 10 per cent CrO_3 after drying to a constant weight at 120 °C in flushing, dry air. The sample containing the CrO_3 lost less weight than did the silica sample even though some CrO_3 certainly gave up oxygen as the temperature was increased to 800 °C. This difference can be accounted for only by a difference in bound water content existing at 120 °C.

Further evidence for stabilization of chromium on silica-alumina was obtained from reduction profiles of the chromium catalyst by Holm and Clark.[5] These profiles were determined from changes in the reduction rate of the chromium as the temperature of the sample in contact with hydrogen was increased at a steady rate from 145–550 °C. Hydrogen was circulated continuously over the catalyst, passing through a drying agent on each cycle. Reduction rates, $\Delta p/\Delta t$, were determined by measuring pressure (P) drops in small intervals, and they were plotted against temperature. The heating rate selected was 1·5 °C/min. The profiles shown in Figure 13.1 represent the rates of reduction of samples containing different amounts of chromium from 0·25–3·19 per cent. It is evident from the figures that reduction starts at a lower temperature the higher the chromium content. These results show that the lower the chromium content, the more resistant the chromium is to reduction, and thus the stronger the interaction. Agglomeration of chromium at higher levels may be one explanation for this ease of reduction.

Reduction tests were also carried out at constant temperature in the presence of both hydrogen and carbon monoxide. Both the hydrogen and carbon monoxide were circulated through the catalyst bed and then an adsorbent to remove water in the case of hydrogen, or carbon dioxide in the case of carbon monoxide. Typical results are shown in Figure 13.2.

At 315 °C reduction with carbon monoxide was very rapid and had virtually ceased after 20 minutes, though not complete as the results at 510 °C show. Reduction with hydrogen at 315 °C was slow and was still continuing after 2 hours. Reduction had just about ceased after 4 hours. At 510 °C rates and extent of reduction with carbon monoxide and hydrogen were nearly identical for the first 15 minutes. Reduction with hydrogen had virtually stopped in 45 minutes but reduction with carbon monoxide continued slowly for about

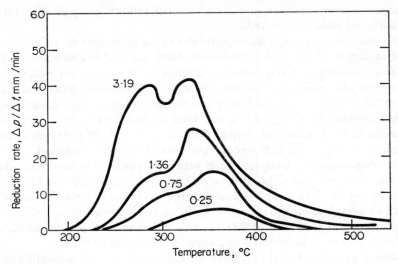

Figure 13.1   Reduction profiles for air-activated catalysts with varying chromium contents

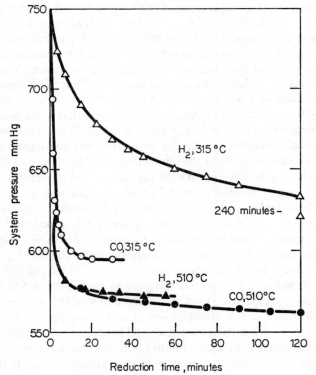

Figure 13.2 H_2 and CO reduction of 2·5 per cent Cr silica-alumina catalyst activated at 540 °C. Consumption of reducing gas against time at constant temperature

two hours. In both cases reduction at 510 °C was far more extensive than at 315 °C, which is indicative of non-uniformity of the supported oxide. Other agents, such as process solvent or diluent and reactant, also cause reduction of the catalyst at reaction conditions.

13.2 DETAILED STUDY OF CO-TREATED CATALYSTS

Further, more detailed studies of the reaction of carbon monoxide with the chromium-silica catalyst were made.[4] It is possible to visually follow the changes in oxidation which result when the catalyst is treated with hydrogen or carbon monoxide. When the orange, activated catalyst was reduced in hydrogen water was formed which in turn reacted with the CrO_3 allowing formation of considerable Cr_2O_3 and yielded a green catalyst. Reduction in carbon monoxide produces only carbon dioxide and does not affect the Cr—O—Si bonding. The colour of carbon monoxide reduced catalyst, flushed and cooled in nitrogen, was blue. When the blue catalyst was evacuated at room temperature the colour changed to light green. Readmission of nitrogen immediately gave the blue colour again. However, when carbon monoxide was admitted to the evacuated catalyst it became purple in colour. Exposure of carbon monoxide reduced catalyst to dry air resulted in the original orange colour after emission of a brilliant orange light due to chemiluminescence.

When ethylene was admitted to the purple, carbon monoxide complexed catalyst at ambient temperature an immediate change to indigo-blue occurred but no polymerization of the ethylene resulted. However, when the reduced catalyst was exposed to ethylene without first being exposed to carbon monoxide polymerization was rapid. Addition of a carbon monoxide ethylene mixture to an evacuated, carbon monoxide reduced catalyst also resulted in the blue-indigo colour but no polymerization occurred until the carbon monoxide was removed and pure ethylene admitted. If, instead of readmitting ethylene to this evacuated catalyst, water vapour is added, then ethylene corresponding to approximately one mole per mole of chromium was found in the gas phase. Further measurements made with the carbon monoxide reduced catalyst at room temperature indicated that one mole of carbon monoxide was adsorbed for each mole of chromium present.

These results suggest that an active chromium site is not only involved with a growing polymer chain but also orients and feeds incoming ethylene molecules to the growing polymer molecule.

13.3 REACTIVITY AND MECHANISM

The polymerization of ethylene over the chromium oxide catalyst is a complex reaction. Kinetic data obtained from literature and Phillips laboratory show that this is true.[5] A study of the polymerization of ethylene was made by Ayscough, *et al.*, at low pressure in the gas phase and in the absence of solvent which indicated that the kinetics followed a zero-order law between −40 °C and 0 °C. Above 0 °C, they found that the rate changed to a first, and later, to an even higher order. Yermakov and Ivanov, using cyclohexane solvent and working below 90 °C, found the kinetics were complex and dependent on whether the pressure was varied during a single experiment or from experiment to experiment. In the first case, reaction order was close to 1·5; and in the second case, calculated for extreme conditions (40 atm and 3 atm), the order was close to 2. The high order observed with respect to ethylene was explained on the assumption that the number of active sites is a function of the ethylene concentration, which is consistent with assuming that the active sites are formed by interaction of ethylene with the active component of the catalyst. At temperatures above 100 °C, Clark found first-order behaviour for both gas phase and solvent operation, gradually reducing to zero order as the pressure was increased above 500 psig. Not all of the peculiarities of the kinetics can be explained but consideration of some polymerization variables may help to explain some of the results.

There is a definite relationship between catalyst activation temperature (perhaps relating to hydroxyl content) and catalyst activity.[6] As catalyst activation temperature is increased in the range of 500–750 °C there is a corresponding increase in polymerization activity. Below 500 °C activity is low due to the presence of unremoved moisture and high hydroxyl content of the catalyst. As activation temperature is increased additional removal of hydroxyl population is accomplished.

An interesting effect of chromium content was obtained using hydrocarbon diluent and a pressure of 550 psig and a temperature of 110 °C.[4] A series of catalysts were tested which had chromium contents ranging from 1·0 to 0·001 weight per cent of the total catalyst. The results obtained with these catalysts activated at 850 °C are presented in Figure 13.3. As the concentration of chromium was decreased, catalyst charge was increased to compensate for any poisoning of catalyst sites by trace impurities and to keep total rate of production about constant. Although polymer yield based on total catalyst decreased sharply, yield based on chromium increased sharply as the chromium content was decreased to 50 ppm. In view of the extreme dilution of chromium atoms on the surface, a mechanism involving polymer growth on a simple site is strongly indicated. The increase in efficiency per chromium atom by dilution indicates that special locations on the catalyst surface which are limited in number provide the best environment for maximum polymerization rate. As chromium content was increased, efficiency decreased, but overall rate increased, indicating that other sites provide activity but at lower rates. Chromium agglomeration may also be involved at the higher chromium levels. As chromium is diluted, effective concentration of ethylene at the catalyst site may increase because of increased ratio of ethylene adsorption (feeder) sites to catalyst sites. The average number of polymer molecules produced per second per chromium atom was calculated to be 2·8 for the catalyst containing 1 per cent chromium. At this chromium level the number average molecular weight of polymer was about 20 000. If it is assumed that all the chromium atoms created active sites, then the average rate of ethylene monomer addition was 2/ms. This is, of course, conservative since no correc-

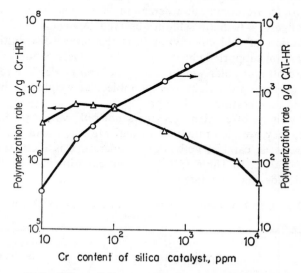

Figure 13.3 Effect of Cr concentration on polymerization rate

tion for inactive chromium or for time elapsed during chain termination and initiation of a new chain is made.

Pretreatment of the chromium oxide catalyst with carbon monoxide increases the activity of the catalyst considerably beyond that obtained with untreated catalyst. This treatment also reduces or eliminates the induction period sometimes encountered in the polymerization reaction. This result can be explained as due to the removal from the catalyst of combined oxygen which forms oxygenated compounds in the presence of ethylene, the oxygenated compounds acting as catalyst poisons.[4]

Certain polar materials are severe poisons for the chromium oxide catalyst. These poisons include oxygen, water vapour, carbon monoxide and more volatile compounds of nitrogen, oxygen, sulphur and halogens. Low concentrations of acetylene and butadiene may also act as poisons but are readily polymerized with this catalyst when used as pure monomers. Most of these polar materials are more tightly associated with the catalyst sites than the olefin, thus interrupting initiation and propagation during polymerization.

The published literature on the effect of process and catalyst variables on the molecular weight of polymers produced is scanty.[5] It was demonstrated that number average and weight average molecular weights approach a limiting value asymptotically as pressure is increased. Weight average-number average molecular weight ratios, which give a measure of the breadth of molecular weight distribution, are usually high in comparison to those obtained in homogeneous systems. In the supported chromium oxide systems, ratios of 13 to 15 are common, whereas in homogeneous systems, values outside the range of 4 to 5 are seldom observed. A logical cause for such a broad distribution of molecular weights is the broad distribution of site energies present on supported chromium oxide catalysts. Studies of the effect of temperature on molecular weight also have been made. The results show that number average and weight average molecular weights decline linearly with increasing temperature.

Molecular weights of ethylene polymers not only vary with temperature and pressure, they are also functions of activation temperature. Molecular weight decreases as catalyst activation temperature is increased over the range of 400–850 °C. In Table 13.1, pertinent experiments are shown for a chromium trioxide-silica catalyst.

Table 13.1   Effect of activation temperature on molecular weight of ethylene polymer

	Catalyst preparation	Relative molecular weight
1.	CrO_3-silica, 850 °C activation	1
2.	CrO_3-silica, 500 °C activation	3·5
3.	Silica calcined at 850 °C, dry-promoted with CrO_3, 500 °C activation	1·2
4.	Silica calcined at 850 °C, wet-promoted with CrO_3, 500 °C activation	2·5

These data show first that polymer molecular weight is increased by decreasing activation temperature from 850 °C to 500 °C. When the silica is first treated at 850 °C in air, then promoted with CrO_3 in the absence of any moisture and activated at 500 °C, the molecular weight corresponds to the 850 °C treatment rather than the 500 °C treatment. But if the silica is wetted after 850 °C treatment, then the molecular weight corresponds more nearly to the 500 °C treatment. The indication is that the population of hydroxyl groups on the surface is one of the factors that controls molecular weight. (Water has been shown to exist only as hydroxyl on silica at a temperature of 500 °C). Surface area and pore volume did not change materially from 500 °C to 850 °C, and the change in molecular weight is not due to the effect of temperature on CrO_3, as can be seen by comparing the second and third experiments in Table 13.1. It thus appears that hydroxyl groups on silica causes an increase in molecular weight.

Molecular weight also changes with concentrations of chromium on the support. For example, the number of average molecular weight increased about 40 per cent as chromium concentration decreased from 0·75 to 0·001 per cent. This increase is in line with the supposition that effective ethylene concentration around the catalyst site increases as chromium concentration decreases at a given gas pressure. A certain amount of ethylene is adsorbed in sites of the support. As chromium is added to the support, some of these sites are covered with chromium, producing new sites, some of which will be coupled with growing polymer chains. Thus, adsorbed monomer concentration is reduced. It may be reduced still further by reaction of adsorbed monomer with growing polymer chains, for if this reaction is appreciably faster than the rate of adsorption of monomer, then the surface will tend to become denuded of monomer.

Theoretical calculations were made by Clark and Bailey[5] for variation of molecular weights and reaction rates with experimental results. Four models were set up for polymerization on solid surfaces. In two of the models, it was assumed that all adsorption sites have the same adsorption energy. The first of these (Rideal mechanism) assumes polymerization occurs by reaction of monomer in the gas phase with adsorbed monomer or adsorbed, growing polymer chains. The second model (Langmuir-Hinshelwood mechanism) assumes polymerization occurs by reaction of an adsorbed monomer molecule with an adjacently adsorbed monomer molecule or growing chain. The other two models are the counterparts of the first two with a variation of adsorption energies among the sites. The individual steps were assumed to be adsorption (initiation), surface reaction (propagation), and desorption (termination). In all models, it was shown that the rate of polymerization and the molecular weights increase without limit as pressure is increased in the Rideal mechanism, whereas in the Langmuir-Hinshelwood mechanism, molecular weights and the rate approach a maximum value asymptotically as pressure is increased. Experimentally, it was found, as mentioned previously, that reaction rate and molecular weights levelled out as pressure is increased, indicating a Langmuir-Hinshelwood mechanism. Models involving a distribution of adsorption energies allow broad molecular weight distributions, that is, high values for weight average-number average molecular weight ratios, as obtained experimentally, whereas models with all adsorption sites of the same energy give narrow distributions with ratios between 1 and 2. It was shown that if termination is assumed to occur by spontaneous transfer (simple desorption, as above) and also by transfer with monomer, then molecular weights reach maximum values asymptotically with increasing pressure for both the Rideal and Langmuir-Hinshelwood mechanisms. However, the rate expression still increases without limit for the Rideal mechanism and reaches a maximum value asymptotically for the Langmuir-Hinshelwood mechanism as pressure is increased.

The number of active sites present in the chromium oxide-silica catalyst was estimated by several different techniques. The first method[4] was based on the polymerization activity of a catalyst containing a low concentration of chromium. It was found that the maximum chromium efficiency in the slurry process occurred at about 100 ppm chromium. However, total efficiency improved up to about 0·5 per cent chromium on silica. To estimate the concentration of active sites on one per cent chromium catalyst the ratio of polymerization rates obtained with one per cent chromium and 100 ppm chromium (5100/590 = 8·64) was multiplied by 100 giving 864 ppm effective chromium. This gave a value of $1·7 \times 10^{-5}$ mol/g active chromium sites. This value should be conservative since diffusion rate and other factors may begin to limit polymerization rate rather than the number of active sites at high chromium loading. Similar data were obtained for the solution process and a value of $3·5 \times 10^{-5}$ mole per gram active chromium sites was determined.

Another method of estimating number of active sites involved injection of a measured amount of catalyst poison into a reactor when polymerization reaction was near maximum and determining the amount of decrease in reaction rate which resulted from the poisoning. Triethylamine, a polar, hydrocarbon soluble, monofunctional molecule, was used as a means of poisoning the catalyst. It was assumed that one active site was poisoned per molecule of triethylamine and that the number of sites poisoned was proportional to the decrease in ethylene consumption. A value of $3·3 \times 10^{-5}$ mole active sites per gram of catalyst was obtained by this method.

A third, unique technique[7] of determining active polymerization sites involved the measurement of adsorbed ethylene and solid polymer on 'conditioned' and 'unconditioned' catalyst (3×10^{19} chromium atoms per gram catalyst) at sub-zero temperatures. The 'conditioned' catalyst was obtained by treating active catalyst (which had been dried and activated in air at 550 °C) with 6·5 moles of ethylene per gram-atom of chromium at 150 °C for one hour. The experiments were conducted in a continuous flow apparatus at sub-atmospheric pressures. The amount of ethylene plus polymer on the catalyst surface during reaction, the amount of ethylene desorbed, and the amount of ethylene plus polymer remaining on the surface after stripping was calculated from flowgrams. The number of active sites present per gram of catalyst is shown in Figure 13.4, and is based on the assumption that one molecule of ethylene occupied each active polymerization site. Adsorption of ethylene on both the 'conditioned' and 'unconditioned' catalyst, curves (A) and (B), were compared under conditions of no reaction (−55 °C) and it was found that both catalysts adsorbed essentially the same amount of ethylene. Thus, the assumption was made that equilibrium adsorption, as obtained on 'unconditioned' catalyst was an approximate measure of the maximum number of occupied sites on active, 'conditioned' catalyst. The difference between this value and the measured number of ethylene molecules on the surface of active catalysts during reaction was taken as the maximum number of sites occupied by adsorbed polymer climbed to 1.7×10^{19} molecule/g of catalyst, or about 3×10^{-5} mol/g. A similar set of experiments and

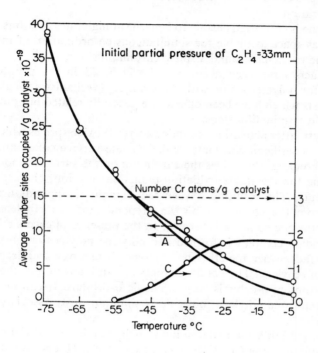

A: BY ETHYLENE ON CONDITIONED CATALYST
B: BY ETHYLENE ON UNCONDITIONED CATALYST (NO REACTION)
C = (B–A): BY ADSORBED POLYMER

Figure 13.4   Average number of sites occupied per gram catalyst

calculations using a catalyst containing 0·1 per cent chromium (2×10^{-5} mol/g) gave a maximum value of 1.6×10^{-5} mol/g for the average number of sites occupied by adsorbed polymers.

These methods gave relatively good agreement.

Certain other relevant facts relating to the polymerization of olefins with the supported chromium catalyst should be stated:

(a) The polyethylene polymer molecule has one vinyl group located at one end of a long unbranched chain of methylene groups.

(b)　Chain termination, reinitiation, and growth is rapid, and molecular weight is primarily a function of reaction temperature.

(c)　Acetone and formaldehyde are found when ethylene is contacted with the chromium catalyst at low temperatures. Other fragments may result from this initial reaction offering a source of hydroxide ions or protons.

(d)　Polymer molecular weight is decreased by addition of hydrogen or addition of α-olefins, which produce short branches and some branched-vinyl unsaturation.

(e)　Acetylene is converted cleanly to benzene at high rates over this catalyst. This indicates space available for simultaneous co-ordination of more than one acetylene molecule around a single chromium site.

Various facts have been accumulated which aid in the proposal of a mechanism for polymerization with this catalyst. Mechanisms for chromium oxide catalysts which have been offered are generally related but differ in the initiation and termination steps.

The catalyst, after activation, contains a Cr(VI) oxide species which can be represented as a silicon chromate or dichromate surface compound having Cr—O—Si bonding which fixes the chromium on the surface. Upon contact with ethylene the inevitable oxidation-reduction reaction, characteristic of Cr(VI), occurs. The resulting oxygenated organic compounds are readily displaced from the chromium by the preponderance of ethylene, and a chromium-ethylene complex is formed. In the presence of excess ethylene, a deep blue-indigo complex is formed and ethylene polymerization proceeds. This colour fades when the ethylene monomer is removed and deepens with the return of ethylene. The colour changes are an indication of the involvement of chromium d-orbitals with ethylenic π-orbitals. It can be assumed that more than one ethylene molecule is simultaneously associated with a chromium site.

The proposed initiation reaction involves interaction of ethylene with the chromium site, forming Cr—CH=CH$_2$ or Cr—CH$_2$CH$_3$ co-ordination. This can be pictured as follows

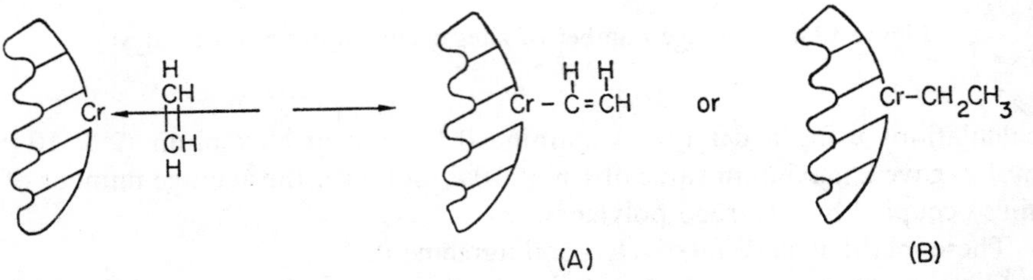

(A)　　　(B)

Since the concentration of ethylene around a site is large, it is readily available to become adsorbed and polarized, ready to insert itself into the chromium-carbon bond formed with the first adsorbed ethylene molecule.

Polymer growth for both types of co-ordination, (A) and (B), would follow as shown

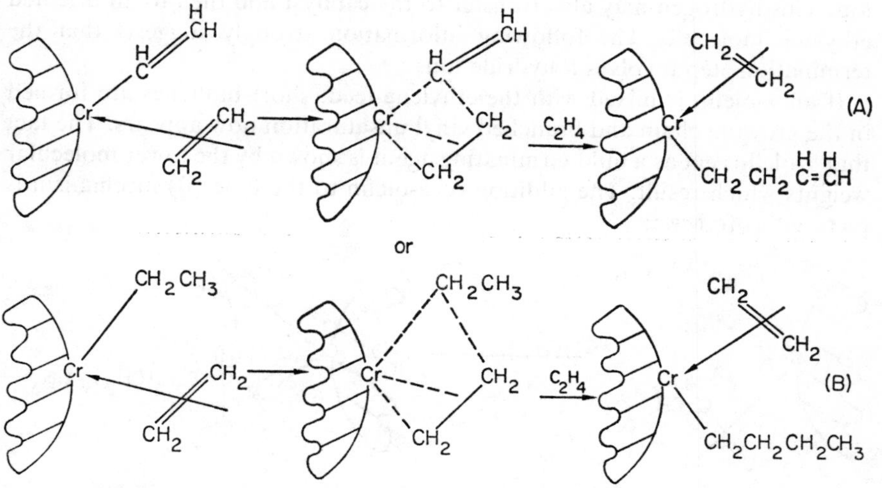

This step could be quite rapid if more than one ethylene molecule was adsorbed and polarized in sequence to adsorption on a site. This is suggested by the fact that three acetylene molecules are oriented to form a six-carbon ring structure on the same chromium sites. Thus, as a molecule of ethylene is inserted into the growing chain another molecule would become associated and polarized at the chromium site.

In either case the polymer would continue to grow by this insertion method. One polymer chain (A) would contain the unsaturated group at the end farthest removed from the catalyst site. The other growing polymer chain (B) would have its olefinic groups formed during the termination step. Termination would be expected to follow the pattern below

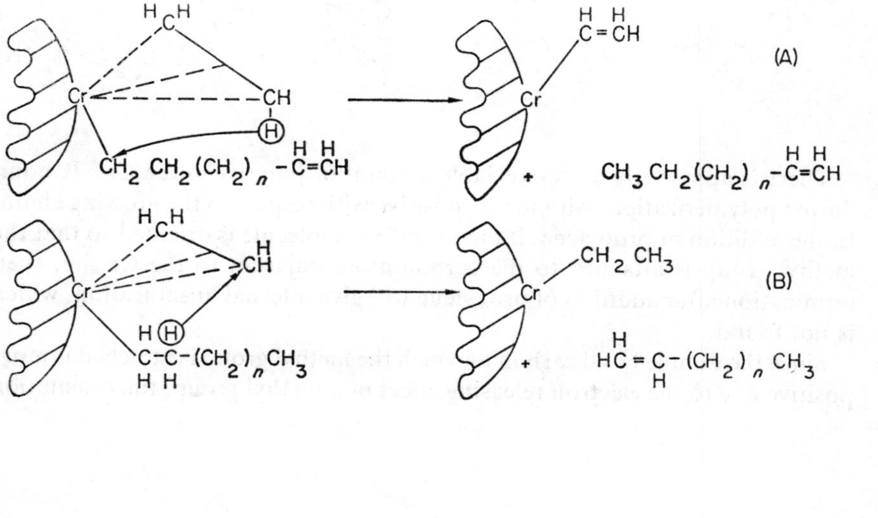

At this stage the transient hydrogen involved in the termination reaction has not been labelled and could be considered as either a proton or hydride ion. This hydrogen may also transfer to the catalyst and then to an oriented ethylene molecule. The following information strongly suggests that the termination step involves a hydride ion.

If an α-olefin is mixed with the ethylene feed, short branches are formed in the growing chain and branched vinyl unsaturation now appears. The fact that α-olefins act as a mild terminating agent is shown by the lower molecular weights which result. The addition of α-olefin to the type (B) mechanism is pictured as follows:

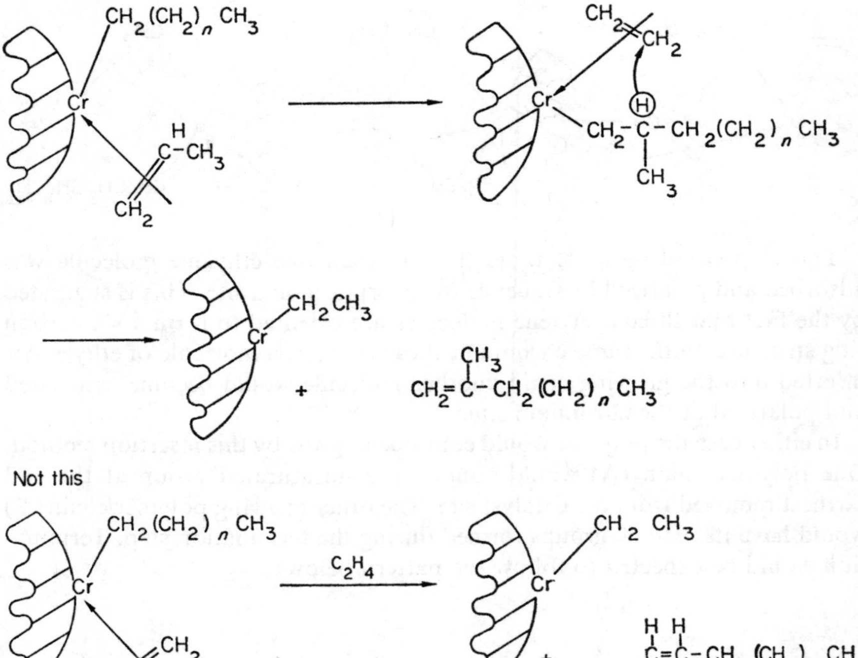

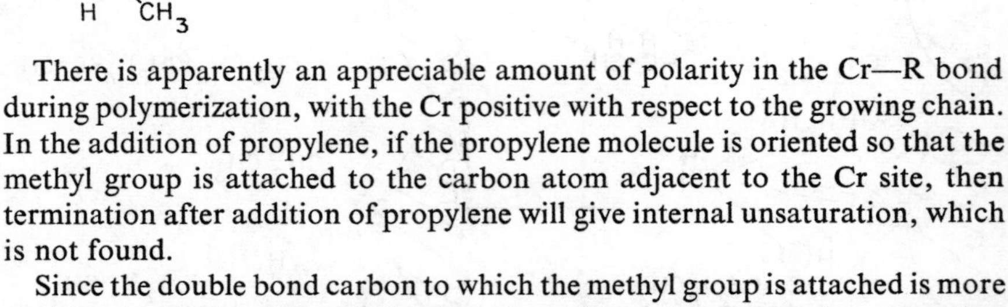

There is apparently an appreciable amount of polarity in the Cr—R bond during polymerization, with the Cr positive with respect to the growing chain. In the addition of propylene, if the propylene molecule is oriented so that the methyl group is attached to the carbon atom adjacent to the Cr site, then termination after addition of propylene will give internal unsaturation, which is not found.

Since the double bond carbon to which the methyl group is attached is more positive due to the electron releasing effect of a methyl group, the orientation required to account for branched vinyl unsaturation indicates that chromium is more positive than the first carbon of the growing chain. This suggests that the transient hydrogen involved in the termination step is a hydride ion.

13.4   REFERENCES

1. J. P. Hogan and R. L. Banks, Belgian Patent 530 617, (24 January 1955) and US Patent 2 825 721, (4 March, 1958) (to Phillips Petroleum Company).
2. E. W. Fawcett, R. O. Gibson and M. W. Perrin, US Patent 2 153 553, (11 April, 1939) (to ICI).
3. Phillips Petroleum Company, Belgian Patent 551 826 (1956).
4. J. P. Hogan, 157th National Meeting, American Chemical Society, Symposium, Division of Polymer Chemistry, on Ethylene Polymerization Catalysis Over Chromium Oxide, 16 April, 1969.
5. A. Clark, *Chemical Reviews*, 3(2), 145–174, (1969).
6. A. Clark and J. P. Hogan, p. 29 in *Polythene*, 2nd edition, eds. A. Renfrew and P. Morgan, Interscience, New York, 1969.
7. A. Clark, J. N. Finch and B. H. Ashe, *Third International Congress on Catalysis*, Vol. II, pp. 1010–1020, 1964.

14

Interaction of Light with Monomers and Polymers

F. C. De Schryver and
G. Smets

Department of Chemistry, University of Louvain

14.1 INTRODUCTION

The main purpose of this chapter is to discuss fundamental photochemical processes which intervene during the photopolymerization of monomers and the photochemical reactions in which polymers can be involved. Its aim is not to give an exhaustive review of the existing literature in this field but rather to afford a comparative and critical survey of existing data. After a short reminder of the basic principles of photochemistry, the subject is treated in three sections. The first of these presents a discussion on photoinitiated polymerization, usually known, although erroneously, as photopolymerization in which only the initiation step results from interaction of light with a photosensitive compound or system. The second section deals with photochemical chain lengthening processes in which the absorption of light is indispensable for each propagation step. Such photochemical step reactions should be considered as true photopolymerizations. The last section discusses the interaction of a polymer with light resulting in either reversible or irreversible transformations.

14.2 PRINCIPLES OF PHOTOCHEMICAL PROCESSES

The first and most important step in any photochemical reaction is the absorption of a quantum of light. By this interaction the molecule is brought into an excited state, electrons from a non-bonding or bonding molecular orbital being promoted to an anti-bonding orbital. This absorption is evidently related to the electronic structure of the chemical compound; the more easily electrons are promoted, the smaller the excitation energy and the larger the wavelength of the light which will be absorbed. As most absorptions obey the Franck-Condon principle,[1] the state obtained in an absorption is

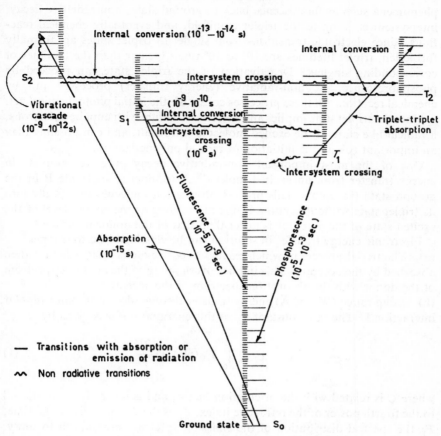

Figure 14.1 Modified Jablonski diagram

not only electronically but also vibrationally excited. Since most molecules have in the ground state a singlet manifold, they will be excited to the singlet excited state (Figure 14.1).

Events following light excitation are strongly dependent on the physical state of the molecule.

In the gaseous state the dissipation of the excess vibrational energy will depend on the total pressure as well as on the partial pressure of the photo-activated gas; therefore reactions from 'hot' ground and excited states are often possible, and different reaction products can be obtained. On the contrary, in the solid and liquid state excess vibrational energy is dissipated very rapidly to the zero vibrational level of S_X^* and, through internal conversion, leads to the lowest vibrational level of the first excited singlet state (S_1^*). The lifetime of the latter (about 10^{-11} s) is determined by several competing

16

phenomena such as fluorescence back to ground state, non-radiative decay, intersystem crossing to the triplet manifold, and eventually chemical reactions. Since electronic transitions from singlet to triplet states are formally forbidden, triplet lifetimes are 10^3 to 10^6 times greater than the lifetimes of corresponding singlets.[2] Deactivation of the triplet state can proceed by phosphorescence, by non-radiative (energy transfer) processes and by chemical reaction. All these processes constitute the initial photochemical act.

Since the triplet state implies an electronic structure with unpaired electrons, a radical-like state of the excited molecule will result, and can possibly play an important role in the initiation of radical polymerization.

One of the most important non-radiative decay processes consists in energy transfer from an excited molecule A (donor) to molecule B in the ground state (acceptor). This process promotes a molecule specifically into its triplet state without exciting it directly so long as the energy level of the excited state of the acceptor is lower than that of the donor molecule.[3]

Electronic energy transfer in solution can be classified into four types:

(a) The trivial process in which excited donor re-emits light, which is then absorbed by the acceptor. This implies overlapping of the emission spectrum of the donor with the absorption spectrum of the acceptor.

(b) Long range (50–150 Å) energy transfer through donor-acceptor dipolar interaction.[4,5] The rate constant k_A of this energy transfer is given by

$$k_A = \frac{C}{Z_D^\circ R^6} \int_0^\infty F_D(\nu)\varepsilon_A(\nu)\frac{\partial V}{\nu^4} \tag{1}$$

where C is related with the orientation factor, and is inversely proportional to the fourth power of the refractive index, Z_D° is the donor radiative lifetime, F_D the spectral distribution of the emission spectrum normalized to unity, $\varepsilon_A(\nu)$ the molar extinction coefficient of the acceptor, and ν is the wave number. In this system no changes of spin in either donor or acceptor are allowed.

(c) Short range (10–15 Å) collisional energy transfer based on an exchange interaction mechanism; this is one of the most important energy transfer mechanisms for the triplet-triplet energy transfer and most efficient if the triplet energy of the donor exceeds that of the acceptor by 3–4 kcal or more. Energy transfer then occurs at a diffusion controlled rate.

(d) Excitation migration in which the excitation is passed around over several donor molecules before being transferred to an acceptor.

Most electronic transitions are either $n \to \pi^*$ or $\pi \to \pi^*$ depending on whether a non-bonding n or π electron is promoted to a π^* antibonding molecular orbital. The most important differences between the two processes are the easier intersystem-crossing from an $n \to \pi^*$ singlet to a triplet state and the longer lifetime of the $\pi \to \pi^*$ singlet state.

The electronic excitation so far described was concerned only with isolated molecules (except for solvation), a large number of compounds however can undergo complexation in the ground state to form charge-transfer complexes. This results in an additional electronic transition, i.e. the charge transfer bond. Excitation of the charge transfer bond leads to a greater charge separation and at the limit eventually to an electron transfer.

$$D + A \rightleftharpoons |D \cdots A| \overset{h\nu}{\rightleftharpoons} |D^{\delta+} \cdots A^{\delta-}| \rightleftharpoons D^{+\cdot} + A^{-\cdot} \qquad (2)$$

$$\downarrow$$

Triplet

14.3 PHOTOINITIATED POLYMERIZATION

As stated in the introduction, only polymerization processes in which each individual propagation step involves the interaction of light with the 'monomer' should be called photopolymerizations. All other processes are to be considered as photoinitiated polymerizations, or photochemical post-polymerizations. Photochemical initiation of polymerization may occur in different ways: (a) through energy transfer of an excited sensitizer molecule to either a monomer or foreign molecule resulting in the formation of a species capable of initiation, (b) through direct excitation of a molecule, monomeric or polymeric, or of a molecular complex, followed by reaction creating an initiating species.

(a) $S \overset{h\nu}{\longrightarrow} S^*$ (b) $A \overset{h\nu}{\longrightarrow} A^*$

$\qquad S^* + A \rightarrow S + A^* \qquad\qquad A^* \longrightarrow A^\circ$

$\qquad A^* \rightarrow A^\circ$

$$A^\circ + M \overset{k_i}{\longrightarrow} AM^\circ$$

$$AM^\circ + nM \overset{k_p}{\longrightarrow} A(M)_n M^\circ$$

Since emphasis in this type of photoprocess lies in initiation, we should consider its implications for the kinetics of polymerization. If the sensitizer (S) only, absorbs light the rate of formation of the initiating species (A°) can in general be written as

$$V_{A^\circ} = \alpha I_0 (1 - e^{-\varepsilon_s l}[S])$$

where I_0 is the incident light intensity, [S] and ε_s respectively the sensitizer concentration and its molar extinction coefficient, l is the path length, while α is the overall quantum yield of formation of the initiating species as measured by actinometry; this takes into account the quantum yield of formation of the excited state, the energy transfer yield and the efficiency of formating initiating species. If the product $\varepsilon_s l/[S]$ is small, this equation becomes

$$V_{A^\circ} = \alpha I_0 \varepsilon l/[S]$$

Not all activated species A° succeed in initiating polymerization; recombination of primary radicals formed in the solvent cage decreases the efficiency (f) to a value lower than one. It has also been found[6] that an increase of viscosity lowered the initiation efficiency of azo-bis-isobutyronitrile in photoinitiated polymerization of styrene. A detailed kinetic analysis of photoinitiated polymerizations and a description of the methods of measuring data have been discussed recently by Oster.[7]

The main feature of this type of initiation is the fact that one can, through variation of the light intensity, control the rate of formation of the initiating species and the molecular weight distribution. Photoinitiation has also been an excellent tool in the study of pre- and post-effects in radical polymerization.[8] From the ratio of the velocity constant in non-steady state conditions, in combination with steady state kinetic data absolute values for propagation and termination rate constants for radical polymerizations can be obtained.

Several systems, used for the initiation of radical or ionic polymerizations, were recently reviewed by Oster,[7] Delzenne,[9] Rabek[10] and Tazuke.[11] Direct irradiation of monomer has been studied in the case of styrene[12] and methylmethacrylate[13] but one should be very careful, as far as polymer formation is concerned, in interpreting these results since the presence of impurities and initial formation of peroxides might deeply influence, even overwhelm the rate of direct photoinitiation.

If direct irradiation results only in the formation of singlet or triplet excited state of the monomer, dimerization into cyclobutane derivatives could be a competitive process.

Polymerization through sensitized exitation of the monomer has been found in gas-phase polymerization sensitized by mercury vapour.[10] The first step is triplet energy transfer from mercury to olefin resulting in the triplet excited state of the monomer which, either through intersystem crossing to vibrationally excited ground state or, through dissociation or hydrogen transfer forms a radical species capable of initiating the polymerization.

Most photoinitiation systems are, however, based on the photolysis or on a photochemical reaction of an added substance; they are therefore progressively modified as the latter are consumed as a function of time.

One such type of compounds are carbonyl-group-containing organic molecules. Their initiating capability is due to the $n\pi^*$ excited state; in the case of dialkylketones excitation is mainly due to Norrish cleavage I. With diarylketones fragmentation does not occur in condensed state but radicals are formed by substrate (SH) hydrogen abstraction by the $n\pi^*$ triplets.

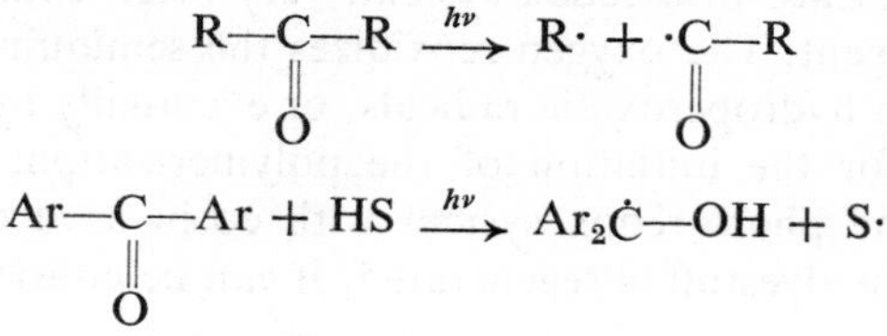

$$R-\underset{\underset{O}{\|}}{C}-R \xrightarrow{h\nu} R\cdot + \cdot\underset{\underset{O}{\|}}{C}-R$$

$$Ar-\underset{\underset{O}{\|}}{C}-Ar + HS \xrightarrow{h\nu} Ar_2\dot{C}-OH + S\cdot$$

It has been shown that only photoreducible ketones do initiate polymerization, ortho-hydroxybenzophenone being inactive. Incorporation of the ketones into the polymer chain was clearly demonstrated in the polymerization of vinylacetate photoinitiated with bromoacetophenone.[14]

Peroxides can be photolyzed into radicals, either by direct photolysis or through sensitization with electronic energy donors such as anthracene, naphthalene[15] or even chlorophyl.[7]

By direct photolysis the radicals have an excess energy of 78 kcal/mol and 56 kcal/mol at 253 nm and 313 nm respectively, while the bond energy of the peroxide O—O linkage varies between 30 kcal/mol and 39 kcal/mol. This situation is evidently much less pronounced in the case of photosensitized reactions. One drawback in the use of peroxides as initiators lies in their propensity to induced decomposition which makes kinetic studies of these systems more complex. This difficulty does not arise in the decomposition of azo compounds.

Decomposition of azo compounds at 320 nm occurs through $n\pi^*$ excitation of an electron of the lone pair of the nitrogen atom. Recent studies by Bartlett[16,17] indicate that the decomposition of acyclic azo-compounds will occur for an important part from singlet excited states during the photoinitiation of polymerization; indeed azo-2-methyl-2-propane does

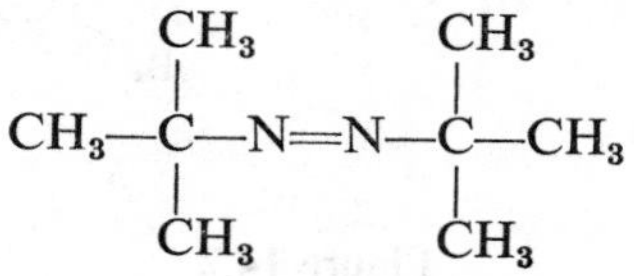

$$CH_3-\underset{\underset{CH_3}{\overset{CH_3}{|}}}{\overset{\overset{CH_3}{|}}{C}}-N{=}N-\underset{\underset{CH_3}{\overset{CH_3}{|}}}{\overset{\overset{CH_3}{|}}{C}}-CH_3$$

quench fluorescence of triphenylene and 9,10-diphenyl-anthracene, while sensitization with benzophenone results in cis/trans-isomerization around the nitrogen-nitrogen bond. Several azo compounds have been used as photoinitiators of the polymerization of styrene,[6] methyl methacrylate[18] and vinyl chloride.[19] The most efficient initiators are those with an electron withdrawing group alpha to the —N=N— linkage. The resulting radical is sufficiently reactive for initiation but is, on the other hand, stabilized by resonance by the presence of the substituent.

Dye sensitization has been developed mainly by Oster and coworkers.[20] Highest conversions were obtained for systems in which the dye is photo-excited in the presence of a redox system, very often composed of oxygen and a reducing agent. The oxygen reoxidizes the semiquinone of the leuco dye, and resulting hydroperoxidic radicals, or eventually hydrogen peroxide are responsible for the initiation of the polymerization. The mechanism proposed[21] for the photo-redox system with eosin as dye is described in Figure 14.2. As the dyestuff is regenerated, it can be considered as a photo-

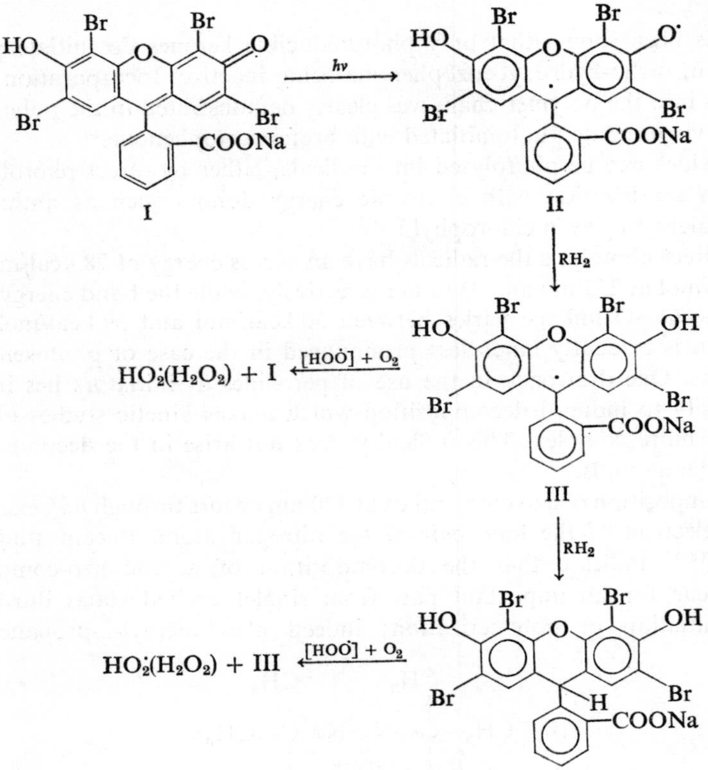

Figure 14.2

sensitizer. It should however be pointed out that in certain circumstances another initiation mechanism is operative.

Dyes such as eosin and methylene-blue are known to form singlet oxygen[22] through triplet-triplet annihilation of the triplet excited dye and triplet oxygen ground state.

$$Dye^\circ \xrightarrow{h\nu} Dye^1 \longrightarrow Dye^3$$
$$Dye^3 + O_2^3 \longrightarrow Dye^\circ + O_2^1$$

This singlet oxygen can then react with monomer or another component of the reaction mixture and form hydro- or endoperoxides capable of initiation on their decomposition. It was indeed shown that a system containing 2,5-diphenyloxazole, eosin and oxygen could initiate the polymerization of acrylamide in glycol-water mixture on irradiation with visible light.[23]

All above mentioned photoinitiated polymerizations are based on reactions in which radicals are produced and are responsible for the initiation of polymerization. Recently photoinduced ionic polymerizations have been reported.[11,31] Direct ionization of organic molecules requires 7–12 eV and consequently is only accessible in the vacuum ultra-violet region. However, formation of charge-transfer complexes can greatly lower the energy required for charge separation. As stated in the introduction, this charge separation is even more important in the excited state of the charge transfer complex. If moreover one of the components of the complex is a suitable monomer, polymerization can be initiated by the radical ion formed. Since the propagation rate constant for cationic polymerization is several orders of magnitude higher than that for radical polymerization, cationic polymerization can occur. This type of photoinitiation has been reported for systems in which both components are monomeric, for example N-vinyl carbazole-acrylonitrile, and for systems in which only one component is monomeric, for example styrene-tetracyanobenzene. Irradiation of the N-vinylcarbazole-acrylonitrile system[24,25] results in a mixture of homo-poly-N-vinylcarbazole and copolymer; the first results from cationic polymerization of N-vinylcarbazole as demonstrated by its inhibition on adding ammonia while the latter is formed by radical polymerization and can be avoided in the presence of radical chain inhibitors. No evidence was found for anionic propagation. It is suggested therefore, that initiation involves formation of a radical cation derived from N-vinylcarbazole. Acrylonitrile can be replaced in this system by acetonitrile; the complex is still able to photoinduce the polymerization of N-vinylcarbazole. Cationic photoinduced polymerization was also observed for other N-vinylcarbazole-electron acceptor systems, for example with nitrobenzene,[27] but side reactions occur with tetranitromethane[28] and with chloranil[29,30]

Hayashi recently studied the photoinduced polymerization of the styrene and α-methylstyrene-tetracyanobenzene system.[26] Based on copolymer composition and inhibition of the polymerization by trace amounts of water a cationic polymerization mechanism was proposed. As expected in these systems, the polymer acts also as a possible but weaker electron donor than the monomer itself; consequently a competition in complex formation with monomer and polymer exists and complicates the reaction mechanism. Charge-transfer complexes, in which one of the partners is a monomer, do not only photoinitiate ionic polymerization; irradiation of acrylonitrile-isobutylvinylether[32] and triphenyl-phosphine-methylmethacrylate[33,34] enhances radical polymerization, initiation occurring by excitation of the complex formed between both reactants.

14.4 PHOTOPOLYMERIZATION

As stated above the concept of photopolymerization should be limited to those systems in which through a repetitive photochemical act polymer chain lengthening occurs, i.e. for which each propagation step is a photochemical reaction.

Although this idea has been applied successfully in cross-linking of photo-sensitive polymers it has only recently been used to obtain soluble macromolecular substances.

14.4.1 Photopolymerization in the solid state

One of the most important photochemical reactions is certainly the photo-cyclomerization in which a molecule containing a π-bond adds on electronic

Table 14.1

Monomer	λ^{a}_{max}, nm	$[\eta]^{b}_{red}$.
1. 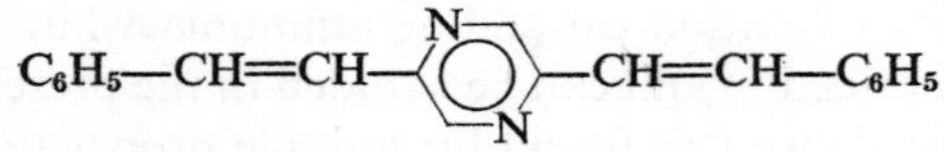 $C_6H_5—CH=CH—\langle\text{pyrazine}\rangle—CH=CH—C_6H_5$	384	1–10
2. 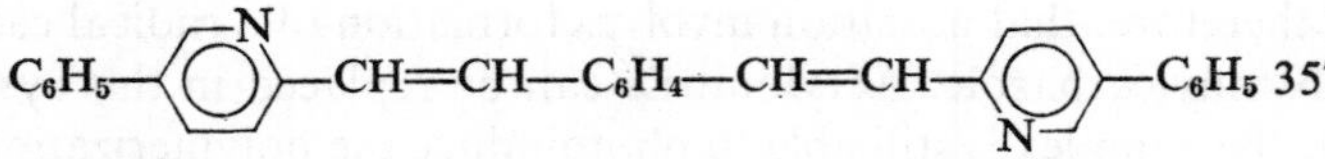 $C_6H_5—\langle\text{pyridine}\rangle—CH=CH—C_6H_4—CH=CH—\langle\text{pyridine}\rangle—C_6H_5$	357	0·3–1·2
3. 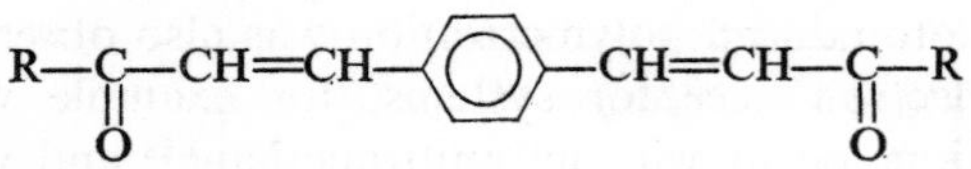 $R—\overset{O}{\underset{\parallel}{C}}—CH=CH—\langle\text{benzene}\rangle—CH=CH—\overset{O}{\underset{\parallel}{C}}—R$		
where R represents:		
a. OH	363	0·12
b. OCH_3	320	6·95–25
c. OC_2H_5	320	0·16–1·41
d. OC_3H_7 n.	321	0·26
e. OC_3H_7 is.	321	0·45
f. OC_6H_5	328	0·90
g. NH_2	353	1·5

a. All λ measured in dioxane except 3_a and 3_g in sulphuric acid. All $\pi — \pi^*$ transitions, $\varepsilon > 10^4$.
b. Measured in concentrated sulphuric acid at 30 °C except 3_f which is in aqueous sodium hydroxide.

excitation to another molecule containing a π-bond; this reaction is an easy, symmetry allowed, process if it occurs in the singlet excited manifold of one of the molecules,[35] but many cyclodimerizations are known to happen between a molecule in its triplet excited state[36] with another in its ground state. Although mentioned in 1958 by Koelsch[37] it was only recently through the very interesting results of M. Hasegawa and coworkers[38] that the feasibility of true photopolymerization in the solid state to obtain soluble high polymers was demonstrated.

The monomers studied by the Japanese workers are reported in Table 14.1 and are all linear symmetric diolefins of the type

$$R_1\text{—CH}\text{=}\text{CH—}R_2\text{—CH}\text{=}\text{CH—}R_1$$

The general scheme of these polymerization reactions is based on the analogy with the solid state photodimerization of cinnamic derivatives[39] and of stilbazoles.[40]

$C_6H_5\text{—CH}\text{=}\text{CH—R} \rightarrow$ Several isomeric cyclobutane derivatives

$R = \text{CO—Aryl, COO—Alk,}$

Although the absolute configuration of the cyclobutane rings built into the polymer chain has not been established it was assumed on the basis of solution properties of poly-2,5-distyrylpyrazine that a 1,2- or 1,3-*cis*-cyclobutane linkage with restricted rotation was present.[41] Hydrolysis of polyphenylene diacrylic derivatives followed by intramolecular anhydride formation shows the influence of the nature of acrylic derivatives on the main cyclobutane isomer[39] as demonstrated by infrared analysis.

All high polymers obtained on these topochemical reactions are highly crystalline and are only soluble in strong acids or bases. Several factors are important in these photopolymerizations; the polymerization temperature, as such unimportant in usual photoinitiated polymerization, does determine the polymerizability and the degree of crystallinity of lower melting monomers. The crystal structure is of major importance on account of the topochemical nature of the reaction; some crystalline varieties of monomers being inactive from the point of view of the polymerization. Molten monomers and solutions of these monomers do not polymerize through cyclobutane forming mechanisms which is easily understandable as an easier deactivation process for example *cis/trans*-isomerization is possible. Some monomers described by Hasegawa where R_1 is 9-anthryl or 1-naphthyl are not active because of intramolecular energy transfer.

Kinetic studies[42] showed that the photopolymerization of 2,5-distyrylpyrazine is also a true stepwise polyaddition reaction. This was proved by the influence of irradiation time on the reduced viscosity of the polymer, and by the broad molecular weight distribution which corresponds formerly to a step polyaddition and not to a photoinitiated polymerization. In these last systems a gaussian molecular weight distribution is usually obtained.

14.4.2 Solution photopolymerization

It was recently shown that photopolymerization could also be realized in solution.[43] N,N'-alkalene-bis-maleimides react on irradiation either following path (a) Figure 14.3 or path (b) Figure 14.3 depending on the nature and length of the link.[44]

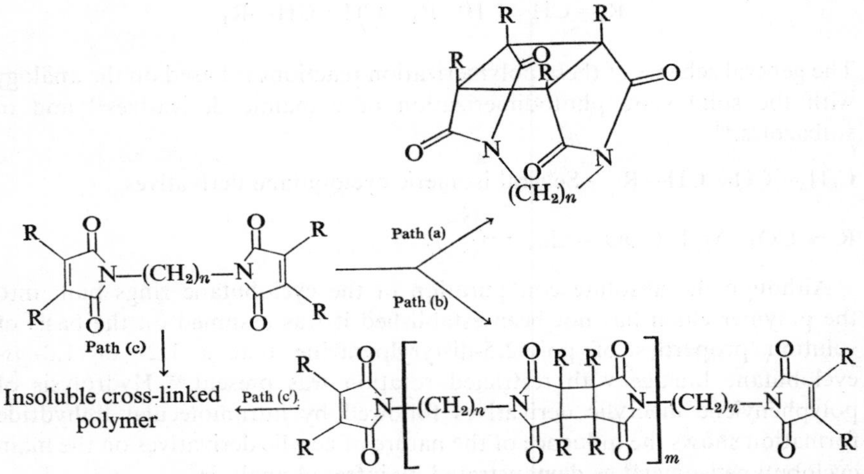

Figure 14.3 Photochemical behaviour of N,N'-polymethylene-bis-maleimides

If properly substituted by halogen no vinyl-polymerization of the double bond will occur and no cross-linked materials will be formed. High polymers were obtained for alkylene chains with nine or eleven methylene units (Table 14.2). The rate of polymerization is enhanced in the presence of a sensitizer such as benzophenone. This is the first example of a sensitized photopolymerization.

The repeating cyclobutane unit has an exo-structure as shown by infrared and proton magnetic resonance data on low molecular weight oligomers of the undecamethylene-N,N-monochloromaleimide. The polymerization follows a stepwise addition mechanism as can be seen from the influence of reaction time on the molecular weight (Table 14.2). Polymers obtained through this repetitive dimerization of maleimide groups are soluble in dichloromethane if the methylene chain contains an uneven number of methylene groups; they form transparent, flexible films. If the methylene chain between

Table 14.2

	Monomer	Irradiation time in hours	$[\eta]\ 20°$ in dimethylformamide
$(CH_2)_7$	R_1, R_2, R_3, R_4 $R_1 = R_2 = R_3 = R_4 = Cl$	121	0·25
$(CH_2)_9$	$R_1 = R_2 = R_3 = R_4 = Cl$	120	0·45
$(CH_2)_{11}$	$R_1 = R_2 = R_3 = R_4 = Cl$	110	0·65
$(CH_2)_{11}$		74	0·11

the two maleimide groups contains an even number of methylene units only lower molecular weight oligomers are obtained on account of their precipitation in the course of the reaction; they are only soluble in hot dimethylformamide or hot N-methylpyrrolidone. Solution photopolymerization is not restricted to photopolycyclomerization. Polyalcohols and polyamides were obtained[45,46] by repetitive photoreductive polyaddition of bisbenzophenones and bisazomethines as shown in reaction schemes, Figure 14.4.

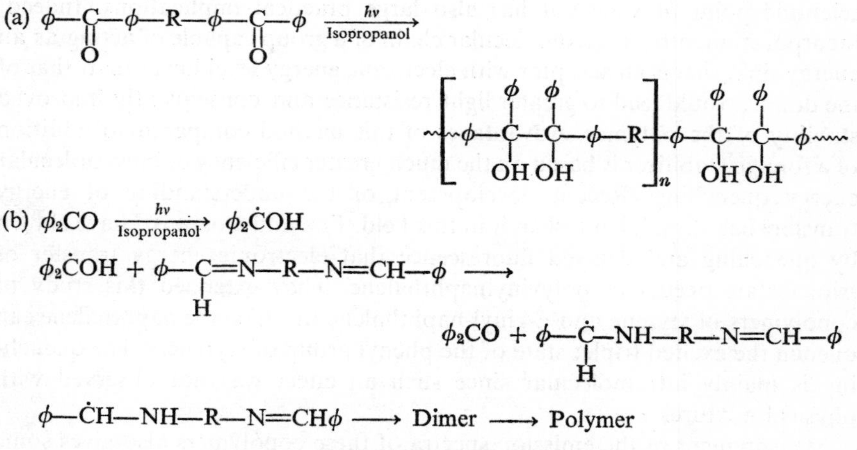

Figure 14.4   Reductive photopolymerization

Formation of high molecular weight compounds is strongly dependent on the monomer concentration and on the solubility of monomer and polymer. Both these factors are relevant as can be seen from Table 14.3.

Other photochemical reactions with high yields can and will certainly be applied to bifunctional systems in order to obtain macromolecules that are not easily accessible by conventional methods. (*Since this article was prepared,*

Table 14.3   Reductive photopolyaddition of bis-benzophenones.

Monomer	[M] mol l^{-1}	Isopropanol	Mñ
p,p'-Dibenzoyldiphenyl-methane	0·53	1/5	30 000
Terephthalophenone	0·035	1/5	4 500
Isophthalophenone	0·047	1/7	3 000
p,p'-Dibenzoyl-1,8-di-phenyloctane	0·028	1/7	11 000

the use of singlet and triplet states in this type of photopolymerization has been demonstrated.)[120]

14.5 PHOTOCHEMICAL PROCESSES IN HIGH POLYMERS

Two different types of photochemical process can be considered in macro-molecules: energy absorption and transfer, or absorption followed by a reversible or irreversible reaction.

The first type of photochemical process is not only important from a scientific point of view but has also large practical implications. Indeed, incorporation into a macromolecular chain of a group capable of acting as an energy sink, that is an acceptor with electronic energy level lower than that of the donor, would lead to greater light resistance and consequently improved stability of the polymer. Advantages of this method compared to addition of a foreign stabilizer is based on the much greater efficiency of intramolecular energy quenching. Recent development of the understanding of energy transfer[3] has stimulated research in this field. Fox and Cozzens[47] have shown by quenching and delayed fluorescence that electronic energy transfer of triplet state occurs in polyvinylnaphthalene. They extended this study to copolymers of styrene and α-vinyl-naphthalene in which the naphthalene can quench the excited triplet state of the phenyl group of styrene.[48] The quench-ing is mainly intramolecular since such an effect was not observed with physical mixtures.

Measurement of the emission spectra of these copolymers also gives some idea about the distribution of the comonomers. Two types of emission are observed in copolymers containing more than 5 per cent of the naphthyl derivative: the phosphorescence of the naphthyl group and its delayed fluorescence due to triplet-triplet annihilation. The ratio between these two intensities is constant from 5 to 100 per cent naphthyl group containing polymers; below 5 per cent however the intensity of the phosphorescence increases. This indicates that in this region the naphthyl group containing sequences are short and no triplet-triplet annihilation takes place. Geuskens[49] showed that polyvinylbenzophenone films containing naphthalene effectively transferred energy from the triplet excited state of the benzophenone to naphthalene as indicated by the phosphorescence of naphthalene polymer films.

In contrast benzophenone-naphthalene solid mixtures have a noticeably greater critical radius at which energy transfer is still possible. Similar results were obtained for polyphenylvinylketone. If a sufficiently high concentration of quencher is used $(5 \times 10^{-2} \text{ mol})$ chain scission is strongly inhibited.[5] This indicates that scission is due to a reaction starting from the triplet excited state of the ketone. The fact that the critical radius of polyphenyl-vinylketone was smaller than that of polyvinylbenzophenone shows that migration of electronic energy along the polymer chain is less important in polyphenylvinylketone. Energy transfer from polyvinylnaphthalene to benzophenone was observed as a decrease of excimer fluorescence of the polymer, but no phosphorescence of the acceptor could be detected. This probably indicates an energy transfer of the benzophenone triplet to poly-vinylnaphthalene and loss of energy by triplet-triplet annihilation. The rate constant of energy transfer for this system is similar to that found for the polystyrene-1,1,4,4-tetraphenylbutadiene[51] couple. If anthracene is used as acceptor of singlet energy instead of benzophenone, fluorescence of anthracene replaces excimer fluorescence. This is emission from a complex between a singlet-excited state and ground state molecule. Since geometric restrictions for complex formation are severe[52] excimer fluorescence can give important information on the shape of polymers in solution and polymer composition.

Pure polyvinylnaphthalene only emits excimer fluorescence indicating that on excitation of the naphthalene group migration along the chain occurs until geometric conditions are fulfilled for excimer formation. A relation of excimer fluorescence in polystyrene solutions to the average density of polymer segments has been assumed.

It is clear from the above discussion that polymers containing photoactive groups can be used as macro sensitizers. *Cis/trans*-isomerization of 1,3-pentadiene can be affected by polyphenylvinylketone.[56]

The two most important reversible processes photochemically induced in polymers are the *cis/trans*-isomerization of double bond containing macromolecules and the photochromism of spiropyran containing polymers. Photoisomerization of *cis*-polybutadiene to its *trans*-isomer was studied by Golub.[55] The interconversion was 'sensitized' by organic bromides, disulphides or mercaptans. The mechanism of most of these isomerizations however is not based on sensitization by energy transfer but on radical addition followed by radical elimination resulting in the thermodynamically most stable isomer. This became clear from *cis/trans*-isomerization studies on model compounds.[57] A slightly different mechanism was proposed by Bishop[58] involving initial formation of a weak charge transfer complex between the sulphur containing compounds and the double bonds of the polymer, breaking of the complex through irradiation followed by *cis/trans*-isomerization, a mechanism resembling the bromine catalysed *cis/trans*-isomerization of olefins. It should however be possible to obtain rubbers with a varying *cis/trans*-ratio using sensitizers with different triplet energies.[59] *Cis/trans*-isomerization by direct irradiation of polybutadiene and of *cis*- or *trans*-1,4-polyisoprene does occur in films to reach an equilibrium value around 70/30 with a quantum yield of 0·041 for polyisoprene; it is however accompanied by irreversible processes.[60,61] The study in polymer systems of another reversible photo-reaction, photochromism, has recently been initiated.[62] As in *cis/trans*-isomerization, this study can give important information on the structure of the polymer chain.

An indolinobenzospiran group incorporated in the main chain or as a pendant substituent is excited, probably in the chromene part of the molecule,[63] and recent data of Bach and Calvert[64] suggest that singlet and triplet excited states can give rise to colour formation.

Substances which undergo a reversible change in absorption spectrum, on irradiation with light of a suitable wavelength are called photochromic compounds. Where the photochromic group becomes a part of a polymer molecule, changes of photochromic behaviour might be expected, resulting from polar and steric effects due to the proximity of the polymer chain, and from internal viscosity phenomena by which the segmental motions are restricted. It is evident that such differential effects will only be perceptible if the photochromic phenomenon involves an appreciable movement of one part of the molecule with respect to the other. Such a condition is usually not fulfilled in photochromic processes such as valence tautomerisms, photo-tropic tautomerisms or photochemical dissociation processes; on the contrary *cis/trans*-isomerizations may offer interesting possibilities as shown by Lovrien on phototropic macromolecules containing amino-azobenzene side-groups.[65] We have concentrated our efforts to spirobenzopyran derivatives, where the photochromism consists in the scission of the C—O-pyran bond followed by a rotation of one part of the molecule so as to approach planarity, Figure 14.5.

The open ring merocyanines are characterized by a very strong solvato-chromism[66,69] affecting their maximum absorptions wavelength as well as their decoloration rates,[70,72] moreover they are very sensitive to steric effects, especially in the benzthiazol series.[73,74] These were the main reasons for choosing benzopyrylospiranes for incorporation in macromolecules.[73,62] The next question to be answered is in how far a photochromic polymer differs from a mixture of the corresponding polymer and an appropriate photochromic substance.[75,76] Photochromic polymers were synthesized with the photosensitive group attached as a side group on the main chain and as a part of the polymeric backbone. In general it can be said that, with the exception of photochromic polytyrosine,[77] the photochromic behaviour of a benzopyrylospiran group incorporated in a polymer molecule is not very different in solution from that of the corresponding model substances. The differences are most pronounced in less polar media and depend on the nature of the comonomer; they affect mainly the reaction kinetics, the polymer decolorizing slower than the low molecular weight compound. Spectrophotometric data show the existence of two or more coloured isomers, for

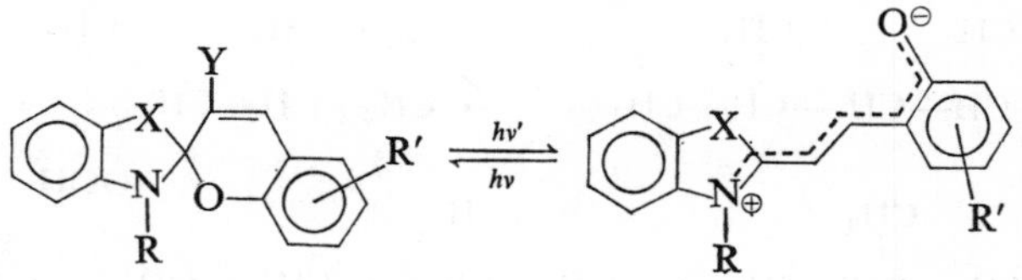

when X = C(Me)$_2$ Y = H
     X = S   Y = Me

hv' = visible light
hv = ultra-violet light

Figure 14.5 Photochromism

which interconversion is considerably slowed down by the polymeric structure. Similar differences are shown between polymeric films and model substances dispersed in the corresponding matrix.[78,79] On the contrary, photochromic polytyrosine shows a very low solvatochromism compared to its model compound, and it is assumed that this difference is related to the helicoïdal structure of polypeptides. Interaction of light with polymer films containing photoactive groups can also result in irreversible reactions. One very important industrial process is photodegradation. Recent studies of polypropylene photodegradation led to the suggested sequence of steps shown in Figure 14.6. Either during the moulding process (in the presence of air), or through photooxidation, hydroperoxides (1) are formed which are photochemically labile, decomposing to ketone containing polymers.[80-81] Degradation follows by both Norrish-type (I) and (II) fragmentations.[82]

Recent mechanisms were also put forward for the photodegradation of polyethylene by Guillet[83,84] and Trozolo.[85]

Photodegradation also starts by absorption of light by a carbonyl group incorporated in polyethylene during moulding or extrusion. Based on model copolymers of carbon monoxide and ethylene, Guillet and Hartley showed that Norrish (I) and (II) fragmentations occur, the latter being the most important at ambient temperature. Chain breaking could be quenched by energy transfer to cyclooctadiene-1,3 but only to a certain extent (45 per cent) indicating that chain scission through Norrish fragmentations occurred partly from the excited singlet state. Since certain conformational conditions have to be fulfilled in the intramolecular Norrish (II) reaction this process is inhibited in the region of the glass transition.

In air[85] most of the triplet excited state would be quenched by oxygen resulting in singlet excited oxygen capable of oxidizing the double bonds formed in the Norrish degradation from the singlet excited state. As pointed out by Trozolo, stabilization based on quenching of the n-π* triplet excited

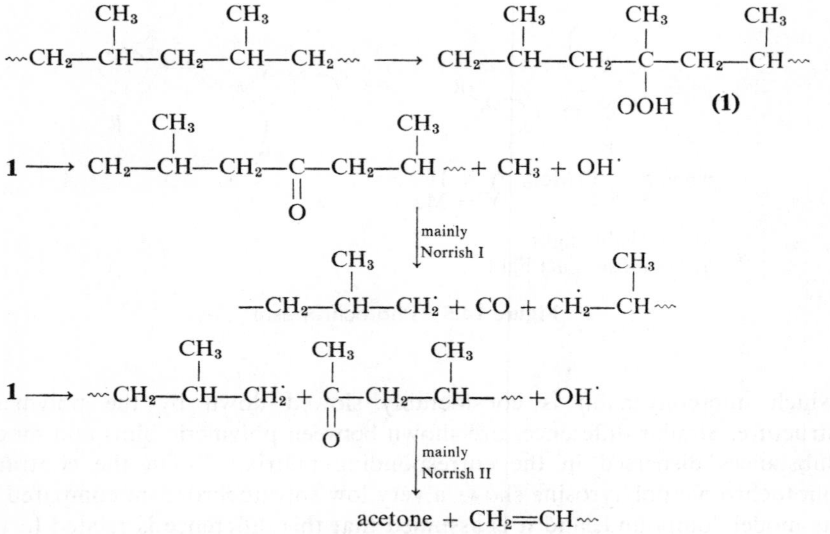

Figure 14.6 Photodegradation

state is not only ineffective, as indeed found by Guillet, but harmful since the quencher can transfer this energy to oxygen resulting in promoted oxidation.

As mentioned earlier, photoisomerization is accompanied by irreversible processes such as formation of cyclopropylrings in the main chain of poly-isoprene (2)[61] ($\phi = 0.018$) and formation of cyclobutane and cyclopentane units in poly-1,2-butadiene (3) and poly-3,2-isoprene (4).

Photodegradation of other synthetic polymers or copolymers of methyl-methacrylate[86,87] polymethylsiloxane[88] and the photodegradation of natural polymers as rubber,[89] cellulose[90] and protein[91] also have been studied recently in solution or in film. Most of these photodegradations include mainly chain scission at the weakest link. Protein denaturation does however occur by intramolecular photodimerization of purine bases and by photo-hydration. Photolysis of aromatic polyesters results in hydroxylation of the aromatic nucleus[92] while a light induced rearrangement of polycarbonates

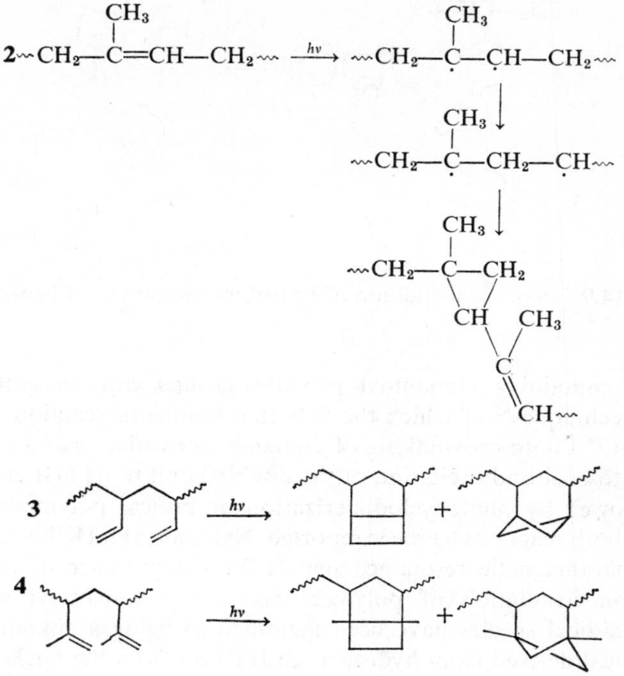

Figure 14.7 Irreversible phototransformations of polyisoprene (2), poly-1,2-buta-diene (3) and poly-3,2-isoprene (4)

accompanied by partial degradation, was reported by Bellus.[93] This rearrangement is analogous to the photo-Fries rearrangement. Schulz[94] recently reported an irreversible phototransformation of polyvinylbenzophenone through photocycloaddition to furan and benzofuran.

One of the most thoroughly studied irreversible photoreactions on polymers which will be considered in more detail is the photo-crosslinking of cinnamic derivatives. Their photocyclodimerization was reported a long time ago[95] but only recently it was found that, depending on the crystal structure of the monomer, two types of dimeric species are formed in the solid state.[96]

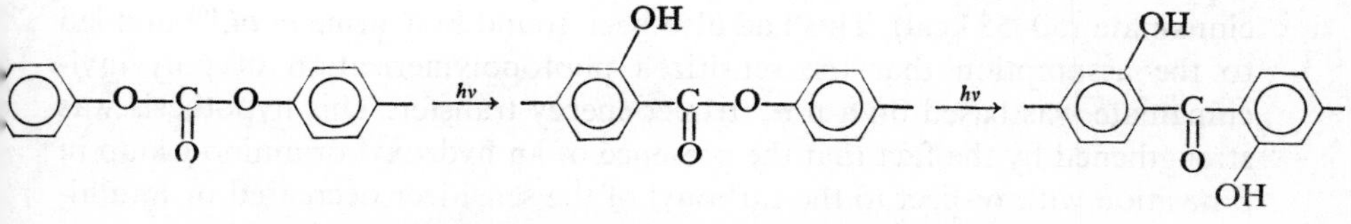

Figure 14.8 Light-induced rearrangement of polycarbonate

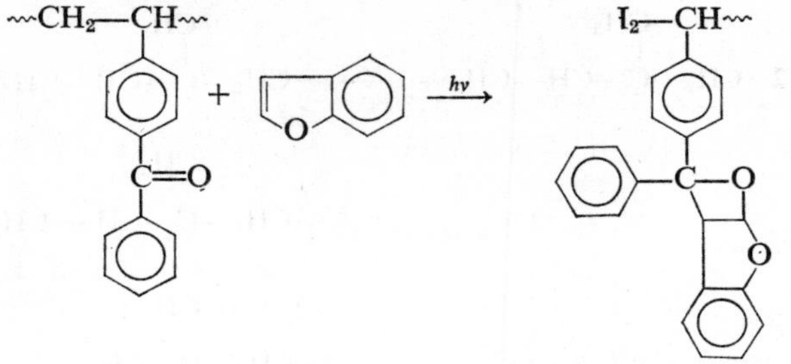

Figure 14.9   Photocycloaddition of benzofuran to polyvinyl benzophenone

Polymers containing cinnamoyl pendant groups can be synthetized by different techniques[97] of which the Schotten-Baumann reaction is the most convenient.[98] Photo-crosslinking of cinnamic derivatives can occur by direct or by photosensitized excitation of the double bond to its first excited triplet state, followed by photocyclodimerization or radical polymerization. Evidence for both reactions has been reported. Nakamura and Kikuchi[99] obtained electron paramagnetic resonance signals for the presence of two different radicals on irradiation of polymers containing cinnamoyl substituents. The two radical species have been identified as being a cinnamoyl radical and a radical derived from hydrogen abstraction from the back-bone of the polymer chain. On the other hand, the structural identification of the truxillic dimers obtained by Srinivasan and Sonntag[100] after hydrolysis of bridged polyvinylcinnamate, confirms the interpretation of Curme, Natale and Kelly,[101] based on spectroscopic data.

One should bear in mind that the triplet state has a radical character and that therefore the cinnamoyl triplet is liable, in a side reaction, to abstract hydrogen from the main chain. Photodimerization is certainly the main process for photo-crosslinking of cinnamic acid derivatives, as was made clear by the elegant work of Tsuda,[102] Nakamura and Kikuchi[103] and Moreau.[104] The latter showed that fluorescence of 2-benzoyl-methylene-1-methyl naphthothiazoline was not affected by the presence of polyvinylcinnamate but that the quantum yield of phosphorescence decreased from 0·85 to 0·15. He came to the conclusion that to be active for cross-linking polyvinylcinnamate a sensitizer should have a high intersystem crossing efficiency, a triplet life-time greater than 10^{-2} s and a triplet energy above that of the cinnamate (50–55 kcal). This had also been found by Curme et al.[101] and led to the assumption that the sensitized photopolymerization of polyvinylcinnamate was based on a n-π^* triplet energy transfer. This hypothesis was strengthened by the fact that the presence of an hydroxyl or amino group in β position with respect to the carbonyl of the sensitizer decreased or annihilated the energy transfer on account of an intramolecular photo-process and, or, a lower energy level of the π-π^* triplet. Nakamura and Kikuchi[103] found that the O—O transition of the phosphorescence of cinnamic acid must be located at 20 000 cm^{-1} and showed that sensitizers (Table 14.4), effective for photopolymerization, had a phosphorescence transition in the same region.

Table 14.4

	Sensitizer	E_T kcal/mol	ISC[a]	k_S/k_0 [b]
a.	Michler's ketone	61	1	640
b.	Naphthalene	61	0·75	3
c.	Benzophenone	69·5	1	20
d.	Chrysene	56	0·67	18
e.	p-Dimethylamino-nitrobenzene	55		137
f.	2,6-Dibromo-4-nitro-1-dimethyla-mino-benzene	55		797
g.	4-Nitro-1-naphthylamine	49·4		0
h.	N-acetyl-4-nitro-1-naphthylamine	52·5		1100

a. ISC intersystem crossing efficiency.
b. k_S/k_0 relative reactivity, i.e. the ratio of the sensitized to unsensitized reaction.

Tsuda[102] pointed out the relation between phosphorescence intensity and sensitizer activity (Table 14.4 e and f). Incorporation of a 'heavy atom' enhances intersystem crossing and also the efficiency of the sensitizer (797 instead of 137); on the contrary compounds with a triplet energy lower than that of the cinnamic derivatives do not sensitize the dimerization reaction (Table 14.4 g). This is not due solely to the presence of an aminogroup; indeed p-nitro-aniline shows a relative reactivity k_S/k_0 equal to 410. On the other hand, acylation of the amino function increases the triplet energy level, enough to make the acetyl derivative a very effective sensitizer. Quantum yield determinations[105] on the direct photopolymerization of polyvinylcinnamate indicate an efficiency of 0·5, while the photosensitized reactions have lower quantum yields, e.g. Michler's ketone 0·35, p-nitroaniline 0·25. These data indicate incomplete energy transfer and, or, partial singlet reaction in the direct irradiation.

The use of a sensitizer does however increase the spectral sensitivity range of the system.

Photopolymerization of this type is evidently not restricted to polyvinyl-cinnamate. Other side groups of the polymer chain capable of cyclodimerization are shown in Table 14.5. The syntheses and some of the properties of these polymers were reviewed recently[106] but no detailed mechanistic studies as for the cinnamic derivatives are available. That cyclodimerization is in fact the major polymerization step in these systems is based on analogy with dimerizations of the model compounds. Special attention should be paid to the stilbazole and anthracene derivatives since photodimerization of these compounds probably takes place from the singlet excited state.

Cross-linking can also be obtained by photoinitiated reactions. Photolabile groups that build in the polymer give, on direct irradiation, or sensitization, reactive species such as nitrenes, carbenes or radicals. Such transformations have recently been reviewed by G. Delzenne.[106]

Table 14.5

Side group	Photopolymerization Unsensitized	Sensitized	Reference
—O—CO—(CH=CH)$_n$—C$_6$H$_5$ Cinnamates + higher vinylogues	+	+	107, 108, 109
—Ar—CO—(CH=CH)$_n$—C$_6$H$_5$ Chalcone + higher vinylogues	+	+	o.a. 110, 111, 112
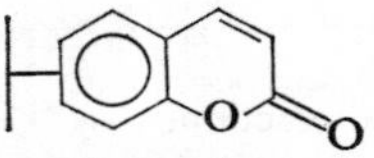 Coumarine	+	+	113
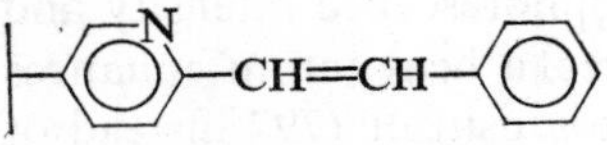 Stilbazole derivatives	+		114, 115
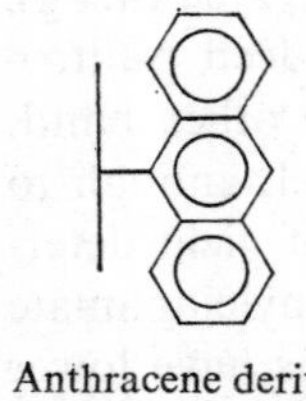 Anthracene derivatives	+		116, 108

We will consider as a typical example, polymers containing azide groups. Photolysis of polymers containing arylazides as pendant groups result in nitrene formation. Dissociation is thought to occur[117] from a vibrationally excited state of the azide after rapid intersystem crossing from the n → π* or π → π* excited state to the ground state. The nitrenes can then insert within carbon-hydrogen bonds, abstract hydrogen or form azo-bonds between the polymers. The mechanistic picture of the photosensitized decomposition of azides is not so well-defined. It has been shown recently that aryl-azide decomposition can be sensitized by energy transfer of the n → π* triplet excited state of aromatic ketones.[118]

However, singlet-singlet energy transfer from polyaromatic compounds such as naphthalene, pyrene and anthracene also results in arylazide decomposition[118,120] and anthracene fluorescence is quenched by arylazides. Sensitized decomposition of arylazides is then possible from singlet and triplet excited manifolds of the azide but the exact nature of resulting active species is still in doubt.

14.6 CONCLUSIONS

The interaction of light with some functions of the monomer or polymer molecule may result in transformations or reactions leading to macromolecules that are difficult to synthesize by ground state reactions and, therefore, makes initiation of polymerization possible in reaction conditions in which ordinary initiating systems fail.

Although most photochemical reactions are specific as far as the reaction site is concerned, it should be stated that since states with high electronic and or vibrational energy content are involved, side reactions such as hydrogen abstraction can occur.

A better understanding and application of photochemical processes in small molecules has been and will be of great help in the development of photochemical processes in relation to polymer chemistry.

14.7 REFERENCES

1. J. G. Calvert and J. N. Pitts, *Photochemistry*, Wiley, New York, 1966.
2. N. J. Turro; *Molecular Photochemistry*, p. 47, W. A. Benjamin, New York, 1965.
3. A. A. Lamola, *Energy transfer and organic Photochemistry; Technique of Organic chemistry* Vol. XIV, Interscience, New York, 1969.
4. Th. Förster, *Discussions of Faraday Soc.*, **27**, 7, (1959).
5. Th. Förster, *Delocalized Excitation and Excitation Transfer*, *Modern Quantum Chemistry* Vol. 3, ed. O. Suranglu, Academic Press, New York, 1965.
6. F. De Schryver and G. Smets, *J. Polym. Sci.*, **A-1**, 2201, (1966).
7. G. Oster and N. L. Young, *Chem. Rev.* **68**, 125, (1968).
8. G. M. Burnett, p. 1107 in *Investigation on Rates and Mechanisms of Reactions*, eds S. L. Fries, E. S. Lewis and A. Weissberger, Interscience, New York, 1963.
9. G. Delzenne, *European Polymer Journal, Supplement*, **1969**, 55.
10. J. F. Rabek, *Photochemistry and Photobiology*, **7**, 5, (1968).
11. S. Tazuke, *Adv. Polymer Sci.*, **6**, 321, (1969).
12. R. G. Norrish and J. P. Simmons, *Proc. Roy. Soc. (London)*, **A215**, 4, (1959).
13. V. A. Krongauz, *Teor. i Eksperim, Khim. Akad. Nauk Ukr. U.S.S.R. I*, 47, (1965).
14. S. Sukegawa, K. Muzaki and T. Ozowa, *Kobunski Kagaku*, **18**, 700, (1961).
15. C. Luner and M. Szwarc, *J. Chem. Phys.*, **23**, 1978, (1955).
16. P. D. Bartlett and N. A. Porter, *J. Am. Chem. Soc.*, **90**, 5317, (1968).
17. P. D. Bartlett and P. S. Engel, *J. Am. Chem. Soc.*, **90**, 2960, (1968).

18. H. Migama, *Bull. Chem. Soc. Japan*, **29**, 715, (1956).
19. G. M. Burnett and W. W. Wright, *Proc. Roy. Soc.*, **A221**, 28, (1954).
20. See reference cited in reference 7.
21. G. Delzenne, S. Toppet and G. Smets, *J. Polymer Sci.*, **48**, 347, (1960).
22. C. S. Foote, *Accounts Chem. Research*, **1**, 104, (1968).
23. F. C. De Schryver and M. O'Kelly de Galway, unpublished results.
24. S. Tazuke and S. Okamura, *J. Polym. Sci.*, **B6**, 173, (1968).
25. S. Tazuke and S. Okamura, *J. Polym. Sci.* **A-1 6**, 2907, (1968).
26. K. Hayashi, personal communications.
27. S. Tazuke, M. Asai, S. Ikeda, S. Okamura, *J. Polym. Sci.*, **B, 5** 453, (1967).
28. D. H. Iles, A. Ledwith, *Chem. Comm.*, **1969**, 346.
29. L. P. Ellinger, *Polymer*, **5**, 559, (1964).
30. L. P. Ellinger, *Polymer*, **6**, 549, (1965).
31. N. G. Gaylord, *Polymer Preprints*, **10**, 277, (1969).
32. S. Tazuke and S. Okamura, *J. Polym. Sci.*, **A-1, 7**, 715, (1969).
33. T. J. Mao and R. J. Eldred, *J. Polym. Sci.*, **A-1, 5**, 1741, (1967).
34. R. J. Eldred, *J. Polym. Sci.*, **A-1, 7**, 265, (1967).
35. R. B. Woodward and R. Hoffmann, *The conservation of arbital symmetry*, p. 65, Verlag Chemie, 1970.
36. R. Steinmetz, *Fortschritte Chem. Forsch.*, **7**, 445, (1967).
37. C. Koelsh and W. Gumprecht, *J. Org. Chem.*, **23**, 1603, (1958).
38. M. Hasegawa and Y. Suzuki, *J. Polym. Sci.* **B, 5**, 813, (1967).
39. F. Suzuki, Y. Suzuki, M. Nakanishi and M. Hasegawa, *J. Polym. Sci.*, **A, 7**, 2319, (1969).
40. M. Hasegawa, Y. Suzuki, F. Suzuki and M. Nakanishi, *J. Polym. Sci.*, **A7**, 743, (1969).
41. S. Fujishige and M. Hasegawa, *J. Polym. Sci.*, **A7**, 2037, (1969).
42. M. Nakanishi, Y. Suzuki, F. Suzuki and M. Hasegawa, *J. Polym. Sci.*, **A-1, 7**, 753, (1969).
43. F. De Schryver, J. Feast and G. Smets, *J. Polym. Sci.*, **A-1, 8**, 1939, (1970).
44. F. De Schryver, I. Bhardwaj and J. Put, *Ang. Chem. (Int. Ed.)*, **8**, 213, (1969).
45. F. De Schryver, Tran van Tien and G. Smets, unpublished results.
46. F. De Schryver, H. Leblanc and G. Smets, unpublished results.
47. R. F. Cozzens and R. B. Fox, *J. Chem. Phys.*, **50**, 1532, (1969).
48. R. B. Fox and R. F. Cozzens, *Macromolecules*, **2**, 181, (1969).
49. C. David, W. Dumarteau and G. Geuskens, *European Polymer Journal*, **6**, 537, (1970).
50. C. David, W. Dumarteau and G. Geuskens, *European Polymer Journal*, in press.
51. L. J. Borsile, *Trans. Far. Soc.*, **60**, 1702, (1964).
52. F. Hirayama, *J. Chem. Physics*, **42**, 1256, (1965).
53. T. Nishihara and M. Kaneko, *Macromol. Chem.*, **124**, 84, (1969).
54. G. S. Hammond, N. J. Turro and P. A. Leemakers, *J. Phys. Chem.*, **66**, 1144, (1962).
55. M. Golub, *J. Polymer Sci.*, **25**, 373, (1957).
56. M. Golub, *J. Am. Chem. Soc.*, **82**, 5093, (1960).
57. C. Sivertz, *J. Phys. Chem.*, **63**, 34, (1959).
58. W. A. Bishop, *J. Polymer Sci.*, **55**, 827, (1961).
59. G. S. Hammond *et al.*, *J. Am. Chem. Soc.*, **86**, 3197, (1964).
60. M. Golub and C. L. Stephens, *J. Polym. Sci.*, **A-1, 6**, 763, (1968).
61. M. Golub and C. L. Stephens, *J. Polym. Sci.*, **C, 16**, 765, (1966).
62. P. H. Vandeweyer and G. Smets, *J. Polym. Sci.*, **C, 22**, 231, (1968).

63. N. W. Tyer Jun. and R. Becker, *J. Am. Chem. Soc.*, **91**, 1289, 1295, (1970).
64. H. Bach and J. Calvert, *J. Am. Chem. Soc.*, **91**, 2608, (1970).
65. R. Lovrien, *Amer. Chem. Soc. Div. Polym. Chem. Preprints*, **4**, 715, (1963).
66. K. Dimroth, C. Reichardt, T. Siepmann and F. Bohlmann, *Ann. Chem.*, **661**, 1, (1963).
67. L. G. S. Brooker, G. H. Keyes and D. W. Heseltine, *J. Am. Chem. Soc.*, **73**, 5350, (1951).
68. A. I. Kiprianov and W. J. Petrukin, *Zh. Obshch. Khim.*, **10**, 613, (1940).
69. A. I. Kiprianov and E. S. Timoshenko, *Zh. Obshch. Khim.*, **17**, 1468, (1947).
70. H. Kokelenberg, Ph.D. Thesis Louvain University 1967.
71. J. B. Flannery Jun., *J. Am. Chem. Soc.*, **90**, 5660, (1968).
72. P. H. Vandeweyer and G. Smets, *Tetrahedron*, **25**, 3251, (1969).
73. A. Hinnen, C. Audic and R. Gautron, *Bull. Soc. Chim. France*, **1968**, 3190.
74. J. Hoefnagels, Ph.D. thesis Louvain University 1969.
75. Z. G. Gardlund, *J. Polym. Sci.*, **B, 6**, 57, (1968).
76. S. J. Laverty and Z. G. Gardlund, *J. Polym. Sci.*, **B, 7**, 161, (1969).
77. P. Vandeweyer and G. Smets, *J. Polym. Sci.*, in press.
78. J. Verborgt, doctoral thesis Louvain University, October 1970.
79. G. Smets, *Proceed Hungarian acad. Sciences*, in press.
80. D. J. Carlsson and D. M. Miles, *Macromolecules*, **2**, 587, (1969).
81. D. J. Carlsson, Y. Kato and D. M. Miles, *Macromolecules*, **1**, 459, (1968).
82. D. J. Carlsson and D. M. Miles, *Macromolecules*, **2**, 597, (1969).
83. G. Hartley and J. Guillet, *Macromolecules*, **1**, 165, (1968).
84. M. Heskins and J. Guillet, *Macromolecules*, **1**, 97, (1968).
85. A. Trezolo and E. Window, *Macromolecules*, **1**, 98, (1968).
86. H. Atwater and B. Mc. Cene, *J. Polymer Sci.*, **B, 7**, 477, (1969).
87. N. Grassie, B. Torrance and J. Colford, *J. Polym. Sci.*, **A-1, 7**, 1425, (1969).
88. A. Delman, M. Ludy and B. Simmon, *J. Polym. Sci.*, **A-1**, 3375, (1969).
89. F. Rabeck, *Photochemistry and Photobiology*, **7**, 5, (1968).
90. R. Desien and J. Shiells, *Makromol. Chem.*, **122**, 134, (1969).
91. J. Burr, *Advance in Photochemistry*, **6**, 193, (1968).
92. J. Pacifici and J. Straley, *J. Polym. Sci.*, **B7**, 7, (1969).
93. O. Bellus, P. Hrollovic, Z. Manasek and P. Slawa, *J. Polym. Sci.*, **C, 16**, 267, (1967).
94. R. Schulz, C. Rihe, H. Adler, *European Polymer Journal, Supplement*, **1969**, 309.
95. J. Bertram and R. Küsten, *J. Prakt. Chem.*, **51**, 325, (1895).
96. M. Cohen, G. Schmidt and F. Sonntag, *J. Chem. Soc.*, **1964**, 2000.
97. Esterification, US Patent, 1973 493, Eastman Kodak; Transesterification, M. Tsuda and Y. Minoshima, *Kogyo Kagaku Zushi*, **67**, 472, (1964).
98. M. Tsuda, *Makromol. Chem.*, **72**, 183, (1964).
99. K. Nakamura and S. Kikuchi; *Bull. Chem. Soc. Japan*, **40**, 2684, (1967).
100. F. Sonntag and R. Srinivasan, preprints technical conference of photopolymers, New York, 1967.
101. H. Curme, C. Natale and D. Kelly; *J. Phys. Chem.*, **71**, 767, (1967).
102. M. Tsuda, *Bull. Chem. Soc. Japan*, **42**, 905, (1969).
103. K. Nakamura and S. Kikuchi, *Bull. Chem. Soc. Japan*, **41**, 1977, (1968).
104. N. M. Moreau, *Polymer Preprints ACS*, **10**, 362, (1969).
105. Y. Kirsh, K. Lyalikov and K. Kalminsk; *Russ. J. Phys. Chem.*, **39**, 1002, (1965).
106. G. Delzenne, *European Polymer Journal, Supplement*, **1969**, 55.
107. UK Patent Spec. BP$_a$ 949 919, Eastman Kodak Cy. (23 October 1962).

108. UK Patent Spec. 951 928, Eastman Kodak Cy. (23 October 1960).
109. French Patent 1351 1542, Kodak-Pathé (27 December 1962).
110. C. C. Unruh, *J. Appl. Polym. Sci.*, **2**, 358, (1958).
111. US Patent 27.08.65 Eastman Kodak Cy. (12 September 1951).
112. H. Calvayracn and Y. Gola, *Bull. Soc. Chim. France*, **1968**, 1076.
113. G. A. Delzenne and U. Laridon, *Preprints 36e Congres de Chimie Industrielle*, Bruxelles (1966).
114. J. L. K. Williams and D. G. Borden, *Makromol. Chem.*, **203**, 713, (1964).
115. Belgian Patent 679 251, Gevaert-Agfa (8 May 1966).
116. C. A. Schröter and P. Riegger; *Kunststoffe*, **44**, 278, (1954).
117. A. Reiser, F. W. Willets; G. C. Terry, V. Williams and R. Marley, *Trans. Faraday Soc.*, **64**, 3265, (1968).
118. J. S. Swenton, T. J. Ikeler and P. H. Williams, *J. Am. Chem. Soc.*, **92**, 3103, (1970).
119. G. A. Delzenne, U. Laridon and H. Peeters, personal communications.
120. F. C. De Schryver, *Pure Appl. Chem.* **34**, 213, (1973); F. C. De Schryver, N. Boens and G. Smets, *J. Polym. Sci.*, *A-1*, **10**, 1687, (1972); J. G. Higgins and D. A. McCombs, *Chem. Tech.*, **1972**, 176; F. C. De Schryver, L. Anand, G. Smets and J. Switten, *J. Polym. Sci.*, *B*, **9**, 777, (1971); D. J. Andrews and W. J. Feast, IUPAC Symposium, Leuven, 1972; F. C. De Schryver, T. Tran Van and G. Smets, *J. Polym. Sci.*, *B*, **9**, 425, (1971); J. Higgins, A. H. Johannes, J. F. Jones, R. Schultz, D. A. McCombs and C. S. Menon, *J. Polym. Sci.*, *A-1*, **8**, 1987, (1970); D. E. Pearson and P. D. Thiemann, *J. Polym. Sci.*, *A-1*, **8**, 2103, (1970); D. A. McCombs, C. S. Menon and J. Higgins, *J. Polym. Sci.*, *A-1*, **9**, 1261 (1971); D. A. McCombs, C. S. Menon and J. Higgins, *J. Polym. Sci.*, *A-1*, **9**, 1799, (1971); Y. Musa and M. P. Stevens, *J. Polym. Sci.*, *A-1*, **10**, 319, (1972) and references cited therein; N. Kardush and M. P. Stevens, *J. Polym. Sci.*, *A-1*, **10**, 1093, (1972); H. H. Bössler and R. C. Schultz, *Makromolekulare Chemie*, **158**, 113, (1972); M. Hasegawa, Y. Suzuki and T. Tamaki, *Bull. Chem. Soc Japan*, **43**, 3020, (1970); H. Kanetsuna, M. Hasegawa, S. Mitsuhashi, T. Kurita, K. Sasaki, K. Moeda, H. Obata and T. Hatakeyama, *J. Polym. Sci.*, *A-2*, **8**, 1027, (1970); H. Nakanishi, N. Nakano and M. Hasegawa, *J. Polym. Sci.*, *B*, **8**, 755, (1970); Y. Sasada, H. Shimanoushi, H. Nakanishi and M. Hasegawa, *Bull. Chem. Soc. Japan*, **44**, 1262, (1971); T. Tamaki, Y. Suzuki and M. Hasegawa, *Bull. Chem. Soc. Japan*, **45**, 1988, (1972); R. H. Baughman, *J. Appl. Phys.*, **42**, 4579, (1971); H. Nakanishi, K. Ueno and M. Hasegawa, *Chem. Letters*, **1972**, 301; F. Nakanishi and M. Hasegawa, *J. Polym. Sci.*, *A-1*, **8**, 2151, (1970); N. J. Leonard, R. S. McCreadie, M. W. Logne and R. L. Cundall, *J. Amer. Chem. Soc.* **95**, 2320, (1973); J. K. Frank and I. C. Paul, *J. Amer. Chem. Soc.*, **95**, 2324, (1973).

15

Configuration and Conformation in High Polymers

Pier Luigi Luisi

Swiss Federal Institute of Technology, Zurich, Switzerland

Francesco Ciardelli

Institute of Industrial Organic Chemistry of the University, Pisa, Italy

15.1 INTRODUCTION

In this chapter we will use the terms configuration and conformation according to the classical organic chemistry nomenclature.[1-4] Configurational isomerism is then due to the presence in the molecule of one or more dissymmetric centres, in the simplest case asymmetric carbon atoms, each of which can have (R) and (S) absolute configuration; and, or, to the presence of double bonds which can give *cis-* and *trans-*geometrical isomers. In the case of synthetic polymers, typical problems of configurational isomerism are, for instance, related to the distribution of (R) and (S) asymmetric carbon atoms along the main chain. These asymmetric carbon atoms can originate either from dissymmetric monomers (e.g. propylene oxide), or from non-dissymmetric monomers (e.g. vinyl monomers), the polymerization of which yields at least one asymmetric carbon atom per monomeric unit. Configurational isomerism in macromolecules arising from the presence of asymmetric carbon atoms in the lateral chains, for example in the polymerization of dissymmetric vinyl monomers, is also possible and nowadays widely investigated.[5-11]

By conformations (rotamers, rotational isomers, or conformers), in keeping with the classical organic chemistry nomenclature,[1-4] we mean the different arrangements in space produced by the rotation (not breaking) of bonds of the molecules with respect to each other.

Typical conformational problems in polymers are, for instance, related to the question of whether a chain molecule is in the form of random coils or in the form of more or less regular spiralled conformations.

This precisation of nomenclature may seem trivial. Unfortunately, however, much confusion is present in macromolecular chemistry because of the indiscriminate use of the terms conformation and configuration. Some workers, for instance, commonly use the term 'configurational statistics'[12–14] or 'configurational transitions'[12] to indicate what, according to the organic chemistry nomenclature, should be referred to as 'conformational statistics', 'conformational transitions'. The expression 'configurational statistics' in dealing with conformational problems of macromolecules was also used by Volkenstein in the early years,[15] but later abandoned by him and his school. The situation is even more confused in the field of biological macromolecules where the terms 'conformer', 'conformational change' and 'configurational transition' are often used in a very misleading context. Of course, it is obvious that the discrimination between conformation and configuration, as used in classical organic chemistry, may become critical in some instances. However, we think that in the case of synthetic polymers there is no reason to abandon the organic chemistry nomenclature.

In dealing with configuration and rotational isomerism in this chapter, we will assume that the macromolecules have a known chemical structure (for instance, a regular head-to-tail enchainment in the case of vinyl polymers). Operationally, indeed, before asking questions on configurational isomerism, the problem of chemical structure should already have been solved. Once the chain configuration is known, one can finally begin to investigate conformational properties. Clearly, however, the three levels of investigation of a polymer structure (chemical structure, configuration and conformation) are closely correlated and in practice they often overlap. This three-step differentiation is, however, very useful from an heuristic point of view, and we will discuss first configurational problems assuming a known chemical structure, and secondly conformational problems, assuming a known configuration. In particular, we will begin to consider the relationship between chain configuration and monomer chemical structure (section 15.2). Afterwards we will consider the physical methods used to investigate the configurational order in the chains (section 15.2.2) and finally the influence of the configurational isomerism on some physical properties (section 15.2.3). Clearly, the actual configuration of the macromolecules of a given polymer depends upon the stereo-specificity of the polymerization process and so, in order to have a more complete picture of the question, one should also consider the mechanism of the stereo-specific polymerization and the influence of different experimental conditions on it. Such a discussion is, however, out of the scope of the present chapter, as it has been presented in other parts of this book.

As far as conformational problems are concerned, we will examine the relationship between conformational equilibria in the solid state and in solution, emphasizing the particular difficulties encountered in the latter case both from the experimental and conceptual view-point (section 15.3 and section 15.5). A brief review of the conformations in the crystalline state which are known at the present time is reported in section 15.4, with emphasis on the influence of the monomeric unit structure and configuration isomerism on the conformation of some typical stereo-regular polymers. Finally, the relationships between rotational isomerism and some physical properties are considered (section 15.6) with particular emphasis on optical activity (section 15.7).

15.2 CONFIGURATION

In the case of high polymers the real configurational problem is to know the law of succession along the chain of the configuration of the dissymmetric centre(s) present in each monomeric unit. In the cases in which the above law produces a steric order in the main chain, the polymer has been called stereo-regular or tactic.[16-17]

The first nomenclature for tactic polymers was introduced for vinyl polymers, the macromolecules of which consist of monomeric units having a tertiary carbon atom on the main chain, which is asymmetric* since the chain has a finite length.

Natta[18] called 'isotactic' a polymer consisting of macromolecules having a chain backbone containing long sequences of monomeric units in which the carbon atoms carrying the substituents have identical absolute configuration. Syndiotactic[19] macromolecules by contrast contain backbone asymmetric carbon atoms with alternate absolute configuration.

The detailed analysis of the configuration of the main chain asymmetric carbon atoms of the macromolecules of isotactic vinyl polymers leads to the conclusion that an ideal isotactic chain with equal end groups has a meso-structure,[22,23] i.e. in the first half of the chain the asymmetric carbon atoms have opposite absolute configuration with respect to those of the second half.

Accordingly, the true low molecular weight model of the isotactic macro-molecule, for instance in the case of a compound containing two asymmetric centres, is the meso-diastereoisomer.[24]

The definition of an isotactic macromolecule as that in which all the main chain asymmetric carbon atoms have the same absolute configuration holds however in the case of polypropyleneoxide, poly-α-aminoacids and poly-1,3-pentadiene.

In the conventional representation popularized by Natta, the atoms of the main chain lie in a plane and form a zig-zag conformation, and an isotactic

---

* It has been proposed[20] to call 'pseudoasymmetric' the tertiary carbon atoms of vinyl polymer chains, as two of the neighbour groups are ($-CH_2-$). In our opinion this is not very convenient considering the use of the above term in classical organic chemistry.[21] A single real 'pseudoasymmetric' carbon atom can be present in a macromolecule only in a few special cases.[22]

vinyl macromolecule (1) is that in which all the side chains are on the same side with respect to such a plane. In a syndiotactic vinyl macromolecule (2) the lateral chains are alternately above and below such a plane.[25]

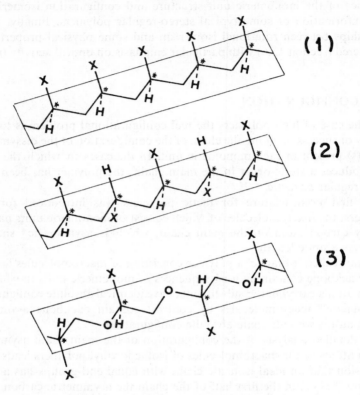

The extension of the above representation to non-vinyl polymers might generate confusion. In fact the monomeric units of some of them, for example polypropyleneoxide and poly-α-aminoacids, contain an odd number of atoms in the chain backbone; therefore in an isotactic macromolecule of this type (3) the side chains are alternately above and below the plane containing the extended main chain. This discrepancy is not observed in Fischer projection formulae (4) and (5) shown below.

A very useful definition has been introduced by Corradini, who proposed[26] that the two bonds (a) and (b) of the chain adjacent to the asymmetric carbon atom can be distinguished from a configurational view-point as (+) or (−) bonds. As bonds astride an asymmetric carbon atom are always opposite in sign by the above definition, if (a) is (+), (b) must be (−), and vice versa. Two monomeric units are identical, from this point of view, if the corresponding bonds are characterized by the same set of (+) and (−) signs. An

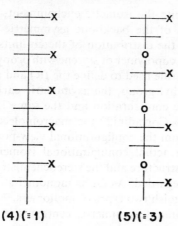

(4)(≡1)   **(5)(≡3)**

isotactic 'vinyl' macromolecule having DP = n can be described by $(+-)_n$ (6), and a syndiotactic polymer by $[(++)(--)]_{n/2}$ (7). This definition can be extended to other types of polymers with more complex tacticity and it is, in our opinion, especially useful for copolymers of two vinyl monomers.

For the last, the term 'co-stereosymmetric' has been used[27] to indicate macromolecules in which both co-monomers units are enchained in an

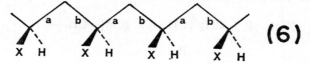

(6)

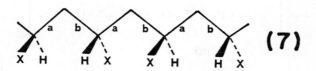

(7)

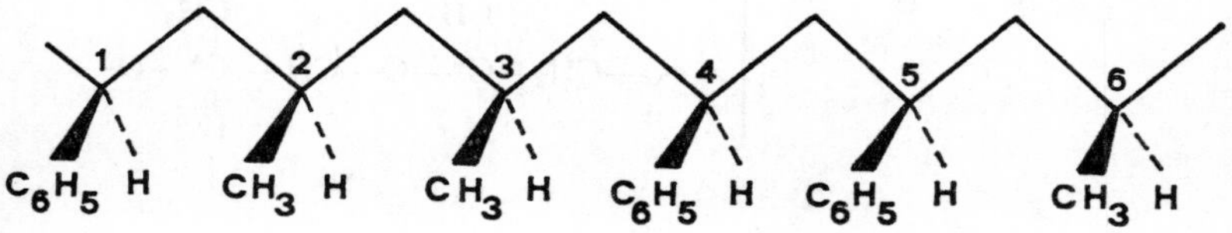

(8)

isotactic fashion. Despite the isotactic enchainment, the succession of the absolute configuration of the backbone asymmetric carbon atoms can be disordered because of the distribution of the co-units. This concept can be easily exemplified for a copolymer of styrene with propylene (**8**). As the phenyl group, in the sequence rule used to define the (R) and (S) absolute configuration, precedes the methyl group, the asymmetric carbon atoms (2) and (3) have opposite absolute configuration and the same is true for numbers (4) and (5). According to Corradini,[26] the macromolecule (**8**) is however unequivocally defined, from the configurational view-point, as $(+-)_n$.

In dealing with the actual configurational isomerism of polymers, the effect of the monomer structure and the stereochemistry of the polymerization process cannot be disregarded. As far as monomer structure is concerned, it is convenient to distinguish two types of monomers. The first includes monomers which do not contain asymmetric centres, the second includes chiral monomers.

In the case of achiral monomers, such as vinyl monomers $(CH_2=CHR)$ and 1-substituted 1,3-dienes $(CH_2=CH-CH=CHR)$, the formation of stereo-regular polymers demonstrate that the polymerization is stereo-specific, the steric control being in general exerted by the catalytic complex.[28] Since in each monomer addition a new asymmetric carbon atom is formed (**9**), both iso- and syndiotactic macromolecules can in principle be obtained.

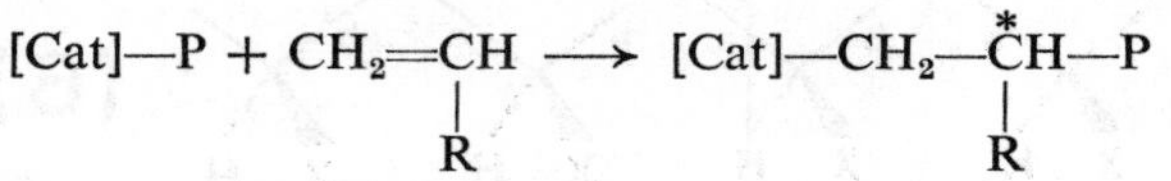

$$[Cat]-P + CH_2=\underset{\underset{R}{|}}{CH} \longrightarrow [Cat]-CH_2-\underset{\underset{R}{|}}{\overset{*}{CH}}-P$$

(**9**) (P = polymer growing chain)

In the case of chiral monomers like propyleneoxide, the asymmetric carbon atom, which will be present in the polymer backbone, is already present in

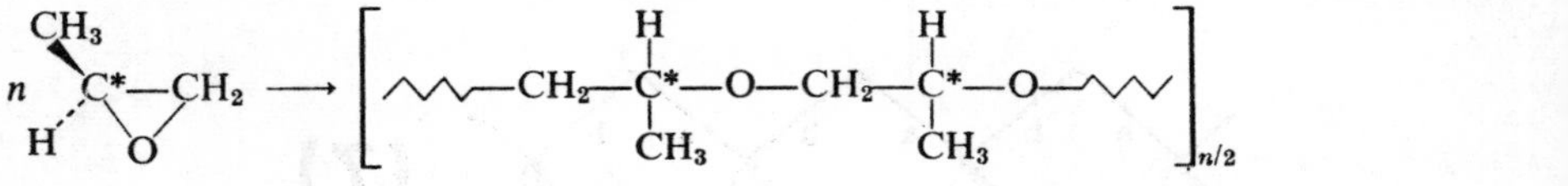

(**10**)

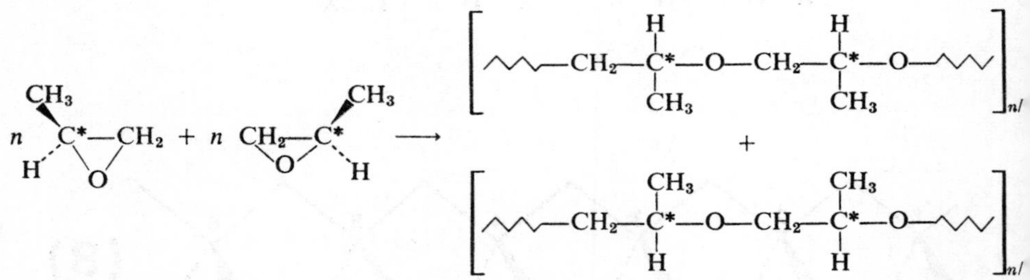

(**11**)

the monomer. Therefore isotactic macromolecules (**10**) can be obtained simply by polymerizing a monomer antipode under conditions of no racemization.[29] As is easily understandable, no syndiotactic polymers from monomers of this type have yet been obtained. The polymerization of racemic propylenoxide can yield either atactic or isotactic polymers depending on the catalyst. For instance, isotactic polymers have been obtained by stereo-selective catalysts[30] which give macromolecules (**11**) each of which predominantly consists of a single antipode.

The case of polymers from chiral vinyl monomers, that is those bearing a dissymmetric centre in the side chain (**12**), is more complicated from the configurational point of view. In fact, the monomeric unit has at least two asymmetric centres, one in the chain backbone and the other in the side chain.

$$
\begin{array}{ccc}
CH_2{=}CH & & {\sim}{-}CH_2{-}\overset{*}{CH}{-}{\sim} \\
| & & | \\
H{-}C^*{-}R_2 & \longrightarrow & H{-}C^*{-}R_2 \\
| & & | \\
R_1 & & R_1
\end{array}
\qquad (12)
$$

If the starting monomer is optically pure the problem can be treated as that of non-chiral vinyl monomers, taking into account that all units have side chain asymmetric centres with the same absolute configuration.

The complication arises when the monomer is non-optically pure or racemic; in this case the molar fraction and the distribution of the two monomeric units having asymmetric carbon atoms with opposite configurations in the side chains of each isotactic macromolecule must be considered. The problem could in principle be treated from a statistical view-point assuming random distribution of both types of monomeric units.[31] However, some catalysts are able to give stereo-selective polymerization of racemic vinyl monomers.[32]

The stereo-selectivity might be, in principle, considered as independent of the stereo-specificity. Actually, however, the two processes do not appear to be completely independent; for a deeper discussion on this point the reader is referred to the specialized literature.[6-8]

15.2.1 Evaluation of stereo-regularity type and degree

It is well known that synthetic high polymers in general consist of molecular species varying in molecular weight over a considerable range. In the case of vinyl polymers especially, the same is true as far as stereo-regularity is concerned. In fact, from the same polymerization process macromolecules with different type and degree of stereo-regularity may be and actually have been obtained. Moreover when type and degree of tacticity are the same, the distribution of the configurational irregularities can be different.

Therefore, in the case of stereo-regularity, as for the degree of polymerization, only average information can be obtained.

For at least partially crystalline polymers, X-ray examination allows the unequivocal determination of stereo-regularity type.[26] However, X-ray data are not very useful from a quantitative point of view as the crystallinity is limited in general to a portion of the polymeric sample and is markedly affected by thermic treatment, molecular weight and distribution of configurational irregularities. Finally, several tactic polymers are known, which are not able to crystallize.[33]

High resolution NMR appears to be at present the most powerful technique for determining the configuration of macromolecular chains. Relationships between magnetic shielding of nuclei and molecular symmetry can be used for this purpose. For instance in vinyl polymers, which are the most investigated by NMR, the two protons of the methylene groups have the same chemical shift in syndiotactic chains, whereas differentiated bands are observed in isotactic sequences.[24] When the above differentiation is possible the methylene group has been called heterosteric,[34] whereas a homosteric methylene group is that in which the two protons are equivalent. The same classification applies to $-CX_2-$ groups in general. These symmetry classifications are independent of molecular conformation and, with the exception of few cases, protons of a heterosteric methylene group have different shieldings and then give two different signals in the NMR spectrum. The symmetry properties of a methylene group included in tactic sequences having different length (diads, triads, tetrads, and pentads) and their relationships with magnetic shieldings have been discussed in detail.[24] This detailed analysis combined with the improved resolution power of the NMR spectrometers now available, has allowed determination of stereo-regularity type and degree of a large number of vinyl polymers.[24] The main limitation of the application of NMR to polymer solutions is due to the fact that intramolecular interactions restrict segmental motions and hence the widths of the bands from dissolved macromolecules are larger than those from small molecules. An increase of temperature[35] can help in this problem more than a reduction of the concentration, which limits the sensitivity. The difficulty arising from the presence in many polymers of insufficiently differentiated protons can be overcome by investigating polymers obtained from monomers deuterated in a suitable position[36] and, more recently, by the use of spectrometers working with higher magnetic fields.[24] Particularly promising is the study of other nuclei, such as ^{13}C.[24,37]

IR spectra have also been used in many cases for determining stereoregularity. In fact in vibrational spectra of polymers, stereo-regularity bands have been observed both in the solid state and in solution.[38-40]

The type of stereo-regularity can be identified by symmetry considerations and by comparison of the polymer spectra with those of suitable low molecular weight model compounds.[40,42-45]

As a relative stereo-regularity index, the ratio between the optical density of one of the above bands with respect to the optical density of bands independent of stereo-regularity has been reported.[41] A better quantitative evaluation of the stereo-regularity degree can be obtained if standard samples are available having a known degree of tacticity, obtained by independent methods.

Considerable help can be also given by the use of partially deuterated samples. For more details we refer to recent reviews on the application of IR spectroscopy to polymer chemistry.[42-45]

In the case of crystalline polymers, a relative stereo-regularity index can be obtained from solvent extraction data. This method has been used extensively for polypropylene,[46] but reliable results can be obtained only by carrying out solvent extraction under similar conditions and on polymer samples having similar average molecular weight as well as molecular weight distribution.[47,48] The fractionation by solvent extraction is in any case a very useful technique for obtaining, from the crude heterogeneous polymer, fractions which are more uniform with respect to molecular weight and stereo-regularity. By this method it has been possible to demonstrate that during a polymerization process, especially by heterogeneous catalysis, macromolecules with different stereo-regularity degree and even stereo-regularity type can be formed.[49]

The relationships between tacticity and optical activity are very peculiar and two types of polymers must be considered from this point of view. In the case of polymers obtained from chiral monomers, such as polypropyleneoxide and poly-α-aminoacids, the optical activity can be a direct measurement of stereo-regularity type and degree. As synthesized from optically active monomer, an isotactic polymer will have the highest optical rotation, the syndiotactic polymer will be optically inactive due to intramolecular compensation, and the heterotactic polymer will show in general very low optical rotation. A similar situation occurs in tactic polymers of 1-substituted or 1,4-disubstituted-1,3-dienes prepared with optically active catalysts.[23] As mentioned above, an isotactic polymer obtained from a racemic monomer through a stereo-selective polymerization process, is optically inactive, being an equimolar mixture of homopolymers of the two antipodes. This appears to be the case with polypropyleneoxide; in fact, crystalline polymers of the racemic monomer have been obtained[50] and separated in fractions having optical activity of opposite sign. The above considerations refer to the determination of the type of tacticity. The problem of the quantitative evaluation of the degree of stereo-regularity by optical activity measurements can be solved if a linear relationship exists between polymer optical rotation and chain configurational purity, which in the above polymers corresponds to the optical purity of the polymerized monomer. A limitation to this evaluation derives from the fact that in general the optical purity of the monomer included in

17

the macromolecular chains (and hence the maximum rotatory power of the polymer) is not exactly known.

An attempt has been made by Price and coworkers[51] to determine the relationship between molar rotatory power and number of configurational inversions in polypropyleneoxide by ozonolysis. This method, however, allows the determination of the number of head-to-head, tail-to-tail enchainments of the monomeric units (presence of —O—O— groups) and not the number of configurational inversions which could be present in macromolecules having a regular head-to-tail enchainment.

No direct relationship between optical rotation and stereo-regularity exists in polymers from vinyl and related monomers. In fact, as previously discussed, a perfectly isotactic or syndiotactic macromolecule of this type has mirror symmetry and the corresponding polymer cannot be optically active.

The possibility of molecular dissymmetry can arise in atactic polymers or tactic polymers having different end-groups; however as, in general, both monomer and catalyst are non-dissymmetric or at least non-optically active, intermolecular compensation eliminates the possibility of optical activity. The optical rotation of polymers of this type prepared by polymerization with optically active catalysts is rather low and decreases with increasing molecular weight. In general, it can be entirely accounted for by considering the presence in the polymer of groups originally present in the catalyst.[52,53]

An appreciable specific optical rotation in vinyl polymers similar to or larger than that of low molecular weight structural models can be obtained by starting with optically active monomers.

However, even in this case appreciable differences of optical rotation among polymer samples obtained from the same monomers and having different stereo-regularity degree and type have been observed only when the side chain asymmetric centre is not too far from the main chain. For instance, in the case of poly-α-olefins (13), the mentioned effect has been clearly

Table 15.1 Influence of stereo-regularity on the optical rotation of poly-(S)-4-methyl-1-hexene[8]

Sample	Intrinsic[a] viscosity dl/g	Relative IR[b] stereo-regularity index	X-ray examination	$[\phi]_D^{25}$ [c]
(1)	0·26	0·62	Amorphous	+178
(2)	0·22	0·72	Moderately crystalline	+227
(3)	0·08	0·83	Crystalline	+266
(4)	0·48	0·95	Highly crystalline	+292
(5)	5·15	0·96	Highly crystalline	+296

a. In tetralin at 120 °C.
b. $D_{B\,995}/D_{B\,964}$.
c. In hydrocarbon solution, per mole of monomeric unit.

Table 15.2 Influence of stereo-regularity[a] on the optical rotation of poly-(−)-menthylmethacrylate[55] and poly-(+)-α-methylbenzylmethacrylate[56]

Polymer	Isotactic per cent	Stereo-regularity Heterotactic per cent	Syndiotactic per cent	$[\alpha]_D^{15}$ [b]
Poly-(−)-menthyl- methacrylate	69·2	22·7	8·1	−82
	55·7	28·0	16·3	−82·6
	12·9	46·9	40·2	−91·7
Poly-(+)-α-methyl- benzylmethacrylate	78·4	16·6	5·0	+116·5
	15·6	30·1	54·3	+104·2
	10·6	28·2	61·2	+104·0

a. Determined by NMR on polymethylmethacrylate derived from the optically active polymer.
b. In aromatic hydrocarbons.

observed only when the side chain asymmetric carbon atom is in the α or in the β position with respect to the main chain.[54] The data for (13) ($x = 1$) are reported in Table 15.1. In the case of optically active polymethacrylates (14),[55,56] in which the closest side chain asymmetric carbon can be in the γ-position with respect to the main chain, only a slight effect of tacticity on optical rotation has been observed (Table 15.2).

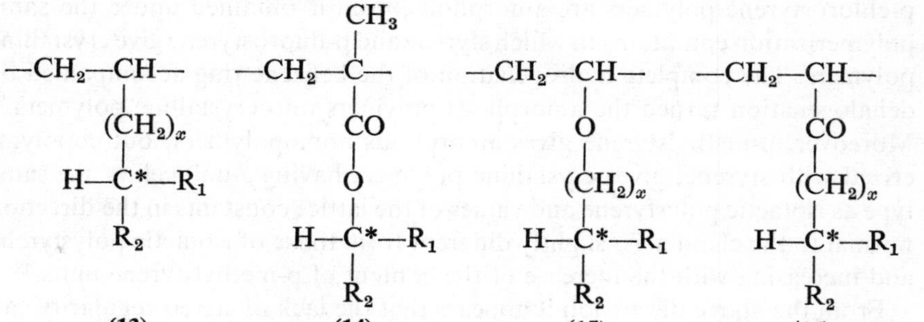

$$
\begin{array}{cccc}
& \mathrm{CH_3} & & \\
& | & & \\
-\mathrm{CH_2-CH-} & -\mathrm{CH_2-C-} & -\mathrm{CH_2-CH-} & -\mathrm{CH_2-CH-} \\
| & | & | & | \\
(\mathrm{CH_2})_x & \mathrm{CO} & \mathrm{O} & \mathrm{CO} \\
| & | & | & | \\
\mathrm{H-C^*-R_1} & \mathrm{O} & (\mathrm{CH_2})_x & (\mathrm{CH_2})_x \\
| & | & | & | \\
\mathrm{R_2} & \mathrm{H-C^*-R_1} & \mathrm{H-C^*-R_1} & \mathrm{H-C^*-R_1} \\
& | & | & | \\
& \mathrm{R_2} & \mathrm{R_2} & \mathrm{R_2} \\
(13) & (14) & (15) & (16)
\end{array}
$$

In the case of polyalkylvinylethers (15)[57] and polyalkylvinylketones (16)[58] an appreciable dependence of rotatory power on polymer tacticity has been observed when $x = 0$. The interpretation of these data is complicated by the fact that optical activity changes with varying stereo-regularity and is also connected to differences in conformational equilibria positions, as discussed later.

In the presence of stereo-selective catalysts,[8] furthermore, highly stereo-regular polymers having no optical activity or very low optical activity can be obtained from racemic monomers or from monomers with very low optical purity, respectively.

Finally, in the polymers from optically active vinyl monomers, there is no linear relationship, in many cases, between polymer optical rotation and optical purity of polymerized monomer.[9,59]

15.2.2 Influence of stereo-regularity on polymer properties

It has been long recognized that stereochemical regularity, in addition to chemical regularity, is a necessary requisite for polymer crystallizability.[60] Actually, even partially stereo-regular macromolecules can give a partial, but observable crystallization and, in the end, the percentage of crystalline polymer will be related to the degree of stereo-regularity.

However, few cases are known in which configurational irregularities along the main chain do not prevent crystallizability. For instance poly-vinylalcohol, obtained by hydrolysis of atactic and amorphous polyvinyl-acetate, is crystalline, as CH_2 and $CH(OH)$ groups are near enough to the same size to fit in similar crystalline lattices.[61] Another particular case is that of polymers of linear α-olefins in which the crystalline melting point decreases from polypropylene to poly-1-pentene; higher members are rubbery, but poly-octadecene is again crystalline owing to crystallization of long lateral groups.[62] Furthermore, several tactic polymers are not able to crystallize owing to their particular chemical structure. A typical example is given by the polymers of styrene derivatives. For instance among halostyrene polymers, p-chlorostyrene polymers are amorphous, even if obtained under the same polymerization conditions in which styrene and p-fluorostyrene give crystalline polymers. The complete hydrogenation of the benzene ring accompanied by dehalogenation turned the amorphous polymers into crystalline polymers.[63] Moreover, p-methylstyrene gives amorphous homopolymers, but copolym-erized with styrene, gives crystalline polymers having a unit cell of the same type as isotactic polystyrene and values of the lattice constants in the direction normal to the chain axis, slightly different from those of isotactic polystyrene and increasing with the increase of the content of p-methylstyrene units.[33]

From the above discussion it appears that the lack of stereo-regularity can prevent the crystallization of macromolecules even if they have regular chemical structure.

The degree of stereo-regularity also affects the melting range of crystalline polymers.[64,65] In fact, the disappearance of a polymer crystalline phase occurs at lower temperatures in less stereo-regular samples. This is probably a con-sequence of the lower perfection of crystalline order in polymer consisting of macromolecules with higher number of configurational inversions. However, many other factors affect the melting temperature of polymers and the separation of the different contributions is still under investigation.[64]

The influence of tacticity on glass-transition temperature is remarkable.[66,67] This conclusion is well-documented despite the fact that individual polymers may exhibit multiple glass-transition temperatures which may be difficult to detect and to correlate to tacticity.

Stereo-regularity has a pronounced effect on other properties in the solid state, particularly on mechanical properties. This is the primary reason for the widespread use of stereo-regular polymers as technologically important materials. Understandably, the discussion of the relationship between technological properties and type and degree of stereo-regularity in vinyl polymers is one of the most important topics in macromolecular chemistry nowadays. Such a discussion is, however, beyond the scope of the present paper; for some more details the reader is referred, for instance, to the recent review by Johnson and Porter.[68]

Solubility is also affected by stereo-regularity and increases when the latter decreases. This property has been used to achieve separation of iso-tactic chains from stereo-block or atactic material by solvent extraction of the crude polymer, as discussed in the previous section. The different solu-bility of samples of the same polymer having different tacticity degree is probably connected to the considerably lower stability of the crystal lattice when a relatively small content of chain segments is present with the 'wrong' steric configuration.[20] A very particular example of the influence of the configurational isomerism on the solubility is given by the isotactic polymers of 4-methyl-1-hexene. In fact, the polymer obtained from a single enantiomer is more soluble than that obtained from the corresponding racemic monomer. It is interesting to mention that the latter behaves as the equimolar mixture of the homopolymers of the two antipodes.[8,69]

Solution properties are, in general, affected both by stereo-regularity and conformation and it is very difficult to discriminate between them, as dis-cussed in the next section.

It is worth mentioning the effect of stereo-regularity on chemical reactivity. However, in this case it is also very difficult to separate the effect of stereo-regularity from that of conformation. In any case, a review of this topic is beyond the scope of the present chapter. An exhaustive picture of the chem-ical reactivity of polymer chains is given in the book by Fettes.[70] Moreover, van Beylen[71] has recently reviewed some special cases in which reactions on tactic polymers have been studied from the stereochemical view-point.

15.3 BASIC PROBLEMS IN THE INVESTIGATION OF CONFORMATIONAL EQUILIBRIA

Conformational analysis problems are usually dealt with in the framework of the rotational isomeric state approximation, as introduced by Volken-stein,[15,72] and later extensively developed by his group[73-76] and by other authors.[12]

According to such an approximation, only the conformations of bonds corresponding to energy minima are taken into account to describe the conformation of the molecule. In this way, the number of rotamers for the molecule is a discrete number, which is usually in simple relationship with the structure of the molecule. For instance, in the case of open chain paraffins and related compounds, only staggered conformations of C—C bonds are considered, and if the rotation takes place around n bonds, we have a total of 3^n conformers. A discussion of the limit and validity of the rotational isomeric scheme can be found in the book by Birshtein and Ptitsyn,[73] in that by Flory,[12] and in some recent reviews.[77,78]

The sum of the relative statistical weights of all the f conformers of a molecule is usually referred to as the 'conformational partition function' of the molecule, and takes the form

$$Q = \sum_{i=1}^{f} g_i$$

where $g_i = \exp\left(-\Delta E_i / RT\right)$ and ΔE_i is the energy difference between the ith rotamer and the reference energy state (usually the energy of the most stable rotamer, for which the relative statistical weight g is taken equal to one). The main problem in conformational analysis is the determination of Q, that is the number of the conformers of a molecule and their relative energy. Once that this is known, most of the properties of the molecule can be obtained through classical thermodynamical statistical treatments.[12,15,73] The thermodynamic treatment in terms of Q is, of course, valid when the system is in equilibrium, i.e. when all the considered rotamers can rapidly interconvert one into the other. This can be realized only if the potential energy barriers hindering the rotation in molecules[79] can easily be overcome under operational conditions. Only in this case can we properly speak of conformers, and can the system be treated in terms of the Maxwell-Boltzmann distribution.

The bond rotation cannot take place in the solid state and conformational analysis problems are simplified. Since, for stereo-regular macromolecules, it is commonly assumed that the conformations in the crystalline state correspond to conformational energy minima,[26,73] we can obtain experimentally, through X-ray analysis, the preferential conformation of the monomeric unit in the chain.

The determination of the most stable conformation of a molecule can be carried out also on the basis of semi-empirical interaction potentials.[12,13,80—87] These calculations of energy minima are nowadays of increasing reliability. In most cases, a very good agreement between the calculated conformations and those experimentally found by X-ray analysis has been obtained.

The knowledge of the most stable conformation of a molecule in the solid state, however, does not solve the problem of its conformation in solution.

In solution when the rotation of the bonds in the chain is possible, the conformational equilibrium will be characterized by the fast interconversion of an extremely large number of chain conformers. This is the case for poly-α-olefins and for many other types of synthetic polymers so far investigated. Proteins and some polypeptides, on the contrary, are supposed to maintain the same conformation in solution as in the solid state, because of the presence of non-covalent bondings.[73,82,88,89] This point indicates probably the most important difference between the conformational properties of biopolymers and non-biological synthetic polymers. It is the reason, for instance, why the sharp helix-coil transitions present in some polypeptides[89–92] have not been found in non-biological synthetic polymers.

15.4 THE CONFORMATION OF MACROMOLECULES IN THE CRYSTALLINE STATE

Table 15.3 shows the conformation in the crystalline state of some isotactic vinyl polymers. Many other data could be listed and discussed both for stereo-regular vinyl polymers and for other polymers. The few examples reported in Table 15.3 are, however, sufficient to illustrate, in our opinion, the most important points concerning the conformation of stereo-regular macromolecules in the crystalline state.

Most of the isotactic vinyl polymers are characterized by helical conformations, in which, according to the postulate of equivalence,[26] the main chain bonds of all the monomeric units have the same pair of internal rotation angles* (ψ_1, ψ_2). The ideal helical conformation in the crystalline state of a macromolecule of this kind can be simply described as $(\psi_1, \psi_2)_n$, where n is the polymerization degree. In the case of 3_1 helices, we have $(0,120)_n$ for a right-handed helix and $(-120,0)_n$ for a left-handed helix.[26,73]

The angles of internal rotation 0 and ± 120 correspond to *trans* (t) and *gauche* ($g+$ and $g-$) staggered conformations respectively, i.e. they correspond to the minima of energy characterizing the bond rotation in low molecular weight paraffins.[12,73]

In some cases, however, there are deviations from the ideal t and g states, and helix conformations other than 3_1 can be assumed by isotactic vinyl macromolecules in the crystalline state. The cases of poly-4-methyl-1-pentene, poly-3-methyl-1-butene, poly-m-methylstyrene are illustrative in this respect. Such deviations from the ideal t and g states are present when the side chain contains a further substituent in the α or β position with respect to the main chain. However, the phenyl group in the α position does not cause deviations

* For the definition of internal rotation angle, and the meaning of ψ_1 and ψ_2, see footnote of Table 15.3.

Table 15.3   Conformations of some isotactic vinyl polymers in the crystalline state

Polymers	Type of helix	Angles of internal rotation$^{a(\circ)}$ ψ_1, ψ_2	References
Polypropylene Poly-1-butene Poly-5-methyl-1-hexene Poly-5-methyl-1-heptene Poly-1,2-butadiene Polystyrene Poly-3-phenyl-1-propene	3_1	0, 120	33, 93–102
Poly-4-methyl-1-pentene Poly-4-methyl-1-hexene	7_2	−13, 110	33, 100, 103
Poly-3-methyl-1-butene^b	4_1	−24, 96	100, 104
Poly-α-vinylnaphthalene Poly-o-methylstyrene	4_1	0, 90	33, 105, 106
Poly-m-methylstyrene	11_3	−16, 104	33, 107
Polyisopropylacrylate Polysecbutylacrylate Polyvinylmethylether Polyvinyl-n-butylether	3_1	0,120	33, 108, 109

a. The angles of internal rotation of the main chain bonds are, in this paper, evaluated, according to Birshtein and Ptitsyn,[73] i.e. assuming as zero angle the *trans*-conformation of bonds and moving clockwise. The $g+$ conformation corresponds to 240° (−120°), the $g-$ to 120°.

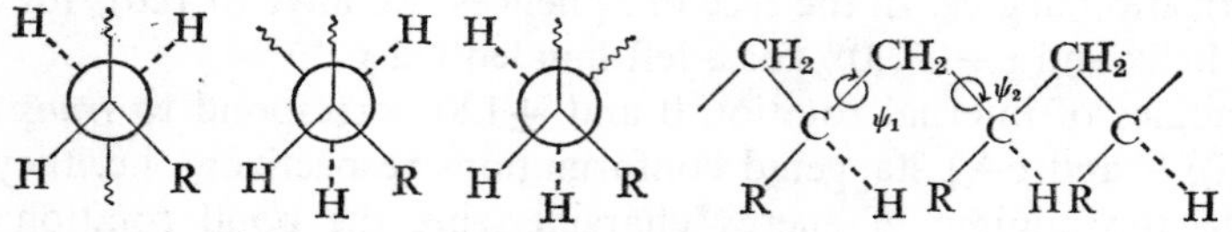

$$\psi_1 = 0(t) \quad \psi_1 = +120(g-) \quad \psi_1 = -120(g+)$$

Other authors, for instance, Natta, Corradini and coworkers use a different evaluation of the internal rotation angles and of the $g+$ and $g-$ conformations of bonds.[26,33]

b. Recently, Corradini and coworkers reinvestigated the crystal structure of isotactic poly-3-methyl-1-butene, proposing somewhat different conformational parameters.[110]

from the ideal t and g conformations of bonds (see, for instance, poly-3-phenyl-1-propene). Isotactic poly-α-vinylnaphthalene and poly-o-methylstyrene are characterized by 4_1 helices with a very large deviation from the ideal t and g states.

Also isotactic polyaldehydes[111–115] crystallize in 4_1 helices. It is of interest to compare polyaldehydes with the corresponding poly-α-olefins having a

—CH_2— group replacing the oxygen in the main chain. In the case of poly-α-olefins, in fact, the conformation is governed by repulsive forces between the aliphatic groups, which lead to staggered conformations of bonds, and hence in most cases to 3_1 helices. In the case of polyaldehydes, however, attractive forces between oxygen atoms and aliphatic groups lead to partially eclipsed conformations of bonds.

The comparison between polyethylene chains and polyoxymethylene chains is also of some interest. As it is well known, polyethylene assumes a fully *trans*-conformation in the solid state.[116] Polyoxymethylene assumes, on the contrary, a helix conformation.[117] The reason for such a difference has been the subject of many investigations.[118-120] It is now assumed that the large dipole moment of the polyoxymethylene chains is determining the helical conformation; if the molecule would assume the planar zig-zag conformation, the dipole vectors should be oriented parallel. In the case of the 9_5 (or 2_1) helix, found for polyoxymethylene, the dipole-moment vectors are oriented alternately antiparallel which leads to a much more favoured dipole-dipole interaction.[121] Also polyethyleneoxide assumes a helical conformation, though a very loose one.[122] Also the molecular structure of polythiomethylene (—CH_2S—)$_n$ has been found to be a 17_9 helix.[123] Polypropylene oxide crystallizes, however, in the form of planar *trans*-chains.[124,125] In the case of polymethyl- and polyisobutylvinylether, the regular 3_1 helical conformation has been found.[26] Isotactic polyvinyl-tert-butylether is, however, in the form of a 4_1 helix,[126] while isotactic polyisopropylvinylether is a 17_5 helix.[127] The conformation in the crystalline state of isotactic polyacrylates are generally characterized by 3_1 helical conformations.[33,108,109]

Syndiotactic polymers of type (—CH_2—$CH(R)$—) crystallize either in the form of planar *trans*-chains ($\psi_1 = \psi_2 = 0$), for example poly-1,2-butadiene,[128] or in the form of a helix which corresponds to the alternation of (ψ_1, ψ_1) and (ψ_2, ψ_2) angles. For instance, the conformation in the crystalline state of syndiotactic polypropylene can be described in terms of unit pair conformations $[(g+, g+)(t, t)]_n$ (left-hand helix) or $[(t, t)(g-, g-)]_n$ (right-handed helix) as the following scheme shows

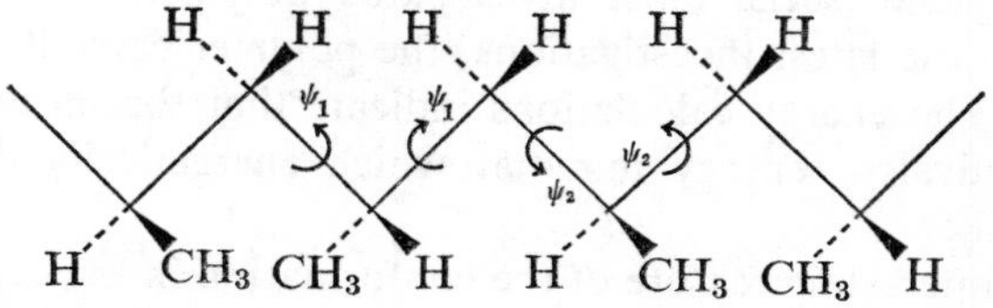

Chain conformation		ψ_1	ψ_1	ψ_2	ψ_2	
zig-zag planar	$\cdots\cdots t$	t	t	t	t	$\cdots\cdots$
left-handed helix	$\cdots\cdots g+$	$g+$	$g+$	t	t	$\cdots\cdots$
right-handed helix	$\cdots\cdots t$	t	t	$g-$	$g-$	$\cdots\cdots$

Note, that the sequence of internal rotation angles $\cdots [(g+, g+)(t, t)] \cdots$ (or, of course, $[(g-, g-)(t, t)]$ for the chain having mirror image configuration with respect to that shown above)[26] can be seen also in terms of the sequence $\cdots [(g+, t)(t, g+)] \cdots$, i.e. with an alternation of monomeric units left-handed and right-handed spiralled according to the internal rotation angles found in the isotactic polypropylene. The fact that the experimentally found[33] conformation in the crystalline state of syndiotactic polypropylene corresponds to helices instead of zig-zag planar main chain conformations, was predicted by Corradini, Natta and Ganis on the basis of conformational analysis.[129,130] However, the existence of a crystalline modification of syndiotactic polypropylene having planar *trans*-chain conformation has been reported.[131] Recently, the crystal structure of syndiotactic polypropylene has been reinvestigated by Corradini, Natta and coworkers,[132] who have confirmed the above data. It is of some interest to consider the reason for the different conformation in the crystalline state of syndiotactic polypropylene and poly-1,2-butadiene. Borisova and Birshtein[74] calculated that the side groups of syndiotactic poly-1,2-butadiene are weakly interacting with each other and with the H atoms of the main chain, so that a fully *trans*-conformation of the chain bonds is allowed. In the case of syndiotactic polypropylene, the helical conformation is favoured, as the energy of the sequences $(t, t)(g+, g+)(t, t)$ is about 1 kcal/mol lower than that of the fully *trans*-sequence of bonds $(t, t)(t, t)(t, t)$.[73,74] Syndiotactic polyacetaldehyde also does not crystallize in a planar zig-zag conformation, but with an alternation of *trans*- and partially eclipsed conformation of bonds.[133] Diisotactic polymers[33,134,135] assume helical conformations in the solid state which are of the type 3_1, 4_1, 7_2 or 10_3. Erythro- and threo-isomers assume, however, different conformations. For instance, in the case of 1-chloro-2-butoxyethylene, the erythro-diisotactic macromolecules assume 4_1 helices, while the threo-isomer crystallizes in the form of 3_1 helices.

Not much is known about the crystalline conformations of polymers of the type $(-CH_2-CR_2-)$ or $(-CH_2-CRR_1-)$. In this case staggered conformations of bonds may lead to high energy interactions.[73]

Polyisobutylene is the most investigated polymer of this type.[83,136-140] According to the latest investigations, the polymer crystallizes in helices of the type 8_3. The energy calculations indicate that the monomeric unit has four non-equivalent energy minima, which energetically differ very little from each other.

The most interesting feature of the conformation in the crystalline state of polyisobutylene is, however, the fact that one of the two monomer unit C—C bond angles is considerably distorted with respect to the ideal tetrahedral value, being close to 123°.[138] This is due to the strong interaction between vicinal $-C(CH_3)_2-$ groups, and it may be taken as an example of the limits of our usual approximation to consider conformational isomerism on the basis of fixed configurational parameters (bond angles, bond lengths etc.); in this case, in fact, the macromolecule assesses itself in the energy minima not only by rotational isomerism around C—C bonds, but also modifying the same bond angles.[138]

The discussion of the crystal structures of polyamides and polyesters is out of the scope of this paper. It is just worth mentioning here that the conformation in the crystalline state (and in solution) of those compounds is strongly determined by the planarity of the amide or ester group. Poly-(S)-lactic acid can be reported as an example of a polyester which is crystalline and assumes a helical conformation in the solid state.[141] A case in which the presence of the planar amide group determines the actual conformation in the crystalline state is also given by poly-n-butyl-isocyanate,[142] which assumes a 8_3 helical conformation.

The latest book by Flory[12] gives a useful description of the features which govern the chain conformations in the case of polyesters and polyamides. This author also discusses the case of polyphosphates and polydimethylsiloxane.

For a discussion of the conformation in the crystalline state of polymers having double bonds in the main chain, the reader is referred to the recent review by Corradini.[26] This author also discusses in this paper the relationship between chain conformation and the mode of packing in the crystal lattice, and some phenomena of isomorphism and polyisomorphism arising in the solid state in the case of some stereo-regular polymers.

15.5 STATISTICAL APPROACH TO THE INVESTIGATION OF THE CONFORMATIONAL ISOMERISM OF MACRO-MOLECULES IN SOLUTION

From the data reported in the preceding section, we can conclude that stereo-regular polymers preferentially assume helical conformations in the crystalline state. We can also conclude, that intermolecular interactions, for instance those arising during the packing of molecules in the crystal lattice, do not play any important role in determining the chain conformation, otherwise the chain conformation in the crystalline state would be dependent upon the size and nature of the lateral group and not upon the short range interaction between side chain and main chain.[26] This means that the conformation in the crystalline state is the one having the minimum of intramolecular energy.[12,26,73,121] This statement is of great importance, as it allows us to make the further assumption, that the most probable conformations for the monomeric units of the chain in solution are those corresponding to those found in the crystalline state. Stretching this approximation a little further, we can assume that only such conformations are allowed in an ideal solution.[73] This assumption refers, however, only to the main chain bond conformations. A larger number of conformations in solution may, in fact, be allowed for the monomeric unit in the case of polymers having two or more carbon atoms in the lateral chains. This effect is irrelevant for the point we want to stress here, and from now on in this section, when speaking of monomeric unit conformation, we will refer only to the conformation of the main chain bonds. In this case, for instance, of isotactic vinyl polymers, we will consider in solution only $(tg+)$ and $(tg-)$ conformations for the monomeric unit, as those are the only two allowed for the monomeric unit in the crystalline state. However, the fact that only these two conformations are allowed in solution for the monomeric unit does not mean that the macromolecule in solution must be seen as a 'rigid' helix, as in the solid state. In fact, in solution we have to consider that bonds can rotate with respect to each other, because of the relatively low energy barriers hindering bond rotation.[12,15,73] If all the bonds of the chain can rotate from one minimum of energy to another one, and if, for instance in the case of isotactic vinyl polymers, only $(tg-)$ and $(tg+)$ conformations are allowed for the monomeric units, we are led to conclude that both senses of spiralization must be present in the same macromolecule in solution. This can then be visualized as an alternation of right-handed and left-handed spiralled sections, continuously changing their length and continuously inter-converting due to the very fast rotation of the main chain bonds. The macromolecule in solution is therefore a 'random-coil' in which the chain segments possess a regular spiralled conformation. Macromolecules in solution can be characterized by 'average' conformational properties, as, for instance, the average length of right-handed and left-handed sections, and the average number of conformational reversals (junctions between sections spiralled in opposite screw sense).

The conformational reversals, for instance $\cdots tg - tg+ \cdots$ or $\cdots tg - g + t \cdots$ or $\cdots g - t\ tg+ \cdots$ etc, correspond to local deviations from the regular helical main chain conformation, and hence they produce in general a local enhancement of energy.[73,143,144] Indeed, in the case of isotactic vinyl polymers, for instance, the average energy excess, ΔU, associated with a conformational reversal is in the range of 600–2000 cal/mol.[73,78,143,144] However, this increase of energy is in general compensated by entropic effects; since the conformational reversals can fluctuate along the chain, each conformational reversal provides each chain with a 'stabilization energy' of the order of $RT \ln (n)$, if n is the polymerization degree. These considerations on conformational reversals also suggest, that in order to express statistically the conformational properties of the macromolecules in solution, one has to consider the mutual influence of each monomeric unit on the vicinal ones. The chain molecule in solution is indeed a monodimensional co-operative system[12,73,143,144] where the term co-operative indicates the mutual interdependence of the conformation of neighbouring monomeric units.

The parameter that analytically expresses the extent of co-operativity is ΔU. The larger ΔU, the larger will be the tendency for an isotactic chain to exist in long spiralled sequences in which the monomeric units have the same conformation; the lower ΔU, the greater will be the tendency to produce conformational reversals, and the lower will be the value of the average length of the spiralled sections. Statistical treatments that express analytically the conformational properties of isotactic macromolecules in solution as a function of the co-operativity parameter ΔU have been the subject of many investigations.[12,73,143,144,146–150] The end-to-end distance, the dipole moment and the average length of the spiralled sections, have, in fact, been expressed in terms of ΔU for a number of polymer structures.[73] For some typical isotactic poly-α-olefins, ΔU has been numerically evaluated through conformational partition function of pairs of contiguous monomeric units. In this way, it has been possible to obtain a rough evaluation of the conformational properties of the macromolecules in solution, and how those properties are affected by the monomeric unit structure.[78] The calculated characteristic ratio $(\bar{h}^2/Nl^2)$* for typical isotactic poly-α-olefins in a solution is in the range 10–20; the regularly spiralled sections consist on average of 5–10 monomeric units.

Table 15.4 reports the calculated conformational properties of some typical isotactic poly-α-olefins. By increasing the size of the lateral chain, the stiffness of the macromolecule generally increases, i.e. increases the characteristic ratio $\bar{h}^2/Nl^2$ and the average length of the regularly spiralled sections. When a branching is introduced in the α or β position of the lateral chain, ΔU generally increases, with a corresponding further increase of the stiffness of the chain in solution. Consider, however, that only a rough qualitative meaning can be attributed to these figures, given the uncertainty of reliability of the approximations involved in the calculations. The limit and validity of such an approach, as well as the critical comparison with the available experimental data, are presented elsewhere.[78]

In the case of syndiotactic polymers having conformations in the crystalline state of the kind $(\psi_1, \psi_1; \psi_2, \psi_2)$, each monomeric unit in the crystalline state has a fixed conformation, either (ψ_1, ψ_1) or (ψ_2, ψ_2). In solution, however, each monomeric unit can rapidly interconvert between the two states, and so, for the syndiotactic macromolecules in solution, we have basically the same model previously outlined in the case of isotactic chains. In this case, however, together with the co-operativity parameter ΔU, we have to consider the energy difference between the two conformational states (ψ_1, ψ_1) and

---

* $\bar{h}^2$ is the actual quadratic end-to-end distance, N is the number of main chain bonds and l the C—C bond length. Since Nl^2 is the end-to-end distance of the 'freely jointed' chain,[12] the characteristic ratio gives a measure of the flexibility of the macromolecule in solution. The closer $\bar{h}^2/Nl^2$ is to one, the higher is the chain flexibility.

(ψ_2, ψ_2). More details on this point can be found in the book by Birshtein and Ptitsyn.[73]

In most of the statistical treatments of stereo-regular macromolecules in solution, the chains are generally supposed to be ideally stereo-regular and of infinite length. The investigations of the conformational properties of the macromolecules of isotactic vinyl polymers in solution having configurational

Table 15.4 Conformational properties of some isotactic poly-α-olefins in solution[a]

Polymer	ΔU	ν	h^2/Nl^2 [b,c]	
			calculated	experimental
Polypropylene	500	3·2	6·0	5·7 at 145 °C[152,d]
			4·2 at 400 K	
Poly-1-butene	820	5·0	10·5	9·4 at 45 °C[154]
Poly-1-pentene	860	5·2	11·0	10 at 31·5 °C[156]
				9 at 62·4 °C[153]
Poly-3-methyl-1-butene	1500	13·0	30·0	—
Poly-4-methyl-1-pentene	1200	8·0	19·0	—
Poly-5-methyl-1-hexene	850	5·0	11·0	—

a. Calculated at 300 K, unless otherwise specified, with $\Delta E_1 = 500$ cal/mol, $\Delta E_2 = 1500$ cal/mol, $\Delta E_3 = 2500$ cal/mol. Note that the temperature of the experimental data is not the same as the calculated values. We did not consider it worthwhile carrying out calculations at each particular temperature, given the rough qualitative meaning that we wish to attribute to the calculations.

b. h^2 indicates in this work the unperturbed mean square end-to-end distance. Other authors use the symbols $\langle R_0^2 \rangle$ or $\langle r_0^2 \rangle$, which we want to avoid in order not to produce confusion with the gyration radius, indicated with $\langle \bar{r} \rangle$ by many authors. Note also that in some other authors' papers the characteristic ratio is given in terms of the poly-merization degree instead of the number of C—C bonds in the main chain.

c. The experimental values reported in this Table have been taken from the book by Flory[12] and from the review by Crescenzi.[195] For more details about the experimental and theoretical problems connected with average dimension problems in polymers see the review by Kurata and Stockmayer.[151]

d. A value ranging between 4·6 and 5·8 has been obtained by Nakajima and Saijyo[145] in the temperature range of 125–183 °C.

irregularities in the chain have been carried out recently by Flory and his group.[14] Also, statistical treatments that depart from the simplification of infinite chain length have been presented.[12,157]

So far, we have considered only stereo-regular macromolecules, as this is the case that mostly attracted the attention of researchers in the field. However, the same kind of statistical approach can be applied to many other kinds of polymers in solution. Polyethylene is the case most intensely investi-gated.[13,75,157,158] Recently, the cases of polytetrafluoroethylene,[159,160] poly-oxymethylene,[161] polyoxyethylene,[162] polydimethylsiloxane,[163] polyisoprene,[164] polyamides and polyesters[165,166] have also been investigated. A critical discussion of the conformational equilibria proposed for all these macro-molecules in solution cannot be given in this paper. The reader is referred to the latest book by Flory.[12]

15.6 EXPERIMENTAL INVESTIGATIONS ON THE CONFORMATION OF MACROMOLECULES IN SOLUTION

There are in principle two ways to look at the experimental investigation of the conformation of macromolecules in solution. On the one hand, the experiments should give confirmation of the model previously outlined, which has been obtained throughout statistical treatments. For instance, in the case of stereo-regular vinyl polymers, the experiments should provide evidence of the presence of the rotational isomerism and of the short range helical order in the macromolecules in solution.

On the other hand, the experiments should gather information on the influence of the conformational equilibria on the physical properties. How-ever, these two levels of investigation in practice always overlap and a clear classification on the available experimental data according to such a criterion is not possible.

The existence of the rotational isomerism in macromolecules is confirmed by a number of spectroscopic data both in the state of solution and in the bulk. In the case of polyethylene, it has been demonstrated by the temperature dependence of IR absorption bands.[167] Mizushima, Shimanouchi and co-workers have been able to detect the rotational isomerism in polyvinylchloride and polyisoprene hydrochloride,[168-170] and Myake detected it in the case of polyethylene terephthalate[171] and polyethyleneglycol.[172] A detailed discussion of the IR spectral bands in relation to the configuration and conformation of stereo-regular polymers can be found in the papers by Zerbi and coworkers,[39,40] in the review by Miyazawa,[43] and in the books by Elliott[44] and Hummel.[45]

Also viscometric methods have been employed to obtain evidence of the chain flexibility of macromolecules in solution.[173] More recently, the deter-mination of intramolecular motion in macromolecules has been attempted by polarized luminescence.[174-180] By this method, the problem of the segmental motion of macromolecules in solution can be examined kinetically, as frequencies of bond rotation in the range $10^7–10^{10}$ s^{-1} can be directly deter-mined. Acrylic and methacrylic polymers with different tacticities and polyethers, have been particularly investigated by fluorescence,[175-177] and it is indeed a shared belief that this technique will become of increasing utility to study the conformation of polymers in solution.

In the first part of this paper we briefly discussed the use of high resolution NMR spectroscopy to investigate the configuration of polymers. In principle, this technique could be useful to obtain information on the conformation of the chain as well, since the nuclei in different environments (as we have in the different rotational isomers) tend to produce different chemical shifts.[24] At room temperature, however, the rapid interconversion between rotamers may cause complete averaging of the signals deriving from the different conformers. In order to obtain information on the conformational equilibria by NMR spectroscopy, one has to operate at low temperature, or to measure the average spectral parameters at different temperatures. Using this second technique, several low molecular weight model compounds of vinyl polymers, and in particular the various stereoisomers of 2,4-disubstituted pentanes and 2,4,6-tri-substituted heptanes, have been investigated.[181–185]

Lately, NMR spectroscopy has been utilized to investigate the dynamics of chain segments, as the relaxation time of protons is particularly sensitive to local motion. In particular, the nuclear magnetic relaxation of protons in polyethyleneoxide has been interpreted by Liu and Ullmann[186] in terms of jumps between rotameric states of unequal statistical weight. Herman and Weill reinvestigated the case of polyethyleneoxide, extending the study to polymethyleneoxide and polydioxalane.[187]

IR spectroscopic studies have been useful to detect the existence of the short range helical order in the main chain bonds of synthetic macromolecules in solution. Volchek,[188] for instance, demonstrated for polypropylene that the IR band at 974 cm^{-1}, characteristic of isotactic samples and due to intramolecular short range interaction, is present both in the solid state and in solution, though with different intensity. Analogous data have been collected for isotactic polystyrene,[114,189] annealed poly-p-deuterostyrene,[160] melted poly-p-deuterostyrene,[160] and melted poly-p-fluorostyrene.[114,189]

An indication of the presence of helical conformation in the case of molten isotactic propylene comes from the work of Zerbi and coworkers.[40] The data presented in this paper provide a better understanding of the IR spectrum of liquid isotactic polypropylene and provide experimental evidence that segments of isotactic polypropylene which are still helical, exist in the molten state and consist of at least five monomer units.

The presence of helical conformations of the main chain in stereo-regular macromolecules in solution has also been demonstrated by Pino and coworkers using isotactic vinyl polymers having asymmetric carbon atoms in the lateral chains.[22,54,190–192] As will be discussed in more detail in the next section, the experimental value of optical activity for a number of different polymer structures can, in fact, be interpreted only by assuming helical conformations in solution. The study of the temperature coefficient of optical activity also provided evidence that the bond rotation accompanying optical activity changes with temperature must be fast, at least compared with the time necessary for optical activity measurements.

Further support for the existence of highly spiralled conformations of the main chain of isotactic vinyl macromolecules comes from the study of optical activity of copolymers obtained from a co-monomer mixture of optically active and non-dissymmetric vinyl monomers, for instance (S)-4-methyl-1-hexene and 4-methyl-1-pentene.[193] The results indicate that the 4-methyl-pentene units are also contributing to the optical activity of the polymer in solution. This can be explained assuming that the (S)-4-methyl-1-hexene monomeric units, which tend to assess themselves preferentially in left-handed spiralled sections,[22,24,190-192] induce the same kind of preferential spiralization in the 4-methyl-1-pentene units, present in the same chain. In the case of copolymers of (R)-3,7-dimethyl-1-octene with styrene[194] more direct evidence of this induced dissymmetry has been given. In fact, in this case there is a negative Cotton effect at about 260 nm, due to the phenyl group, which clearly demonstrates that the styrene units become dissymmetric in the copolymer chains; this dissymmetry is indeed easily explained on the basis of the preferential right-handed screw sense spiralization[54] of the sections of (R)-3,7-dimethyl-1-octene units.

Concerning the influence of conformation on the physical properties, much emphasis has been given to the investigation of the end-to-end distance and dipole moment of polymer chains in solution. In particular, since the statistical calculations predict that the conformational equilibria of isotactic and syndiotactic vinyl polymers in solution should be different,[73] much attention has been paid to the comparison of the average dimensions in solution of stereo-regular vinyl polymers of different tacticity.

We will not dwell upon the experimental methods to determine the average dimensions. For this, the reader is referred, for instance, to the useful book by Morawetz[20] and to the recent review by Crescenzi.[195]

The results of these experiments indicate that $\bar{h}^2/Nl^2$ is somewhat larger in the case of isotactic polymers than in the case of atactic polymers. Polystyrene,[196] poly-1-butene[154] polymethacrylates[197] are the cases usually quoted (see Table 15.5). The differences are, however, only slightly larger than the average error in the measurements. Also the differences in the end-to-end distance of polymers having different structures are rather small (as shown in Table 15.5), and the general conclusion can be drawn that the characteristic ratio is not a very sensitive tool for the investigation of conformational equilibria of macromolecules in solution.

The dipole moment characteristic ratio $\bar{\mu}^2/\mu_0^2$, where $\bar{\mu}^2$ is the quadratic dipole moment per monomeric unit of the vinyl polymer and μ_0 the dipole moment of the hydrogenated monomer, is not more sensitive than $\bar{h}^2/Nl^2$ to the conformational equilibria of chains in solution. It is commonly accepted that $\bar{\mu}^2/\mu_0^2$ of isotactic samples is higher than that of the atactic ones, at least in the case of polystyrene,[205] polymethylmethacrylate,[206,207] poly-tertbutylmethacrylate[208] and polyvinylisobutylether.[209] However, the differences are not larger than 10–15 per cent (see Table 15.6). The dipole moments of syndiotactic samples seem to be identical within the experimental

Table 15.5   Characteristic ratio ($\overline{h^2}/Nl^2$) for some stereo-
regular macromolecules in solution[a]

Polymer	$\overline{h^2}/Nl^2$	References
Polypropylene		
isotactic	5·2, 7·4	151, 152
atactic	7·0	198
Poly-1-butene		
isotactic	9·4	154, 153
atactic	7·2	199
Polystyrene		
isotactic	10·6	151
atactic	10·0	200
Polymethylmethacrylate		
isotactic	9·6	201, 202
syndiotactic	7·4	202
atactic	7·5, 11·5	201, 200
Polyisopropylacrylate		
isotactic	9·7, 15·2	203, 204
syndiotactic	7·2	203
atactic	7·1, 14·8	203, 204

a. See footnote Table 15.6.

error to those of atactic samples.[206,207,210] Also the dipole moment of isotactic
optically active vinyl ethers do not show any remarkable difference with
respect to the non-optically active polymers of similar structure,[211] though the
chain conformation equilibria are supposed to be remarkably different in
the two cases.[22,190]

Recently, much attention has been given to the temperature dependence
of the average dimensions, both from a theoretical and experimental point of
view. The cases of poly-1-butene, poly-1-pentene,[14] polyisobutylene,[13,212-215]
polyethylene[158,215-217] and polydimethylsiloxane[218] have been particularly
investigated.

Along with spectroscopy and average dimension studies, polarizability,
that is the tendency of the electron cloud of a molecule to become displaced
under the influence of an electric field, can be correlated to the conformation of the macromolecules in solution. Volkenstein, Gotlib and Tsvet-
kow[15,76,220-222] gave a valid contribution to this field, developing analytical
expressions for the anisotropic polarizability of different polymer chains in
solution. Also, the low angle X-ray scattering technique has been widely
employed for the study of conformation of synthetic polymers and polypeptides.[223-228] In the case of polymethylmethacrylate in benzene solution,
the differences in the X-ray patterns between isotactic and syndiotactic
samples have been ascribed to the differences in conformational equilibria
in the two cases.[223] For polymethylmethacrylate in dilute solution, there is

Table 15.6 Dipole moment of some stereo-regular macromolecules in solution[a]

Polymer	$\bar{\mu}^2/\mu_0^2$	References
Polyisobutylvinylether		
isotactic	0·90, 0·66, 0·67	209, 219, 211
atactic	0·77, 0·65, 0·70	
Polyisopropylvinylether		
isotactic	0·39	211
atactic	0·61	
Poly-(S)-1-methylpropylvinylether		
isotactic	0·52	211
atactic	0·56	
Poly-(S)-2-methylbutylvinylether		
isotactic	0·73	211
atactic	0·76	
Polymethylmethacrylate		
isotactic	0·76, 0·65	
syndiotactic	0·70, 0·67	207
atactic	0·66, 0·70	
Polystyrene		
isotactic	0·53	205
atactic	0·36	

a. Given the uncertainty of the determination of the stereo-regularity in all these polymer samples, the terms isotactic and atactic as reported in literature and as here used, must be more properly intended as referring to the most and least stereo-regular fraction, respectively. The lack of an exact measurement of the stereo-regularity index, makes it critical to compare, in terms of average dimensions, samples prepared by different authors, and, or, with different catalytic systems.

another phenomenon that correlates conformation and configuration properties, i.e. the formation of a stereocomplex between isotactic and syndiotactic sections.[224–232]

Particularly interesting are the conformational properties of polyisocyanates. In fact, whereas the chains of stereo-regular vinyl polymers, about which we have been mostly concerned so far, have a high flexibility in solution, a series of physicochemical investigations on poly-n-butyl-isocyanate[233–236] and poly-n-hexyl-isocyanate[237] seem to indicate that these polymers may have a much more rigid chain in solution. This is, in principle, not surprising, since in this case we have amide bonds in the main chain, and the rotation around them may be forbidden, at room temperature at least, as in the case of some polypeptides. For more details about the properties of polyisocyanates, see the recent papers by Goodman and coworkers.[82,238]

A physical property that has also been widely investigated in terms of the conformational equilibria in the chains, is the polymer stretching in bulk. Clearly, stretching depends upon the chain flexibility, i.e. upon the tendency of the chain to undergo conformational transitions under the application of an external force. This problem is discussed in the already mentioned book by Birshtein and Ptitsyn[73] both from the theoretical and experimental point of view. The presence of the rotational isomerism under stretching has been detected by Russian authors and by Mizushima and coworkers in the case of several structures, with the use of polarized light.[169,188,239,240] Mention should also be made of the importance that particular conformational equilibria have on physical properties of polyelectrolytes in solution. This point is reviewed by Crescenzi.[195]

The relationship between configuration and conformation in relation to the electric properties of polymeric material in the solid state has been extensively discussed by Boguslavskii and Vannikov.[241]

15.7   OPTICAL ACTIVITY AND CONFORMATION OF MACROMOLECULES IN SOLUTION

Optical activity is probably the most sensitive physical property to the conformational equilibrium of a molecule. As such, and by virtue of the fact that optical activity measurements can usually be carried out easily and with a high degree of reproducibility, this technique has received much attention both in the case of polymers and low molecular weight compounds.

In particular, the different conformations of an asymmetric molecule will, in general, correspond to different perturbations of the dissymmetric chromophor, so that the average molar optical rotation $[\phi]_\lambda^T$ will be given by

$$[\phi]_\lambda^T = \frac{\sum \phi_i p_i}{\sum p_i}$$

where ϕ_i and p_i are respectively the molar rotation and the relative statistical weight of the i-th conformer at the wavelength λ and temperature T. Once the ϕ_i values are known by some independent method, the previous equation could be used, at least in some simple cases, to determine the energy parameters p_i (practically, the conformational equilibrium of the molecule) through the experimental values of $[\phi]_\lambda^T$. Unfortunately, exact determination of the terms ϕ_i is not possible and this represents nowadays the most serious difficulty in utilizing optical activity for conformational analysis studies. At the present time, in fact, we have to use semi-empirical methods to calculate the factors ϕ_i, and the information gathered on the conformation equilibria is consequently only qualitative.

As an example of a procedure which has been used to obtain information on the conformational equilibria through optical activity studies, Table 15.7 shows the comparison between $[\phi]_D^T$, per mole of monomeric unit, of some isotactic optically active polymers, and the $[\phi]_D^T$ of low molecular weight structural models. In these cases, the enhancement of $[\phi]_D^T$ found in polymers with respect to the low molecular weight models has been explained assuming

Table 15.7 Molar optical rotation of the most stereo-regular fractions of some vinyl polymers, and of some low-molecular weight model compounds

Polymers[a]			Models[a]			
Monomeric unit structure	$[\phi]_D^{25c}$	λ_0	Structure	n	$[\phi]_D^{25}$	λ_0
CH_3—C—$(CH_2)_n$—CH ; C_2H_5 ; CH_2	+160	167[b]	CH_3 ; H—C—$(CH_2)_n$—CH ; C_2H_5 ; CH_3	0	−11.4	176[b]
	+290	165[b]		1	+21.3	170[b]
	+68	169[b]		2	+11.7	—
CH_3—C—$(CH_2)_n$—O—CH ; C_2H_5 ; CH_2	+312	~190[d]	CH_3 ; H—C—$(CH_2)_n$—O—CH_2 ; C_2H_5 ; CH_3	0	+34.5	~190[d]
	+6.5			1	+1.1	
CH_3—C(O=C)—$(CH_2)_n$—CH ; C_2H_5 ; CH_2	−118	292[e]	CH_3 ; H—C—$(CH_2)_n$—C(O=C)—CH_2 ; C_2H_5 ; CH_3	0	+34.8	283[e]
	−43	290[e]		1	+11.5	282[e]

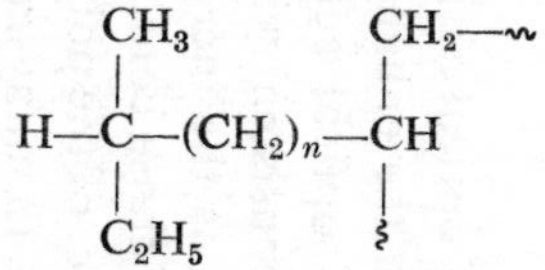

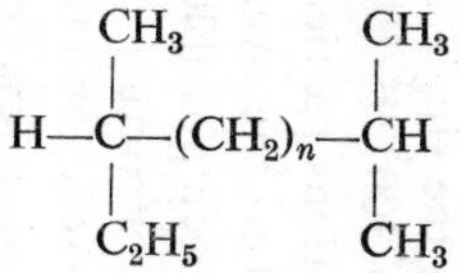

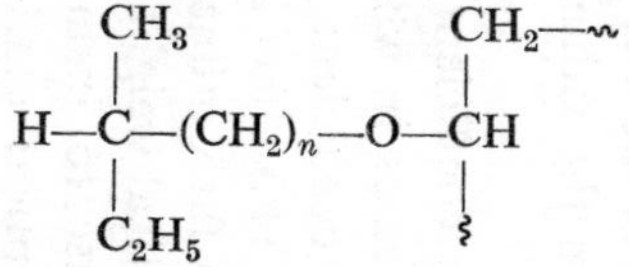

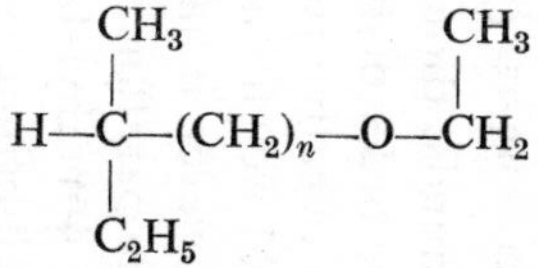

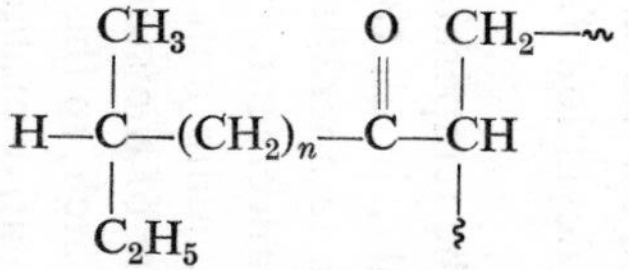

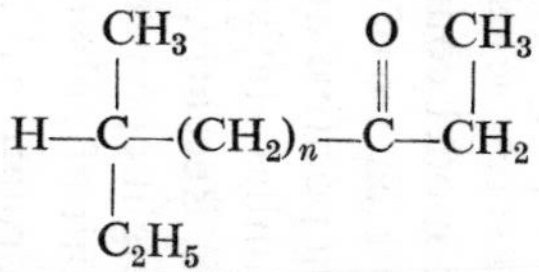

a. For more details on the models and polymers, see references 22, 26, 58, 192.
b. Wavelength corresponding to the longest wavelength optically active transition, evaluated from the one term Drude equation, $[\phi]_\lambda^T = K/[\lambda^2 - \lambda_0^2]$.
c. Referred to one monomeric unit, in aromatic hydrocarbon solution.
d. Wavelength corresponding to the nearest UV absorption maximum.
e. Wavelength corresponding to the nearest optically active electronic transition, evaluated from the experimental ORD curves.

that, in the monomeric unit, the conformational equilibrium is displaced towards a few conformations, which happen to have high optical rotation of the same sign.[22,54,190–192,242]

In particular, in the case of isotactic optically active poly-α-olefins, using the empirical method of Brewster[243] to predict the 'allowed' conformations and their molar rotation, it has been possible to calculate the average molar rotation of the monomeric unit included in a left-handed or in a right-handed spiralled section.[54] As reported in Table 15.8, the experimental molar rotation is close to the molar rotation calculated for the monomeric unit included in the left-handed screw sense, and the conclusion was drawn, that the macromolecules of those poly-α-olefins in solution are overwhelmingly spiralled in the left-handed screw sense.[22,54,190–192] On the contrary, in the case of low molecular weight model compounds, though the single conformations have molar rotations which are of the order of $100–300°$,[243,244] many allowed conformations having positive and negative rotations participate in the equilibrium with very similar statistical weight, and so the average $[\phi]_D^T$ is a relatively low number. Recently, however, it has been possible to synthesize some low molecular weight open chain paraffins, the structure of which is similar to the monomeric unit of poly-(S)-4-methyl-1-hexene, and for which a displacement of the conformational equilibrium towards a few conformations having high rotation of the same sign had been predicted[245] by the empirical method of Brewster.[243] The experimental $[\phi]_D^T$ values in this case have been found to be in the range $100–200$,[245–247] i.e. of the order of that found in polymers listed in Table 15.7. These data confirm that the enhancement of $[\phi]_D^T$ found in polymers is due to conformational effects.

Indeed, there is, in principle, another way to explain the enhancement of optical activity found in polymers. That is, to invoke, instead of conformational effects, some kind of interaction between the chromophores in the chain, so that the optically active electronic transition takes place at a different wavelength in polymers than in low molecular weight compounds. In the case of the polymers reported in Table 15.7, this possibility has been, however, excluded on the basis of the fact that λ_0 in polymers does not show any appreciable deviation from the values found in low molecular weight compounds.

The study of the temperature coefficient of optical activity is also of some interest. Generally, in all the cases in which the enhancement of $[\phi]_D$ has been found, the temperature coefficient of optical activity also shows a considerable enhancement.[22,54] This effect is consistent with the feature of the conformational equilibrium so far outlined, as the increase of temperature causes an increase of the statistical weight of those conformations which are less stable and which, in the case, for instance, of the polymers of Table 15.8, have high and negative molar optical rotation. In many cases no enhancement of optical activity has been found in polymers with respect to low molecular

Table 15.8 Conformational properties in solution of some isotactic optically active poly-α-olefins having the structure

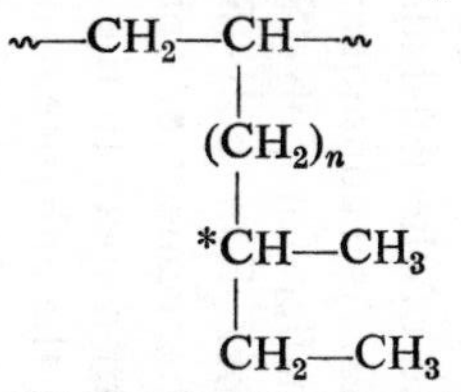

$$\text{---CH}_2\text{---CH---} \quad (\text{CH}_2)_n \quad *\text{CH---CH}_3 \quad \text{CH}_2\text{---CH}_3$$

Polymer	Number of allowed conformations[a]		Rough evaluation of ΔE[b] cal/mol	Calculation of the monomeric unit molar rotation			$[\phi]_D$ (exp)[d]
	n_l	n_d		$M_1°$	$M_d°$	Average[a]	
$n = 0$ Poly-(S)-3-methyl-1-pentene	2	1	~400	+180	−240	+40	+160
$n = 1$ Poly-(S)-4-methyl-1-hexene	2	1	~400	+240	−300	+60	+280
$n = 2$ Poly-(S)-5-methyl-1-heptene	5	4	~100	+228	−225	+22	+68
$n = 3$ Poly-(S)-6-methyl-1-octene	11	10	~50	+240	−192	+34	+20

a. Evaluated according to Brewster;[243] n_l and n_d are respectively the number of conformations allowed to a monomeric unit inserted in a left-handed and a right-handed spiralled section.

b. Assuming, in first approximation, that $\Delta E = RT \ln n_l/n_d$, at 300 °K. A more precise evaluation of ΔE, considering the actual conformational partition functions of the monomeric units left-handed and right-handed spiralled, has been calculated elsewhere,[78] but it is not essential here.

c. M_l and M_d are the molar rotation, calculated according to Brewster, of the left-handed and right-handed spiralled monomeric unit, respectively. The average is calculated, according to Brewster, assuming that the allowed conformations have the same statistical weight.

d. In aromatic hydrocarbon solution, referred to one monomeric unit, at 25 °C. For more details about the experimental values, see reference 54.

weight model compounds.[22] This is not surprising, since the enhancement of optical activity of polymers with respect to low molecular weight compounds is by no means a general rule but it can take place only with particular monomer unit structures. This point can be better understood in the examples of poly-α-olefins reported in Table 15.8, considering the number of conformations allowed to the monomeric unit included in left-handed or right-handed spiralled sequences respectively, in the case of the homologous poly-α-olefins. The difference in the number of allowed conformations in the two cases derives from the different conformational rigidity of the lateral chain.[22,54] Such a determination of n_l and n_d allows us to calculate the energy difference ΔE, per mole of monomeric unit, between the two main chain regularly spiralled conformations assuming in first approximation that $\Delta E = RT \ln n_l/n_d$.[73]

The ΔE value so obtained is about 400 cal/mol in the case of poly-(S)-4-methyl-1-hexene and poly-(S)-3-methyl-1-pentene, and much smaller in the case of poly-(S)-5-methyl-1-heptene. Since the higher ΔE the higher is the tendency of the macromolecule to spiralize in the more favoured screw sense,[248] it is clear that in the last case the prevalence of the left-handed screw sense is much smaller than in the two previous cases. As a consequence, the absolute value of the optical activity is much smaller for poly-(S)-5-methyl-1-heptene than for poly-(S)-4-methyl-1-hexene. For poly-(S)-6-methyl-1-octene, for which ΔE is close to zero, no enhancement of optical activity has been found in the polymer in solution with respect to the low molecular weight structural models.[54] In general, then, the enhancement of optical activity in isotactic vinyl polymers in solution is possible when one of the two screw senses is largely prevailing in solution, and when the monomeric units, belonging to this prevailing screw sense, have high molar rotations of the same sign. In turn, the prevalence of one sense of spirallization is possible when there is a sizeable energy difference between the two regularly spiralled conformations.

The conformational properties of optically active isotactic vinyl polymers, so far outlined in qualitative terms, have received much attention from a statistical mechanical point of view.[73,76,144,248,249] The statistical treatment most commonly considered is in terms of two energy parameters; ΔE, previously defined, and ΔU, the energy excess due to conformational reversals, defined in the previous section. All the conformational properties of the macromolecules in solution, such as the end-to-end distance, the average length of regularly spiralled sequences, the probability of conformational reversals and the molar fraction of monomeric units both left-handed and right-handed spiralled, can be expressed analytically in terms of ΔE and ΔU.[73,248]

The most characteristic general property of optically active vinyl polymers in solution, as obtained through this theoretical approach, is expressed in

Figure 15.1, which gives the molar fraction of monomeric units spiralled in the more favoured screw sense as a function of ΔE and ΔU. For ΔU in the usual range of 500–1500 cal/mol,[73,145] ΔE values as small as 200 cal/mol are sufficient to induce a chain spiralization in the most favourite sense of the order of 80–90 per cent at room temperature.[145,243]

This peculiar conformational feature, deriving from the large degree of co-operativity present in these macromolecules, gives support from a thermodynamical point of view to the hypothesis that the enhancement of optical

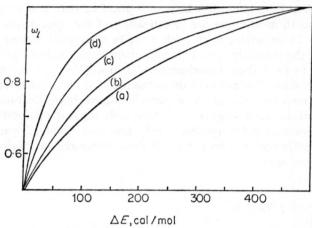

Figure 15.1 Dependence of the molar fraction ω_1 of monomeric units spiralled in the more favoured screw sense, on the energy parameters ΔE and ΔU. Curve (a) $\Delta U = 500$; curve (b) $\Delta U = 1000$; curve (c) $\Delta U = 1500$; curve (d) $\Delta U = 2000$ cal/mol

activity found in some polymers is due to conformational factors alone. Recently, utilizing this relatively simple model in terms of ΔE and ΔU, it has been possible to explain semiquantitatively some of the properties of optically active isotactic vinyl polymers in solution, e.g. the temperature coefficient of optical activity,[145,242] the dependence of $[\phi]_D^T$ on molecular weight[250] and the conformational equilibrium in copolymers obtained from mixtures of enantiomeric α-olefins at various degrees optical purity.[11] Such a statistical treatment in terms of ΔE and ΔU allows predictions to be made of some physical properties of the macromolecules in solution. For instance, it has been calculated that the end-to-end distance of some optically active isotactic poly-α-olefins is remarkably larger than that of non-optically active vinyl polymers having analogous structure.[145,242] Also, it has been possible to investigate the temperature dependence of some characteristic conformational properties, like the length of the regularly spiralled sections, or the molar fraction of monomeric unit spiralled in each of the two main chain screw senses.[145,242] For instance, according to these calculations, in the case of poly-(S)-4-methyl-1-hexene, the macromolecules at room temperature are composed by alternated left-handed and right-handed sections containing on average about 35 and 2 monomeric units respectively, values which become 11 and 2 respectively at 400 K. The percentage of spiralization in the left-handed screw sense, which is about 95 at 300 K, is reduced to about 85 at 400 K, whereas in the same temperature range the characteristic ratio $\bar{h}^2/Nl^2$ goes from 55 to 15.[145,242] The limits of such a model, and the corresponding reliability of these figures, are discussed elsewhere.[78]

Concerning the relationship between optical activity and conformation, it is also worth mentioning the investigation of the optical rotation in the solid state. In principle, from such a study one could obtain the molar rotation of the regularly spiralled main chain conformations present in the solid state. In turn, the comparison of $[\phi]_D^T$ in solution and in the solid state could give a direct indication of the difference of the conformational equilibria in the two cases. So far the experiments have been performed, in the case of synthetic stereo-regular polymers, only for isotactic optically active poly-α-olefins and polyvinylethers.[251,252] The data obtained are consistent with the conformational properties of these stereo-regular macromolecules outlined previously.

15.8   CONCLUSIONS

The determination of the chain configuration in polymers is important from a two-fold view-point. On the one hand, it can provide useful information on the stereo-specificity (and, or, stereo-selectivity) of the polymerization process; on the other hand, the knowledge of the chain configuration of a given polymer sample is the necessary prerequisite to approach in a meaningful way the problem of the relationship between structure and physical properties in polymers.

The experimental determination of the chain configuration is not, in general, an easy matter. Due to the incomplete stereo-specificity of the polymerization process, the polymer samples, even after an exhaustive fractionation, are not homogeneous with respect to stereo-regularity. Even with a very high stereo-regularity degree (for instance 98 per cent), an extremely large number of chain configurational isomers are, in principle, possible.

In the case of polymers, therefore, configurational isomerism is viewed in terms of an average degree of steric order, and this is a basic difference with respect to configurational isomerism problems in classical organic chemistry.

X-ray analysis and NMR spectroscopy are the methods which are usually employed to obtain information on the stereo-regularity type and, or, stereo-regularity degree. These are absolute methods. X-ray analysis is, however, limited to crystallizable samples, and usually it can provide information only on the stereo-regularity type. High resolution NMR spectroscopy does not go beyond the analysis of sequences of four to five monomeric units, and hence, for the present time at least, it leaves unsolved the problem of the long range configurational order. Furthermore, the NMR spectra of polymers containing side chains with two or more —CH_2— or —CH_3 groups, is generally too complicated to be resolved in a reliable way.

IR spectroscopy, solubility properties, melting point, crystallinity, can provide a qualitative criterion of relative stereo-regularity for different samples of the same polymer, but none of these methods is sufficient *per se* to obtain reliable information on a large variety of polymer samples. Optical activity can be of some utility in obtaining information on the stereo-regularity type and degree of polymers obtained from dissymmetric monomers. But no general relationship between optical activity and stereo-regularity can be obtained for all optically active polymers.

In conclusion, though there are many methods to obtain a qualitative indication of the degree of stereo-regularity of a given polymer sample, the problem of the quantitative determination is possible so far for only a very limited number of cases.

As far as conformational problems are concerned, X-ray analysis is the most powerful technique to obtain information on the preferential conformation of the monomeric units of a given macromolecule in the solid state. There are no such sensitive techniques to investigate the conformation of macromolecules in solution, where generally the problem is complicated by the fact that all main chain and side chain bonds can rotate, providing the macromolecule with an extremely large number of chain rotational isomers. The problem of the conformational equilibria of macromolecules in solution is approached with mechanical statistical methods, which express the chain average conformational properties in terms of energy differences between the conformational states of the monomeric unit. The experimental determination of these energy parameters, and eventually the experimental check of the chain conformational properties obtained through them, are, however, very difficult problems. Statistical treatments are available, at the present time, for a large variety of polymers; however the quantitative evaluation of the conformational properties (for instance the average length of sequences of monomeric units in the same conformation, or the average number of conformational reversals, or the molar fraction of monomeric units present in each of the possible conformations) is possible only in a very limited number of cases and only in a rather indirect way. For instance in the case of isotactic poly-α-olefins, combining statistical algebra with the classical conformational analysis valid for low molecular weight paraffins, and assuming as energy parameters those known for low molecular weight structural compounds, a rough quantitative evaluation of the above mentioned conformational properties is possible. A check of the reliability of such a procedure, is given by the fact that the conformational properties of optically active isotactic poly-α-olefins, calculated in this way, are consistent with the data of optical activity and with the temperature coefficient of optical activity, experimentally found for these polymers in solution.

Further progress on the knowledge of conformational equilibria of polymers in solution is to be expected only if reliable information on the conformational analysis of low molecular weight compounds becomes available, and if, at the same time, a better understanding of the polymer-solvent interaction is furnished.

Acknowledgement

P. L. L. expresses his gratitude to the Swiss National Fund and F. C. to C.N.R. (Roma-Italy) for their support of this work.

15.9 REFERENCES

1. W. Klyne and W. Prelog, *Experentia*, **16**, 521, (1960).
2. E. L. Eliel, *Stereochemistry of Carbon Compounds*, McGraw-Hill, 1962.
3. M. Hanack, *Conformation Theory*, Acad. Press, 1965.
4. K. Mislow, *Introduction to Stereochemistry*, Benjamin, 1965.
5. P. Pino, F. Ciardelli, G. P. Lorenzi and G. Natta, *J. Am. Chem. Soc.* **84**, 1487, (1962).
6. P. Pino, G. Montagnoli, F. Ciardelli and E. Benedetti, *Makromol. Chem.*, **93**, 158, (1966).
7. G. Montagnoli, D. Pini, A. Lucherini, F. Ciardelli and P. Pino, *Macromolecules*, **2**, 684, (1969).
8. F. Ciardelli, G. Montagnoli, D. Pini, O. Pieroni, C. Carlini and E. Benedetti, *Makromol. Chem.*, **147**, 53, (1971).
9. E. Chiellini, G. Montagnoli and P. Pino, *Polymer Letters*, **7**, 121, (1969).
10. K. Matsuzaki and N. Tateno, *J. Polymer Sci.*, Part C, **23**, 733, (1968).
11. P. L. Luisi and O. Bonsignori, *J. Phys. Chem.*, **56**, 4298, (1972).
12. P. J. Flory, *Statistical Mechanics of Chain Molecules*, Interscience, 1969.
13. A. Abe, R. L. Jernigan and P. J. Flory, *J. Am. Chem. Soc.*, **88**, 631, (1966).
14. P. J. Flory, J. E. Mark and A. Abe, *J. Am. Chem. Soc.*, **88**, 639, (1966).
15. M. W. Volkenstein, *Configurational Statistics of Polymer Chains*, Interscience, 1963.
16. M. L. Huggins, G. Natta, V. Desreux and H. Mark, *J. Polymer Sci.*, **56**, 153, (1962).
17. M. Farina, M. Peraldo and G. Natta, *Angew. Chem.*, **77**, 149, (1965).
18. G. Natta, *Rend. Accad. Nazl. Lincei*, **4**, 61, (1955).
19. G. Natta and P. Corradini, *J. Polymer Sci.*, **20**, 251, (1956).
20. H. Morawetz, *High Polymers*, Vol. XXI, Wiley, (1965).
21. H. Gilman, *Organic Chemistry. An Advanced Treatise*. Vol. I, p. 235, Wiley, New York, 1953.
22. P. Pino, *Adv. Polym. Sci.*, **4**, 393, (1965).
23. M. Farina and G. Bressan, in *Progress in Stereochemistry*, Vol. 3, p. 181, ed. A. D. Ketley, Dekker, New York, 1968.

24. F. A. Bovey, *Polymer Conformation and Configuration*, Academic Press, New York, 1969.
25. G. Natta, P. Pino and G. Mazzanti, *Chim. Ind. (Milan)*, **37**, 927, (1955).
26. P. Corradini, in *The Stereochemistry of Macromolecules*, Vol. 3, p. 1, 1968.
27. H. W. Coover, Jun., R. L. McConnell, F. B. Joyner, D. F. Sloneker and R. L. Combs, *J. Polymer Sci.* A1, **4**, 2563, (1966).
28. J. Boor, Jun., *Macromol. Rev.*, **2**, 115, (1967).
29. a. C. C. Price, M. Osgan, R. E. Hughes, and C. Shambelan, *J. Am. Chem. Soc.*, **78**, 690, (1956); b. C. C. Price and M. Osgan, *J. Am. Chem. Soc.*, **78**, 4787, (1956).
30. T. Tsuruta, in *Progress in Stereochemistry* Vol. 2, p. 177, ed. A. D. Ketley, Dekker, New York, 1967.
31. P. L. Luisi, G. Montagnoli and M. Zandomeneghi, *Gazz. Chim. Ital.*, **97**, 222, (1967).
32. P. Pino, F. Ciardelli and G. Montagnoli, *J. Polymer Sci.*, Part C, **16**, 3265, (1968).
33. G. Natta, *Makromol. Chemie*, **35**, 93, (1960).
34. H. L. Frisch, C. L. Mellows and F. A. Bovey, *J. Chem. Phys.*, **45**, 1565, (1966).
35. K. C. Ramey and J. Messick, *J. Polymer Sci.*, A4, 155, (1966).
36. F. Heatley and A. Zambelli, *Macromolecules*, **2**, 618, (1969).
37. J. Schaefer, *Macromolecules*, **2**, 533, (1969).
38. G. Zerbi, F. Ciampelli and V. Zamboni, *Chim. Ind. (Milan)*, **46**, 1, (1964).
39. G. Zerbi, F. Ciampelli and V. Zamboni, *J. Polymer Sci.*, Part C, **7**, 141, (1964).
40. G. Zerbi, M. Gussoni and F. Ciampelli, *Spectrochim. Acta*, **23A** (2), 301, (1967).
41. F. Ciardelli, E. Benedetti and P. Pieroni, *Makromol. Chem.*, **103**, 1, (1967).
42. C. E. H. Bawn and A. Ledwith, *Quart. Rev.*, **16**, 427, (1962).
43. T. Miyazawa, in *The Stereochemistry of Macromolecules*, Vol. 3, p. 147, 1968.
44. A. Elliott, *IR Spectra and Structure of Organic Long-Chain Polymers*, Edward Arnold, 1969.
45. D. O. Hummel, IR Spectra of Polymers, *Polymer Reviews*, **1966, 14**,
46. G. Natta, P. Pino, P. Corradini, F. Danusso, E. Mantica, G. Mazzanti and G. Moraglio, *J. Am. Chem. Soc.*, **77**, 1708, (1955).
47. G. Natta, P. Pino and G. Mazzanti, *Polymer Letters*, **2**, 443, (1964).
48. O. Fuchs, *Makromol. Chem.*, **58**, 247, (1962).
49. G. Natta, M. Pegoraro and M. Peraldo, *Ricerca Scientifica (Rome)*, **28**, 1473, (1958).
50. T. Tsuruta, S. Inoue and I. Tsukuma, *Makromol. Chem.*, **84**, 298, (1965).
51. C. C. Price, R. Spector and A. L. Tumolo, *J. Polymer Sci.*, A-1, **5**, 407, (1967).
52. D. Braun and W. Kern, *J. Polymer Sci.*, Part C, **4**, 197, (1964).
53. P. Pino, F. Ciardelli and G. P. Lorenzi, *J. Polymer Sci.*, Part C, **4**, 21, (1964).
54. P. Pino, F. Ciardelli, G. P. Lorenzi and G. Montagnoli, *Makromol. Chem.*, **61**, 207, (1963).
55. K. Matsuzaki, A. Ishida and N. Tateno, *J. Polymer Sci.*, Part C, **16**, 2111, (1967).
56. H. Yuki, K. Ohta, K. Ono and S. Murahashi, *J. Polymer Sci.*, A1, **6**, 829, (1968).
57. G. P. Lorenzi, E. Benedetti and E. Chiellini, *Chim. Ind. (Milan)* **46**, 1474, (1964).

58. O. Pieroni, F. Ciardelli, C. Botteghi, L. Lardicci, P. Salvadori and P. Pino, J. Polymer Sci., Part C, **22**, 993, (1969).
59. P. Pino, F. Ciardelli, G. Montagnoli and O. Pieroni, *Polymer Letters*, **5**, 307, (1967).
60. A. Sharples, *Introduction to Polymer Crystallization*, Edward Arnold, London, 1966.
61. C. W. Bunn and H. S. Peiser, *Nature*, **159**, 161, (1947).
62. T. W. Campbell, and A. C. Haven, Jun., *J. Appl. Polymer Sci.*, **1**, 73, (1959).
63. G. Natta and D. Sianesi, *Rend. Accad. Nazl. Lincei*, (8), **26**, 418, (1959).
64. F. Danusso, p. 239 in *Chimica delle Macromolecole*, ed. CNR, Roma, 1963.
65. A. Ledwith, *Ind. Chemist.*, **37**, 71, (1961).
66. J. A. Shetler, *J. Polymer Sci.*, **B1**, 209, (1963).
67. F. E. Karasz and W. J. Macknight, *Macromolecules*, **1**, 537, (1968).
68. J. F. Johnson and R. S. Porter in *The Stereochemistry of Macromolecules*, Vol. 3, p. 213, ed. A. D. Ketley, Dekker, New York, 1968.
69. A. Lucherini, Thesis, University of Pisa, February 1968.
70. *Chemical Reactions of Polymers*, ed. E. M. Fettes, Wiley, New York, 1965.
71. M. M. Van Beylen, in *The Stereochemistry of Macromolecules*, Vol. 3, p. 333, ed. A. D. Ketley, Dekker, New York, 1968.
72. M. W. Volkenstein, *Dokl. Akad. Nauk. SSSR*, **78**, 879, (1951).
73. T. M. Birshein and O. B. Ptitsyn, *Conformation of Macromolecules*, Interscience, 1966.
74. N. P. Borisova and T. M. Birshtein, *Vysokomol. Soed.*, **5**, 279, (1963).
75. N. P. Borisova, *Vysokomol. Soed.*, **6**, 135, (1964).
76. T. M. Birshtein, V. A. Zubkov and M. V. Volkenstein, *J. Polymer Sci.*, Part A-2, **8**, 177, (1970).
77. R. A. Pethrick and E. Wyn-Jones, *Quart. Rev.*, **23**, 301, (1969).
78. P. L. Luisi, *Polymer*, **13**, 232, (1972).
79. J. P. Lowe, *Progress in Physical Organic Chemistry*, **6**, 1, (1968).
80. R. A. Scott and M. A. Sheraga, *Biopolymers* **4**, 237, (1966).
81. T. M. Birshtein and O. B. Ptitsyn, *J. Polymer Sci*, **C16**, 4617, (1969).
82. M. Goodman, A. S. Verdini, N. S. Choi and Y. Masuda, *Topics in Stereochemistry*, **5**, 69, (1970).
83. P. de Santis, E. Giglio, A. M. Liquori and A. Ripamonti, *Nuovo Cimento*, **26**, 616, (1962).
84. D. A. Braut, *Macromolecules*, **1**, 291, (1968).
85. M. Goodman and G. C. C. Nik, *Macromolecules*, **1**, 223, (1968).
86. S. Lifson, *J. Chem. Phys. Physicochim. Biol.*, **65** (1), 40, (1968).
87. S. Lifson and A. Warshel, *J. Chem. Phys.*, **49**, (11), 5116, (1968).
88. D. Poland and H. A. Sheraga, *Theory of Helix-Coil Statistics in Biopolymers*, Academic Press, 1970.
89. J. Applequist and P. Doty, in *Polyamino Acids, Polypeptides and Proteins*, ed. M. A. Stahmann, Wisconsin Press, 1962.
90. P. Doty, A. M. Holtzer, J. H. Bradbury and E. R. Blout, *J. Am. Chem. Soc.*, **76**, 4493, (1954).
91. E. Katchalski, I. Z. Steinberg, *Ann. Rev. Phys. Chem.*, **12**, 433, (1961).
92. B. H. Zimm and N. R. Kallenbach, *Ann. Rev. Phys. Chem.*, **13**, 171, (1962).
93. R. L. Miller and L. E. Nielsen, *J. Polymer Sci.*, **55**, 643, (1961).
94. G. Natta, P. Corradini and M. Cesari, *Atti Accad. Nazl. Lincei, Rend. Classe Sci. Fis., Mat. Nat.*, (8), **21**, 365, (1956).
95. G. Natta and P. Corradini, *Nuovo Cimento* (*Suppl.* 1), **15**, 40, (1960).
96. Z. W. Wilchinsky, *J. Appl. Phys.*, **31**, 1969, (1960).

97. Z. Mencik, *Chem. Prymisl.*, **10**, 377, (1960).
98. G. Natta, P. Corradini and I. W. Bassi, *Makromol. Chem.*, **21**, 240, (1956).
99. G. Natta, P. Corradini and I. W. Bassi, *Nuovo Cimento*, (*Suppl.* 1), **15**, 52, (1960).
100. G. Natta, P. Corradini and I. W. Bassi, *Atti Accad. Nazl. Lincei, Rend. Classe Sci. Fis. Mat. Nat.*, [8], **19**, 404, (1955).
101. G. Natta, P. Corradini and I. W. Bassi, *Atti Accad. Nazl. Lincei, Rend. Classe Sci. Fis. Mat.* [8], **23**, 363, (1957).
102. G. Natta, P. Corradini and I. W. Bassi, *Nuovo Cimento* (*Suppl.* 1), **15**, 68, (1960).
103. P. Corradini and P. Ganis, *J. Polymer Sci.*, **43**, 311, (1960).
104. G. Natta, P. Corradini and I. W. Bassi, *Makromol. Chem.*, **33**, 247, (1959).
105. P. Corradini and P. Ganis, *Nuovo Cimento* (*Suppl.* 1), **15**, 104, (1960).
106. P. Corradini and P. Ganis, *Nuovo Cimento* (*Suppl.* 1), **15**, 96, (1960).
107. F. C. Frank, A. Keller and A. O. 'Connor, *Phil. Mag.*, **4**, 200, (1959).
108. G. Dall'Asta and I. W. Bassi, *Chim. Ind.* (*Milan*), **43**, 999, (1961).
109. T. Makimoto, T. Tsuruta and J. Furukawa, *Makromol. Chem.*, **50**, 116, (1961).
110. P. Corradini, P. Ganis and V. Petraccone, *Eur. Polymer J.*, **6**, 281, (1970).
111. G. Natta, P. Corradini and I. W. Bassi, *J. Polymer Sci.*, **51**, 505, (1961).
112. G. Natta, G. Mazzanti, P. Corradini and I. W. Bassi, *Makromol. Chem.*, **37**, 156, (1960).
113. G. Natta, G. Mazzanti, P. Corradini, P. Chini and I. W. Bassi, *Atti Accad. Nazl. Lincei, Rend. Classe Sci. Fis. Mat. Nat.*, [8], **28**, 8, (1960).
114. G. Natta, G. Mazzanti, P. Corradini, A. Valvassori and I. W. Bassi, *Atti Accad. Nazl. Lincei, Rend. Classe Sci. Fis. Mat. Nat.*, [8], **28**, 18, (1960).
115. G. Natta, *Chim. Ind.* (*Milan*), **42**, 1207, (1960).
116. R. A. V. Raff and J. B. Allison, *Polyethylene*, Interscience, New York, 1956.
117. H. Tadokoro, T. Yasumoto, S. Murahashi and I. Nitta, *J. Polymer Sci.*, **44**, 266, (1960).
118. P. De Santis, E. Giglio, A. M. Liquori and A. Ripamonti, *J. Polymer Sci.*, **A1**, 965, (1963).
119. H. Tadokoro, M. Kobayashi, K. Mori and R. Chûjô, *Rept. Progr. Polymer Phys. Japan*, **8**, 45, (1965).
120. T. Uchida, Y. Kurita, N. Koizumi and M. Kubo, *J. Polymer Sci.* **21**, 313, (1956).
121. H. Tadokoro, *Macromol. Reviews*, **1**, 119, (1967).
122. C. S. Fuller, *Chem. Rev.* **26**, 143, (1940).
123. G. Carazzolo, and M. Mammi, *J. Polymer Sci.*, **B2**, 1057, (1964).
124. G. Natta, P. Corradini and G. Dall'Asta, *Atti Accad. Nazl. Lincei, Rend. Classe Sci. Fis. Mat. Nat.*, [8], **20**, 408, (1956).
125. E. Stanley and M. Litt, *J. Polymer Sci.*, **43**, 453, (1960).
126. I. W. Bassi, G. Dall'Asta, U. Campigli and E. Strepparola, *Makromol. Chem.*, **60**, 202, (1963).
127. G. Dall'Asta and N. Otto, *Chim. Ind.* (*Milan*), **42**, 1234, (1960).
128. G. Natta and P. Corradini, *J. Polymer Sci.*, **20**, 251, (1956).
129. G. Natta, P. Corradini and P. Ganis, *Makromol. Chem.*, **39**, 238, (1960).
130. G. Natta, P. Corradini and P. Ganis, *Rend. Accad. Nazl. Lincei Sci. Natural. Fis. Mat.*, **33**, 200, (1962).
131. G. Natta, M. Peraldo and G. Allegra, *Makromol. Chem.*, **75**, 215, (1964).
132. P. Corradini, G. Natta, P. Ganis and P. A. Temussi, *J. Polymer Sci.*, Part C, **16**, 2477, (1967).

133. M. Letort, and A. Richard, *J. Chim. Phys.*, **57**, 752, (1960).
134. G. Natta, M. Peraldo, M. Farina and G. Bressan, *Makromol. Chem.*, **55**, 139, (1962).
135. G. Natta, M. Farina, M. Peraldo, P. Corradini, G. Bressan and P. Ganis, *Atti Accad. Nazl. Lincei, Rend. Classe Sci. Fis. Mat. Nat.* **28**, 442, (1960).
136. R. Riande, J. G. Fatou and G. G. López-Heredia, *Abstracts Intern. Symp. Macromol., Leiden 1970*, p. 77, Inter Scientias, 1970.
137. C. W. Bunn, and D. R. Holmes, *Discussions Faraday Soc.*, **25**, 95, (1958).
138. G. Allegra, E. Benedetti and C. Pedone, *Macromolecules*, **3**, 727, (1970).
139. G. Wasai, T. Saegusa and J. Furukawa, *Makromol. Chem.*, **86**, 1, (1965).
140. A. M. Liquori, *Acta Cryst.*, **8**, 345, (1955).
141. P. de Santis and J. Kovacs, *Biopolymers*, **6**, 299, (1968).
142. U. Shmueli, W. Traub and K. Rosenheck, *J. Polymer Sci.* Part A-2 **7**, 515, (1969).
143. P. Corradini and G. Allegra, *Atti Accad. Nazl. Lincei, Classe Sci. Fis. Mat. Nat.*, [8] **30**, 516, (1961).
144. G. Allegra, P. Ganis and P. Corradini, *Makromol. Chem.*, **61**, 225, (1963).
145. A. Nakajima and A. Saijo, *J. Polymer Sci.*, **A-2, 6**, 735, (1968).
146. K. Nagai, *J. Chem. Phys.*, **37**, 490, (1962).
147. S. Lifson, *J. Chem. Phys.*, **30**, 964, (1959).
148. T. M. Birshtein, *Vysokomol. Soed.*, **1**, 798, (1959).
149. P. J. Flory, *J. Am. Chem. Soc.*, **89**, 1798, (1967).
150. T. M. Birshtein and O. B. Ptitsyn, *Zh. Tekhn. Fis.*, **29**, 1048, (1959).
151. M. Kurata and W. H. Stockmayer, *Fortschr. Hochpolymer. Forsch.*, **3**, 196, (1963).
152. J. B. Kinsinger and R. E. Hughes, *J. Phys. Chem.*, **67**, 1922, (1963).
153. J. E. Mark and P. J. Flory, *J. Am. Chem. Soc.*, **87**, 1423, (1965).
154. W. R. Krigbaum, J. E. Kurz and P. Smith, *J. Phys. Chem.*, **65**, 1984, (1961).
155. G. Allegra, P. Corradini and P. Ganis, *Makromol. Chemie*, **90**, 60, (1966).
156. G. Moraglio and J. Brzezinski, *J. Polymer Sci.*, **B2**, 1105, (1964).
157. P. J. Flory and R. L. Jernigan, *J. Chem. Phys.*, **42**, 3509, (1965).
158. C. A. J. Hoeve, *J. Chem. Phys.*, **35**, 1266, (1961).
159. T. W. Bates and W. H. Stockmayer, *Macromolecules*, **1**, 12, (1968).
160. H. Tadokoro, S. Nozakura, T. Kitazawa, Y. Yasuhara and S. Murahashi, *Bull. Chem. Soc. Japan*, **32**, 313, (1959).
161. P. J. Flory and J. E. Mark, *Makromol. Chem.*, **75**, 11, (1964).
162. J. E. Mark and P. J. Flory, *J. Am. Chem. Soc.*, **88**, 3720, (1966).
163. P. J. Flory, V. Crescenzi and J. E. Mark, *J. Am. Chem. Soc.*, **86**, 146, (1964).
164. J. E. Mark, *J. Am. Chem. Soc.*, **89**, 6829, (1967).
165. D. A. Brandt and P. J. Flory, *J. Am. Chem. Soc.*, **87**, 2791, (1965).
166. L. A. LaPlanche and M. T. Rogers, *J. Am. Chem. Soc.*, **86**, 337, (1964).
167. I. I. Novak, *Zh. Tekhn. Fis.*, **25**, 1854, (1955).
168. S. Mizushima, T. Shimanouchi, K. Nakamura, M. Hayashi and S. Tsuchiya, *J. Chem. Phys.*, **26**, 970, (1957).
169. T. Shimanouchi, S. Tsuchiya and S. Mizushima, *J. Chem. Phys.*, **30**, 1365, (1959).
170. K. Jimura, and M. Takeda, *J. Polymer Sci.*, **51**, 551, (1961).
171. A. Miyake, *J. Polymer Sci.*, **38**, 497, (1959).
172. A. Miyake, *J. Am. Chem. Soc.*, **82**, 3040, (1960).
173. A. M. Liquori, G. Barone, V. Crescenzi, F. Quadrifoglio and V. Vitagliano, *J. Macromol. Chem.*, **1**, 291, (1966).

174. E. V. Anufrieva, Yu. Ya. Gottlib, M. G. Krakovjak, S. S. Skorokhodov and T. V. Sheveleva, *Abstracts Intern. Symp. Macromol., Leiden 1970*, p. 483, Inter Scientias, 1970.
175. E. Dubois-Violette, F. Geny, L. Monnerie and O. Parodi, *J. Chim. Phys.*, **66**, 1865, (1969).
176. L. Monnerie and S. Gorin, *J. Chim. Phys.*, **67**, 422, (1970).
177. B. Valeur, C. Noel and L. Monnerie, *Abstracts Intern. Symp. Macromol. Leiden 1970*, p. 485, Inter Scientias, 1970.
178. M. T. Vala, J. Haebig and S. A. Rice, *J. Chem. Phys.*, **43**, 886, (1965).
179. S. Yanari, F. A. Bovey and L. Lumry, *Nature*, **200**, 242, (1963).
180. L. J. Basile, *J. Chem. Phys.*, **36**, 2204, (1962).
181. S. Satoh, *J. Polymer Sci.*, **A2**, 5221, (1964).
182. F. A. Bovey, F. P. Hood III, E. W. Anderson and L. C. Snyder, *J. Chem. Phys.*, **42**, 3900, (1965).
183. T. Shimanouchi, M. Tasumi and Y. Abe, *Makromol. Chem.*, **86**, 43, (1965).
184. Y. Abe, M. Tasumi, T. Shimanouchi, S. Satoh and R. Chŷjô, *J. Polymer Sci.*, **A4**, 1413, (1966).
185. D. Doskocilova and B. Schneider, *J. Polymer Sci.*, **B3**, 213, (1965).
186. K. J. Liu and R. Ulmann, *J. Chem. Phys.*, **48**, 1158, (1968).
187. G. Hermann and G. Weill, *Abstracts IUPAC, Intern. Symposium Macromol., Leiden 1970*, p. 885, Inter Scientias, 1970.
188. B. Z. Volchek, *Vysokomol. Soed.*, **5**, 260, (1963).
189. M. Takeda, K. Jimura, A. Yamoda and Y. Ymanura, *Bull. Chem. Soc. Japan*, **33**, 1219, (1960).
190. P. Pino, P. Salvadori, E. Chiellini and P. L. Luisi, *J. Pure Appl. Chem.*, **16**, 469, (1968).
191. P. Pino and P. L. Luisi, *J. Chim. Phys.*, **65**, 130, (1968).
192. P. Pino, F. Ciardelli and M. Zandomeneghi, *Annual Review Phys. Chem.*, **21**, 561, (1970).
193. C. Carlini, F. Ciardelli and P. Pino, *Makromol. Chem.*, **119**, 244, (1968).
194. P. Pino, C. Carlini, E. Chiellini, F. Ciardelli and P. Salvadori, *J. Am. Chem. Soc.*, **90**, 5025, (1968).
195. V. Crescenzi, in *The Stereochemistry of Macromolecules*, Vol. 3, p. 243, 1968.
196. W. R. Krigbaum, D. K. Carpenter and S. Newman, *J. Phys. Chem.*, **62**, 1586, (1958).
197. S. Krause and E. Cohn-Ginsberg, *Polymer*, **3**, 565, (1962).
198. F. Danusso and G. Moraglio, *Rend. Accad. Nazl. Lincei, Ser. VIII*, **25**, 509, (1958).
199. G. Moraglio, and C. Giannotti, *J. Polymer Sci.*, **27**, 374, (1959).
200. H. Bauman, *J. Polymer Sci.*, **B3**, 1069, (1965).
201. E. Hamori, L. R. Prusinowski, P. G. Sparks and R. E. Hughes, *J. Phys. Chem.*, **69**, 1101, (1965).
202. G. V. Schulz, W. Wunderlich and R. Kirste, *Makromol. Chem.*, **74**, 22, (1964).
203. J. E. Mark, R. A. Wessling and R. E. Hughes, *J. Phys. Chem.*, **70**, 1895, (1966).
204. R. A. Wessling, J. E. Mark, E. Hamori and R. E. Hughes, *J. Phys. Chem.*, **70**, 1903, (1966).
205. W. R. Krigbaum and A. Roig, *J. Chem. Phys.*, **31**, 544, (1959).
206. H. A. Pohl, R. Bacskai and W. P. Purcell, *J. Phys. Chem.*, **64**, 1701, (1960).
207. R. Salovey, *J. Polymer Sci.*, **50**, S. 7, (1961).
208. G. P. Mikhailov, and L. L. Burshtein, *Vysokomol. Soed.*, **6**, 1713, (1964).

209. M. Takeda, Y. Imamura, S. Okamura and S. Higashimura, *J. Chem. Phys.*, **33**, 631, (1960).
210. T. I. Borisova, L. L. Burshtein and G. P. Mikhailov, *Vysokomol. Soedin.*, **4**, 1479, (1962).
211. P. L. Luisi, E. Chiellini, P. F. Franchini and M. Orienti, *Makromol. Chem.*, **112**, 197, (1968).
212. T. G. Fox and P. J. Flory, *J. Am. Chem. Soc.*, **73**, 1909, (1951).
213. N. Kuwahara, M. Kaneko, Y. Miyake and J. Furuichi, *J. Phys. Soc. Japan*, **17**, 568, (1962).
214. C. E. Bawn and R. D. Patel, *Trans. Faraday Soc.*, **52**, 1669, (1956).
215. A. Ciferri, C. A. J. Hoeve and P. J. Flory, *J. Am. Chem. Soc.*, **83**, 1015, (1961).
216. P. J. Flory, A. Ciferri and R. Chiang, *J. Am. Chem. Soc.*, **83**, 1023, (1961).
217. K. Nagai and T. Ishikawa, *J. Chem. Phys.*, **37**, 496, (1962).
218. J. E. Mark and P. J. Flory, *J. Am. Chem. Soc.*, **86**, 138, (1964).
219. H. A. Pohl and H. H. Zabusky, *J. Phys. Chem.*, **66**, 1390, (1962).
220. V. N. Tsvetkov, Soviet Phys.-Usp. **6**, 639 (1963).
221. Y. Y. Gotlib, *Soviet Phys.-Tech. Phys.*, **2**, 637, (1957).
222. Y. Y. Gotlib, *Soviet Phys.-Tech. Phys.*, **3**, 749, (1958).
223. R. Kirste and W. Wunderlich, *Makromol. Chem.*, **73**, 240, (1964).
224. R. Kirste and W. Wunderlich, *J. Polymer Sci.*, **B3**, 851, (1964).
225. G. W. Brody and R. Salovey, *J. Am. Chem. Soc.*, **86**, 3499, (1964).
226. V. Luzzati, M. Cesari, G. Spach, F. Mason and J. M. Vincent, *J. Mol. Biol.*, **3**, 566, (1961).
227. G. W. Brody, R. Salovey and J. M. Reddy, *Biopolymers*, **3**, 573, (1965).
228. P. Saludjian, and V. Luzzati, *J. Mol. Biol.*, **15**, 681, (1966).
229. W. H. Watanabe, C. F. Ryan, P. C. Fleisher and B. S. Garrett, *J. Phys. Chem.*, **65**, 896, (1961).
230. C. F. Ryan and P. C. Fleisher, *J. Phys. Chem.*, **69**, 3384, (1965).
231. M. D'Alagni, P. DeSantis, A. M. Liquori and M. Savino, *J. Polymer Sci.*, **B2**, 925, (1964).
232. A. M. Liquori, G. Anzuino, V. M. Coiro, M. D'Alagni, P. De Santis and M. Savino, *Nature*, **206**, 358, (1965).
233. W. Burchard, *Makromol. Chem.*, **67**, 182, (1963).
234. H. Yu, A. J. Bur and L. J. Fetters, *J. Chem. Phys.*, **44**, 2568, (1966).
235. A. J. Bur and D. E. Roberts, *J. Chem. Phys.*, **51**, 406, (1969).
236. L. J. Fetters and H. Yu, *Polymer. Prepr. Amer. Chem. Soc., Div. Polymer Chem.*, **7**, (2), 443, (1966).
237. N. S. Schneider, S. Furusaki and R. W. Lenz, *J. Polym. Sci.*, Part A, **3**, 933, (1965).
238. M. Goodman, and Shih-Chung Chen, *Macromolecules*, **3**, 398, (1970).
239. B. Z. Volchek and Zh. N. Roberman, *Vysokomol. Soed.*, **2**, 1157, (1960).
240. V. N. Nikitin, M. V. Volkenstein and B. Z. Volchek, *Zh. Tekh. Fis.*, **25**, 2486, (1955).
241. L. I. Boguslavskii and A. V. Vannikov, *Organic Semiconductors and Biopolymers*, Plenum Press, 1970.
242. P. L. Luisi and P. Pino, *J. Phys. Chem.*, **72**, 2400, (1968).
243. J. H. Brewster, *J. Am. Chem. Soc.*, **81** 5475, (1959).
244. J. H. Brewster, *Topics in Stereochemistry*, **2**, 1, (1967).
245. S. Pucci, M. Aglietto, P. L. Luisi and P. Pino, *J. Am. Chem. Soc.*, **89**, 2787, (1967).
246. S. Pucci, M. Aglietto, P. L. Luisi and P. Pino, pp. 203–218 in *Conformational Analysis*, Vol. 21, ed. G. Chiurdoglu, Academic Press, 1971.

247. S. Pucci, M. Aglietto and P. L. Luisi, *Gaz. Chim. Ital.*, **100**, 159, (1970).
248. T. M. Birshtein and P. L. Luisi, *Vysokomol. Soed.*, **6**, 1238, (1964).
249. A. Abe, *J. Am. Chem. Soc.*, **90**, 2205, (1968).
250. P. L. Luisi, *Polymer Letters*, **6**, 177, (1968).
251. O. Bonsignori and G. P. Lorenzi, *J. Polymer Sci.*, Part A-2, **8**, 1639, (1970).
252. I. W. Bassi, O. Bonsignori, G. P. Lorenzi, P. Pino, P. Corradini and P. A. Temussi, *J. Polymer Sci.*, Part A-2, **9**, 193, (1971).

16
Thermodynamics of Addition Polymerization Processes

K. J. Ivin

The Queen's University of Belfast

16.1 INTRODUCTION

α-Methylstyrene polymerizes at 0 °C yet does not polymerize at 100 °C,[1] except at high pressure.[2,3] α-Methoxystyrene copolymerizes readily with styrene, yet does not homopolymerize at any temperature.[4] Acetaldehyde does not polymerize at 20 °C unless high pressure is applied.[5] These are but three of the many observations in polymer chemistry which have a thermodynamic explanation. An understanding of the thermodynamics of polymerization processes is therefore of some importance to the preparative chemist in search of new polymers.

Many of the early synthetic polymers were obtained accidentally during the preparation of unsaturated compounds. Later, as polymer chemistry advanced, vast numbers of both unsaturated and ring compounds were tested for polymerizability and certain rules were formulated from experience. It was found, for example, that six-membered ring compounds were difficult or impossible to polymerize, and that too many substituents had a bad effect on polymerizability. This chapter, which might be sub-titled 'Structure and Polymerizability', will be mainly concerned with the thermodynamic basis for these rules and the means whereby an 'unpolymerizable' monomer may sometimes be rendered polymerizable by changing the conditions. In this chapter the term 'polymerizable' will generally be used in its thermodynamic sense and will refer to the formation of long chain polymer as distinct from oligomers.

16.2 TYPES OF EQUILIBRIA IN POLYMER-MONOMER SYSTEMS

Three types of interwoven equilibria may be distinguished in addition polymerization systems:

1. equilibria between polymeric species,
2. equilibria between species which may be described as monomeric,
3. equilibria between monomeric and polymeric species.

Type 1 equilibria arise from the multitude of chain lengths which polymeric species may possess. A polymer in which all these species are present in their equilibrium proportions is said to possess the most probable, or thermodynamic, distribution of molecular weights,[6] characterized by $\bar{M}_w/\bar{M}_n = 2$, where $\bar{M}_w$ and $\bar{M}_n$ are the weight-average and number-average molecular weights respectively. To a good approximation the heat content of a long-chain polymer per mole of repeat unit is independent of $\bar{M}_n$ so that such equilibria are governed solely by entropy (statistical) considerations. Ability to achieve a thermodynamic distribution experimentally for a given polymer depends on the availability of a facile exchange mechanism such as ester interchange,[7] reversible propagation without termination,[8,9] or other exchange mechanisms as in the case of polytetrahydrofuran.[10] In free-radical polymerizations a thermodynamic distribution may be produced directly for certain mechanisms at low conversion.[6]

The most important case of type 2 is the equilibrium between oligomeric ring species, such as the siloxanes, which all open to give the same linear or macrocyclic polymer. This case will be considered in detail in paragraph 16.7.3. It suffices to note here that when the rings are above a certain size the heat of polymerization is zero (or a small constant value) and the equilibria are again governed solely by entropy considerations.

For type 3 equilibria, in contrast to types 1 and 2, the heat of reaction plays an important, often dominant role in determining the position of equilibrium. It is with these equilibria that we shall be mainly concerned in this chapter.

16.3 ADDITION POLYMERIZATION CONSIDERED AS AN AGGREGATION PROCESS

For any chemical reaction at constant temperature and pressure the criterion for equilibrium is that the free energy change for the reaction shall be zero. Addition polymerization reactions differ, however, from all other types of reaction in that a large number of molecules come together to form a single molecule. This gives to such reactions some of the features more usually associated with physical aggregation processes, in particular the existence of a sharp temperature, defined as a 'ceiling temperature,' above which the formation of long-chain polymer does not occur. The comparison with the freezing of a liquid is shown in Table 16.1. The essential feature common to both types

Table 16.1 Comparison between physical and chemical aggregation processes

Aggregation process	Physical (freezing)	Chemical (polymerization)
Non-aggregated form	Liquid	Monomer
Aggregated form	Solid	Polymer
Temperature at which $\Delta G = 0$	Melting point (T_m)	Ceiling temperature (T_c)

Table 16.2 Some observed ceiling temperatures

Monomer	Concentration (mol l⁻¹)	Solvent[a]	Polymer structure	T_c (°C)	Reference
Styrene	1.2×10^{-4}	PhH	—CH₂—CHPh—	110	17
α-Methylstyrene	pure	THF	—CH₂—CMePh—	61	2
	0.76			0	1, 18, 19
Methyl atropate	pure	PhMe	—CH₂—C(COOMe)Ph—	-8	20
	1.0			-40	20
Formaldehyde	0.06	CH₂Cl₂	—O—CH₂—	30	21
Acetaldehyde	pure	—	—O—CHMe—	-31	22
Tetrahydrofuran	pure	—	—O—(CH₂)₄—	80	23
1,3-Dioxolane	1.0	CH₂Cl₂	—O—CH₂—O—(CH₂)₂—	1	24
Phthalaldehyde	1.0	CH₂Cl₂	(benzene ring fused acetal: —CH—O—...—O—CH—)	-43	25
Thioacetone	pure	—	—S—CMe₂—	95	26
n-Hexyl isocyanate	2	DMF	—NHex^n—CO—	-22	27

a. THF = tetrahydrofuran, DMF = dimethylformamide.

of aggregation is that at the sharp temperature the free energy change for the process is zero. In both cases the transition temperature is depressed by the addition of a diluent, and is independent of the size of the aggregate except when this is very small.

The first observation of a ceiling temperature was made in 1938 by Snow and Frey,[11] working on the copolymerization of alkenes with sulphur dioxide. They found, for example, that with a 1:1 liquid mixture of isobutene and sulphur dioxide the reaction to form poly(isobutene sulphone) could be started and stopped at will by lowering and raising the temperature below and above 4 °C in the presence of a suitable initiator. The thermodynamic explanation of this effect was given ten years later.[12,13]

Scattered observations may be found in the literature before 1938 which have their origins in the same effect. Thus it was known that formaldehyde vapour at 300 mmHg pressure would polymerize only at temperatures below 100 °C.[14] The reverse effect, that of a 'floor temperature', although rather rare, is exemplified by the classic case of sulphur, which polymerizes to a highly viscous liquid only above 159 °C.[15] In this case both ΔH and ΔS for polymerization are positive and ΔG becomes negative when the temperature is raised above the floor temperature.

The complete generality of the ceiling temperature effect is illustrated in Table 16.2. A more complete list of monomers may be found in the Polymer Handbook.[16] Before discussing the effect of chemical structure we will summarize the methods for the determination of ceiling temperature and consider the effect of other variables.

16.4 METHODS FOR DETERMINING CEILING TEMPERATURES

The very term ceiling temperature carries the implication that it can be found by varying the temperature while keeping the reactant composition constant. This indeed is so and the onset of polymerization may be observed directly if the polymer is insoluble, as in the case of poly(isobutene sulphone),[11] or from dilatometric measurements[28] as illustrated in Figure 16.1. Alternatively, the yield, the rate or degree of polymerization may be measured at several temperatures below T_c and the results extrapolated to zero yield, rate R_p, or degree of polymerization, as shown in Figure 16.2. This method has the advantage that the chain length may remain quite high even up to a temperature within a few degrees of the ceiling temperature. The extrapolated value for T_c then corresponds to the limiting value for long chains. The limiting slope $(dR_p/dT)_{T_c}$ is proportional to the concentration of chain centres and to the heat of polymerization.[13] The 'blurring' of the transition at T_c as the result of the formation of short-chain polymer is usually very slight, although it must be expected in systems where chain transfer is dominant.[29]

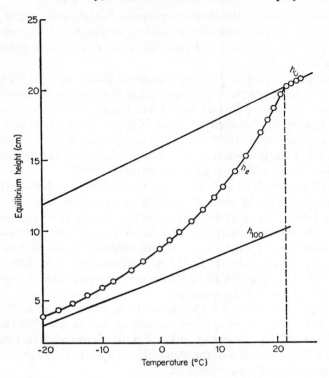

Figure 16.1   Determination of T_c by direct dilatometry: case of α-methylstyrene in tetrahydrofuran, initiated by 'living' tetramer dianions.[28] [M] = 2·31 mol l⁻¹, T_c = 21·8 °C. The upper straight line indicates the height h_0 corresponding to no polymerization; the lower straight line indicates the height h_{100} corresponding to 100 per cent polymerization. The points denote equilibrium heights h_e in the dilatometer; those below 21·8 °C are ceiling temperatures for solutions containing [M] < 2·31 mol l⁻¹ and finite polymer concentrations

Another way to determine ceiling temperatures is to hold the temperature constant, to allow the polymerization to proceed to equilibrium and then to determine the composition of the equilibrium mixture. This method has been applied, for example, to methyl methacrylate,[30] α-methylstyrene[28] and 1,3-dioxolane,[31] using titration, gas liquid chromatography and NMR respectively to analyse for monomer. In general it is necessary also to determine the polymer concentration for a complete thermodynamic analysis (see section 16.5). Alternatively the initial rate or degree of polymerization may be measured at a series of initial monomer concentrations and the results extrapolated to zero rate or degree of polymerization.[32]

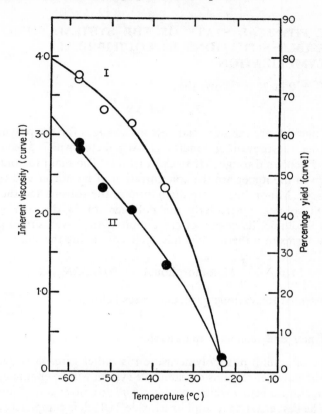

Figure 16.2 Determination of T_c by extrapolation of a plot of rate or degree of polymerization against temperature: case of n-hexyl isocyanate in dimethylform-amide, initiated by sodium cyanide.[27] Curve I: percentage yield. Curve II: inherent viscosity of polymer at 0·1 per cent concentration in benzene

The only difference in these two approaches lies in the choice of independent variable: reactant composition or temperature; thermodynamically there is no distinction. In practice it is found that some polymer-monomer equilibria are very mobile and equilibrium can be readily attained from both sides. This is particularly so for 'living' anionic polymerization systems and also for the polymerization of aldehydes. In such cases direct measurement of T_c or the equilibrium monomer concentration $[M_e]$ is a practical proposition. Other polymer-monomer equilibria are less easy to attain, particularly where the polymerization proceeds by a radical mechanism; in such cases the extrapolation procedures are generally more satisfactory.

16.5 THE PHYSICAL STATE OF THE SYSTEM; SOME POSSIBLE SITUATIONS IN EQUILIBRIUM POLYMERIZATION

Since $\Delta G = 0$ at T_c it follows that

$$T_c = \Delta H / \Delta S \tag{1}$$

where ΔH and ΔS are the heat and entropy changes for the polymerization process under the prevailing conditions. Any factor which affects ΔG will also affect T_c either through ΔH or ΔS or both. The main factors affecting ΔG for a given monomer are the concentration of monomer and the hydrostatic pressure. Minor factors are the nature of the solvent and the presence of dissolved solutes, particularly the polymer; or in the case where the polymer is insoluble, its degree of crystallinity. Any given situation may be analysed in terms of a thermodynamic cycle containing the step

$$M(liq) \rightarrow \frac{1}{n} M_n(amorphous) \qquad \Delta H_{lc}, \Delta S_{lc}, \Delta G_{lc}$$

This is illustrated for various particular cases below.

16.5.1 Polymer and monomer immiscible

This type of equilibrium polymerization is rather rare. Provided the free energy of the polymer remains constant the system will be characterized by a ceiling temperature below which there is *complete* conversion of monomer to polymer. The best example is sulphur trioxide[33] which is completely converted to a crystalline polymer below 30·4 °C. Another appears to be n-butyraldehyde.[34]

Although rare, the simplicity of this type of system makes it worth considering the various free energy relationships which may exist between solid and liquid monomer on the one hand and polymer on the other. These are illustrated in Figure 16.3 where the polymer is assumed to be amorphous and of a constant tacticity; the curves for crystalline polymer would be lower. All the curves have negative slopes (corresponding to positive entropies) which increase numerically with temperature. The properties of these systems would be as follows:

(a)   liquid monomer polymerizable above T_f,
(b)   solid and liquid monomer polymerizable above T_f,
(c)   liquid and solid monomer polymerizable below T_c,
(d)   solid monomer polymerizable below T_c,
(e)   monomer polymerizable only between T_f and T_c,
(f)   monomer unpolymerizable, except in glassy state below T_c.[35]

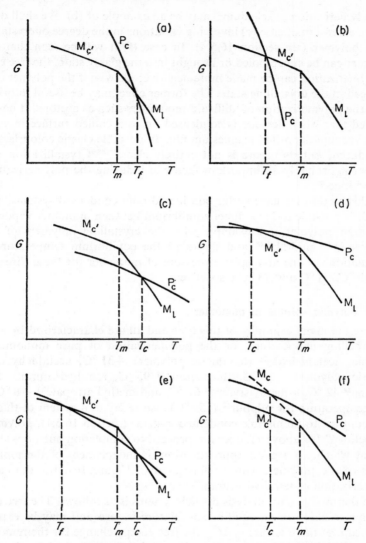

Figure 16.3 Some possible relationships between the free energy of crystalline monomer $M_{c'}$, liquid monomer M_l (per mole), and amorphous polymer P_c (per mole of repeat unit). T_m denotes the melting point of the monomer. T_f and T_c denote floor and ceiling temperatures respectively. M_c denotes supercooled or glassy monomer

The polymerization of selenium[36] may be an example of (b). We shall discuss later an extension of case (c) involving variations in the degree of crystallinity of the polymer (see section 16.5.4). In case (f) it will be seen that if the monomer can be supercooled or brought into the glassy state, the free energy of polymerization can be made more negative; likewise if the polymer can be produced in the crystalline state. The former effect may be useful in bringing about the polymerization of 'difficult' monomers such as acetone. It has been reported that when acetone is condensed on to a chilled surface in contact with a vacuum-deposited magnesium film an unstable elastic colourless solid is produced, but its nature is not entirely clear.[37,38] Crystallization of the polymer is probably an important factor in inducing the polymerization of 2-piperidone.[39]

In this section we have so far considered only condensed systems. Sometimes it is possible to have direct equilibrium between monomer vapour and condensed polymer, for example with the crystalline polymers of formaldehyde,[40,41] chloral[42,43] and fluoral;[43] the equilibrium temperatures for 'sublimation' to one atmosphere pressure of monomer for these three cases are 119 °C, 98 °C and 73 °C respectively.

16.5.2 Polymer soluble in monomer

There are many examples of this type and all are characterized by a fairly sharp ceiling temperature for the polymerization of pure monomer, for example, acetaldehyde[22] (to atactic polymer) −31 °C, acetaldehyde[22] (to isotactic polymer) −39 °C, thioacetone[26] 95 °C, tetrahydrofuran[23] 80 °C, oxepane[44] 42 °C, α-methylstyrene[2] 61 °C, and methyl atropate[20] −8 °C. The floor temperature for sulphur[15] (159 °C) also refers to a system of this type.

In contrast to immiscible condensed systems (section 16.5.1), polymerization below T_c (or above T_f) does not proceed to completion but stops short at a point which, to a good approximation, is independent of the molecular weight of the polymer. This is illustrated in Figure 16.4 for the cationic polymerization of tetrahydrofuran.[23]

The thermodynamic analysis of such systems is as follows. The free energy of polymerization in an equilibrium mixture is zero and may be expressed as the sum of three terms: $-\Delta \bar{G}_1$, the free energy change for the removal of 1 mole of liquid monomer from the mixture; ΔG_{lc}, as already defined (p. 520); and $\Delta \bar{G}_2$, the free energy change for the addition of 1 mole (expressed as repeat units) of amorphous polymer to the mixture. Therefore

$$-\Delta \bar{G}_1 + \Delta G_{lc} + \Delta \bar{G}_2 = 0 \qquad (2)$$

Expressions for $\Delta \bar{G}_1$ and $\Delta \bar{G}_2$ may be inserted from the Flory-Huggins equation[6] for the free energy of mixing of polymer and solvent monomer, giving equation (3),

$$\Delta G_{lc}/RT = \ln \phi_1 - \frac{1}{n} \ln \phi_2 - \frac{1}{n} + 1 + \chi(\phi_2 - \phi_1) \qquad (3)$$

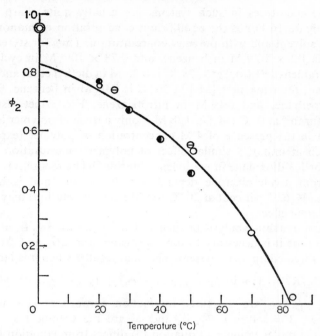

Figure 16.4 Equilibrium volume fraction of polymer ϕ_2 in the bulk polymerization of tetrahydrofuran.[23] Results from ○ Sims,[45] ⊖ Bawn, Bell and Ledwith,[46] ◑ Rozenberg and coworkers.[47] The line drawn is theoretical and corresponds to chosen values of χ, ΔH_{lc} and ΔS_{lc} (see text)

where ϕ_1 and ϕ_2 are the volume fractions of monomer and polymer, and χ is the polymer–monomer interaction parameter. For large n (degree of polymerization) and not too small ϕ_2 this reduces to

$$\Delta G_{lc}/RT = 1 + \ln \phi_1 + \chi(\phi_2 - \phi_1) \tag{4}$$

whence for given values of ΔH_{lc}, ΔS_{lc} and χ, ϕ_2 may be calculated as a function of temperature. The line in Figure 16.4, for example, was calculated[23] taking $\chi = 0.3$ (assumed independent of concentration and temperature), $\Delta H_{lc} = -12.4 \text{ kJ mol}^{-1}$ and $\Delta S_{lc} = -40.8 \text{ J K}^{-1} \text{ mol}^{-1}$ (both assumed independent of temperature).

16.5.3 Polymer and monomer in a solvent; effect of concentration

Numerous equilibrium systems of this type have been studied, in particular for the following monomers: styrene;[17] $CH_2{=}CMeR$ where $R = CN$,[48] $COOMe$,[30] Ph,[1,18,19] 2-naphthyl,[19] p-biphenyl;[19] $CH_2{=}CPhR$ where $R = COOMe$,[20] CN;[20] 1,3-dioxolane[24] and 1,3-dioxepane;[49] chloral,[42] 1,2-dichloropropionaldehyde,[50] isobutyraldehyde[51] and phthalaldehyde.[25]

Ceiling temperatures in such systems are usually quite sharp (see, for example, Figure 16.1) but the equilibrium concentration of monomer $[M_e]$ varies with solvent and with polymer concentration. Thus for styrene[17] $[M_e]$ at 110 °C is $1\cdot2 \times 10^{-4}$ M in benzene, and $0\cdot78 \times 10^{-4}$ M in cyclohexane; for isobutyraldehyde[51] $[M_e]$ at -78 °C is $0\cdot35$ M in diethyl ether and $0\cdot087$ M in n-heptane; for trioxane[52] $[M_e]$ at 30 °C is $0\cdot05$ M in benzene, $0\cdot13$ M in 1,2-dichloroethane, and $0\cdot19$ M in nitrobenzene. For α-methylstyrene in tetrahydrofuran[28] at 0 °C $[M_e]$ is $0\cdot78$ M if only a trace of polymer is present, but $0\cdot34$ M in the presence of 4 M concentration of polymer (expressed in terms of repeat units); a similar effect of polymer concentration has been observed for 1,3-dioxolane in methylene chloride.[31] The fact that the yield of polymer from δ-valerolactone depends on the solvent and is highest in nitrobenzene[50] ($62\cdot5$ per cent at 20 °C for 20 per cent solution) may well be a thermodynamic effect.

The thermodynamic analysis of such systems again follows from equation (2) but this time it is necessary to use expressions for $\Delta \bar{G}_1$ and $\Delta \bar{G}_2$ appropriate to a three-component system. The final result[28] when n is large is

$$\Delta G_{lc}/RT = 1 + \ln \phi_1 + \chi_{12}(\phi_2 - \phi_1) + (\chi_{13} - \chi_{32}V_1/V_3)\phi_3 \qquad (5)$$

where χ_{12}, χ_{13}, χ_{32} are the three interaction parameters (1 = monomer, 2 = polymer, 3 = solvent), ϕ_1, ϕ_2 and ϕ_3 are the volume fractions, and V_1, V_3 are the molar volumes. Equation (5) differs from equation (4) by the inclusion of the last term, the magnitude of which is determined mainly by the difference in monomer-solvent and polymer-solvent interaction parameters. The former generally shows the stronger dependence on solvent and hence is mainly responsible for the variation of equilibrium monomer concentration with solvent. Using a simplified form of equation (5) Bywater and Worsfold[17] were able to explain quantitatively the solvent effect for styrene (see above). When the variation of ϕ_1 with ϕ_2 is not too large it can be shown from equation (5) that ϕ_1 will be a linear function of ϕ_2, as found for α-methylstyrene in tetrahydrofuran[28] and dioxane.[53]

It will be evident that a thermodynamic analysis in terms of equation (5) can only be carried out properly if the three interaction parameters are known or can be estimated. Even then the treatment will be approximate since the parameters are liable to vary with composition and temperature. For these reasons recourse has frequently been made to a less exact treatment in terms of equation (6).

$$T_c = \Delta H_{ss}/\Delta S_{ss} = \Delta H_{ss}^0/(\Delta S_{ss}^0 + R \ln [M]) \qquad (6)$$

where ΔH_{ss} and ΔS_{ss} are the heat and entropy changes for the polymerization of monomer in solution to polymer in solution at the ceiling temperature for a monomer concentration $[M]$. ΔH_{ss}^0 and ΔS_{ss}^0 are thus defined in terms of a standard state of 1 mol l^{-1} of monomer. Variations of the activity coefficient of the monomer with temperature or with composition are thus absorbed into the values of ΔH_{ss}^0 and ΔS_{ss}^0 instead of being expressed through the interaction terms in equation (5). These variations are generally sufficiently small or monotonously continuous that equation (6) provides a very good representation of the experimental data, as illustrated in Figure 16.5 for isobutyraldehyde.[51] The slope of the plot of log [M] against $1/T_c$ (equivalent to a plot of log $[M_e]$ against $1/T$) gives an average value of ΔH_{ss}^0 which may be combined with appropriate heats of solution of monomer and polymer[51] to obtain ΔH_{lc}.

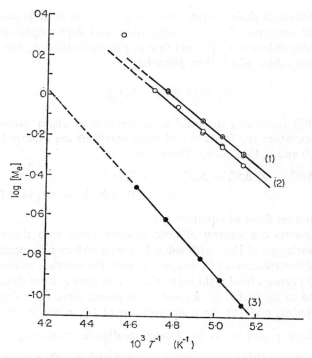

Figure 16.5 Temperature dependence of equilibrium monomer concentration of isobutyraldehyde in (1) tetrahydrofuran, (2) diethyl ether, and (3) n-pentane[51]

Values of ΔH_{lc} or ΔH_{ss}^0 obtained by the application of equations (4), (5) and (6) are generally in good agreement with calorimetric values, for example for styrene, α-methylstyrene, methyl and ethyl methacrylate, methacrylonitrile, tetrahydrofuran, caprolactam and for the copolymerization of sulphur dioxide with but-1-ene, cis- and trans-but-2-ene, isobutene and hexadec-1-ene. Details may be found in the Polymer Handbook.[16]

Values of ΔS_{lc} or ΔS_{ss}^0 obtained by the application of equations (4), (5) and (6) are in reasonable agreement with third law values in the relatively few cases where comparison can be made. Details may be found in the Polymer Handbook.[16] An absolute comparison is not strictly possible since the imperfectly ordered polymer has a small residual entropy at the absolute zero.

At this point we may consider how equation (6) may be derived from kinetic considerations. Near the ceiling temperature the propagation reaction is reversible and we may write

$$P_n^* + M \underset{k_d}{\overset{k_p}{\rightleftharpoons}} P_{n+1}^* \qquad K = k_p/k_d \qquad (7)$$

where P_n^* denotes a chain carrier containing n monomer units and k_p and k_d are the rate constants for the propagation and depropagation reactions, assumed independent of n. The net rate of polymerization R_p for reasonably large average values of n is then given by

$$R_p = (k_p[M] - k_d) \sum_{}^{n} [P_n^*] \qquad (8)$$

where $\sum^{n} [P_n^*]$ represents the total concentration of chain carriers. At the ceiling temperature the forward and back reactions are exactly balanced so that $R_p = 0$ and $k_p[M_e] = k_d$. Therefore

$$\Delta H_{ss}^0 - T_c \Delta S_{ss}^0 = \Delta G_{ss}^0 = -RT_c \ln K$$
$$= -RT_c \ln k_p/k_d = RT_c \ln [M_e] \qquad (9)$$

which is another form of equation (6).

Several points are worthy of note in connection with these equations. First, the variation of $[M_e]$ with solvent or with polymer concentration must be reflected in variations of k_p and, or, k_d with the reaction medium. Second, equation (8) cannot hold right up to the ceiling temperature since the chains will become so short that (a) k_p and k_d no longer have their limiting values for long chains, and (b) the concentration of P_1^*, a species which cannot undergo depropagation, is no longer negligible compared with $\sum^{n} [P_n^*]$. For longer chains $[P_1^*]$ is negligible compared with $\sum^{n} [P_n^*]$ even though it will generally be the largest term in the series. As stated earlier (section 16.4), these restrictions do not present any serious obstacle to the practical determination of ceiling temperatures from rate measurements. Third, equation (8) expresses the fact that the ceiling temperature must be independent of the nature and concentration of catalyst since these affect only $\sum^{n} [P_n^*]$. Fourth, since the average degree of polymerization is given by $\overline{DP} = R_p/f(R_i)$, where $f(R_i)$ is a function of the rate of initiation, $\overline{DP}$ must also extrapolate to zero at the same ceiling temperature, independent of the nature or concentration of catalyst. These last two points are illustrated in Figures 16.6 and 16.7,

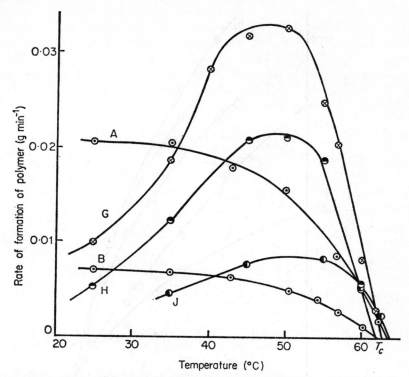

Figure 16.6 Rate of formation of poly (but-1-ene sulphone) from a 10:1 molar mixture of sulphur dioxide and but-1-ene.[54] A and B, photochemical initiation at two different intensities; G and H, initiation by silver nitrate at two different concentrations; J, initiation by benzoyl peroxide

where the curves for a given monomer composition all extrapolate to $T_c =$ 63 °C.

16.5.4 1:1 Copolymerization of miscible monomers to give insoluble polymer

This somewhat unusual type of system is found in the copolymerization of sulphur dioxide, S, with certain alkenes, M, such as isobutene[55] and 3-methylbut-1-ene.[56] Its special interest lies in the fact that this is one of the few cases where monomer activities have been directly determined and where the degree of crystallinity of the polymer is known to play a variable role.

These reactions may be written as

$$M(liq) + S(liq) \rightarrow \frac{1}{n} (MS)_n (amorphous) \qquad \Delta H_{lc}, \Delta S_{lc}, \Delta G_{lc}$$

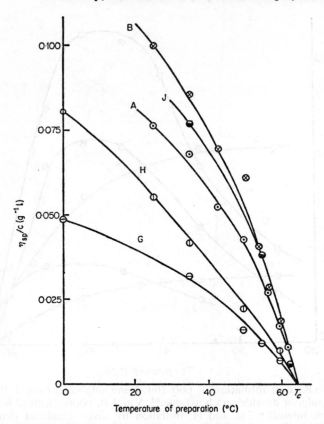

Figure 16.7   Specific viscosity of poly (but-1-ene sulphone) in acetone ($c = 8$ g l^{-1}) plotted against temperature of preparation, with constant percentage conversion.[32] Conditions as in Figure 16.6

where ΔH_{lc} etc. refer to the conversion of the liquid monomers (in their unmixed state) to amorphous polymer. The free energy of polymerization in an equilibrium mixture is zero and may be expressed as the sum of three terms: $-\Delta \bar{G}_M$, the free energy for the removal of 1 mole of liquid M from the mixture; $-\Delta \bar{G}_S$, the free energy for the removal of 1 mole of liquid S from the mixture; ΔG_{lc} as defined above; and assuming that there is no free energy change on adding solid polymer to the liquid monomer mixture. Therefore

$$-\Delta \bar{G}_M - \Delta \bar{G}_S + \Delta G_{lc} = 0 \tag{10}$$

Therefore

$$\Delta H_{lc} - T_c \Delta S_{lc} = \Delta \bar{G}_M + \Delta \bar{G}_S = RT_c \ln a_M a_S \tag{11}$$

where a_M and a_S are the activities of the two monomers in the mixture. Therefore

$$\frac{1}{T_c} = \frac{\Delta S_{lc}}{\Delta H_{lc}} + \frac{R}{\Delta H_{lc}} \ln a_M a_S \qquad (12)$$

As expected from this equation T_c rises to a maximum as the mole fraction x_M is increased and then falls off again. The maximum occurs at 9 °C ($x_M = 0.25$) for isobutene[55] and 40 °C ($x_M = 0.3$) for 3-methylbut-1-ene.[56] On either side of this maximum there exist pairs of compositions with the same

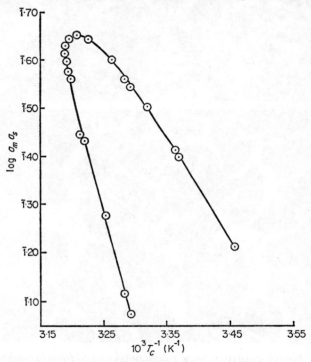

Figure 16.8 Test of equation (12) for M = 3-methylbut-1-ene, S = sulphur dioxide.[56] Compositions range from $x_M = 0.025$ (end of lower arm) to $x_M = 0.951$ (end of upper arm). Activities determined from vapour pressure measurements[57]

activity product $a_M a_S$ and therefore one might have expected also with same ceiling temperature. This, however, turns out not to be the case, as shown in Figure 16.8. At a given T_c the two values of $RT_c \ln a_M a_S$ differ by up to 3 kJ mol^{-1}. The explanation of this discrepancy is that the polymer is not precipitated as an amorphous powder over the entire composition range. The product obtained in SO$_2$-rich mixtures (lower arm of Figure 16.8), when spread on a microscope slide and allowed to evaporate, has the appearance

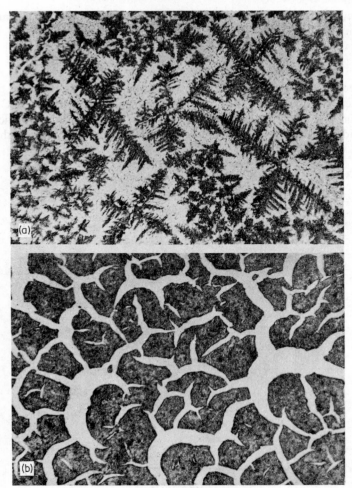

Figure 16.9 Poly (3-methylbut-1-ene sulphone) prepared from (a) sulphur-dioxide-rich mixture and (b) alkene-rich mixture[58]

shown in Figure 16.9a; the product obtained in M-rich mixtures (upper arm of Figure 16.8), when treated in the same way, has the appearance shown in Figure 16.9b. The increased driving force associated with the lower arm of Figure 16.8 thus stems, at least in part, from the increased crystallinity of the product formed from SO_2-rich mixtures. The degree of crystallinity of the product is thus affected by the polarity of the medium, which presumably acts through its effect on the relative proportion of different conformations about the main-chain bonds.

16.5.5 Effect of pressure (see also Chapter 6)

The effect of pressure on polymer–monomer equilibria has been studied for the monomers listed in Table 16.3. For a pure monomer the variation of ceiling temperature T_c with pressure P may be expressed in the form of the Clapeyron-Clausius equation

$$d(\ln T_c)/dP = \Delta V/\Delta H \qquad (13)$$

Table 16.3 Monomer–polymer equilibria which have been studied as a function of pressure

Monomer	Reference	$-\Delta V$ (ml mol^{-1})	$-\Delta H$ (kJ mol^{-1})	$\Delta V/\Delta H$ (ml kJ^{-1})
α-Methylstyrene	2, 3	14·1	33·9	0·42
Chloral	42	22	33·5	0·66
n-Butyraldehyde	34	8	21·1	0·38
Tetrahydrofuran	59	9·5	18·0	0·53
Sulphur	60	6·15	−13·2	−0·47
Para-substituted α-methylstyrenes	61			
n-Perfluorohept-1-ene	62		qualitative or	
Acetaldehyde	5		semi-quantitative studies	
γ-Butyrolactone	63			
δ-Valerolactam	63			

where ΔV and ΔH are the volume and heat changes for the polymerization under the prevailing conditions. For pure α-methylstyrene[2,3] and tetrahydrofuran[59] log T_c has been shown to be a linear function of P (Figure 16.10), the slope being consistent with the known values of ΔV and ΔH. Chloral and n-butyraldehyde have been studied in solution, and the equilibrium monomer concentration measured as a function of pressure as shown in Figure 16.11. The slope of $RT\ln[M_e]$ against P is equal to ΔV. For n-butyraldehyde the average value was -8 ± 1 ml mol^{-1} which agrees with that calculated from the molar volumes of the monomer and polymer at atmospheric pressure, after allowing for a considerable difference in the compressibilities of monomer and polymer.[34] The floor temperature for sulphur ought to decrease with increasing pressure, since ΔV is negative and ΔH is positive. Instead it increases from 159 °C at 1 atm to 180 °C at 300 atm and then decreases.[60] This anomaly has not been explained. Although para-substituted α-methylstyrenes polymerize on applying high pressure the corresponding ortho-substituted α-methylstyrenes do not.[65] Thus the adverse effect on T_c caused by the introduction of the ortho-substituent cannot be offset by the application of pressure.

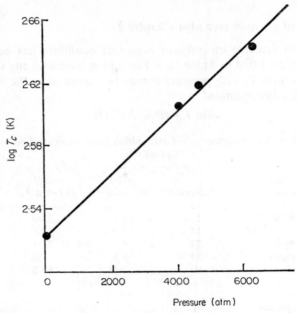

Figure 16.10   Variation of T_c with pressure for pure α-methylstyrene (plotted[64] from the results of Kilroe and Weale[2])

The effect of pressure on melting points is generally a good deal larger than the effect on ceiling temperatures[64] so that for the case shown in Figure 16.5c application of pressure will tend to reduce $T_c - T_m$ and may even reverse its sign. In principle it would be possible by the application of pressure to go from an unpolymerizable liquid to a polymerizable solid.

Since practically all addition polymerizations are accompanied by a decrease in both volume and enthalpy the most favourable thermodynamic conditions for polymerization are high pressure and low temperature. Usually one or other of these variables has been applied to encourage 'difficult' monomers to polymerize, but seldom both together. It is clearly a mistake to raise both the pressure and temperature since the two effects on ΔG are then working in opposition (see Chapter 6).

16.6   EFFECT OF MONOMER STRUCTURE ON POLYMERIZABILITY OF UNSATURATED COMPOUNDS

16.6.1 Ethylene derivatives

Ethylene itself can be readily polymerized either at high pressure and moderately high temperature to give weakly branched polymer or at lower

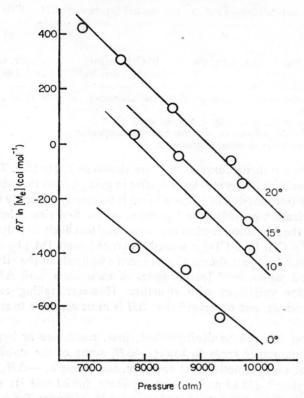

Figure 16.11 Variation of equilibrium concentration of n-butyraldehyde with pressure.[34] Solvent, n-hexane. Temperatures indicated in °C

pressures and temperatures in the presence of Ziegler-Natta or oxide catalysts to give linear polymer. Most mono-substituted ethylenes can also be readily polymerized, and the back reaction is unimportant unless the reaction is carried out at exceptionally low concentration and high temperature, e.g. 10^{-3} M and 120 °C for styrene.[17]

Reversibility is more in evidence for 1,1-disubstituted ethylenes and the ability of the monomer to polymerize is then critically dependent on the nature of the substituents, especially on their bulkiness. Some results on substituted methyl acrylates illustrate this point in Table 16.4. Clearly the free energy of polymerization at a given temperature becomes less negative on going from left to right across the table and it is probably ΔH rather than ΔS which is affected by structure.[13]

Another instructive series of monomers is that derived from α-methyl-styrene, with the general formula $CH_2{=}CMeAr$. The equilibrium monomer

Table 16.4   Polymerizability of substituted methyl acrylates,[66] $CH_2{=}CRCOOMe$

Group	A	B	C	D
R	H, Me	Et, n-Pr, Ph	PhCH$_2$, n-Bu, iso-Bu, tert-Bu[67]	iso-Pr, sec-Bu cyclohexyl

Group A   polymerize both radically at 65 °C and anionically at −78 °C to give high polymer.
Group B   give high polymer at −78 °C but not at 65 °C.
Group C   give only low polymer or oligomer at either temperature.
Group D   give no polymer at either temperature.

concentrations in tetrahydrofuran at 0 °C are shown in Table 16.5. The free energy of polymerization becomes less negative in going across the table from left to right. The introduction of a second ring is thermodynamically favourable to polymerization provided the 2-position on the first ring is left clear. A substituent at the 2-position renders the compound less likely to polymerize. Replacement of α-CH$_3$ by α-CD$_3$ in α-methylstyrene lowers [M$_e$] by a factor of 3 or 4, the effect being attributed to a reduced amplitude of the vibrations in the deuterated monomer.[69] In this series of monomers both ΔH_{ss}^0 and ΔS_{ss}^0 show minor variations with structure. However, taking ethylenic monomers as a whole, one observes[13] that ΔH is more sensitive to structure than is ΔS.

Two effects on ΔH can be distinguished: first, resonance or hyperconjugation in the monomer tends to lower $-\Delta H$, some of the stabilization energy being lost on polymerization; compare, for example, $-\Delta H_{lc}$ values for methyl acrylate[70] (78 kJ mol^{-1}) and styrene[71] (69 kJ mol^{-1}); second, steric hindrance in the polymer tends to lower $-\Delta H$; compare, for example, $-\Delta H_{lc}$ values for α-methylstyrene[72] (39 kJ mol^{-1}) and styrene (69 kJ mol^{-1}), and for methyl methacrylate[73] (57 kJ mol^{-1}) and methyl acrylate (78 kJ mol^{-1}).

Table 16.5   Polymerizability of derivatives of α-methylstyrene,[19] $CH_2{=}CMeAr$
Solvent, tetrahydrofuran at 0 °C

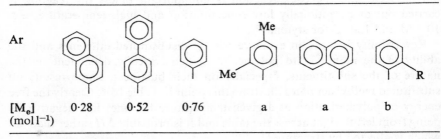

Ar					
[M$_e$] (mol l^{-1})	0·28	0·52	0·76	a	a	b

a. Not polymerizable at −80 °C.
b. Not polymerizable at 6000 atm and 120 °C for R = Me, Cl,[65] but polymerizable below −30 °C (1 atm) for R = MeO.[68]

One or two other structural effects are worthy of note. A methoxy group in place of a methyl group has a decidedly adverse effect on polymerizability. Thus α-methoxystyrene is not polymerizable[4] at any temperature down to −78 °C; and, unlike isobutene, α-methylvinyl methyl ether, $CH_2{=}CMeOMe$, has a ceiling temperature below room temperature.[74] In contrast a nitrile group in place of an ester group has a beneficial effect on polymerizability. Thus atroponitrile, $CH_2{=}CPhCN$, and para-substituted atroponitriles have ceiling temperatures much above those of the corresponding esters.[20,75] 1,1-Diphenylethylene is unpolymerizable as also are all tri- and tetra-substituted ethylenes except in certain cases where the substituents are fluorine atoms. Unpolymerizable monomers will, however, frequently copolymerize with other monomers (see section 16.8).

16.6.2 Aldehydes and ketones

Formaldehyde is readily polymerized cationically in solution if the concentration exceeds the equilibrium value (0·06 M in CH_2Cl_2 at 30 °C),

Table 16.6 Thermodynamic data for the polymerization of a series of aldehydes in tetrahydrofuran[77]

Monomer	$-\Delta H_{ss}^0$ (kJ mol^{-1})	$-\Delta S_{ss}^0$ (J K^{-1} mol^{-1})	T_c (°C) for 1 M solution
CCl_3CHO	14·6	51·9	11
$MeCCl_2CHO$	17·1	69·1	−24
$Me_2CClCHO$	19·7	90·4	−54
Me_3CCHO	—	—	< −78

and in the gas phase if the pressure exceeds the equilibrium value[40] (1 atm at 119 °C). The situation is complicated both in solution and in the gas phase by the concurrent equilibrium with cyclic oligomers,[40] particularly the trimer and tetramer. Hydrated forms are also present in solution.[76]

Substitution of the first hydrogen in formaldehyde by methyl or ethyl to give acetaldehyde and propionaldehyde lowers T_c to −31 °C for the formation of atactic polymer from the pure monomer, and to −39 °C for the formation of isotactic polymer.[22] Further substitution at the α-position to give isobutyraldehyde, Me_2CHCHO, lowers T_c to −63 °C for 1 M solution in tetrahydrofuran.[51] Pivalic aldehyde, Me_3CCHO, is unpolymerizable[77] at −78 °C. Although substitution by methyl groups has an adverse effect on polymerizability, substitution by halogen atoms has a beneficial effect. The series of monomers from chloral to pivalic aldehyde, shown in Table 16.6, is worthy of note.[77] Here the entropy effect outweighs the heat effect, unlike the case of the ethylenic monomers discussed above (section 16.6.1). Likewise gaseous fluoral, CF_3CHO, has a T_c of 73 °C at 1 atm pressure, much higher than the hypothetical value of $-10\,°C$ estimated for acetaldehyde[43] under the same conditions. Again, T_c for liquid 1,2-dichloropropionaldehyde,[50] $CH_2ClCHClCHO$, is in the region of $0\,°C$, which is higher than that for propionaldehyde $(-31\,°C$, see above).

The cationic polymerization of phthalaldehyde in methylene chloride (Table 16.2) is the only known example[25] of a cyclopolymerization with a readily measurable equilibrium monomer concentration $(1·0\,M$ at $-43\,°C)$.

Ketones are generally not polymerizable although it has been claimed[37] and suggested[35] that acetone may polymerize in the supercooled state (see section 16.5.1). In view of the possible commercial importance of a polymer of acetone the thermodynamics of its polymerization have received close attention.[26,78] It is estimated that the ceiling temperature for pure acetone is in the region of $-150\,°C$ to $-220\,°C$. Certainly it resists all attempts[38] to polymerize it at $-78\,°C$; but the effect of pressure has not been tried. Thioacetone on the other hand has a ceiling temperature of about $100\,°C$ and polymerizes spontaneously at room temperature to a high molecular weight product which is unstable and degrades to the cyclic trimer.[26,79,80] In contrast trioxane[52] polymerizes in solution at room temperature while trithiane can be polymerized in bulk at $220\,°C$, though not in solution with the same catalysts.[81] Thioacetophenone also is reported to polymerize both in bulk and in toluene solution.[233]

Fluorinated thioketones and aldehydes such as F_2CS, $(CF_3)_2CS$ and RCFS can also be polymerized.[82,83] H_2CS and RCHS do not polymerize.[84]

16.6.3 Isocyanates

Certain isocyanates can be polymerized in the presence of sodium cyanide[27] or other catalysts,[85] such as $Nipyr_2Et_2$, to give 1-nylons. A ceiling temperature effect is very clearly shown in the case of n-hexyl isocyanate[27] (Figure 16.2). The non-polymerizability of many other isocyanates, particularly those substituted close to the nitrogen atom, is attributable to a lowering of the ceiling temperature. For example isopropyl[27] and 1-phenylethyl[86] isocyanates will not polymerize but phenyl[85] and 2-phenylpropyl[86] isocyanates will polymerize.

16.6.4 The 1:1 copolymerization of alkenes with sulphur dioxide

Many ethylene derivatives M copolymerize with sulphur dioxide S to give 1:1 copolymers regardless of the composition of the feed. Thermodynamically such systems may be treated like homopolymerizations, and the effect of structure of the organic monomer on the ceiling temperature is therefore directly comparable with such effects in homopolymerizations.

Some ceiling temperatures for polysulphone formation are shown in Table 16.7. A small part of the variations is due to the physical effects

Table 16.7 Ceiling temperatures (°C) for 1:1 polysulphone formation at a standard concentration product of $[M][S] = 27 \text{ mol}^2 \text{l}^{-2}$ (excess of sulphur dioxide[87,88,89] except where indicated)[a]

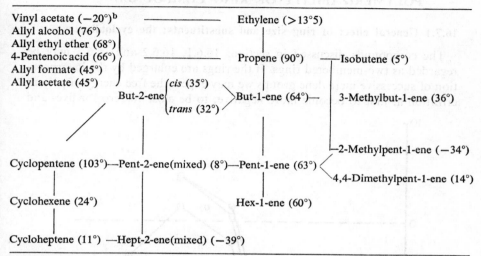

Vinyl acetate $(-20°)$[b] — Ethylene $(>13·5)$
Allyl alcohol $(76°)$
Allyl ethyl ether $(68°)$
4-Pentenoic acid $(66°)$ — Propene $(90°)$ — Isobutene $(5°)$
Allyl formate $(45°)$
Allyl acetate $(45°)$
But-2-ene $\begin{cases} cis\ (35°) \\ trans\ (32°) \end{cases}$ But-1-ene $(64°)$ — 3-Methylbut-1-ene $(36°)$

—2-Methylpent-1-ene $(-34°)$
Cyclopentene $(103°)$—Pent-2-ene(mixed) $(8°)$—Pent-1-ene $(63°)$ 4,4-Dimethylpent-1-ene $(14°)$

Cyclohexene $(24°)$ Hex-1-ene $(60°)$

Cycloheptene $(11°)$ —Hept-2-ene(mixed) $(-39°)$

a. The lines between the monomers M indicate structural relationships.
b. $[M][S] = 1$.

described in sections 16.5.1, 16.5.2 and 16.5.3, but in the main they reflect an effect of structure on ΔG_{lc}. The following rules may be formulated:

(a) The introduction of any substituent tends to lower T_c.

(b) Substituents at the double bond cause a much greater lowering of T_c than substituents away from the double bond. Compare, for example, isobutene (5 °C), cis-but-2-ene (35 °C) and but-1-ene (64 °C) with propene (90 °C).

(c) Electronegative substituents have a much greater effect than alkyl substituents. Compare, for example, vinyl acetate (-20 °C) with propene (90 °C), or allyl acetate (45 °C) with but-1-ene (64 °C).

The failure of alkenes such as 2-ethylbut-1-ene, 2-ethylhex-1-ene, 2,4,4-trimethylpent-1-ene and 4-methylpent-2-ene to form polysulphones at any temperature down to -80 °C at normal pressures may thus be interpreted as a lowering of T_c to below -80 °C by the increased substitution.[87] It is possible that polysulphones could be made from these alkenes by γ-irradiation of the mixture of solid monomers at still lower temperatures, as for vinyl acetate[89,90] at -150 °C, or at high pressures.

It appears from the limited data available that structural effects on T_c run broadly parallel with effects on ΔH.[87]

16.7 EFFECT OF MONOMER STRUCTURE ON POLYMERIZABILITY OF RING COMPOUNDS

16.7.1 General effect of ring size and substituents; the cycloalkanes

The compounds discussed in sections 16.6.1, 16.6.2 and 16.6.3 may be regarded as two-membered rings. If the rings are enlarged by the incorporation of successive methylene groups we may expect the free energy of polymerization at first to become more negative, to be close to zero for five- and

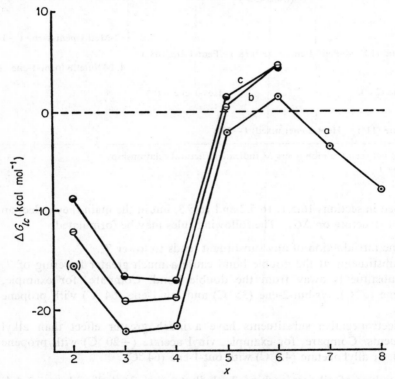

Figure 16.12 Free energy of polymerization of cycloalkanes as a function of the number of atoms in the ring, x. (a) unsubstituted; (b) methyl substituted; (c) 1,1-dimethyl substituted. Adapted from reference 13, using more recently available data.[16] 1 kcal $\equiv$ 4.184 kJ

six-membered rings, then to become negative again and finally to approach a value close to zero. This statement is based on the data for cycloalkanes plotted in Figure 16.12. The trend in ΔG values follows the trend in ΔH values. For three- and four-membered rings the strain energy arises from distortion of the normal bond angles. For five- and six-membered rings the strain energy is rather small and the sign of ΔG is then very sensitive to small changes in the physical conditions and chemical structure. For seven- and eight-membered rings ΔH and ΔG are again negative, but now the strain energy results from crowding within the ring. For larger rings ΔG must pass through a shallow minimum and approach a negative value, close to zero, determined mainly by the entropy change. This last situation will be considered further in section 16.7.3. It is also evident from Figure 16.12 that, as with unsaturated compounds, successive alkyl substitutions at the same carbon atom make ΔG more positive. In fact it is difficult to polymerize cycloalkanes to high polymers even when ΔG is negative; 1,1-dimethyl-cyclopropane is one of the few such compounds which will give reasonably high polymer.[91]

Before passing on to heterocyclic compounds it should be mentioned that changes in stabilization energy may sometimes play a role in determining the heat of polymerization of a cyclic compound. Thus it has been calculated that the hypothetical polymerization of cyclooctatetrene to linear conjugated polyacetylene would be exothermic by no less than 170 kJ mol^{-1} as a result of the gain in resonance energy.[92] The polymerization of cyclic $(PNCl_2)_4$ to linear polymer probably derives part of its driving force from this source. Such polymerizations have been termed resonance-induced.[92] The opposite situation occurs with sulphur, S_8, where polymerization is endothermic, indicating stabilization of the ring relative to the chain.[93]

16.7.2 Heterocyclic compounds

For heterocyclic compounds the general shape of the relationship between ΔG and ring size may be expected to be similar to that for the cycloalkanes, particularly if the hetero-atom does not differ too much from carbon in size and bond angles, for example in oxygen and nitrogen compounds, but not in sulphur compounds.[94] Ionic ring-opening mechanisms are generally available for heterocyclic compounds so that non-polymerizability is nearly always synonymous with a positive free energy change for polymerization. Cyclic sulphones provide one exception to this statement.[95] The considerable body of information on the polymerizability of heterocyclic compounds will now be considered in the light of the foregoing discussion.

Nearly all three- and four-membered heterocyclic rings can be polymerized, whether substituted or not; likewise for all eight- and higher-membered rings. One exception is the four-membered ring, $[(CH_2)_3NR_2]^+Br^-$, which does not polymerize when $R = Pr^n$ or Bu^n but does when $R = Me$.[96]

The polymerizability of some five-, six- and seven-membered ring compounds is summarized in Tables 16.8, 16.9 and 16.10. For ease of comparison some of the information for unsubstituted cyclic compounds is given in another form in Table 16.11. All the seven-membered rings listed in Table

Table 16.8 Polymerizability of five-membered rings (hydrogen atoms not shown)

Ring	Polymerized	Not polymerized	References
C_2N_2O	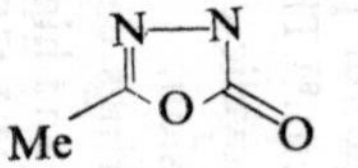 (note a.) R = Me, Ph, H	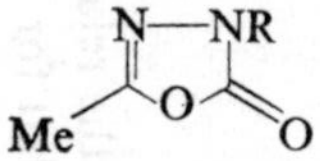 R = Me, Ph	97
C_3NO	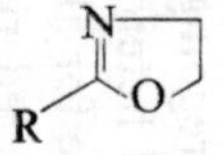 R = Me, Ph, H	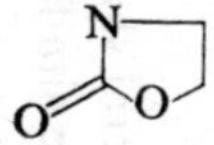 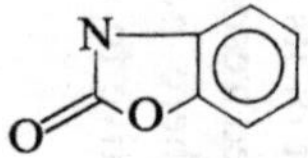	98,99, 254
C_3NS	(note b.)	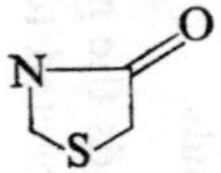 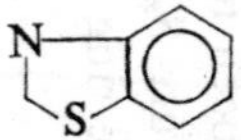 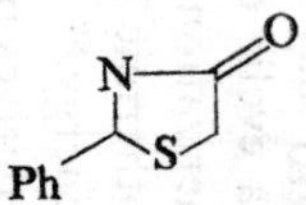 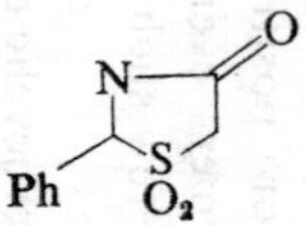	99, 100
C_3N_2	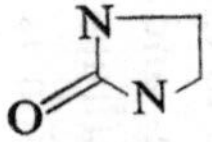	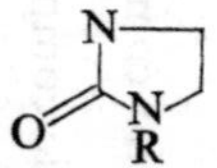 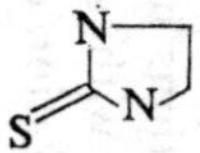 R = Me, MeCO	99
C_3OS	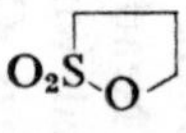	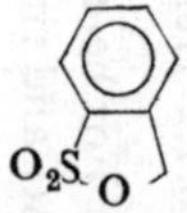	101

(contd.)

		References
C_3O_2	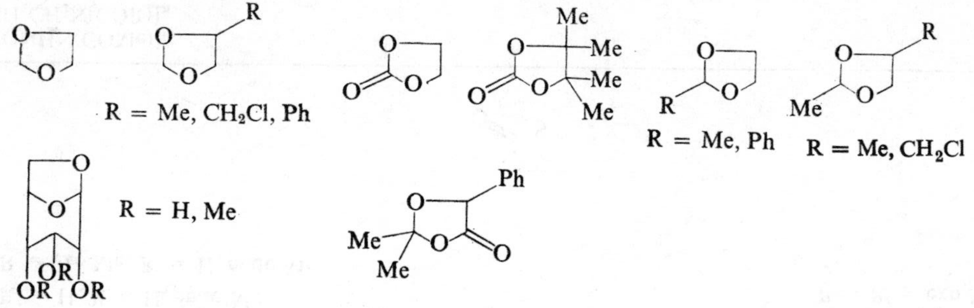 R = Me, CH$_2$Cl, Ph; R = H, Me; R = Me, Ph; R = Me, CH$_2$Cl (cyclic carbonate/dioxolane structures with Me, Ph substituents)	24, 31, 99, 102–108 234–236
C_3S_2	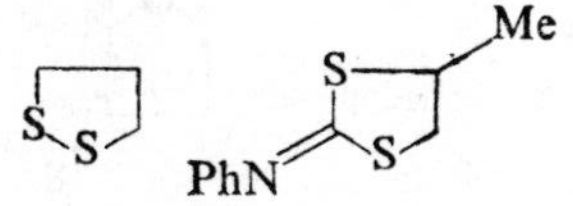 (dithiolane structures) R = Me	109, 110 237
C_4N	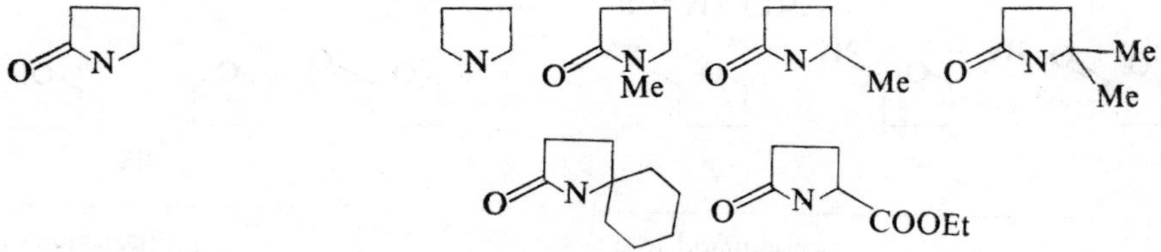 (pyrrolidinone/lactam structures) Me, Me; COOEt; N-Me	100, 111–113
	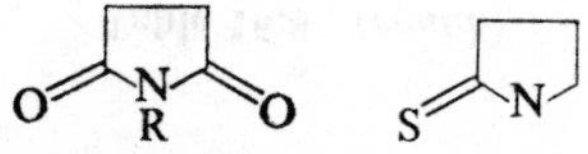 R = H, Me, Am^n; S	99, 100, 107

(contd.)

Table 16.8 *(contd.)*

Ring	Polymerized	Not polymerized	References
C$_4$O	Me (on tetrahydrofuran ring)	Me, Me (disubstituted tetrahydrofuran)	23, 45, 46, 59, 63, 99, 107, 114–121, 238–240
	(lactone, bicyclic) R' = H; R = H, endo Me; R = exo Me; R' = H, endo Me	R = Me, CH$_2$Cl (lactone)	
		(fused bicyclic lactone)	
		(γ-butyrolactone with Me)	
		R = R' = exo Me	
C$_4$S		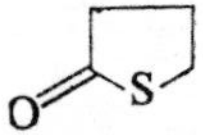 (thiolactone)	122

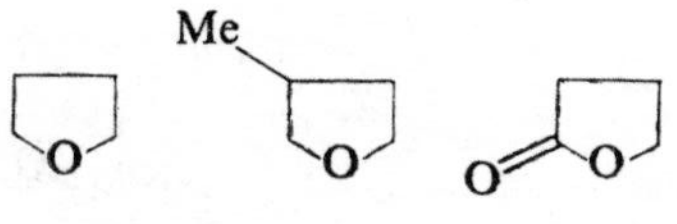

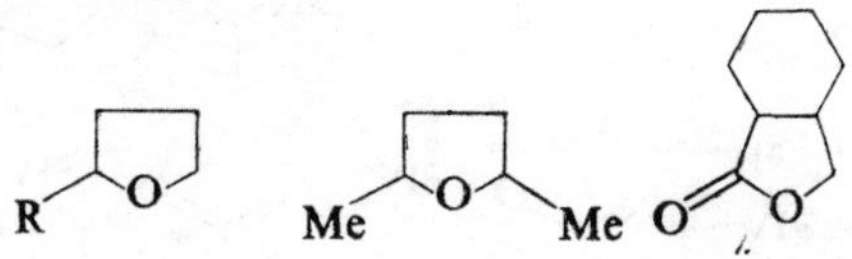

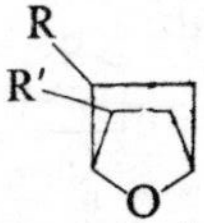

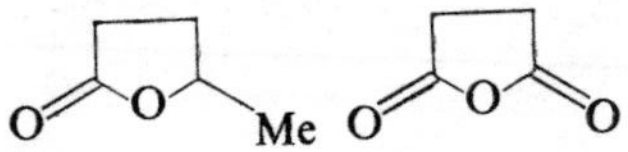

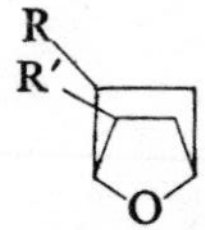

a. Polymer is [CONHN(COMe)]$_n$.
b. Polymer is [CH$_2$CH$_2$N(COR)]$_n$.

Table 16.9 Polymerizability of six-membered rings (hydrogen atoms not shown)

Ring	Polymerized	Not polymerized	References
C_3N_2O			99
C_3N_3		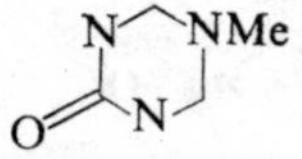	99
C_3O_3	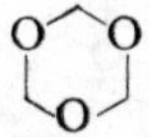	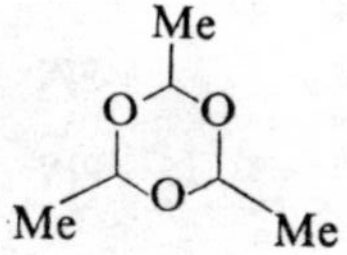	123–126
C_3S_3	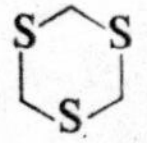	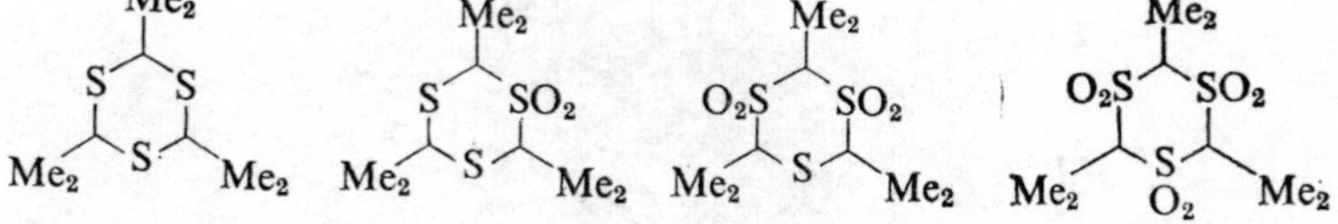	26, 81, 127, 241
C_4NO	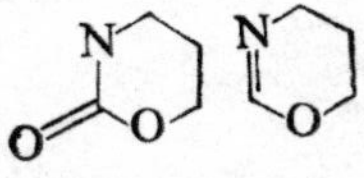	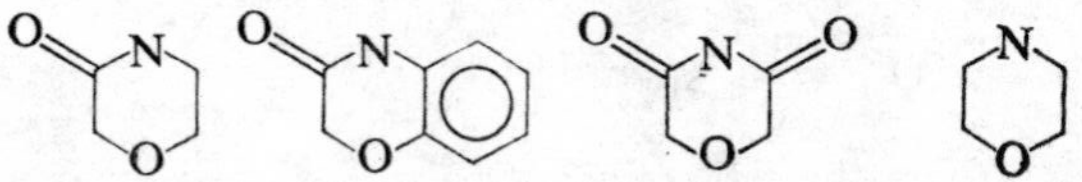	99, 100, 242, 255
C_4NS		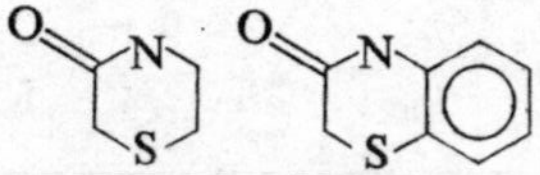	100

(*contd.*)

Table 16.9 (contd.)

Ring	Polymerized	Not polymerized	References
C_4N_2	(structures)	(structures) R = Me, Ph; with Me, Pr^i; R = Me, Ph	99, 100, 128
C_4OS		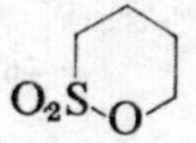 (structures) O$_2$S	101
C_4O_2	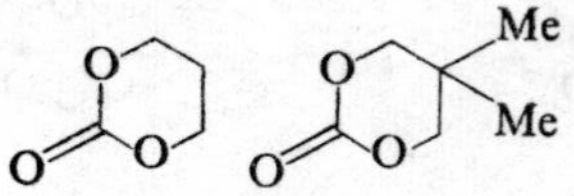 (structures) Me, Me; Me, Me; Me	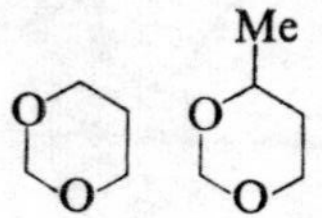 (structures) Me; Et; R = Me$_2$ Ph	49, 99, 103, 107, 129
			110, 132
	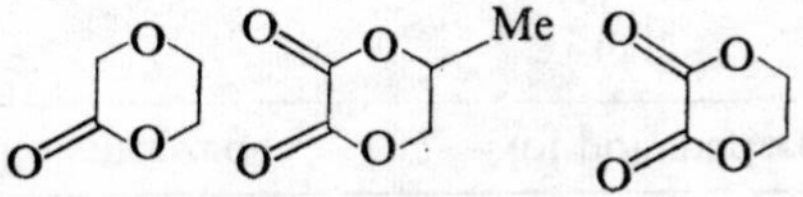 (structures) Me	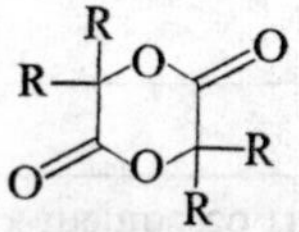 (structures) R, R; R, R R = Me, Ph	99, 107, 129–131

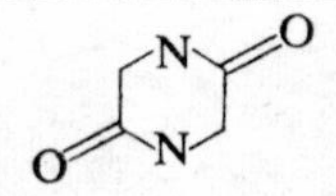

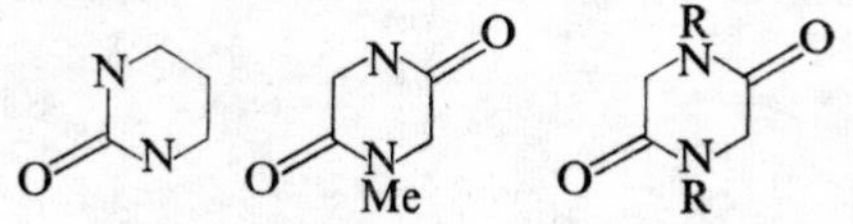

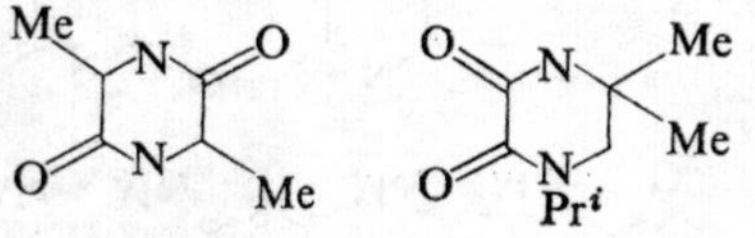

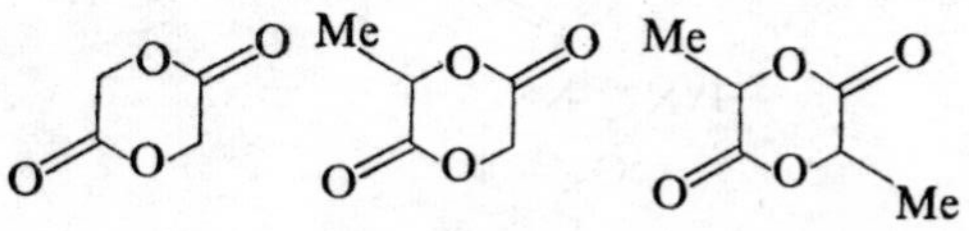

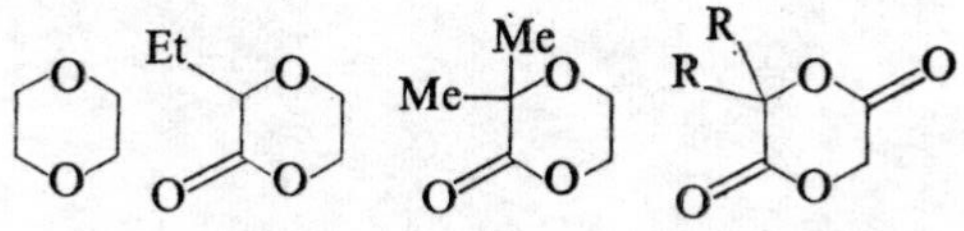

(contd.)

			References
C_4S_2	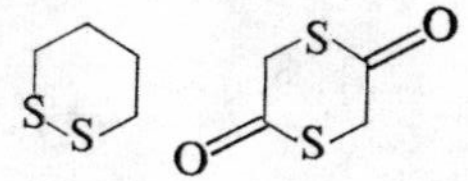		110, 132
C_5N	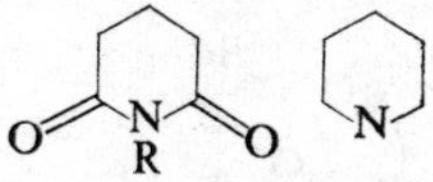 R = H, Et, Ph		39, 63, 94, 99, 100, 107
C_5O	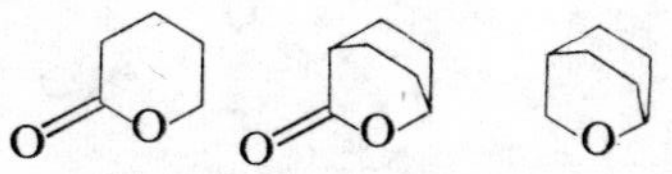	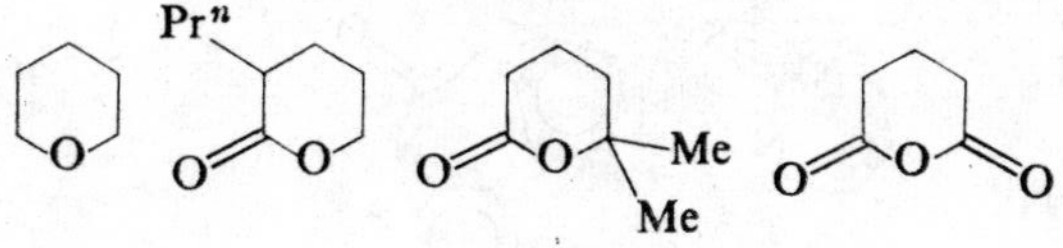	50, 94, 99, 107, 131, 133–136,
	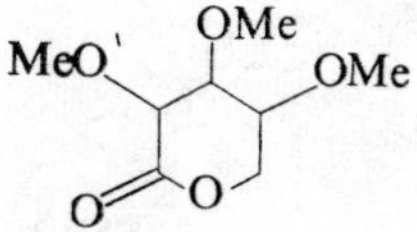		243
C_5S	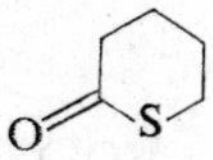		122, 242

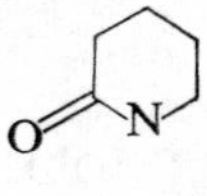

Table 16.10 Polymerizability of seven-rings (hydrogen atoms not shown)

Ring	Polymerized	Not polymerized	References
C_4OS_2			137
C_5NS			100
C_5N_2			99, 100
C_5O_2			49, 102, 103, 129, 138, 139
C_5S_2			110

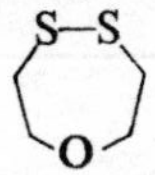

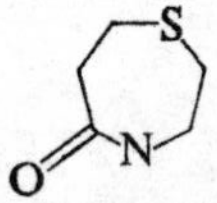

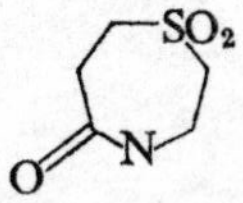

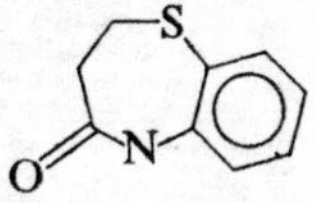

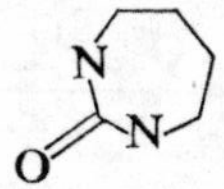

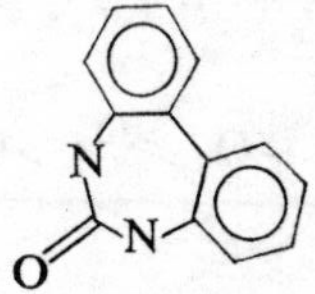

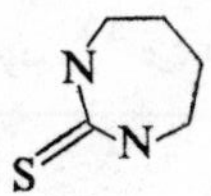

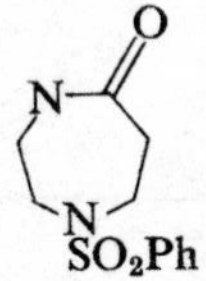

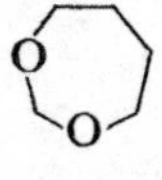

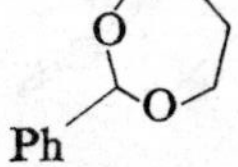

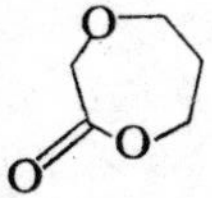

(*contd.*)

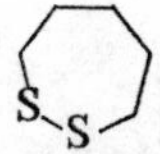

C_6N

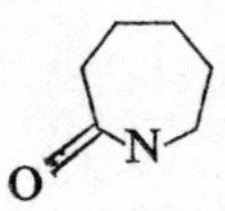

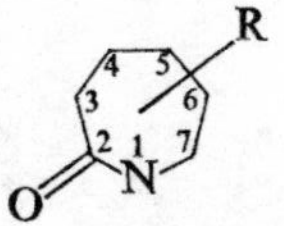

100, 113, 140–149, 256

100, 107, 146, 147, 150

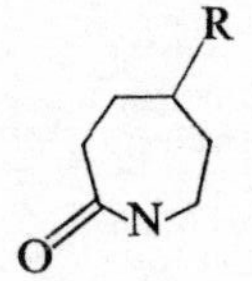

$R = Ph, PhCH_2, Hep^n$

$R = Me, Ph, CH_2OH, EtSCH_2$

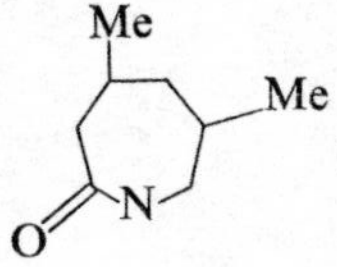

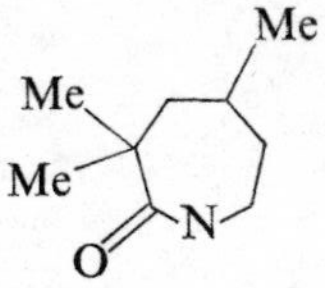

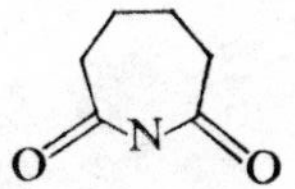

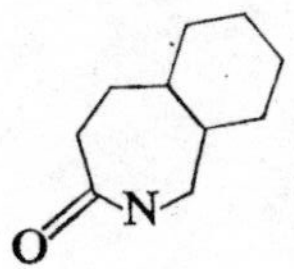

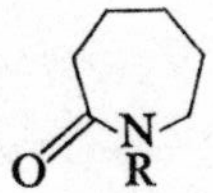

$R = H, Me$

$R = 3\text{-Me}, 4\text{-Me}, 5\text{-Me}, 5\text{-Et}, 5\text{-Pr}^n, 5\text{-Pr}^i, 5\text{-cyclohex}, 6\text{-Me}, 7\text{-Me}$

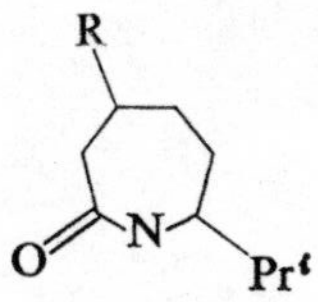

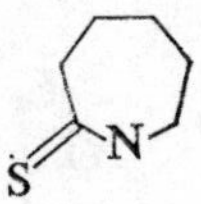

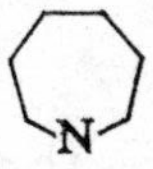

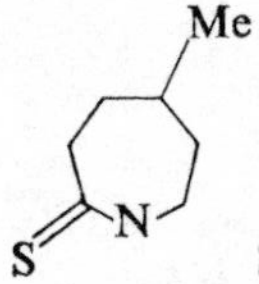

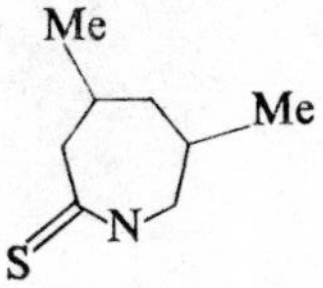

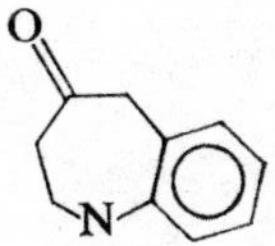

(contd.)

Table 16.10 (contd.)

Ring	Polymerized	Not polymerized	References
C_6O	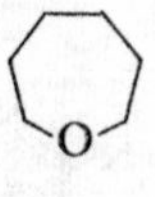 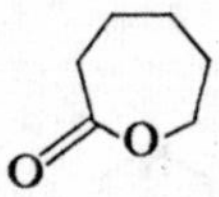 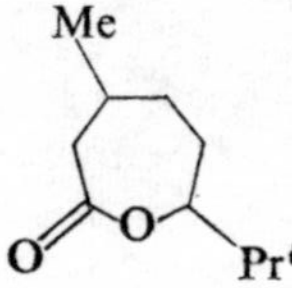 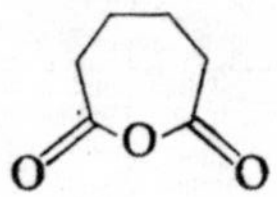 (oxepane, O) and (ε-caprolactone, C=O)	(methyl/Pr^t-substituted lactone) and (adipic anhydride, O=C–O–C=O)	44, 50, 99, 107, 151
C_6S	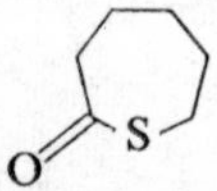 (thiepanone, S, C=O)		122

Table 16.11   Polymerizability ($+$ or $-$) of some unsubstituted rings

Group	Type of ring		Name of five-membered ring	Ring size 5	6	7
1	1,3-C_nNO	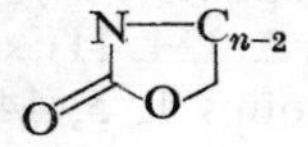	2-Oxazolidinone		+	
2	1,3-C_nN_2	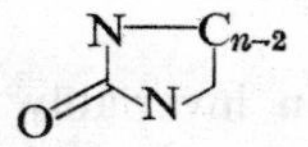	2-Imidazolidinone	+	−	+
3	1,2-C_nOS	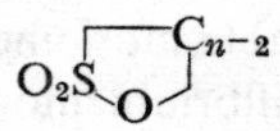	1,2-Oxathiolane-2,2-dioxide	+		
4	1,3-C_nO_2	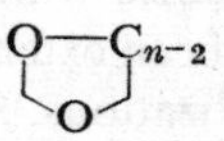	1,3-Dioxolane	−	+	
5		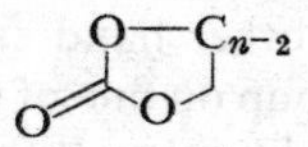	Ethylene carbonate	−	+	
6	1,2-C_nS_2	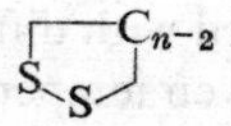	1,2-Dithiolane	+	+	+
7	C_nN	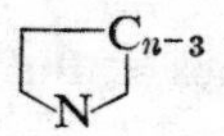	Pyrrolidine	−	−	+
8		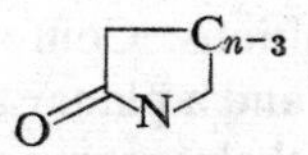	2-Pyrrolidinone	+	+	+
9		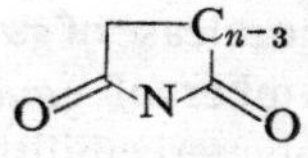	Succinimide	−	−	+
10		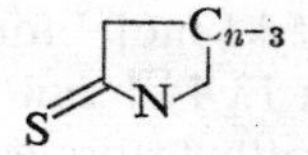	2-Pyrrolidinethione	−		+
11	C_nO	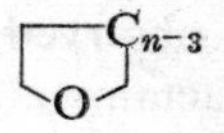	Tetrahydrofuran (oxolane)	+	−	+
12		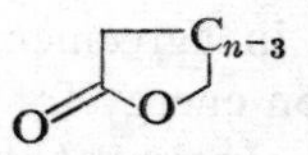	Oxolane-2-one (γ-butyrolactone)	+	+	+
13		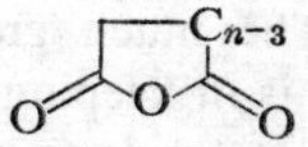	Oxolane-2,5-dione (succinic anhydride)	−		+
14	C_nS	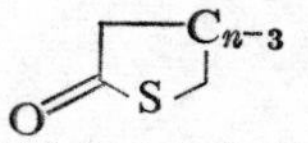	Thiolan-2-one (γ-thiobutyrolactone)	−	+	+

16.11 are polymerizable, although inspection of Table 16.10 shows that some unsubstituted seven-membered rings do not polymerize. For the unsubstituted five- and six-membered rings sometimes both will polymerize (groups 6, 8, 12), sometimes neither will polymerize (groups 7, 9, 13), sometimes only the five-membered ring will polymerize (groups 2, 3, 4, 11), and sometimes only the six-membered ring will polymerize (groups 1, 5, 14). This reflects the sensitivity of the sign of ΔG to small changes in the nature of the five- and six-membered rings.

For a given basic structure, substitution invariably leads to decreased likelihood of polymerization. Ample evidence for this may be found in Tables 16.8, 16.9 and 16.10. For polymerizable monomers it is always found that substitution leads to a decreased equilibrium monomer concentration and hence to a less negative ΔG, in accordance with the expectations from Figure 16.12. The substituted lactams have been extensively investigated and provide the best examples of these generalizations.[100,148,152] Substitution of 2-pyrrolidinone in any position by any substituent renders the compound unpolymerizable. For caprolactam on the other hand substitution must be either on the nitrogen atom or by a bulky group on one of the carbon atoms to have such a drastic effect. Differences can however be detected in the percentage polymerization which can be attained with different substituents. The equilibrium monomer concentration at a given temperature rises in the order of substituents:[148]

$$Me < Et < Pr^n < Pr^i < \text{cyclohex} < Bu^t < \text{gem-Me}_2$$

and a methyl group in the fourth or sixth positions has a bigger effect than a methyl group in the fifth and seventh positions. Conformational analysis, based on a chair structure for the monomer and a planar zig-zag structure for the polymer, allows one to interpret the enthalpy component of the driving force in the polymerization as arising from the release of *gauche* interactions.[152] Substitution reduces the change in the number of *gauche* interactions on polymerization and so leads to reduced polymerizability. The contribution of these interactions to $-\Delta H$, allowing $3 \cdot 35$ kJ mol^{-1} for each gauche interaction, decreases in the order: caprolactam $13 \cdot 4$ kJ mol^{-1}, 3- and 7-methylcaprolactams $10 \cdot 0$ kJ mol^{-1}, 4-, 5- and 6-methylcaprolactams $6 \cdot 7$ kJ mol^{-1}, 4,6-dimethylcaprolactam 0 kJ mol^{-1}. The observed order of equilibrium conversions for 7-, 4- and 6-methylcaprolactams and the non-polymerizability of 4,6-dimethylcaprolactam fit in with these calculations, although the result for the 5-methyl compound is not in agreement (see above). The decrease in the calculated *gauche* interaction energy for the polymerization of caprolactam ($13 \cdot 4$ kJ mol^{-1}) is in good agreement with the observed[153] heat of polymerization ($13 \cdot 9$ kJ mol^{-1}). The much greater sensitivity of 2-pyrrolidinone to substituents (see above) is in keeping with its very small value of $-\Delta H$ (2–3 kJ mol^{-1}).[154] Liquid 2-piperidone can be polymerized at 45–93 °C and it appears that the driving force may stem largely from the crystallization of the polymer.[39]

16.7.3 Ring-chain equilibria involving rings of different sizes

When tetraoxane, the cyclic tetramer of formaldehyde, is polymerized in 1,2-dichloroethane at 50 °C, the concentration of tetraoxane falls to an equilibrium value of less than 0·008 M, but at the same time an equilibrium concentration of 0·2 M trioxane is built up.[124] At concentrations of tetraoxane below 0·008 M no linear polymer is formed but cyclic trimer is rapidly produced.[155] Such behaviour is typical of a whole range of cyclic compounds. The establishment of equilibrium between cyclic compounds containing different numbers of repeat units usually occurs via linear polymer or oligomer, but sometimes a direct ring expansion mechanism is available, as in the polymerization of cycloalkenes on the catalyst WCl_4—ROH—$RAlCl_2$ to give a macrocyclic polyene.[156] For cyclopentene, cyclic structures with 2–7 repeat units have been identified. Other examples include isocyanates[27,157,158] $[RNCO]_{2,3}$, phosphonitriles[159] $[PNCl_2]_{3-7}$, caprolactam[160] $[(CH_2)_5CONH]_{2-5}$ ethylene terephthalate[161] $[CH_2CH_2OCOC_6H_4COO]_{3-5}$, decamethylene adipate[162] $[(CH_2)_{10}OCO(CH_2)_4COO]_x$, sulphur[84,163] S_{2-10}, phosphates[84,163] $[RPO_3]_{3-8}$, silthianes[84,163] $[Me_2SiS]_{2,3}$, but above all the siloxanes $[RMeSiO]_x$, where R = H,[164] Me,[165-169] Et,[164] Pr^n,[164] $CF_3CH_2CH_2$[164] and Ph.[170] Further references may be found in the reviews of Allcock[84,163] and Carmichael.[171]

In a remarkable study on the dimethylsiloxane system, Brown and Slusarczuk[168] prepared an equilibrium mixture from the cyclic tetramer in the presence of potassium hydroxide and from the equilibrate were able not only to isolate individual cyclics up to $x = 25$ by vapour phase chromatography but also were able to demonstrate by gel permeation chromatography that these formed the first members of a continuous population extending up to at least $x = 400$. The total macrocyclic population represented 8–10 per cent of the total polymer in systems equilibrated at 200–250 g l^{-1}, and 2–3 per cent of the total polymer in commercial silicone oils, gums and rubbers. The data so obtained provided the first, and still the most extensive, test of the Jacobson-Stockmayer[172] theory of ring-chain equilibria.

The basic postulates of this theory are that:

1. The distribution of ring sizes is determined only by entropy considerations, i.e. all rings are strainless.
2. The end-to-end distances of linear chains obey Gaussian statistics; the random flight model is assumed and no allowance is made for the excluded volume effect.
3. The probability of ring formation is governed by the frequency with which the ends coincide.

4.   The probability of bond formation is independent of whether the ends form part of the same chain or are ends of separate chains.

The theory predicts that:

1.   No linear polymer should be formed below a critical concentration, expressed as moles of repeat units per litre.

2.   The concentration of rings, $c\text{-}M_x$, in equilibrium with linear polymer, M_y, should be independent of dilution.

3.   The equilibrium constant K_x, defined by the equation $M_y \rightleftharpoons M_{y-x} + c\text{-}M_x$, should be proportional to $x^{-2\cdot5}$ and independent of temperature. In contrast to the thermodynamic distribution of linear polymers, this means that the *weight* distribution of macrocyclic constituents at equilibrium is predicted to be a monotonically decreasing function of ring size.

All these predictions are in reasonable agreement with experiment provided x is sufficiently large; Figure 16.13 shows that for polydimethylsiloxane K_x is nearly a linear function of $x^{-2\cdot5}$ for $x > 15$. Deviations at lower values of x may be attributed to the failure of the first postulate. K_x (and hence $[c\text{-}M_x]_e$) is also dependent on the nature of the medium for small values of x, as may be expected from earlier considerations (see section 16.5.3).

As a check on the validity of the first postulate we may examine heats of polymerization ΔH as a function of ring size. For the cycloalkanes $(CH_2)_x$, $-\Delta H$ has fallen to $8\cdot4\,\mathrm{kJ\,mol^{-1}}$ for $x = 17$ (liquid monomer going to amorphous polymer),[16] while for the series $(PNCl_2)_x$, $-\Delta H$ falls to zero for $x = 7$ (fourteen-atom ring).[159] For the siloxanes it appears from Figure 16.13 that the ring size must be somewhat bigger (thirty-atom ring) before it becomes strainless relative to the open chain.

The postulate of a random flight model is a very rough approximation: bond angles have well-defined values and there is restricted rotation about the bonds arising from both short-range and long-range interactions. This explains why the theoretical proportionality constant between K_x and $x^{-2\cdot5}$ differs markedly from that observed. Jacobson and Stockmayer[162] expressed this deviation in terms of an *effective* length of a chain link of $4\cdot5 \pm 0\cdot2$ Å. More recently Flory and Semlyen[169] have revised the theory, replacing the random flight model with the rotational-isomeric-states model. This enables the proportionality constant to be expressed in terms of the parameters describing the proportion of conformers about each main-chain bond. These parameters may be obtained from a study of the solution properties of linear polydimethylsiloxane. The predicted absolute values of K_x are then in good agreement with experiment for large values of x (Figure 16.13). The weak minimum in the experimental curve at lower values of x appears to be genuine and is found for other polysiloxanes.[164] It is to be noted that the experimental and calculated curves agree at one particular small value of x corresponding closely to that expected for a strainless ring.

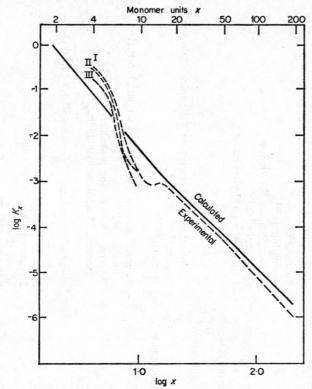

Figure 16.13 Macrocyclic equilibrium constants for polydimethylsiloxane. $M_y \rightleftharpoons M_{y-x} + c\text{-}M_x$, K_x. x = number of repeat units (Me_2SiO). Experimental curves: I^{168}, II^{173}, III^{167}. Calculated curve: Flory and Semlyen[169]

16.8 COPOLYMERIZATIONS INVOLVING REVERSIBLE ADDITION

The copolymerization of two monomers M_1 and M_2 is generally considered in terms of four propagation steps:

$$M_1^* + M_1 \rightarrow M_1^* \qquad k_{11} \qquad r_1 = \frac{k_{11}}{k_{12}}$$
$$M_1^* + M_2 \rightarrow M_2^* \qquad k_{12}$$
$$M_2^* + M_2 \rightarrow M_2^* \qquad k_{22} \qquad r_2 = \frac{k_{22}}{k_{21}}$$
$$M_2^* + M_1 \rightarrow M_1^* \qquad k_{21}$$

where M_1^*, M_2^* represent growing polymer chains possessing M_1 and M_2 end-units respectively. The copolymer composition, expressed as the molar ratio $d[M_1]/d[M_2]$, is then given by the Lewis-Mayo equation (14).

$$\frac{d[M_1]}{d[M_2]} = \frac{[M_1]}{[M_2]}\left[\frac{r_1[M_1] + [M_2]}{r_2[M_2] + [M_1]}\right] \qquad (14)$$

Table 16.12 Monomers M_2 which do not homopolymerize but which can be copolymerized with monomer M_1

M_2	M_1	
	$d[M_1]/d[M_2]$ always > 1	$d[M_1]/d[M_2]$ sometimes approx. 1
Oxygen		Styrene,[175] α-methylstyrene,[175] methyl methacrylate[175]
Carbon monoxide	Vinyl monomers[176]	Formaldehyde[177]
Carbon dioxide	N-phenylethylenimine[244]	Oxiranes[178-180,245]
Carbon oxysulphide		Oxiranes,[181] aziridines[182]
Carbon disulphide		Thiirane[253]
Phenyl isothiocyanate		Methylthiirane[237]
Sulphur dioxide	3-Methyl-3-chloromethyloxetane[183]	Alkenes,[95] dienes,[95] styrenes[a,95]
Tetrahydropyran		Epichlorohydrin[184]
1,3-Dioxane	3,3-Bis(chloromethyl)oxetane,[185] trioxane[250]	3,3-Bis(chloromethyl)oxetane[184-187]
1,4-Dioxane	3,3-Bis(chloromethyl)oxetane[187]	
2-Methyltetrahydrofuran		3,3-Bis(chloromethyl)oxetane[184-186] epichlorohydrin[148]
Benzaldehyde	Ethylene,[188] vinyl monomers[188]	Styrene,[246-248] dienes[246]
Hexafluoroacetone	Styrene etc.[b,191]	Oxiranes[189,190]
α-Trifluoromethyl vinyl acetate		Styrene,[4] acrylonitrile,[4] methyl acrylate,[4] methyl methacrylate[4]
α-Methoxystyrene		Styrene,[c,192,252] o- and p-methoxy-styrenes,[193] isoprene,[194] 2,3-dimethylbutadiene[195]
1,1-Diphenylethylene	Styrene[d,196]	
Trans-crotonitrile		Styrene,[197] butadiene,[198] isoprene,[198] 2,3-dimethyl-butadiene[198]
Trans-stilbene		Styrene,[199] acrylonitrile[199]
Trans-1,2-di(2-pyridyl)-ethylene	Butadiene[199]	
Trans-1,2-dibenzoylethylene		Styrene,[174] n-butyl vinyl ether[174]
Trans-1,2-diacetylethylene		Styrene[174]
α-Stilbazole		2-Vinylpyridine[200]
2,3-Dimethylbut-2-ene	Thiocarbonyl fluoride[201]	

a. $M_2M_1^* + M_2$ does not occur at 25 °C but does at −78 °C.
b. No details of composition.
c. $M_2^* + M_1$ reversible at 25 °C.

This equation has been found satisfactory for a wide variety of monomer pairs,[6] but it is clear that it will fail if one of the reverse reactions becomes important compared with its forward reaction, as may happen when the monomer to be added is in the region of its ceiling temperature. If the conditions are such that M_2 is far above its ceiling temperature, the addition of M_2 to M_2^* may be very rapidly reversed whilst the addition of M_2 to M_1^* is not reversed. Under these circumstances the *effective* value of k_{22} and hence of r_2 will be zero and equation (14) reduces to equation (15).

$$\frac{d[M_1]}{d[M_2]} = 1 + r_1 \frac{[M_1]}{[M_2]} \tag{15}$$

Figure 16.14 shows an example of this type of behaviour for a monomer, α-methoxystyrene, which itself is unpolymerizable, presumably because its

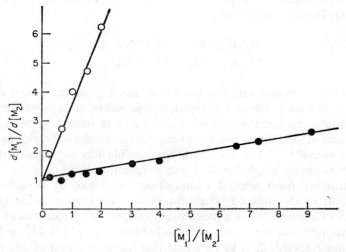

Figure 16.14 Radical copolymerization of α-methoxystyrene (M_2) with methyl acrylate (filled circles) and methyl methacrylate (open circles)[4] at 60 °C

ceiling temperature is so far below room temperature.[4] At low values of $[M_1]/[M_2]$ the copolymer tends to the alternating 1:1 composition; the slope gives $r_1 = 0.17$ for methyl acrylate and $r_1 = 2.5$ for methyl methacrylate (radical copolymerization at 60 °C). If r_1 is less than 0.01 the composition of the copolymer will be close to 1:1 over a wide range of feed composition, as for example in the radical copolymerization of styrene with *trans*-1,2-dibenzoylethylene.[174]

In Table 16.12 are summarized most of the known cases of monomers which do not themselves polymerize but which readily copolymerize, sometimes to the extent of forming alternating copolymers. The ability to form such copolymers is sometimes critically dependent on the nature of the catalyst and solvent. For example, in the copolymerization of 3,3-bis(chloromethyl)-oxetane with 2-methyltetrahydrofuran an alternating copolymer can be achieved with Et_3Al/H_2O as catalyst, but not with BF_3/Et_2O.[186] Another example is the anionic copolymerization of isoprene with 1,1-diphenylethylene,[194] for which r_1 has a constant value of 0·12, independent of cation, in tetrahydrofuran as solvent at 0 °C, but varies from 37 for Li^+ as cation to 0·05 for K^+ as cation when the solvent is benzene at 40 °C. Again, *trans*-stilbene can be copolymerized with butadiene in tetrahydrofuran as solvent, but not in benzene.[198] Particularly worthy of note is the fact that polycarbonates can be made by the 1:1 copolymerization of carbon dioxide with oxiranes, using Et_2Zn/H_2O as catalyst.[178-180]

It will be seen from the footnotes in Table 16.12 that in some cases the ability of M_2 to add to M_1^* depends both on the penultimate unit and on the temperature; in other words the processes

$$M_1M_1^* + M_2 \rightarrow M_1M_1M_2^*$$
$$M_2M_1^* + M_2 \rightarrow M_2M_1M_2^*$$

sometimes have different rate constants and may become reversible as the temperature is raised. General theoretical treatments of copolymerization which take into account the reversal of one or more steps have been given by Lowry and others. These treatments are concerned not only with copolymer composition[202-204] but also with the equilibrium sequence distribution,[205-209] molecular weight[29] and ceiling temperature.[210, 211]

The transition from normal composition behaviour as described by equation (14) to the limiting behaviour as described by equation (15) is illustrated in Figure 16.15 for a theoretical example with conditions as given in the caption[202] and for Lowry's[202] mechanism I in which $M_1^* + M_2 \rightarrow M_1M_2^*$ is not reversed. It is to be noted that the proportion of M_2 in the copolymer may exceed 0·5 even above the ceiling temperature for M_2. This is because there is a finite probability that $M_2M_2^*$ will add M_1 before it can shed the terminal M_2 unit; such M_2 units then become locked into place in the polymer chain. This mechanism applies to both radical[212] and anionic[213] copolymerizations of methyl methacrylate (M_1) with methyl atropate (M_2) at 60 °C, and to its radical copolymerization with α-methylstyrene (M_2)[214] at 100 °C; also to the radical part of the copolymerization of acrylonitrile or methacrylonitrile with α-methylstyrene at 40 °C, catalysed by $AlEt_{1.5}Cl_{1.5}$.[215]

If both $M_2^* + M_2 \rightarrow M_2M_2^*$ and $M_1^* + M_2 \rightarrow M_1M_2^*$ are reversed, regardless of the detailed structure of M_1^* and M_2^*, the polymer formed under conditions of high temperature and high M_2 concentration will be the homopolymer of M_1. This is Hazell and Ivin's mechanism IV and applies to the

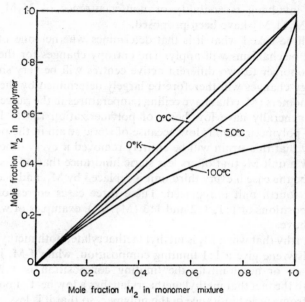

Figure 16.15 Copolymer composition [202] as affected by reversibility of the process $M_2^* + M_2 \underset{k_{-22}}{\overset{k_{22}}{\rightleftharpoons}} M_2 M_2^*$. Conditions: (a) $M_1^* + M_2 \rightarrow M_1 M_2^*$ assumed not to be reversed; (b) $r_1 = r_2 = 1$; (c) $T_c = 50\,°C$ for $[M_2] = 1$ mol l^{-1}; (d) $E_{22} = 8\cdot4$ kJ mol^{-1}, $E_{-22} = 50\cdot2$ kJ mol^{-1}

formation of a polysulphone from a mixture of cyclopentene (M_1) and isobutene (M_2) with sulphur dioxide.[203]

Yet another possibility is that M_2 may add to $M_1 M_2^*$ but not to $M_2 M_2^*$, steric strain only preventing the occurrence of the forward reaction when it would lead to the presence of three consecutive M_2 units in the chain. In this case the limiting composition of the copolymer at high temperature and high M_2 concentration is 1:2 ($M_1:M_2$). This is Lowry's[202] mechanism II and applies to the radical copolymerization of styrene (M_1) with methyl methacrylate (M_2)[216] at 132 °C, to the radical copolymerization of styrene,[212,217] methyl acrylate[212] and acrylonitrile[212] (M_1) with methyl atropate (M_2) at 60 °C, and to the cationic copolymerization of vinyl mesitylene (M_1) with α-methylstyrene (M_2)[216] at 0 °C. For the radical copolymerization of α-methylstyrene (M_2) with acrylonitrile,[218–221] methacrylonitrile,[218,219] fumaronitrile[218,222] and styrene (M_1)[221] the limiting proportion of M_2 may be even higher than 0·67, possible 0·75; likewise for the copolymerization of 3,3-bis(chloromethyl)oxetane (M_1) with tetrahydrofuran (M_2) at 120 °C, catalysed by tri-isobutyl aluminium,[223] and for the copolymerization of styrene (M_1) with o-phthalaldehyde (M_2) at 0 °C, catalysed by BF$_3$, Et$_2$O.

Mechanisms involving reversibility or non-occurrence of $M_1M_2M_2M_2^* + M_2 \rightarrow M_1M_2M_2M_2M_2^*$ have been proposed.

It may well be asked what it is that determines whether one or other of the proposed mechanisms will apply. The entropy changes for the addition of a given monomer to two different active centres will be very similar and the free energy changes will therefore be largely determined by the enthalpy changes. Monomers (M_2) that have ceiling temperatures in the range $-100\,°C$ to $+150\,°C$ generally have low heats of polymerization (20–50 kJ mol^{-1}). If the heat of polymerization is low because of steric strain in the polymer of M_2 it is likely that this strain will be entirely removed if every other M_2 unit is replaced by a unit M_1 that offers less steric hindrance than M_2; sometimes this will still be the case if every third unit is replaced by M_1, and occasionally if only every fourth unit is replaced. These three cases correspond to the limiting compositions of $1:1$, $1:2$ and $1:3$ ($M_1:M_2$), examples of which have been given above.

It is noteworthy that when M_1 is methyl methacrylate, both methyl atropate and α-methylstyrene give a $1:1$ limiting composition, while if M_1 is styrene, methyl acrylate or acrylonitrile the limiting composition is $1:2$. This is consistent with the fact that methyl methacrylate has a low heat of polymerization, arising from steric hindrance in the polymer, so that it is less effective in removing strain than the other M_1 comonomers.

Although a certain amount of data exists on heats of copolymerization[16] none of it is directly relevant to the systems discussed above. It may be predicted that in these systems $-\Delta H_{12} > -\Delta H_{22}$, and that the overall heat of copolymerization per monomer unit will be numerically greater than the weighted average heat of polymerization of the separate monomers. This is indeed the case in the systems styrene/methyl methacrylate and vinyl acetate/methyl methacrylate,[224] where the two cross-propagation reactions overall are about 10 kJ mol^{-1} more exothermic than the two homopropagation reactions overall.

If the heat of polymerization of M_2 is low, not because of steric strain in the polymer, but for other reasons such as resonance stabilization of the monomer, then the exothermicity of the reaction $M_1^* + M_2 \rightarrow M_1M_2^*$ will be much the same as for $M_2^* + M_2 \rightarrow M_2M_2^*$ and both reactions may be expected to become reversible under the same conditions. In these circumstances only homopolymer of M_1 will be produced at high temperature and high M_2 concentration. Copolymerizations involving aldehydes, isocyanates or ring compounds (M_2) may be expected to fall into this category, but the polysulphone reaction (see above) seems to be the only one in which this prediction has been put to the test.[203] Thus acetaldehyde (M_2) copolymerizes with 3,3-bis(chloromethyl)oxetane[225] at $-5\,°C$ but is not expected to copolymerize at much higher temperatures. Surprisingly tetrahydrofuran (M_2) copolymerizes with 3,3-bis(chloromethyl)oxetane[223] at $120\,°C$ to the extent of 75 per cent even though the temperature is 40 °C above the ceiling tempera-ture for pure tetrahydrofuran. Depropagation effects have been observed in copolymerizations involving 1,3-dioxolane and 1,3,5-trioxane.[226,227]

There have been very few attempts[203,214,216,221,223-251] to study experimentally the transition from 'normal' copolymerization behaviour through 'abnormal' behaviour to 'limiting' behaviour as illustrated theoretically in Figure 16.15 for one particular mechanism. The reason for this is that it is not generally obvious from a graph of copolymer composition against feed composition at a single temperature that some of the propagation steps may be undergoing reversal.[214] The experimental precision and limitations are such that most data can be expressed in terms of equation (14), and failing that by a modifica-tion of the mechanism in which certain propagation constants are assigned a rate constant of zero. Caution is therefore needed in interpreting reactivity ratios quoted for monomer pairs in which one monomer is either unpolym-erizable or within 40 °C of its ceiling temperature under the prevailing conditions. This applies particularly to five-, six- and seven-membered ring compounds[227-229] and to monomers such as α-methylstyrene.[230] Care is also needed in distinguishing between true and apparent values of r_2 close to zero. Examples of true values close to zero are found in the anionic copolymeriza-tion of β-cyanopropionaldehyde (M_1) with methyl isocyanate (M_2)[231] at -78 °C, and in the radical copolymerization of vinyl acetate (M_1) with crotonaldehyde (M_2).[232] Examples of apparent values close to zero have already been given.

16.9 REFERENCES

1. H. W. McCormick, *J. Polymer Sci.*, **25**, 488, (1957).
2. J. G. Kilroe and K. E. Weale, *J. Chem. Soc.*, **1960**, 3849.
3. D. J. Stein, P. Wittmer and J. Tölle, *Angew. Makromol. Chem.*, **8**, 61, (1969).
4. H. Lüssi, *Makromol. Chem.*, **103**, 68, (1967).
5. A. Novak and E. Whalley, *Canad. J. Chem.*, **37**, 1710, (1959).
6. P. J. Flory, *Principles of Polymer Chemistry*, Cornell University Press, Ithaca, 1953.
7. P. J. Flory, *J. Amer. Chem. Soc.*, **64**, 2205, (1942).
8. W. B. Brown and M. Szwarc, *Trans. Faraday Soc.*, **54**, 416, (1958).
9. M. Szwarc, *Adv. Polymer Sci.*, **4**, 457, (1967).
10. M. P. Dreyfuss and P. Dreyfuss, *Amer. Chem. Soc. Polymer Preprints*, **11**, 203, (1970).
11. R. D. Snow and F. E. Frey, *Ind. Eng. Chem.*, **30**, 176, (1938).
12. F. S. Dainton and K. J. Ivin, *Nature*, **162**, 705, (1948).
13. F. S. Dainton and K. J. Ivin, *Quart. Rev. Chem. Soc. (London)*, **12**, 61, (1958).
14. J. E. Carruthers and R. G. W. Norrish, *Trans. Faraday Soc.*, **32**, 195, (1936).
15. L. Rotinjanz, *Z. Physik. Chem.*, **62**, 609, (1908); R. F. Bacon and R. Fanelli, *J. Amer. Chem. Soc.*, **65**, 639, (1943).
16. *Polymer Handbook*, ed. J. Brandrup and E. H. Immergut, Interscience, New York, 1966; second edition 1973.

17. S. Bywater and D. J. Worsfold, *J. Polymer Sci.*, **58**, 571, (1962).
18. D. J. Worsfold and S. Bywater, *J. Polymer Sci.*, **26**, 299, (1957).
19. H. Hopff and H. Lüssi, *Makromol. Chem.*, **62**, 31, (1963).
20. H. Hopff, H. Lüssi and L. Borla, *Makromol. Chem.*, **81**, 268, (1965).
21. W. Kern and V. Jaacks, *J. Polymer Sci.*, **48**, 399, (1960).
22. A. M. North and D. Richardson, *Polymer, London*, **6**, 333, (1965).
23. K. J. Ivin and J. Léonard, *Polymer, London*, **6**, 621, (1965).
24. P. H. Plesch and P. H. Westermann, *J. Polymer Sci.*, **C16**, 3837, (1968).
25. C. Aso, S. Tagami and T. Kunitake, *J. Polymer Sci.*, **A-1, 7**, 497, (1969).
26. V. C. E. Burnop and K. G. Latham, *Polymer, London*, **8**, 589, (1967).
27. V. E. Shashoua, W. Sweeny and R. F. Tietz, *J. Amer. Chem. Soc.*, **82**, 866, (1960).
28. K. J. Ivin and J. Léonard, *European. Polymer J.*, **6**, 331, (1970).
29. G. G. Lowry, *Amer. Chem. Soc., Polymer Preprints*, **11**, 189, (1970).
30. S. Bywater, *Trans. Faraday Soc.*, **51**, 1267, (1955).
31. L. I. Kuzub, M. A. Markevich, A. A. Berlin and N. S. Yenikolopyan, *Polymer Sci., U.S.S.R.*, **10**, 2332, (1968); *Vysokomol. Soedinenya*, **A10**, 2007, (1968).
32. F. S. Dainton and K. J. Ivin, *Proc. Roy. Soc.*, **A212**, 207, (1952).
33. D. C. Abercromby, R. A. Hyne and P. F. Tiley, *J. Chem. Soc.*, **1963**, 5832.
34. Y. Ohtsuka and C. Walling, *J. Amer. Chem. Soc.*, **88**, 4167, (1966).
35. D. Heikens and H. Geilen, *Polymer, London*, **3**, 591, (1962).
36. A. Eisenberg and A. V. Tobolsky, *J. Polymer Sci.*, **46**, 19, (1960).
37. V. A. Kargin, V. A. Kabanov, V. P. Zubov and I. M. Papisov, *Dokl. Akad. Nauk S.S.S.R.*, **134**, 1098, (1960).
38. V. C. E. Burnop, *Polymer, London*, **6**, 411, (1965).
39. N. Yoda and A. Miyake, *J. Polymer Sci.*, **43**, 117, (1960).
40. W. K. Busfield and D. Merigold, *Makromol. Chem.*, **138**, 65, (1970).
41. Y. Isawa and T. Imoto, *J. Chem. Soc. Japan, Pure Chem. Sect.* [*Nippon Kagaku Zasshi;* **84**, 29, (1963).
42. W. K. Busfield and E. Whalley, *Trans. Faraday Soc.*, **59**, 679, (1963).
43. W. K. Busfield, *Polymer, London*, **7**, 541, (1966).
44. F. P. Jones, Ph.D. thesis, University of Keele, 1970.
45. D. Sims, *J. Chem. Soc.*, **1964**, 864.
46. C. E. H. Bawn, R. M. Bell and A. Ledwith, *Polymer, London*, **6**, 95, (1965).
47. B. A. Rozenberg, O. M. Chekhuta, E. B. Ludwig, A. R. Gantmakher and S. S. Medvedev, *Vysokomol. Soedinenya*, **6**, 2030, (1964).
48. S. Bywater, *Canad. J. Chem.*, **35**, 552, (1957).
49. P. H. Plesch and P. H. Westermann, *Polymer, London*, **10**, 105, (1969).
50. H. Sumimoto and T. Nakagawa, *J. Polymer Sci.*, **B7**, 739, (1969).
51. I. Mita, I. Imai and H. Kambe, *Makromol. Chem.*, **137**, 169, (1970).
52. T. Miki, T. Higashimura and S. Okamura, *J. Polymer. Sci.*, **A-1, 8**, 157, (1970).
53. J. Léonard and S. L. Malhotra, *J. Polymer Sci.*, **A-1, 9**, 1983, (1971).
54. F. S. Dainton and K. J. Ivin, *Discuss. Faraday Soc.*, **14**, 199 (1953).
55. R. E. Cook, K. J. Ivin and J. H. O'Donnell, *Trans. Faraday Soc.*, **61**, 1887, (1965).
56. B. H. G. Brady and J. H. O'Donnell, *Trans. Faraday Soc.*, **64**, 29, (1968).
57. B. H. G. Brady and J. H. O'Donnell, *Trans. Faraday Soc.*, **64**, 23, (1968).
58. B. H. G. Brady and J. H. O'Donnell, *European Polymer J.*, **4**, 537, (1968).
59. M. Rahman and K. E. Weale, *Polymer, London*, **11**, 122, (1970).
60. G. C. Vezzoli, F. Dachille and R. Roy, *J. Polymer Sci.*, **A-1, 7**, 1557, (1969).
61. V. V. Korshak, A. M. Polyakova and I. M. Stoletova, *Izvest. Akad. Nauk S.S.S.R., Otdel. Khim. Nauk*, **1959**, 1471.

62. L. A. Wall and D. W. Brown, *J. Polymer Sci.*, **C4**, 1151, (1963).
63. F. Korte and W. Glet, *J. Polymer Sci.*, **B4**, 685, (1966).
64. K. J. Ivin, *Pure & Applied Chem.*, **4**, 271, (1962).
65. V. V. Korshak, A. M. Polyakova and I. M. Stoletova, *Izvest. Akad. Nauk S.S.S.R., Otdel. Khim. Nauk*, **1959**, 1477; *Chem. Abstr.*, **54**, 1368, (1960).
66. K. Chikanishi and T. Tsuruta, *Makromol. Chem.*, **73**, 231, (1964); **81**, 198, 211, (1965).
67. J. W. C. Crawford, *J. Chem. Soc.*, **1953**, 2658.
68. Y. Okamoto, H. Takano and H. Yuki, *Polymer J.*, **1**, 403, (1970).
69. L. J. Fetters, W. J. Pummer and L. A. Wall, *J. Polymer Sci.*, **A-1**, 3003, (1966).
70. L. K. J. Tong and W. O. Kenyon, *J. Amer. Chem. Soc.*, **69**, 2245, (1947).
71. R. M. Joshi, *J. Polymer Sci.*, **56**, 313, (1962).
72. B. J. Cottam, J. M. G. Cowie and S. Bywater, *Makromol. Chem.*, **86**, 116, (1965).
73. K. G. McCurdy and K. J. Laidler, *Canad. J. Chem.*, **42**, 818, (1964).
74. M. Goodman and You-Ling Fan, *J. Amer. Chem. Soc.*, **86**, 4922, (1964).
75. M. Sonntag and W. Funke, *Makromol. Chem.*, **137**, 23, (1970).
76. J. B. Thompson and W. M. D. Bryant, *Amer. Chem. Soc. Polymer Preprints*, **11**, 204, (1970).
77. I. Mita, I. Imai and H. Kambe, *Makromol. Chem.*, **137**, 155, (1970).
78. D. R. Waywell, *J. Polymer Sci.*, **B8**, 327, (1970).
79. W. J. Bailey and H. Chu, *Amer. Chem. Soc. Polymer Preprints*, **6**, 145, (1965).
80. R. D. Lipscomb and W. H. Sharkey, *J. Polymer Sci.*, **A-1**, **8**, 2187, (1970).
81. E. Gipstein, E. Wellisch and O. J. Sweeting, *J. Polymer Sci.*, **B1**, 237, (1963).
82. W. J. Middleton, H. W. Jackson, R. E. Putnam, H. C. Walter, D. G. Pye and W. H. Sharkey, *J. Polymer Sci.*, **A**, **3**, 4115, (1965).
83. A. L. Barney, J. M. Bruce, J. N. Coker, H. W. Jacobson and W. H. Sharkey, *J. Polymer Sci.*, **A-1**, **4**, 2617, (1966).
84. H. R. Allcock, *Heteroatom Ring Systems and Polymers*, Academic Press, New York, 1967.
85. T. Kashiwagi, M. Hidai, Y. Uchida and A. Misono, *J. Polymer Sci.*, **B8**, 173, (1970).
86. M. Goodman and Shih-Chung Chen, *Macromolecules*, **3**, 398, (1970).
87. R. E. Cook, F. S. Dainton and K. J. Ivin, *J. Polymer Sci.*, **26**, 351, (1957).
88. J. E. Hazell and K. J. Ivin, *Trans. Faraday Soc.*, **58**, 176, (1962).
89. Z. Kuri, T. Yoshimura, N. Sakurai and M. Hirutani, *Kogyo Kagaku Zasshi*, **68**, 1117, (1965).
90. Z. Kuri and T. Yoshimura, *J. Polymer Sci.*, **B1**, 107, (1963).
91. H. Pines, W. D. Huntsman and V. N. Ipatieff, *J. Amer. Chem. Soc.*, **75**, 2315, (1953).
92. R. J. Orr, *Polymer, London*, **5**, 187, (1964).
93. F. Fairbrother, G. Gee and G. T. Merrall, *J. Polymer Sci.*, **16**, 459, (1955).
94. P. A. Small, *Trans. Faraday Soc.*, **51**, 1717, (1955).
95. K. J. Ivin and J. B. Rose, p. 335 in *Advances in Macromolecular Chemistry*, Vol. 1., ed. W. M. Pasika, Academic Press, London, 1968.
96. C. F. Gibbs and C. S. Marvel, *J. Amer. Chem. Soc.*, **57**, 1137, (1935).
97. T. Endo, *Bull. Chem. Soc. Japan*, **44**, 870, (1971).
98. T. Kagiya, S. Narisawa, T. Maeda and K. Fukui, *Kogyo Kagaku Zasshi*, **69**, 732, (1969); *J. Polymer Sci.*, **B4**, 441, (1966).
99. H. K. Hall and A. K. Schneider, *J. Amer. Chem. Soc.*, **80**, 6409, (1958).
100. H. K. Hall, *J. Amer. Chem. Soc.*, **80**, 6404, (1958).
101. S. Hashimoto and T. Yamashita, *Kobunshi Kagaku*, **27**, 400, (1970).

102. J. W. Hill and W. H. Carothers, *J. Amer. Chem. Soc.*, **57**, 925, (1935).
103. A. A. Strepikheev and A. V. Volokhina, *Doklady Akad. Nauk S.S.S.R.*, **99**, 407, (1954).
104. G. A. Clegg and T. P. Melia, *Polymer, London*, **10**, 912, (1969).
105. M. Okada, Y. Yamashita and Y. Ishii, *Makromol. Chem.*, **80**, 196, (1964).
106. Y. Yamashita, M. Okada, K. Sujama and H. Kasahara, *Makromol. Chem.*, **114**, 146, (1968).
107. W. H. Carothers, Collected papers of, ed. H. Mark and G. S. Whitby, Interscience, New York, 1940.
108. I. Maruyama, M. Nakaniwa, T. Saegusa and J. Furukawa, *J. Chem. Soc. Japan, Ind. Chem. Sect.*, **68**, 1149, (1965).
109. R. B. Whitney and M. Calvin, *J. Chem. Phys.*, **23**, 1750, (1955).
110. J. G. Affleck and G. Dougherty, *J. Org. Chem.*, **15**, 865, (1950).
111. W. O. Ney, W. R. Nummy and C. E. Barnes, United States Patent 2638463, (1953).
112. W. O. Ney and M. Crowther, United States Patent 2739959, (1956).
113. M. P. Kozina and S. M. Skuratov, *Doklady Akad. Nauk S.S.S.R.*, **127**, 561, (1959).
114. H. Meerwein, *Angew. Chem.*, **59**, 168, (1947).
115. H. Meerwein, D. Delfs and H. Morschel, *Angew. Chem.*, **72**, 927, (1960).
116. M. P. Dreyfuss and P. Dreyfuss, *J. Polymer Sci.*, A4, 2179, (1966).
117. T. Saegusa and S. Matsumoto, *J. Macromol. Sci., Chem.*, A4, 873, (1970).
118. G. A. Clegg, D. R. Gee, T. P. Melia and A. Tyson, *Polymer, London*, **9**, 501, (1968).
119. J. B. Rose and J. Stuart-Webb, unpublished results quoted in *The Chemistry of Cationic Polymerization*, ed. P. H. Plesch, Pergamon, Oxford, 1963.
120. R. Chiang and J. H. Rhodes, *J. Polymer Sci.*, B7, 643, (1969).
121. C. L. Hamermesh and V. E. Haury, *J. Org. Chem.*, **26**, 4748, (1961).
122. C. G. Overberger and J. K. Weise, *J. Amer. Chem. Soc.*, **90**, 3533, (1968).
123. W. Kern, H. Cherdron and V. Jaacks, *Angew. Chem.*, **73**, 177, (1961).
124. T. Miki, T. Higashimura and S. Okamura, *J. Polymer Sci.*, A-1, **8**, 157, (1970).
125. G. A. Clegg, T. P. Melia and A. Tyson, *Polymer, London*, **9**, 75, (1968).
126. G. A. Clegg and T. P. Melia, *Makromol. Chem.*, **123**, 194, (1969).
127. J. B. Lando and V. Stannett, *J. Polymer Sci.*, B2, 375, (1964).
128. A. B. Meggy, *J. Chem. Soc.*, **1956**, 1444.
129. H. K. Hall, *Amer. Chem. Soc., Polymer Preprints*, **6**, 535, (1965).
130. W. Dittrich and R. C. Schulz, *Angew. Makromol. Chem.*, **15**, 109, (1971).
131. E. Hollo, *Ber.*, **61**, 895, (1928).
132. A. Schöberl and G. Wiehler, *Ann.*, **595**, 101, (1955).
133. F. Fichter and A. Beisswenger, *Ber.*, **36**, 1200, (1903).
134. K. Saotome and Y. Kodaira, *Makromol. Chem.*, **82**, 41, (1965).
135. H. Batzer and G. Fritz, *Makromol. Chem.*, **14**, 179, (1954).
136. H. D. K. Drew and W. N. Haworth, *J. Chem. Soc.*, **1927**, 775.
137. F. S. Dainton, J. Davies, P. P. Manning and S. Zahir, *Trans. Faraday Soc.*, **53**, 813, (1957).
138. G. A. Clegg and T. P. Melia, *Polymer, London*, **11**, 245, (1970).
139. M. H. Palomaa and V. Toukola, *Ber.*, **66B**, 1629, (1933).
140. D. D. Coffman, N. L. Cox, E. L. Martin, W. E. Mochel and F. J. van Natta, *J. Polymer Sci.*, **3**, 85, (1948).
141. W. E. Hanford and R. M. Joyce, *J. Polymer Sci.*, **3**, 167, (1948).
142. H. R. Mighton, United States Patent 2647105, (1953).
143. F. N. S. Carver and B. L. Hollingsworth, *Makromol. Chem.*, **95**, 135, (1966).

144. V. P. Kolesov, I. E. Paukov and S. M. Skuratov, *Zh. Fiz. Khim.*, **36**, 770, (1962); *Russ. J. Phys. Chem.*, **36**, 401, (1962).
145. A. V. Tobolsky and A. Eisenberg, *J. Amer. Chem. Soc.*, **81**, 2302, (1959).
146. A. Schäffler and W. Ziegenbein, *Ber.*, **88**, 1374, (1955).
147. W. Ziegenbein, A. Schäffler and R. Kaufhold, *Ber.*, **88**, 1906, (1955).
148. L. E. Wolinski and H. R. Mighton, *J. Polymer Sci.*, **49**, 217, (1961).
149. M. Imoto, H. Sakurai and T. Kono, *J. Polymer Sci.*, **50**, 467, (1961).
150. N. L. Cox and W. E. Hanford, United States Patent, 2276164, (1942); Swiss Patent, 270546, (1951); 276924, (1952).
151. R. Gehm, *Angew. Makromol. Chem.*, **18**, 159, (1971).
152. R. C. P. Cubbon, *Makromol. Chem.*, **80**, 44, (1964).
153. A. K. Bonetskaya and S. M. Skuratov, *Vysokomol. Soedinenya*, **A11**, 532, (1969); *Polymer Sci. U.S.S.R.*, **11**, 604, (1969).
154. O. Riedel and P. Wittmer, *Makromol. Chem.*, **97**, 1, (1966).
155. T. Miki, H. Higashimura and S. Okamura, *J. Polymer Sci.*, **A-1, 5**, 2997, (1967).
156. K. W. Scott, N. Calderon, E. A. Ofstead, W. A. Judy and J. P. Ward, *Adv. Chem. Ser.*, **1969**, No. 91, 399.
157. T. Kashiwagi, M. Hidai, Y. Uchida and A. Misono, *J. Polymer Sci.*, **B8**, 173, (1970).
158. K. C. Frisch and L. P. Rumao, *J. Macromol. Sci.*, *Rev. Macromol. Sci.*, **C5**, 103, (1970).
159. J. K. Jacques, M. F. Mole and N. L. Paddock, *J. Chem. Soc.*, **1965**, 2112.
160. J. A. Semlyen and G. R. Walker, *Polymer, London*, **10**, 597, (1969).
161. G. R. Walker and J. A. Semlyen, *Polymer, London*, **11**, 472, (1970).
162. H. Jacobson, C. O. Beckmann and W. H. Stockmayer, *J. Chem. Phys.*, **18**, 1607, (1950).
163. H. R. Allcock, *J. Macromol. Sci.*, *Rev. Macromol. Chem.*, **C4**, 149, (1970).
164. P. V. Wright and J. A. Semlyen, *Polymer, London*, **11**, 462, (1970).
165. D. W. Scott, *J. Amer. Chem. Soc.*, **68**, 2294, (1946).
166. Z. Laita and M. Jelinek, *J. Polymer Sci.*, **B4**, 739, (1965).
167. J. B. Carmichael and P. Winger, *J. Polymer Sci.*, **A-1, 3**, 971, (1965).
168. J. F. Brown and G. M. J. Slusarczuk, *J. Amer. Chem. Soc.*, **87**, 931, (1965).
169. P. J. Flory and J. A. Semlyen, *J. Amer. Chem. Soc.*, **88**, 3209, (1966).
170. M. S. Beevers and J. A. Semlyen, *Polymer, London*, **12**, 373, (1971).
171. J. B. Carmichael, *J. Macromol. Chem.*, **1**, 207, (1966).
172. H. Jacobson and W. H. Stockmayer, *J. Chem. Phys.*, **18**, 1600, (1950).
173. H. A. Hartung and S. M. Camiolo, *Amer. Chem. Soc. Abstracts*, 141st meeting, Washington 1962.
174. T. Nishimura, T. Yogo, C. Azuma and N. Ogata, *Polymer J.*, **1**, 493, (1970).
175. R. Kerber and V. Serini, *Makromol. Chem.*, **140**, 1, (1970).
176. M. Otsuka, Y. Yasuhara, K. Takemoto and M. Imoto, *Makromol. Chem.*, **103**, 291, (1967).
177. M. Modena, M. Ragazzini and E. Gallimella, *J. Polymer Sci.*, **B1**, 567, (1963).
178. S. Inoue, H. Koinuma and T. Tsuruta, *J. Polymer Sci.*, **B7**, 287, (1969).
179. S. Inoue, H. Koinuma and T. Tsuruta, *Makromol. Chem.*, **130**, 210, (1969).
180. S. Inoue, H. Koinuma, Y. Yokoo and T. Tsuruta, *Makromol. Chem.*, **143**, 97, (1971).
181. S. W. Osborn and E. Broderick, United States Patent, 3213108, (1965).
182. H. Yokota and M. Kondo, *J. Polymer Sci.*, **A-1, 9**, 13, (1971).
183. N. M. Geller, V. A. Kropachev and B. A. Dolgoplosk, *Vysokomol. Soedinenya*, **A9**, 575, (1967).

184. A. Ishigaki, T. Shono and Y. Hachihama, *Makromol. Chem.*, **79**, 170, (1964).
185. T. Tsuda, T. Nomura and Y. Yamashita, *Makromol. Chem.*, **86**, 301, (1965).
186. T. Tsuda and Y. Yamashita, *Makromol. Chem.*, **99**, 297, (1966).
187. J. Furukawa, *Polymer, London*, **3**, 487, (1962).
188. E. G. Howard and P. B. Sargeant, *J. Macromol. Sci. (Chem.)*, **A1**, 1011, (1967).
189. N. L. Madison, *Amer. Chem. Soc., Polymer Preprints*, **7**, 1099, (1966).
190. F. S. Fawcett and E. G. Howard, United States Patent, 3316216, (1967).
191. H. C. Haas and N. W. Schuler, *J. Polymer Sci.*, **A2**, 1641, (1964).
192. E. Ureta, J. Smid and M. Szwarc, *J. Polymer Sci.*, **A-1**, **4**, 2219, (1966).
193. H. Yuki and Y. Okamoto, *Polymer J.*, **1**, 13, (1970).
194. H. Yuki and Y. Okamoto, *Bull. Chem. Soc. Japan*, **42**, 1644, (1969).
195. H. Yuki, K. Hatada and T. Inoue, *J. Polymer Sci.*, **A-1**, **6**, 3333, (1968).
196. D. G. L. James and T. Ogawa, *J. Polymer Sci.*, **B2**, 991, (1964).
197. H. Yuki, M. Kato and Y. Okamoto, *Bull. Chem. Soc. Japan*, **41**, 1940, (1968).
198. H. Yuki, Y. Okamoto, K. Tsubota and K. Kosai, *Polymer J.*, **1**, 147, (1970).
199. C. S. Marvel, A. T. Tweedie and J. Economy, *J. Org. Chem.*, **21**, 1420, (1956).
200. G. Natta, P. Longi and U. Nordio, *Makromol. Chem.*, **83**, 161, (1965).
201. A. L. Barney, J. M. Bruce, J. N. Coker, H. W. Jacobson and W. H. Sharkey, *J. Polymer Sci.*, **A-1**, **4**, 2617, (1966).
202. G. G. Lowry, *J. Polymer Sci.*, **42**, 463, (1960).
203. J. E. Hazell and K. J. Ivin, *Trans. Faraday Soc.*, **58**, 342, (1962); **61**, 2330, (1965).
204. A. A. Durgaryan, *Vysokomol. Soedinenya*, **8**, 790, (1966).
205. T. Alfrey and A. V. Tobolsky, *J. Polymer Sci.*, **38**, 269, (1959).
206. J. A. Howell, M. Izu and K. F. O'Driscoll, *J. Polymer Sci.*, **A-1**, **8**, 699, (1970).
207. M. Izu and K. F. O'Driscoll, *Polymer J.*, **1**, 27, (1970).
208. M. H. Theil, *Macromolecules*, **2**, 137, (1969).
209. M. H. Theil, *Amer. Chem. Soc., Polymer Preprints*, **11**, 173, (1970).
210. H. Sawada, *J. Polymer Sci.*, **3**, 2483, (1965).
211. H. Sawada, *J. Polymer Sci.*, **5**, 1383, (1967).
212. H. Lüssi, *Makromol. Chem.*, **103**, 62, (1967).
213. H. Yuki, K. Hatada, J. Ohshima and T. Komatsu, *Polymer J.*, **2**, 812, (1971).
214. P. Wittmer, *Makromol. Chem.*, **103**, 188, (1967).
215. N. G. Gaylord and B. K. Patnaik, *Makromol. Chem.*, **146**, 125, (1971).
216. K. J. Ivin and R. H. Spensley, *J. Macromol. Sci. (Chem)*, **A1**, 653, (1967).
217. K. Chikanishi and T. Tsuruta, *Makromol. Chem.*, **81**, 198, 211, (1965).
218. G. E. Ham, *J. Polymer Sci.*, **45**, 183, (1960).
219. R. G. Fordyce, E. C. Chapin and G. E. Ham, *J. Amer. Chem. Soc.*, **70**, 2489, (1948).
220. G. E. Ham, *J. Polymer Sci.*, **14**, 87, (1954).
221. K. F. O'Driscoll and F. P. Gasparro, *J. Macromol. Sci., (Chem.)* **A1**, 643, (1967).
222. R. G. Fordyce and G. E. Ham, *J. Amer. Chem. Soc.*, **73**, 1186, (1951).
223. I. Penczek and S. Penczek, *J. Polymer Sci.*, **B5**, 367, (1967).
224. M. Suzuki, H. Miyama and S. Fujimoto, *J. Polymer Sci.*, **31**, 212, (1958).
225. T. Oota and S. Masuda, *Kogyo Kagaku Zasshi*, **69**, 721, (1966).
226. Y. Yamashita, T. Asakura, M. Okada and K. Ito, *Makromol. Chem.*, **129**, 1, (1969).
227. T. M. Frunze and V. V. Kurashev, *Russ. Chem. Reviews*, **37**, 681, (1968).
228. R. A. Patsiga, *J. Macromol. Sci.—Reviews Macromol. Chem.*, **C1**, 223, (1967).
229. Y. Yamashita, T. Tsuda, M. Okada and S. Iwatsuki, *J. Polymer Sci.*, **A-1**, **4**, 2121, (1966).

230. T. Masuda and T. Higashimura, *Polymer J.*, **2**, 29, (1971).
231. K. Hashimoto and H. Sumitomo, *J. Polymer Sci.*, A-1, **9**, 107, (1971).
232. M. Georgiewa and W. Kabaivanov, *Angew. Makromol. Chem.*, **15**, 233, (1971).
233. T. Kunitake, M. Yasumatsu and C. Aso, *J. Polymer Sci.*, A-1, **9**, 3675, (1971).
234. B. Krummenacher and H. G. Elias, *Makromol. Chem.*, **150**, 271, (1971).
235. P. C. Wollwage and P. A. Seib, *J. Polymer Sci.*, A-1, **9**, 2877, (1971).
236. C. C. Tu and C. Schuerch, *J. Polymer Sci.*, B-1, 163, (1963).
237. G. Belonovskaya, Z. Tchernova and B. Dolgoplosk, *European Polymer J.*, **8**, 35, (1972).
238. J. Kops and H. Spanggaard, *Makromol. Chem.*, **151**, 21, (1972).
239. K. Weissermel and E. Nölken, *Makromol. Chem.*, **68**, 140, (1963).
240. T. Saegusa, M. Matoi, S. Matsumoto and H. Fujii, *Macromolecules*, **5**, 233, 236, (1972).
241. O. G. von Ettinghausen and E. Kendrick, *Polymer, London*, **7**, 469, (1966).
242. G. Nabi, *Pakistan J. Sci.*, **20**, 29, (1968).
243. T. Saegusa, T. Hodaka and H. Fujii, *Polymer J.*, **2**, 670, (1971).
244. T. Kagiya and T. Matsuda, *Polymer J.*, **2**, 398, (1971).
245. S. Inoue, H. Koinuma and T. Tsuruta, *Polymer J.*, **2**, 220, (1971).
246. C. Aso, S. Tagami and T. Kunitake, *Kobunshi Kagaku*, **23**, 63, (1966).
247. A. A. Durgaryan and A. V. Agumyan, *Vyokomol. Soedinenya*, **5**, 1755, (1963).
248. R. Raff, J. L. Cook and B. V. Etting, *J. Polymer Sci.*, A-3, 3511, (1965).
249. C. Aso, S. Tagami and T. Kunitake, *J. Polymer Sci.*, A-1, **8**, 1323, (1970).
250. D. Fleischer, R. C. Schulz and B. Turcsányi, *Makromol. Chem.*, **152**, 305, (1972).
251. Y. Inaki, S. I. Nozakura and S. Murahashi, *J. Macromol. Sci. (Chem.)*, A-6, 313, (1972).
252. J. P. Fischer, *Makromol. Chem.*, **155**, 227, (1972).
253. G. A. Razuvaev, V. S. Etlis and L. N. Gribov, *Zhur. Obshchei Khim.*, **33**, 1366, (1963).
254. T. Saegusa, H. Ikeda and H. Fujii, *Polymer J.*, **3**, 35, 176, (1972).
255. T. Saegusa, Y. Nagura and S. Kobayashi, *Macromolecules*, **6**, 495, (1973).
256. T. Kodaíra, J. Stehlíček and J. Šekenda, *European Polymer J.*, **6**, 1451, (1970).

Author Index

The first number refers to the page on which the reference is quoted, the number in parenthesis gives the reference number on that page, and the italic number gives the page on which the reference is quoted in full.

Abe, A., 408(64) *429*; 472(14) *506*; 472, 492 (13) *506*; 502(249) *513*
Abe, Y., 444(183, 184) *511*
Abercromby, D. C., 520(33) *560*
Accaseina, F., 251(25) *306*
Acree, F. S., 192(31) *225*; 330(67a) *348*
Adler, H., 462(94) *469*
Affleck, J. G., 541(110) *562*
Aglietto, M., 500(245, 246) *512*; 500 (247) *513*
Akhren, I. S., 264(68) *307*
Alcock, W. G., 36(13) *51*
Alexander, A. E., 218(103) *227*
Alfrey, T., 123(6) *140*; 125, 128(15) *141*; 205(66, 67) *226*; 556(205) *564*
Alfrey Jun., T., 372(70) *381*
Allcock, H. R., 536(84) *561*; 551(163) *563*
Allegra, G., 394, 411(42) *428*; 399, 404 (49a) *429*; 423(92) *430*; 488(131) *509*; 488(138) *510*; 490(143, 144) *510*
Allen, R. H., 205(66) *226*
Allen, V. R., 371(67) *381*
Allison, J. B., 487(116) *509*
Altarea, T. A., 371(67) *381*
Altier, M. W., 198(54) *226*; 216(100) *227*
Anderson, A. W., 387(29) *428*
Anderson, E. W., 494(182) *511*
Anderson, J. D., 359(25) *380*
Andruzzi, F., 261, 262(55) *307*
Angove, S. N., 215(96) *227*
Anneser, E., 340(102a) *349*
Anufrieva, E. V., 493, 494(174) *511*
Anzuino, G., 497(232) *512*
Appel, B., 235, 237(6) *243*
Appenrodt., J., 345(113b) *349*

Applequist, J., 485(89) *508*
Araki, K., 192, 210(104) *227*
Arest-Yakubovich, A. A., 367(47, 48, 49) *381*
Arlman, E. J., 389, 403(32) *428*; 393, 402(38) *428*; 414(76) *429*
Armstrong, D. R., 392(34a) *428*
Arnett, E. M., 95(76) *115*
Arrialdi, A. C., 193(36) *225*
Asai, M., 453(27) *468*
Asakura, T., 559(226) *564*
Ashby, C. E., 387(29) *428*
Ashe, B. H., 440(7) *445*
Ashworth, J., 93, 94, 95(72) *115*; 94(73) *115*
Aso, C., 279(106) *308*; 516, 536(25) *560*
Atherton, J. N., 144, 153(17) *156*
Atherton, N., 343(111b) *349*
Atwater, H., 462(86) *469*
Audie, C., 460(73) *469*
Ayres, D. C., 338(94b) *349*
Ayres, P. W., 129(104) *349*; 131(65) *348*
Ayscough, P. B., 36(11) *51*
Azorlosa, J. L., 180(11) *224*
Azuma, C., 554(174) *563*

Baader, H., 301(139) *309*
Baba, H., 372(72) *381*
Babitskii, B. D., 207(76) *226*; 416(80) *430*
Bach, H., 460(64) *469*
Bacon, R. F., 517, 522(15) *559*
Bacskai, R., 495, 496(206) *511*
Baer, M., 376(86) *382*
Bafus, D. A., 322(35a) *347*
Bahsteter, F., 375(84) *382*
Baikova, R. I., 172(46) *174*

Bailey, J. T., 214(93) *227*
Bailey, W. J., 536(79) *561*
Baker, R., 235, 237(6) *243*
Balas, J. G., 421(91) *430*
Ballard, D. G. H., 398(47a) *429*
Bamford, C. H., 52, 57, 71(3) *113*; 53
   (4a, 7) *113*; 53, 55, 73(5) *113*; 53, 63,
   78, 80(6) *113*; 53, 69, 71(8) *113*; 53,
   56, 59, 64, 70(9) *113*; 53, 54(10) *113*;
   53, 64, 66, 69(12) *113*; 54, 57, 66(13)
   113; 54, 70, 89(14) *113*; 55, 69, 70,
   71(15) *113*; 55, 67, 75, 76, (16) *113*;
   55, 67, 69(17, 19) *113*; 55(19a) *114*;
   56, 64, 68, 71, 80, 84(20) *114*; 58(21)
   114; 58, 92(22) *114*; 57, 61, 93(23) *114*;
   56, 59, 67, 85(24) *114*; 60(25, 26) *114*;
   63, 64(30) *114*; 67(33a, 35) *114*; 71
   (36) *114*; 73, 76(36a) *114*; 76(42) *114*;
   86(55) *115*; 87(57) *115*; 87, 88(58)
   115; 87, 91(60) *115*; 88(62) *115*; 88,
   93(63) *115*; 89(64) *115*; 89, 91(65) *115*;
   93(68, 69) *115*; 93, 94, 95(70) *115*; 93,
   94(71, 72) *115*; 94(73) *115*; 96, 99, 101,
   103 (78) *115*; 99, 112(80) *115*; 104,
   105, 107, 108(85) *115*; 106(87, 88) *116*;
   124(12) *140*; 127(19, 20) *141*; 142(3)
   156; 155(42, 45) *157*; 370(61) *381*
Bank, S., 335(85b) *348*
Banks, R. L., 431(1) *445*
Barb, W. G., 76(42) *114*; 123(7) *140*;
   142(3) *156*; 370(61) *381*
Barnes, C. E., 541(111) *562*
Barney, A. L., 536(83) *561*; 554(201)
   564
Barone, G., 493, 494(173) *510*
Bartlett, P. D., 317(13) *346*; 332(52)
   347; 451(16, 17) *467*
Basile, L. J., 494(180) *511*
Basolo, F., 67(34) *114*; 75(37, 39, 40, 41)
   114; 82(47) *114*
Basova, R. V., 367(47, 48) *381*
Bassi, I. W., 486(98, 99, 100, 101, 102,
   104) *509*; 486, 487(108) *509*; 486(111,
   112, 113) *509*; 486, 494(114, 115, 126)
   509; 504(252) *513*
Bates, R. B., 336(88) *348*
Bates, T. M., 248, 278(17) *306*
Bates, T. W., 492(159) *510*
Battenberg, E., 267(84) *307*
Batzer, H., 545(135) *562*
Bauld, N. L., 329(61) *348*
Bauman, H., 496(200) *511*

Baur, H-J., 151(33) *157*
Bawn, C. E. H., 12(4) *29*; 249, 261, 263,
   264, 271, 279, 291, 303(22) *306*; 263,
   271(63) *307*; 263(67) *307*; 265(71)
   307; 266(77) *307*; 299(135) *309*; 384
   (7) *427*; 478, 479(42) *507*; 496(214)
   512; 523(46) *560*
Beard, J. M., 273(96) *308*
Beason, L. R., 214(93) *227*
Bebb, R. L., 329(56) *347*
Becher, F., 179(7) *224*
Becker, R., 460(63) *469*
Becker, R. S., 343(108b) *349*
Beckmann, C. O., 551(162) *563*
Beevers, M. S., 551(170) *563*
Begley, J. W., 397(46) *429*
Behar, J. V., 168(31) *174*
Beisswenger, A., 545(133) *562*
Bell, R. M., 263(63) *307*; 265(71) *307*;
   523(46) *560*
Bello, A., 362(33) *380*
Bellus, O., 463(93) *469*
Belousova, M. I., 343(110a) *349*
Belt, F., 418, 421(83) *430*
Belt, R. F., 421(90) *430*
Belyavskii, A. B., 52(1, 2) *113*
Bemis, A. G., 341(103) *349*
Benedetti, E., 471(6) *506*; 471, 477, 481
   (8) *506*; 479(41) *507*; 481(57) *508*;
   489, 494(138) *510*
Bengough, W. I., 144(11) *156*
Benkeser, R. A., 329(63) *348*
Benoit, H., 151(30) *157*; 345(115) *349*;
   372(74) *381*
Benson, R. E., 318(25b) *347*
Benson, S. W., 33(1) *50*; 144(10) *156*;
   146, 147(23) *157*; 155(39) *157*
Berces, T., 36(8) *50*
Berding, C., 97, 103(77) *115*
Bereznoj, G. D., 216(102) *227*
Bergmann, E., 352(6) *380*
Berlin, A. A., 167(25) *174*; 518(31) *560*
Bertoli, V., 246(13) *306*
Bertram, J., 463(95) *469*
Best, J. V. G., 248, 278(17) *306*
Best, R. J., 331(69d) *348*
Bethel, D., 230(3) *243*
Betz, W., 365(42) *381*
Bevington, J. C., 155(41) *157*
Bevza, T. I., 103(82) *115*
Bey, A. E., 208(78) *226*
Beynon, K. I., 123(5) *140*

Bhadani, S. N., 322(34b) *347*; 359(24) *380*

Bhardwaj, I., 456(44) *468*

Bhattacharyya, D. N., 328(51) *347*; 363 (36) *380*

Bhattacharyya, D. W., 247, 279(15) *306*

Bickel, A. F., 385(19) *428*

Bier, G., 386(25) *428*

Bingham, J., 88(62) *115*

Birshtein, T. M., 483, 484, 485, 488, 489, 490, 491(73) *508*; 483(74) *508*; 483, 502(76) *508*; 484(81) *508*; 491(148, 150) *510*; 502(248) *513*

Bishop, W. A., 459(58) *468*

Blackledge, J., 344(112) *349*

Block, H., 88(62) *115*; 148, 154(27) *157*

Blout, E. R., 485(90) *508*

Bobalek, E. G., 206(69) *226*

Body, R. H., 315(16) *346*

Boguslavskii, L. I., 498(241) *512*

Bohlmann, F., 460(66) *469*

Bond, J., 193, 218(45) *225*

Bonetskaya, A. K., 550(153) *563*

Bonin, M. D., 267, 278, 291(80) *307*

Bonsignori, O., 471(11) *506*; 504(251, 252) *513*

Boor Jun., J., 384(9) *427*; 399, 402, 404, 405(50) *429*; 384, 398, 406, 408(10) *428*; 384, 398, 408(12) *428*; 388, 402 (28) *428*; 476(28) *507*

Booth, C., 214(93) *227*

Borden, D. G., 466(114) *470*

Bordwell, F. G., 310(2) *346*

Borgwardt, U., 147(22) *156*

Borisov, A. E., 334(82a, 82b) *348*

Borisova, N. P., 483, 488(74) *508*; 483 (75) *508*

Borisova, T. I., 496(210) *512*

Borkenshire, J. L., 36(9) *50*

Borla, L., 516, 522, 535(20) *560*

Borsile, L. J., 459(51) *468*

Borzel, P., 290(122) *308*

Bostick, E. E., 377(88) *382*

Botteghi, C., 481(58) *508*

Boudreau, R. J., 356(19) *380*

Bovey, F. A., 24(10) *30*; 368(52) *381*; 473, 478, 495(24) *507*; 478(34) *507*; 494(179, 182) *511*

Bowyer, P. M., 263, 271, 273, 278, 291 298, 303(62) *307*

Boyle Jun., W. J., 310(2) *346*

Bradford, E. B., 192(30) *225*; 205(67) *226*; 206(70) *226*

Bradley, J. N., 155(38) *157*

Bradley, R. S., 158(1) *173*

Brady, B. H. G., 527, 529(56) *560*; 529 (57) *560*; 530(58) *560*

Braendlin, H. P., 385(19) *428*

Brandbury, J. H., 485(90) *508*

Brandrup, J., 368(53) *381*; 517, 526, 538, 552(16) *559*

Brandt, D. A., 492, 494(165) *510*

Brauman, J. I., 312, 313(8) *346*

Braun, D., 365(42) *381*; 450(52) *507*

Braut, D. A., 484(84) *508*

Breil, H., 383, 384(1) *427*

Breitenbach, J. W , 183, 199, 209(16) *225*

Breitschaft, S , 75(39) *114*

Brenner, N., 77(44) *114*

Breslow, D. S., 373(75) *381*; 393(40) *428*

Breslow, R., 318(26) *347*

Bressan, G., 473, 479(23) *506*; 488, 494 (134, 135) *510*

Brewster, J. H , 500(243, 244) *512*

Briggs, E. R., 77(46) *114*

Briton, R. K., 40(15, 19) *51*

Broaddus, C. D., 321(33) *347*; 329(64) *348*

Broderick, E., 554(181) *563*

Brodnyan, J. G., 181(12) *224*, 193(44) *225*

Brody, G. W., 496(22) *512*; 496, 497 (227) *512*

Brooks, B. W., 179(2) *228*; 370(60) *381*

Brower, F. M., 362(33) *380*

Brown, D. W., 167(24) *174*; 172(44) *174*; 531(62) *561*

Brown, H. C., 340(98) *349*

Brown, J. F., 551(168) *563*

Brown, R. E., 331(72) *348*

Brown, R. W., 204(65) *226*; 220(108) *227*

Brown, T. L., 322(35b) *347*

Brown, W. B., 515(8) *559*

Bruce, J. M., 536(83) *561*; 554(201) *564*

Brumsby, S., 106(87, 88) *116*

Bryant, J. T., 33(6) *50*

Bryant, W. M. D. 535(76) *561*

Brzezinski, J., 492(156) *510*

Buckton, G., 330(67b) *348*

Buls, V. W., 394(41) *428*

Bunn, C. W., 482(61) *508*; 488, 494 (137) *510*

Bur, A. J., 497(234, 235) *512*

Burchard, W., 497(233) *512*

Burkhart, R. D., 144(20) *156*

Burleigh, J. E., 178(5) *228*; 214(92, 93) 227

Burmistrova, M. F., 254(39) *306*

Burnett, G. M., 142(2) *156*; 147(25) *157*; 155(43) *157*; 214(92a) *227*; 450 (8) *467*; 451(19) *468*

Burnop, V. C. E., 516, 522(26) *560*; 522 (38) *560*

Burr, J., 462(91) *469*

Burshtein, L. L., 495(208) *511*; 496(210) *512*

Busfield, W. K., 171(36) *174*; 522, 535, (40) *560*; 522, 523, 531(42) *560*; 522 536(43) *560*

Bywater, S., 353(11) *380*; 361(30) *380*; 364(37, 38) *380*; 365(41) *381*; 366(43) *381*; 367(46) *381*; 371(66) *381*; 516, 523, 524(17) *560*; 516, 523(18) *560*; 523(30, 48) *560*; 534(72) *561*

Cadogan, J. I. G., 272(92) *308*

Cala, J. A., 193(44) *225*; 215(95) *227*; 216(100) *227*

Calderon, N., 18(9) *30*; 551(156) *563*

Caldwell, R. A., 319, 320(28) *347*; 339 (55) *347*

Callen, J. E., 77(44) *114*

Calvayracn, H., 466(112) *470*

Calvert, J., 460(64) *469*

Calvert, J. G., 446(1) *467*

Calvert, M. L., 267, 278, 291(80) *307*

Calvin, M., 541(109) *562*

Cambini, M., 14(6) *29*

Cameron, C. G., 214(92a) *224*

Camiolo, S. M., 553(173) *563*

Campbell, T. W., 482(62) *508*

Campigli, U., 487(126) *509*

Canterino, P. J., 213(91) *227*

Cantow, M. J., 211(86) *227*

Capon, B., 233, 240(5) *243*

Carrazzolo, G., 487(123) *509*

Carlini, C., 471, 477, 481(8) *508*; 495 (193, 194) *511*

Carlsson, D. J., 461(80, 81, 82) *469*

Carmichael, J. B., 551(167, 171) *563*

Carnighan, R. H., 336(88) *348*

Carothers, W. H., 541(102, 107) *562*

Carpenter, D. K., 496(196) *511*

Carrick, W. L., 387(30) *428*

Carruthers, J. E., 517(14) *559*

Carruthers, R. A., 263(67) *307*

Carver, F. N. S., 547(143) *562*

Casey, B. A., 55(19) *113*

Castanza, A. J., 193(47) *225*

Cekeda, J., 208(78) *226*

Cesari, M., 486(94) *508*; 496, 497(226) *512*

Chadron, H., 194, 219(42) *225*

Chalmers, W., 261(58) *307*

Chan, L., 326(45) *347*; 327(46) *347*

Chandhur, J., 342(109) *349*

Chang, P., 326(44a) *347*

Chapin, E. C., 557(219) *564*

Chaser, A. G., 387(30) *428*

Chatt, J., 391(34) *428*

Chekhuta, O. M., 523(47) *560*

Chen, E., 343(108a, 108b) *349*

Chen, E. S., 179(7) *224*

Cherdron, H., 543(123) *562*

Chernikova, A. Ya, 228(7) *228*

Chesnokova, N. N., 207(76) *226*; 416 (80) *430*

Chiang, R., 497(216) *512*; 542(120) *562*

Chiellini, E., 471, 482(9) *506*; 481(57) *509*; 495(190, 194) *511*; 496(211) *512*

Chikanishi, K., 534(66) *561*; 557(217) *564*

Chini, P., 486(113) *509*

Chiurdoglu, G., 500(246) *512*

Chong, C.-H., 192, 210(33) *225*

Chu, H., 536(79) *561*

Chujo, R., 487(119) *509*

Ciampelli, F., 478(38) *507*; 478, 493(39, 40) *507*

Ciardelli, F., 407(61, 62, 63) *429*; 471 (5) *506*; 471, 477(6, 7) *506*; 471, 480, 481(8) *506*; 477(32) *507*; 479(41) *507*; 480(53) *507*; 481, 494, 500(54) *507*; 481(58) *508*; 482(59) *508*; 495(192, 193, 194) *511*

Ciferri, A., 496(215, 216) *512*

Clark, A., 435, 438, 439(5) *445*; 436(6) *445*; 440(7) *445*

Clarke, K. J., 408(64) *429*

Clegg, G. A., 541(104) *562*; 542(118) *562*; 543(125, 126) *562*; 546(138) *562*

Cline, C. W., 211(86) *227*

Clippinger, E., 323(39) *347*

Closs, G. L., 340(100) *349*

Closs, L. E., 340(100) *349*
Coates, G. E., 352, 359(10) *380*; 426 (73) *429*
Coetzec, J. F., 310(1) *346*
Coffman, D. D., 547(140) *562*
Cohen, M., 463(96) *469*
Cohn, E. S., 370(64) *381*
Cohn-Ginsberg, E., 495(197) *511*
Coiro, V. H., 497(232) *512*
Coker, J. N., 536(83) *561*; 554(201) *564*
Colford, J., 462(87) *469*
Collins, C. J., 242(10) *243*
Colonius, H., 352(5) *380*
Combs, R. L., 475(27) *507*
Conant, J. B., 313, 329(6) *346*
Connor, A. O., 486(107) *509*
Cook, R. E., 527, 529(55) *560*; 537(87) *561*
Cooper, W., 410, 422(66) *429*; 413 418, 420(71) *429*
Coover Jun., H. W., 401(56) *429*; 475 (27) *507*
Cope, A. C., 319(27b) *347*
Corradini, P., 394, 411(42) *428*; 473(19) *506*; 474, 476, 478, 489(26) *507*; 479 (46) *507*; 486(94, 95) *508*; 486(98, 99, 100, 101, 102, 103, 104, 105, 106, 110, 111, 112, 113, 114, 115) *509*; 487(124, 128) *509*; 488(129, 132) *509*; 488(135) *510*; 490(143) *510*; 490 491(144) *510*; 492(155) *510*; 504(252) *513*
Cossee, P., 390, 392, 402(31) *428*; 392 (35, 36) *428*; 393, 402(38) *428*; 398, 402, 415(47) *429*
Cottam, B. J., 534(72) *561*
Cowan, D. O., 333(79a) *348*
Cowell, G. W., 11(2) *29*
Cowie, J. M. G., 534(72) *561*
Cox, N. L., 547(140) *562*; 547(150) *563*
Cozzens, R. F., 458(47, 48) *468*
Cram, D. J., 311(3) *346*; 315(18) *346*; 320(31) *347*; 322(37) *347*; 323(38a, 38b) *347*; 334(84) *348*; 336(92) *349*
Crawford, J. W. C., 534(67) *561*
Crescentini, L., 189, 195, 218(21b) *225*
Crescenzi, V., 492, 494(163) *510*; 493, 494(173) *510*; 495(195) *511*
Crosato-Arnaldi, A., 173(47) *174*
Crössman, F., 331(70) *348*
Crowe, P. A., 87(58) *115*; 87, 91(60) *115*
Crowther, M., 541(112) *562*
Crump, J. W., 334(81) *348*

Cubbon, R. C. P., 550(152) *563*
Curme, H., 464(101) *469*
Curtin, D. Y., 332(75a, 75b) *348*; 334 (81) *348*
Cvetanovic, R. J., 40, 43(18) *51*
Cyhjo, R., 494(184) *511*

Dachille, F., 531(60) *560*
Dāhl, P., 193, 195(39) *225*
Dainton, F. S., 15(8) *30*; 297(127) *309*; 360(27) *380*; 517(12, 13) *559*; 518(32) *560*; 527(54) *560*; 537(87) *561*; 546 (137) *562*
D'Alagni, M., 497(231, 232) *512*
Dall'Asta, G., 18(9) *30*; 486(108) *509*; 487, 494(124, 126, 127) *509*
Dalton, F. L., 192(31) *225*
Danusso, F., 479(46) *507*; 482(64) *508*; 496(198) *511*
Darcy, L. E., 251, 254, 303(30) *306*
Dart, E. C., 336(87) *348*
Das, A. K., 111(91) *116*
Das, B., 207(73) *226*; 420(88) *430*
Dauby, R., 207(75) *226*
David, C., 458(49) *468*
Davidson, J. B., 331(69b, 69d) *348*
Davies, J., 546(137) *562*
Davison, W. H., 267, 278(79) *307*
Dawans, F., 207(75) *226*; 417(82) *430*; 426(95) *430*
Day, J. P., 75(40) *114*
De Bruijn, P. M., 386(21) *428*
Degler, G., 425(93) *430*
DeGraff, A. W., 228(6) *228*
De la Mare, H. E., 421(91) *430*
Delbecq, C. J., 202(58e) *226*
Delfs, D., 270(88) *308*; 542(115) *562*
Dellsperger, W., 163(15) *174*
Delman, A., 462(88) *469*
Delzenne, G., 450(9) *467*; 452(21) *458*; 465(106) *469*
Delzenne, G. A., 458, 467(120) *470*; 466(113) *470*
DeMore, W. B., 155(39) *157*
Denbigh, K. C., 201(58b) *226*
Dennison, J. T., 250(24) *306*
Denyer, R., 53, 63, 78, 80(6) *113*; 53, 64, 66, 69(12) *113*; 53, 69, 71(8) *113*; 63, 65(30) *114*; 67(33a) *114*
DeSantis, P., 484(83) *508*; 487, 494(118) *509*; 489, 494(141) *510*; 497(231, 232) *512*

De Schryver, F. C., 450, 451(6) *467*; 456(43, 44) *468*; 457(45, 46) *468*; 453(23) *468*
Desien, R., 462(90) *469*
Desreux, V., 471(16) *506*
Desyatova, N. V., 367(47, 48) *381*
Devlin, T. R. E., 15(8) *30*
Dewar, M. J. S., 257(48) *307*
De Wilde, M. C., 123(10) *140*
Diaz, A., 235, 237(6) *243*
Dillon, R. L., 315(15) *346*
Dimroth, K., 460(66) *469*
Di Pietro, C., 86(53) *115*
Dislich, H., 353(15) *380*; 359(23) *380*
Dittrich, W., 544(130) *562*
Dixon, P. S., 42(22) *51*
Dobson, G. R., 86(51, 52) *114*
Dodge, J. S., 206, 224(68a) *226*
Doering, W. von E., 317(14) *346*; 340 (101) *349*
Dolgoplosk, B. A., 103(82) *115*; 554 (183) *563*
Doskocilova, D., 494(185) *511*
Doty, P., 485(89, 90) *508*
Douek, M., 318(26) *347*
Dougherty, G., 541(110) *562*
Dreher, E., 286(115) *308*
Drew, H. D. K., 545(136) *562*
Dreyfuss, M. P., 267(83) *307*; 271(90) *308*; 271, 301(91) *308*; 515(10) *559*; 542(116) *562*
Dreyfuss, P., 267(83) *307*; 271(90) *308*; 271, 301(91) *308*; 515(10) *559*; 542 (116) *562*
Dubois-Violette, E., 493, 494(175) *511*
Duck, E. J., 220(105) *227*
Duck, E. W., 214(94) *227*; 385(18) *428*
Duffy, M., 161(10) *174*
Dulmage, W. J., 345(113a) *349*
Dumarteau, W., 458(49) *468*
Duncan, F. J., 57, 61, 93(23) *114*
Dunkelberger, D. L., 370(64) *381*
Dunlop, Co. Ltd., 208, 223(79, 80) *226*
Dunn, A. S., 228(1) *228*; 192(33) *225*
Durand, J. P., 417(82) *430*
Durgaryan, A. A., 556(204) *564*
Dwyer, F. P., 111(89) *116*
Dyson, R. W., 88, 93(63) *115*; 93(68, 69) *115*

Eastman Kodak, 464(97) *469*, 466(107) *469*; 466(108, 111) *470*

Eastmond, G. C., 53, 63, 78, 80(6) *113*; 53(7) *113*; 53, 69, 71(8) *113*; 53, 54 (10) *113*; 54, 70, 89(14) *113*; 55, 69, 70, 71(15) *113*; 55, 67, 75, 76(16) *113*; 55(19a) *114*; 56, 64, 68, 71, 80, 85(20) *114*; 58(21) *114*; 60(25, 26) *114*; 63, 65(30) *114*; 86(55) *115*; 88, 93(63) *115*; 93(68, 69) *115*; 93, 94, 95(70) *115*; 93(71) *115*
Eaves, D. E., 425(93) *430*
Eckard, A. D., 246, 263(14) *306*
Economy, J., 554(199) *564*
Edelhauser, H., 203(59) *226*
Egle, G., 378(92) *382*
Eguchi, W., 202(58d) *226*
Ehring, R. J., 259(50) *307*
Ehrlich. P., 164(19) *174*; 172(40) *174*
Eigen, M., 337(91) *349*
Eirich, F., 386(24) *428*
Eisenberg, A., 372(73) *381*; 522(36) *560*; 547(145) *563*
Eldred, R. J., 453(33, 34) *468*
Eley, D. D., 249, 262(21) *306*; 261(57) *307*; 263(59) *307*; 294, 303(125) *309*
Elgood, B. G., 210(85) *227*
Eliel, E. L., 471(2) *506*
Ellinger, L. P., 249(19) *306*; 265(73, 74) *307*; 453(29, 30) *468*
Elliot, A., 478, 479, 493(44) *507*
Elliott, J. J., 289(121) *308*
Ellis, F. R., 374(80) *382*; 375(83) *382*
El Sayed, M. F. A., 86(51) *114*
Emery, P., 67(35) *114*
Endo, T., 540(97) *561*
Endrenyi, L., 40, 42(17) *51*
Engel, P. S., 451(17) *467*
Enikolopyan, N. S., 167(25) *174*; 173 (50, 51) *174*
Evans, A. G., 246(11) *306*; 273(98) *308*
Evans, P., 147(25) *157*
Ewald, A. H., 161(6, 7) *173*; 163(12) *174*; 168(30) *174*
Ewart, R. H., 186(18) *225*

Fagherazzi, G., 423(92) *430*
Fahla, I., 212(89) *227*
Fainberg, A. H., 323(39) *347*
Fairbrother, F., 539(93) *561*
Fanelli, R., 517, 522(15) *559*
Farina, M., 471(17) *506*; 473, 479(23) *506*; 488, 494(134, 135) *510*
Fatou, J. G., 488(136) *510*

Fawcett, E. W., 431(2) *445*
Fawcett, F. S., 554(190) *564*
Fearn, J. E., 172(44) *174*
Feast, J., 456(43) *468*
Fellman, R. P., 368(56) *381*
Ferrar, A. N., 99, 100, 112(80) *115*; 104, 105, 106, 107, 108, 109(85) *115*
Fetters, L. J., 353, 357, 377(14) *380*; 497 (234, 236) *512*; 534(69) *561*
Fettes, E., 483(70) *508*
Fettis, G. C., 36(7) *50*
Fichter, F., 545(133) *562*
Figini, R. V., 365(40) *381*
Figueruelo, J. E., 368(55) *381*
Fildes, F. J. T., 53(7) *113*; 71(36) *114*; 86(55) 115
Finaz, G., 377(91) *382*
Finch, C. A., 52, 57, 59, 71(3) *113*; 53 (4a) *113*; 53, 56, 59, 64, 70(9) *113*; 56, 57, 66(13) *113*
Finch, J. N., 440(7) *445*
Findlay, D. E., 208(78) *226*
Fink, 190, 193(38) *225*
Finnegan, R. A., 329(60) *348*
Fiore, L., 415(78) *430*
Fischer, H., 313, 319(5) *346*; 317(21) *346*; 336(89a) *348*
Fischer, H. P., 322(37) *347*
Fischer, J. P., 144(14) *156*
Fitch, R. M., 190(25) *225*
Fitzsimmonds, C., 245(7) *305*; 249, 261 263, 264, 271, 278, 291, 298, 303(22) *306*; 263, 271(63) *307*
Flautt, T. J., 329(64) *348*
Fleisher, P. C., 497(229, 230) *512*
Florin, R. E., 202(58e) *226*
Flory, P. J., 362(32) *380*; 472, 484, 491 (12) *506*; 472, 484, 492, 496(13) *506*; 472, 492(14) *506*; 491(149) *510*; 492, 494(153, 157, 161, 162, 163, 165) *510*; 496(212, 216, 218) *512*; 515(6, 7) *559*; 551, 552(169) *563*
Follett, A. E., 220(107) *227*
Foote, C. S., 452(22) *468*
Fordham, J. W. L., 183(15) *225*
Fordyce, R. G., 557(219, 222) *564*
Forman, L. E., 367(50) *381*
Förster, Th., 448(4, 5) *467*
Fox, R. B., 458(47, 48) *468*
Fox, T. G., 496(212) *512*
Franchini, P. F., 496(211) *512*
Frank, F. C., 486(107) *509*

Frank, R. L., 213(91) *227*
Freedman, H. H., 334(83) *348*
French, D. M., 190, 192, 210(24) *225*
Frey, F. E., 517(11) *559*
Frey, M., 151(30) *157*
Freydlina, R. K., 52(1, 2) *113*
Freyss, D., 345(115) *349*; 372(74) *381*
Friedlander, H. N., 385(15, 17) *428*
Friedman, S., 332(52) *347*
Fries, S. L., 450(8) *467*
Frisch, H. L., 478(34) *507*
Frisch, K. C., 15(7) *29*; 245(6) *305*; 551 (158) *563*
Fritz, G., 545(135) *562*
Fritze, H., 183, 199, 209(16) *225*
Frunze, T. M., 559(227) *564*
Fryling, C. F., 220(107) *227*
Fuchs, O., 479(48) *507*
Fujishige, S., 455(41) *468*
Fukui, K., 540(98) *561*
Fuller, C. S., 487(122) *509*
Funke, W., 535(75) *561*
Funt, B. L., 359(24) *380*
Fuoss, R. M., 251(25) *306*
Furuichi, J., 496(213) *512*
Furukawa, J., 127(16) *141*; 356(22) *380*; 372(69) *381*; 486, 487(109) *509*; 488 (139) *510*; 541(108) *562*; 554(187) *564*
Furusaki, S., 497(237) *512*
Futrell, J. H., 33(2) *50*; 34(6a) *50*

Gabant, J. A., 398(48) *429*
Gaj, B. J., 329(59) *348*
Gallazzi, M. C., 410, 416, 424(67) *429*
Gallimella, E., 554(177) *563*
Gallot, Y., 377(91) *382*
Gandini, A., 251(32, 33, 35) *306*; 251, 281(34) *306*
Ganis, P., 486(103, 105, 106, 110) *509*; 488, 494(129, 130, 132) *509*; 488, 494 (135) *510*; 490, 494(144) *510*
Gantmakher, A. R., 367(47, 48) *381*
Gantmakhers, A. R., 268(85) *307*; 272 (94) *308*; 523(47) *560*
Gardlund, Z. G., 460(75, 76) *469*
Gardon, J. L., 178(4) *228*; 191, 198(27) *225*; 198, 199(50) *225*; 198(20) *225*; 193(53) *226*
Garratt, P. J., 318(25a) *347*
Garrett, B. S., 497(229) *512*
Garst, J. F., 341(104) *349*; 345(65) *348*

Gautron, R., 460(73) *469*
Gaylord, N., 125(13) *140*
Gaylord, N. G., 453(31) *468*; 556(215) *564*
Gechele, G. B., 189, 195, 218(21b) *225*
Gee, D. R., 542(118) *562*
Gee, G., 539(93) *561*
Gehm, R., 548(151) *563*
Geilen, H., 520(35) *560*
Geller, N. M., 554(183) *563*
Geny, F., 493, 494(175) *511*
Georgiewa, M., 559(232) *565*
Gerber, G., 378(93) *382*
Gerlach, K., 86(50) *114*
Gerrens, H., 190, 193(38) *225*; 192(116) *228*; 201(57) *226*
Gershberg, D. B., 202(58c) *226*
Gerters, R. L., 322(35b) *347*
Gervasi, J. A., 192, 210(104) *227*
Geuskens, G., 458(49) *468*
Gevaert-Agfa, 466(115) *470*
Giannini, U., 398(47a) *429*; 423(92) *430*
Giannoti, C., 496(199) *511*
Gibbs, C. F., 539(96) *561*
Gibbs, W. E., 216(97) *227*
Gibson, R. O., 431(2) *445*
Giglio, E., 484(83) *508*; 487, 494(118) *509*
Gilbert, J. M., 312(9) *346*
Gilchrist, A., 385(20) *428*
Giles, R. D., 33(5) *50*
Gilman, H., 331(72, 73) *348*; 473(21) *506*
Gilman, H. J., 329(59) *348*
Gippin, M., 383(4) *427*; 421(86) *430*
Gipstein, E., 536(81) *561*
Giusti, D., 261, 262(55) *307*
Glaze, W. H., 332(76) *348*; 375(85) *382*
Gleason, E. H., 194, 219(41) *225*
Glet, W., 531(63) *561*
Glusker, D. L., 368(58) *381*
Goering, H. L., 322(36) *347*
Goff, A. L., 168(29) *174*
Gola, Y., 466(112) *470*
Gold, H., 267(84) *307*
Gold, V., 158(2) *173*; 230(3) *243*
Gol'Danskii, V. I., 173(51) *174*
Goldberg, E. J., 371(68) *381*
Golden, D. M., 33(1) *50*
Goldfarb, Y. Y., 288(118) *308*
Goldfinger, G., 123(6) *140*
Golinkin, H. S., 162(11) *174*

Golub, M., 459(55, 56) *468*; 460(60, 61) *468*
Gonikberg, M. G., 168(29) *174*; 172(46) *174*
Goode, W. E., 368(56) *381*
Goodman, M., 408(64) *429*; 484, 485 (82) *508*; 484(85) *508*; 497(238) *512*; 535(74) *561*; 536(86) *561*
Goodrich, B. F., 222(112) *227*
Gorin, S., 493, 494(176) *511*
Gosser, L., 323(38a, 38b) *347*
Gotoh, A., 274(100) *308*
Gottlib, Yu. Ya., 493, 494(174) *511*; 496(221, 222) *512*
Graham, E. S., 215(96) *227*
Graham, N. B., 177(2) *224*
Graham, R. K., 370(64) *381*
Grancio, M. R., 222(112a) *227*
Granger, M. R., 319, 320(28) *347*
Granlio, M. R., 205, 206(68) *226*
Grassie, N., 462(87) *469*
Greber, G., 378(92) *382*
Green, M. L. H., 426(73) *429*
Gregg, R. A., 77(45) *114*
Gregory, N. L., 77(44) *114*
Greth, G., 180, 212(8) *224*
Grevels, F. W., 85(48) *114*
Griffin, W. C. 179(6) *224*
Gross, M. D., 193(47) *225*
Grotewold, J., 33(3) *50*
Grübel, H., 53(4) *113*; 87(61) *115*
Grunwald, E., 324(40) *347*
Guarise, G. B., 164(16) *174*; 173(47, 48) *174*
Guillet, J., 461(83, 84) *469*
Guillot, J., 123(9) *140*
Guinn, H. W., 263(64) *307*
Gulbekian, E. V., 210(85) *227*
Gumbolt, A., 386(26) *428*
Gumprecht, W., 455(37) *468*
Gussoni, M., 478, 493(40) *507*
Guttman, J. Y., 395(43) *428*
Guzman, G. M., 362(33) *380*

Haas, H., 91(67) *115*
Haas, H. C., 554(191) *564*
Habmann, V., 272(92) *308*
Hachihama, Y., 554(184) *564*
Haebig, J., 494(178) *511*
Haines, L. I. B., 67(32, 33) *114*
Hall, H. K., 540(99, 100) *561*; 544(129) *562*

Hallowell, A. T., 331(69a) *348*
Ham, G. E., 11(3) *29*; 123(5, 11) *140*; 153(37) *157*; 245, 294(5) *305*; 267(81) *307*; 557(218, 219, 220, 228) *564*
Hamann, S. D., 158(1) *173*; 161(7) *173*; 163(13) *174*; 170(32) *174*
Hamanoue, K., 167(23) *174*; 168, 172 (26) *174*
Hamermesh, C. L., 542(121) *562*
Hammond, G. S., 460(59) *468*
Hammons, J. H., 312(8) *346*; 314(11a) *346*
Hamori, E., 496(201, 204) *511*
Hanack, M., 471(3) *506*
Hanford, W. E., 547(141) *562*; 547(150) *563*
Hank, R., 425(93) *430*
Harada, M., 202(58d) *226*
Haren, H. J. V., 392(36) *428*
Hargitay, B., 403(59) *429*
Hargreaves, K., 53, 54(10) *113*; 53, 54, 75(11) *113*; 55(19a) *114*; 55, 67, 69(17) *113*; 56, 59, 67, 85(24) *114*
Harkins, W. D., 185(17) *225*
Harries, C., 352(4) *380*
Harrod, J. F., 417(81) 430
Hartley, G., 461(83) *469*
Hartmann, P., 87(56) *115*
Hartung, H. A., 553(173) *563*
Hasegawa, M., 455(38, 39, 40, 41, 42) *468*
Hashimoto, K., 559(231) *565*
Hashimoto, S., 541(101) *561*
Hashimoto, Y., 248, 280(18) *306*
Haszeldine, R. N., 45(33, 34, 35, 36, 37, 38, 39) *51*
Hata, G., 61(28) *114*
Hatada, K., 554(195) *564*; 556(213) *564*
Haury, V. E., 542(121) *562*
Hauser, C. R., 339(96) *349*
Haworth, W. N., 545(136) *562*
Hay, P. M. 189, 194(22) *225*
Hayashi, K., 267, 278, 291(80) *307*; 278 (103) *308*; 453(26) *468*
Hayashi, M., 493, 494(168) *511*
Hazell, J. E., 537(88) *561*; 556, 557, 559 (203) *564*
Hearn, J., 203(60a) *226*
Heatley, F., 478(36) *507*
Hechenbleikner, I., 329(58) *347*
Heck, R. F., 75(38, 39) *114*
Heiberger, C. A., 211(86) *227*

Heikens, D., 520(35) *560*
Helin, A. F., 180, 190, 217(23) *225*; 190, 194, 203(61) *226*; 217(10) *224*
Henglein, A., 147(22) *156*
Henrici-Olive, G., 155(44) *157*
Henry, P. M., 75(39) *114*
Henry, M. C., 86(53) *115*
Herbig, K., 340(102a) *349*
Herman, D. F., 385(14) *428*
Hermann, G., 494(187) *511*
Herschbach, D. R., 77(43) *114*
Heskins, M., 461(84) *469*
Heven, Jun., A. C., 482(62) *508*
Hiatt, R., 33(1) *50*
Hidai, M., 536(85) *561*; 551(157) *563*
Higashimura, S., 495(209) *512*
Higashimura, T., 251(27) *306*; 254(40) *306*; 261, 277(53) *307*; 274(100) *308*; 277(102) *308*; 279(105) *308*; 524(52) *560*; 559(230) *565*
Higasi, K., 372(72) *381*
Higgins, J. P. J., 172(43) *174*
Higgins, T. L., 394(41) *428*
Higginson, W. C. E., 352(8) *380*
Hilbers, C. W., 413(74) *429*
Hilditch, G., 215(96) *227*
Hill, J. W., 541(102) *462*
Hine, J., 339(97) *349*
Hinlicky, J. A., 284(112) *308*
Hinnen, A., 460(73) *469*
Hiratsuka, S., 151(31) *157*
Hirayama, F., 459(52) *468*
Hirota, M., 266(76) *307*
Hirota, N., 343(111a) *349*
Hirutani, M., 537(89) *561*
Hitzke, J., 172(42) *174*
Hobbs, J., 87(57) *115*; 87, 91(60) *115*
Hochstein, F. A., 319(27b) *347*
Hoefnagels, J., 460(74) *469*
Hoeg, D. F., 401(57) *429*
Hoeve, C. A. J., 492, 494, 496(158) *510*
Hoffmann, R., 455(35) *468*
Hogan, J. P., 431(1) *445*; 433, 436, 437, 440(4) *445*; 436(6) *445*
Hogen-Esch, T. E., 301(137) *309*; 324 (42, 43) *347*; 328(54) *347*
Hogg, A. M., 40(16) *51*
Holden, H. W., 177(2) *224*
Holdworth, S. D., 201(58a) *226*
Hollingsworth, B. L., 547(143) *562*
Hollis, C. E., 216(99) *227*
Hollo, E., 544, 545(131) *562*

20

Holmes, D. R., 488(137) *510*
Holmes, J., 272(95) *308*
Holtzer, A. M., 485(90) *308*
Holzkamp, E., 383, 384(1) *427*
Hood, III, F. P., 494(182) *511*
Hopff, H., 212(89) *227*; 516(19, 20) *560*
Hopgood, D., 67(33) *114*
Horiuti, J., 340(99) *349*
Hornung, K. M., 268(87) *308*
Hornyak, J., 173(49) *174*
Hoshi, Y., 211(87) *227*
House, H. O., 330(66) *348*
Howard, E. G., 554(188, 190) *564*
Howell, J. A., 556(206) *564*
Howland, L. H., 204(64, 65) *226*; 220 (108) *227*
Hoyland, J. R., 138(23) *141*
Hrollovic, P., 463(93) *469*
Hsieh, H. L., 375(85) *382*
Huch, A., 386, 399(13) *428*
Huch, C., 386, 399(13) *428*
Hudson, Jr., B. E., 289(121) *308*
Huggins, M. L., 386(27) *428*; 473(16) *506*
Hughes, E. O., 55, 59, 68, 72, 76, 82(18) *113*
Hughes, J. 154(47) *157*
Hughes, R. E., 477(29a) *507*; 492, 494 (152) *510*; 496(201, 203, 204) *511*
Huibers, D. Th. A., 211(86) *227*
Huisgen, R., 340(102a, 102b) *349*
Hulse, G. E., 373(75) *381*
Hummel, D. O., 478, 479, 493(45) *507*
Humphlett, W. J., 339(96) *349*
Hunter, D. H., 301(138) *309*; 334(84) *348*
Hunter, W. H., 256(42) *306*
Huntsman, W. D., 539(91) *561*
Husar, A., 153(36) *157*
Hush, N. S., 344(112) *349*
Hyde, F. J., 208(77) *226*
Hyne, J. B., 162(11) *174*
Hyne, R. A., 520(33) *560*

Ikeda, K., 251(27) *306*
Ikeda, S., 453(27) *468*
Ikeler, T. J., 466, 467(118) *470*
Iles, D. H., 453(28) *468*
Imai, I., 523, 525(51) *560*; 535(77) *561*
Imaizumi, Y., 279(106) *308*
Imamura, Y., 495(209) *512*

Immergut, E. H., 123(5) *140*; 368(53) *381*; 517, 526, 538, 552(16) *559*
Imoto, M., 554(176) *563*
Imoto, T., 522(41) *560*; 547(149) *563*
Impastato, F. J., 332(77a) *348*
Inagaki, H., 369(59) *381*
Ingold, C. K., 337(90) *349*
Inone, S., 479(50) *507*
Inoue, S., 356(22) *380*; 554(178, 179, 180) *563*
Inoue, T., 554(195) *564*
Ipatieff, V. N., 539(91) *561*
Irwin, R. S., 40, 43(18) *51*
Isack, F. L., 249, 262(21) *306*
Isawa, Y., 522(41) *560*
Isemura, T., 178(3) *224*
Ishi, Y., 541(105) *562*
Ishida, A., 481(55) *507*
Ishida, S., 283(110) *308*
Ishigaki, A., 554(184) *564*
Ishii, Y., 149(29) *157*; 267(82) *307*
Ishikawa, T., 496(217) *512*
Ito, K., 146, 152(24) *157*; 559(226) *564*
Itoh, M., 144(15) *156*
Ivin, K. T., 360(27) *380*; 516(23) *560*; 517(12, 13) *559*; 517(28) *560*; 518(32) *560*; 527(54) *560*; 529(55) *560*; 537 (87, 88) *561*; 539(95) *561*; 556, 557 (203) *564*; 557 (216) *564*
Iwatsuki, S., 559(229) *564*
Izu, M., 556(206, 207) *564*

Jaacks, V., 301(139) *309*; 516(21) *560*; 543(123) *562*
Jackson, H. W., 536(82) *561*
Jacobson, H., 551(162, 172) *563*
Jacobson, H. W., 536(83) *561*; 554(201) *564*
Jacques, J. K., 551(159) *563*
Jagur-Grodzinski, J., 322(34b) *347*; 342 (109) *349*; 345(106) *349*
James, D. G. L., 41(20, 21) *51*; 544(196) *564*
Jelinek, M., 551(166) *563*
Jenkins, A. D., 55(19) *113*; 73, 76(36a) *114*; 76(42) *114*; 77(46) *114*; 122(4) *140*; 124(12) *140*; 127(19, 20) *141*; 140(25) *141*; 142(3) *156*; 155(42, 45) *157*; 370(61) *381*
Jenner, G., 172(42) *174*
Jensen, F. R., 333(79b) *348*

Jernigan, R. L., 472, 496(13) *506*; 492, 494(157) *510*
Jimura, K., 493, 494(170) *510*; 494(189) *511*
Johari, D. P., 48(42) *51*
Johnson, A. F., 374(79) *382*
Johnson, J. F., 483(68) *508*
Johnston, R., 73, 76(36a) *114*; 124(12) *140*; 155(45) *157*
Jones, D. M., 298(132) *309*
Jones, F. P., 522(44) *560*
Jones, R. G., 331(73) *348*
Jordan, D. O., 246(12) *306*; 399(52) *429*
Joshi, R. M., 11(3) *29*; 534(71) *561*
Jovanovic, S., 199(55) *226*
Joyce, R. M., 547(141) *562*
Joyner, F. B., 475(27) *507*
Judy, W. A., 551(156) *563*
Jun, M-J., 86(54) *115*

Kabaivanov, W., 559(232) *565*
Kabanov, V. A., 522(37) *560*
Kaeriyama, K., 104, 105, 108, 110(84) *115*
Kagiya, T., 540(98) *561*
Kaizerman, S., 198(54) *226*
Kajima, H., 202(58d) *226*
Kallenbach, N. R., 485(92) *508*
Kalminsk, K. 465(105) *469*
Kambe, H., 523, 525(51) *560*; 535(77) *561*
Kammereck, R. F., 373(76) *381*
Kaneko, M., 496(213) *512*
Kangas, L. R., 75(39) *114*
Kani, M., 149(29) *157*
Kanoh, N., 261(53) *307*; 274(100) *308*
Kapustyan, V. M., 173(50) *174*
Karabatsos, G. J., 299(134) *309*
Karasz, F. E., 482(67) *508*
Kargin, V. A., 522(37) *560*
Kasahara, H., 541(106) *562*
Kashiwagi, T., 536(85) *561*; 551(157) *563*
Kastning, E. G., 97, 103(77) *115*
Kataoka, Y., 103, 107, 108(83) *115*
Katchalski, E., 485(91) *508*
Kato, M., 554(197) *564*
Kato, Y., 461(81) *469*
Katz, T. J., 318(24) *346*; 318(25a) *347*
Kaufhold, R., 547(147) *563*
Kawabami, M., 410(68) *429*
Kawabata, N., 127(16) *141*; 372(69) *381*

Kazakevich, A. G., 173(51) *174*
Kearney, J. J., 210(104) *227*
Kebarle, P., 40(16) *51*
Keller, A., 486(107) *509*
Kelley, E. L., 181(12) *224*; 193(44) *225*
Kelly, D., 464(101) *469*
Kennedy, J. P., 245(4, 7) *305*; 245, 282, 285, 288, 290(10) *306*; 253, 256, 259, 263(36) *306*; 259, 263(52) *307*; 282 (108) *308*; 286(116) *308*; 288(117) *308*; 289(119, 121) *308*; 290(122) *308*; 297(128) *309*; 298(129, 130) *309*; 304 (141) *309*; 367(51) *381*
Kent, L., 339(95) *349*
Kenyon, W. O., 534(70) *561*
Kerber, R., 554(175) *563*
Kern, W., 192, 219(42) *225*; 301(139) *309*; 365(42) *381*; 480(51) *507*; 516 (21) *560*; 543(123) *562*
Kerr, J. A., 36(12) *51*; 42, 46(29) *51*; 155(40) *157*
Ketley, A. D., 273(97) *308*; 285(114) *308*; 384(11) *428*; 477(30) *507*; 483 (68, 71) *508*
Keyerson, S. J., 299(134) *309*
Kikuchi, S., 464(99) *469*; 464, 465(103) *469*
Kilroe, J. G., 169, 170(34) *174*; 514(2) *559*
Kinsinger, J. B., 492, 494(152) *510*
Kinsler, D., 210(85) *227*
Kiprianov, A. I., 460(69) *469*
Kirmse, W., 340(101) *349*
Kirsh, Y., 465(105) *469*
Kirste, R., 496(202) *511*; 496(223) *512*; 496, 497(224) *512*
Kitazawa, T., 493, 494(160) *510*
Klager, K., 319(27a) *347*
Kleiner, H., 331(70) *348*
Klimisch, R. L., 340(98) *349*
Klinedinst, Jr., P. E., 250(23) *306*
Klyne, W., 471(1) *506*
Knox, J. H., 36(7) *50*
Kobayashi, M., 487(119) *509*
Kochi, J. K., 155(48) *157*
Kodaira, Y., 545(134) *562*
Kodak-Pathe, 466(109) *470*
Koehl, Jun., W. J., 332(75a, 75b) *348*
Koelsh, C., 455(37) *468*
Koerner von Gustorf, E., 85(48) *114*; 86 (53, 54) *115*
Kohla, H., 143(8) *156*

Kohler, J., 278(104) *308*
Kohnlein, E., 190, 193(38) *225*; 192 (116) *228*
Koinuma, H., 554(178, 179, 180) *563*
Koizumi, N., 487(120) *509*
Kokelenberg, H., 460(70) *469*
Kolesov, V. P., 547(144) *563*
Kolthoff, I. M., 203(62) *226*
Komatsu, T., 556(213) *564*
Konar, R. S., 192, 211, 222(111) *227*
Kondo, M., 554(182) *563*
Konishi, A., 372(71) *381*
Konizer, G., 327(47) *347*
Kono, T., 547(149) *563*
Kooymann, E. C., 385(19) *428*
Kormer, V. A., 207(76) *226*; 416(80) *430*
Korshak, V. V., 256(44) *306*; 531(61) *560*; 531(65) *561*
Korshak, Yu. V., 103(82) *115*
Korte, F., 531(63) *561*
Kosower, E. M., 317(22) *346*
Kovacs, J., 489(141) *510*
Kozina, M. P., 541(113) *562*
Krackeler, J. J., 201(58) *226*
Krakovjak, M. G., 493, 494(174) *511*
Kramar, V., 330(66) *348*
Kraus, C. A., 363(35) *380*
Krause, S., 495(197) *511*
Kreider, R. W., 181(12) *224*
Krentsel, B. A., 288(118) *308*; 385(16) *428*
Krieger, I. M., 205(68a) *226*
Krieghoff, N. G., 333(79a) *348*
Krighaum, W. R., 492, 494, 496(154) *510*; 495(196, 205) *511*
Krishan, T., 190, 216(101) *227*
Krongauz, V. A., 450(13) *467*
Kropachev, V. A., 554(183) *563*
Kubo, M., 487(120) *511*
Kuchner, K., 183, 199, 209(16) *225*
Kuhn, R., 337(89b) *348*
Kuhn, S. J., 263(64) *307*
Kunitake, T., 263(65, 66) *307*; 279(106) *308*; 516(25) *560*
Kuntz, I., 271(89) *308*
Kurashev, V. V., 559(227) *564*
Kurata, M., 492, 494(151) *510*
Kuri, Z., 537(89, 90) *561*
Kurita, Y., 487(120) *509*
Kursanov, D. N., 264(68) *307*
Kurz, J. E., 492, 494(154) *510*

Kusabayashi, S., 266(76) *307*
Küsten, R., 463(95) *469*
Kuwahara, N., 496(213) *512*
Kuzub, L. I., 518, 541(31) *560*
Kwok, J. C., 87(59) *115*

Laasko, T. M., 345(113a) *349*
Labes, M. M., 266(75) *307*
Ladd, J. A., 322(35b) *347*
Lagally, P., 143(8) *156*
Laidlaw, W. G., 162(11) *174*
Laidler, K. J., 534(73) *561*
Laita, Z., 551(166) *563*
Lal, J., 255(41) *306*
LaLancette, E. A., 318(25b) *347*
Lamb, J. A., 164, 167(18) *174*
Lamb, R. C., 341(104) *349*; 345(65) *348*
Lamola, A. A., 448, 458(3) *467*
Landfield, H., 194(43) *225*; 194, 222 (117) *228*; 198, 213(51) *225*
Lando, J. B., 543(127) *562*
Langer, Jun., A. W., 245(4) *305* 329(49) *347*; 329 (62) *348*; 353(12) *380*
Laplanche, L. A., 493, 494(166) *510*
Lardicci, L., 481(58) *508*
Laridon, U., 458, 467(120) *470*; 466 (113) *470*
Latham, K. G., 516, 522(26) *560*
Laverty, S. J., 460(76) *469*
Lawler, R. G., 319(29) *347*
Lawry, P. S., 55(19) *113*
Lebedev, N. N., 257(44) *306*
Leblanc, H., 457(46) *468*
Ledwith, A., 11(2) *29*; 239(7) *243*; 245 (7) *305*; 246, 263(14) *306*; 249, 263, 264, 271, 278, 291, 298(22) *306*; 261, 262(54) *307*; 263(60, 67) *307*; 263, 280, 301(61) *307*; 263, 271, 273, 278, 291, 301, 303(62) *307*; 263, 271(63) *307*; 264(70) *307*; 265(71) *307*; 266(77) *307*; 298(131) *309*; 299(135) *309*; 453(28) *468*; 478, 479(42) *507*; 482(65) *508*; 523(46) *560*
Lee, B. E., 12(4) *29*
Lee, C. C., 298(133) *308*
Lee, C. L., 247, 279(15) *306*; 345(116) *349*; 363(36) *380*
Lee, P. I., 193, 218(45, 46) *225*
Leffler, J. E., 338(94a) *349*
Lehr, M. H., 384(8) *427*

Lehr, M. L., 412(69) *429*
Lehr, M. M., 410(65) *429*
Le Noble, W. J., 158(3) *173*; 161(10) *174*; 171(38) *174*
Lenz, R. W., 497(237) *512*
Leonard, J., 516, 522(23) *560*; 517(28) *560*; 524(53) *560*
Le Roy, D. J., 40, 42(17) *51*
Letort, M., 488(133) *510*
Levy, L. K., 317(14) *346*
Levy, M., 352(9) *380*; 370(63) *381*; 376 (87) *382*
Lewis, E. S., 450(8) *467*
Lewis, F. M., 374(77) *382*
Liebman, S., 401(57) *429*
Lifson, S., 484(86, 87) *508*; 491, 494 (147) *510*
Lim, D., 330(71) *348*
Lin, Y. T., 301(138) *308*
Lind, D. J., 96, 99, 101, 102, 103(78) *115*; 103(81) *115*
Lipscomb, R. D., 536(80) *561*
Liquori, A. M., 484(83) *508*; 487, 494 (118) *508*; 488, 494(140) *510*; 493, 494 (173) *510*; 497(231 232) *512*
Lissi, E. A., 33(3) *50*
Liston, T. V., 329(63) *348*
Litt, M., 192, 210(32) *225*; 192, 210, 222(111a) *227*; 487(125) *511*
Liu, K. J., 494(186) *511*
Livshitz, I. A., 216(98) *227*
Llano, E., 368(55) *381*
Loback, M. I., 416(80) *430*
Lochmann, L., 330(71) *348*
Lockett, E. C., 278, 291(123) *308*
Logan, T. J., 329(64) *348*
LoGiudice, F., 423(92) *430*
Löhr, G., 364(37) *380*; 365(40) *381*
Longbottom, H. M., 193, 218(46) *226*
Longfield, E., 202(58c) *226*
Longi, P., 554(200) *564*
Longworth, W. R., 273(99) *308*
Looy, H. M. van., 392, 398, 402(37) *428*; 398(48) *429*
Lopez-Heredia, G. G., 488(136) *510*
Lorenzi, G. P., 407(61, 62, 63) *429*; 471 (5) *506*; 480(53) *507*; 481(54, 57) *507*; 504(251, 252) *513*
Lovelock, J. E., 77(44) *114*; 343(108a) *349*
Lovrien, R., 460(65) *469*
Lowe, J. P., 484(79) *508*

Lowry, G. G., 517(29) *560*; 556(202) *564*
Lucherini, A., 471, 477(7) *506*; 483(69) *508*
Ludlum, D. B., 178(4) *224*; 387(29) *428*
Ludwig, E. B., 523(47) *560*
Ludy, M., 462(88) *469*
Luisi, P. L., 471(11) *506*; 477(31) *507*; 484, 491(78) *508*; 495(190, 191) *511*; 496(211) *511*; 500(242, 245), 246) *512*; 500(247) *513*; 502(248) *513*; 503 (250) *513*
Lumry, L., 494(179) *511*
Luner, C., 451(15) *467*
Lussi, H., 514(4) *559*; 516(19) *560*; 516, 522, 535(20) *560*; 556(212) *564*
Luzzati, V., 496, 497(226, 228) *512*
Lyalikov, K., 465(105) *469*
Lyudvig, E. B., 268(85) *307*; 272(94) *308*

MacCallum, D., 41(20) *51*
MacCarthy, B., 251, 254, 303(31) *306*
Mack, W., 340(102b) *349*
Macknight, W. J., 482(67) *508*.
Maclean, C., 413(74) *429*
Macosko, C., 163(14) *174*
Madigan, J. C., 204, 221(64) *226*
Madison, N. L., 554(189) *564*
Maeda, T., 540(98) *561*
Maercker, A., 327(50) *347*
Magagnini, P. L., 289(119) *308*
Makimoto, T., 368(54) *381*; 486, 487 (109) *509*
Malhotra, S. L., 524(53) *560*
Mallan, J. M., 329(56) *347*
Maltman, W. R., 53, 54, 73(5) *113*; 54, 70, 89(14) *113*; 55, 70, 71(15) *113*; 56, 64, 70, 71, 80, 85(20) *114*; 71(36) *114*
Mammi, M., 487(123) *509*
Manasek, Z., 463(93) *469*
Mandelcorn, L., 40(14) *51*
Mankovitch, A. M., 178(5) *224*
Manning, P. P., 546(137) *562*
Mantell, G. J., 180, 190, 217(23) *225*, 190, 194, 203(61) *226*; 217(10) *224*
Mantica, E., 479(46) *507*
Manyasek, Z., 192, 216(29) *225*
Mao, T. J., 452(33) *468*
Marcus, R. A., 259(50) *307*
Marek, M., 257(45) *306*
Mares, F., 320(30) *347*
Margaritova, M., 190, 216(101) *227*

Margerison, D., 322(35a) *347*
Mark, H., 386(24) *428*; 473(16) *506*; 541 (107) *562*
Mark, H. F., 123(5) *140*
Mark, J. E., 471(14) *506*; 492, 494(153, 161, 162, 163, 164) *510*; 496(203, 204) *511*; 496(218) *512*
Markevich, M. A., 518, 541(31) *560*
Marley, R., 466(117) *470*
Martin, E. L., 547(140) *562*
Martin, H., 383, 384(1) *427*
Maruyama, I., 541(108) *562*
Marvel, C. S., 208(71) *226*; 539(96) *561*; 544(199) *564*
Mason, F., 496, 497(226) *512*
Masuda, S., 558(225) *564*
Masuda, T., 559(230) *565*
Mateo, J. L., 228(3) *228*
Matlack, A. S., 373(75) *381*; 393(40) *428*
Matsuda, M., 345(106) *349*
Matsuguma, Y., 263(65, 66) *307*; 279 (106) *308*
Matsumoto, S., 248, 280(18) *306*; 542 (117) *562*
Matsuzaki, K., 471(10) *506*; 481(55) *507*
Matthews, F. E., 352(3) *380*
Mayo, F. R., 77(45, 46) *114*; 120(3) *140*; 123(8) *140*; 139(24) *141*; 374(77) *382*
Mazzanti, G., 386(22) *428*; 474(25) *507*; 479(46, 47) *507*; 486(112, 113, 114) *509*
McCene, B., 462(86) *469*
McClelland, B. J., 356, 357(21) *380*
McConnell, R. L., 475(27) *507*
McCormack, H. W., 362(33) *380*; 514, 516, 523(1) *559*
McCurdy, K. G., 534(73) *561*
McEwan, W. K., 312(7) *346*
McGrath, J. E., 255(41) *306*
McIntosh, A., 210(84) *227*
McKinley, S. V., 334(83) *348*
McMillan, D. F., 33(1) *50*
McNees, R. S., 329(60) *348*
Medvedev, S. S., 179(7) *224*; 186, 199 (19) *225*; 268(85) *307*; 272(94) *308*; 367(47, 48, 49) *381*; 523(47) *560*
Meerwein, H., 267(84) *307*; 270(88) *308*; 542(114, 115) *562*
Meggy, A. B., 544(128) *562*
Mehdi, S. A., 169(33) *174*

Meier, M., 82(47) *114*
Melia, T. P., 541(104) *562*; 542(118, 125, 126, 138) *562*
Mellor, D. P., 390(33) *428*
Mellows, C. L., 478(34) *507*
Melville, H. W., 144(11) *156*; 147(25) *157*; 155(41) *157*
Mencik, Z., 486(97) *509*
Mendelsohn, M. A., 95(76) *115*
Merigold, D., 522, 535(40) *560*
Merrall, G. T., 539(93) *561*
Merrett, F. M., 164(20) *174*
Merz, E., 123(6) *140*
Mesrobian, R. B., 189(21a) *225*
Messick, J., 478(35) *507*
Metzger, G., 161, 168(9) *173*
Meyersen, K., 371(67) *381*
Mialto, M., 403(59) *429*
Mibawa, H., 266(76) *307*
Michael, A., 345(113b) *349*
Michel, R. M., 283(111) *308*
Michelin, 221(109) *227*
Middleton, W. J., 536(82) *561*
Migama, H., 451(18) *468*
Mighton, H. R., 547(142) *562*; 547(148) *563*
Mikhailov, G. P., 495, 496(208) *511*; 496(210) *512*
Miki, T., 524(52) *560*; 543(124) *562*
Miles, D. M., 461(80, 81, 82) *469*
Milkovitch, R., 352(9) *380*
Miller, A. G., 266(75) *307*
Miller, I. K., 203(62) *226*
Miller, J. R., 193(47) *225*
Miller, R. L., 486(93) *508*
Miller, W. L., 267(80) *307*
Millet, M., 172(42) *174*
Millrine, W. P., 251, 254, 303(30, 31) *306*
Minaghi, M., 14(6) *29*
Minckler, Jr., L. S., 286(116) *308*
Mino, G., 194, 219(41) *225*
Minoshima, Y., 464(97) *469*
Mislow, K., 471(4) *506*
Misono, A., 536(85) *561*; 551(157) *563*
Mita, I., 523, 525(51) *560*; 535(77) *561*
Miyake, A., 493, 494(171, 172) *510*; 522(39) *560*
Miyake, Y., 496(213) *512*
Miyamoto, T., 369(59) *381*
Miyazawa, T., 478, 479, 493(43) *507*
Mizushima, S., 493, 494(168, 169) *510*

Mochel, W. E., 223(114) *227*; 547(140) *562*

Modena, M., 554(177) *563*

Mole, M. F., 551(159) *563*

Monnerie, L., 493, 494(175, 176, 177) *511*

Montagnoli, G., 471, 477(6, 7) *506*; 471, 477, 481, 483(8) *506*; 471, 482(9) *506*; 477(31, 32) *507*; 481, 494, 495(54) *507*; 482(59) *508*

Montecatini, 383, 416, 418(3) *427*

Moore, D. E., 192(90) *227*

Moore, J. C., 361(31) *380*

Moore, J. E., 368(56) *381*

Moraglio, G., 492(156) *510*; 496(198, 199) *510*; 479(46) *507*

Morawetz, H., 473, 483, 495(20) *506*

Moreau, N. M., 464(104) *469*

Morgan, P., 436(6) *445*

Mori, K., 487(119) *509*

Moriata, H., 367(45) *381*

Mørk, P. C., 193, 195(39) *225*; 198, 212 (52) *226*

Morris, C. E., 218(103) *227*

Morris, C. E. M., 181(14) *224*

Morrow, R. C., 375(83) *382*

Morsal, H., 269(88) *308*

Morschel, H., 542(115) *562*

Morton, A. A. 329(58) *347*; 331(69a, 69b, 69c, 69d) *348*

Morton, M., 194(43) *225*; 194, 222(117) *228*; 198, 213(51) *225*; 198(54) *226*; 207(73) *226*; 215(95) *227*; 216(97, 100) *227*; 373(76) *381*; 374(80) *382*; 377 (88) *382*; 420(88) *430*

Motoyama, T., 192(35) *225*

Motroni, G., 18(9) *30*

Moyer, P. H., 412(69) *429*

Msi, N., 299(134) *309*

Muck, D. L., 321(33) *347*

Mucke, J., 144(14) *156*

Murshashi, S., 481(56) *507*; 487(117) *509*; 492, 494(160) *510*

Murata, N., 103, 107, 108(83) *115*

Murphy, P., 55, 67, 75, 76(16) *113*

Muzaki, K., 451(14) *467*

Naarman, H., 96, 103(77) *115*

Naegele, W., 290(122) *308*

Nagai, K., 491(146) *510*; 496(217) *512*

Nagata, S., 202(58d) *226*

Naidus, H., 201(58) *226*

Nakagawa, T., 178(3) *224*; 523(50) *560*

Nakajima, A., 490, 491, 494(145) *510*

Nakamaye, K. L., 333(79b) *348*

Nakamura, K., 464(99) *469*; 464, 465 (103) *469*; 493, 494(168) *510*

Nakamura, T., 267(78) *307*

Nakane, R., 257(47) *307*

Nakaniwa, M., 541(108) *562*

Nakayama, Y., 368(54) *381*

Nakinishi, M., 455(39, 40, 42) *468*

Napper, D. H., 192(34) *225*

Narisawa, S., 540(98) *561*

Natale, C., 464(101) *469*

Natsuume, T., 266(76) *307*

Natta, F. J. van, 547(140) *562*

Natta, G., 383, 384(2) *427*; 384(6) *427*; 386, 387(22) *428*; 394, 411(42) *428*; 399(49) *429*; 407(61) *429*; 410, 416, 424(67) *429*; 413(72) *429*; 415(77) *429*; 415(78) *430*; 471(5) *506*; 473(16, 17, 18, 19) *506*; 474(25) *507*; 478, 482, 486, 487(33) *507*; 479(46, 47, 49) *507*; 482(63) *508*; 486(94, 95) *508*; 486 (99, 100, 101, 102, 104, 111, 112, 113, 114, 115, 124) *509*; 487(128) *509*; 488(129, 130, 131, 132) *509*; 488(134, 135) *510*

Nechvatal, A., 36(9) *50*

Neklutin, V. C., 220(108) *227*

Nelson, N. K., 385(14) *428*

Nenitzescu, C. D., 386, 399(13) *428*

Nesmeyanov, A. N., 334(82a, 82b) *348*

Neumann Jr., R. C., 168(31) *174*

Newey, H. A., 331(69b, 69c) *348*

Newman, S., 495(196) *511*

Newmann, M. G., 33(3) *50*

Newport, J. P., 322(35a) *347*

Ney, W. O., 541(111, 112) *562*

Nicholson, A. E., 164, 166(21) *174*

Nielsen, L. E., 486(93) *508*

Nik, G. C. C., 484(85) *508*

Nikitin, V. N., 498(240) *512*

Nishijima, Y., 151(31) *157*

Nishikawa, Y., 95(74, 75) *115*; 96(79) *115*

Nishimura, T., 554(174) *563*

Nitta, I., 487(117) *509*

Noel, C., 493, 494(177) *511*

Nomura, M., 202(58d) *226*

Nomura, T., 554(185) *564*

Nordio, V., 554(200) *564*

Noro, K., 177(1) *224*

Norrish, R. G., 450(12) *467*

Norrish, R. G. W., 143(7) *156*; 164(20) *174*; 164, 166(21) *174*; 517(14) *559*

North, A. M., 12(4) *29*; 142(1, 4) *156*; 143(5) *156*; 143, 146(6) *156*; 144(9, 10) *156*; 144, 146(13) *156*; 144, 149(16) *156*; 144, 153(17, 18, 19) *156*; 146, 147(23) *157*; 148, 154(27) *157*; 148(28) *157*; 151(32) *157*; 153(35) *157*; 155(43, 46) *157*

Novak, A., 514(5) *559*

Novak, I. I., 493, 494(167) *510*

Noyes, R. M., 146(21) *157*

Nozakura, S., 493, 494(160) *510*

Nummy, W. R., 541(111) *562*

Nunomoto, S., 14(6) *29*

Oae, S., 317(19) *346*

O'Deen, L. A., 42(26) *51*

Odian, G., 3(1) *29*

Odintzova, P. P., 181(13) *224*

O'Donnell, J. H., 527, 529(55, 56) *560*; 529(57) *560*; 530(58) *560*

O'Donnell, J. T., 189(21a) *225*

O'Driscoll, K. F., 142(1) *156*; 153(36) *157*; 155(46) *157*; 324(41) *347*; 353, 355, 364(13) *380*; 356(19) *380*; 356 (206, 207) *564*

Ofstead, E. A., 551(156) *563*

Ogata, N., 554(174) *563*

Ogawa, T., 41(21) *51*; 544(196) *564*

Ohno, A., 317(19) *346*

Ohshina, J., *556*(213) *564*

Ohta, K., 481(56) *507*

Ohtsuka, S., 211(87) *227*

Ohtsuka, Y., 171(37) *174*; 520(34) *560*

Oita, K., 385(15, 17) *428*

Okada, M., 541(105, 106) *562*; 559(226, 229) *564*

Okamoto, Y., 534(68) *561*; 554(193, 194, 197, 198) *564*

Okamura, S., 192(35) *225*; 249, 279(20) *306*; 251(27) *306*; 254(40) *306*; 261, 277(53) *307*; 267, 278, 291(80) *307*; 277(102) *308*; 278(103) *308*; 453(24, 25, 27, 32) *468*; 495(209) *512*; 524(52) *560*; 551(155) *563*

O'Kelly de Galway, M., 453(23) *468*

Olah, C. A., 229(1) *243*; 230(2) 243; 245 (2) *305*; 263(64) *307*

Olive, G. H. 395(44) *428*; 398(44a) *428*

Olive S., 155(44) *157*; 395(44) *428*; 398 (44a) *428*

Onishi, T., 267(78) *307*

Ono, K., 481(56) *507*

Onyon, P. F., 76(42) *114*; 142(3) *156*; 370(61) *381*

Oosterhof, H. A., 210(83) *227*

Oota, T., 558(225) *564*

O'Reilly, D. E., 421(90) *430*

Orienti, M., 496(211) *509*

Orr, R. J., 539(92) *561*

Osborn, A. R., 171(39) *174*

Osborn, S. W., 554(181) *563*

Osborne, A. G., 91(66) *115*

Osgan, M., 477(29a, 29b) *507*

Oster, G., 144(12) *156*; 450(7) 467

Oster, G. K., 144(12) *156*

Osugi, J., 168, 172(26) *174*

O'Sullivan, W. I., 339(96) *349*

Otsu, T., 95(74, 75) *115*; 96(79) *115*; 138(22) *141*; 282(108) *308*

Otsuka, M., 554(176) *563*

Otsuka, S., 410(68) *429*

Ott, N., 340(102a) *349*

Ottewill, R. H., 203(60a) *226*

Otto, N., 487(127) *509*

Ottolenghi, A., 370(62) *381*

Overberger, C. G., 259(50) *307*; 283(110) *308* 356(20) *380*; 542(22) *562*

Owen, G. E., 42, 43(24) *51*

Owens, F. H., 368(56) *381*

Ozowa, T., 451(14) *467*

Pacifici, J., 462(92) *469*

Paddock, N. L., 551(159) *563*

Padgett II, W. M., 317(20) *346*

Palit, S. R., 192, 211, 222(111) *227*

Palomaa, M. H., 546(139) *562*

Palvarini, A., 416(79) *430*

Papir, Y. S., 205, 206, 224(68a) *226*

Papisov, I. M., 522(37) *560*

Paprotny, J., 89(64) *115*; 89, 90(65) *115*

Parker, A. J., 312(4) *346*

Parodi, O., 493, 494(175) *511*

Parrod, J., 370(65) *381*; 377(91) *382*

Parsonage, M., 36(12) *51*; 42, 46(29) *51*

Parts, A. G., 181(14) *224*; 192(34) *225*; 218(103) *227*

Pasika, W. M., 539(95) *561*

Pasquon, I., 384(6) *427*; 399(49) *428*

Patel, R. D., 496(214) *512*

Patit, F., 386(23) *428*

Patnaik, B. K., 556(215) *564*
Paton, R. M., 272(92) *308*
Patraik, B., 125(13) *140*
Patsiga, R., 192, 210(32) *225*
Patsiga, R. A., 559(228) *564*
Paukov, I. E., 547(144) *563*
Peaker, F. W., 144(9) *156*
Pearson, J. M., 42(23) *51*; 42, 43(24) *51*
Pearson, R. G., 75(40) *114*; 82(47) *114*;
  315(15) *346*
Pederson, C. J., 327(48) *347*
Pedone, C., 488(138) *510*
Peeters, H., 458, 467(120) *470*
Peggion, E., 193(37) *225*
Pegoraro, M., 479(49) *507*
Peisach, J., 161(5) *173*
Peiser, H. S., 482(61) *508*
Pellon, J., 166(22) *174*; 168(27) *174*
Penczek, I., 557(223) *564*
Penczek, S., 557(223) *564*
Penfold, J., 249, 261, 263, 264, 271, 278,
  291, 298, 303(22) *306*
Pennella, F., 397(46) *429*
Pepper, D. C., 245(2, 3) *305*; 251, 279
  (26) *306*; 251, 254, 279(28) *306*; 251,
  254, 281(29) *306*; 251, 254, 303(30, 31)
  306
Peraldo, M., 471(17) *506*; 479(49) *507*;
  488, 494(131) *509*; 488, 494(134, 135)
  510
Perkins, P. G., 392(34a) *428*
Perrin, M. W., 161(8) *173*; 431(2) *445*
Peterson, J. H., 223(114) *227*
Petit, R., 272(95) *308*
Petraccone, V., 486(110) *509*
Petrick, R. A., 484(77) *508*
Petrov, E. S., 343(110a, 110b) *349*
Pfeil, E., 267(84) *307*
Phillips Petroleum Company, 221(110)
  227; 431(1) *445*; 432(3) *445*
Phillips, P. J., 148(28) *157*
Phillips, R., 211(86) *227*
Pichot, C., 123(9) *140*
Pierce, J. B., 332(77b) *348*
Pieroni, O., 471, 477, 481(8) *506*; 481
  (58, 59) *508*
Pieroni, P., 479(41) *507*
Piirma, I., 207(73) *226*; 215(95) *227*
Pines, H., 539(91) *561*
Pini, D., 471, 477(7) *506*; 471, 477, 481
  (8) *506*
Pinner, S. H., 267, 278(79) *307*

Pino, P., 11(194) *511*; 28(11) *30*; 407
  (61, 62, 63) *429*; 423(92) *430*; 471(9)
  506; 471(5) *506*; 471, 477(6, 7) *506*;
  473, 494, 495, 500(22) *506*; 473(25)
  507; 477(32) *507*; 479(46, 47) *507*;
  480(53) *507*; 481, 494, 500(54) *507*;
  581(58) *508*; 482(59) *508*; 494(190,
  191, 192) *511*; 495(193) *511*; 500
  (242, 245) *512*; 500(246) *512*; 504
  (252) *513*
Pittman, C. W., 229(1) *243*
Pitts, J. N., 446(1) *467*
Pizzoli, M., 189(21b) *225*
Plesch, P. H., 245(1) *306*; 245, 251, 254,
  279, 303(8) *307*; 246(13) *307*; 251, 281
  (32, 33, 34, 35) *307*; 273(96, 99) *308*;
  289(119) *308*; 294(124) *308*; 516(24)
  560; 523(49) *560*
Pochlein, C. W., 228(8) *228*
Poddubnyi, I. Ya., 207(76) *226*
Poe, A. J., 67(32, 33) *114*
Pohl, H. A., 495, 496(206) *511*; 497
  (219) *512*
Poland, D., 485(88) *508*
Polyakova, A. M., 531(61) *560*; 531(65)
  561
Poole, Jun., C. P., 418, 421(83) *430*
Porri, L., 410, 416, 424(67) *429*; 415(78)
  430; 416(79) *430*
Porter, G., 146(21) *156*; 374(78) *382*
Porter, N. A., 451(16) *467*
Porter, R. S., 483(68) *508*
Pospisil, J., 330(71) *348*
Postlethwaite, D., 142(1) *156*; 144, 153
  (18, 19) *156*; 155(46) *157*
Powell, J., 420(87) *430*
Prati, G., 144(12) *156*
Pregaglia, G. F., 14(6) *29*
Prelog, W., 471(1) *506*
Prenosil, M. B., 190(25) *225*
Price, C. C., 125(14) *140*; 124, 128(15)
  141; 477(29a, 29b) *507*; 480(51)
  507
Prince, M., 173(49) *174*
Prischepa, N. D., 288(118) *308*
Pritchard, G. O., 33(4, 6) *50*
Prusinowski, L. R., 496(201) *511*
Ptitsyn, O. B., 483, 484, 485, 488, 489,
  490, 491(73) *508*; 484(81) *508*; 491,
  494(150) *510*
Pucci, S., 500(245, 246) *512*; 500(247)
  513

Pudjaatmaka, A. H., 313(8) *346*
Pummer, W. J., 534(69) *561*
Purcell, W. P., 495, 496(206) *511*
Purma, I., 420(88) *430*
Put, J., 456(44) *468*
Putnam, R. E., 536(82) *561*
Pye, D. G., 536(82) *561*

Quadrifoglio, F., 493, 494(173) *510*
Quick, L. M., 33(5) *50*; 36(13) *51*

Rabeck, F., 462(89) *469*
Rabek, J. F., 450(10) *467*
Raff, R. A. V., 487(116) *509*
Ragazzini, M., 554(177) *563*
Rahman, M., 170(35) *174*; 531(59) *560*
Ramey, K. C., 478(35) *507*
Ramsey, J. B., 250(24) *306*
Ranauto, H. J., 179(7) *224*
Rangnes, P., 193, 195(39) *225*
Rao, D. V. R., 111(91) *116*
Rayner, M. G., 55(19) *113*
Read, G. A., 144, 146(13) *156*; 144, 149 (16) *156*
Reader, B. E. L., 210(84) *227*
Reddy, J. M., 496, 497(227) *512*
Reed, G. A., 106(86) *116*
Reegen, S. L., 15(7) *29*; 245(6) *305*
Reich, J., 368(57) *381*
Reichardt, C., 460(66) *469*
Reijen, L. L. V., 392(35, 36) *428*
Reilly, D. E., 418, 421(83) *430*
Reilly, P. J., 251, 254, 279(28) *306*; 251, 254, 281(29) *306*
Reis, H., 97, 103(77) *115*
Reiser, A., 466(117) *470*
Rembaum, A., 372(72, 73) *381*; 375(83) *382*
Rembaum, A. A., 377(88) *382*
Rempp, P., 345(115) *349*
Rempp, R., 370(65) *381*; 372(74) *381*; 377(91) *382*
Renfrew, A., 436(6) *445*
Reppe, W., 319(27a) *347*
Rewicki, D., 313, 319(5) *346*; 317(21) *346*; 336(89a, 89b) *348*
Reynolds, R. J. W., 57, 61, 93(23) *114*
Reynolds, W. B., 213(91) *227*
Rezabek, A., 192, 216(29) *225*
Rhodes, J. H., 542(120) *562*

Riande, R., 488(136) *510*
Richard, A., 488(133) *510*
Richards, A. W., 261(57) *307*; 263(59) *307*
Richards, D. H., 361(29) *380*; 377(89) *382*
Richardson, D., 359(24) *380*; 516(22) *560*
Richardson, G. M., 331(69a) *348*
Richardson, W. S., 172(41) *174*
Rice, S. A., 494(178) *511*
Riedel, O., 550(154) *563*
Riegger, P., 466(116) *470*
Rihe, C., 463(94) *469*
Rinehart, R. E., 207(72, 74) *226*; 222 (113) *227*
Ringsdorf, H., 283(110) *308*
Ripamonti, A., 484(83) *508*; 487(118) *509*
Ritchie, C. D., 310(1) *346*; 314(12a, 12b) *346*
Robb, J. C., 144(9) *156*
Robb, I. D., 190(88) *227*
Roberman, Zh. N., 498(239) *512*
Roberts, D. E., 497(235) *512*
Roberts, J. D., 327(50) *347*; 333(78) *349*
Roberts, R. C., 344(117) *349*
Robinson, G. C., 250(23) *306*; 323(39) *347*; 362(34) *380*
Robinson, V. J., 60(25) *114*
Rochester, C. M., 249, 262(21) *306*
Rodriguez, L., 403(59) *429*
Rodriguez, L. A., 398(48) *429*
Rodriguez, L. A. M., 392, 398, 402(37) *428*
Roe, C. P., 190, 209(26) *225*
Rogers, C. E., 375(82) *382*
Rogers, M. T., 493, 494(166) *510*
Roig, A., 368(55) *381*; 495(205) *511*
Romatowaki, J., 196(48, 49) *225*; 199 (55) *226*
Romeyn, H., 207(72) *226*; 222(113) *227*
Rose, J. B., 539(95) *561*; 542(119) *562*
Rosenheck, K., 489(142) *510*
Ross, S., 179(7) *224*
Rotinjanz, L., 517, 522(15) *559*
Rowe, Jun., C. A., 335(85a, 85b) *348*
Roy, R., 531(60) *560*
Rozenberg, B. A., 268(85) *307*; 272(94) *308*; 523(47) *560*
Ruchardt, C., 272(92) *308*
Rumao, L. P., 551(158) *563*

Russell, G. A., 337(93) *349*; 341(103) *349*
Rutherford, P. P., 273(96) *308*
Ryan, C. F., 497(229, 230) *512*

Sadron, C., 370(65) *381*
Saegusa, T., 248, 280(18) *306*; 488(139) *510*; 541(108) *562*; 542(117) *562*
Saijo, A., 490, 491, 494(145) *510*
Sakai, 267(82) *307*
Sakamoto, Y., 340(99) *349*
Sakurai, H., 547(149) *563*
Sakurai, N., 537(89) *561*
Salatiello, P. P., 194(43) *225*; 194, 222 (117) *228*; 198, 213(51) *225*
Salovey, R., 495(207) *511*; 496, 497(225, 227) *512*
Saludjian, P., 496, 497(228) *512*
Salvadori, P., 481(58) *508*; 495(190, 194) *511*
Sambhi, M., 266(77) *307*
Sandel, V. R., 334(83) *348*
Sangster, J. M., 42, 43(25) *51*; 42(27, 30) *51*; 280(107) *308*
Saotone, K., 545(134) *562*
Sargeant, P. B., 554(188) *564*
Sargent, G. D., 240(9) *243*
Sargeson, A. M., 111(89) *116*
Satoh, S., 494(181, 184) *511*
Sattlemeyer, R., 216(102) *227*
Sauer, J., 340(102a) *349*
Savino, M., 497(231, 232) *512*
Sawada, H., 556(210, 211) *564*
Schaefer, Ann., O., 331(70) *348*
Schaefer, J., 478(37) *507*
Schafer, O., 352(5) *380*
Schaffler, A., 547(146, 147) *563*
Scheck, G. O., 86(54) *115*
Schildknecht, C. E., 289(120) *308*
Schlinder, A., 368(57) *381*
Schissler, D. C., 421(91) *430*
Schlenk, W., 345(113b) *349*; 352(6) *380*
Schleyer, P. von R., 230(2) *243*
Schlichting, O., 319(27a) *347*
Schlick, S., 356(87) *382*
Schlosser, M., 328(57) *347*
Schmidt, G., 463(96) *469*
Schmidt, M., 386(26) *428*
Schnabel, W., 147(22) *156*
Schneider, A. K., 540, 541(99) *561*
Schneider, B., 494(185) *511*
Schneider, N. S., 497(237) *512*

Schreiber, H., 370(65) *381*
Schrisesheim, A., 335(85a, 85b) *348*
Schroter, C. A., 466(116) *470*
Schuler, N. W., 554(191) *564*
Schulz, G. V., 144(14) *156*; 196(48, 49) *225*; 199(55) *226*; 365(40) *381*; 496 (202) *511*
Schulz, R., 463(94) *469*
Schulz, R. C., 194, 219(42) *225*; 544(130) *562*
Schwager, I., 317(20) *346*
Scilly, N. F., 377(89) *382*
Scott, D. A., 322(37) *347*
Scott, D. W., 551(165) *563*
Scott, H., 266(75) *307*; 418, 421(83) *430*; 421(90) *430*
Scott, K. W., 551(156) *563*
Scott, R. A., 484(80) *508*
Seddon, J. D., 57, 61, 93(23) *114*
Selman, C. M., 332(76) *348*
Semlyen, J. A., 551(160, 161, 164, 169, 170) *563*
Sepulchre, M., 407(60) *429*
Serini, V., 554(175) *563*
Shambelan, C., 477(29a) *507*
Sharkey, W. H., 536(80, 82, 83) *561*; 554(201) *564*
Sharples, A., 482(60) *508*
Shashoua, V. E., 516(27) *560*
Shatenstein, A. I., 314(10a, 10b) *346*; 343(110a, 110b) *349*
Shaw, B. L., 391(34) *428*; 420(87) *430*
Shaw, J. N., 203(60a) *226*
Sheldon, R. A., 155(48) *157*
Sheline, R. K., 86(51) *114*; 91(67) *115*
Sheraga, H. A., 484(80) *508*; 485(88) *508*
Sherrington, D. C., 246, 263(14) *306*; 249, 261, 263, 264, 271, 278, 291(22) *306*; 261, 262(54) *307*; 263, 271, 273, 278, 291, 300, 301, 303(62) *307*; 263, 271(63) *307*; 265(72) *307*
Shetler, J. A., 482(66) *508*
Sheveleva, T. V., 493, 494(174) *511*
Shiells, J., 462(90) *469*
Shih-Chung Chen, 497(238) *512*; 536 (86) *561*
Shimanouchi, T., 493, 494(168, 169) *510*; 494(183, 184) *511*
Shimomura, T., 365(39) *380*
Shimonura, T., 328(53) *347*
Shinkawa, A., 298(130) *309*
Shinoda, K., 178(3) *224*

Shinohara. M. 366(44) *381*
Shiraishi, M., 177(1) *224*
Shirota, Y., 266(76) *307*
Shitova, A. A., 181(13) *224*
Shono, T., 554(184) *564*
Shostakovsky, M. F., 254(39) *306*
Shmueli, U., 489(142) *510*
Sidebottom, H. W., 33(1) *50*; 36(12) *51*; 42(29, 31) *51*; 48(42) *51*
Sienesi, D., 482(63) *508*
Siepmann, T., 460(66) *469*
Sigwalt, P., 407(60) *429*
Simmon, B., 462(88) *469*
Simmons, J. P., 450(12) *467*
Sims, D., 523(45) *560*
Sinn, H., 386(23) *428*
Sinn, H. J., 419(85) *430*
Sivertz, C., 459(57) *468*
Skorokhodov, S. S., 493, 494(174) *511*
Skuratov, S. M., 541(113) *562*; 547 (144) *563*; 550(153) *563*
Slates, R. V., 326(44a, 44b) *347*
Slawa, P., 463(93) *469*
Sloneker, D. F., 475(27) *507*
Slusarczuk, G. M. J., 551(168) *563*
Small, P. A., 15(8) *29*; 539(94) *561*
Smets, G., 123(10) *140*; 450(6) *467*; 452 (21) *468*; 456(43) *468*; 457(45, 46) *468*; 460(62) *469*; 460(72, 77, 79) *469*
Smid, J., 247, 279(15) *306*; 324(42, 43) *347*; 326(45) *347*; 327(46, 47) *347*; 328(51, 53, 54) *347*; 345(116) *349*; 353, 355, 364(13) *380*; 364(36) *380*; 365(39) *380*; 366(44) *381*; 375(84) *382*; 554(192) *564*
Smid, J. J., 301(137) *309*
Smith, E. G., 93, 94, 95(72) *115*; 94(73) *115*
Smith, H. P., 207(72) *226*; 222(113) *227*
Smith, H. S., 204, 221(64) *226*
Smith, J. J. 387(30) *428*
Smith, P., 492, 494(154) *510*
Smith, P. V., 213(91) *227*
Smith, R. R., 143(7) *156*
Smith, W. V., 186(18) *225*; 192, 209(28) *225*
Snow, R. D., 517(11) *559*
Snyder, L. C., 494(182) *511*
Snyder, W. H., 368(56) *381*
Sokolov, V. N., 207(76) *226*
Solovykh, D. A., 367(47, 48) *381*
Soja, M., 267(48) *307*

Sonntag, F., 463(96) *469*; 464(100) *469*
Sonntag, M., 535(75) *561*
Soutar, I., 151(32) *157*
Spach, G., 370(63) *381*; 497(226) *512*
Sparks, P. G., 496(201) *511*
Spassky, N., 407(60) *429*
Spector, R., 480(51) *507*
Spensley, R. H., 557, 559(216) *564*
Spolsky, R., 220(106) *227*
Sprick, K. J., 190(25) *225*
Squires, R. G., 290(122) *308*; 304(141) *309*
Srinivasan, R., 464(100) *469*
Stables, C. E., 336(88) *348*
Stahmann, M., 485(89) *508*
Stanley, E., 487(125) *509*
Stannett, V., 192, 210(32) *225*; 192, 210 (104) *227*; 192, 211, 222(111a) *227*; 278 (104) *308*; 543(127) *562*
Staudinger, H., 286(115) *308*; 352(7) *380*
Steacie, E. W. R., 36(11) *51*; 40(14) *51*
Steele, B. R., 45(34, 35, 36, 38) *51*
Stein, D. J., 514, 531(3) *559*
Steinberg, I. Z., 485(91) *508*
Steiner, E. C., 313(9) *346*
Steinmetz, R., 455(36) *468*
Stephens, C. L., 460(60, 61) *468*
Sterns, R. S., 367(50) *381*
Stewart, J. J. P., 392(34a) *428*
Stewart, R. A., 215(96) *227*
Stiddard, M. H. B., 91(66) *115*
Stiles, E., 368(58) *381*
Stiles, M., 332(52) *347*
Stockmayer, W. H., 492(159) *510*; 492, 494(151) *510*; 551(172) *563*; 551, 552 (162) *563*
Stoke, M. A., 408(64) *429*
Stoletova, I. M., 531(61) *560*; 531(65) *561*
Stolz, I. W., 86(51) *114*
Stotskaya, L. L., 385(16) *428*
Straley, J., 462(92) *469*
Strange, E. H., 352(3) *380*
Streitwieser, A., 158, 162(3) *173*
Streitwieser, Jun., A., 312, 313(8) *346*; 312, 314(11a) *346*; 314(12a) *346*; 317, 318(20) *346*; 318(23) *346*; 319, 320(28) *347*; 319(29) *347*; 320(30, 32) *347*; 335(86) *348*; 339(55) *347*; 342(105) *349*
Strepikheev, A. A., 541(103) *562*
Strepparola, E., 487(126) *509*

Strohmeier, W., 53(4) *113*; 65(31) *114*; 86(49, 50) *114*; 87(56, 61) *115*
Stryker, H. K., 180, 190, 217(23) *225*; 190, 194, 203(61) *226*; 217(10) *224*
Stuart-Webb, J., 542(119) *562*
Stutchbury, J. E., 161(7) *173*
Sujama, K., 541(106) *562*
Sukegawa, S., 451(14) *467*
Sumimoto, H., 523, 524(50) *560*; 559 (231) *565*
Suranglu, O., 448(5) *467*
Suzuki, F., 455(39, 40, 42) *468*
Suzuki, Y., 455(38, 39, 40, 42) *468*
Swain, C. G., 339(95) *349*
Swamer, F. W., 339(96) *349*
Sweeney, W., 516(27) *560*
Sweeting, O. J., 536(81) *561*
Swenton, J. S., 466, 467(118) *470*
Syamal, A., 111(90) *116*
Symcox, R. O., 164(19) *174*
Szwarc, M., 42(22, 23) *51*; 42, 43(24) *51*; 247, 279(15) *306*; 277, 292(101) *308*; 322, 326(34a) *347*; 322(34b) *347*; 326 (44a, 44b) *347*; 328(51, 53) *347*; 342 (107a, 107b, 109) *349*; 344(117) *349*; 345(106, 114) *349*; 350(1) *380*; 351(2) *380*; 352(9) *380*; 354(16) *380*; 360(28) *380*; 363(36) *380*; 365(39) *380*; 366 (44) *381*; 370(63) *381*; 375(84) *382*; 451(15) *467*; 515(8, 9) *559*; 554(192) *564*

Tabata, Y., 267(81) *307*
Tachibana, T., 168, 172(26) *174*
Tadokoro, H., 487(117, 119) *509*; 487, 489(121) *509*; 487, 489(160) *510*
Taft, R. W., 158, 162(3) *173*
Taft, Jun., R. W., 152(34) *157*
Tagaki, W., 317(19) *346*
Tagami, S., 516(25) *560*
Takahashi, A., 125(13) *140*
Takano, H., 534(68) *561*
Takeda, M., 493, 494(170) *510*; 494(189) *511*; 494(209) *512*
Takemoto, K., 554(176) *563*
Talamini, G., 173(47) *174*; 193(36, 37) *225*
Talalay, L., 204(63) *226*
Tamamushi, B., 178(3) *224*
Tamura, T., 267(78) *117*
Tan, C. C., 272(92) *308*
Tanaka, I., 359(26) *380*

Tanaka, M., 103, 107, 108(83) *115*
Tanner, D. D., 172(45) *174*
Tarkowski, H. L., 192(30) *225*; 206(70) *226*
Tarnawiecki, H., 183, 199, 209(16) *225*
Tasumi, M., 494(183, 184) *511*
Tateno, N., 471(10) *506*; 481(55) *507*
Taylor, P. A., 192, 210, (33) *225*
Taylor, R. P., 155(41) *157*
Tazuke, S., 249, 279(20) *306*; 301(136) *309*; 450, 453(11) *467*; 450, 453(24) *468*; 453(25, 27, 32) *468*
Tedder, J. M., 33(1) *50*; 36(9) *50*; 36(12) *51*; 42(28, 29, 31, 32) *51*; 46, 48(40) *51*; 47(41) *51*; 48(42, 43) *51*; 49(44) *51*
Temussi, P. A., 488(132) *509*; 504(252) *513*
Teplitzky, D. R., 169(32) *174*
Teramoto, A., 151(31) *157*
Terry, G. C., 466(117) *470*
Terry, J. O., 33(2) *50*; 34(6a) *50*
Testa, F., 180(9) *224*; 193(37) *225*
Teyssie, Ph., 207(75) *226*; 417(82) *430*; 426(95) *430*
Thal, A., 345(113b) *349*
Theil, M. H., 556(208, 209) *564*
Thomas, R. M., 286(116) *308*; 297(128) *309*
Thomas, W. M., 194, 219(41) *225*
Thommarson, R. L., 33(4) *50*
Thompson, J. B., 535(76) *561*
Thomson, C., 272(92) *308*
Thorat, P. L., 214(92a) *227*
Thorsteinson, E. M., 75(39) *114*
Thynne, J. C. J., 42, 43(25) *51*; 42(27, 30) *51*
Tietz, R. F., 516(27) *560*
Tiley, P. F., 520(33) *560*
Timoshanko, E. S., 460(69) *469*
Tipper, C. F. H., 285(113) *308*
Tirpitz, W. von., 419(85) *430*
Tobolsky, A. V., 268, 280(86) *308*; 356 (19) *380*; 367(45) *381*; 375(82, 83) *382*; 522(36) *560*; 547(145) *563*; 556 (205) *564*
Toepel, T., 319(27a) *347*
Tolle, J., 514, 531(3) *559*
Tolle, K. J., 365(39) *380*
Toney, M. K., 341(104) *349*
Tong, L. K. J., 534(70) *561*
Toohey, A. C., 164, 168(17) *174*
Toukola, V., 546(139) *562*

Topchiev, A. V., 385(16) *428*
Toppet, S., 452(21) *468*
Tornquist, E., 245(7) *305*; 298(129) *308*
Tornquist, E. G. M., 367(51) *381*
Torrance, B., 462(87) *469*
Tran van Tien, 457(45) *468*
Traub, W., 489(142) *510*
Treloar, F. E., 246(12) *306*
Trezolo, A., 461, 462(85) *469*
Tromsforff, E., 143(8) *156*
Trotman-Dickenson, A. F., 36(8) *50*;
    36(10) *51*
Tsuchiya, S., 493, 494(168, 169) *510*
Tsuda, M., 464(97, 98) *469*; 464, 465
    (102) *469*
Tsuda, T., 554(185, 186) *564*; 559(229)
    564
Tsukuma, I., 479(50) *507*
Tsuruta, T., 127(16) *141*; 142(1) *156*;
    155(46) *157*; 279(105) *308*; 324(41)
    347; 353, 355, 364(13) *386*; 356(22)
    380; 368(54) *381*; 372(69) *381*; 477
    (30) *507*; 479(50) *507*; 486(109) *509*;
    534(66) *561*; 554(178, 179, 180) *563*;
    557(217) *564*
Tsutumi, S., 60(27) *114*
Tsvetkov, V. N., 496(220) *512*
Tumolo, A. L., 480(51) *507*
Turro, N. J., 448(2) *467*
Tweedie, A. T., 554(199) *564*
Tyer, Jun, N. W., 460(63) *469*
Tyson, A., 542(118) *562*; 543(125) *562*

Uchida, T., 487(120) *509*
Uchida, Y., 536(85) *561*; 551(157) *563*
Uehara, K., 103, 107, 108(83) *115*
Uelzmann, H., 421(89) *430*
Ueno, K., 267, 278, 291(80) *307*; 278
    (103) *308*
Ugelstad, J., 193, 195(39) *225*, 198, 212
    (52) *226*
Ulmann, R., 494(186) *511*
Unruh, C. C., 466(110) *470*
Uranek, C. A., 178(5) *228*; 214(92, 93)
    227
Ureta, E., 554(192) *564*
Uschold, R. E., 314(12a,b) *346*

Vala, M. T., 494(178) *511*
Valeur, B., 493, 494(177) *511*

Valvassori, A., 486, 494(114) *509*
Van Eeylen, M. M., 483(71) *508*
Van Den Hul, H. J., 203(60b) *226*
Van Der Hoff, B. M. E., 209(81) *226*;
    209(82) *227*
Vanderhoff, J. W., 192(30) *225*; 203(60b)
    226; 205(67) *226*; 206(70) *226*
Vandeweyer, P. H., 460(62) *468*; 460
    (72, 77) *469*
Van Helden, R., 385(19) *428*
Vannikov, A. V., 498(241) *512*
Vaughan, G., 425(93) *430*
Verborgt, J., 461(78) *469*
Vezzoli, G. C., 531(60) *560*
Vialle, J., 123(9) *140*
Vianello, G., 180(9) *224*
Vidotto, G., 191(36) *225*
Vincent, J. M., 496, 497(226) *512*
Vinogradov, P. A., 181(13) *224*
Vitagliano, V., 493, 494(173) *510*
Vitkuske, J. F., 205(67) *226*
Vofsi, D., 268, 280(86) *308*
Vogl, O., 13(5) *29*
Volchek, B. Z., 494(188) *511*; 498(240,
    239) *512*
Volkenau, N. A., 334(82a) *348*
Volkenstein, M. V., 498(240) *512*
Volkenstein, M. W., 471(15) *506*; 483
    (72) *508*; 483, 502(76) *508*
Volkov, V. I., 370(65) *381*
Volokhina, A. V., 541(103) *562*
Vol'Pin, M. E., 264(68) *307*
Vrieze, K., 413(74) *429*

Wade, K., 426(73) *439*
Wahl, P., 151(30) *157*
Walborsky, H. M., 332, 333(77a, 77b,
    80) *348*
Walker, D. A., 285(113) *308*
Walker, G. R., 551(160, 161) *563*
Walker, H. W., 223(114) *227*
Wall, F. T., 119(2) *140*; 202(58e) *226*
Wall, L. A., 127(17) *141*; 167(24) *174*;
    531(62) *561*; 534(69) *561*
Wallace, L. R., 417(81) *430*
Walling, C., 77(46) *114*; 120(3) *140*;
    161(5) *173*; 161, 168(9) *173*; 166(22)
    174; 168(27) *174*; 171(37) *174*; 172(45)
    174; 520(34) *560*
Walter, H. C., 536(82) *561*

Walton, J. C., 36(12) *51*; 42(28, 29, 31, 32) *51*; 46, 48(40) *51*; 47(41) *51*; 48 (42, 43) *51*; 49(44) *51*
Wanless, G. C., 286(116) *308*; 288(117) *308*
Ward, J. P., 551(156) *563*
Warshel, A., 484(87) *508*
Wasai, G., 488(139) *510*
Watanabe, W. H., 497(229) *512*
Watanuma, S., 95(75) *115*
Watkins, K. W., 42(26) *51*
Watanabe, T., 254(40) *306*; 257(47) *307*
Watson, R. A., 36(12) *51*
Watts, C. E., 336(87) *348*
Wawersik, H., 67(34) *114*
Wayne, R. P., 87(57) *115*; 87, 91(60) *115*
Waywell, D. R., 536(78) *561*
Weale, K. E., 158, 167(4) *173*; 163(14, 15) *174*; 164, 168(17) *174*; 164, 167(18) *174*; 169(33) *174*; 169, 170(34) *174*; 170(35) *174*; 172(43) *174*; 514(2) *559*; 531(59) *560*
Webster, O. W., 315(17) *346*
Wehrly, J. R., 208(77) *226*
Weightman, J. A., 249, 261, 263, 264, 271, 278, 291, 298, 303(22) *306*
Weill, G., 494(187) *511*
Weiner, H. 194(40) *225*
Weise J. K., 542, 545, 548(122) *562*
Weiss, M. D., 77(44) *114*
Weissberger, A., 450(8) *467*
Weissman, S. I., 343(111b) *349*
Wellisch, E., 536(81) *561*
Wentworth, W. E., 343(108a) *349*
Werner, H. G., 204, 221(64) *226*; 220 (108) *227*
Wertz, W., 153(36) *157*
Wessling, R. A., 201(56) *226*; 496(203, 204) *511*
Westermann, P. H., 516(24) *560*; 523 (49) *560*
Weyenberg, D. R., 208(78) *226*
Whalley, E., 158(2) *173*; 171(36, 39) *174*; 514(5) *549*; 522(42) *560*
Wheland, G. W., 313, 329(6) *346*
Whitby, G. S., 193(47) *225*; 541(107) *562*
White, F. L., 215(96) *227*
Whitmore, F., 253(37) *306*
Whitney, R. B., 541(109) *562*
Whittle, D., 60(26) *114*; 93(69, 71) *115*; 93, 94, 95(70) *115*

Whittle, E., 33(5) *50*; 36(13) *51*
Wichterle, O., 257(45) *306*
Wiehler, G., 544(132) *562*
Wilchinsky, Z. W., 486(96) *508*
Wiles, D. M., 365(41) *381*
Wilke, G., 383(5) *427*
Wilkinson, B. W., 192(30) *227*
Willets, F. W., 466(117) *470*
Willey, F., 322(37) *347*
Willfgang, G., 267(84) *307*
Williams, D. J., 205, 206(68) *226*; 206 (69) *226*; 222(112a) *227*
Williams, Ff., 248, 278(17) *306*; 267, 278, 291(80) *307*; 298(130) *308*
Williams, F. J., 377(89) *382*
Williams, H., 256(43) *306*
Williams, H. L. 183(15) *225*; 220(106) *227*
Williams, J. L. K., 466(114) *470*
Williams, J. L. R., 345(113a) *349*
Williams, P. H., 466, 467(118) *470*
Williams, R. L., 361(29) *380*
Williams, V., 466(117) *470*
Wilson, J. E., 180, 212(8) *224*
Wilson, K. R., 77(43) *114*
Window, E., 461, 462(85) *469*
Winger, P., 551(167) *563*
Winstein, S., 235, 237(6) *243*; 250(23) *306*; 323(36, 39) *347*; 336(87) *348*; 362(34) *380*
Winter, H., 419(85) *430*
Winton, K. D. R., 47(41) *51*; 48(43) *51*; 49(44) *51*
Wislicenus, J., 231(56) *307*
Witanowski, M., 333(78) *348*
Witt, H. S., 207(72) *226*; 222(113) *227*
Wittmer, P., 514, 531(3) *559*; 550(154) *563*; 556(214) *564*
Wojeieki, J., 75(37) *114*
Wolfstirn, K. B., 77(46) *114*
Wolinski, L. E., 547(148) *563*
Wong, K. H., 327(47) *347*
Wood, N. F., 298(132) *308*
Woodbrey, J. C., 172(40) *174*
Wooding, N. S., 352(8) *380*
Woods, G. F., 317(13) *346*
Woods, H. J., 298(131) *308*
Woods, M. E., 205, 206, 224(68a) *226*
Woodward, A. E., 189(21a) *225*
Woodward, F. E., 213(91) *227*
Woodward, R. B., 455(35) *468*
Worrall, R., 267, 278(79) *307*

Worsfold, D. J., 279(107) *308*; 361(30)
    380; 364(38) *380*; 366(43) *381*; 367
    (46) *381*; 371(66) *381*; 374(79) *382*;
    516, 523, 524(17) *560*; 516, 523(18)
    560
Wright, P. V., 551(164) *563*
Wright, W. W., 451(19) *468*
Wunderlich, J. A., 390(33) *428*
Wunderlich, W., 496(202) *511*; 496(223)
    512; 496, 497(224) *512*
Wyman, D. P., 371(67) *381*
Wyn-Jones, E., 484(77) *508*

Yakoleva, E. A., 343(110b) *349*
Yamashita, T., 540(101) *561*; 541(105,
    106) *562*
Yamashita, Y., 554(185, 186) *564*; 559
    (226, 229) *564*
Yamamoto, N., 356(20) *380*
Yamamoto, T., 138(21, 22) *141*
Yamazaki, N., 359(26) *380*
Yamoda, A., 494(189) *511*
Yampol'Skii, P. A., 173(51) *174*
Yanari, S., 494(179) *511*
Yasuhara, Y., 492, 494(160) *510*; 554
    (176) *563*
Yasumoto, T., 487(117) *509*
Yats, L. D., 205(66) *226*
Yee, K. C., 310(2) *346*
Yen, S-P. S., 377(90) *382*
Yenikolopyan, N. S., 518(31) *560*
Ymanura, Y., 494(189) *511*
Yobe, R. V., 256(42) *306*
Yoda, N., 522(39) *560*
Yogo, T., 554(174) *563*
Yokoo, Y., 554(180) *563*
Yokota, H., 554(182) *563*
Yokota, K., 144(15) *156*; 149(29) *157*
Yokoyama, N., 40(19) *51*
Yoncoskie, B., 368(58) *381*
Yoshikawa, S., 211(87) *227*
Yoshimura, T., 537(89, 90) *561*
Yoshisato, E., 60(27) *114*
You-Ling, Fan., 535(74) *561*

Young, A. E., 332, 333(77a, 80) *348*
Young, L. J., 119(1) *140*; 123(5) *140*
Young, N. L., 450, 451(7) *467*
Young, W. G., 322(36) *347*
Young, W. R., 321(32) *347*
Youngman, E. A., 384(9) *427*; 399, 402,
    404, 405(50) *429*
Yu, H., 497(234, 236) *512*
Yuki, H., 481(56) *511*; 554(193, 194,
    195, 197) *564*; 554, 556(198) *564*; 556
    (213) *564*

Zabusky, H. H., 497(219) *512*
Zagorbinina, V. N., 168(29) *174*
Zahir, S., 546(137) *562*
Zambelli, A., 384(6) *427*; 399(49) *429*;
    478(36) *507*
Zamboni, V., 478(38) *507*; 478, 493(39)
    507
Zandomeneghi, M., 477(31) *507*; 494
    (192) *511*
Zanini, G., 415(78) *430*; 416(79) *430*
Zazda, A., 305(142) *309*
Zerbi, G., 478(38) *507*; 478, 493(39, 40)
    507
Zharov, A. A., 167(25) *174*; 173(50, 51)
    174
Zhulin, V. M., 168(29) *174*; 172(46) *174*
Ziegenbeim, W., 268(87) *308*
Ziegenbein, W., 547(146, 147) *563*
Ziegler, K., 331(70) *348*; 352(5) *380*;
    353(15) *380*; 359(23) *380*; 383, 384(1)
    427
Ziegler, M. L., 91(67) *115*
Zilkha, A., 370(62) *381*
Zimm, B. H., 485(92) *508*
Zlamal, Z., 245, 294(5) *305*; 305(142)
    309
Zoeller, Jun., J. H., 329(61) *348*
Zubkov, V. A., 483, 502(76) *508*
Zubov, V. P., 522(37) *560*
Zucchini, U., 398(470) *429*
Zvereva, T. M., 272(94) *308*
Zwolinski, B. J., 11(3) *29*

Subject Index

Absolute velocity constants, 118
Acenaphthylene, 261
Acetaldehyde, 13, 536
  ceiling temperature, 516, 522
  copolymerization of, 558
  effect of pressure on polymerization, 514
  equilibrium polymerization, 531
  polymerizability, 535
Acetone, anion, 315
  ceiling temperature, 536
  polymerizability, 536
  polymerization, 522
Acetophenone, electron affinity, 343
trans-2-Acetoxycyclohexyl-p-toluenesul-
  phonate, acetolysis, 233
Acetyl acetonates, reactivity in polymerization, 95
Acetyl acetone, with manganese carbonyl, 89
Acetylene, anion, 320
  in Phillips process, 437
Acetylenes, polymerizability, 11
Achiral monomers, 476
Acrolein, emulsion polymerization, 219
  data, 194
Acrylamide, anionic polymerization, 373
  effect of viscosity on polymerization, 144
Acrylic anhydride, inducement of cyclopolymerization by pressure, 172
Acrylic polymers, relaxation, 148
Acrylonitrile, 8
  copolymerization, 554, 557
    with isobutyl vinyl ether, 453
  emulsion polymerization, 219
    data, 194
  polymerization at high pressure, 172
  transfer reactions, 370
  with metal chelates, 101

Acylium salts, 253
  for cyclic ether polymerization, 270
Addition polymerization, as aggregation
  process, 515–517
Addition processes, 2
After effects, 88
Alcohols, pK_a values, 329
Aldehydes, chloro-, ceiling temperature, 535
  entropy of polymerization, 535
  heat of polymerization, 535
  polymerizability, 535
  polymerization, 519
Allyl acetate, copolymerization with
  sulphur dioxide, 537
  polymerization under pressure, 168
Allyl alcohol, copolymerization with
  sulphur dioxide, 537
Allyl ethyl ether, copolymerization with
  sulphur dioxide, 537
Allyl formate, copolymerization with
  sulphur dioxide, 537
π-Allyl nickel halides, 410, 416
π-Allyl rhodium complexes, 207
Allylic anions, 334
  protonation, 336
N,N'-Alkalene-bis-maleimides, photol-
  ysis, 456
Alkali metal alkoxides, 365
Alkali metal alkyls, as anionic initiators, 352
Alkanes, hydrogen abstraction from, 36
Alkenes, copolymerization, 554
  with sulphur dioxide, 517, 527, 536
Alkoxy anions, 341
Alkyl lithiums, polymeric nature, 329
Alkyl methacrylates,
  effect of ester groups on chain motion, 147
  effect of viscosity on polymerization, 144

Alkyl methacrylates (*contd.*)
  dependence of termination rate constant on viscosity, 144
Alkyl migrations, 242
Alkyl radicals, addition to ethylene, 42
  combination rates, 33
  cross combination ratios, 34
Alkyl styrenes, cationic polymerization, 289
Alkyl vinyl ethers, 9
  cationic initiation, 264, 278
  cationic polymerization, 261, 263
    by iodine, 262
  polymerization, 275
  transfer in cationic polymerization, 298
Alkylene sulphide, emulsion polymerization, recipe, 223
Alternating tendency, 120, 123, 125
Alternation, regular, 125
Aluminium alkyls, 252, 256
  role in Ziegler-Natta polymerization, 419
Aluminium chloride, 247, 252, 256
Amines, pK_a values, 329
  transfer reactions of, 121
Aminium salts, 253
  stable, 264
2-Amino-3-methyl butane, deamination, 242
1-Amino propane, deamination, 242
Ammonium ion, 248
Ammonium persulphate, 181
Anchimeric assistance, 233
Anionic copolymerization, 373–379
Anionic initiators, monofunctional, 353
Anionic polymerization, 350–382
  definition, 350
  degree of polymerization, 365
  design features, 372–373
  effect of ionic species, 362
  historical aspects, 352
  initiation, 352–359
    by alkali metals, 353
    by electrolysis, 359
    by ether cleavage, 358
    by living polymers, 357
  molecular weight distribution, 361
  polymer structure, 366
  practical example, 350
  propagation, 359–369
    kinetics, 362

sequential monomer addition, 376
  specific end groups, 369
  temperature effects, 364
  termination, 369–372
    by air, 369
    by impurities, 371
    by isomerization, 370
    by proton donors, 369
    by reactive groups, 378
    by transfer, 369
    producing reactive end groups, 371
Anionic systems, living, 519
Anions, coupling of, 371
  localized, inductive stabilization, 319
Anisotropic polarizability, 496
2-*p*-Anisyl-3-butyl-*p*-toluenesulphonate, solvolysis, 240
threo-3-*p*-Anisyl-2-butyl-*p*-bromobenzene sulphonate, special salt effect, 237
Anthracene, electron affinity, 343
  energy transfer, 467
  photodimerization of derivatives, 466
Anti-aromatic anions, 318
Anti-Markovnikov addition, 43
Antimony pentachloride, 252, 256
Atroponitrile, 523
  polymerizability, 535
Aromatic alkylation, 243
Aryl diazonium salts, 253
  for cyclic ether polymerization, 271
Arylazide group, photolysis, 466
Auto-ionization, 257
Average energy excess, 490
Aziridines, copolymerization, 554
Azo compounds, in emulsion polymerization, 183
  photolysis, 451
Azo-bisisobutyronitrile, comparison with molybdenum carbonyl, 73
bis-Azomethines, photoreductive polyaddition, 457
Azo-2-methyl-2-propane, 451

'B' strain, 232
Back-biting, 20, 295
Backbone rotation, relation to termination rate, 148
Back-donation, 391
Benzaldehyde, electron affinity, 343
Benzene, hydrogen transfer, 129

Benzofuran, photocyclo addition reaction, 464
Benzophenone, as photosensitizer, 456
  sodium reduction, 342
bis-Benzophenones, photoreductive polyaddition, 457, 458
Benzopyrylospiranes, 460
Benzyl anion, 370
Benzyl potassium, 352
Benzyl sodium, 328
Binary copolymerization, termination reactions, 152
p-Biphenyl-2-propene, 523
Block copolymerization, 375
Block copolymers, 358, 371
  AB, 376
  ABA, 376
  ABC, 376
  from metal carbonyl initiation systems, 92
  properties, 378
Boron trifluoride, 247, 252, 256
Branching, 2, 94
Bromine atoms, hydrogen abstraction by, 30
dl-erythro-3-Bromobutanol, reaction with hydrogen bromide, 233
dl-threo-3-Bromobutanol, reaction with hydrogen bromide, 233
Bromonium ions, 233
Bromotrichloromethane, photochemical addition to olefins, 46
Brönsted catalysis law, 314
Butadiene, 416
  copolymerization, 554, 556
  dimerization, 425
  emulsion copolymerization, 213
    with acrylonitrile, 215
    with methacrylic acid, 204
    with styrene, *see* butadiene/styrene,
  emulsion polymerization, 212, 213
    data, 194
    molecular weight control, 213
    molecular weight distribution, 214
    recipe, 222
  heat of polymerization, 12
  in Phillips process, 437
  initiation, by Mn(acac)$_3$, 96
  Φ_M value, 198
  polymerization, 1–4, 421
    by complex catalysts, 423, 426
    by rhodium catalysts, 207

structure of anionic polymers, 366
Butadiene/styrene, anionic copolymerization, 374
  emulsion copolymerization, 184, 215
  inhibition, 180
  molecular weight control, 228
  rate coefficients 228
  recipes, 219–221
  redox initiation, 183
1-Butene, 304, 404
  copolymerization with sulphur dioxide, 537
  Phillips polymerization, 431
cis-2-Butene, copolymerization with sulphur dioxide, 537
trans-2-Butene, copolymerization with sulphur dioxide, 537
t-Butoxy radicals, hydrogen abstraction by, 36
Butyl acrylate, effect of viscosity on polymerization, 144
n-Butyl acrylate, Φ_M value, 198
t-Butyl benzene, hydrogen transfer, 129
t-Butyl chloride, reactivity in solvolysis, 233
Butyl lithium, effect of N,N,N',N'tetra methylethylene diamine, 327
  with THF, 353
n-Butyl lithium, 352
  complex formation, 353
sec-Butyl lithium, 353
  optically active, 332
t-Butyl lithium, 353
t-Butyl radicals, combination, 33
  disproportionation-combination ratio, 34
  ionization polymerization, 231
Butyl rubbers, 260
n-Butyl vinyl ether, copolymerization, 554
t-Butyl vinyl ether, cationic polymerization, temperature effect, 291
n-Butyraldehyde, equilibrium polymerization, 531
  equilibrium monomer concentration, effect of pressure, 533
  polymerization, 520
  pressure effect on ceiling temperature, 171
γ-Butyrolactone, equilibrium polymerization, 531

Caprolactam, 17, 550, 551
  heat of polymerization, 525
  3-methyl, 550
  4-methyl, 550
  5-methyl, 550
  6-methyl, 550
  7-methyl, 550
  4,6-di-methyl, 550
  ring-opening polymerization, 373
  substituent effects, 550
Carbanions, 310–349
  addition to double bonds, 339
  configurational stability, 332–335
  delocalization, 315, 317
  delocalized, reactivity, 335–338
  effect of hydrogen bonding solvents, 311
  field effects, 319
  hybridization, 315
  molecular orbital calculations, 319
  preparation, 328–332
    by direct metalation, 330
    by ether cleavage, 331
    by halogen-metal interconversion, 330, 331
    by metal-metal exchange, 330
    by metalation, 328
  reactions, 338–342
  reactions with organic halides, α-elimination, 339
    β-elimination, 339
    metal-halogen exchange, 339
    substitution, 339
  reactions with oxygen, 341
  reactivity, effect of counterion, 331
  stabilization, 315
  stability, use of pK_a as index, 311–321
  stereochemistry, 332–335
  structure, relation to pK_a, 314
Carbenes, generation, 340
Carbon atom hybridization, effect of strain, 321
Carbon dioxide, copolymerization of, 554, 556
Carbon-hydrogen bond, effect of s-character, 321
Carbon monoxide, 554
Carbon oxysulphide, copolymerization of, 554
Carbonation, 333, 345
Carbonium ions, 229–243, 244, 245, 248
  effect of added salts, 236

elimination, 239
formation, 229–233
  by addition of cations to olefins, 229
  by electron impact, 229
  by heterolytic cleavages, 229
  by oxidation of radicals, 229
ideal structure, 232
in stable salts, 270
non-classical, 242, 299
propagating, 275
reactions, 239–240, 242
rearrangement, 240–242
solvent effects on, 235–239
stability, 229–233
stabilization, 229
  hyperconjugative, 239
stable, 253
strain, 232
  energy, 235
substitution, 239
Carbonyl compounds, photoinitiation by, 450
  polymerizability, 13
Carboxonium ion, 244
  salts, 253
    as initiators, 269
N-Carboxyacidanhydrides, 16
Cascade systems, 202
Cationic copolymerization, 248
Cationic intermediates, detection, 246
Cationic polymerization, 244
  active intermediates, 245–252
  additive effects, 247
  bond rearrangements, 283
  chain breaking reactions, 292–305
  cocatalysis, by alkyl halides, 259
  cocatalysts, 257
    poisoning, 259
  conductance effects, 248
  free ions, 250
  impurity effects, 247
  inhibition, 247
  initiation, 245, 252–274
    by aprotonic acids, 256
    by charge transfer, 265
    by iodine, 261
    by protonic acids, 253
    by radiation, 267
    by stable carbonium ion salts, 263
    solvent effect on, 273
    temperature effect on, 273
  ion pairs, 250

Cationic polymerization (*contd.*)
mechanism and reactivity, 244–309
propagation, 245, 274–292
by free ions, 276
by ion pairs, 276
solvent effect on, 290
temperature effect on, 290
with isomerization, 282
radiation induced, 278
solvent effects, 248, 249
termination, 245, 292, 302
transfer, 292, 294
to catalyst, 296
to cocatalyst, 296
to monomer, 245
to polymer, 295
Cationic systems, pre-initiation equilibrium, 252
Ceiling temperature, 360, 515–517, 524
determination, 517–519
pressure dependence of, 170, 531
table, 516
Cellulose, photodegradation, 462
Chain configurational isomers, 504
Chain end mobility, observation, 150
Chain reaction polymerization, 6, 7
features, 8
Chain reaction processes, 2
Chain transfer, 94, 119, 120, 138
effect of pressure, 165, 167
effect on molecular weight, 370
in anionic systems, 370
to hydrocarbons, 129
Characteristic ratio, 491
table, 496
Charge-transfer, 139
Charge-transfer complexes, 125, 453
Chemically controlled reactions, 143
Chiral monomers, 476, 479
Chiral vinyl monomers, stereospecific polymerization, 477
Chloral, ceiling temperature, pressure effect on, 171
initiation, by acetyl acetonates, 95
polymerization, 522
equilibrium, 531
in solution, 523
polymerizability, 535
Chloranil, 265, 453
Chlorine atoms, abstraction by, 36
3-Chloro-3-methyl-1-butene, cationic polymerization, 289

3,3-*bis*(Chloromethyl)oxetane, copolymerization, 554, 557, 558
Chloroprene, as crosslinking agent, 93
emulsion polymerization, 215, 216
data, 194
recipe, 223
p-Chlorostyrene, 377
Chlorosulphuric acid, 269
Chromium oxide, 394, 431
on silica, active site estimation, 440
mechanism of ethylene polymerization, 442
reaction, 433
reaction with support, 432
treatment with carbon monoxide, 437
Chromium oxide catalysts, poisons, 437
reactivity and mechanism, 431–445
reduction, 433
Chromium (π-allyl)$_3$ catalysts, 420
Chromium triacetylacetonate, 416
Cinnamates, photocrosslinking, 463
photodimerization, 455
Cinnamic acid, phosphorescence, 465
Clapeyron-Clausius equation, 531
Cobalt carbonyl, 410
inhibition of MMA polymerization, 74
Cobalt (II) chloride, 418
Cobalt complexes, in diene polymerization, 418
Collision volume, 147
Comb copolymers, 376, 377
Complexes, π-bonded, 390
Condensation polymerization, 2, 378
Configuration, of polymers, 471–513
Configuration isomerism, 471, 504
statistical approach, 489–493
Conformation, definition, 471
of polymers, 471–513
Conformational analysis, 483
Conformational equilibria, problems, 483–485
Conformational partition function, 484
Conformational reversals, 490, 502
Conformational statistics, 472
Co-operative systems, monodimensional, 490
Co-operativity, 503
Co-operativity parameter, 491
Co-ordinated anionic polymerization, 399

Copolymerization, 119, 126, 127
  1 : 1, 527
  anionic, 373–379
    random, 373
    effect of reversible propagation, 553–
      559
    composition of polymers, 557
  equation, 119
  ideal, 119
  of non-homopolymerizable mono-
    mers, 554
  salt effects, 125
Copolymers, equimolar, 125
  from Phillips catalysis, 431
  graft, 2
  linear, 1
Copper acetylacetonate, as initiator, 96
Coulombic interaction, 326
Coupling agents, 377
  difunctional, 377
Coupling reactions, for block copolymers, 377
Critical micelle concentration, 176
Crosslinking, 2
  role of termination by combination,
    94
trans-Crotonitrile, copolymerization of,
  554
Cumene hydroperoxide, 181
Cumylmethyl ether, 353, 359
β-Cyanopropionaldehyde, reactivity
  ratios, 559
Cyclic crown ethers, 327
Cyclic ethers, initiation, by acetylacetonates, 95
  polymerizability, 18
  polymerization, propagation, 275
  ring-opening polymerization, 267
Cyclic monomers, cationic propagation,
  279
  rate constants, 280
Cyclic olefins, polymerization, 18
Cyclic siloxanes, cationic emulsion
  polymerization, 208
  ring-opening polymerization, 373
Cyclic sulphides, emulsion polymerization, 208
  polymerizability, 18
Cycloalkanes, polymerizability, 538
  polymerization, 551
    free energy of, 538
    heat of, 535

  1,1 dimethyl substituted, 538
  methyl substituted, 538
Cyclobutadiene, dianion, 318
Cyclobutene, 9
Cyclodimerizable groups, table, 466
Cyclodimerization, 455
Cycloheptatrienyl anion, 318
Cycloheptatrienyl cation, 229, 263, 270
Cyclopheptene, sulphur dioxide co-
  polymerization, 537
Cyclohexane, hydrogen transfer, 129
Cyclohexanone, with manganese car-
  bonyl, 89
Cyclohexene, copolymerization with sul-
  phur dioxide, 537
  metalation, 321
Cyclononatetraene anion, 318
1,3-Cyclooctadiene, as donor with metal
  chelates, 103
Cyclooctatraene, 539
  dianion, 318
Cyclopentadienyl anion, 318, 370
Cyclopentane derivatives, reactivity in
  solvolysis, 232
Cyclopentene, 551
  copolymerization with sulphur diox-
    ide, 537
  metalation, 321
  mixed polysulphone, 557
  ring-opening polymerization, 18
Cyclopolymerization, 283
  effect of pressure on, 172
Cyclopropane, protonated, 242
  ring-opened polymerization, 285
Cyclopropanes, protonated, 298
  reactivity in solvolysis, 232

Decamethyl adipate, 551
trans-1,2-Diacetyl ethylene, copolymerization of, 554
Dialkylketones, 450
Diallyl cyanamide, pressure effect on
  cyclopolymerization of, 172
1-2-Diaminopropane, 105
Diarylketones, 451
Diazoethane, 11
trans-1,2-Dibenzoyl ethylene, 555
  copolymerization of, 554
1,6 Dibromohexane, 377
Di-t-butyl peroxide, effect of pressure on
  decomposition, 161
1,2-Dichloropropionaldehyde, 523, 536

Dielectric constant, effect on ion pair formation, 364
effect on reaction rate, 126
Dielectric relaxation, of polymers, 148
Diene polymers, isomerism, 25–27
Dienes, 1,3- 1-substituted, 479
1,4-substituted, 479
anionic copolymerization with styrene, 374
conjugated, emulsion copolymerization, 212
polymerizability, 12
transfer in cationic polymerizations, 304
copolymerization, 554
emulsion polymerization, 212
polymerization, 1,2-, 424
1,4-, 424
by cobalt catalysts, 416
by complex catalysts, 421, 422
by π-allyl catalysts, 416
by transition metal complexes, 409–427
structures of anionic polymers, 366, 367
Ziegler-Natta polymerization, 409–427
mechanism, 412
nature of catalyst, 411
stereoregulation, 414
Diffusion, ball and chain model, 147
Diffusion controlled reaction, 143
Dilithiostilbene, 357
α,α-Dimethyl allyl chloride, 322
γ,γ-Dimethyl allyl chloride, 236
2,3,Dimethyl butadiene, copolymerization of, 554
emulsion polymerization, 212, 215
data, 194
Dimethyl-2-butene, 240
2,3-Dimethyl-2-butene, copolymerization of, 554
1,1-Dimethyl cyclopropane, 539
Dimethyl dichlorosiloxane, 377
(R,S)-3,7,-Dimethyl-1-octene, 407
4,4-Dimethyl-1-pentene, copolymerization with sulphur dioxide, 537
Dimethyl siloxane, equilibrium polymerization, 551
Dimethyl sulphoxide, 311
1,3-Dioxane, copolymerization of, 554
1,4-Dioxane, copolymerization of, 554

effect on initiation of polymerization by metal carbonyls, 63
1,3-Dioxepane, solution polymerization, 523
1,3-Dioxolane, 16
ceiling temperature, 516
depropagation, 559
equilibrium polymerization, 518
solution polymerization, 523
1,3-Diphenylallyl anion, 370
1,1-Diphenyl ethylene, 372
carbonium ion, 246
copolymerization of, 554
radical anion dimerization, 345
α,α′ Diphenyl-β-picryl hydrazyl, 247
Dipole moment, characteristic ratio, 495
Dipole reorientation frequency, comparison with termination rate constants, 149, 150
trans-1,2-Di-(2-pyridyl)ethylene, copolymerization of, 554
Dissymetric monomers, 471
o-Divinyl benzene, 279
Dye sensitization, 452

Electron affinity, of aromatic carbonyl compounds, 343
of aromatic hydrocarbons, 343
Electron transfer, 121, 264, 272, 342, 354
from transition metals to halides, 59
Emulsifiers, 176
anionic, 177
classification, 176
non-ionic, 177
properties, 178
Emulsion polymerization, 175–228
additives, 184
antifreeze agents, 184
auxiliary stabilizers, 180, 204
chain transfer agents, 184
continuous, 200
description, 185
effect of electrolytes, 184, 204
effect of monomer concentration, 198
free radicals/polymer particle, 195
general features, 176
initiators, 181
inverse, 206
kinetics, 186–200
latices, 202
Medvedev theory, 186

Emulsion polymerization (*contd.*)
  molecular weights, 188, 199
    effect of laminar flow on, 201
  non-radical, 207
  nucleation, 189
  particle distribution, 204
  particle number, 189
  particle size, 204
  particle surface, 199
  recipes, 219–224
  redox systems, 182
  semi-continuous, 200
  Smith-Ewart kinetics, 186, 199, 202,
    205, 209, 211
  stereospecificity, 228
  surface chemistry of particles, 203
  termination, 204
Emulsion stabilizer, choice, 179
Emulsions, effect of pH, 180
Energy absorption, in polymers, 458
Energy transfer, 448, 449
  in polymers, 458
  long range, 448
  short-range, 448
  singlet-singlet, 467
Entropy, of polymerization, 524, 534
  of solvation, 327
  residual, 526
  translational, 324
Eosin, as photosensitizer, 452
Epichlorohydrin, copolymerization, 554
Equilibrium polymerization, 518, 520–
  532
Erythro-isomers, conformation, 488
Ester interchange, 515
ε_T values (for solvent polarity), 236
Ethane, anion, 320
Ethyl acetate, effect on initiation by
  metal carbonyls, 63
Ethyl benzene, hydrogen transfer, 129
2-Ethyl-1-butene, 537
2-Ethyl-1-hexene, 537
2-Ethyl-hexyl vinyl ether, 263
Ethyl methacrylate, heat of polymerization, 525
Ethyl radicals, addition to ethylene, 42
  addition to polar olefins, 41
  combination, 33
  cross combination ratios, 34
  disproportionation-combination
    ratio, 34
  ionization potential, 231

Ethyl *p*-toluene sulphonate, acetolysis,
  235
Ethylene, 8, 384
  addition to, radical reactivities, 42
  anion, 320
  bonding, to titanium, 392
    to transition metals, 390
  copolymerization, with 1-butene, 404
    with sulphur dioxide, 537
  derivatives, polymerizability, 532
  disubstituted, stereoregular, 25
  1,1-disubstituted,   polymerizability,
    533
  1,2-disubstituted, polymerizability, 10
  emulsion polymerization, 217
    data, 194
  free radical polymerization, 295
  heat of polymerization, 12
  monosubstituted, tacticity, 22
  polymerizability, 532
  polymerization by Phillips catalysts,
    392, 431
    effect of activation temperature on
      molecular weight, 438
    initiation, 442
    propagation, 443
    termination, 443
  polymerization, effect of pressure on,
    164
  teleomerization via metal carbonyls,
    52
  tetra-substituted,   polymerizability,
    535
  tri-substituted, polymerizability, 535
  Ziegler-Natta polymerization, 384–
    409
Ethyleneimine, 16
Ethylene oxide, 16, 371
  copolymerization, 554
  ring-opening polymerization, 373
Ethylene sulphide, emulsion polymerization, 208
Ethylene terephthalate, 551
Excitation migration, 448
External return, 236

Fischer projection, 474
Floor temperature, 517
Flory-Huggins equation, 522
Fluoral, polymerization, 535, 522
Fluorescence, 448
  delayed, 458

Fluorescence (*contd.*)
  depolarization, 150
    comparison of measurements with
      termination rate constants, 151
    tacticity, determination by, 493
Fluoropolymers, 180
Fluorosulphuric acid, 269
Formaldehyde, 13
  ceiling temperature, 516
  copolymerization, 554
  equilibrium monomer concentration,
    535
  equilibrium pressure, 535
  polymerization, 517, 522
Franck-Condon principle, 446
Free energy of polymerization, 515, 520,
  522
Free ions, *see* Ions, free
Friedel-Crafts halides, 257, 268, 270
  in cyclic ether polymerization, 272
Friedel-Crafts, polymerization, 419
  reaction, 243
Fumaronitrile, copolymers, 557

Gel permeation chromatography, 361
Gegen ions, 362
Glass transition, 155
  temperature, effect of tacticity, 482,
    see also individual monomers
Graft copolymers, from metal carbonyl
  systems, 92
Grafting reactions, for block copoly-
  mers, 377
Group migration polymerization, 289

HLB number, 179
Halide ion migration polymerization,
  289
Halides, activity, 77
  electron affinity, 77
  transfer reactions of, 121
Haloforms, deprotonation, 340
Hammett σ-function, 132, 138
Heat of polymerization, 515, 517, 524,
  531, 534, 558
  table, 12
  variation with ring size, 552
Helical conformations, 485, 494
Heptafluoropropyl radicals, addition
  tofluoro-olefins, 47
  orientation of, 49

Heptanes, 2,4,6-tri substituted, 494
2-Heptene, sulphur dioxide copolymerization, 537
Heteroatom anions, 337
Heteroatom carbon acids, pK_a^W values,
  315
Heterocyclic compounds, polymerizability of, 539
Hexafluoroacetone, copolymerization of,
  554
Hexafluoroazomethane, photolysis, 41
Hexakis-(phenyl isocyano)-metal derivatives as initiators, 73
1-Hexene, 304
  copolymerization with sulphur diox-
    ide, 537
  Phillips polymerization, 431
1-Hexenyl anion, 336
n-Hexyl isocyanate, ceiling temperature,
  516
  determination, 519
  polymerizability, 536
Homopolymers, linear, 1
Hoyland, electronegativity scheme, 139
  parameters, 138
Hückel π-electron distribution, 336
Hydride ion, abstraction, 243, 264, 271
  involvement in termination reactions,
    445
  shift, 242
    1,2-, 288
    1,3-, 288
  transfer, 280, 301, 303
Hydride ion shift polymerization, 286
Hydrocarbon acids, pK_a values, 312
Hydrocarbons, pK_a^W values, 316
  transfer reactions of, 121
Hydrogen, transfer agent, 399
Hydrogen abstraction, from halogeno-
  methanes, 36
  from hydrogen, 36
Hydrogen peroxide, 181
Hydrophile-lyophile balance, 179
Hydroxyl radical, 181

'I'-strain, 232
Immonium ion, 244
Indanyl cation, 246
Indene, cationic polymerization, 263
Indolinobenzospiran group, 460
Induced dipole-dipole, 312
Inductive effects, 319

Infra-red spectroscopy for polymer conformation, 494
Inhibition, 121
Initiation, by organometallic derivatives/ organic halides, 52–116
of chain reaction polymerization, 6
Initiators, difunctional, 370
dissociation at high pressures, 168
monofunctional, 376
Internal conversion, 447
Internal return, 236
Intersystem crossing, 448
Ion-pair dissociation constant, temperature effects, 291
Ion-pair equilibria, 326
solvent effects, 327
Ion-pair free ion equilibria, 250, 291
Ion-pairing, effective reactivity, 322–328
Ion-pairs, 235–239, 273, 310, 322, 362
contact, 238, 245, 250, 324
substituent effects, 326
reactivity in styrene polymerization, 328
role in electrophillic substitution, 323
salt effects, 374
solvent separated, 238, 245, 250, 324, 362
solvolysis, 238
spectral studies, 324
Ionic polymerization, effect of pressure on, 169
photoinitiation, 450
Ionic species, propagation by, 364
Ionization potential, 342
Ions, free, 245, 250, 273, 362
reactivity in styrene polymerization, 328
Ions, triple, 363
Iron carbonyl, photo-adducts with vinyl monomers, 86
Isobutene, 9, 248, 254
copolymerization with sulphur dioxide, 517, 527, 537
heat of polymerization, 12
mixed polysulphone, 557
polymerizability, 535
transfer reactions in cationic polymerization, 297
Isobutyl methacrylate, dependence of termination rate constant on viscosity, 145
Isobutyl vinyl ether, cationic polym-

erization, 298
solvent effect, 292
temperature effect, 291
Isobutyraldehyde, ceiling temperature, 524
equilibrium monomer concentration, 525
polymerization, 535
solution polymerizability, 523
Isocyanates, 551
polymerizability, 536
Isoprene, block copolymer, 379
copolymerization, 554
with 1,1 diphenyl ethylene, 556
effect of pressure on dimerization, 161
emulsion polymerization, 212, 215
heat of polymerization, 12
initiation, by lithium butyl, 353
by manganese acetylacetonate, 96
structure of anionic polymers, 366
Isopropyl benzene, hydrogen transfer, 129
Isopropyl isocyanate, polymerizability, 536

Jablonski diagram, 447
Jacobson-Stockmayer theory, 551

Ketenes, polymerizability, 13
Ketones, polymerizability, 535

LCAO molecular orbital theory, 372
Lactams, heat of polymerization, 550
substituted, 550
Latices, narrow distribution, 205
Lewis acid, 247
halides, 269
Lithium soaps, 205
Living polymers, see Polymers, living
Localization energy, 138

Macromolecules, configurational isomerism, 471
co-stereosymmetric, 475
crystalline conformation, 485
solution conformation, determination, 493–498
relation to optical activity, 498–504
Macroradicals, combination, 143
disproportionation, 143
Maleic anhydride, 265
Maleimides, 9

Manganese(III)acetyl acetonate, 96
  derivatives, 96
  enhancement of activity, 104
  initiation by, monomer orders in, 97
  monomer selectivity, 101
Manganese carbonyl, after effects in
    photoinitiation by, 89
  evolution of carbon dioxide, 69
  irradiation in presence of electron
    donors, 91
  photo-initiation by, 87
    quantum yield, 88
  thermal initiation by, 66
Manganese (III) chelates, kinetic and
    thermodynamic parameters for ini-
    tiation, 99
Manganese 1,1,1-trifluoroacetylacetone,
    in methylmethacrylate polymeri-
    zation, 99
  monomer selectivity, 101
  retardation by, 100
Markovnikov addition, 43
Maxwell-Boltzmann distribution, 484
Mayo equation, in anionic systems, 374
Mercaptans, transfer reactions, 121
Mercuration, 333
Mercury, photosensitization by, 450
Metal allyls, π-bonded, 413
  σ-bonded, 413
Metal carbonyl systems, active end
    group synthesis by, 92
  block copolymers from, 92
  dependence of rate on halide con-
    centration, 57
  effect of acetic anhydride, 63
  effect of water, 61
  graft copolymers from, 92
  halide component activity, 57
  initiation, enthalpy of activation of,
    67
    entropy of activation of, 67
    kinetics, 55, 64
  monomer selectivity, 68
  network synthesis by, 93
  practical applications, 91–95
  reaction mechanisms, 59–72
  reactivity of solvents, 65
  S_N1 mechanism, 66
  S_N2 mechanism, 62
  solvent effects, 63
Metal carbonyls, evolution of carbon
    dioxide from, 68

  photochemistry, 81
  photoinitiation by, 85–91
Metal chelates, activation by 1–2
  diamino propane, 107
  complex formation, 108
  exciplex formation, 113
  initiation by, 95
    effects of donors, 103
    photochemical, 112
  monomer complexes, kinetics of for-
    mation, 109
  monomer selectivity, 103, 108
Metalation reactions, 353
  kinetic control, 329
  thermodynamic control, 329
Methacrylonitrile, copolymers, 557
  equilibrium polymerization, 523
  heat of polymerization, 525
Methide shift, 289
4-Methoxy-1-methylbutyl-p-bromo-
    benzene sulphonate, acetolysis, 234
4-Methoxypentyl derivatives, acetolysis,
    234
Methoxy radicals, hydrogen abstraction
    by, 36
α-Methoxy styrene, ceiling temperature,
    555
  copolymerization, 514, 554, 555
  polymerizability, 535
o-Methoxy styrene, copolymerization,
    554
p-Methoxy styrene, 248, 254
  cationic polymerization, 261, 263
  copolymerization, 554
Methyl acrylate, 8
  copolymerization, 554, 558
    with α-methoxy styrene, 555
  emulsion polymerization, 201, 218
  ΔH_{lc} value, 534
  heat of polymerization, 12
  substituted, polymerizability, 534
Methyl affinity, 127
Methyl atropate, ceiling temperature,
    516, 522
  copolymerization, 556, 557
  equilibrium polymerization, 523
3-Methyl 1-butene, copolymerization
    with sulphur dioxide, 527, 529, 537
  Lewis acid polymerization, 286
  Ziegler-Natta polymerization, 286
3-Methyl butyl vinyl ether, pressure
    effect on polymerization, 169

3-Methyl-3-chloromethyl oxetane, copolymerization, 554
4-Methyl-1-hexene, isotactic polymer
   from, 483
(R,S) 4-Methyl-1-hexene, 408
   stereoselective polymerization, 407
(S) 4-Methyl-1-hexene, 495
Methyl isocyanate, copolymerization,
   reactivity ratios, 559
Methyl methacrylate, 9
   as crosslinking agent, 93
   copolymer with citraconic-anhydride,
      dipole reorientation, 149
   copolymerization, 554, 556
      effect of pressure, 164
      effect of viscosity, 144
      with citraconic anhydride, 149,
         153
      with α-methoxy styrene, 555
      with styrene, 96, 356, 373, 558
      with vinyl acetate, 153, 558
   effect of pressure on stereoregularity,
      172
   emulsion polymerization, 191, 218
      data, 193
      kinetics, 197
      recipe, 222
   equilibrium polymerization, 518
      in solution, 523
   glass transition temperatures, 368
   ΔH_{lc} value, 534
   heat of polymerization, 12, 525
   in water, 190
   initiation, by manganese acetylace-
      tonate, 96
   irradiation, 450
   Φ_M value, 198
   photoinitiation, by azo compounds,
      451
      by chelates, 112
   polymerization, anionic, 368
      by chelate-donor systems, 104
      by metal carbonyls, 57
      by molybdenum carbonyl/tetra-
         bromomethane systems, 81
      effect of viscosity, 144
      free radical, 368
      stationary rates, 85
   rates of initiation by metal carbonyls,
      76
   stereo-block polymer from, 368
   strength of bond to nickel, 82

   termination, dependence on viscosity,
      145
   effect of temperature, 155
   with metal chelates, 101
Methyl pentadiene, anion, 336
2-Methyl-1-pentene, copolymerization
   with sulphur dioxide, 537
3-Methyl-1-pentene, 407
(R,S) 3-Methyl-1-pentene, 407
4-Methyl-1-pentene, 288, 495
   Phillips polymerization, 431
4-Methyl-2-pentene, 537
4-Methyl pent-3-enyl-p-toluene sulphonate, acetolysis, 235
Methyl radicals, addition, orientation
   of, 49
   to cis-2-butene, 40
   to $trans$-2-butene, 40
   to ethylene, 39, 40, 42, 43
   to fluoro olefins, 49
   to hydrocarbon olefins, 40
   to tetrafluoroethylene, 43
   combination, 33
   cross combination ratios, 34
   disproportionation-combination
      ratio, 34
   halogen substituted, orientation of
      addition, 49
   hydrogen abstraction by, 36
   ionization potential, 231
α-Methyl stilbene, cis-$trans$ interaction,
   334
α-Methyl styrene, 9, 524
   cationic polymerization, 254
      effect of pressure, 169
   ceiling temperature, 360, 516, 522
      determination, 518
      effect of pressure, 170, 532
      in solution, 524
   copolymerization, 554, 556, 557
      reactivity ratios, 559
   effect of pressure on polymerization,
      514
   electroinitiation, 359
   equilibrium polymerization, 518, 531
      in solution, 523
   ΔH_{lc} value, 534
   heat of polymerization, 12, 525
   substituted, polymerizability, 533
   o-substituted, 531
   p-substituted, 531
      equilibrium polymerization, 531

α-Methyl styrene (*contd.*)
  tetramer, 345, 361
  volume of polymerization, 171
  with tetra cyanobenzene, 453
β-Methyl styrene, anion, 334
  with tin chloride, 286
2-Methyl-tetrahydrofuran, copolymerization, 554
α-Methyl vinyl methyl ether, polymerizability, 535
2-Methyl-5-vinyl pyridine, 218
  emulsion polymerization, 191
    data, 194
Methylene blue, photosensitization by, 452
Micelles, 176
Michael addition, 352
Molecular weight, distribution, 29, 515
  number average, 28
  pressure dependence of, 163
  viscosity average, 29
  weight average, 29
Monomers, activity, 527
  in anionic polymerization, 358
  effect of structure on polymerizability, 532–537
  equilibria, 514
  equilibrium concentration, 550
  halogen substituted, 373
  interaction with light, 446–470
  π-complexes, 423
  strain effects in, 11
Molybdenum carbonyl, as initiator, 78
  comparison with AIBN, 73
  esr of mixture with carbon tetrachloride and methyl methacrylate, 78
Multi-block copolymerization, 376, 377

1-Naphthaldehyde, electron affinity, 343
2-Naphthaldehyde, electron affinity, 343
Naphthalene, electron affinity, 343
  energy transfer, 467
  sodium reduction, 342
Naphthyl group, phosphorescence, 458
2-Naphthyl-2-propene, 523
Negative activation energy, 365
Neighbouring group participation, 233–235
Neopentyl chloride, solvolysis, 241
Neopentylamine, deamination, 241
Nickel carbonyl, 410

  as initiator, 75
Nitriles, N-alkylation, 243
Nitrobenzene, 453
Nitromethane, anion, 315
2,5-Norbornadiene, transannular polymerization, 284
anti-Norbon-2-en-7-yl tosylate, acetolysis, 235
Norborn-7-yl tosylate, acetolysis, 235
Non-radiative decay, 448
Norrish cleavage, type I, 450, 461
  type II, 461
Nuclear magnetic resonance spectroscopy, for polymer configuration, 478
  for polymer conformation, 493
  for polymer stereoregularity, 504
Nucleation time, 188
Number average degree of polymerization, 118

Octamethyl cyclotetrasiloxane, 377
Olefins, *see also* Alkenes,
  cis-opening, 400
  propagation rate constants for cationic polymerization, 278
  trans-opening, 401
α-Olefins, polymerization by Phillips catalysts, 431
  Ziegler-Natta polymerization, 384–409
Optical activity, temperature coefficient, 500
Organic halides, activity in polymerization initiated by metal carbonyls, 76–78
Organic peroxides, 181
  in emulsion polymerization, 181
Organolithium compounds, addition to benzophenones, 339
Organometallic initiators, 53–116, 383–430
  activities, 73–76
  homogeneous, 384
  structure, 416
  thermal, 53–55
Ostwald's dilution law, 251
Oxepane, ceiling temperature, 522
Oxiranes, copolymerization, 554
Oxonium ions, 244, 248
  propagation by, 275
  salts, 253, 268

Oxygen, 247
  copolymerization of, 554
  singlet, 453

Palladium salts, as initiators, 420
Patterns treatment, 127, 135, 140
  α-values, correlation with σ-values, 135
  α,β values for monomers, 134
Penultimate unit effects, 374
1,3-pentadiene, 410
  cis-trans isomerization, 459
  emulsion polymerization, 212, 216
Pentadienylic anions, protonation, 336
Pentafluoroethyl radicals, orientation of addition, 49
Pentafluoro-sulphur radicals, addition to ethylene, 42
Pentane, 2,4 disubstituted, 494
1-pentene, 304
  copolymerization with sulphur dioxide, 537
  Phillips polymerization, 431
2-Pentene, copolymerization with sulphur dioxide, 537
4-Pentenoic acid, copolymerization with sulphur dioxide, 537
n-Perfluoro-1-heptene, equilibrium polymerization, 531
n-Perfluoropenta-1-4-diene, pressure effect on cyclopolymerization, 172
Peroxides, photolysis, 451
Persulphates, 181
  emulsion polymerization initiators, 209
Phase changes, in high pressure polymerization, 172
Phase separation, in high pressure polymerization, 172
9-Phenanthraldehyde, electron affinity, 343
Phenanthrene, electron affinity, 343
Phenol-formaldehyde polymer, 4
Phenonium ion, 240
Phenyl diazomethane, 11
4-Phenyl-1,3-dioxolane, copolymerization, 554
1-Phenyl ethyl isocyanate, polymerizability, 536
Phenyl isocyanate, polymerizability, 536
Phenyl lithium, 329
Phenyl nitro-methyl anion, 337

2-Phenyl-propyl isocyanate, polymerizability, 536
Phenyl sodium, 328
Phillips catalysts, 392, 431
  nature of active site, 435
  preparation, 431
  reduction with carbon monoxide, 433
  reduction with hydrogen, 433
Phillips petroleum company, 431
Phillips process, see also individual monomers
  description, gas phase, 432
    slurry, 432
    solution, 432
  effect of activating temperature, 436
  effect of chromium concentration, 437
    on molecular weights, 439
  kinetics, 435
  models, 439
  molecular weight distribution, 438
  polymer molecular weights, 438
  reaction rates, 439
Phosphates, 551
Phosphonitriles, 551
Photochemical chain lengthening, 446
Photochemical processes, principles, 446–449
Photochromic compounds, 460
Photochromic polymers, 460
Photochromic process, 460
Photo-cyclomerization, 454
Photo-Fries rearrangement, 463
Photoinitiation systems, 450
Photopolymerization, 446, 449, 454–458
  in solid state, 454
  in solution, 456
Photosensitizers, triplet energies, 465
Phthalaldehyde, ceiling temperature, 516
  equilibrium monomer concentration, 536
  solution polymerization, 523
Picric acid, 312
β-Pinene, polymerization by Friedel Crafts halides, 285
2-Piperidone, 550
Pivalic aldehyde, polymerizability, 535
pK_a, choice of solvent, 313
  determination, from equilibria, 313
    from kinetics, 313
  measurement, 313
  values, 358
    of hydrocarbon acids, 312

pK_a^W, values, of heteroatom carbon acids, 315
of hydrocarbons, 316
Poisson molecular weight distribution, 365
anionic polymerization, 361
Polar character, effect on reactivity, 130
Polarized luminescence, 493
Poly (acetaldehyde), isomerism, 27
syndiotactic, 488
Poly (acetals), 4
isomerism, 27
Poly (acetylene), 539
Poly(acrylamide), fluorescence depolarization, 151
Poly (acrylate) latex, recipe, 222
Poly (acrylonitrile), 8
radical, rate constants, 128
reactivity, 122
termination, 155
Poly (alcohols), 457
Poly (aldehydes), isotactic, 486
Poly (alkenamers), 19
Poly(alkyl methacrylates), effect of alcohol residue on relaxation, 149
Poly (alkyl vinyl ethers), 9
Poly (alkylidenes), 11
Poly (amides), 4, 457
conformation, 489
conformation equilibria, 492
N-halogenated, as network precursors, 93
Poly (-α-amino acids), 474, 479
isotactic, 473
Poly (benzylidene), 11
Poly (butadiene), 1,2-, 411, 415, 420
crystalline conformation, 487
irreversible photo transformation, 463
isotactic conformation, 486
syndiotactic conformation, 488
Poly(butadiene), 1,4- *cis*, 415, 416
trans, 414
cis-trans isomerism, 459
effect of pressure on structure, 172
Poly (1-butene), 496
characteristic ratio, 496
isotactic, conformation, 486
solution conformation, 492
Poly(1-butene sulphone), rate of formation, 527
specific viscosity, 527

Poly (*sec*-butyl acrylate), isotactic conformation, 486
Poly (*n*-butyl isocyanate), 497
conformation, 489
Poly (butyl methacrylate), fluorescence depolarization, 151
radical, termination rate constant, 147
Poly (*n*-butyl vinyl ether), 486
Poly (*t*-butyl vinyl ether), conformation, 487
Poly (ε-caprolactam), 17
Poly (carbonates), 5, 556
photo-rearrangement, 462
Poly (N-carboxyacidanhydrides), 16
Poly (1-chloro-2-butoxy ethylene), *erythro* and *threo* isomer conformation, 488
Poly (*p*-chloro styrene), amorphous, 482
dipole reorientation, 150
termination rate constant, 150
Poly (cyclobutene), 9
Poly (*p*-deuterostyrene), 494
Poly (3, 7-dimethyl-1-octene), 408
Poly (dimethyl siloxane), 496, 552
conformation, 489
equilibria, 492
Poly (diolefins), 1,4- *cis*, 411
1,4- *trans*, 411
Poly (1,3-dioxolane), 16
NMR, 494
Poly (2,5-distryrylpyrazine), 455
Poly (electrolytes), 498
Poly (esters), 4
conformation, 489
equilibria, 492
isomerism, 27
Poly (ethylene), 8, 496
conformation, 487
equilibria, 492
high density, 398, 431
low pressure, 431
phase separation, 162
pressure diminution of branching, 171
Poly(ethylene glycol), rotational isomerism, 493
Poly (ethylene imine), 16
Poly (ethylene oxide), 16, 177
chain motion, 147
helical conformation, 487
NMR, 494
radical, effect of chain length on termination rate constant, 146

Poly(ethylene terephthalate) rotation isomerism, 493
Poly (ethylidene), 11
Poly(*p*-fluoro styrene),conformation,494
  crystalline, 482
Poly (*n*-hexyl isocyanate), 497
Poly (iso-butene), 9, 259, 298, 496
  conformation, 488
Poly (isobutene sulphone), 517
Poly(isobutyl vinyl ether), conformation, 487
  dipole moment, 497
Poly (isopropyl acrylate), characteristic ratio, 496
  isotactic conformation, 486
Poly(isopropyl vinyl ether), conformation, 487
  dipole moment, 497
Poly (isoprene), *1,4- trans*, 414
  3,4- 411
  conformation equilibria, 492
  effect of pressure on structure, 172
  hydroxyl terminated, 371
  irreversible phototransformation, 643
  1-4, photochemical isomerization, 460
Poly (isoprene hydrochloride), rotational isomerism, 493
Poly((S)-lactic acid), conformation, 489
Poly (maleimides), 9
Poly ((−) menthyl methacrylate), effect of stereoregularity on optical rotation, 481
Poly (methacrylates), optically active, 481
Poly (methacrylonitrile), radical, rate constants, 128
Poly (methyl acrylate), 8
  dipole reorientation, 149
  energy barriers to backbone rotation, 148
  fluorescence depolarization, 151
  radical, rate constants, 128
  termination rate constant, 149
Poly ((+)-α-methyl benzyl methacrylate)
  effect of stereoregularity on optical rotation, 481
Poly (3-methyl-1-butene), isotactic, 485
  conformation, 486
  solution conformation, 492
Poly (3-Methyl-1-butene sulphone), 530
Poly((S)-2-methyl-butyl vinyl ether),

dipole moment, 497
Poly (5-methyl-1-heptene), isotactic conformation, 486
Poly ((S)-5-methyl-1-heptene), 502
  conformation properties, 501
Poly (4-methyl-1-hexene), 409
  isotactic conformation, 486
Poly ((S)-4-methyl-1-hexene), 502
  conformation, 501
  effect of stereoregularity on optical rotation, 480
  spiralization, 504
Poly (5-methyl-1-hexene), isotactic, conformation, 486
  solution conformation, 492
Poly (methyl methacrylate), 9
  anion, 370
  chain motion, 147
  characteristic ratio, 496
  dipole moment, 497
  dipole reorientation, 149
  effect of molecular weight on termination rate constant, 146
  energy barriers to backbone rotation, 148
  fluorescence depolarization, 151
  helical configuration, 368
  isotactic, 368
  photodegradation, 462
  radical, comparison of termination rate constant with other poly (methacrylate) radicals, 148
    disproportionation, 155
    rate constants, 128
  replication processes, 369
  stereocomplex, 497
  syndiotactic, 368
    in emulsion polymerization, 228
  termination rate constant, 149
Poly ((S)-6-methyl-1-octene), 502
  conformation, 501
Poly ((S)-3 methyl-1-pentene), 502
  conformation, 501
Poly (4-methyl-1-pentene), isotactic, 485
  conformation, 486
  solution conformation, 492
Poly((S)-1-methyl propyl vinyl ether), dipole moment, 497
Poly (methyl siloxane), photodegradation, 462
Poly (α-methyl styrene), 9
Poly (*o*-methyl styrene), isotactic, 486

conformation, 486
Poly (*m*-methyl sytrene), isotactic, 485
  conformation, 486
Poly (*p*-methyl styrene), amorphous, 482
Poly (methyl vinyl ether), conformation, 487
  isotactic conformation, 486
Poly (methylene oxide), 16
  NMR, 494
Poly (nonyl methacrylyl) radical, termination rate constant, 147
Poly (nortricyclene), 284
Poly (octadecene), crystalline, 482
Poly (α-olefins), 480, 505
  characteristic ratio, 491, 492
  conformation, 502
  conformation equilibria, 485
  crystallinity, 482
  isotactic, optically active, 500, 501
    solution conformation, 492
Poly (oxymethylene), conformation, 487
  equilibria, 492
Poly (1,3-pentadiene), 473
Poly (1-pentene), 482, 496
  isotactic, solution conformation, 492
Poly (peptides), 461
  as network precursors, 93
  helix-coil transitions, 485
Poly (3-phenyl-1-propene), isotactic, 486
  conformation, 486
Poly (phenyl vinyl ketone), 459
Poly (phenylene oxide), 5
Poly (phosphates), conformation, 489
Poly (propiolactone), 16
Poly (propylene), 8, 482
  characteristic ratio, 496
  conformation, from infra-red, 494
  isotactic, conformation, 486
    solution conformation, 492
  stereoregularity, 479
  syndiotactic, 404
    crystalline conformation, 487, 488
Poly (propylene oxide), 16, 474, 479
  conformation, 487
  isotactic, 473
Poly (propylene sulphide), 16
Poly (siloxanes), 4
Poly (styrene), 8
  carboxylic acid ended, 371
  characteristic ratio, 496
  conformation, from infra-red spectra, 494

crystalline, 482
  dipole moment, 497
  fluorescence depolarization, 151
  isotactic, 482
    conformation, 486
  latices, particle diameters, 191
  radical, half life, 195
    rate constants, 128
    reactivity, 122
    termination, 155
    transfer constants, 77
Poly (styryl sodium), in tetra hydro-pyran 366
Poly (sulphides), 4
Poly (sulphones), 4
  ceiling temperatures, 536, 537
Poly (tetrahydrofuran), 515
Poly (tetrafluoroethylene), 9
  conformation equilibria, 492
Poly (tetramethylene oxide), 16
Poly (thiomethylene), conformation, 487
Poly (trimethylene oxide), 16
Poly (urethanes), 4
Poly (δ-valerolactone), 17
Poly (vinyl acetate), 9, 482
  dipole reorientation, 149
  radical, rate constants, 128
  relaxation, 149
  termination rate constant, 149
Poly (vinyl alcohol), acetylated, 177
  crystalline, 482
Poly (vinyl bromide), dipole reorientation, 150
  termination rate constant, 150
Poly (N-vinyl carbazole), 453
  dipole reorientation, 150
  termination rate constant, 150
Poly (vinyl chloride), 8
  dipole reorientation, 150
  rotational isomerism, 493
  termination rate constant, 150
Poly (vinyl cinnamate), quantum yield of photopolymerization, 465
Poly (vinyl benzophenone), 464
  film, 458
Poly (vinyl ethers), optically active, 481
Poly (vinyl ketones), optically active, 481
Poly (vinylnapthalene), 458
  excimer fluorescence, 459
  isotactic, 486
    conformation, 486

Poly (vinyltrichloroacetate), as network precursor, 93
Poly (vinylidene chloride), 9
Polymer-monomer equilibria, 514, 519
 immiscible systems, 520
 in solvents, 523
 involving gaseous monomer, 522
 polymer soluble in monomer, 522
Polymer networks, 93
Polymer radicals, propagation, 117–120
 reactivity, 117–141
 transfer reactions, 120, 121
Polymer stretching, effect of conformation, 497
Polymeric anions, isomerization, 370
 spectra, 370
Polymeric species, equilibria, 514
 with monomeric species, 514
Polymerizability, effect of monomer structure, 532–537
 of unsaturated compounds, 7–15
Polymerization, effect of pressure, 158–174
 inhibition by emulsifiers, 180
 photoinitiated, 449–454
 reactivity and structure, 1–29
 resonance induced, 539
 thermodynamics of addition, 514–565
Polymers, configuration, 471–513
 conformation, 471–513
 crystallinity of, effect on Free Energy, 527
 dipole moment, 495
 end-to-end distance, 495, 502
 effect of stereoregularity on properties, 482
 electrical properties, 498
 growth from crystal edges, 403
 helical conformations, 485
 interaction with light, 446–470
 isotactic, 473
 light resistance, 458
 living, 277, 351
  termination by air, 371
 microstructure, 366
  pressure dependence of, 171
 molecular weights, 28, 29
 nature, 1
 photochemical processes, 458–467
 photodegradation of, 461
 stereoregular, 365, 473
 strain effects, 11

syndiotactic, 473
 crystalline conformation, 491
 crystallization, 487
 tacticity, 473
 ultrasonic studies, 151
Potassium benzophenone radical anion, 357
Potassium persulphate, reaction with iron sulphate, 181, 183, 228
Pre-initiation equilibrium, 273
Pressure, effect on equilibrium constants, 159–163
 effect on polymerization, 514
 effect on rate constants, 159–163, 160
 physical effects influencing reactivity, 162
Propagation, 6, 118
 reversible, 515, 520
  effect on rate of polymerization, 526
Propiolactone, 16
Propionaldehyde, polymerizability, 535, 536
2-Propyl radical, disproportionation-combination ratio, 34
 ionization potential, 231
n-Propyl radical, disproportionation-combination ratio, 34
 ionization potential, 231
Propylene, 8, 304, 384
 copolymerization, with sulphur dioxide, 537
 deuterium labelled, 400
 heat of polymerization, 12
 high pressure polymerization of, 167
 isotactic placement, 400
 solution polymerization, 401
 syndiotactic placement, 405
Propylene oxide, 16, 27
 copolymerization, 554
 stereospecific polymerization, 476
Propylene sulphide, 16
 emulsion polymerization, 208
 ring-opening polymerization, 373
Protein, denaturation, 462
 photodegradation, 462
Proton transfer, 294, 298
Protonic acids, for cyclic ether polymerization, 269
Pseudo-cationic polymerization, 251, 281
Pyrene, electron affinity, 343
 energy transfer, 467

2-Pyrrolidinone, 550
Pyrylium derivatives, cations, 263

Quaternary ammonium ion, 268
Quaternary ammonium salts, as sur-
    factants, 177
Q,e parameters, 126
Q,e scheme, 125, 126, 128, 134, 138, 139,
    372
Q-e-e* scheme, 127
Quinones, 247

Racemization, by S_N1 reactions, 236
Radical-anions, 342–346, 355
    as reducing agents, 345
    dimerization, 345
    disproportionation, 344
    generation, 342
    ion-pair equilibira, 343
Radical polymerization, photoinitiation,
    450
    under high pressure, 163–166
Radicals, addition reactions of, 37–50
    addition to vinyl monomers, effects of
        pressure, 166
    combination reactions of, 32–34, 155
    complexed, 124
    complexes with salts, 124
    disproportionation reactions of, 34,
        35, 155
    electronegativity, 138
    ionization potentials, 231
    polar parameters, 127
    polarity of, 125
    reactivity of, 31–50, 122, 125, 128, 136
        data interpretation, 125–139
        data treatment by computer
            methods, 138
        penultimate unit effects, 123, 124
        steric effects, 123
    recombination rates, 345
    transfer reactions of, 35–37
Randomizing agents, 375
Rate constants, pressure dependence, 162
Reactive intermediates, in chain polym-
    erization, 14
Reactivity patterns, variation with sub-
    strate, 133
Reactivity ratios, 119, 120, 123, 138
    pressure dependence, 167
Relative stereoregularity index, 479
Retardation, 121

Rhenium carbonyl, photoinitiation by,
    87
    after effect in, 88
Rhodium chloride, 207, 420
Rhodium salts, as initiators, 420
Ring-chain equilibria, 551
    theory of, 551
Ring compounds, five membered, polym-
        erization of, 540–542
    polymerizability of, 15–19, 538–553
    seven membered, polymerizability of,
        546–548
    six membered, polymerizability of,
        543–545
Ring-opening polymerization, 373
    strain relief, 285
    transfer, 301
Rotational isomeric scheme, 484
Rubber, photodegradation, 462
    synthetic, 372
Ruthenium salts, as initiators, 420

Saytzeff orientation, 240
Seed latices, 189
Seed polymerization, 206
Segmental-controlled reaction, 147
Segmental motion, 478
Segmental rearrangement, 142
    diffusion, 145
    slow, 144
Selenium, polymerization, 522
Self diffusion, 145
Self-initiation, 57, 70
Self retardation, 70
Sequences, alternating, 1
    block, 1
    random, 1
Shotten-Baumann reaction, 464
σ-bond participation in cationic frag-
    mentation, 242
Silica, as Phillips catalyst support, 431
Silica/alumina, as Phillips catalyst sup-
    port, 431
Siloxanes, 551
    macrocyclic polymer from, 515
Silthianes, 551
Smith-Ewart kinetics, *see* Emulsion
    polymerization
Sodium naphthalene, 356
    initiation of styrene polymerization,
        357
Sodium persulphate, 181

Sodium tetraphenyl boride, 365
Solid state polymerization, at high pressure, 173
Solvato-chromism, 460
Solvent, cage, 327
  hydrogen bonding, 310
  polarity, scale of measurement, 235
Solvent-solvent dispersion interactions, 310
Sorbitol, 177
Special salt effect, 237, 238
Spreading coefficient, 179
Stabilization energy, 125, 490
Star copolymer, 376, 377
Stationary state, 118, 246
Step reaction polymerization, 2–6
  features, 8
Stereo-block copolymerization, 376
Stereo-elective polymerization, 27, 406, 408
Stereoregular polymerization, 27
Stereoregularity of polymers, effect on mechanical properties, 483
  effect on reactivity, 483
  evaluation, 477
    from infra-red spectra, 478
    from optical activity, 479
Stereoselective polymerization, 27, 406, 477
α-Stilbazole, copolymerization, 554
Stilbazoles, photodimerization, 455, 466
Stilbene, *cis-trans* interconversion, 334
trans-Stilbene, copolymerization, 554, 556
  dianion, 355
cis-Stilbenyl lithium, isomerization, 334
Styrene, 8, 122, 248, 514
  ABA block polymers, properties, 379
  anionic polymerization, 350
    lithium butyl initiation, 353
  as crosslinking agent, 93
  cationic polymerization, 261, 279
    effect of pressure on, 169
    transfer, 296
  ceiling temperature, 516
    in solution, 523
  copolymerization, 554
    with 1-2 dibenzoylethylene, 555
    with (R)-3,7-dimethyl-1-octene, 495
    with methyl methacrylate, 557
    with *o*-phthalaldehyde, 557
    with propylene, 476

    with α-vinyl naphthalene, 458
  electrolysis in tetrahydrofuran, 359
  emulsion polymerization, 201, 208
    data, 192
    effect of conversion, 202
    kinetics, 197, 228
  freezing point, pressure dependence of, 162
  ΔH_{lc} value, 534
  heat of polymerization, 12, 525
  initiation, by acetyl acetonates, 96
    by manganese (III) acetylacetonate, 96
    micellar, 191
  irradiation, 450
  latices, 206
  α-methyl, *see* α-methyl styrene,
  β-methyl, *see* β-methyl styrene,
  α-methoxy, *see* α-methoxy styrene,
  o-methoxy, *see* *o*-methoxy styrene,
  p-methoxy, *see* *p*-methoxy styrene,
  molecular weight, pressure dependence of, 163
  narrow distribution copolymer latices, recipe, 224
  Φ_M value, 198
  photoinitiation, 450
    by azo compounds, 451
    by chelates, 112
  polymerizability, 533
  polymerization, at high pressure, 164, 173
    by acids, 251
    effect of pressure on, 163
    in tetrahydrofuran, 363
    in water, 190
    volume of activation, 165, 166
    with chelate-donor system, 104
  radical anion, dimerization, 345
    reactions, 355
  strength of bond to nickel, 82
  termination, 94
  thermal polymerization, pressure acceleration, 164, 165
  transfer, pressure effect, 168
  with divinyl benzene, latices, 206
  with iodine, 281
  with metal chelates, 101
  with perchloric acid, 254, 281, 302
  with tetracyanobenzene, 453
  with tin tetrachloride, 256
Sulphate radical ion, 181

Sulphonium ion, 244, 248
Sulphur, 551
  equilibrium polymerization, 531
  floor temperature, 522
  polymerization, 517, 539
Sulphur dioxide, copolymerization of, 554
  with alkenes, 517, 525, 527, 536
  with 3-methyl-1-butene, 529
Sulphur trioxide, 520
Super acids, 230
Surfactants, cationic, 177

Tacticity, relation to optical activity, 479
Taft polar factor (σ^*), correlation with termination rate constants, 152
Taft steric factor (E_s), 152
Termination rate constant, molecular weight effects, 146
Termination reactions, 6, 95, 118
  activation energy, 144
  diffusive processes, 143–155
  effect of chain structure, 142–157
  effect of copolymer composition, 152
  effect of temperature, 154
  rate determining step, 143
Terpenes, isomerization polymerization 285
1,2,4,5,Tetra-chloromethyl benzene, 377
Tetracyanoethylene, 253, 265
Tetrafluoroethylene, 9
  heat of polymerization, 12
Tetrahydrofuran, cationic polymerization, 263
  ceiling temperature, 516, 522
    pressure dependence, 170, 531
  copolymerization, 557, 558
  equilibrium polymerization, 531
  heat of polymerization, 525
  ring-opening polymerization, 267
  volume of polymerization, 171
Tetrahydropyran, copolymerization of, 554
Tetrakis-(triphenyl phosphite) nickel (0), 81
  ligand exchange with methyl methacrylate, 81
  ligand exchange with styrene, 81
  reaction parameters with methyl methacrylate, 82
  reaction parameters with styrene, 82
Tetramethylene oxide, 16

Tetranitro-methane, 265, 453
Tetraoxane, equilibrium monomer concentration, 551
Thermodynamics, of addition polymerization, 514–565
Thermoplastic elastomers, 379
Thioacetone, 13
  ceiling temperature, 516, 522, 536
  cyclic trimer, 536
Thioaldehydes, fluorinated, polymerizability, 536
  polymerizability, 536
Thiocarbonyl fluoride, copolymerization, 554
Thioketones, fluorinated, polymerizability, 536
Threo-isomers, conformations, 488
Tin tetrachloride, 252
Titanium tetrachloride, 247, 252
Titanium trichloride, crystalline modifications, 402
  α-form, 402, 411, 414
  β-form, 402, 411, 413, 415
  γ-form, 411
  δ-form, 402, 411
  structures, 394
Toluene, as anionic transfer agent, 370
Transannular polymerization, 284
Transfer constant, 120
Transfer reactions, penultimate unit effects, 123
Transition metal-alkyl σ-bonds, 389
Transition metal complexes, bonding, 390
Transition metals, oxidation state in catalysis, 399
Translation controlled reaction, 147
Translational diffusion, 142, 145
Trialkyl aluminium, 415
Trialkyl oxonium ions, 268
Tribromo-methyl radical, 81
Trichloroethylene, 265
Trichloromethyl radicals, addition, orientation of, 49
  to ethylene, 42
  to olefins, 46
  to unsymmetric olefins, 48
  combination, 33
  hydrogen abstraction by, 36
  initiation by, 60
Trifluoromethyl radicals, addition, orientation of, 49

Trifluoromethyl radicals (*contd.*)
　to ethylene, 42, 43
　to fluoro olefins, 42
　to tetrafluoro ethylene, 43
　to unsymmetric olefins, 45
　cross combination ratios, 34
　hydrogen abstraction by, 36
α-Trifluoromethyl vinyl acetate, co-
　　polymerization of, 554
Trimethylene oxide, 16
3,5,5,Trimethyl-hexyl methacrylate, de-
　　pendence of termination constant
　　on viscosity, 145
Trimethylorthoformate, 302
2,4,4,Trimethyl-1-pentene, 537
1,3,5,Trinitrobenzene, 265
Trioxane, 16
　ceiling temperature, 524
　depropagation, 559
　polymerizability, 536
Triphenyl methane, 271
　cation, 263, 270, 280
Triphenylene, electron affinity, 343
Triphenyl phosphine, 74
　with methyl methacrylate, 453
Trithiane, polymerizability, 536
Trypticenyl anion, 319

Ultrasonic measurements on polymers,
　151

δ-Valerolactam, equilibrium polymeri-
　　zation, 531
δ-Valerolactone, 17
　polymerization, 524
Vanadium triacetylacetonate, 415
Vinyl acetate, 9, 122
　copolymerization, with crotonalde-
　　hyde, reactivity ratios, 559
　　with sulphur dioxide, 537
　effect of viscosity on polymerization,
　　144
　emulsion copolymerization, 210
　emulsion polymerization, 209
　　data, 192
　　γ-ray initiation of, 210
　　retardation, 210
　heat of polymerization, 12
　ϕ_M value, 198
　photoinitiation, 451

polymerization, with chelate-donor
　　system, 104
　with metal chelates, 101
9-Vinylanthracene, polymerization by
　　Lewis acids, 283
Vinyl bromide, effect of temperature on
　　kT, 155
Vinyl caproate, emulsion polymerization
　　data, 192
Vinyl carbanions, 334
N-Vinyl carbazole, cationic copolym-
　　erization, 249
　cationic initiation, 278
　cationic polymerization, 261
　　photoinduced, 453
　　temperature effect, 291
　　transfer reactions, 300
　effect of temperature on kT, 155
　initiation, by aminium salts, 265
　　by stable cations, 263
　with acrylonitrile, 453
　with electron acceptors, 265
　with perchlorates, 303
　with TCNE, 266
Vinyl chloride, 8
　emulsion copolymerization, 211
　emulsion polymerization, 211
　　data, 193
　　recipe, 223
　heat of polymerization, 12
　photoinitiation by azo compounds,
　　451
　polymerization at high pressure, 173
Vinyl cyclohexane, 287
Vinyl cyclopropane, cationic polym-
　　erization, 285
Vinyl ethers, termination of cationic
　　polymerization, 303
　polymerization by acids, 254
Vinyl mesitylene, copolymerization, 557
Vinyl monomers, copolymerization, 554
　initiation, by acetyl acetonates, 95
　　cationic, 253
　cationic propagation, 276
1-Vinyl naphthalene, copolymerization
　　with styrene, 375
Vinyl polymers, chain branching, 19
　erythro-di-isotactic configuration of,
　　24
　threo-di-isotactic configuration of, 24
　di-syndiotactic, 24
　helical conformations, 494

Vinyl polymers (*contd.*)
  isomerism, 19–25
  isotactic, 22, 473, 475
    conformation, 485
    optical activity, 502
    molar optical rotation, 499
  sequence isomerism, 20
  stereoisomerism, 21
  structural isomerism, 19
  syndiotactic, 22, 474
2-Vinyl pyridine, copolymerization, 554
Vinyl stearate, emulsion polymerization
  data, 192
Vinylidene chloride, 9, 372
  emulsion polymerization, 189, 216
    data, 194
  heat of polymerization, 12
Volume of activation, 160
Volume of polymerization, 531

Walling-Mayo hypothesis, 44, 45

Xanthylium derivatives, cations, 263
X-rays, for conformational analysis, 505

for stereoregularity, 504

Y-values of solvent polarity, 236

Z-values, of solvent polarity, 236
Ziegler-Natta catalysts, 253, 282
  polymerization, bimetallic mecha-
    nism, 386
    Boor-Youngman model, 404
    Cossee-Arlman model, 402
    Cossee monometallic mechanisms,
      388
    early mechanistic theories, 385
    evidence for existing mechanisms,
      398
    heterogeneous catalysis, 397
    heterogeneous mechanism, 394
    homogeneous catalysis, 395, 398
    initiation, 385
    monometallic mechanism, 388
    origin of stereoregulation, 400
    propagation, 253
    recent developments, 393
    uniform site theory, 395
Zinc chloride, as donor, 103